COLD SPRING HARBOR SYMPOSIA ON QUANTITATIVE BIOLOGY

VOLUME LXXXI

COLD SPRING HARBOR SYMPOSIA ON QUANTITATIVE BIOLOGY

VOLUME LXXXI

Targeting Cancer

symposium.cshlp.org

Symposium Organizers: Scott Lowe (*Memorial Sloan Kettering Cancer Center*),
Kornelia Polyak (*Dana-Farber Cancer Institute*),
David Stewart and Bruce Stillman (*Cold Spring Harbor Laboratory*),
and Eileen White (*The Rutgers Cancer Institute of New Jersey*)

Editors: David Stewart and Bruce Stillman (*Cold Spring Harbor Laboratory*)

COLD SPRING HARBOR LABORATORY PRESS
2016

COLD SPRING HARBOR SYMPOSIA ON QUANTITATIVE BIOLOGY VOLUME LXXXI

International Standard Book Number 978-1-621822-09-7 (cloth)
International Standard Book Number 978-1-621822-10-3 (paper)
International Standard Serial Number 0091-7451
Library of Congress Catalog Card Number 34-8174

Printed in the United States of America

COLD SPRING HARBOR SYMPOSIA ON QUANTITATIVE BIOLOGY
Founded in 1933 by
REGINALD G. HARRIS
Director of the Biological Laboratory 1924 to 1936

Previous Symposia Volumes

I (1933) Surface Phenomena
II (1934) Aspects of Growth
III (1935) Photochemical Reactions
IV (1936) Excitation Phenomena
V (1937) Internal Secretions
VI (1938) Protein Chemistry
VII (1939) Biological Oxidations
VIII (1940) Permeability and the Nature of Cell Membranes
IX (1941) Genes and Chromosomes: Structure and Organization
X (1942) The Relation of Hormones to Development
XI (1946) Heredity and Variation in Microorganisms
XII (1947) Nucleic Acids and Nucleoproteins
XIII (1948) Biological Applications of Tracer Elements
XIV (1949) Amino Acids and Proteins
XV (1950) Origin and Evolution of Man
XVI (1951) Genes and Mutations
XVII (1952) The Neuron
XVIII (1953) Viruses
XIX (1954) The Mammalian Fetus: Physiological Aspects of Development
XX (1955) Population Genetics: The Nature and Causes of Genetic Variability in Population
XXI (1956) Genetic Mechanisms: Structure and Function
XXII (1957) Population Studies: Animal Ecology and Demography
XXIII (1958) Exchange of Genetic Material: Mechanism and Consequences
XXIV (1959) Genetics and Twentieth Century Darwinism
XXV (1960) Biological Clocks
XXVI (1961) Cellular Regulatory Mechanisms
XXVII (1962) Basic Mechanisms in Animal Virus Biology
XXVIII (1963) Synthesis and Structure of Macromolecules
XXIX (1964) Human Genetics
XXX (1965) Sensory Receptors
XXXI (1966) The Genetic Code
XXXII (1967) Antibodies
XXXIII (1968) Replication of DNA in Microorganisms
XXXIV (1969) The Mechanism of Protein Synthesis
XXXV (1970) Transcription of Genetic Material
XXXVI (1971) Structure and Function of Proteins at the Three-dimensional Level
XXXVII (1972) The Mechanism of Muscle Contraction
XXXVIII (1973) Chromosome Structure and Function
XXXIX (1974) Tumor Viruses
XL (1975) The Synapse
XLI (1976) Origins of Lymphocyte Diversity
XLII (1977) Chromatin
XLIII (1978) DNA: Replication and Recombination
XLIV (1979) Viral Oncogenes
XLV (1980) Movable Genetic Elements
XLVI (1981) Organization of the Cytoplasm
XLVII (1982) Structures of DNA
XLVIII (1983) Molecular Neurobiology
XLIX (1984) Recombination at the DNA Level
L (1985) Molecular Biology of Development
LI (1986) Molecular Biology of Homo sapiens
LII (1987) Evolution of Catalytic Function
LIII (1988) Molecular Biology of Signal Transduction
LIV (1989) Immunological Recognition
LV (1990) The Brain
LVI (1991) The Cell Cycle
LVII (1992) The Cell Surface
LVIII (1993) DNA and Chromosomes
LVIX (1994) The Molecular Genetics of Cancer
LX (1995) Protein Kinesis: The Dynamics of Protein Trafficking and Stability
LXI (1996) Function & Dysfunction in the Nervous System
LXII (1997) Pattern Formation during Development
LXIII (1998) Mechanisms of Transcription
LXIV (1999) Signaling and Gene Expression in the Immune System
LXV (2000) Biological Responses to DNA Damage
LXVI (2001) The Ribosome
LXVII (2002) The Cardiovascular System
LXVIII (2003) The Genome of Homo sapiens
LXIX (2004) Epigenetics
LXX (2005) Molecular Approaches to Controlling Cancer
LXXI (2006) Regulatory RNAs
LXXII (2007) Clocks and Rhythms
LXXIII (2008) Control and Regulation of Stem Cells
LXXIV (2009) Evolution: The Molecular Landscape
LXXV (2010) Nuclear Organization and Function
LXXVI (2011) Metabolism and Disease
LXXVII (2012) The Biology of Plants
LXXVIII (2013) Immunity and Tolerance
LXXIX (2014) Cognition
LXXX (2015) 21st Century Genetics: Genes at Work

Front cover (*paperback*): Robert Delaunay, *Circular Forms* (*Formes circulaires*), 1930, oil on canvas, 50 3/4 × 76 3/4 inches (128.9 × 194.9 cm), Solomon R. Guggenheim Museum, New York, Solomon R. Guggenheim Founding Collection 49.1184.

For a complete catalog of all Cold Spring Harbor Laboratory Press publications, visit our website www.cshlpress.org.

Online access: Please visit our companion website at symposium.cshlp.org. For access issues, please contact Cold Spring Harbor Laboratory Press at subscriptions@cshl.edu.

Symposium Participants

ABBRUZZESE, GENEVIEVE, Massachusetts Institute of Technology, Cambridge, Massachusetts
AKRE, MONICA, University of Minnesota, Minneapolis, Minnesota
ALAGESAN, BRINDA, Cold Spring Harbor Laboratory, Cold Spring Harbor, New York
ALBANESE, CHRISTINA, Rensselaer Polytechnic Institute, Troy, New York
ALBRENGUES, JEAN, Cold Spring Harbor Laboratory, Cold Spring Harbor, New York
ALDERTON, GEMMA, *Nature Reviews Cancer,* London, United Kingdom
ALEXANDROVA, EVGUENIA, Stony Brook University, Stony Brook, New York
ALONSO CURBELO, DIRENA, Memorial Sloan Kettering Cancer Center, New York, New York
ALVANIA, REBECCA, Rockefeller University Press, New York, New York
ALZRIGAT, MOHAMMAD, Uppsala University, Uppsala, Sweden
AMBRICO, ALEXANDRA, Cold Spring Harbor Laboratory, Cold Spring Harbor, New York
AMELIO, ANTONIO, University of North Carolina at Chapel Hill, Chapel Hill, North Carolina
AMON, ANGELIKA, Massachusetts Institute of Technology, Cambridge, Massachusetts
ANCZUKOW, OLGA, Cold Spring Harbor Laboratory, Cold Spring Harbor, New York
ANTAL, CORINA, Salk Institute, La Jolla, California
ARANDA, VICTORIA, *Nature,* New York, New York
ARDITO-ABRAHAM, CHRISTINE, Lustgarten Foundation, Bethpage, New York
ARMSTRONG, SCOTT, Memorial Sloan Kettering Cancer Center, New York, New York
ARUN, GAYATRI, Cold Spring Harbor Laboratory, Cold Spring Harbor, New York
ATRETKHANY, KAMAR-SULU, Engelhardt Institute of Molecular Biology, Moscow, Russia
AYENI, DEBORAH, Yale University, New Haven, Connecticut
AZZOPARDI, STEPHANIE, Weill Cornell Medical College, New York, New York
BAKER, LINDSEY, Cold Spring Harbor Laboratory, Cold Spring Harbor, New York
BAKHOUM, SAMUEL, Memorial Sloan Kettering Cancer Center, New York, New York
BENEVOLENSKAYA, ELIZAVETA, University of Illinois at Chicago, Chicago
BERGERS, GABRIELE, University of California, San Francisco, San Francisco, California
BEVERLY, LEVI, University of Louisville, Louisville, Kentucky
BHOLA, PATRICK, Dana-Farber Cancer Institute, Boston, Massachusetts
BIFFI, GIULIA, Cold Spring Harbor Laboratory, Cold Spring Harbor, New York
BIKFALVI, ANDREAS, INSERM and University Bordeaux, Pessac, France
BISWAS, ROMI, National Institutes of Health, Bethesda, Maryland
BLAIN, STACY, State University of New York Downstate Medical Center, Brooklyn, New York
BLASCO, MARÍA, Spanish National Cancer Research Center, Madrid, Spain
BOSBACH, BENEDIKT, Memorial Sloan Kettering Cancer Center, New York, New York
BOSE, ROHIT, Memorial Sloan Kettering Cancer Center, New York, New York
BOYE, KEVIN, INSERM and Bordeaux University, Pessac, France
BRADY, COLLEEN, *Cancer Cell,* Cambridge, Massachusetts
BRISARD, DAPHNE, Northwestern University, Chicago, Illinois
BRUGGE, JOAN, Harvard Medical School, Boston, Massachusetts
BRUZAS, EMILIS, Cold Spring Harbor Laboratory, Cold Spring Harbor, New York
BUDMAN, DANIEL, Northwell Health, Lake Success, New York
BURGER, MEGAN, Massachusetts Institute of Technology, Cambridge, Massachusetts
BURGOS-BARRAGAN, GUILLERMO, University of Cambridge, Cambridge, United Kingdom
CALDAS, CARLOS, Cancer Research UK Cambridge Institute, Cambridge, United Kingdom
CAMPBELL, ROBERT, Brown University, Providence, Rhode Island
CANNER, DAVID, Massachusetts Institute of Technology, Cambridge, Massachusetts
CAO, ZHEN, Memorial Sloan Kettering Cancer Center, New York, New York
CARBINI, MARIANA, Italy
CARUGO, ALESSANDRO, University of Texas MD Anderson Cancer Center, Houston, Texas
CARVALHO, TIAGO, Rockefeller University, New York, New York
CASANOVA, IRENE, Cold Spring Harbor Laboratory, Cold Spring Harbor, New York
CASPER, KAREN, Washington Technology School, Roseville, Minnesota
CHA, ADRIEL, Stanford University, Stanford, California
CHAKRABORTY, ABHISHEK, Dana-Farber Cancer Institute, Boston, Massachusetts
CHALLA, SRIDEVI, H. Lee Moffitt Cancer Center, Tampa, Florida
CHANG, KUNG-CHI, Cold Spring Harbor Laboratory, Cold Spring Harbor, New York
CHEN, CHONG, Sichuan University, Chengdu, China
CHEN, HSIAO-CHI, National Yang-Ming University, Taipei, Taiwan
CHEN, LIANG, Duke University, Durham, North Carolina
CHEN, SIDI, Yale University, West Haven, Connecticut
CHEN, YUNCHING, National Tsing Hua University, Hsinchu, Taiwan
CHEON, DONG-JOO, Albany Medical College, Albany, New York
CHEUNG-ONG, KAHLIN, Onkaido Therapeutics, Cambridge, Massachusetts
CHI, MAOYEN, Cold Spring Harbor Laboratory, Cold Spring Harbor, New York
CHIANG, CHENG-MING, UT Southwestern Medical Center, Dallas, Texas
CHIO, CHRISTINE, Cold Spring Harbor Laboratory, Cold Spring Harbor, New York
CHOI, SOYOUNG, Weill Cornell Medical College, New York, New York
CHOWDHURY, A.M. MASUDUL AZAD, Doshisha University, Kyotanabe, Japan
CHUANG, CHEN-HUA, Stanford University, Mountain View, California
CHUNG, TAE MUN, Cold Spring Harbor Laboratory, Cold Spring Harbor, New York
CICHOWSKI, KAREN, Harvard Medical School/Brigham and Women's Hospital, Boston, Massachusetts
CLARK, SUSAN, Garvan Institute of Medical Research, Sydney, Australia
COHEN, PAULA, Cornell University, Ithaca, New York
COLAVITO, SIERRA, University of Wisconsin-La Crosse, La Crosse, Wisconsin
COLLADO, MANUEL, Health Research Institute of Santiago Idis, Santiago de Compostela, Spain
DA SILVA-ALVAREZ, SABELA, Health Research Institute of Santiago Idis, Santiago de Compostela, Spain
DAHLHOFF, MAIK, Ludwig Maximilian University of Munich, Munich, Germany
DANG, CHI VAN, University of Pennsylvania, Philadelphia, Pennsylvania
DAS, KAKOLI, Duke-NUS Medical School, Singapore, Singapore
DAS, SANJEEV, National Institute of Immunology, New Delhi, India
DAVIS, NICHOLAS, Duke University, Durham, North Carolina

Delfino, Tess, Genentech, South San Francisco, California
DeMare, Laura, Cold Spring Harbor Laboratory Press, Woodbury, New York
Deswal, Sumit, Research Institute of Molecular Pathology, Vienna, Austria
Detchokul, Sujitra, University of Melbourne, Heidelberg, Australia
Devoe, Craig, Northwell Health, Lake Success, New York
Diermeier, Sarah, Cold Spring Harbor Laboratory, Cold Spring Harbor, New York
Donen, Maury, Manitoba Civil Service Superannuation Board, Winnipeg, Manitoba, Canada
Dorniak, Piotr, University of Texas MD Anderson Cancer Center, Houston, Texas
Dow, Luke, Weill Cornell Medicine, New York, New York
Drake, Justin, Rutgers Cancer Institute of New Jersey, New Brunswick, New Jersey
Drosos, Yiannis, St. Jude Children's Research Hospital, Memphis, Tennessee
Du, Yi-Chieh Nancy, Weill Cornell Medical College, New York, New York
Dutta, Aditya, Columbia University Medical Center, New York, New York
Eberlein, Cath, Cancer Research UK Manchester Institute, Manchester, United Kingdom
Egeblad, Mikala, Cold Spring Harbor Laboratory, Cold Spring Harbor, New York
Elyada, Ela, Cold Spring Harbor Laboratory, Cold Spring Harbor, New York
Engelman, Jeffrey, Massachusetts General Hospital, Boston, Massachusetts
Engle, Dannielle, Cold Spring Harbor Laboratory, Cold Spring Harbor, New York
Enos, Miriam, Brigham and Women's Hospital, Boston, Massachusetts
Er, Ekrem Emrah, Memorial Sloan Kettering Cancer Center, New York, New York
Evan, Gerard, University of Cambridge, Cambridge, United Kingdom
Evans, Ronald, Salk Institute for Biological Studies, La Jolla, California
Fearon, Douglas, Cold Spring Harbor Laboratory, Cold Spring Harbor, New York
Feigin, Michael, Cold Spring Harbor Laboratory, Cold Spring Harbor, New York
Fein, Miriam, Cold Spring Harbor Laboratory, Cold Spring Harbor, New York
Fein Levy, Carolyn, Cohen Children's Medical Center, New Hyde Park, New York
Feinberg, Andrew, Johns Hopkins University School of Medicine, Baltimore, Maryland
Feldhahn, Niklas, Imperial College London, London, United Kingdom
Fernandez, Mario, H. Lee Moffitt Cancer Center, Tampa, Florida
Filant, Justyna, University of Texas MD Anderson Cancer Center, Houston, Texas
Fillmore, Christine, Boston Children's Hospital, Boston, Massachusetts
Fiorani, Simona, *Nature Communications,* London, United Kingdom
Fisher, Daniel, Centre National de la Recherche Scientifique, Montpellier, France
Fonseca, Cecilia, University of São Paulo, São Paulo, Brazil
Frauman, Albert, University of Melbourne, Heidelberg, Australia
Fujiwara, Hiroaki, University of Tokyo, Tokyo, Japan
Galli, Giorgio, Novartis Institute for Biomedical Research, Basel, Switzerland
Ganesan, Rajkumar, Boehringer Ingelheim, Ridgefield, Connecticut
Gartel, Andrei, University of Illinois College of Medicine, Chicago, Illinois
Genovese, Giannicola, University of Texas MD Anderson Cancer Center, Houston, Texas
Ghysdael, Jacques, Institute Curie, Orsay, France
Giannakou, Andreas, Pfizer Inc., Pearl River, New York
Gimotty, Phyllis, University of Pennsylvania, Philadelphia, Pennsylvania
Gorrini, Chiara, Princess Margaret Hospital, Toronto, Ontario, Canada
Gotterer, Jessica, Cold Spring Harbor Laboratory/Northwell Health, Cold Spring Harbor, New York
Gouronnec, Alizée, Cold Spring Harbor Laboratory, Cold Spring Harbor, New York
Grabole, Nils, F. Hoffmann-La Roche, Roche Innovation Center Basel, Basel, Switzerland
Gray, Veronica, ORISE Postbac Fellow at FDA, Silver Spring, Maryland
Grodzicker, Terri, Cold Spring Harbor Laboratory, Cold Spring Harbor, New York
Guha, Udayan, Center for Cancer Research/National Cancer Institute/National Institutes of Health, Bethesda, Maryland
Gunes, Cagatay, Ulm University, Ulm, Germany
Haber, Daniel, Massachusetts General Hospital, Charlestown, Massachusetts
Hahn, William, Dana-Farber Cancer Institute, Boston, Massachusetts
Hall, Matthew, National Institutes of Health Center for Advancing Translational Sciences, Rockville, Maryland
Hammell, Amy, Bristol-Myers Squibb, Princeton, New Jersey
Han, Teng, Weill Cornell Medicine, New York, New York
Han, Xiaoqing, University of Miami, Miami, Florida
Handly-Santana, Abram, Cold Spring Harbor Laboratory, Cold Spring Harbor, New York
Hanemann, Clemens, Plymouth University, Plymouth, United Kingdom
Hao, Yuan, Cold Spring Harbor Laboratory, Cold Spring Harbor, New York
Hart, Traver, University of Texas MD Anderson Cancer Center, Houston, Texas
Hastings, Katherine, Yale University, New Haven, Connecticut
Hatzi, Katerina, Memorial Sloan Kettering Cancer Institute, New York, New York
He, Bing, The Children's Hospital of Philadelphia, Philadelphia, Pennsylvania
Hebert, Jess, Massachusetts Institute of Technology, Cambridge, Massachusetts
Hedegger, Kathrin, Ludwig Maximilian University at Munich, Munich, Germany
Henderson, Christina, Moderna Therapeutics, Inc., Cambridge, Massachusetts
Herr, Winship, University of Lausanne, Lausanne, Switzerland
Hickson, Ian, The Panum Institute, Copenhagen, Denmark
Ho, Yu-Jui, Cold Spring Harbor Laboratory, Cold Spring Harbor, New York
Hodges, Maria, Genome Medicine, London, United Kingdom
Hoesl, Christine, Ludwig Maximilian University at Munich, Munich, Germany
Hogg, Simon, Peter MacCallum Cancer Centre, Melbourne
Hoh, Hong Huat, Okinawa Institute of Science and Technology, Onnason, Okinawa, Japan
Hopkins, Nancy, Massachusetts Institute of Technology, Cambridge, Massachusetts
Houlihan, Shauna, Sloan Kettering Institute, New York, New York
Hsu, Hsin-Ling, National Health Research Institutes, Zhunan Town, Miaoli, Taiwan
Hu, Bomiao, Yale University, New Haven, Connecticut
Hu, Wenwei, Rutgers University-Cancer Institute of New Jersey, New Brunswick, New Jersey
Hu, Yuxuan, The Children's Hospital of Philadelphia, Philadelphia, Pennsylvania
Huang, Chun-Hao, Memorial Sloan Kettering Cancer Center, New York, New York
Huang, Sidong, McGill University, Montreal, Quebec, Canada
Huang, Xinyan, University of New York, New York, New York
Hwang, Chang-Il, Cold Spring Harbor Laboratory, Cold Spring Harbor, New York
Iggo, Richard, University of Bordeaux, Bordeaux, France
Iyer, Swathi, University of Kansas Medical Center, Kansas City, Kansas
Jacks, Tyler, David H. Koch Institute for Integrative Cancer Research at MIT, Cambridge, Massachusetts
Jahchan, Nadine, ORIC Pharmaceuticals, South San Francisco, California
Jailkhani, Noor, Massachusetts Institute of Technology, Cambridge, Massachusetts
Jain, Saket, University of Alberta, Alberta, Edmonton, Canada
Jaiswal, Arushi, University of Toronto, Toronto, Ontario, Canada

JAKIMO, ALAN, Hofstra University, Hempstead, New York
JEN, JAYU, National Cheng Kung University, Tainan, Taiwan
JIANG, YING, University of Virginia, Charlottesville, Virginia
JIN, KE, University of Miami, Miami, Florida
JOHNSON, RON, National Cancer Institute, Bethesda, Maryland
JONKERS, JOS, Netherlands Cancer Institute, Amsterdam, The Netherlands
JOSHI, NIKHIL, Massachusetts Institute of Technology, Cambridge, Massachusetts
JUDE, JULIAN, Research Institute of Molecular Pathology (IMP), Vienna, Austria
JUNE, CARL, University of Pennsylvania, Philadelphia, Pennsylvania
KADOCH, CIGALL, Dana-Farber/Harvard Cancer Center, Boston, Massachusetts
KAELIN, WILLIAM, Howard Hughes Medical Institute/Dana-Farber Cancer Institute, Boston, Massachusetts
KALDIS, PHILIPP, Institute of Molecular and Cell Biology (IMCB), Singapore, Singapore
KALLURI, RAGHU, University of Texas MD Anderson Cancer Center, Houston, Texas
KANG, YIBIN, Princeton University, Princeton, New Jersey
KARAKASHEVA, BAGRYANA, Sabanci University, Istanbul, Turkey
KASTENHUBER, EDWARD, Memorial Sloan Kettering Cancer Center, New York, New York
KATO, JUN-YA, Nara Institute of Science and Technology, Nara, Japan
KATO, NORIKO, Nara Institute of Science and Technology, Nara, Japan
KHAN, SHOWKHIN, Cold Spring Harbor Laboratory, Cold Spring Harbor, New York
KIBERSTIS, PAULA, *Science* Magazine, Newbury, Massachusetts
KIM, EEJUNG, Dana-Farber Cancer Institute, Cambridge, Massachusetts
KIM, JAE-YOUNG, Moffitt Cancer Center, Tampa, Florida
KIM, JIHUN, Asan Medical Center, Seoul, South Korea
KIM, KYUNGTAE, National Cancer Center, Goyang, South Korea
KIM, YOUNG JIN, Cold Spring Harbor Laboratory, Cold Spring Harbor, New York
KIM, YOUNGJOO, SUNY College at Old Westbury, Old Westbury, New York
KOBAYASHI, AKIRA, Doshisha University, Kyotanabe, Japan
KODIGEPALLI, MADHAV KARTHIK, The Ohio State University, Columbus, Ohio
KOEGL, MANFRED, Boehringer Ingelheim RCV GmbH & Co KG, Vienna, Austria
KOLLET, JUTTA, Miltenyi Biotec GmbH, Bergisch Gladbach, Germany
KORIMERLA, NAVYATEJA, Cold Spring Harbor Laboratory, Cold Spring Harbor, New York
KRUCHER, NANCY, Pace University, Pleasantville, New York
KUNDU, SAMRAT, University of Texas MD Anderson Cancer Center, Houston, Texas
KUZMIN, ELENA, McGill University, Goodman Cancer Centre, Montreal, Quebec, Canada
KVAJO, MIRNA, *Cell,* Cambridge, Massachusetts
LAI, SHAO-CHIANG (MICHAEL), Eastern Virginia Medical School, Norfolk, Virginia
LAUGHNEY, ASHLEY, Memorial Sloan Kettering Cancer Center, New York, New York
LAVI, SARA, Tel Aviv University, Tel Aviv, Israel
LEBLEU, VALERIE, MD Anderson Cancer Center, Houston, Texas
LEE, DA-HYE, Korea Advanced Institute of Science & Technology (KAIST), Daejeon, South Korea
LEE, HO, National Cancer Center, Goyang-si, South Korea
LEE, JAE, Kyungpook National University School of Medicine, Daegu, South Korea
LEE, JE, Cold Spring Harbor Laboratory, Cold Spring Harbor, New York
LEE, MATT, Cold Spring Harbor Laboratory, Cold Spring Harbor, New York
LEIBMAN, NICOLE
LEIBOLD, JOSEF, Memorial Sloan Kettering Cancer Center, New York, New York
LETAI, ANTHONY, Dana-Farber Cancer Institute, Boston, Massachusetts
LI, AMY, Massachusetts Institute of Technology, Cambridge, Massachusetts
LI, ANDREW, Dana-Farber Cancer Institute, Boston, Massachusetts
LI, HUIPENG, Genome Institute of Singapore, Singapore, Singapore
LI, RONG, University of Texas Health Science Center at San Antonio, San Antonio, Texas
LI, WEI, Pennsylvania State University, Hershey College of Medicine, Hershey, Pennsylvania
LI, XIANG, Memorial Sloan Kettering Cancer Center, New York, New York
LI, YI, Baylor College of Medicine, Houston, Texas
LIAO, SHENG-YOU, National Cheng Kung University, Tainan, Taiwan
LIN, YI-JANG, Harvard University, Boston, Massachusetts
LIOT, CAROLINE, New York University Medical Center, New York, New York
LIU, HUI, Harvard Medical School/Beth Israel Deaconess Medical Center, Boston, Massachusetts
LIU, JUAN, Rutgers Cancer Institute of New Jersey, New Brunswick, New Jersey
LIU, YU, Sichuan University, Chengdu, China
LIU, ZHIMIN, Stony Brook University, Stony Brook, New York
LIVSHITS, GEULAH, Memorial Sloan Kettering Cancer Center, New York, New York
LOCKWOOD, WILLIAM, British Columbia Cancer Research Centre, Vancouver, British Columbia, Canada
LOIZOU, EVANGELIA, Memorial Sloan Kettering Cancer Center, New York, New York
LOPES, EDUARDO, Universidade de São Paulo, São Paulo, Brazil
LOVE, CASSANDRA, Duke University, Durham, North Carolina
LOWE, SCOTT, Memorial Sloan Kettering Cancer Center, New York, New York
LOZANO, GUILLERMINA (GIGI), University of Texas MD Anderson Cancer Center, Houston, Texas
LU, RICHARD, Cincinnati Children's Hospital Medical Center, Cincinnati, Ohio
LU, TSAILING, University Health Network (UHN), Toronto, Ontario, Canada
LUGA, VALBONA, Weill Cornell Medicine, New York, New York
LUJAMBIO, AMAIA, Icahn School of Medicine at Mount Sinai, New York, New York
LUO, WEIBO, UT Southwestern Medical Center, Dallas, Texas
LYDEN, DAVID, Weill Cornell Medical College of Cornell University, New York, New York
LYONS, SCOTT, Cold Spring Harbor Laboratory, Cold Spring Harbor, New York
MAIA-SILVA, DIOGO, Faculdade de Medicina Universidade de Lisboa, Lisbon, Portugal
MAIORINO, LAURA, Cold Spring Harbor Laboratory, Cold Spring Harbor, New York
MAITY, TAPAN, National Institutes of Health, Bethesda, Maryland
MAK, TAK, University of Toronto, Toronto, Ontario, Canada
MAKI, ROBERT, Northwell Health, Lake Success, New York
MANCHADO, EUSEBIO, Novartis, Basel, Switzerland
MANSUKHANI, ALKA, New York University School of Medicine, New York, New York
MARCH, NIKKI, Cancer Research UK Manchester Institute, Manchester, United Kingdom
MARCHENKO, NATALIA, Stony Brook University, Stony Brook, New York
MARDIS, ELAINE, Nationwide Children's Hospital Research Institute, Columbus, Ohio
MAROTO, MIGUEL, Adaptimmune, Abingdon, United Kingdom
MARTINEZ, LUIS, Stony Brook University, Stony Brook, New York
MASCARENO, MANYA, State University of New York, College at Old Westbury, Old Westbury, New York
MATHEW, GRINU, Cold Spring Harbor Laboratory, Cold Spring Harbor, New York
MAURER, CARLO, Columbia University, New York, New York
MAYER, ANDREAS, Harvard Medical School, Boston, Massachusetts
MAYLE, ALLISON, Memorial Sloan Kettering Cancer Center, New York, New York
MCKENNA, ELIZABETH, *Cancer Discovery,* Philadelphia, Pennsylvania
MCKINNEY, KRISTINE, Moderna Therapeutics, Cambridge, Massachusetts
MCMAHON, MARTIN, University of Utah, Huntsman Cancer Institute, Salt Lake City, Utah
MELNICK, ARI, Weill Cornell Medical College, New York, New York
MENSSEN, ANTJE, Ludwig Maximilian University at Munich, Munich, Germany

MERTENS, CLAUDIA, The Rockefeller University, New York, New York
MIETHING, CORNELIUS, Universitätsklinikum Freiburg, Freiburg, Germany
MIIKKULAINEN, PETRA, University of Turku, Turku, Finland
MILLS, ALEA, Cold Spring Harbor Laboratory, Cold Spring Harbor, New York
MIN, JIMIN, Seoul National University, Seoul, South Korea
MOCK, BEVERLY, National Cancer Institute, National Institutes of Health, Bethesda, Maryland
MOLIK, DAVID, Cold Spring Harbor Laboratory, Cold Spring Harbor, New York
MOLINA, JENNIFER, MD Anderson Cancer Center, Houston, Texas
MOLL, UTE, Stony Brook University, Stony Brook, New York
MONSERRAT-SANCHEZ, JOSEP, The Francis Crick Institute, London, United Kingdom
MOON, HYEONG-GON, Seoul National University College of Medicine, Seoul, South Korea
MORRIS IV, JOHN, Memorial Sloan Kettering Cancer Center, New York, New York
MORRISON, SEAN, UT Southwestern Medical Center, Dallas, Texas
MOU, HAIWEI, University of Massachusetts Medical School, Worcester, Massachusetts
MU, DAVID, Eastern Virginia Medical School, Norfolk, Virginia
MUHAR, MATTHIAS, Research Institute for Molecular Pathology, Wien, Austria
MULEY, ASHLESHA, Weill Cornell Medicine, New York, New York
MURANEN, TARU, Beth Israel Deaconess Medical Center, Boston, Massachusetts
MUZUMDAR, MANDAR, Koch Institute at Massachusetts Institute of Technology, Cambridge, Massachusetts
NACIRI, IKRAME, University Paris Diderot/CNRS, Paris, France
NAKANISHI, SHIMA, H. Lee Moffitt Cancer Center, Tampa, Florida
NAKATSUKA, TAKUMA, The University of Tokyo, Bunkyo-ku, Tokyo, Japan
NAZ, SARWAT, National Cancer Institute, National Institutes of Health, Bethesda, Maryland
NEDOSPASOV, SERGEI, Engelhardt Institute of Molecular Biology, Moscow, Russia
NEEL, BENJAMIN, New York University School of Medicine, New York, New York
NELSON, SARAH, University of Colorado Denver, Aurora, Colorado
NEMAJEROVA, ALICE, Stony Brook University, Stony Brook, New York
NG, SHENG RONG, Koch Institute at Massachusetts Institute of Technology, Cambridge, Massachusetts
NGO, BRYAN, Weill Cornell Medical College, New York City, New York
NOLAN, EMMA, Walter and Eliza Hall Institute, Melbourne, Australia
NOWAK, DAWID, Cold Spring Harbor Laboratory, Cold Spring Harbor, New York
OLIVE, KENNETH, Columbia University, New York, New York
OLSEN, SARAH NAOMI, Brigham and Women's Hospital, Boston, Massachusetts
ONI, TOBILOBA, Cold Spring Harbor Laboratory, Cold Spring Harbor, New York
O'ROURKE, KEVIN, Memorial Sloan Kettering Cancer Center, New York, New York
OSER, MATTHEW, Dana-Farber Cancer Institute, Boston, Massachusetts
OUDIN, MADELEINE, Massachusetts Institute of Technology, Cambridge, Massachusetts
OZDUMAN, KORAY, Acibadem University, School of Medicine, Istanbul, Turkey
PADDOCK, MARCIA, Weill Cornell Medicine/New York-Presbyterian, New York, New York
PAL, DEBJANI, Cold Spring Harbor Laboratory, Cold Spring Harbor, New York
PAPAGIANNAKOPOULOS, THALES, New York University Medical School, New York, New York
PARADA, LUIS, Memorial Sloan Kettering Cancer Center, New York, New York
PARDEE, TIMOTHY, Wake Forest University School of Medicine, Winston-Salem, North Carolina
PARK, WOO-YONG, Korea Advanced Institute of Science and Technology, Daejeon, South Korea
PARK, YOUNGKYU, Cold Spring Harbor Laboratory, Cold Spring Harbor, New York
PARRILLA-MONGE, LAURA, Stony Brook University, Stony Brook, New York
PASHINE, ACHAL, Bristol-Myers Squibb, Lawrenceville, New Jersey
PAUL, DORU, Northwell Health, Lake Success, New York
PENG, DAVID, MD Anderson Cancer Center, Houston, Texas
PHELPS, CODY, Eastern Virginia Medical School, Norfolk, Virginia
PISKOL, ROBERT, Genentech, South San Francisco, California
PISKOUNOVA, ELENA, UT Southwestern, Dallas, Texas
PISTERZI, PAOLA, Cold Spring Harbor Laboratory, Cold Spring Harbor, New York
POLITI, KATERINA, Yale University, New Haven, Connecticut
POLLOCK, MILA, Cold Spring Harbor Laboratory, Cold Spring Harbor, New York
POLYAK, KORNELIA, Dana-Farber Cancer Institute, Boston, Massachusetts
POLYANSKAYA, SOFYA, Cold Spring Harbor Laboratory, Cold Spring Harbor, New York
POMMIER, ARNAUD, Cold Spring Harbor Laboratory, Cold Spring Harbor, New York
POWERS, SCOTT, Cold Spring Harbor Laboratory, Woodbury, New York
PREMSRIRUT, PREM, Mirimus, Inc., Woodbury, New York
PRUMMER, MICHAEL, ETH Zürich, Zürich, Switzerland
PULVIRENTI, TEODORO, *Nature Cell Biology*, New York, New York
QIAO, SHUXI, Mass General Hospital and Harvard Medical School, Boston, Massachusetts
RAGOUSSIS, IOANNIS, McGill University, Montreal, Quebec, Canada
RAJU, PRAVEEN, Weill Cornell Medical College, New York, New York
RAMSEY, MATTHEW, Brigham and Women's Hospital, Harvard Medical School, Boston, Massachusetts
RANJAN, ATUL, University of Kansas Medical Center, Kansas City, Kansas
RANTANEN, KRISTA, University of Turku, Turku, Finland
RAO, ANGAD, National University of Singapore, Singapore, Singapore
RAO, MANISHA, Stony Brook University, Stony Brook, New York
RAPTIS, GEORGE, Northwell Health, Lake Success, New York
RAVICHANDRAN, PRIYADARSHINI, Cold Spring Harbor Laboratory, Cold Spring Harbor, New York
REMPEL, RACHEL, Duke University Medical Center, Durham, North Carolina
RENNHACK, JONATHAN, Michigan State University, East Lansing, Michigan
RITHO, JOAN, MD Anderson Cancer Center, Houston, Texas
ROE, JAE SEOK, Cold Spring Harbor Laboratory, Cold Spring Harbor, New York
ROITMAN, LIOR, Weizmann Institute, Rechovot, Israel
ROMESSER, PAUL, Memorial Sloan Kettering Cancer Center, New York, New York
ROSE, JOHNATHON, MD Anderson Cancer Center, Houston, Texas
ROSEN, MONICA, Weill Cornell Medicine, New York, New York
ROSEN, NEAL, Memorial Sloan Kettering Cancer Center, New York, New York
ROUNBEHLER, ROBERT, H. Lee Moffitt Cancer Center, Tampa, Florida
RUSCETTI, MARCUS, Memorial Sloan Kettering Cancer Center, New York, New York
RYAN, MOLLY, Yale University, New Haven, Connecticut
SACCHI, NICOLETTA, Roswell Park Cancer Institute, Buffalo, New York
SACHAN, NALANI, Cold Spring Harbor Laboratory, Cold Spring Harbor, New York
SADEK, JOULIANA, Weill Cornell Medicine, New York, New York
SADELAIN, MICHEL, Memorial Sloan Kettering Cancer Center, New York, New York
SAFFRAN, DOUG, Warp Drive Bio, Inc., Cambridge, Massachusetts
SAID, SUZANNE, Agensys Inc., Santa Monica, California
SALAMI, FARIMAH, St. Jude Children's Research Hospital, Memphis, Tennessee
SALOTTI, JACQUELINE, National Cancer Institute, National Institutes of Health, Frederick, Maryland
SANCHEZ-RIVERA, FRANCISCO, Memorial Sloan Kettering Cancer Center, New York, New York
SAROSIEK, KRISTOPHER, Dana-Farber Cancer Institute, Boston, Massachusetts
SASAKI, TOMOAKI, Yale University, New Haven, Connecticut

SAWYERS, CHARLES, Memorial Sloan Kettering Cancer Center, New York, New York
SCAGLIONI, PIER, UT Southwestern Medical Center, Dallas, Texas
SCARBOROUGH, HANNAH, University of Colorado Denver, Denver, Colorado
SCHMIDT, LEAH, Koch Institute at Massachusetts Institute of Technology, Cambridge, Massachusetts
SCHOENFELD, ALAN, Adelphi University, Garden City, New York
SCOTT, ANDREW, North Dakota State University, Fargo, North Dakota
SEMENOVA, EKATERINA, The Netherlands Cancer Institute (NKI), Amsterdam, The Netherlands
SERRANO, MANUEL, Spanish National Cancer Center (CNIO), Madrid, Spain
SEVER, RICHARD, Cold Spring Harbor Laboratory Press, Woodbury, New York
SHAND, TIFFANY, University of Virginia, Charlottesville, Virginia
SHAW, REUBEN, Salk Institute for Biological Studies, La Jolla, California
SHELTZER, JASON, Cold Spring Harbor Laboratory, Cold Spring Harbor, New York
SHEN, ROGER, Academia Sinica, Taipei, Taiwan
SHI, JUNWEI, Cold Spring Harbor Laboratory, Cold Spring Harbor, New York
SHI, QING, Baylor College of Medicine, Houston, Texas
SHIELDS, MARIO, Cold Spring Harbor Laboratory, Cold Spring Harbor, New York
SHIROLE, NITIN, Cold Spring Harbor Laboratory, Cold Spring Harbor, Alabama
SHU, SHAOKUN, Dana-Farber Cancer Institute, Boston, Massachusetts
SIERRA GONZALEZ, PABLO, Beatson Institute for Cancer Research, Glasgow, United Kingdom
SIMON, M. CELESTE, University of Pennsylvania Medical School, Philadelphia, Pennsylvania
SIMÓN CARRASCO, LUCÍA, Spanish National Cancer Research Center, Madrid, Spain
SINDHWANI, SHREY, University of Toronto, Toronto, Ontario, Canada
SINGH, KAMINI, Memorial Sloan Kettering Cancer Center, New York, New York
SINGH, MALLIKA, ORIC Pharmaceuticals, South San Francisco, California
SMITH, MATTHEW, H. Lee Moffitt Cancer Center, Tampa, Florida
SOLIT, DAVID, Memorial Sloan Kettering Cancer Center, New York, New York
SOLOMON, LARRY, AbbVie, Inc, North Chicago, Illinois
SONG, DONGYAN, Cold Spring Harbor Laboratory, Cold Spring Harbor, New York
SONG, JI-YING, The Netherlands Cancer Institute (NKI), Amsterdam, The Netherlands
SORDELLA, RAFFAELLA, Cold Spring Harbor Laboratory, Cold Spring Harbor, New York
SPECTOR, DAVID, Cold Spring Harbor Laboratory, Cold Spring Harbor, New York
SRIDEVI, PRIYA, Cold Spring Harbor Laboratory, Cold Spring Harbor, New York
STAHLHUT, CARLOS, Cold Spring Harbor Laboratory, Cold Spring Harbor, New York
STEINBERG, BETTIE, The Feinstein Institute, Northwell Health, Manhasset, New York
STEWART, DAVID, Cold Spring Harbor Laboratory, Cold Spring Harbor, New York
STILLMAN, BRUCE, Cold Spring Harbor Laboratory, Cold Spring Harbor, New York
STOMMEL, JAYNE, National Cancer Institute, Bethesda, Maryland
STRASSER, SAMANTHA DALE, Massachusetts Institute of Technology, Cambridge, Massachusetts
STRATTON, MICHAEL, The Wellcome Trust Sanger Institute, Hinxton, Cambridge, United Kingdom
STROHECKER, ANNE, The Ohio State University, Columbus, Ohio
SÜLTMANN, HOLGER, German Cancer Research Center/National Center for Tumor Diseases, Heidelberg, Germany
SUNDARESAN, VARSHA, University of Florida, Gainesville, Florida
SUSSMAN, HILLARY, Cold Spring Harbor Laboratory Press, Woodbury, New York
SYMONS, MARC, Feinstein Institute, Manhasset, New York
TALOS, FLAMINIA, Stony Brook University, Stony Brook, New York
TAMMELA, TUOMAS, Massachusetts Institute of Technology, Cambridge, Massachusetts
TAN, KAI, University of Pennsylvania, Philadelphia, Pennsylvania
TANAKA, YASUO, The University of Tokyo, Tokyo, Japan
TANG, HAIDONG, UT Southwestern Medical Center, Dallas, Texas
TANG, XUMING, The University of Hong Kong, Hong Kong, Hong Kong
TATEISHI, KEISUKE, The University of Tokyo, Tokyo, Japan
TEESALU, TAMBET, University of Tartu, Tartu, Estonia
TELERMAN, ADAM, Institut Gustave Roussy, Villejuif, France
THOMPSON, CRAIG, Memorial Sloan Kettering Cancer Center, New York, New York
THOMSON, TIMOTHY, CSIC, Barcelona, Spain
TOMASZEWSKI, MONICA, Thermofisher, Pittsburgh, Pennsylvania
TONELLI, CLAUDIA, Cold Spring Harbor Laboratory, Cold Spring Harbor, New York
TRAN, NGOC, Cold Spring Harbor Laboratory, Cold Spring Harbor, New York
TRAN QUANG, CHRISTINE, Institut Curie, Orsay, France
TROTMAN, LLOYD, Cold Spring Harbor Laboratory, Cold Spring Harbor, New York
TRUINI, ANNA, Yale University, New Haven, Connecticut
TRUITT, MORGAN, The Salk Institute for Biological Studies, La Jolla, California
TUVESON, DAVID, Cold Spring Harbor Laboratory, Cold Spring Harbor, New York
TZENG, HONG-TAI, National Cheng Kung University, Tainan City, Taiwan
UNNI, ARUN, Weill Cornell Medicine, New York, New York
VAKILI, JALAL, Servier Laboratories, Suresnes Cedex, France
VAKOC, CHRISTOPHER, Cold Spring Harbor Laboratory, Cold Spring Harbor, New York
VANNER, ROBERT, Princess Margaret Cancer Centre, Toronto, Ontario, Canada
VARMUS, HAROLD, Weill Cornell Medical College, New York, New York
VENUGOPALAN, ABHILASH, National Cancer Institute, Bethesda, Maryland
VERZI, MICHAEL, Rutgers, The State University of New Jersey, Piscataway, New Jersey
VIJAY, VINDHYA, University of Florida, Gainesville, Florida
VOUSDEN, KAREN, Cancer Research UK Beatson Institute, Glasgow, United Kingdom
WAGENBLAST, ELVIN, University Health Network (UHN), Toronto, Ontario, Canada
WANG, HONG, St. Jude Children's Research Hospital, Memphis, Tennessee
WANG, JIN, Stony Brook University, South Setauket, New York
WANG, JINHUA, New York University Medical School, New York, New York
WANG, LU-HAI, National Health Research Institutes, Miaoli County, Taiwan
WANG, WEN-CHI, National Yang-Ming University, Taipei, Taiwan
WANG, YA TING, Academia Sinica, Taipei, Taiwan
WANG, YANG, Sichuan University, Chengdu, China
WATRUD, KAITLIN, Cold Spring Harbor Laboratory, Cold Spring Harbor, New York
WEAVER, VALERIE, University of California, San Francisco, San Francisco, California
WEINBERG, ROBERT, Whitehead Institute for Biomedical Research, Cambridge, Massachusetts
WEISBROD, STU, Iguana Healthcare Partners, LLC, New York, New York
WEISS, JOSH, Weill Cornell, New York, New York
WEISSMAN, IRVING, Stanford University, Stanford, California
WESTCOTT, PETER, Massachusetts Institute of Technology, Cambridge, Massachusetts
WHITE, EILEEN, The Rutgers Cancer Institute of New Jersey, New Brunswick, New Jersey
WIGLER, MICHAEL, Cold Spring Harbor Laboratory, Cold Spring Harbor, New York
WILKINSON, JOHN, North Dakota State University, Fargo, North Dakota
WILKINSON, JOHN, University of Michigan, Ann Arbor, Michigan
WITKOWSKI, JAN, Cold Spring Harbor Laboratory, Cold Spring Harbor, New York

Wu, Gang, CAS-MPG Partner Institute for Computational Biology, Shanghai, China
Wu, Matthew, Stony Brook University, Stony Brook, New York
Xu, Dazhong, New York Medical College, Valhalla, New York
Xu, Longyong, Baylor College of Medicine, Houston, Texas
Xue, Wen, University of Massachusetts Medical School, Cambridge, Massachusetts
Yan, Chunhua, National Cancer Institute, Rockville, Maryland
Yan, Yan, Research Program Units, Helsinki, Finland
Yang, Dian, Stanford University, Stanford, California
Yang, Guang, Jounce Therapeutics, Cambridge, Massachusetts
Yang, Li, National Cancer Institute, National Institutes of Health, Bethesda, Maryland
Yang, T. Jonathan, Memorial Sloan Kettering Cancer Center, New York, New York
Yao, Maojin, University of Virginia, Charlottesville, Virginia
Yao, Min, University of Kansas Medical Center, Kansas City, Kansas
Ye, Xin, Whitehead Institute for Biomedical Research, Cambridge, Massachusetts
Yeh, Edward, MD Anderson Cancer Center, Houston, Texas
Yeo, Syn, University of Cincinnati, Cincinnati, Ohio
Yordanov, Georgi, Cold Spring Harbor Laboratory, Cold Spring Harbor, New York
Yu, Allen, Cold Spring Harbor Laboratory, Cold Spring Harbor, New York
Yu, Xiaofei, The Rockefeller University, New York, New York
Yuen, Kobe, Stowers Institute for Medical Research, Kansas City, Missouri
Zender, Lars, University of Tübingen, Tübingen, Germany
Zhang, Hongbing, Chinese Academy of Medical Sciences, Beijing, China
Zhang, Hongxiang, Third Rock Ventures, Boston, Massachusetts
Zhang, Ping, University of Oxford, Oxford, United Kingdom
Zhang, Siwei, Cold Spring Harbor Laboratory, Cold Spring Harbor, New York
Zhang, Xu, Zhejiang University, Hangzhou, China
Zhao, Na, Baylor College of Medicine, Houston, Texas
Zhao, Shaying, University of Georgia, Athens, Georgia
Zhao, Xiaoyu, Stony Brook University, Stony Brook, New York
Zhao, Zhen, Memorial Sloan Kettering Cancer Center, New York, New York
Zheng, Yuxiang, Weill Cornell Medical College, New York, New York

Row 1: P. Raju, J. Weiss; J. Sheltzer, C. Gunes; S. Nedospasov, L. Pollock
Row 2: H. Varmus, J. Witkowski; W. Herr, S. Weisbrod
Row 3: G. Burgos-Barragan, S. Said, J. Sadek; R. Alvania; V. Sundaresan, V. Vijay
Row 4: U. Pedmale, D. Stewart, D. Spector, M. Spector, P. Sherwood

Row 1: P. Westcott, M. Truitt; M. Koegl, D. Fisher; D. Tuveson, C. Caldas
Row 2: B. Steinberg, M. Symons; J. Jonkers, L. Zender; G. Evan, M. McMahon
Row 3: P. Kaldis, E. White; M. Muhar, B. Bosbach; A. Menssen, M. Koegl
Row 4: R. Evans; C. Caldas, M. Stratton; B. Neel, C. June
Row 5: J. Witkowski, W. Kaelin; D. Lyden, B. Neel; J. Brugge, R. Sever

Row 1: P. Scaglioni, G. Genovese; S. Lowe; M. Kvajo, S. Morrison
Row 2: D. Saffran, C. Fonseca; S. Challa, S. Nelson
Row 3: J. Kollet, R. Johnson; K. Vousden, G. Lozano; P. Raju, L. Beverly
Row 4: Lowe Lab

Row 1: L. Parada; C. Thompson; V. Weaver
Row 2: A.M. Laughney, N. Grabole; C. Vakoc, P. Kiberstis
Row 3: C. Sawyers, S. Armstrong; Y. Tanaka; M. Stratton, R. Weinberg
Row 4: R. Alvania, M. Kvajo; H. Varmus, S. Nedospasov, I. Weissman; C. Hanemann, E. Benevolonskaya

Row 1: P. Miikkulainen, K. Rantanen, A. Menssen; L. Maiorino, E. Bruzas
Row 2: M.C. Simon, C.V. Dang; E. Kuzmin, T. Hunt; T. Papagiannakopoulos, K. Politi
Row 3: S. Clark, M. Kvajo; U. Moll, G. Lozano, R. Iggo
Row 4: N. Kato, J.-y. Kato; P. Kaldis, L. Simón Carrasco, C. Fearon; A. Letai

Foreword

Cancer is a deadly disease that will afflict one out of every two to three people in the developed world in their lifetime. Recent years have produced transformative discoveries revealing the mechanisms behind the development and progression of cancer and the implementation of targeted cancer therapies in the clinic based on these discoveries. We have defined the mutational landscape of many cancers and with that the incipient driver mutations. Sophisticated and elegant models now exist for many types of cancer, and we understand in great detail the changes in signaling and gene expression that promote cancer and have begun to unravel the antitumor immune response. Cell death and survival mechanisms used by cancer cells have been identified and therapeutically exploited, and we have revealed processes that dictate cancer cell growth, repair, and proliferation and have the means to inhibit these processes. We have also seen advances in the mechanistic understanding, implementation, and deployment of cytotoxic cancer therapy. Collectively these advances are improving both the quality of life and overall survival of cancer patients.

The decision to focus the 81st Cold Spring Harbor Laboratory Symposium on Targeting Cancer reflected the enormous research progress achieved in recent years and provided a broad synthesis of the current state of the field, setting the stage for future discoveries and applications. Implications of how the underlying science can drive improvements in diagnostic, prognostic, and therapeutic approaches were a major theme throughout the Symposium. Previous Symposia that have included significant aspects of cancer research have occurred on a roughly five- to 10-year cycle, notably but not limited to Genes and Mutations (1951); Cellular Regulatory Mechanisms (1961); Transcription of Genetic Material (1970); Tumor Viruses (1974); Viral Oncogenes (1979); The Cell Cycle (1991); The Molecular Genetics of Cancer (1994); Biological Responses to DNA Damage (2000); and Molecular Approaches to Controlling Cancer (2005).

Major themes and topics highlighted at the 2016 Symposium included Cancer Genes and Genomes (genome stability, chromatin, epigenetic regulation); Cancer Pathways (signaling pathways, networks); Tumor Cell Biology (cells of origin, metabolism, autophagy, senescence); Cancer Growth & Progression (microenvironment, stroma/niche, tumor evolution, metastasis); Innate & Adaptive Immune Responses (inflammation, immunotherapies); and Enabling Technologies (single-cell sequencing, imaging, genetic screens, genome editing, organoids). Opening night speakers included Elaine Mardis (Nationwide Children's Hospital Research Institute) on translating cancer genomics into therapeutic target identification and vaccine design, Craig Thompson (Memorial Sloan Kettering Cancer Center), who spoke on the emerging treatment paradigm exploiting cancer metabolism, Carl June (University of Pennsylvania) on the use of genetic editing to generate synthetic lethal T cells, and Tyler Jacks (David H. Koch Institute for Integrative Cancer Research at MIT), who addressed engineering the cancer genome. Charles Sawyers (HHMI/Memorial Sloan Kettering Cancer Center) delivered an outstanding Dorcas Cummings Lecture on "Reflections on Precision Medicine and Cancer Moonshots" for Laboratory friends, neighbors, and Symposium participants prior to the annual Symposium dinners. Joan Brugge (Harvard Medical School) provided a masterful Summary at the conclusion of the Symposium immediately prior to the final banquet.

This Symposium was attended by almost 490 scientists from universities around the country and internationally, and the program included 55 invited presentations and more than 200 poster presentations. To disseminate the latest results and discussion of the Symposium to a wider audience, attendees were able to share many of the Symposium talks with their colleagues who were unable to attend using the Leading Strand video archive. A collection of interviews by Gemma Alderton (*Nature Reviews Cancer),* Paula Kiberstis (*Science*), Mirna Kvajo (Cell Press), Elizabeth McKenna (*Cancer Discovery*/ AACR), Richard Sever (CSHL Press), and Jan Witkowski (CSHL Banbury Center) with leading experts in the field were arranged during the Symposium and distributed as free video from the Cold Spring Harbor Symposium interviews website.

We thank Val Pakaluk, Mary Smith, Ed Campodonico, and his staff in the Meetings & Courses Program for their assistance in organizing and running the Symposium, and John Inglis and his staff at Cold Spring Harbor Laboratory Press, particularly Inez Sialiano, Jan Argentine, Kathleen Bubbeo, and Denise Weiss. Major support was provided by the CSHL-Northwell Health Partnership, with additional support provided by Genomic Health and Fluidigm. Financial support from the corporate sponsors of our meetings program is essential for these Symposia to remain a success, and we are most grateful for their support.

Symposium Organizers
Scott Lowe, Memorial Sloan Kettering Cancer Center
Kornelia Polyak, Dana-Farber Cancer Institute
David Stewart, Cold Spring Harbor Laboratory
Bruce Stillman, Cold Spring Harbor Laboratory
Eileen White, The Rutgers Cancer Institute of New Jersey

Editors
David Stewart, Cold Spring Harbor Laboratory
Bruce Stillman, Cold Spring Harbor Laboratory

Sponsors

Contributions from the following companies provide core support for the Cold Spring Harbor meetings program.

Corporate Benefactor

Regeneron

Corporate Sponsors

Agilent Technologies
Bristol-Myers Squibb Company
Calico Labs
Genentech, Inc.
Life Technologies (part of Thermo Fisher Scientific)
Merck
Monsanto Company
New England Biolabs
Pfizer

Corporate Affiliate

Ionis Pharmaceuticals

Contents

Normal and Neoplastic Stem Cells

MELISSA N. MCCRACKEN,[1,2,3,5] BENSON M. GEORGE,[1,2,3,5] KEVIN S. KAO,[1,2] KRISTOPHER D. MARJON,[1,2,3] TAL RAVEH,[1] AND IRVING L. WEISSMAN[1,2,3,4]

[1]*Institute for Stem Cell Biology and Regenerative Medicine, Stanford University School of Medicine, Stanford, California 94305*

[2]*Ludwig Center for Cancer Stem Cell Research and Medicine, Stanford University School of Medicine, Stanford, California 94305*

[3]*Stanford Cancer Institute, Stanford University School of Medicine, Stanford, California 94305*

[4]*Department of Pathology, Stanford University Medical Center, Stanford, California 94305*

Correspondence: irv@stanford.edu

A stem cell is broadly defined as a cell that retains the capacity to self-renew, a feature that confers the ability to continuously make identical daughter cells or additional cells that will differentiate into downstream progeny. This highly regulated genetic program to retain "stemness" is under active investigation. Research in our laboratory has explored similarities and differences in embryonic, tissue-specific, and neoplastic stem cells and their terminally differentiated counterparts. In this review, we will focus on the contributions of our laboratory, in particular on the studies that identified the mouse hematopoietic stem cell (HSC) and the human leukemic stem cell. These studies have led to significant improvements in both preclinical and clinical research, including improved clinical bone marrow transplantation protocols, isolation of nonleukemic HSCs, a cancer immunotherapy currently in clinical trials, and development of a HSC reporter mouse. These studies and the current follow-up research by us and others will continue to identify the properties, function, and regulation of both normal and neoplastic stem cells.

Over the last 40 years our research has focused on understanding normal and neoplastic stem cells. This review will cover the fundamental discoveries in defining similarities and differences between malignant cells and their healthy counterparts to further understand how to treat and cure hematologic malignancies.

The defining trait of stem cells, unlike most terminally differentiated cells within the body, is their ability to self-renew. Throughout an animal's life span these tissue-specific stem cells divide and self-renew, whereas daughter cells replace and replenish dead and/or dying terminally differentiated cells. One of the earliest demonstrations of this phenomenon was in bone marrow transplantation (BMT), when animals that were lethally irradiated were rescued by transplantation of healthy nonirradiated bone marrow (Lorenz et al. 1951; Ford et al. 1956). Early studies in our laboratory expounded on these findings with experiments focused on the hematopoietic system in mice. We worked to identify key surface markers that would allow for the isolation and transplantation of pure hematopoietic stem cells (HSCs). Studies demonstrated that only purified HSCs are capable of long-term self-renewal and complete replacement and replenishment of all blood subsets in normal homeostasis and after transplantation. By isolating the long-term HSC, we provided a method for how to replace the hematopoietic compartment after chemotherapy, radiation, or antibody-based bone marrow depletion without complications that are observed in whole bone marrow transplantation (Czechowicz and Weissman 2011; Muller et al. 2012; Chhabra et al. 2016). As discussed in further detail below, transplantation of pure HSCs removes potentially deleterious contaminating cells such as hematological and nonhematological malignant cells or alloreactive T cells from donors.

In cancer, the ability to self-renew is unlocked by a fraction of malignant cells, which are then able to self-renew and maintain a stem-like population, similar to HSCs. These self-renewing malignant cells are termed cancer stem cells (CSCs) or leukemic stem cells (LSCs). Initial studies for identifying a LSC began with our work on chronic myelogenous leukemia (CML) and acute myeloid leukemia (AML) (Miyamoto et al. 2000; Jamieson et al. 2004a,b; Abrahamsson et al. 2009; Majeti et al. 2009a). We studied primary patient samples to determine key differences between healthy HSCs and their progeny compared with LSCs and blast cells. Several key findings included the identification of LSC markers, identification of how malignant mutations are acquired and retained within the LSC pool, and the identification of CD47 as a "do not eat me" cell surface marker used by all malignancies. In AML, up-regulation of CD47 was observed on almost all cells from a large portion of primary patient samples (Majeti et al. 2009b). Later studies

[5]These authors contributed equally to this work.

Published by Cold Spring Harbor Laboratory Press; doi: 10.1101/sqb.2016.81.030965

found that solid tumors also increase their CD47 expression when compared with normal counterpart cells (Willingham et al. 2012). Through in vitro and in vivo studies we determined that the antibody blockade of CD47 increased phagocytosis and aided in the removal of cancerous cells.

In this review, we will cover seminal discoveries made by our laboratory and others on identifying the HSC, how this led to the discovery of the LSC, and how understanding similarities and differences between these two populations can identify new therapeutic targets.

ELUCIDATING THE HEMATOPOIETIC HIERARCHY

The HSC is the most well-characterized stem cell in both mice and humans. Studying them has provided a wealth of information in the fields of developmental biology and stem cell biology. Additionally, HSCs have been by far the most clinically translated multipotent cell type, with the first successful bone marrow transplant being reported in 1959 (Thomas et al. 1959). Today, HSCs remain at the forefront of regenerative medicine, as they are being leveraged to battle cancer, correct genetic disorders, tame autoimmune diseases, and tackle a host of other pathologies. However, these advances hinge on basic science discoveries that led to the isolation of the HSC as well as its downstream progenitor cells, which give rise to more than 10 distinct mature cell types.

Interest around HSCs began growing in the mid-20th century when Jacobson et al. (1951) showed that mice could be protected from lethal irradiation by shielding the spleen. This early work alluded to a humoral factor secreted by the spleen that could stimulate reconstitution of the damaged bone marrow. Later studies utilizing bone marrow transplantation and tracking clonogenicity via chromosomal abnormalities shed light on the possibility of a self-renewing cell type, rather than a soluble factor, that could give rise to blood cells (Lorenz et al. 1951; Ford et al. 1956). Upon this finding, physicians began to consider bone marrow transplantation as a therapy for radiation toxicity (Thomas et al. 1957). Soon the notion that a single purified population with the ability to self-renew and give rise to multiple differentiated cell types began to gain traction. However, strong evidence for this biological phenomenon was not brought forward until the colony forming assays established by Till and McCulloch (Till 1961; Becker et al. 1963; Wu et al. 1967). Since the work of these pioneering scientists and many others, our laboratory has focused on isolating pure hematopoietic stem and progenitor cells. Much of this review will focus on our contributions to the field of stem cell biology in the context of normal hematopoiesis and cancer.

To study and characterize the HSCs required purifying this population from heterogeneous bone marrow. Two critical innovations that enabled this task were the development of antibodies toward distinct cell surface antigens and the coupling of these antibodies to fluorescent molecules for fluorescence-activated cell sorting (FACS). In pursuit of identifying early B-lymphocyte progenitors, Muller-Sieburg et al. (1986) used a two-color FACS strategy to isolate B-cell precursors and a phenotypic pluripotent stem cell population. The authors utilized the Whitlock–Witte culture system where potential progenitors are cocultured with pre-established bone marrow–derived stromal cells (Whitlock and Witte 1982). With this system, the authors isolated $B220^-$ cells, knowing that pre-B cells expressed B220, and cultured them to test their capacity to generate long-term cultures yielding mature B cells. After successfully demonstrating that $B220^-$ cells could generate long-term cultures whereas $B220^+$ cells could not, the authors then tried to further refine their precursor population using antibodies against antigens on differentiated cell types, known as lineage markers.

Antibodies against CD4, CD8, Gr1, Mac1, and B220 allowed the authors to negatively select for differentiated T-cell, B-cell, granulocyte, and monocyte populations. Additionally, they utilized Thy1.1, a marker highly expressed on T cells, for positive selection. Reports showed that this marker was expressed in bone marrow cells and that $Thy1.1^+$ cells gave rise to splenic colonies when injected into irradiated mice, which was considered a criterion for hematopoietic multipotency (Basch and Berman 1982). After isolating cells negative for lineage markers with low expression of Thy1.1, referred to as $Thy1.1^{lo}$ Lin^-, Muller-Sieburg et al. (1986) showed these cells gave rise to T, B, myeloid, and erthyroid cells via in vitro and in vivo methods. This population, which represented 0.1%–0.2% of bone marrow, also rescued lethally irradiated animals by fully reconstituting their hematopoietic system.

In a landmark paper, Spangrude et al. (1988) were able to further purify this pluripotent hematopoietic stem population using an antibody to Sca-1. Using $Thy1.1^{lo}$ Lin^- $Sca\text{-}1^+$ (TLS) cells, the authors saw a 1000-fold enrichment for splenic colony forming units compared with whole bone marrow; thus, these cells were enriched for stem cell activity. Intrathymic injection of this purified population resulted in thymic colonies with as few as four cells, whereas previous reports had required 5000–8000 whole bone marrow cells to recapitulate this finding. After intravenous injection into a lethally irradiated animal, 40 TLS cells contributed to 50% of leukocyte chimerism at 6 weeks. The donor cells were positive for lymphoid and myeloid markers. Finally, approximately 30 cells could be injected to provide 50% survival of lethally irradiated animals, compared with 1.3×10^4 to 3.3×10^4 whole bone marrow cells. These data represented the most purified population of cells to achieve two of the hallmarks of an HSC: multilineage reconstitution and hematopoietic rescue of lethally irradiated animal. The final hallmark, long-term self-renewal, was shown by Smith et al. (1991). In their experiments, mice were transplanted with a single TLS cell and sacrificed 8 weeks later to isolate the donor TLS cells. These donor cells were then secondarily transferred into another lethally irradiated animal. The second transplants resulted in multilineage chimerism, which could only be possible if the initial single-cell transplant resulted in self-renewal of new HSCs.

The question of whether these TLS cells were the purest HSCs was called into question by many. However, their heterogeneity was not shown until they were fractioned by rhodamine-123 (Rh-123) staining and shown to have different functional capacity in vivo (Spangrude and Johnson 1990). Positive staining with Rh-123, which preferentially binds mitochondria, is indicative of a cell which is in an active state. The authors found that cells with lower Rh-123 staining, or quiescent cells, had a 20-fold-higher capability of creating splenic colonies than the higher Rh-123 population. Our laboratory was able to further refine the phenotypic HSCs based on observations that mutations in the loci for steel factor, also known as stem cell factor, and its receptor, c-Kit, in mice resulted in reduced hematopoietic activity and anemia (Russell 1979). Using antibodies against c-Kit, Ikuta and Weissman (1992) isolated the c-Kit$^+$ subset of TLS bone marrow cells and showed that low-dose transplantation of this population was capable of long-term multilineage hematopoietic reconstitution, whereas the negative fraction resulted in no donor chimerism 10 and 15 weeks after injection.

Identifying c-Kit as a marker of stemness was a major advancement; however, our laboratory, as well as others, continued to strive to identify the long-term HSC (LT-HSC). Although TLS cells produced multilineage reconstitution and rescued lethally irradiated mice, Uchida et al. (1993) estimated that only 1 out of 22 of these phenotypic HSCs initiated blood cell production in vivo. Soon after, the identification of CD34 as a negative marker of the LT-HSC helped refine the population and also provided an interesting deviation from human HSC biology, as our group had previously reported CD34 as a marker for human multipotent hematopoietic precursors (Baum et al. 1992; Osawa et al. 1996). CD34$^+$ c-Kit$^+$ Sca1$^+$ Lin$^-$ cells could provide early hematopoietic reconstitution, but it was transient, whereas the CD34$^{-/lo}$ fraction provided sustained reconstitution.

Continuing to immunophenotype the LT-HSC has also allowed our group to uncover downstream progenitors. The use of Flk2 has enabled us to identify three fractions within the TLS c-Kit$^+$ subset (Christensen and Weissman 2001). Thy1.1^{lo} Flk2$^-$ cells were shown to be LT-HSC, whereas Thy1.1^{lo} Flk2$^+$ cells were identified as short-term HSCs (ST-HSCs), and Thy1.1$^-$ Flk2$^+$ were further downstream from the ST-HSCs. As the expression of Flk2 increased and Thy1.1 decreased, there was a progressive loss of self-renewal capacity, indicating progressive differentiation. Combining Flk2 with the lineage marker antibodies allowed identification of LT-HSCs by using three-color sorting, with the other two channels being used for Sca1 and c-Kit. By minimizing the number of channels needed to isolate these cells, additional surface markers could be used to further characterize the population. Kiel et al. (2005) were also able to minimize the number of markers needed to identify LT-HSCs and multipotent hematopoietic progenitors (MPPs) with an approach involving the signaling lymphocyte activation molecule (SLAM) family members. This work, combined with our previous findings and that of others mentioned earlier, characterized the LT-HSC as Lin$^-$ c-Kit$^+$ Sca1$^+$ Flk2$^-$ CD34$^-$ CD150$^+$.

Identifying the LT-HSC and its immediate downstream progenitors has provided us with the tools to further characterize the hematopoietic hierarchy. (Kondo et al. (1997) identified the first oligopotent progenitor cell by isolating a murine population that was lymphoid-restricted. This work was initiated by the observation that IL-7 played a pivotal role in both T- and B-cell development. The authors looked for expression of IL-7Rα in Lin$^-$ cells and isolated a Lin$^-$ IL-7Rα$^+$ Thy1.1$^-$ Sca-1^{lo} c-Kit^{lo} population. When these cells were transplanted into lethally irradiated congenic mice, the only donor-derived cells were B220$^+$ or CD3$^+$. Donor chimerism in these lineages was highest at 5 weeks but steadily decreased over time, indicating the inability of the common lymphoid progenitor (CLP) to self-renew. Additionally, when the CLP was transplanted into immunodeficient mice, natural killer (NK) cells could be found in the spleen. These findings confidently indicated that the CLPs were irreversibly restricted to producing all the mature lymphoid subsets while lacking the ability to differentiate into myeloid cells.

Our group was later able to isolate the common myeloid progenitor (CMP) population, which together with the CLP cells represented a definitive branch point in early hematopoiesis and the transition from multipotency to oligopotency. Given that CLP cells were IL-7Rα$^+$, Akashi et al. (2000) intuitively searched the IL-7Rα$^-$ population to identify cells with the ability to differentiate into the myeloid and erythroid subsets. The IL-7Rα$^-$ fraction of Lin$^-$ c-Kit$^+$ cells were capable of substantial myeloerythroid colony formation activity; however, within this population the Sca-1$^+$ cells contained HSC activity and thus needed to be removed to fully purify a lineage-restricted oligopotent cell. The subsequent IL-7Rα$^-$ Lin$^-$ c-Kit$^+$ Sca-1$^-$ could be further subdivided by FcγR and CD34. The authors demonstrated that the FcγR^{lo} CD34$^+$ cells were identical to HSCs in colony forming assays. However, the FcγR^{hi} CD34$^+$ cells formed only granulocytes or macrophages in vitro and in vivo, whereas the FcγR^{lo} CD34$^-$ cells exclusively formed megakaryocytes and erythrocytes. Based on these findings, the authors coined the FcγR^{hi} CD34$^+$ population as granulocyte/macrophage lineage-restricted progenitors (GMPs) and the FcγR^{lo} CD34$^-$ as megakaryocyte/erythrocyte lineage-restricted progenitors (MEPs).

Over time, our laboratory has further helped isolate, characterize, and define the HSC and its downstream progenitors previously shown in Seita and Weissman (2010). We have continued to build upon these findings within the mouse system to ultimately understand properties that govern stemness, HSC maintenance, and the HSC niche.

Hoxb5 AS THE DEFINITIVE MARKER FOR LONG-TERM HSCs IN MICE

Since their first isolation (Spangrude et al. 1988), our characterization of the immunologic profile of HSCs has become increasingly precise with the identification of specific cell surface markers (Osawa et al. 1996; Christensen and Weissman 2001; Kiel et al. 2005; Fathman et al. 2014). However, our present definition of HSCs still remains limited in its ability to clearly distinguish phenotypic subsets within the HSC population—namely, the long-term and short-term HSCs. In addition, the need for a complex antibody cocktail to discriminate HSCs remains incompatible with microscopy, limiting our abilities to visualize the anatomical position of the HSC and its interactions with neighboring cells within the bone marrow. As a result, in recent years, numerous groups have generated fluorescent reporter mice to overcome these limitations and better characterize HSCs and their interactions within the bone marrow microenvironment (Hills et al. 2011; Gazit et al. 2014; Acar et al. 2015).

Through unbiased multistep screening, our group has managed to identify a single gene, homeobox B5 (*Hoxb5*, *Hox2.1*), whose expression in the bone marrow is exclusive to the HSC transcriptome (Chen et al. 2016). Utilizing the Gene Expression Commons database (Seita et al. 2012) to compare gene expression across 28 distinct hematopoietic populations and eight bone and stromal populations, we were able to obtain a list of 45 candidate HSC-specific genes. To ensure that our candidates could be detected both by flow cytometry and microscopy, we utilized a previously generated Bmi-1-eGFP knock-in reporter mouse (Hosen et al. 2007), in which eGFP expression is detectable in the HSC population. We performed RNA-seq on hematopoietic cells from this animal to estimate the minimal fragments per kilobase of transcript per million mapped reads (FPKM) value necessary for detection. Based off of this, we concluded that a minimal FPKM value of ~20 would be necessary to detect a single copy of eGFP. However, as this criterion would have excluded all potential candidates, we designed a targeting construct containing three copies of mCherry to amplify fluorescent signal, lowering our estimated detection threshold to ~7. Combining this strategy with RNA-seq of immunophenotypic HSCs (pHSCs) ($Lin^-c\text{-}Kit^{+}Sca\text{-}1^+Flk\text{-}2^-CD34^{-/lo}CD150^+$), multipotent progenitors subset A ($Lin^-c\text{-}Kit^+Sca\text{-}1^+Flk\text{-}2^-CD34^+CD150^+$), and multipotent progenitors subset B ($Lin^-c\text{-}Kit^+Sca\text{-}1^+Flk\text{-}2^-CD34^+CD150^-$), from C57/B6J mice, we were able to quantify the FPKM value of each candidate gene and narrow down our list to three candidates—*Hoxb5*, *Rnf208*, and *Smtnl1*. Given that numerous groups have reported heterogeneity within the phenotypic HSC (pHSC) population, we rationalized that the ideal pan-HSC marker would possess a bimodal expression pattern, with differences in gene expression potentially implicating diversity within the population. Through single-cell quantitative polymerase chain reaction (qPCR), we managed to identify *Hoxb5* as the only candidate within the HSC transcriptome that satisfied all the criteria of our multistep screen.

By engineering an in-frame triple-mCherry *Hoxb5* knock-in mouse (Hoxb5-tri-mCherry), we have managed to design a model system whereby *Hoxb5*-expressing HSCs can be isolated and visualized via endogenous fluorescent signal. To validate the specificity of this reporter, we isolated bone marrow from 12-wk-old reporter mice and evaluated the mCherry expression in various progenitor and differentiated cell populations. From this we observed mCherry expression in 21.8 ± 0.90% of pHSCs, 1.88 ± 0.64% of MPPs, and no expression elsewhere in the bone marrow (Chen et al. 2016).

The two hallmarks of a HSC are its ability to self-renew and its ability to give rise to all lineages of blood cells. As such, the gold standard of evaluating whether or not a cell is truly a HSC is the transplantation assay, with successful engraftment imparting multilineage reconstitution. Within our current understanding of HSC biology, there exist two subtypes of HSCs, short-term (ST) and long-term (LT), with the latter able to successfully undergo serial transplantation. To validate our finding of *Hoxb5* as a marker for HSCs, we isolated $mCherry^+$ and $mCherry^-$ cells within the pHSC population from the bone marrow of 12-wk-old reporter mice and transplanted them into lethally irradiated recipients. As hypothesized, the $mCherry^+$ cells were able to endow multilineage reconstitution of the hematopoietic system, while the $mCherry^-$ cells mostly conferred lymphoid reconstitution, indicative of a lymphoid-biased HSC or MPP phenotype. To further evaluate the biology of our $mCherry^+$ pHSCs, after 16 weeks we isolated whole bone marrow from the primary recipients and transplanted them into lethally irradiated hosts. Strikingly, in our secondary transplantation setting, only the whole bone marrow from $mCherry^+$ primary recipients was able to give robust multilineage reconstitution, validating *Hoxb5* as a marker of LT-HSCs.

By antibody-based methods (Kiel et al. 2005; Mendez-Ferrer et al. 2010) and microscopy-based detection of HSCs in fluorescent reporter mice (Acar et al. 2015), numerous groups over the years have previously reported that HSCs reside near blood vessels. To validate these findings, we utilized a solvent-based clearing technique—CUBIC—to diminish heme-derived autofluorescence within bone marrow plugs (Susaki et al. 2014) and clarify the tissue for whole-mount imaging. Strikingly, we have managed to reveal a near-homogenous perivascular niche for LT-HSCs, with >94% of LT-HSCs being directly attached to vascular endothelial (VE)-cadherin$^+$ cells.

With our discovery of *Hoxb5* as a marker for murine LT-HSCs, numerous secondary questions must be answered. Namely, does *Hoxb5* have a functional role in regulating the capacity for self-renewal or multipotency in LT-HSCs, and if so are variable levels in gene expression indicative of progression in the hematopoietic hierarchy? In addition, at what point in development do hematopoietic precursors or HSCs begin expressing *Hoxb5*, and how do these cells function in native hematopoiesis? Lastly, does an equivalent genetic marker exist in human HSCs, and if so how can we leverage this to study normal hematopoiesis? This knowledge could be instrumental in understanding dysregulated hematopoiesis, particularly in the setting of malignancy.

ACCUMULATION OF MUTATIONS WITHIN THE NORMAL HSC AND PROGENITOR COMPARTMENT LEADS TO LEUKEMIA

Most leukemia requires several mutations before transformation into a blast crisis. With the Majeti laboratory, we sought to understand the origin of these mutations by studying primary patient AML samples (Jan et al. 2012; Corces-Zimmerman et al. 2014). Knowing that the blast phase requires several mutations, we focused on how these were acquired and in what order they occurred by using advanced sorting and sequencing techniques.

First, we performed exome sequencing to identify known AML mutations. We then performed targeted resequencing and found that the mutations were present in a fraction of cells, which demonstrated that these were hemizygous mutations in a nonhomogenous population. When this was performed on phenotypic HSCs from leukemia patients, some but not all mutations were present in this subset, signifying that there were preleukemic cells within the HSC compartment (Jan et al. 2012).

This led us to perform single-cell sequencing to identify the order of clonal evolution in AML. In the HSC compartment, single or biallelic loss of TET2, a driver mutation, is observed at a high frequency. However, other driver mutations, like FLT3-ITD, which are seen in the primary leukemia blasts, were not observed in the HSC fraction. This suggests that the order of mutations acquired is not stochastic in AML. Preleukemic clones, which are phenotypically HSCs but harbor leukemia-associated mutations, likely maintain mutations in the HSC pool through self-renewal. These clones then either outcompete healthy counterpart HSCs or are maintained within the HSC fraction at a similar frequency. Our single-cell mapping confirmed this and showed that additional mutations are slowly acquired in these preleukemic clones, eventually a mutation within the MPP is acquired and results in malignant transformation. In our samples, we were able to find additional evidence of clonal evolution in the HSC compartment by detecting other early mutation events in several individuals (e.g., mutation A-loss of TET2, A+B-loss of other TET2 allele, A+B+C-minor mutation). This allowed us to make a roadmap from normal HSC to leukemia (Jan et al. 2012; Corces-Zimmerman et al. 2014).

This slow progression and maintenance of minor mutations within the HSC pool provides insight into both HSC and LSC biology. For example, because HSCs can self-renew and are maintained in vivo, the accumulation of mutations can occur slowly over time, which may be one reason why AML can affect an aging population (Estey and Dohner 2006; Estey 2006; Jan and Majeti 2013; Beerman et al. 2014). For some individuals, these mutations may provide a competitive advantage and begin to outcompete their remaining healthy HSCs (Corces-Zimmerman et al. 2014). This clonal progression also provides insight into how to treat AML (or other blood malignancies) effectively. In some instances, a relapse may occur not because of residual disease but because of the accumulation of an alternate mutation in the progenitor population. In these instances, targeted therapies that will remove all blast, LSC, and preleukemic clones may be necessary (Majeti and Weissman 2011). We hypothesize that this sequential accumulation of mutations may also occur in other malignancies, such as solid tumors, and that their accumulation of mutation events can also be mapped in future studies to understand the progression of disease.

PHENOTYPIC AND BIOLOGICAL DIFFERENCES BETWEEN THE HSC AND LSC

AML was reported to be propagated by an LSC that was most likely an early progenitor, such as HSCs (Lapidot et al. 1994; Bonnet and Dick 1997). As described above, my laboratory has isolated and identified HSCs and early progenitors and demonstrated the functional differences of these populations (Muller-Sieburg et al. 1986; Kondo et al. 1997; Morrison et al. 1997; Akashi et al. 2000; Manz et al. 2002; Majeti et al. 2007; Mori et al. 2015). Using bone marrow from AML-1/ETO$^+$ translocation leukemias, we determined what fraction gives rise to AML. We found that the CD34^{+-}CD38^{lo}CD90$^-$Lin$^-$ MPP fraction propagates AML, whereas the HSC population, which still maintained the translocation, did not propagate the disease (Miyamoto et al. 2000). This suggested that the LSCs were more differentiated than the HSCs, as the two populations could give rise to two different phenotypes when transplanted into mice. These findings supported the idea that the LSCs are unique but maintain similar properties to the HSCs, as we had previously proposed (Reya et al. 2001).

To investigate the differences in more detail we relied on the fundamental findings that leukemia was self-perpetuating independently of the HSC population. These findings suggest that there is a pathway that is enriched in the progenitor LSC population that allows for the propagation of leukemia. We proposed that the wnt β-catenin pathway, and other pathways, were critical in the progression of LSCs and CSCs (Reya et al. 2001). We found that wnt3a and the β-catenin pathway regulate not only cell cycle but also self-renewal of HSCs, emphasizing the importance of this pathway in normal hematopoiesis (Reya et al. 2003). We explored this pathway in more detail when we investigated the progression of CML to blast crisis. Interestingly, the GMP population from CML patients in blast crisis had increased levels of nuclear β-catenin and had self-renewing capacity when compared with normal GMP populations (Jamieson et al. 2004a). In the chronic phase, BCR-ABL transcripts are also increased in the HSCs compared with myeloid progenitors; however, during the crisis phase, BCR-ABL was increased predominately in the myeloid progenitors, especially in the GMP population. During blast crisis the

GMP-LSC is able to promote disease progression through the β-catenin pathway. We found that some of these patients had constitutively active β-catenin because of a splice variant, GSK3β that is expressed only in the GMP population (Abrahamsson et al. 2009). GSK3β is a critical factor of the destruction complex for β-catenin, and the splice variant of GSK3β lacked the capacity to phosphorylate β-catenin, allowing β-catenin to escape degradation and promote downstream signaling pathways. Additionally, we showed that when full-length GSK3β is re-expressed in the CML cells, β-catenin is decreased and leukemic engraftment is inhibited. This highlights the importance of GSK3β in the propagation of the LSC in CML, which propagates blast crisis from a progenitor downstream from the HSC. These data indicate that the HSC can harbor the mutation, and the mutations within the HSC may increase competiveness over non-mutant-bearing HSCs within the bone marrow. In all of the leukemias that we have evaluated, although we can detect mutations within the HSC population, they are completely distinct from the LSC, which is a downstream progenitor. Taken together, these findings indicate that although the mutation can be observed within HSCs, it is not until a later step that the leukemia arises owing to other contributing factors.

As discussed previously, in collaboration with the Majeti laboratory we determined what the progression is from HSCs to LSCs, to help identify potentially what the key differences are in the LSCs from the HSCs (Jan et al. 2012). Additionally, we found that there are mutations to the genetic landscape within the preleukemic HSCs that prime them for full leukemic transformation, which is observed in the LSCs' mutational load (Corces-Zimmerman et al. 2014). Further analysis between HSCs and LSCs led us to identify dysregulated pathways in LSCs compared with HSCs (Majeti et al. 2009a). This type of analysis also led to the discovery that we could phenotypically isolate normal HSCs from preleukemic HSCs with the cell surface marker TIM3 (Jan et al. 2011) and identified a role for CD47 in both HSCs and LSCs (Jaiswal et al. 2009; Majeti et al. 2009b).

We have identified key mutations, signaling pathways, and surface markers for LSCs. This profile has allowed us and others to determine what the differences are between LSCs and HSCs. In short, we have demonstrated that LSCs are not HSCs but are downstream progenitors to HSCs that have acquired additional mutations. These additional mutations arise progressively in downstream progenitors, which influences survival signaling pathways and the up-regulation of the major "do not eat me" signal, CD47.

IDENTIFYING CD47 AS A CELL SURFACE MOLECULE FOR INNATE IMMUNE EVASION

We found that CD47 is transiently induced on normal HSCs providing protection to HSCs during inflammation-mediated mobilization (Jaiswal et al. 2009). This mechanism of temporary up-regulation in CD47 expression serves an important physiological purpose but is hijacked by cancer cells as a method to evade the immune system.

CD47 is a ubiquitously expressed "do not eat me" signal, whose receptor, SIRPα, is expressed by macrophages. By comparing CD47 expression on LSCs with normal HSCs we discovered that CD47 is constitutively up-regulated in both mouse and human myeloid leukemia. We therefore hypothesized that the overexpression of CD47 on leukemic cells was a potential mechanism for allowing them to evade macrophage phagocytosis and increase their pathogenicity. By overexpressing CD47 in a mouse leukemia cell line, we were able to confirm this hypothesis, enhancing the engraftment of these cells in mice and demonstrating a direct correlation between the level of CD47 and the degree of pathology in the host. Conversely, by blocking CD47 on these cells with an anti-CD47 antibody, we were able to restore the ability of macrophages to phagocytose the leukemic cells (Jaiswal et al. 2009).

When we extended our studies to multiple types of solid tumors we discovered that numerous cancer subtypes up-regulated CD47 on their surface. This was true for ovarian cancer, breast cancer, colon cancer, bladder carcinoma, hepatocellular carcinoma, prostate cancer, and glioblastoma (Willingham et al. 2012). By screening with in vitro phagocytosis assays in which we cocultured dissociated cells from human primary tumor samples with macrophages, we found that the addition of blocking anti-human CD47 (hCD47) mAbs significantly increased the efficiency of cancer cells phagocytosis by either human or mouse macrophages. When tested efficacy in NOD-scid-γ (NSG) mouse xenotransplantation models, CD47 blockade increased the survival of the mice over time, preventing metastasis and inhibiting growth of larger tumors, while being potentially curative for smaller tumors (Willingham et al. 2012). Thus, we concluded that CD47 blockade could be a potent therapeutic strategy against many types of cancer.

Interestingly, despite the ubiquitous expression of CD47, macrophages did not engulf healthy cells from corresponding tissues even in the presence of high-dose anti-CD47 antibodies. Therefore, we postulated that there must be a prophagocytic "eat me" signal on cells that are destined for elimination by macrophages, and that such a recognition signal would be present on malignant cells in a tissue but not on their healthy counterparts. We identified calreticulin as one such signal, which is present on cancer stem cells and not on normal stem cells and acts as a robust "eat me" signal (Chao et al. 2010b).

Following this discovery, we began to investigate how calreticulin is trafficked to the surface of cancer cells. Calreticulin is an endoplasmic reticulum (ER) protein, which upon Toll-like receptor (TLR) activation is secreted by macrophages. It has been shown that TLR activation leads to Bruton's tyrosine kinase (BTK) signaling, which in turn phosphorylates calreticulin. Phosphorylated calreticulin is then cleaved to separate the KDEL domain, allowing macrophages to secrete soluble calreticulin. Soluble calreticulin then binds to tumor cells and to

LRP1, a receptor for apoptotic cells, on macrophages and serves both as a recognition signal and as a tethering bridge between the macrophage and the tumor cell (Feng et al. 2015). Thus, calreticulin functions as a molecular "sensor" secreted by macrophages to label their targets for elimination. When a cancer cell is labeled with calreticulin but also expresses high levels of the "do not eat me" signal CD47—the antiphagocytic effect of the CD47–SIRPα signaling axis can neutralize the macrophage, allowing the cancer cell to evade the innate immune system. However, with anti-CD47 blockade, this dynamic can be shifted, priming macrophages to clear malignant cells that they have recognized; this demonstrated the use of anti-CD47 blockade as an avenue for therapeutic intervention in enhancing the removal of malignant cells through immunotherapies.

Importantly, the effect of anti-CD47 blockade can also be used to help initiate an adaptive immune response. In a follow-up study, we found that anti-CD47 antibody-mediated phagocytosis of cancer cells by macrophages can lead to priming of $CD8^+$ T cells—measured by proliferation and activation. However, in our model these macrophages did not prime $CD4^+$ T cells, demonstrating potential differential priming of T-cell responses by macrophages after anti-CD47 mediated phagocytosis. Priming an effective $CD8^+$ T- cell response can help lead to durable long-term antitumor immunity. Therefore, the utilization of anti-CD47 as a modified "vaccination" strategy serves as a promising new avenue for the initiation of an antitumor T-cell response (Tseng et al. 2013).

These preclinical studies present an opportunity to utilize this finding for the development of novel cancer immunotherapy. We developed a humanized anti-CD47 antibody (Hu5F9-G4) with potent efficacy and favorable toxicokinetic properties as a candidate therapy. Administration of this antibody induced potent macrophage-mediated phagocytosis of primary human AML cells in vitro and completely eradicated human AML in xenotransplanted mouse models leading to long-term disease-free survival. Moreover, Hu5F9-G4 synergized with rituximab to eliminate non-Hodgkin's lymphoma (NHL) engraftment and cures xenografted mice (Chao et al. 2010a). Finally, toxicokinetic studies in nonhuman primates showed that Hu5F9-G4 could be safely administered intravenously at doses able to achieve potentially therapeutic serum levels in primates (Liu et al. 2015). Thus, Hu5F9-G4 has been entered into clinical trials in patients with AML and solid tumors (NCI trials NCT02678338 and NCT02216409; additional trials are anticipated).

In summary, our recent discovery of CD47 as an up-regulated "do not eat me" signal on cancer is one mechanism by which they evade the immune system. These key studies have resulted in the development of a new cancer immunotherapy, utilizing blockade of CD47 with monoclonal antibodies and disrupting the CD47-SIRPα axis, resulting in cancer cell phagocytosis. We have also shown that there exist "eat me" signals, like soluble calreticulin secreted from macrophages. These discoveries have opened a major avenue of research exploring the mechanisms of recognition and clearance of target cells by phagocytes. Future studies will continue to understand the interplay of additional pro- and anti-phagocytic pathways in various biological systems, with a fundamental aim to develop and identify ways to enhance programmed cell removal for effective anti-cancer therapy.

CONCLUSION

Normal and neoplastic stem cells share overlapping genetic programs for survival in vivo. In cancer, cells must acquire several additional key features to survive and transform into a malignancy. Our laboratory has studied several of these, but the work described here is focused on two key features—self-renewal and immune evasion.

For self-renewal, we identified the LSC population that was able to give rise to leukemia when transplanted into mice in vivo unlike blast cells that had little to no transplant capacity. These studies parallel initial BMT that utilized whole bone marrow and then was later refined to pHSCs. Through phenotypic sorting and transplantation of leukemic subpopulations, these experiments identified the tumorigenic cell population (LSC). Follow-up studies demonstrated that in the same primary patient sample, the HSC never contained the LSC population because HSCs only contained a fraction of the genetic mutations and the full oncogenic transformation occurred in downstream progeny.

For immune evasion, we looked again at LSCs and identified CD47 as a surface marker that was up-regulated. Similar to mobilized pHSCs that up-regulate CD47 to prevent being phagocytosed, the LSC population had constitutively high CD47 expression as a mechanism to avoid immune recognition and phagocytosis by macrophages. Our work then led us to develop an innate focused immunotherapy by targeting CD47 with a monoclonal antibody.

There are several similarities between normal and neoplastic stem cells as highlighted in the work above. By understanding what drives normal stem cells, we can begin to ask whether cancer acquires or maintains similar properties through mutual or distinct pathways. Therefore, the stem cell and/or cancer studies in our laboratory can provide mutual insight into the biology of both cell types and may lead to new approaches to improve cancer therapies.

ACKNOWLEDGMENTS

M.N.M is supported by the A.P. Giannini Foundation. I.L.W. is supported in part by the Virginia and D.K. Ludwig Fund for Cancer Research, and by a Stand Up To Cancer–Lustgarten Foundation Pancreatic Cancer Convergence Dream Team Translational Research Grant (grant number: SU2C-AACR-DT14-14). Stand Up To Cancer is a program of the Entertainment Industry Foundation administered by the American Association for Cancer Research.

REFERENCES

Abrahamsson AE, Geron I, Gotlib J, Dao KH, Barroga CF, Newton IG, Giles FJ, Durocher J, Creusot RS, Karimi M, et al. 2009. Glycogen synthase kinase 3β missplicing contributes to leukemia stem cell generation. *Proc Natl Acad Sci* **106:** 3925–3929.

Acar M, Kocherlakota KS, Murphy MM, Peyer JG, Oguro H, Inra CN, Jaiyeola C, Zhao Z, Luby-Phelps K, Morrison SJ. 2015. Deep imaging of bone marrow shows non-dividing stem cells are mainly perisinusoidal. *Nature* **526:** 126–130.

Akashi K, Traver D, Miyamoto T, Weissman IL. 2000. A clonogenic common myeloid progenitor that gives rise to all myeloid lineages. *Nature* **404:** 193–197.

Basch RS, Berman JW. 1982. Thy-1 determinants are present on many murine hematopoietic cells other than T cells. *Eur J Immunol* **12:** 359–364.

Baum CM, Weissman IL, Tsukamoto AS, Buckle AM, Peault B. 1992. Isolation of a candidate human hematopoietic stem-cell population. *Proc Natl Acad Sci* **89:** 2804–2808.

Becker AJ, McCulloch EA, Till JE. 1963. Cytological demonstration of the clonal nature of spleen colonies derived from transplanted mouse marrow cells. *Nature* **197:** 452–454.

Beerman I, Seita J, Inlay MA, Weissman IL, Rossi DJ. 2014. Quiescent hematopoietic stem cells accumulate DNA damage during aging that is repaired upon entry into cell cycle. *Cell Stem Cell* **15:** 37–50.

Bonnet D, Dick JE. 1997. Human acute myeloid leukemia is organized as a hierarchy that originates from a primitive hematopoietic cell. *Nat Med* **3:** 730–737.

Chao MP, Alizadeh AA, Tang C, Myklebust JH, Varghese B, Gill S, Jan M, Cha AC, Chan CK, Tan BT, et al. 2010a. Anti-CD47 antibody synergizes with rituximab to promote phagocytosis and eradicate non-Hodgkin lymphoma. *Cell* **142:** 699–713.

Chao MP, Jaiswal S, Weissman-Tsukamoto R, Alizadeh AA, Gentles AJ, Volkmer J, Weiskopf K, Willingham SB, Raveh T, Park CY, et al. 2010b. Calreticulin is the dominant pro-phagocytic signal on multiple human cancers and is counterbalanced by CD47. *Sci Transl Med* **2:** 63ra94.

Chen JY, Miyanishi M, Wang SK, Yamazaki S, Sinha R, Kao KS, Seita J, Sahoo D, Nakauchi H, Weissman IL. 2016. Hoxb5 marks long-term haematopoietic stem cells and reveals a homogenous perivascular niche. *Nature* **530:** 223–227.

Chhabra A, Ring AM, Weiskopf K, Schnorr PJ, Gordon S, Le AC, Kwon HS, Ring NG, Volkmer J, Ho PY, et al. 2016. Hematopoietic stem cell transplantation in immunocompetent hosts without radiation or chemotherapy. *Sci Transl Med* **8:** 351ra105.

Christensen JL, Weissman IL. 2001. Flk-2 is a marker in hematopoietic stem cell differentiation: A simple method to isolate long-term stem cells. *Proc Natl Acad Sci* **98:** 14541–14546.

Corces-Zimmerman MR, Hong WJ, Weissman IL, Medeiros BC, Majeti R. 2014. Preleukemic mutations in human acute myeloid leukemia affect epigenetic regulators and persist in remission. *Proc Natl Acad Sci* **111:** 2548–2553.

Czechowicz A, Weissman IL. 2011. Purified hematopoietic stem cell transplantation: The next generation of blood and immune replacement. *Hematol Oncol Clin N Am* **25:** 75–87.

Estey EH. 2006. General approach to, and perspectives on clinical research in, older patients with newly diagnosed acute myeloid leukemia. *Semin Hematol* **43:** 89–95.

Estey E, Dohner H. 2006. Acute myeloid leukaemia. *Lancet* **368:** 1894–1907.

Fathman JW, Fernhoff NB, Seita J, Chao C, Scarfone VM, Weissman IL, Inlay MA. 2014. Upregulation of CD11A on hematopoietic stem cells denotes the loss of long-term reconstitution potential. *Stem Cell Rep* **3:** 707–715.

Feng M, Chen JY, Weissman-Tsukamoto R, Volkmer JP, Ho PY, McKenna KM, Cheshier S, Zhang M, Guo N, Gip P, et al. 2015. Macrophages eat cancer cells using their own calreticulin as a guide: Roles of TLR and Btk. *Proc Natl Acad Sci* **112:** 2145–2150.

Ford CE, Hamerton JL, Barnes DW, Loutit JF. 1956. Cytological identification of radiation-chimaeras. *Nature* **177:** 452–454.

Gazit R, Mandal PK, Ebina W, Ben-Zvi A, Nombela-Arrieta C, Silberstein LE, Rossi DJ. 2014. Fgd5 identifies hematopoietic stem cells in the murine bone marrow. *J Exp Med* **211:** 1315–1331.

Hills D, Gribi R, Ure J, Buza-Vidas N, Luc S, Jacobsen SE, Medvinsky A. 2011. Hoxb4-YFP reporter mouse model: A novel tool for tracking HSC development and studying the role of Hoxb4 in hematopoiesis. *Blood* **117:** 3521–3528.

Hosen N, Yamane T, Muijtjens M, Pham K, Clarke MF, Weissman IL. 2007. Bmi-1-green fluorescent protein-knock-in mice reveal the dynamic regulation of bmi-1 expression in normal and leukemic hematopoietic cells. *Stem Cells* **25:** 1635–1644.

Ikuta K, Weissman IL. 1992. Evidence that hematopoietic stem cells express mouse c-kit but do not depend on steel factor for their generation. *Proc Natl Acad Sci* **89:** 1502–1506.

Jacobson LO, Simmons EL, Marks EK, Eldredge JH. 1951. Recovery from radiation injury. *Science* **113:** 510–511.

Jaiswal S, Jamieson CH, Pang WW, Park CY, Chao MP, Majeti R, Traver D, van Rooijen N, Weissman IL. 2009. CD47 is upregulated on circulating hematopoietic stem cells and leukemia cells to avoid phagocytosis. *Cell* **138:** 271–285.

Jamieson CH, Ailles LE, Dylla SJ, Muijtjens M, Jones C, Zehnder JL, Gotlib J, Li K, Manz MG, Keating A, et al. 2004a. Granulocyte-macrophage progenitors as candidate leukemic stem cells in blast-crisis CML. *N Engl J Med* **351:** 657–667.

Jamieson CH, Weissman IL, Passegue E. 2004b. Chronic versus acute myelogenous leukemia: A question of self-renewal. *Cancer Cell* **6:** 531–533.

Jan M, Majeti R. 2013. Clonal evolution of acute leukemia genomes. *Oncogene* **32:** 135–140.

Jan M, Chao MP, Cha AC, Alizadeh AA, Gentles AJ, Weissman IL, Majeti R. 2011. Prospective separation of normal and leukemic stem cells based on differential expression of TIM3, a human acute myeloid leukemia stem cell marker. *Proc Natl Acad Sci* **108:** 5009–5014.

Jan M, Snyder TM, Corces-Zimmerman MR, Vyas P, Weissman IL, Quake SR, Majeti R. 2012. Clonal evolution of preleukemic hematopoietic stem cells precedes human acute myeloid leukemia. *Sci Transl Med* **4:** 149ra118.

Kiel MJ, Yilmaz OH, Iwashita T, Yilmaz OH, Terhorst C, Morrison SJ. 2005. SLAM family receptors distinguish hematopoietic stem and progenitor cells and reveal endothelial niches for stem cells. *Cell* **121:** 1109–1121.

Kondo M, Weissman IL, Akashi K. 1997. Identification of clonogenic common lymphoid progenitors in mouse bone marrow. *Cell* **91:** 661–672.

Lapidot T, Sirard C, Vormoor J, Murdoch B, Hoang T, Caceres-Cortes J, Minden M, Paterson B, Caligiuri MA, Dick JE. 1994. A cell initiating human acute myeloid leukaemia after transplantation into SCID mice. *Nature* **367:** 645–648.

Liu J, Wang L, Zhao F, Tseng S, Narayanan C, Shura L, Willingham S, Howard M, Prohaska S, Volkmer J, et al. 2015. Preclinical development of a humanized anti-CD47 antibody with anti-cancer therapeutic potential. *PLoS One* **10:** e0137345.

Lorenz E, Uphoff D, Reid TR, Shelton E. 1951. Modification of irradiation injury in mice and guinea pigs by bone marrow injections. *J Natl Cancer Inst* **12:** 197–201.

Majeti R, Weissman IL. 2011. Human acute myelogenous leukemia stem cells revisited: There's more than meets the eye. *Cancer Cell* **19:** 9–10.

Majeti R, Park CY, Weissman IL. 2007. Identification of a hierarchy of multipotent hematopoietic progenitors in human cord blood. *Cell Stem Cell* **1:** 635–645.

Majeti R, Becker MW, Tian Q, Lee TL, Yan X, Liu R, Chiang JH, Hood L, Clarke MF, Weissman IL. 2009a. Dysregulated gene expression networks in human acute myelogenous leukemia stem cells. *Proc Natl Acad Sci* **106:** 3396–3401.

Majeti R, Chao MP, Alizadeh AA, Pang WW, Jaiswal S, Gibbs KD Jr, van Rooijen N, Weissman IL. 2009b. CD47 is an adverse prognostic factor and therapeutic antibody target on human acute myeloid leukemia stem cells. *Cell* **138:** 286–299.

Manz MG, Miyamoto T, Akashi K, Weissman IL. 2002. Prospective isolation of human clonogenic common myeloid progenitors. *Proc Natl Acad Sci* **99:** 11872–11877.

Mendez-Ferrer S, Michurina TV, Ferraro F, Mazloom AR, MacArthur BD, Lira SA, Scadden DT, Ma'ayan A, Enikolopov GN, Frenette PS. 2010. Mesenchymal and haematopoietic stem cells form a unique bone marrow niche. *Nature* **466:** 829–834.

Miyamoto T, Weissman IL, Akashi K. 2000. AML1/ETO-expressing nonleukemic stem cells in acute myelogenous leukemia with 8;21 chromosomal translocation. *Proc Natl Acad Sci* **97:** 7521–7526.

Mori Y, Chen JY, Pluvinage JV, Seita J, Weissman IL. 2015. Prospective isolation of human erythroid lineage-committed progenitors. *Proc Natl Acad Sci* **112:** 9638–9643.

Morrison SJ, Wandycz AM, Hemmati HD, Wright DE, Weissman IL. 1997. Identification of a lineage of multipotent hematopoietic progenitors. *Development* **124:** 1929–1939.

Muller AM, Kohrt HE, Cha S, Laport G, Klein J, Guardino AE, Johnston LJ, Stockerl-Goldstein KE, Hanania E, Juttner C, et al. 2012. Long-term outcome of patients with metastatic breast cancer treated with high-dose chemotherapy and transplantation of purified autologous hematopoietic stem cells. *J Am Soc Blood Marrow Transpl* **18:** 125–133.

Muller-Sieburg CE, Whitlock CA, Weissman IL. 1986. Isolation of two early B lymphocyte progenitors from mouse marrow: A committed pre-pre-B cell and a clonogenic Thy-1-lo hematopoietic stem cell. *Cell* **44:** 653–662.

Osawa M, Hanada K, Hamada H, Nakauchi H. 1996. Long-term lymphohematopoietic reconstitution by a single CD34-low/negative hematopoietic stem cell. *Science* **273:** 242–245.

Reya T, Morrison SJ, Clarke MF, Weissman IL. 2001. Stem cells, cancer, and cancer stem cells. *Nature* **414:** 105–111.

Reya T, Duncan AW, Ailles L, Domen J, Scherer DC, Willert K, Hintz L, Nusse R, Weissman IL. 2003. A role for Wnt signalling in self-renewal of haematopoietic stem cells. *Nature* **423:** 409–414.

Russell ES. 1979. Hereditary anemias of the mouse: A review for geneticists. *Adv Genet* **20:** 357–459.

Seita J, Weissman IL. 2010. Hematopoietic stem cell: Self-renewal versus differentiation. *Wiley Interdiscip Rev Syst Biol Med* **2:** 640–653.

Seita J, Sahoo D, Rossi DJ, Bhattacharya D, Serwold T, Inlay MA, Ehrlich LI, Fathman JW, Dill DL, Weissman IL. 2012. Gene Expression Commons: An open platform for absolute gene expression profiling. *PLoS One* **7:** e40321.

Smith LG, Weissman IL, Heimfeld S. 1991. Clonal analysis of hematopoietic stem-cell differentiation in vivo. *Proc Natl Acad Sci* **88:** 2788–2792.

Spangrude GJ, Johnson GR. 1990. Resting and activated subsets of mouse multipotent hematopoietic stem cells. *Proc Natl Acad Sci* **87:** 7433–7437.

Spangrude GJ, Heimfeld S, Weissman IL. 1988. Purification and characterization of mouse hematopoietic stem cells. *Science* **241:** 58–62.

Susaki EA, Tainaka K, Perrin D, Kishino F, Tawara T, Watanabe TM, Yokoyama C, Onoe H, Eguchi M, Yamaguchi S, et al. 2014. Whole-brain imaging with single-cell resolution using chemical cocktails and computational analysis. *Cell* **157:** 726–739.

Thomas ED, Lochte HL Jr, Lu WC, Ferrebee JW. 1957. Intravenous infusion of bone marrow in patients receiving radiation and chemotherapy. *New Engl J Med* **257:** 491–496.

Thomas ED, Lochte HL Jr, Cannon JH, Sahler OD, Ferrebee JW. 1959. Supralethal whole body irradiation and isologous marrow transplantation in man. *J Clin Invest* **38:** 1709–1716.

Till JE. 1961. Radiosensitivity and chromosome numbers in strain L mouse cells in tissue culture. *Radiat Res* **15:** 400–409.

Tseng D, Volkmer JP, Willingham SB, Contreras-Trujillo H, Fathman JW, Fernhoff NB, Seita J, Inlay MA, Weiskopf K, Miyanishi M, et al. 2013. Anti-CD47 antibody-mediated phagocytosis of cancer by macrophages primes an effective antitumor T-cell response. *Proc Natl Acad Sci* **110:** 11103–11108.

Uchida N, Fleming WH, Alpern EJ, Weissman IL. 1993. Heterogeneity of hematopoietic stem cells. *Curr Opin Immunol* **5:** 177–184.

Whitlock CA, Witte ON. 1982. Long-term culture of B lymphocytes and their precursors from murine bone marrow. *Proc Natl Acad Sci* **79:** 3608–3612.

Willingham SB, Volkmer JP, Gentles AJ, Sahoo D, Dalerba P, Mitra SS, Wang J, Contreras-Trujillo H, Martin R, Cohen JD, et al. 2012. The CD47-signal regulatory protein α (SIRPa) interaction is a therapeutic target for human solid tumors. *Proc Natl Acad Sci* **109:** 6662–6667.

Wu AM, Till JE, Siminovitch L, McCulloch EA. 1967. A cytological study of the capacity for differentiation of normal hemopoietic colony-forming cells. *J Cell Physiol* **69:** 177–184.

Targeting the Epithelial-to-Mesenchymal Transition: The Case for Differentiation-Based Therapy

Diwakar R. Pattabiraman[1] and Robert A. Weinberg[1,2,3]
[1]Whitehead Institute for Biomedical Research, Cambridge, Massachusetts 02142
[2]Department of Biology, Massachusetts Institute of Technology, Cambridge, Massachusetts 02142
[3]Ludwig/MIT Center for Molecular Oncology, Cambridge, Massachusetts 02139
Correspondence: weinberg@wi.mit.edu

Although important strides have been made in targeted therapy for certain leukemias and subtypes of breast cancer, the standard of care for most carcinomas still involves chemotherapy, radiotherapy, surgery, or a combination of these. Two processes serve as obstacles to the successful treatment of carcinomas. First, a majority of deaths from these types of cancers occurs as a result of distant metastases and not the primary tumors themselves. Second, subsets of cells that are able to survive conventional therapy drive the aggressive relapse of the tumors, often in forms that are resistant to treatment. A frequently observed feature of malignant carcinomas is the loss of epithelial traits and the gain of certain mesenchymal ones that are programmed by the cell-biological program termed the epithelial-to-mesenchymal transition (EMT). The EMT program can confer (i) an ability to disseminate, (ii) an ability to become stem-like tumor-initiating cells, (iii) an ability to found new tumor colonies at distant anatomical sites, and (iv) an elevated resistance to therapy. These multiple powers of the EMT program explain why it has become an attractive target for therapeutic intervention. Recent work has revealed the variable nature of the EMT, with multiple versions of the program being observed depending on the tissue context and the stage of tumor progression. In this review, we attempt to crystallize emerging concepts in the research on EMT and stemness and discuss the benefits of using a differentiation-based therapeutic strategy for the eradication of stem-like populations that have adopted various versions of the EMT program.

The epithelial-to-mesenchymal transition (EMT) is a cell-biological program that is important in various aspects of development, wound healing, maintenance of stemness, and tumor progression. Initial observations of the loss of epithelial properties and the concomitant gain of mesenchymal traits were made in the laboratory of Elizabeth Hay, who studied the development of cornea, limb, notochord, and lens epithelia (Greenburg and Hay 1982). These findings have opened an entire field of research, which associates EMT programs with a multitude of distinct pathophysiologic processes, including the malignant progression and metastatic dissemination of carcinomas. The transition involves a highly coordinated program whereby epithelial cells loose their characteristics of apical–basal polarity, cell–cell junctions and adherence to the basement membrane and gain, among other traits, certain properties of mesenchymal cells that enable them to migrate and invade. The EMT is known to occur in various different biological contexts, which has led to its classification into three different types (Kalluri and Weinberg 2009). A type I EMT refers to the transition that takes place in the context of development, including processes that give rise to the mesoderm and primary mesenchyme from the primitive streak during gastrulation, as well as those that give rise to the migratory neural crest cells. The type II EMT is utilized during process of tissue regeneration, such as transient fibrosis and wound healing. The version of EMT highlighted in the present discussion is the type III EMT, which is associated with tumor progression.

EMT IN TUMOR PROGRESSION

Over the past decade, the importance of the EMT program in tumor progression has been established by hundreds of studies in a wide variety of carcinomas, including breast, ovarian, pancreatic, prostate, colorectal, and renal types. The role of the EMT in tumor progression has been comprehensively reviewed elsewhere (De Craene and Berx 2013; Tsai and Yang 2013; Nieto et al. 2016) and is not the major focus of the present review. Briefly, the acquisition of mesenchymal traits through induction of an EMT program is thought to enable carcinoma cells to complete many of the steps of the invasion-metastasis cascade—the sequence of steps that begins with the local invasiveness of the neoplastic cells within primary tumors, their intravasation, translocation through the circulation, extravasation into the parenchyma of distant tissues, the founding of micrometastatic deposits, and the

Published by Cold Spring Harbor Laboratory Press; doi: 10.1101/sqb.2016.81.030957

outgrowth of these deposits into macroscopic metastases—the last step being termed "colonization" (Fidler 2003; Yang et al. 2004). Expression of the EMT is maintained during the early and intermediate stages of the cascade, including invasion through tissue around the primary tumor, intravasation, transportation as circulating tumor cells (CTCs) in the vasculature (Yu et al. 2013), and extravasation following their vascular journey (Labelle et al. 2011). Following extravasation, disseminated tumor cells may undergo the reverse of the EMT, termed the mesenchymal-to-epithelial transition (MET) and thereby regain the epithelial traits exhibited by their progenitors within the primary tumor (MET) (Ocana et al. 2012; Tsai et al. 2012); this last step appears to be critical to allow metastatic colonization to proceed efficiently.

ALTERNATIVE VERSIONS OF TUMOR CELL DISSEMINATION

Several reports now show that the escape of carcinoma cells from primary tumors, as enabled by an EMT program, can occur by other mechanisms besides the frequently studied single-cell dissemination. In particular, the process of collective invasion has been widely reported and occurs when cohorts of cells that held together by cell–cell junctions advance through the extracellular matrix (Friedl and Gilmour 2009; Friedl et al. 2012). In fact, three-dimensional reconstruction studies of tumors have shown that single cell dissemination is, by contrast, relatively rare, with cell clusters being the prevalent agents of invasion (Bronsert et al. 2014). The polyclonal nature of certain metastatic colonies is consistent with their having arisen through collective dissemination of tumor cell clusters (Cheung et al. 2016), rather than from disseminated single cells, which would ostensibly give rise to monoclonal metastatic colonies. The expression of E-cadherin—a key marker of the epithelial cell state—throughout the bulk of these invasive masses has led some to question the involvement of EMT in the process of carcinoma invasiveness. However, certain lines of evidence indicate that the cells at the leading edges of invasive cohorts express certain EMT characteristics (Revenu and Gilmour 2009; Westcott et al. 2015; Ye et al. 2015). At present, it has not yet been demonstrated directly that the EMT characteristics expressed by these leader cells, including notably the trait of invasiveness, are critical to the forward migration of the cohorts as a whole.

Additionally, two recent studies using lineage-tracing studies in mouse models of breast and pancreatic cancer have questioned the essential role of the EMT program in the process of metastasis (Fischer et al. 2015; Zheng et al. 2015). Fischer et al. used Fsp1 as a lineage-tracing marker of cells that have undergone an EMT and found that large numbers of Fsp1-negative cells formed metastases in the lung. These results were interpreted as EMT not having been expressed, even transiently, for the formation of metastases. Although Fsp1 is expressed in certain versions of the EMT program (e.g., in renal tubular cells [Okada et al. 1997]), it cannot be used as a reliable marker of activation of an EMT program, especially in the absence of convincing data proving that Fsp1 is actually expressed in a majority of cells undergoing an EMT. A second paper utilized a lineage-traced mouse model of pancreatic ductal adenocarcinoma (PDAC), which exhibited a significant reduction in the percentage of α-smooth muscle actin (αSMA)-expressing cells relative to tumors with an intact Twist locus (Zheng et al. 2015). As described above, use of αSMA expression as a reliable marker of EMT is complicated by the fact that it is rarely induced spontaneously upon activation of EMT during the course of tumor development (Aiello NM and Stanger BZ, unpubl.) in this particular mouse tumor model, undermining the conclusions presented in this report. To summarize, we believe that neither of these papers proved that the EMT did not occur in the metastasizing carcinoma cells, leaving intact the notion that the EMT is indeed necessary for the metastatic dissemination of carcinoma cells. However, these two reports highlight the fact that the EMT is not a single, stereotypical program, and that various versions of this program are expressed under different conditions in a variety of carcinomas. This portrayal of multiple versions of the EMT program will be discussed further below.

EMT AND STEMNESS

Since the initial discovery of the connection between breast carcinomas that have undergone an EMT and the entrance into a stem-like state (Mani et al. 2008; Morel et al. 2008), a number of studies have extended these findings to report similarly the acquisition of stemness following the EMT in multiple tumor types, including colorectal (Fan et al. 2012), ovarian (Long et al. 2015), pancreatic (Rasheed et al. 2010), prostate (Kong et al. 2010), and renal (Zhou et al. 2016), among other types of carcinomas. Additionally, the EMT program has also been shown to be important for the normal mammary epithelial stem cell state (Guo et al. 2012; Nassour et al. 2012).

Although the connection between the EMT and acquisition of stem-like properties is now widely accepted, not all cells that have undergone an EMT exhibit increased stemness. Thus, several studies have reported the loss of stem-like properties in carcinomas upon undergoing an EMT (Celia-Terrassa et al. 2012; Chang et al. 2013; Xie et al. 2014), indicating that the acquisition of mesenchymal traits is not, on its own, conducive to acquisition of increased stemness. In more detail, there is increasing evidence that, rather than operating as a binary switch, the EMT program generates cells residing in a spectrum of multiple phenotypic states lying between the fully epithelial and fully mesenchymal states (Fig. 1), and that cells that have passed through a complete EMT program and become entirely mesenchymal actually lack the ability to function as stem cells, both normal and neoplastic. Instead, there is growing evidence that a cell that has only undergone a partial EMT, thereby expressing both

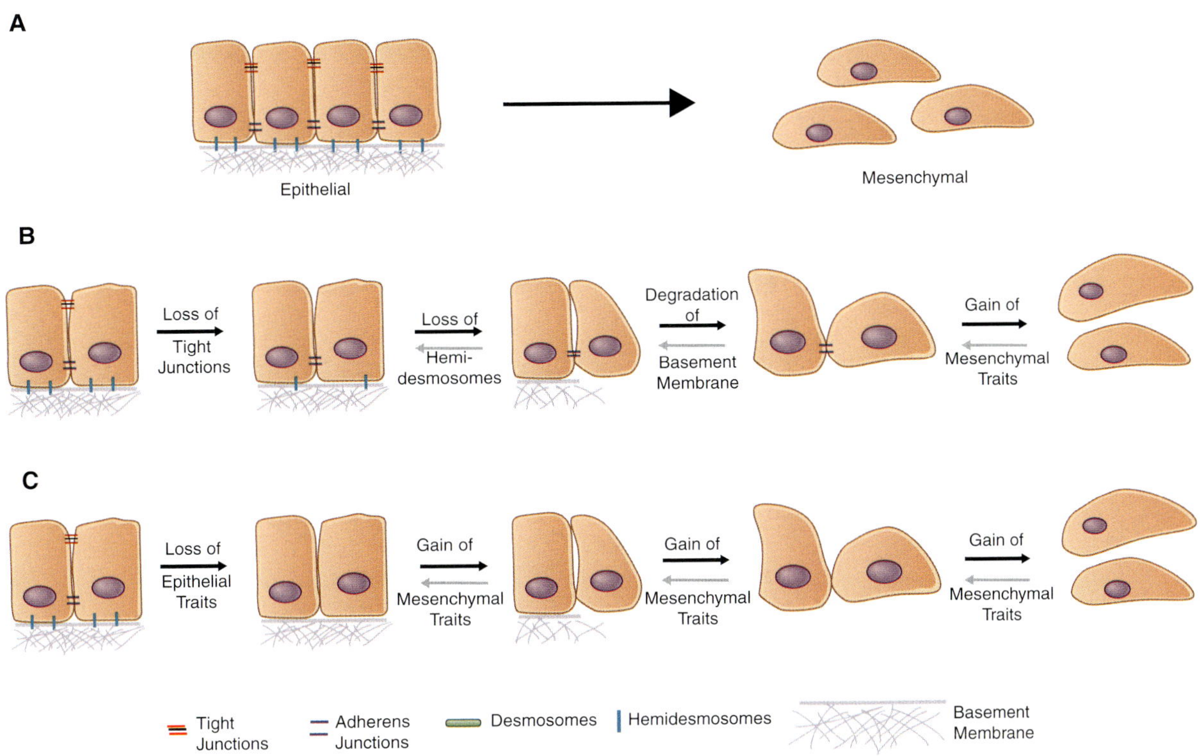

Figure 1. Although we have, in the past, thought of the epithelial-to-mesenchymal transition (EMT) program as a binary switch between two distinct cell states (*A*), more recent evidence points to the EMT representing a spectrum of different cell states anywhere in between the extreme epithelial and mesenchymal states depending on the state that the cell was originally residing in and the nature of the program that manifests. The sequence of events that defines the EMT is still poorly understood—that is, whether (*B*) the loss of epithelial traits occurs over multiple steps, with the gain of mesenchymal traits being the final step in the cascade, or if (*C*) the loss of epithelial traits occurs over a relatively short period of time, followed by a stepwise acquisition of mesenchymal properties.

retained epithelial and acquired mesenchymal traits, is best positioned to acquire stem-like properties (Grosse-Wilde et al. 2015; Jolly et al. 2015a,b; Andriani et al. 2016).

COULD THE EMT ENCOMPASS THE INITIAL LOSS OF EPITHELIAL TRAITS OBSERVED DURING MALIGNANCY?

In fact, the actions of the EMT program are not limited to advanced stages of carcinoma progression. Instead, an EMT can also participate in the disruption of the epithelial state that is a common feature of early stages of malignant transformation (Fig. 2). These early steps are characterized by the loss of polarized organization of epithelial tissue (Lee and Vasioukhin 2008), a loss of cell–cell junctions that compromises the barrier function of epithelial tissue (Martin and Jiang 2009), and the degradation of interactions of epithelial cells with the underlying basement membrane (Barsky et al. 1983).

Loss of cell polarity may be the earliest manifestation of the actions of an EMT program. As an example, the overexpressed HER2 protein, which functions as a driver of a significant proportion of breast cancers, disrupts normal apical–basal epithelial cell polarity by interacting with components of the Par polarity complex, including Par6 and aPKC. Inhibition of these interactions deprives HER2 of its ability to disrupt the acinar organization and polarity of the mammary epithelium, a key event during the initial stages of carcinogenesis (Aranda et al. 2006). Moreover certain proteins that act as tumor suppressors play important roles in maintaining cell polarity—for example, PTEN maintaining apical–basal polarity during epithelial morphogenesis (Martin-Belmonte et al. 2007) and Par3 curbing tumor progression and metastasis (McCaffrey et al. 2012; Li et al. 2014).

These reports indicate the importance of maintaining epithelial polarity as a means of curbing malignancy but do not address whether EMT programs are activated upon loss of polarity. In fact, studies have shown that both the Snail and Zeb family of EMT-inducing transcription factors (EMT-TFs) are capable of repressing the transcription of key polarity genes. Snail is known to repress the transcription of *Crumbs3*, which codes for a key polarity factor, while also disrupting the localization of the Crumbs and Par complexes (Whiteman et al. 2008). Similarly, Zeb1 also represses transcription of Crumbs3, while also repressing *HUGL2, Lgl2*, and *PATJ* in breast and colon cancer cells (Aigner et al. 2007; Spaderna et al. 2008). Loss of Zeb1 enabled a partial restoration of epithelial polarity, demonstrating that the EMT-TFs could be directly responsible for the loss of this epithelial trait. Thus, the loss of polarity that is characteristic of the

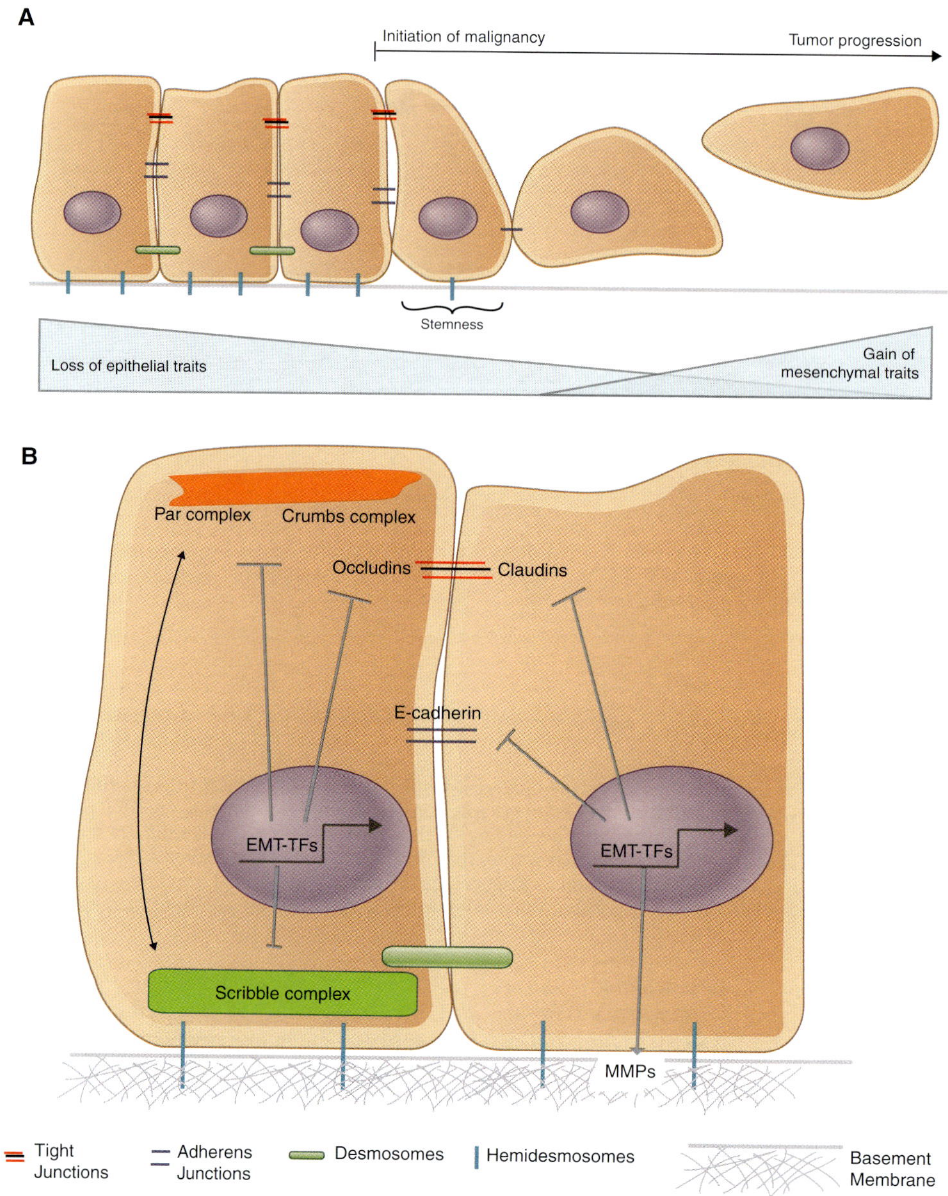

Figure 2. The epithelial-to-mesenchymal transition (EMT) has traditionally been associated, almost exclusively, with the gain of mesenchymal properties. More recent evidence indicates that an overt gain of mesenchymal traits may not be required for the stem-like properties of cells that have undergone an EMT (*A*). Thus, the ability of the EMT-inducing transcription factors to alter apical–basal polarity and repress the transcription of genes required to maintain cell–cell junctions may be equally important functions of EMT-inducing transcription factors (*B*). Similarly, the role of the EMT-inducing transcriptional factors (EMT-TFs) in inducing the transcription of genes whose products are known to degrade components of the basement membrane is also key in the initial stages of the EMT.

initial stages of carcinogenesis may represent an early manifestation of the EMT program.

Similarly, it has been recognized for some time now that loss of adherens and tight junctions that enable cell–cell adhesions can prevent the acquisition of invasive properties of carcinomas (Frixen et al. 1991; Martin and Jiang 2009). Indeed, multiple studies report the importance of retaining adherens and tight junctions to prevent the acquisition of malignant phenotypes in colorectal (Bornholdt et al. 2011), pancreatic (Takai et al. 2005), and breast carcinomas (Vleminckx et al. 1991; Cavallaro and Christofori 2004; Martin et al. 2010), among others. Thus, together with cell polarity, cell–cell junctions represent key traits associated with cells of epithelial tissue and their architectural organization. These traits are often lost at the initial stages of malignancy, and in many cases, must be lost in order for successful malignant transformation. Here again, it is well documented that the EMT-TFs play a significant role in the disruption of cell junctions through the transcriptional repression of *CDH1* (Comijn et al. 2001; Hajra et al. 2002; Bolos et al. 2003; Eger et al. 2005), which codes for the key adherens junction protein

E-cadherin, as well as tight junction proteins such as claudins (Martinez-Estrada et al. 2006) and occludins (Ikenouchi et al. 2003; Ohkubo and Ozawa 2004). Through the repression of these genes, whose products maintain epithelial cell junctions, the EMT program enables an initial advance of carcinoma cell progression toward high-grade malignancy, doing so without the gain of additional overt mesenchymal traits.

Another key property of epithelial tissue is the maintenance of interactions with the basement membrane, which provides structural support while also regulating cell behavior through its ability to control invasion and thereby maintain tissue integrity (Paulsson 1992; Kelley et al. 2014). Snail is known to up-regulate the expression of several matrix metalloproteinase (MMP) genes, including *MMP9*, which degrades the basement membrane and stimulates tumor cell invasion (Jorda et al. 2005; Miyoshi et al. 2005; Kessenbrock et al. 2010). Additionally, other factors such as Zeb2 and TGF-β indirectly activate *MMP2* through activation of the Ets-2 transcription factor (Taki et al. 2006). Hence, a variety of EMT-inducing signals are capable of repressing key epithelial traits that pave the way for malignancy and tumor progression. Importantly, in none of these cited reports was the full spectrum of EMT-associated mesenchymal traits examined, leaving open the possibility that partial EMT programs, which only involve the loss of certain epithelial traits without gain of a full suite of mesenchymal ones, were launched.

DOES AN EMT INVARIABLY ENTAIL THE ACQUISITION OF OVERT MESENCHYMAL TRAITS?

As suggested above, increasing evidence indicates that the term EMT subsumes a collection of cell-biological programs rather than a single, stereotypical one. Initially, it was believed that cells that have undergone an EMT resemble fibroblasts (i.e., acquire properties that make them completely mesenchymal). This prompted various efforts to induce as complete an EMT program as experimentally possible, this being undertaken with the preconception that the more complete the EMT, the more aggressive carcinoma cells would appear.

With the passage of time, this thinking has become more nuanced, and we now realize that the acquisition of a full range of EMT-induced mesenchymal traits and associated loss of all epithelial markers is not necessarily correlated with the acquisition by carcinoma cells of aggressive traits and, as discussed above, stemness. Moreover, because different carcinoma cell types exhibit differing proportions of epithelial and mesenchymal traits (Li and Kang 2016), certain studies have attempted to classify breast cancer cells by assigning them scores that would indicate the extent of their EMT-associated features (Tan et al. 2014). Such studies have highlighted a by-now widely embraced notion: there is no standard, stereotypical version of the EMT program. It is therefore likely that in certain cases, aggressive metastasizing carcinomas may exhibit overt mesenchymal properties that aid its metastatic spread (Trimboli et al. 2008; Bonnomet et al. 2012), whereas in other cases they may not require these properties; moreover, display of these traits may actually be counterproductive for stemness and tumor progression (Celia-Terrassa et al. 2012).

These various reports converge on the increasingly accepted notion that advance partially through an EMT program and resulting residence by carcinoma cells in a partially epithelial, partially mesenchymal state favors tumor progression and metastasis (Lundgren et al. 2009; Jordan et al. 2011; Bednarz-Knoll et al. 2012; Sampson et al. 2014; Grosse-Wilde et al. 2015; Hong et al. 2015; Schliekelman et al. 2015). This suggests, in turn, than an acquisition of overt mesenchymal traits may not be necessary for carcinoma cells to develop aggressive, metastatic properties. Given the prevalence of processes such as collective migration/invasion by tumor cell clusters, as discussed earlier, it is likely that the partial EMT state, which enables some cells to transiently repress certain epithelial properties, is a driving factor for this form of dissemination (Grigore et al. 2016).

REDEFINING THE EMT PROGRAM

Given the currently rapid progress of EMT research and the extensive literature on this topic that has been produced in recent years, we would propose two guidelines that can be used to understand EMT programs in the context of carcinoma progression and metastasis. First, the EMT appears to generate cells residing in a spectrum of states lying in multiple phenotypic states between the highly organized epithelial state characteristic of normal tissue architecture and the extremely mesenchymal state resembling that of fibroblasts. A transition between any two states within this spectrum to a more mesenchymal state may occur under the aegis of an EMT program (Fig. 1). In the majority of instances, carcinoma cells disseminate by means of tumor cell clusters; the forward march of these clusters does not seem to require an overt transition of the bulk of the cells to a truly mesenchymal state. Instead, the majority of cells in these clusters may have lost certain epithelial traits, such as adherence to a basement membrane and an alteration in their form of cell polarity, whereas subpopulations of cells in these invading cell cohorts, specifically those at the leading/invasive edges of these groups, may indeed express, at least transiently, traits typically associated with EMT programs.

Accordingly, we propose that the EMT program can actually be seen as a program presenting in many different forms: In some forms one may observe the loss of epithelial integrity and architecture while still retaining the expression of certain epithelial markers (Khalil and Friedl 2010; Clark and Vignjevic 2015), whereas in other cases one may observe the adoption of certain mesenchymal traits and markers while still retaining the expression of epithelial traits (McCaffrey et al. 2012; Yu et al. 2013). In yet another version of the EMT program, one may observe a more extended expression of the program that involves an almost-complete loss of epithelial traits and a gain of fibroblast-like mesenchymal traits (Bonnomet et al. 2012).

Second, these various versions of the EMT program are united by the fact that members of a small cohort of EMT-TFs are expressed in cells that have begun to activate this program, being increasingly expressed as cells advance progressively toward a fully mesenchymal state. These EMT-TFs include prominently Snail, Twist, Slug, and Zeb1, among others (De Craene and Berx 2013). Depending on the nature of the EMT, these factors may be expressed at levels that are sufficient to repress the transcription of certain genes that regulate key epithelial features while not being sufficient to induce the expression of mesenchymal markers. Thus, *CDH1*, which codes for E-cadherin, can be repressed by a number of EMT-TFs, leading to the destabilization of adherens junctions without the concomitant acquisition of overt mesenchymal markers such as vimentin. In other cases, the coexpression of both epithelial and mesenchymal markers is observed in the same cell, indicating the operation of a different transcriptional program that maintains a hybrid state. Nevertheless, it is clear that the version of the EMT program induced is highly dependent on the specific EMT-TF or combination of TFs that are expressed in a particular context.

TARGETING THE EMT THROUGH DIFFERENTIATION-INDUCING THERAPY

In light of the malignant traits displayed by cancer stem cells (CSCs), an important focus of current research is the identification and targeting of signaling pathways that sustain residence of carcinoma cells in the CSC state. Unfortunately, however, attempts to selectively target carcinoma cells exhibiting exhibit stem-like properties is thwarted by the fact that our understanding of the factors contributing to stemness—including where CSCs lie along the epithelial-mesenchymal spectrum—remains vague at best. Thus, in the absence of knowing which signaling pathways are preferentially activated in the CSCs, the choice of a particular pathway for therapeutic targeting remains arbitrary.

A possibly more attractive therapeutic strategy for dismantling the aggressive traits of CSCs that have acquired EMT-associated traits is a strategy designed to reverse this program by inducing a MET. The induction of an MET in carcinoma cells would ideally involve the forced shedding of mesenchymal traits and a reestablishment of the epithelial cell state, integrity and architecture, which would be accompanied by a loss of stem-like properties.

As mentioned above, it is becoming increasingly clear that as cells undergo an EMT, they slip into a more dedifferentiated state that enhances their ability to resist chemotherapeutic assault (Mani et al. 2008; Morel et al. 2008; Gupta et al. 2009). Hence, the loss of stemness and the process of undergoing differentiation is likely to deprive these cells of traits that fuel their malignant behavior. In fact, the induction of differentiation as a form of therapy was first utilized for the treatment of acute promyelocytic leukemia, where the administration of all-trans retinoic acid was successfully used as a targeted therapy (Warrell et al. 1993). In the case of breast and endometrial carcinomas, the use of histone deacetylase inhibitors (HDACis) induces their differentiation, presumably through the derepression of certain gene promoters (Munster et al. 2001; Uchida et al. 2007; Xu et al. 2007). Along these lines, it appears that processes that serve to maintain the epithelial state of cells act as potent tumor suppressors. This is especially true for features, such as the maintenance of cell polarity (Lee and Vasioukhin 2008), as well as the presence of adherens and tight junctions (Jeanes et al. 2008; Martin and Jiang 2009) between epithelial cells and of hemidesmosomes that maintain interactions with the basement membrane (De Arcangelis et al. 2016). Thus, in the case of carcinomas, it is plausible that processes that maintain the epithelial state are especially relevant for the prevention of malignancy and EMT-induced tumor progression.

As cells undergo an EMT, a series of signaling pathways is activated, while a host of others are down-regulated. The switch in some of these pathways, such as TGF-β and Wnt, is initiated by paracrine signals from the tumor-associated stroma, whereas the maintenance of the mesenchymal state is propagated through autocrine signals that maintain their residence in that state (Scheel et al. 2011). Other factors such as PGE2, secreted by mesenchymal stem cells (MSCs), are also known to induce a mesenchymal/CSC state by creating a favorable niche (Li et al. 2012). NF-κB signaling activated by inflammatory cytokines present in the tumor microenvironment also triggers the activation of NOTCH signaling, which allows expansion of the CSC pool in basal-like breast cancers (Yamamoto et al. 2013). Consequently, it seems that disruption of these paracrine and autocrine signaling axes that maintain residence of tumor cells in a more mesenchymal/stem-like state can force them into a more differentiated epithelial state that would render them more susceptible to conventional therapy. Importantly, such a treatment strategy would restore some of the key traits of epithelial cells, such as the cell–cell junctions that form the foundation of epithelial tissue integrity and morphology.

We recently reported that the elevation of 3′,5′-cyclic adenosine monophosphate (cAMP) levels in certain breast cancer cells can, through the activation of protein kinase A (PKA), induce an MET by enforcing a gene expression program reminiscent of cells that normally exhibit a fully epithelial identity, rendering these cells more susceptible to treatment with conventional chemotherapeutic drugs (Pattabiraman et al. 2016). Activation of such a pathway, which serves to induce and perpetuate residence of cells in an epithelial state, should ideally be an attractive target for differentiation therapy, as it curtails the malignant traits that are associated with tumor progression. Moreover, the continued stimulation of a pathway required for epithelial maintenance is unlikely to have any adverse effects on the already-epithelial, more benign compartment of carcinomas or on the normal resident epithelia of the tissue-of-origin. A caveat of using such an MET-induced differentiation therapy in the case of carcinomas is the observed requirement of an MET to complete the colonization stage of the metastasis cascade. Consequently, the induction of an MET might inadvertently support the process of metastatic colonization at distant sites. Nevertheless, the potential of differentiation therapy makes it an attractive strategy for curtailing the malignant and stem-like properties that accompany the induction of an EMT in carcinomas.

CONCLUSION

Recent findings have provided more insights into the functioning of the EMT program, the nature of its manifestations, and its contribution to tumor progression and resistance to conventional therapy. These developments have furthered our understanding of the spectrum of phenotypic states orchestrated by various versions of the EMT program and provided some evidence for a partial EMT state favoring the CSC state of carcinoma cells. Although the precise mixture of epithelial and mesenchymal traits required for carcinoma cells to enter into a stem-like state is still unknown, there is no longer any doubt that the epithelial state, as characterized by cell–cell junctions, apical–basal polarity, and adherence to a basement membrane, is unfavorable for stemness and tumor progression. Hence, forced entrance of carcinoma cells that previously embarked on an EMT program into a more epithelial state likely represents an attractive strategy that can be applied to eradicate the aggressive properties of a wide range of carcinoma cells.

ACKNOWLEDGMENTS

We thank Dr. Arthur Lambert, Dr. Jordan Krall, Dr. Brian Bierie, and other members of the Weinberg laboratory for helpful discussions and Ms. Meredith Leffler for assistance with the illustrations. D.R.P was supported by a C.J. Martin Overseas Biomedical Fellowship from the National Health and Medical Research Council of Australia (NHMRC APP1071853) and is currently supported by a K99/R00 Pathway to Independence Award (NIH/NCI 1K99CA201574-01A1). The Weinberg laboratory is supported by grants from the National Institutes of Health (NIH; R01 CA078461, P01 CA080111), the Samuel Waxman Cancer Research Foundation, and the Ludwig Center for Molecular Oncology at MIT. R.A.W. is an American Cancer Society Research Professor and a Daniel K. Ludwig Cancer Research Professor.

REFERENCES

Aigner K, Dampier B, Descovich L, Mikula M, Sultan A, Schreiber M, Mikulits W, Brabletz T, Strand D, Obrist P, et al. 2007. The transcription factor ZEB1 (δEF1) promotes tumour cell dedifferentiation by repressing master regulators of epithelial polarity. *Oncogene* **26:** 6979–6988.

Andriani F, Bertolini G, Facchinetti F, Baldoli E, Moro M, Casalini P, Caserini R, Milione M, Leone G, Pelosi G, et al. 2016. Conversion to stem-cell state in response to microenvironmental cues is regulated by balance between epithelial and mesenchymal features in lung cancer cells. *Mol Oncol* **10:** 253–271.

Aranda V, Haire T, Nolan ME, Calarco JP, Rosenberg AZ, Fawcett JP, Pawson T, Muthuswamy SK. 2006. Par6-aPKC uncouples ErbB2 induced disruption of polarized epithelial organization from proliferation control. *Nat Cell Biol* **8:** 1235–1245.

Barsky SH, Siegal GP, Jannotta F, Liotta LA. 1983. Loss of basement membrane components by invasive tumors but not by their benign counterparts. *Lab Invest* **49:** 140–147.

Bednarz-Knoll N, Alix-Panabieres C, Pantel K. 2012. Plasticity of disseminating cancer cells in patients with epithelial malignancies. *Cancer Metastasis Rev* **31:** 673–687.

Bolos V, Peinado H, Perez-Moreno MA, Fraga MF, Esteller M, Cano A. 2003. The transcription factor Slug represses E-cadherin expression and induces epithelial to mesenchymal transitions: A comparison with Snail and E47 repressors. *J Cell Sci* **116:** 499–511.

Bonnomet A, Syne L, Brysse A, Feyereisen E, Thompson EW, Noel A, Foidart JM, Birembaut P, Polette M, Gilles C. 2012. A dynamic in vivo model of epithelial-to-mesenchymal transitions in circulating tumor cells and metastases of breast cancer. *Oncogene* **31:** 3741–3753.

Bornholdt J, Friis S, Godiksen S, Poulsen SS, Santoni-Rugiu E, Bisgaard HC, Lothe IM, Ikdahl T, Tveit KM, Johnson E, et al. 2011. The level of claudin-7 is reduced as an early event in colorectal carcinogenesis. *BMC Cancer* **11:** 65.

Bronsert P, Enderle-Ammour K, Bader M, Timme S, Kuehs M, Csanadi A, Kayser G, Kohler I, Bausch D, Hoeppner J, et al. 2014. Cancer cell invasion and EMT marker expression: A three-dimensional study of the human cancer–host interface. *J Pathol* **234:** 410–422.

Cavallaro U, Christofori G. 2004. Cell adhesion and signalling by cadherins and Ig-CAMs in cancer. *Nat Rev Cancer* **4:** 118–132.

Celia-Terrassa T, Meca-Cortes O, Mateo F, Martinez de Paz A, Rubio N, Arnal-Estape A, Ell BJ, Bermudo R, Diaz A, Guerra-Rebollo M, et al. 2012. Epithelial–mesenchymal transition can suppress major attributes of human epithelial tumor-initiating cells. *J Clin Invest* **122:** 1849–1868.

Chang CC, Hsu WH, Wang CC, Chou CH, Kuo MY, Lin BR, Chen ST, Tai SK, Kuo ML, Yang MH. 2013. Connective tissue growth factor activates pluripotency genes and mesenchymal–epithelial transition in head and neck cancer cells. *Cancer Res* **73:** 4147–4157.

Cheung KJ, Padmanaban V, Silvestri V, Schipper K, Cohen JD, Fairchild AN, Gorin MA, Verdone JE, Pienta KJ, Bader JS, et al. 2016. Polyclonal breast cancer metastases arise from collective dissemination of keratin 14–expressing tumor cell clusters. *Proc Natl Acad Sci* **113:** E854–E863.

Clark AG, Vignjevic DM. 2015. Modes of cancer cell invasion and the role of the microenvironment. *Curr Opin Cell Biol* **36:** 13–22.

Comijn J, Berx G, Vermassen P, Verschueren K, van Grunsven L, Bruyneel E, Mareel M, Huylebroeck D, van Roy F. 2001. The two-handed E box binding zinc finger protein SIP1 downregulates E-cadherin and induces invasion. *Mol Cell* **7:** 1267–1278.

De Arcangelis A, Hamade H, Alpy F, Normand S, Bruyere E, Lefebvre O, Mechine-Neuville A, Siebert S, Pfister V, Lepage P, et al. 2016. Hemidesmosome integrity protects the colon against colitis and colorectal cancer. *Gut*. doi: 10.1136/gutjnl-2015-310847.

De Craene B, Berx G. 2013. Regulatory networks defining EMT during cancer initiation and progression. *Nat Rev Cancer* **13:** 97–110.

Eger A, Aigner K, Sonderegger S, Dampier B, Oehler S, Schreiber M, Berx G, Cano A, Beug H, Foisner R. 2005. DeltaEF1 is a transcriptional repressor of E-cadherin and regulates epithelial plasticity in breast cancer cells. *Oncogene* **24:** 2375–2385.

Fan F, Samuel S, Evans KW, Lu J, Xia L, Zhou Y, Sceusi E, Tozzi F, Ye XC, Mani SA, et al. 2012. Overexpression of snail induces epithelial-mesenchymal transition and a cancer stem cell-like phenotype in human colorectal cancer cells. *Cancer Med* **1:** 5–16.

Fidler IJ. 2003. The pathogenesis of cancer metastasis: The 'seed and soil' hypothesis revisited. *Nat Rev Cancer* **3:** 453–458.

Fischer KR, Durrans A, Lee S, Sheng J, Li F, Wong ST, Choi H, El Rayes T, Ryu S, Troeger J, et al. 2015. Epithelial-to-mesenchymal transition is not required for lung metastasis but contributes to chemoresistance. *Nature* **527:** 472–476.

Friedl P, Gilmour D. 2009. Collective cell migration in morphogenesis, regeneration and cancer. *Nat Rev Mol Cell Biol* **10:** 445–457.

Friedl P, Locker J, Sahai E, Segall JE. 2012. Classifying collective cancer cell invasion. *Nat Cell Biol* **14:** 777–783.

Frixen UH, Behrens J, Sachs M, Eberle G, Voss B, Warda A, Lochner D, Birchmeier W. 1991. E-cadherin-mediated cell-cell adhesion prevents invasiveness of human carcinoma cells. *J Cell Biol* **113:** 173–185.

Greenburg G, Hay ED. 1982. Epithelia suspended in collagen gels can lose polarity and express characteristics of migrating mesenchymal cells. *J Cell Biol* **95:** 333–339.

Grigore AD, Jolly MK, Jia D, Farach-Carson MC, Levine H. 2016. Tumor budding: The name is EMT. Partial EMT. *J Clin Med* **5:** 51.

Grosse-Wilde A, Fouquier d'Herouel A, McIntosh E, Ertaylan G, Skupin A, Kuestner RE, del Sol A, Walters KA, Huang S. 2015. Stemness of the hybrid epithelial/mesenchymal state in breast cancer and its association with poor survival. *PLoS One* **10:** e0126522.

Guo W, Keckesova Z, Donaher JL, Shibue T, Tischler V, Reinhardt F, Itzkovitz S, Noske A, Zurrer-Hardi U, Bell G, et al. 2012. Slug and Sox9 cooperatively determine the mammary stem cell state. *Cell* **148:** 1015–1028.

Gupta PB, Onder TT, Jiang G, Tao K, Kuperwasser C, Weinberg RA, Lander ES. 2009. Identification of selective inhibitors of cancer stem cells by high-throughput screening. *Cell* **138:** 645–659.

Hajra KM, Chen DY, Fearon ER. 2002. The SLUG zinc-finger protein represses E-cadherin in breast cancer. *Cancer Res* **62:** 1613–1618.

Hong T, Watanabe K, Ta CH, Villarreal-Ponce A, Nie Q, Dai X. 2015. An Ovol2-Zeb1 mutual inhibitory circuit governs bidirectional and multi-step transition between epithelial and mesenchymal states. *PLoS Comput Biol* **11:** e1004569.

Ikenouchi J, Matsuda M, Furuse M, Tsukita S. 2003. Regulation of tight junctions during the epithelium-mesenchyme transition: Direct repression of the gene expression of claudins/occludin by Snail. *J Cell Sci* **116:** 1959–1967.

Jeanes A, Gottardi CJ, Yap AS. 2008. Cadherins and cancer: How does cadherin dysfunction promote tumor progression? *Oncogene* **27:** 6920–6929.

Jolly MK, Boareto M, Huang B, Jia D, Lu M, Ben-Jacob E, Onuchic JN, Levine H. 2015a. Implications of the hybrid epithelial/mesenchymal phenotype in metastasis. *Front Oncol* **5:** 155.

Jolly MK, Jia D, Boareto M, Mani SA, Pienta KJ, Ben-Jacob E, Levine H. 2015b. Coupling the modules of EMT and stemness: A tunable 'stemness window' model. *Oncotarget* **6:** 25161–25174.

Jorda M, Olmeda D, Vinyals A, Valero E, Cubillo E, Llorens A, Cano A, Fabra A. 2005. Upregulation of MMP-9 in MDCK epithelial cell line in response to expression of the Snail transcription factor. *J Cell Sci* **118:** 3371–3385.

Jordan NV, Johnson GL, Abell AN. 2011. Tracking the intermediate stages of epithelial-mesenchymal transition in epithelial stem cells and cancer. *Cell Cycle* **10:** 2865–2873.

Kalluri R, Weinberg RA. 2009. The basics of epithelial-mesenchymal transition. *J Clin Invest* **119:** 1420–1428.

Kelley LC, Lohmer LL, Hagedorn EJ, Sherwood DR. 2014. Traversing the basement membrane in vivo: A diversity of strategies. *J Cell Biol* **204:** 291–302.

Kessenbrock K, Plaks V, Werb Z. 2010. Matrix metalloproteinases: Regulators of the tumor microenvironment. *Cell* **141:** 52–67.

Khalil AA, Friedl P. 2010. Determinants of leader cells in collective cell migration. *Integr Biol (Camb)* **2:** 568–574.

Kong D, Banerjee S, Ahmad A, Li Y, Wang Z, Sethi S, Sarkar FH. 2010. Epithelial to mesenchymal transition is mechanistically linked with stem cell signatures in prostate cancer cells. *PLoS One* **5:** e12445.

Labelle M, Begum S, Hynes RO. 2011. Direct signaling between platelets and cancer cells induces an epithelial–mesenchymal-like transition and promotes metastasis. *Cancer Cell* **20:** 576–590.

Lee M, Vasioukhin V. 2008. Cell polarity and cancer–cell and tissue polarity as a non-canonical tumor suppressor. *J Cell Sci* **121:** 1141–1150.

Li W, Kang Y. 2016. Probing the fifty shades of EMT in metastasis. *Trends Cancer* **2:** 65–67.

Li HJ, Reinhardt F, Herschman HR, Weinberg RA. 2012. Cancer-stimulated mesenchymal stem cells create a carcinoma stem cell niche via prostaglandin E2 signaling. *Cancer Discov* **2:** 840–855.

Li J, Liu J, Li P, Mao X, Li W, Yang J, Liu P. 2014. Loss of LKB1 disrupts breast epithelial cell polarity and promotes breast cancer metastasis and invasion. *J Exp Clin Cancer Res* **33:** 70.

Long H, Xiang T, Qi W, Huang J, Chen J, He L, Liang Z, Guo B, Li Y, Xie R, et al. 2015. $CD133^+$ ovarian cancer stem-like cells promote non-stem cancer cell metastasis via CCL5 induced epithelial–mesenchymal transition. *Oncotarget* **6:** 5846–5859.

Lundgren K, Nordenskjold B, Landberg G. 2009. Hypoxia, Snail and incomplete epithelial–mesenchymal transition in breast cancer. *Br J Cancer* **101:** 1769–1781.

Mani SA, Guo W, Liao MJ, Eaton EN, Ayyanan A, Zhou AY, Brooks M, Reinhard F, Zhang CC, Shipitsin M, et al. 2008. The epithelial–mesenchymal transition generates cells with properties of stem cells. *Cell* **133:** 704–715.

Martin TA, Jiang WG. 2009. Loss of tight junction barrier function and its role in cancer metastasis. *Biochim Biophys Acta* **1788:** 872–891.

Martin TA, Mansel RE, Jiang WG. 2010. Loss of occludin leads to the progression of human breast cancer. *Int J Mol Med* **26:** 723–734.

Martin-Belmonte F, Gassama A, Datta A, Yu W, Rescher U, Gerke V, Mostov K. 2007. PTEN-mediated apical segregation of phosphoinositides controls epithelial morphogenesis through Cdc42. *Cell* **128:** 383–397.

Martinez-Estrada OM, Culleres A, Soriano FX, Peinado H, Bolos V, Martinez FO, Reina M, Cano A, Fabre M, Vilaro S. 2006. The transcription factors Slug and Snail act as repressors of Claudin-1 expression in epithelial cells. *Biochem J* **394:** 449–457.

McCaffrey LM, Montalbano J, Mihai C, Macara IG. 2012. Loss of the Par3 polarity protein promotes breast tumorigenesis and metastasis. *Cancer Cell* **22:** 601–614.

Miyoshi A, Kitajima Y, Kido S, Shimonishi T, Matsuyama S, Kitahara K, Miyazaki K. 2005. Snail accelerates cancer invasion by upregulating MMP expression and is associated with poor prognosis of hepatocellular carcinoma. *Br J Cancer* **92:** 252–258.

Morel AP, Lievre M, Thomas C, Hinkal G, Ansieau S, Puisieux A. 2008. Generation of breast cancer stem cells through epithelial-mesenchymal transition. *PLoS One* **3:** e2888.

Munster PN, Troso-Sandoval T, Rosen N, Rifkind R, Marks PA, Richon VM. 2001. The histone deacetylase inhibitor suberoylanilide hydroxamic acid induces differentiation of human breast cancer cells. *Cancer Res* **61:** 8492–8497.

Nassour M, Idoux-Gillet Y, Selmi A, Come C, Faraldo ML, Deugnier MA, Savagner P. 2012. Slug controls stem/progenitor cell growth dynamics during mammary gland morphogenesis. *PLoS One* **7:** e53498.

Nieto MA, Huang RY, Jackson RA, Thiery JP. 2016. EMT: 2016. *Cell* **166:** 21–45.

Ocana OH, Corcoles R, Fabra A, Moreno-Bueno G, Acloque H, Vega S, Barrallo-Gimeno A, Cano A, Nieto MA. 2012. Metastatic colonization requires the repression of the epithelial–mesenchymal transition inducer Prrx1. *Cancer Cell* **22:** 709–724.

Ohkubo T, Ozawa M. 2004. The transcription factor Snail downregulates the tight junction components independently of E-cadherin downregulation. *J Cell Sci* **117:** 1675–1685.

Okada H, Danoff TM, Kalluri R, Neilson EG. 1997. Early role of Fsp1 in epithelial-mesenchymal transformation. *Am J Physiol* **273:** F563–F574.

Pattabiraman DR, Bierie B, Kober KI, Thiru P, Krall JA, Zill C, Reinhardt F, Tam WL, Weinberg RA. 2016. Activation of PKA leads to mesenchymal-to-epithelial transition and loss of tumor-initiating ability. *Science* **351:** aad3680.

Paulsson M. 1992. Basement membrane proteins: Structure, assembly, and cellular interactions. *Crit Rev Biochem Mol Biol* **27:** 93–127.

Rasheed ZA, Yang J, Wang Q, Kowalski J, Freed I, Murter C, Hong SM, Koorstra JB, Rajeshkumar NV, He X, et al. 2010. Prognostic significance of tumorigenic cells with mesenchymal features in pancreatic adenocarcinoma. *J Natl Cancer Inst* **102:** 340–351.

Revenu C, Gilmour D. 2009. EMT 2.0: Shaping epithelia through collective migration. *Curr Opin Genetics Dev* **19:** 338–342.

Sampson VB, David JM, Puig I, Patil PU, de Herreros AG, Thomas GV, Rajasekaran AK. 2014. Wilms' tumor protein induces an epithelial–mesenchymal hybrid differentiation state in clear cell renal cell carcinoma. *PLoS One* **9:** e102041.

Scheel C, Eaton EN, Li SH, Chaffer CL, Reinhardt F, Kah KJ, Bell G, Guo W, Rubin J, Richardson AL, et al. 2011. Paracrine and autocrine signals induce and maintain mesenchymal and stem cell states in the breast. *Cell* **145:** 926–940.

Schliekelman MJ, Taguchi A, Zhu J, Dai X, Rodriguez J, Celiktas M, Zhang Q, Chin A, Wong CH, Wang H, et al. 2015. Molecular portraits of epithelial, mesenchymal, and hybrid states in lung adenocarcinoma and their relevance to survival. *Cancer Res* **75:** 1789–1800.

Spaderna S, Schmalhofer O, Wahlbuhl M, Dimmler A, Bauer K, Sultan A, Hlubek F, Jung A, Strand D, Eger A, et al. 2008. The transcriptional repressor ZEB1 promotes metastasis and loss of cell polarity in cancer. *Cancer Res* **68:** 537–544.

Takai E, Tan X, Tamori Y, Hirota M, Egami H, Ogawa M. 2005. Correlation of translocation of tight junction protein Zonula occludens-1 and activation of epidermal growth factor receptor in the regulation of invasion of pancreatic cancer cells. *Int J Oncol* **27:** 645–651.

Taki M, Verschueren K, Yokoyama K, Nagayama M, Kamata N. 2006. Involvement of Ets-1 transcription factor in inducing matrix metalloproteinase-2 expression by epithelial–mesenchymal transition in human squamous carcinoma cells. *Int J Oncol* **28:** 487–496.

Tan TZ, Miow QH, Miki Y, Noda T, Mori S, Huang RY, Thiery JP. 2014. Epithelial–mesenchymal transition spectrum quantification and its efficacy in deciphering survival and drug responses of cancer patients. *EMBO Mol Med* **6:** 1279–1293.

Trimboli AJ, Fukino K, de Bruin A, Wei G, Shen L, Tanner SM, Creasap N, Rosol TJ, Robinson ML, Eng C, et al. 2008. Direct evidence for epithelial–mesenchymal transitions in breast cancer. *Cancer Res* **68:** 937–945.

Tsai JH, Yang J. 2013. Epithelial-mesenchymal plasticity in carcinoma metastasis. *Genes Dev* **27:** 2192–2206.

Tsai JH, Donaher JL, Murphy DA, Chau S, Yang J. 2012. Spatiotemporal regulation of epithelial–mesenchymal transition is essential for squamous cell carcinoma metastasis. *Cancer Cell* **22:** 725–736.

Uchida H, Maruyama T, Ono M, Ohta K, Kajitani T, Masuda H, Nagashima T, Arase T, Asada H, Yoshimura Y. 2007. Histone deacetylase inhibitors stimulate cell migration in human endometrial adenocarcinoma cells through up-regulation of glycodelin. *Endocrinology* **148:** 896–902.

Vleminckx K, Vakaet L Jr, Mareel M, Fiers W, van Roy F. 1991. Genetic manipulation of E-cadherin expression by epithelial tumor cells reveals an invasion suppressor role. *Cell* **66:** 107–119.

Warrell RP Jr, de The H, Wang ZY, Degos L. 1993. Acute promyelocytic leukemia. *N Engl J Med* **329:** 177–189.

Westcott JM, Prechtl AM, Maine EA, Dang TT, Esparza MA, Sun H, Zhou Y, Xie Y, Pearson GW. 2015. An epigenetically distinct breast cancer cell subpopulation promotes collective invasion. *J Clin Invest* **125:** 1927–1943.

Whiteman EL, Liu CJ, Fearon ER, Margolis B. 2008. The transcription factor snail represses Crumbs3 expression and disrupts apico-basal polarity complexes. *Oncogene* **27:** 3875–3879.

Xie G, Ji A, Yuan Q, Jin Z, Yuan Y, Ren C, Guo Z, Yao Q, Yang K, Lin X, et al. 2014. Tumour-initiating capacity is independent of epithelial-mesenchymal transition status in breast cancer cell lines. *Br J Cancer* **110:** 2514–2523.

Xu WS, Parmigiani RB, Marks PA. 2007. Histone deacetylase inhibitors: Molecular mechanisms of action. *Oncogene* **26:** 5541–5552.

Yamamoto M, Taguchi Y, Ito-Kureha T, Semba K, Yamaguchi N, Inoue J. 2013. NF-κB non-cell-autonomously regulates cancer stem cell populations in the basal-like breast cancer subtype. *Nat Commun* **4:** 2299.

Yang J, Mani SA, Donaher JL, Ramaswamy S, Itzykson RA, Come C, Savagner P, Gitelman I, Richardson A, Weinberg RA. 2004. Twist, a master regulator of morphogenesis, plays an essential role in tumor metastasis. *Cell* **117:** 927–939.

Ye X, Tam WL, Shibue T, Kaygusuz Y, Reinhardt F, Ng Eaton E, Weinberg RA. 2015. Distinct EMT programs control normal mammary stem cells and tumour-initiating cells. *Nature* **525:** 256–260.

Yu M, Bardia A, Wittner BS, Stott SL, Smas ME, Ting DT, Isakoff SJ, Ciciliano JC, Wells MN, Shah AM, et al. 2013. Circulating breast tumor cells exhibit dynamic changes in epithelial and mesenchymal composition. *Science* **339:** 580–584.

Zheng X, Carstens JL, Kim J, Scheible M, Kaye J, Sugimoto H, Wu CC, LeBleu VS, Kalluri R. 2015. Epithelial-to-mesenchymal transition is dispensable for metastasis but induces chemoresistance in pancreatic cancer. *Nature* **527:** 525–530.

Zhou D, Kannappan V, Chen X, Li J, Leng X, Zhang J, Xuan S. 2016. RBP2 induces stem-like cancer cells by promoting EMT and is a prognostic marker for renal cell carcinoma. *Exp Mol Med* **48:** e238.

Trimming the Vascular Tree in Tumors: Metabolic and Immune Adaptations

ELIZABETH ALLEN, RINDERT MISSIAEN, AND GABRIELE BERGERS
KU-Leuven and VIB-Center for Cancer Biology, 3000 Leuven, Belgium
Correspondence: gabriele.bergers@kuleuven.vib.be

Angiogenesis, the formation of new blood vessels, has become a well-established hallmark of cancer. Its functional importance for the manifestation and progression of tumors has been further validated by the beneficial therapeutic effects of angiogenesis inhibitors, most notably ones targeting the vascular endothelial growth factor (VEGF) signaling pathways. However, with the transient and short-lived nature of the patient response, it has become evident that tumors have the ability to adapt to the pressures of vascular growth restriction. Several escape mechanisms have been described that adapt tumors to therapy-induced low-oxygen tension by either reinstating tumor growth by vascular rebound or by altering tumor behavior without the necessity to reinitiate revascularization. We review here two bypass mechanisms that either instigate angiogenic and immune-suppressive polarization of intratumoral innate immune cells to facilitate VEGF-independent angiogenesis or enable metabolic adaptation and reprogramming of endothelial cells and tumor cells to adapt to low-oxygen tension.

Neovascularization, the formation of new blood vessels, has been defined by Hanahan and Weinberg (2011) as one of the six pivotal hallmarks of cancer because it is a necessary process to enable sufficient oxygen supply for an expanding tumor and to dispose of waste products. Vascular sprouting is the most prominent mechanism by which new vessels arise from preexisting capillaries or postcapillary venules. Thereby, endothelial cells become activated by and move toward a gradient of proangiogenic factors, forming expanding sprouts with proliferating endothelial stalk cells and migrating tip cells at the leading edge until tip cells anastomose with cells from neighboring sprouts to connect the newly formed vessel structures. Consequently, vascular growth is terminated and vessels mature and stabilize with the formation of a basement membrane and recruitment and embracement of pericytes (Carmeliet 2005; Adams and Alitalo 2007). Numerous activating and inhibiting factors including the vascular endothelial growth factor (VEGF), fibroblast growth factor (FGF), and angiopoietin tyrosine kinase receptor families and various members of the CXC and CC chemokines orchestrate this multistep process (Bergers and Benjamin 2003; Santoni et al. 2014). However, although blood vessel formation is tightly regulated during physiological conditions, tumors have lost the appropriate balances between positive and negative angiogenic controls. Once tumor angiogenesis is induced, it remains activated leading to a continually and abnormally expanding tumor vasculature (Baluk et al. 2005; Carmeliet 2005). As a consequence, tumor vessels become leaky, aggravating tumor cell intravasation and dissemination and causing a sluggish and irregular blood flow that, together with an expanding tumor mass, increases interstitial pressure and hypoxia, further exacerbating an angiogenic response (Jain 2002; Carmeliet and Jain 2011; Potente et al. 2011).

There is increasing evidence that angiogenesis and inflammation, although being distinct and separable processes, are closely related events that are in part regulated by common chemokines. Indeed, VEGF, one of the most prominent angiogenic factors, which is overexpressed in numerous tumor types, not only promotes angiogenesis but also conveys suppressive effects on different immune cell constituents (Motz and Coukos 2011). VEGF blocks dendritic cell maturation, enhances the expression of the negative checkpoint regulator programmed cell death ligand 1 (PDL1) on dendritic cells, and activates antigen-specific regulatory T cells (Tregs) in a neuropilin-1 (NRP1)-dependent manner (Gabrilovich et al. 1996; Motz and Coukos 2011). Further, VEGF generates an immune-suppressive vasculature by enhancing PDL1 expression and reducing intercellular adhesion molecule 1 (ICAM1) and vascular cell adhesion molecule 1 (VCAM1) on endothelial cells to diminish leukocyte–endothelial interactions and consequently block infiltration of cytotoxic T cells into the tumor (Griffioen et al. 1996; Bouzin et al. 2007; Motz and Coukos 2013).

Recognition of tumor angiogenesis as a hallmark of cancer, together with the functional importance of VEGF in tumor angiogenesis and immune suppression, has led to the development of various VEGF and VEGF receptor inhibitors (Jayson et al. 2016) Consequently, bevacizumab (Avastin, Genentech/Roche), a VEGF ligand-trapping monoclonal antibody, was the first U.S. Federal Drug Administration (FDA)-approved antiangiogenic drug followed by the angiokinase inhibitors sorafenib (Nexavar, Bayer), sunitinib (Sutent, Pfizer), and

Published by Cold Spring Harbor Laboratory Press; doi: 10.1101/sqb.2016.81.030940

axitinib (Inlyta, Pfizer) and others, that target the VEGF receptor (VEGFR) tyrosine kinases and other receptor tyrosine kinases (RTKs) (Ferrara 2004; Jain 2014; Vasudev and Reynolds 2014; McIntyre and Harris 2015). Although these antiangiogenic drugs have produced encouraging and favorable effects in a subset of cancer patients, they have only increased progression-free survival but have had little effect on overall survival. The basis for these treatment limitations lies in part in the development of multiple forms of adaptive resistance to angiogenic inhibitors in which tumors either reinstate growth by neovascularization (Casanovas et al. 2005; Shojaei and Ferrara 2008; Rivera and Bergers 2015), bypass the need for revascularization by co-opting blood vessels, or become more invasive or undergo metabolic reprogramming to withstand the deleterious effects of therapy (Rubenstein et al. 2000; Ebos et al. 2007; Bergers and Hanahan 2008; Paez-Ribes et al. 2009; Lu et al. 2012; Sennino et al. 2012). Importantly, several of these adaptation mechanisms are triggered by hypoxia that is generated by therapy-induced vascular pruning of tumors (Bergers and Hanahan 2008; McIntyre and Harris 2015). The molecular responses to hypoxia are predominantly mediated by a family of hypoxia-inducible factors (HIFs), which are composed of an α and a β subunit, and become stabilized and active under low-oxygen tension (Liao and Johnson 2007; Semenza 2011; Palazon et al. 2014). HIF proteins coordinate an extensive transcriptional program that ensures metabolic and vascular adaptation to low-oxygen tension by producing a plethora of factors implicated in different aspects of cellular homeostasis, including survival, metabolism, invasion, and angiogenesis. In this review, we will discuss two hypoxia-dependent adaptation mechanisms that enable tumors to escape the effects of antiangiogenic therapy: (1) the recruitment and polarization of innate immune cells to recreate a proangiogenic and immune-suppressive environment and, (2) metabolic reprogramming of tumor and host cells to endure tumor propagation in the absence of revascularization.

INNATE IMMUNE CELLS DRIVE RESPONSE AND RELAPSE TO ANTIANGIOGENIC THERAPY

Intratumoral hypoxia drives the mobilization and retention of various bone marrow–derived cells (BMDCs) where they become an additional source of chemokines and cytokines to support tumor progression by promoting angiogenesis and suppressing antitumor immunity (Coussens et al. 2000; Giraudo et al. 2004; Lin et al. 2006; Shojaei and Ferrara 2007b; Du et al. 2008). Under normal conditions, innate immune cells first display immune-stimulatory characteristics to eliminate potential infections when they enter sites of tissue injury but then become immune-suppressive and angiogenic to hinder excessive inflammation and allow tissue repair with the formation of blood vessels and reepithelialization (Kasuya and Tokura 2014). Tumors, however, co-opt and modify the tissue repair program to escape immune-surveillance by converting innate immune cells from an immune-stimulating to an immune-suppressive and angiogenic phenotype (Motz and Coukos 2011).

Although tumors can inherently produce myeloid cell–recruiting factors (e.g., G-CSF, CSF-1, GM-CSF), hypoxia activates many of these molecules and pathways that not only attract innate immune cells but also endorse their angiogenic and immune-suppressive characteristics (Schioppa et al. 2003; Staller et al. 2003; Murdoch et al. 2004; Grunewald et al. 2006; Du et al. 2008; Casazza et al. 2013). VEGF is one of the most prominent hypoxia-regulated angiogenic factors that in addition to affecting endothelial cells can also serve as a mobilizer and chemoattractant for myeloid cells via VEGF receptor 2 (VEGFR2) on monocytes (Dineen et al. 2008). Further, CXCL12 (SDF1α), implicated in the retention of myeloid cells, is induced by HIF-1α (Grunewald et al. 2006; Du et al. 2008), as is its chemokine receptor CXCR4 (Schioppa et al. 2003; Staller et al. 2003). Semaphorin3A (Sema3A) is another hypoxia-induced factor in tumors that is implicated in macrophage recruitment and subsequent angiogenesis. Sema3A mediates Nrp-1-dependent signaling of a PlexinA1/PlexinA4/VEGFR1 holoreceptor complex that leads to VEGFR1 activation in tumor-associated microphages (TAMs) and their subsequent migration into hypoxic regions where they secrete various immune-suppressive and proangiogenic factors. Upon TAM arrival in the hypoxic environment, Nrp1 expression is repressed, which terminates the migratory response of TAMs to Sema3A. Interestingly, hypoxia-dependent Nrp1 repression is facilitated by HIF2α -mediated activation of the nuclear factor κB (NF-κB) pathway. Loss of Nrp-1 on macrophages prevents TAM infiltration in hypoxic regions and thereby maintains an immune-stimulatory phenotype, resulting in delayed tumor growth, which is in turn characterized by impaired vascularization and improved antitumor immunity (Casazza et al. 2013).

Given that hypoxia is a major driver of myeloid-cell recruitment (Du et al. 2008; Mazzieri et al. 2011), it is conceivable that antiangiogenic therapy-induced hypoxia also induces factors that mobilize cells from the bone marrow and attract them to the tumor site. Indeed, several preclinical studies have now shown that tumor-associated innate immune cells sustain angiogenesis during anti-angiogenic therapy, by stimulating VEGF-independent pathways. For example, macrophages induce the expression of various angiogenic molecules, including FGF-1,2, MMP9, and Ang2, in response to angiogenic blockade (Casanovas et al. 2005; Fischer et al. 2007; Rigamonti et al. 2014), and Gr1^{+} immune cells were found to convey resistance to anti-VEGF treatment via secretion of the angiogenic PKR-1/2 ligand Bv8 (Shojaei and Ferrara 2007a,b). Notably, the different intratumoral myeloid cell populations found in tumors (i.e., macrophages, neutrophils, and myeloid-derived suppressor cells [MDSCs]) share redundant functions in angiogenesis and immune modulation and therefore can compensate for each other, leading to an oscillating pattern between immune-cell populations when specific myeloid subpopulations are targeted (Rivera et al. 2015). What these studies also revealed is that VEGF/VEGFR blockade could endorse an angiostatic and immune-stimulating phenotype in all of the intratumoral myeloid cells accompanied with the up-regulation of CXCL14 and other angiostatic chemokines, and that this ability was indeed a requirement for the efficacy of antiangiogenic therapy (Fig. 1A). Because of therapy-induced hypoxia, however, tumors were able to produce factors (e.g., SDF1a and IL-6) that activated phosphoinositide 3-kinase γ (PI3Kγ) signaling in myeloid cells and converted them back into an immune-suppressive and angiogenic state that also disabled their repolarization, which resulted in a proangiogenic tumor relapse (Fig. 1A; Rivera et al. 2015). Congruent with these findings, recent studies have shown that PI3Kγ signals through Akt and mammalian target of rapamycin (mTOR) inhibited NF-κB while stimulating C/EBPβ activation, thereby inducing a transcriptional program that promotes immune suppression (Kaneda et al. 2016). The functional significance of innate immune cells in regulating angiogenesis is further underscored by the observation that the percentage of PI3K-activated myeloid cells in tumors dictated the degree of therapeutic response to angiogenic blockade (Fig. 1B). Congruently, pharmacological inhibition of myeloid PI3Kγ/δ improved and sustained the tumor response to antiangiogenic therapy by converting innate immune cells to an angiostatic and immune-stimulatory state associated with enhanced cytotoxic T-cell infiltration and activity (Fig. 1B; Rivera et al. 2015). Emerging from these studies is the proposition that angiogenesis and inflammation are functionally interregulated and that immune cells play a pivotal role in regulating both processes. It also underscores the tumor's vigilance to conserve an immunosuppressive and angiogenic milieu to overcome growth restrictions and escape immune surveillance.

METABOLIC ENDOTHELIAL CELL ADAPTATION DURING ANGIOGENESIS AND ANGIOGENESIS BLOCKADE

Cancer cells are highly metabolically active, thus they require an adequate vasculature to ensure a constant supply of oxygen and nutrients. In addition, endothelial cells (ECs) undergo metabolic changes when they become activated to form new vessels. Thus targeting metabolic pathways in endothelial cells may be an additional and alternative strategy, because metabolism in these cells not only fuels vascular expansion but also regulates the very formation of blood vessels (Rivera and Bergers 2014). Indeed, such a strategy has proven to inhibit angiogenesis in in vitro sprouting models and induced vessel normalization in tumors (De Bock et al. 2013; Schoors et al. 2015; Cantelmo et al. 2016). Cells use different metabolic programs to convert exogenous molecules to amino acids, nucleotides, and lipids depending on their metabolic needs (Lunt and Vander Heiden 2011; Mihaylova and Shaw 2011). Accordingly, functional differences in angiogenic and quiescent ECs mandate a different metabolic program. Angiogenic ECs require efficient biomass and adenosine triphosphate (ATP)-producing systems to proliferate and migrate (De Bock et al. 2013; Schoors et al. 2015), whereas quiescent ECs, which are exposed to the oxidative environment of the bloodstream, maintain their cellular homeostasis by optimizing the protective mechanisms involved in redox homeostasis and repair mechanisms (Wang et al. 2010; De Bock et al. 2013). Thus, it is conceivable that the angiogenic switch goes hand-in-hand with a metabolic switch, and that this metabolic switch can be an attractive therapeutic target.

In oxygen-rich conditions, the majority of nonproliferating cells rely on oxidative metabolism (Vander Heiden et al. 2009). However, despite the higher ATP yield of oxidative phosphorylation and their exposure to very high oxygen levels, ECs divert the majority of pyruvate to lactate (≈99%) in a process called "aerobic glycolysis" or the "Warburg effect" (Wang et al. 2011; De Bock et al. 2013). ECs rely on glycolysis for >85% of total ATP synthesis, and ECs lining blood vessels show glycolytic flux rates that are much higher than most healthy cells and comparable to those of various cancer cell types (Culic et al. 1997; Vander Heiden et al. 2009; De Bock et al. 2013).

Activation of ECs with growth factors induces proliferative and migratory capacities that require high ATP synthesis rates. EC activation increases the glycolytic flux by 35%–40% (De Bock et al. 2013) and pharmacological inhibition of glycolysis using 3-(3-pyridinyl)-1-(4-pyridinyl)-2-propen-1-one (3PO), which targets 6-phosphofructo-2-kinase/fructose-2,6-bisphosphatase-3 (PFKFB3), lowers glycolytic flux to prevent sprouting, and reduces migration and proliferation of ECs. Consequently, PFKFB3 inhibition targets both stalk and tip cells to impair vessel sprouting in vitro and produces vascular defects in the postnatal mouse retina (De Bock et al. 2013; Xu et al. 2014). In addition, the glycolytic intermediate, glucose-6-phosphate, can be diverted to the pentose phosphate pathway (PPP) to produce ribose-5-phosphate for nucleotide synthesis and the reducing factor nicotinamide adenine dinucleotide phosphate (NADPH). Thus, the PPP has the potential to stimulate biomass production for proliferation or to maintain redox homeostasis (Anastasiou et al. 2011; Lunt and Vander Heiden 2011). Accumulating evidence suggests that the PPP plays a regulatory role in ECs, because inhibition of glucose-6-phoshate dehydrogenase (G6PD), the rate-limiting enzyme of the oxidative branch of the PPP and a NADPH-producing enzyme, increased ROS levels in ECs (Leopold et al. 2001), whereas its overexpression had the opposite effects (Leopold et al. 2003), implying that the PPP in ECs can have a potential protective function and serve as a therapeutic target.

As alternative to glycolysis, certain cell types with high energy demand (e.g., cardiomyocytes, skeletal muscle cells, and certain cancer cells) rely on fatty acid-β-oxidation (FAO) as an energy source or to maintain redox homeostasis (Missiaen et al. 2017). Because ECs show such a high glycolytic flux, it is not surprising that ECs

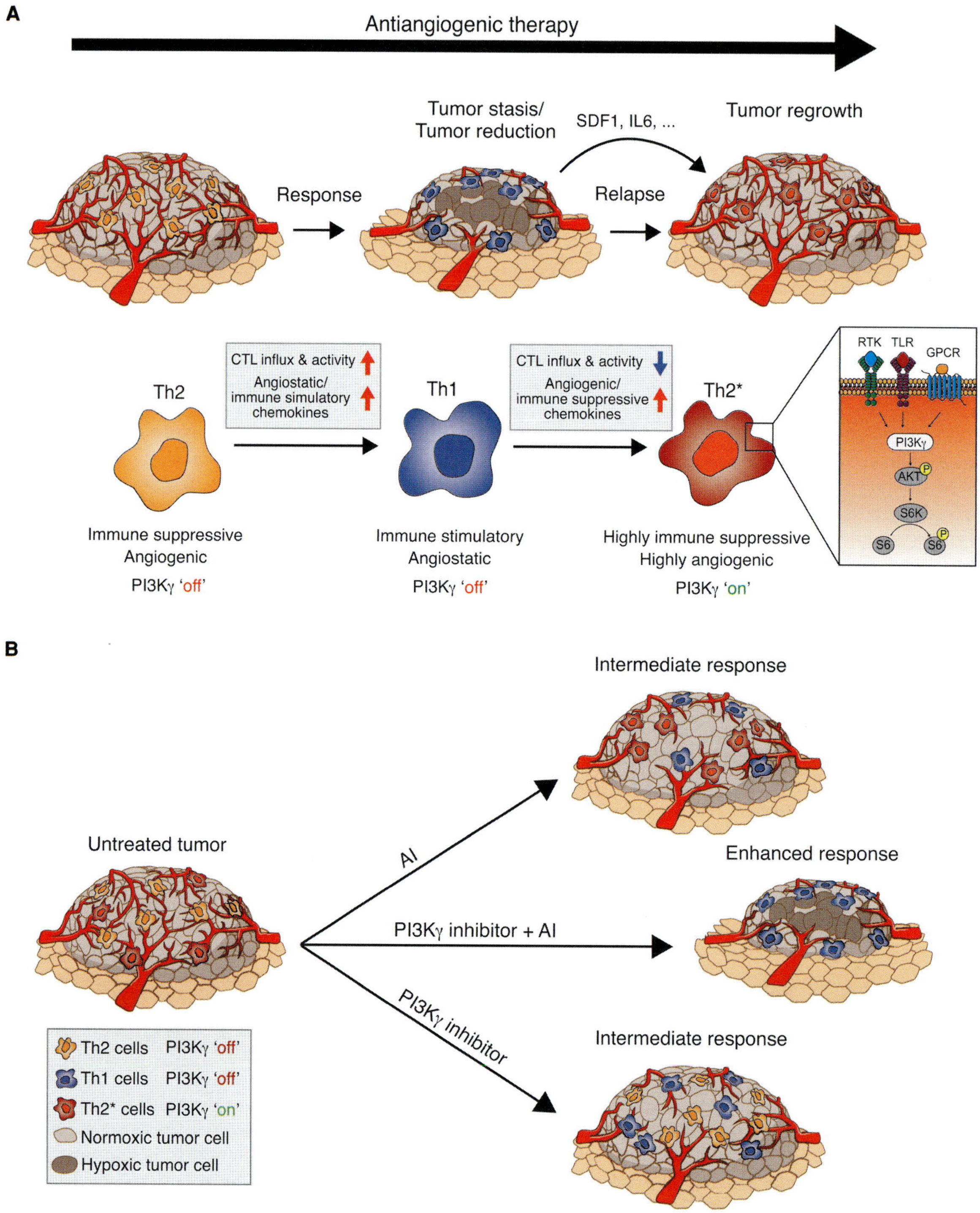

Figure 1. (*A*) Tumors responsive to antiangiogenic inhibition (AI) are infiltrated by Th2-type myeloid cells, which promote an immunosuppressive and proangiogenic phenotype. Treatment with antiangiogenic therapy leads to reduced vessel density and increased hypoxia, and cytokine release transiently shifts the population to a Th1 immune-stimulatory/angiostatic phenotype, leading to cytotoxic T-lymphocyte (CTL) influx, increased apoptosis, and tumor stasis/reduction. Activation of phosphoinositide 3-kinase γ (PI3Kγ) in myeloid cells repolarizes the immune population back to a Th2* phenotype that is now highly immune suppressive and angiogenic and unresponsive to continued antiangiogenic inhibition, leading to vascular rebound and regrowth. RTK, receptor tyrosine kinase; TLR, Toll-like receptor; GPCR, Gαi protein-coupled receptor. (*B*) AI-resistant tumors consist of PI3Kγ-independent, immune suppressive, and angiogenic Th2 cells (orange), and PI3Kγ-dependent, highly immune-suppressive, and angiogenic Th2* cells (red). Single AI treatment results in a shift of Th2 cells to the angiostatic and immune stimulatory Th1 cells (blue), creating an intermediate response. In addition, single PI3Kγ inhibitor treatment shifted Th2* to Th1 cells, leaving the Th2 cell population unaffected and also producing an intermediate response. In contrast, combination AI/PI3Kγ inhibition stimulated a shift of Th2 cells to Th1 cells and eradicated the highly immune-suppressive and angiogenic Th2* cells. This treatment produces an angiostatic and immune-stimulatory effect, resulting in an enhanced response.

have a low oxidative metabolism (De Bock et al. 2013). Surprisingly, pharmacological inhibition of mitochondrial FAO with etomoxir impaired in vitro sprouting in a retinal model, suggesting an important role for FAO in ECs during angiogenesis. This effect was not due to reduced ATP synthesis or affected redox homeostasis; rather FAO inhibition selectively impaired EC proliferation but not migration (Schoors et al. 2015). ^{13}C-palmitate tracing experiments revealed that ECs use fatty acids (FAs) to significantly fuel the tricarboxylic acid (TCA) cycle with carbons. Carbons derived from palmitate were also detected in the intracellular deoxyribonucleotide (dNTP) pool, and this labeling was impaired by FAO inhibition, indicating that FAO plays an essential role in the de novo synthesis of dNTPs in ECs, a requirement for proliferation (Schoors et al. 2015).

In tumors, excessive VEGF and hypoxia induces endothelial dysfunction, resulting in irregularly shaped, leaky, and incompletely perfused tumor blood vessels. VEGF signaling increases glycolysis in ECs by inducing GLUT1 and PFKFB3 expression and inhibition of VEGF signaling resulted in reduced glycolysis (Yeh et al. 2008; De Bock et al. 2013). Interestingly, freshly isolated tumor endothelial cells (TECs) show a higher glycolytic flux than normal endothelial cells (Cantelmo et al. 2016). In contrast to conventional antiangiogenic therapies that aim to inhibit tumor vessel growth, an emerging strategy is to normalize tumor vessels to restore blood vessel perfusion, thereby reducing metastasis and increasing chemotherapeutic delivery. Notably, partial inhibition of the glycolytic flux using low doses of 3PO induced vessel normalization, also a consequence of anti-VEGF treatment (Batchelor et al. 2013). Therefore, it is conceivable that the normalization effect of anti-VEGF treatment could in part be attributed to reduced glycolytic flux in TECs. Unfortunately, studies on the metabolic profile of tumor endothelial cells are limited, and the effect of antiangiogenic therapy on EC metabolism has yet to be investigated.

METABOLIC TUMOR CELL ADAPTATIONS TO ANTIANGIOGENIC THERAPY

As summarized in the preceding text, multiple modes of adaptive or evasive resistance to therapies targeting tumor angiogenesis have been reported, but only heighted invasiveness has been clearly implicated as an adaptive resistance mechanism in glioblastoma (GBM) (Lu and Bergers 2013, and references therein). Intriguingly, Bjerkvig and colleagues explored the metabolic response to antiangiogenic inhibition (AI) therapy using bevacizumab in xenograph models of GBM and linked therapy with increased anaerobic glycolysis and oxidative stress. In addition, they found evidence for increased glutamine consumption that could potentially be used to fuel the TCA cycle (glutaminolysis) and reduce production of reactive oxygen species (Fack et al. 2015). The authors validated their observations by studying a panel of paired human GBM biopsy samples taken before and after bevacizumab treatment that showed increased expression of a key glycolytic enzyme that converts pyruvate to lactate (LDHA), most strikingly up-regulated in the invasive margins of the treated tumors (Fack et al. 2015). It is interesting to note that others have postulated that tumor acidity/high lactate levels may promote local invasion (Estrella et al. 2013). Moreover, they found that the response of GBM to bevacizumab therapy is not mediated by clonal selection mechanisms but is rather an adaptive response. Indraccolo and colleagues studied the response to long-term antiangiogenic therapy and found that it can induce a stable glycolytic phenotype in tumor xenographs derived from cells they had classified as highly or poorly glycolytic, and that the highly glycolytic cells became more rapidly resistant to antiangiogenic therapy. They describe this phenomenon as metabolic evolution; however, it is as yet unclear whether it is the consequence of epigenetic modulation of driver genes, selection of a highly glycolytic population, or an adaptive response to therapy (Curtarello et al. 2015).

The existence of metabolic symbiosis in cancer was reported by Sonveaux, Feron, Dewhirst, and colleagues to arise spontaneously in certain transplant tumor models, and, in a series of publications, they have described a symbiotic state that conveys decreased reliance on circulating glucose and limits damage from secreted lactate (Semenza 2008; Sonveaux et al. 2008; Kennedy et al. 2013). In this model, cancer cells are compartmentalized in conditions of vascular insufficiency into comparatively hypoxic and normoxic compartments based on their proximity to functional blood vessels. In spite of their close proximity to the vasculature, normoxic cells import and catabolize lactate, thereby sparing glucose for hypoxic cells to import and catabolize glycolytically, resulting in the production and secretion of high levels of lactate. This symbiotic relationship was originally described in certain highly metabolically active tissues, including muscle and brain (Brooks 2002). In tumors, the relationship has been extended to interactions between tumor cells and cancer-associated fibroblasts (CAFs), ECs, and macrophages (Nakajima and Van Houten 2013, and references therein).

Three companion reports describe an unanticipated new mode of adaptive resistance to antiangiogenic therapy, metabolic symbiosis, which is induced in response to potent antiangiogenic therapies that cause vascular collapse and consequent hypoxia (Fig. 2; Allen et al. 2016; Jimenez-Valerio et al. 2016; Pisarsky et al. 2016). The reports all show induction and expression patterns of lactate and glucose transporters consistent with the notion of metabolic symbiosis induced in response to different antiangiogenic therapies in preclinical and clinical samples (Fig. 2). The Christofori group showed that gene expression analysis of isolated tumor cells treated with long-term antiangiogenic therapy (nintedanib) revealed a hyperglycolytic signature during the relapse phase, which was characterized by a lack of revascularization and characteristic expression patterns of lactate and glucose transporters. Resistance to nintendanib could be overcome in the PyMT orthotopic preclinical model using the glycol-

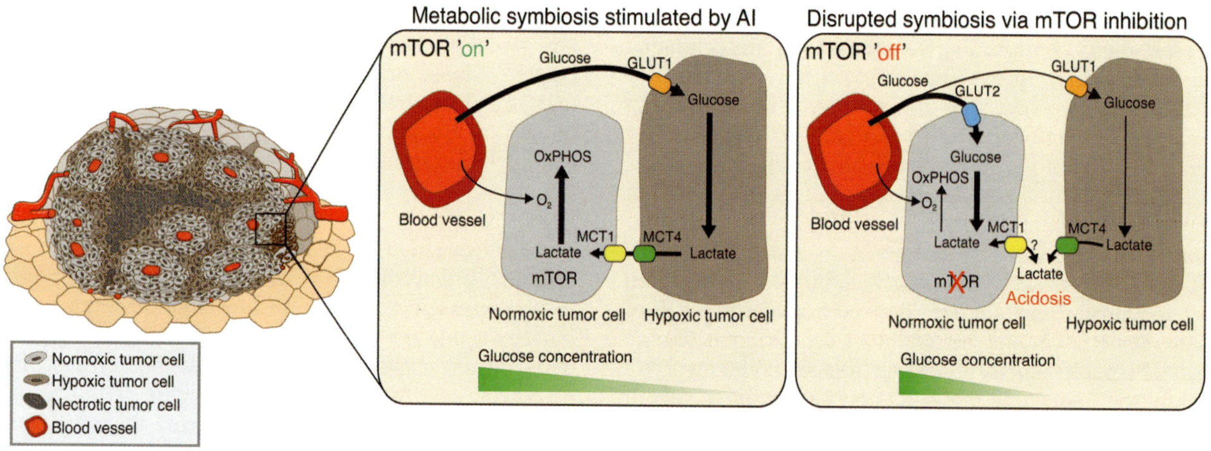

Figure 2. Schematic representation of angiogenesis inhibitor (AI)-induced metabolic symbiosis and the effect of mammalian target of rapamycin (mTOR) inhibition. (*Left*) Treatment with an AI causes vascular collapse to produce normoxic, hypoxic, and necrotic regions, based on their proximity to functional blood vessels. (*Middle*, mTOR"on.") Tumor cells in hypoxic regions up-regulate expression of the glucose transporter GLUT1, increase glycolysis, and export lactate via MCT4. In parallel, mTOR-expressing normoxic tumor cells spare glucose and import lactate via MCT1 to fuel O_2-dependent oxidative phosphorylation (OxPHOs). (*Right*, mTOR"off") Disruption of symbiosis via mTOR inhibition disrupts lactate catabolism in normoxic cells, which can produce increased intracellular lactate concentrations, reduced extracellular lactate clearance, and increased acidosis. Furthermore, mTOR inhibition up-regulates GLUT2 expression in normoxic cells, which can increase glucose consumption by the normoxic cells to reduce its availability for the hypoxic compartment.

ysis inhibitor 3PO, which targets PFKFB3 (Pisarsky et al. 2016). Moreover, symbiosis and tumor progression was blocked in a breast cancer model by genetically suppressing lactate export in antiangiogenic-treated tumors, strengthening the rationale for the development of inhibitors targeting lactate export in the clinic (Pisarsky et al. 2016). In the study from the Hanahan laboratory using a spontaneous mouse model of pancreatic neuroendocrine tumor (PanNET), mTOR signaling was also found to be up-regulated in normoxic tumor regions, and nuclear magnetic resonance (NMR) analysis confirmed lactate catabolism in cultured normoxic tumor cells and tumors in a branched pathway involving glutamine metabolism (glutaminolysis). This pathway is enhanced in antiangiogenic (sunitinib)-treated tumors (Allen et al. 2016), reminiscent of the effects of bevacizumab on GBM (Fack et al. 2015). Moreover, normoxic PanNET cells cultured with lactate and glutamine up-regulate mTOR signaling, which can be reversed with mTOR inhibitors, impairing lactate uptake, or by inhibiting enzymes involved in glutaminolysis. In addition, combination antiangiogenic (sunitinib or axitinib)/mTOR inhibition delays tumor progression and significantly extends survival, in part by blocking lactate catabolism in conjunction with glutamine, and by changing the expression of glucose transporters (Fig. 2; Allen et al. 2016). Furthermore, the Cassanovas group produced evidence corroborating the existence of metabolic symbiosis as a resistance mechanism by studying renal cell carcinoma (RCC) orthoxenografts and paired human RCC biopsies samples taken before and after sunitinib treatment (Jimenez-Valerio et al. 2016). The authors found an up-regulation and redistribution of key glycolytic and lactate transporters in the antiangiogenic-treated tumors consistent with the notion of metabolic symbiosis and also found that mTOR inhibition in concert with antiangiogenic therapy with sunitinib could disrupt the symbiotic patterning and delay tumor progression; critically, the combination was associated disruption of symbiosis in a patient responding to therapy (Jimenez-Valerio et al. 2016). Thus, in the preclinical PanNET model, sunitinib is apparently inducing two distinctive modes of adaptive resistance—metabolic symbiosis, and the perivascular invasive migration of cancer cells at tumor margins (Paez-Ribes et al. 2009). It is presently unclear whether these invading cancer cells also induce symbiosis or whether secreted lactate plays a direct role in the invasion into the surrounding tissue (Estrella et al. 2013).

Noel and colleagues performed a compelling study assessing the metabolic consequences of withdrawing AI therapy in preclinical breast cancer models, which revealed enhanced lipid metabolism concomitant with reduced glycolysis (Sounni et al. 2014), implicating altered lipid metabolism as a potential mechanism of resistance during "rest periods" for patients on antiangiogenic therapy. Both knockdown and pharmacological inhibition of fatty acid synthase (FASN) activity in tumor cells and pharmacological inhibition of FASN with orlistat reversed the tumor rebound and metastasis upon sunitinib withdrawal. In another series of experiments that highlights the role of fatty acid (FA) metabolism in response to hypoxia/antiangiogenic treatment, Harris and colleagues (Bensaad et al. 2014) showed that hypoxia or bevacizumab treatment could induce formation of lipid droplets in breast and glioblastoma xenograph models under the control of HIF1α signaling. Critically, they found that oxidation of these FA stores could provide ATP to tumor cells under conditions of hypoxia or reoxygenation, and reduce levels of toxic reactive oxygen species. In this case, increased FA uptake, rather than de novo synthesis, was essential to this process and controlled by two different fatty acid binding proteins, FABP3 and FABP7. Furthermore, knockdown of these genes impairs tumor growth (Bensaad et al. 2014), providing the rationale for another potential drug target.

CONCLUSION

Although drugs targeting the VEGF pathway did not fulfill initial expectations, their development has been a seminal first step in targeting tumor vessels to impair tumor growth, providing pivotal information about how tumors react to vessel growth restriction. They further have revealed several escape mechanisms that open new avenues for the assessment of combinatorial treatment modalities aimed at targets within these evasive pathways. For example, various antiangiogenic immune therapies are currently tested in the clinic, specifically in combination with checkpoint inhibitors such as CTLA4 or PD1/PDL1 antibodies, showing encouraging results (Garber 2014; Schoenfeld and Dranoff 2011; NCT 02669173, NCT03066427). Ongoing and future studies will reveal which antiangiogenic immunotherapies will more robustly promote an enduring immune-stimulatory milieu and inhibit tumor angiogenesis that could lead to prolonged survival in cancer patients. In line with the concept of targeting evasion mechanisms, targeting endothelial cell metabolism is a new and promising avenue that may circumvent growth factor–mediated resistance, and some of the agents currently used in the clinic, like a next-generation derivative of 3PO, ACT-PFK-158, which targets the glycolytic activator PFKFB3, may have enduring antiangiogenic and vasculature normalization effects. In addition, inhibitors of FAO/CPT1 are currently in the clinic for other indications, most prominently angina, including oxfenicine, perhexiline, and ranolazine (Missiaen et al. 2017, and references therein). Moreover, disruption of induced metabolic pathways might also have value in extending the duration of efficacious responses by traditional antiangiogenic therapies, or potentially antiangiogenic therapy could be used to create a "conditional lethal" metabolic state in tumors, leading to selection and targeting of the newly adapting variants (Bensaad and Harris 2013). Although metabolic symbiosis induced by sunitinib/axitinib can be disrupted by mTOR or glycolysis inhibitors, the significant reductions in tumor burden translate into a relatively modest survival benefit in preclinical studies. In spite of the positive outcome of combination therapy using sunitinib with the rapalog everolimus in the aforementioned clinical study (Jimenez-Valerio et al. 2016), other clinical studies using this combination have shown limited benefit, which may be attributable to high toxicity (Molina et al. 2012). Alternative and potentially attractive strategies may be to use antiangiogenics in combination with compounds that directly inhibit lactate transport, such as AZD3965 and 7ACC2 (Bola et al. 2014; Draoui et al. 2014; Polanski et al. 2014), or glutaminolysis, like CB-839 (Momcilovic et al. 2017). There are increasing numbers of agents in the clinic to target FA metabolism, including those that target FA synthase like orlistat and TVB-2640 (Jones and Infante 2015), and doubtless many more in the pipeline in coming years.

ACKNOWLEDGMENTS

This work was supported by grants from the National Institutes of Health (NIH) (R01 CA201537 and R01 CA188404).

REFERENCES

Adams RH, Alitalo K. 2007. Molecular regulation of angiogenesis and lymphangiogenesis. *Nat Rev Mol Cell Biol* **8:** 464–478.

Allen E, Mieville P, Warren CM, Saghafinia S, Li L, Peng MW, Hanahan D. 2016. Metabolic symbiosis enables adaptive resistance to anti-angiogenic therapy that is dependent on mTOR signaling. *Cell Rep* **15:** 1144–1160.

Anastasiou D, Poulogiannis G, Asara JM, Boxer MB, Jiang JK, Shen M, Bellinger G, Sasaki AT, Locasale JW, Auld DS, et al. 2011. Inhibition of pyruvate kinase M2 by reactive oxygen species contributes to cellular antioxidant responses. *Science* **334:** 1278–1283.

Baluk P, Hashizume H, McDonald DM. 2005. Cellular abnormalities of blood vessels as targets in cancer. *Curr Opin Genet Dev* **15:** 102–111.

Batchelor TT, Gerstner ER, Emblem KE, Duda DG, Kalpathy-Cramer J, Snuderl M, Ancukiewicz M, Polaskova P, Pinho MC, Jennings D, et al. 2013. Improved tumor oxygenation and survival in glioblastoma patients who show increased blood perfusion after cediranib and chemoradiation. *Proc Natl Acad Sci* **110:** 19059–19064.

Bensaad K, Harris AL. 2013. Cancer metabolism as a therapeutic target: Metabolic synthetic lethality. *Oncology (Williston Park)* **27:** 467, 473–474.

Bensaad K, Favaro E, Lewis CA, Peck B, Lord S, Collins JM, Pinnick KE, Wigfield S, Buffa FM, Li JL, et al. 2014. Fatty acid uptake and lipid storage induced by HIF-1α contribute to cell growth and survival after hypoxia-reoxygenation. *Cell Rep* **9:** 349–365.

Bergers G, Benjamin LE. 2003. Tumorigenesis and the angiogenic switch. *Nat Rev Cancer* **3:** 401–410.

Bergers G, Hanahan D. 2008. Modes of resistance to anti-angiogenic therapy. *Nat Rev Cancer* **8:** 592–603.

Bola BM, Chadwick AL, Michopoulos F, Blount KG, Telfer BA, Williams KJ, Smith PD, Critchlow SE, Stratford IJ. 2014. Inhibition of monocarboxylate transporter-1 (MCT1) by AZD3965 enhances radiosensitivity by reducing lactate transport. *Mol Cancer Ther* **13:** 2805–2816.

Bouzin C, Brouet A, De Vriese J, Dewever J, Feron O. 2007. Effects of vascular endothelial growth factor on the lymphocyte-endothelium interactions: Identification of caveolin-1 and nitric oxide as control points of endothelial cell anergy. *J Immunol* **178:** 1505–1511.

Brooks GA. 2002. Lactate shuttles in nature. *Biochem Soc Trans* **30:** 258–264.

Cantelmo AR, Conradi LC, Brajic A, Goveia J, Kalucka J, Pircher A, Chaturvedi P, Hol J, Thienpont B, Teuwen LA, et al. 2016. Inhibition of the glycolytic activator PFKFB3 in endothelium induces tumor vessel normalization, impairs metastasis, and improves chemotherapy. *Cancer Cell* **30:** 968–985.

Carmeliet P. 2005. VEGF as a key mediator of angiogenesis in cancer. *Oncology* **69** (Suppl 3)**:** 4–10.

Carmeliet P, Jain RK. 2011. Molecular mechanisms and clinical applications of angiogenesis. *Nature* **473:** 298–307.

Casanovas O, Hicklin DJ, Bergers G, Hanahan D. 2005. Drug resistance by evasion of anti-angiogenic targeting of VEGF

signaling in late-stage pancreatic islet tumors. *Cancer Cell* **8:** 299–309.
Casazza A, Laoui D, Wenes M, Rizzolio S, Bassani N, Mambretti M, Deschoemaeker S, Van Ginderachter JA, Tamagnone L, Mazzone M. 2013. Impeding macrophage entry into hypoxic tumor areas by Sema3A/Nrp1 signaling blockade inhibits angiogenesis and restores antitumor immunity. *Cancer Cell* **24:** 695–709.
Coussens LM, Tinkle CL, Hanahan D, Werb Z. 2000. MMP-9 supplied by bone marrow–derived cells contributes to skin carcinogenesis. *Cell* **103:** 481–490.
Culic O, Gruwel ML, Schrader J. 1997. Energy turnover of vascular endothelial cells. *Am J Physiol* **273:** C205–C213.
Curtarello M, Zulato E, Nardo G, Valtorta S, Guzzo G, Rossi E, Esposito G, Msaki A, Pasto A, Rasola A, et al. 2015. VEGF-targeted therapy stably modulates the glycolytic phenotype of tumor cells. *Cancer Res* **75:** 120–133.
De Bock K, Georgiadou M, Schoors S, Kuchnio A, Wong BW, Cantelmo AR, Quaegebeur A, Ghesquiere B, Cauwenberghs S, Eelen G, et al. 2013. Role of PFKFB3-driven glycolysis in vessel sprouting. *Cell* **154:** 651–663.
Dineen SP, Lynn KD, Holloway SE, Miller AF, Sullivan JP, Shames DS, Beck AW, Barnett CC, Fleming JB, Brekken RA. 2008. Vascular endothelial growth factor receptor 2 mediates macrophage infiltration into orthotopic pancreatic tumors in mice. *Cancer Res* **68:** 4340–4346.
Draoui N, Schicke O, Seront E, Bouzin C, Sonveaux P, Riant O, Feron O. 2014. Antitumor activity of 7-aminocarboxycoumarin derivatives, a new class of potent inhibitors of lactate influx but not efflux. *Mol Cancer Ther* **13:** 1410–1418.
Du R, Lu KV, Petritsch C, Liu P, Ganss R, Passegue E, Song H, Vandenberg S, Johnson RS, Werb Z, et al. 2008. HIF1α induces the recruitment of bone marrow-derived vascular modulatory cells to regulate tumor angiogenesis and invasion. *Cancer Cell* **13:** 206–220.
Ebos JM, Lee CR, Christensen JG, Mutsaers AJ, Kerbel RS. 2007. Multiple circulating proangiogenic factors induced by sunitinib malate are tumor-independent and correlate with antitumor efficacy. *Proc Natl Acad Sci* **104:** 17069–17074.
Estrella V, Chen T, Lloyd M, Wojtkowiak J, Cornnell HH, Ibrahim-Hashim A, Bailey K, Balagurunathan Y, Rothberg JM, Sloane BF, et al. 2013. Acidity generated by the tumor microenvironment drives local invasion. *Cancer Res* **73:** 1524–1535.
Fack F, Espedal H, Keunen O, Golebiewska A, Obad N, Harter PN, Mittelbronn M, Bahr O, Weyerbrock A, Stuhr L, et al. 2015. Bevacizumab treatment induces metabolic adaptation toward anaerobic metabolism in glioblastomas. *Acta Neuropathol* **129:** 115–131.
Ferrara N. 2004. Vascular endothelial growth factor as a target for anticancer therapy. *Oncologist* **9** (Suppl 1)**:** 2–10.
Fischer C, Jonckx B, Mazzone M, Zacchigna S, Loges S, Pattarini L, Chorianopoulos E, Liesenborghs L, Koch M, De Mol M, et al. 2007. Anti-PlGF inhibits growth of VEGF(R)-inhibitor-resistant tumors without affecting healthy vessels. *Cell* **131:** 463–475.
Gabrilovich DI, Chen HL, Girgis KR, Cunningham HT, Meny GM, Nadaf S, Kavanaugh D, Carbone DP. 1996. Production of vascular endothelial growth factor by human tumors inhibits the functional maturation of dendritic cells. *Nat Med* **2:** 1096–1103.
Garber K. 2014. Promising early results for immunotherapy-antiangiogenesis combination. *J Natl Cancer Inst* **106:** dju392.
Giraudo E, Inoue M, Hanahan D. 2004. An amino-bisphosphonate targets MMP-9-expressing macrophages and angiogenesis to impair cervical carcinogenesis. *J Clin Invest* **114:** 623–633.
Griffioen AW, Damen CA, Martinotti S, Blijham GH, Groenewegen G. 1996. Endothelial intercellular adhesion molecule-1 expression is suppressed in human malignancies: The role of angiogenic factors. *Cancer Res* **56:** 1111–1117.
Grunewald M, Avraham I, Dor Y, Bachar-Lustig E, Itin A, Jung S, Chimenti S, Landsman L, Abramovitch R, Keshet E. 2006. VEGF-induced adult neovascularization: Recruitment, retention, and role of accessory cells. *Cell* **124:** 175–189.
Hanahan D, Weinberg RA. 2011. Hallmarks of cancer: The next generation. *Cell* **144:** 646–674.
Jain RK. 2002. Angiogenesis and lymphangiogenesis in tumors: Insights from intravital microscopy. *Cold Spring Harb Symp Quant Biol* **67:** 239–248.
Jain RK. 2014. Antiangiogenesis strategies revisited: From starving tumors to alleviating hypoxia. *Cancer Cell* **26:** 605–622.
Jayson GC, Kerbel R, Ellis LM, Harris AL. 2016. Anti-angiogenic therapy in oncology: Current status and future directions. *Lancet* **388:** 518–529.
Jimenez-Valerio G, Martinez-Lozano M, Bassani N, Vidal A, Ochoa-de-Olza M, Suarez C, Garcia-Del-Muro X, Carles J, Vinals F, Graupera M, et al. 2016. Resistance to anti-angiogenic therapies by metabolic symbiosis in renal cell carcinoma PDX models and patients. *Cell Rep* **15:** 1134–1143.
Jones SF, Infante JR. 2015. Molecular pathways: Fatty acid synthase. *Clin Cancer Res* **21:** 5434–5438.
Kaneda MM, Messer KS, Ralainirina N, Li H, Leem CJ, Gorjestani S, Woo G, Nguyen AV, Figueiredo CC, Foubert P, et al. 2016. PI3Kγ is a molecular switch that controls immune suppression. *Nature* **539:** 437–442.
Kasuya A, Tokura Y. 2014. Attempts to accelerate wound healing. *J Dermatol Sci* **76:** 169–172.
Kennedy KM, Scarbrough PM, Ribeiro A, Richardson R, Yuan H, Sonveaux P, Landon CD, Chi JT, Pizzo S, Schroeder T, et al. 2013. Catabolism of exogenous lactate reveals it as a legitimate metabolic substrate in breast cancer. *PLoS One* **8:** e75154.
Leopold JA, Cap A, Scribner AW, Stanton RC, Loscalzo J. 2001. Glucose-6-phosphate dehydrogenase deficiency promotes endothelial oxidant stress and decreases endothelial nitric oxide bioavailability. *FASEB J* **15:** 1771–1773.
Leopold JA, Zhang YY, Scribner AW, Stanton RC, Loscalzo J. 2003. Glucose-6-phosphate dehydrogenase overexpression decreases endothelial cell oxidant stress and increases bioavailable nitric oxide. *Arterioscler Thromb Vasc Biol* **23:** 411–417.
Liao D, Johnson RS. 2007. Hypoxia: A key regulator of angiogenesis in cancer. *Cancer Metastasis Rev* **26:** 281–290.
Lin EY, Li JF, Gnatovskiy L, Deng Y, Zhu L, Grzesik DA, Qian H, Xue XN, Pollard JW. 2006. Macrophages regulate the angiogenic switch in a mouse model of breast cancer. *Cancer Res* **66:** 11238–11246.
Lu KV, Bergers G. 2013. Mechanisms of evasive resistance to anti-VEGF therapy in glioblastoma. *CNS Oncol* **2:** 49–65.
Lu KV, Chang JP, Parachoniak CA, Pandika MM, Aghi MK, Meyronet D, Isachenko N, Fouse SD, Phillips JJ, Cheresh DA, et al. 2012. VEGF inhibits tumor cell invasion and mesenchymal transition through a MET/VEGFR2 complex. *Cancer Cell* **22:** 21–35.
Lunt SY, Vander Heiden MG. 2011. Aerobic glycolysis: Meeting the metabolic requirements of cell proliferation. *Annu Rev Cell Dev Biol* **27:** 441–464.
Mazzieri R, Pucci F, Moi D, Zonari E, Ranghetti A, Berti A, Politi LS, Gentner B, Brown JL, Naldini L, et al. 2011. Targeting the ANG2/TIE2 axis inhibits tumor growth and metastasis by impairing angiogenesis and disabling rebounds of proangiogenic myeloid cells. *Cancer Cell* **19:** 512–526.
McIntyre A, Harris AL. 2015. Metabolic and hypoxic adaptation to anti-angiogenic therapy: A target for induced essentiality. *EMBO Mol Med* **7:** 368–379.
Mihaylova MM, Shaw RJ. 2011. The AMPK signalling pathway coordinates cell growth, autophagy and metabolism. *Nat Cell Biol* **13:** 1016–1023.
Missiaen R, Morales-Rodriguez F, Eelen G, Carmeliet P. 2017. Targeting endothelial metabolism for anti-angiogenesis therapy: A pharmacological perspective. *Vascul Pharmacol* **90:** 8–18.
Molina AM, Feldman DR, Voss MH, Ginsberg MS, Baum MS, Brocks DR, Fischer PM, Trinos MJ, Patil S, Motzer RJ.

2012. Phase 1 trial of everolimus plus sunitinib in patients with metastatic renal cell carcinoma. *Cancer* **118:** 1868–1876.

Momcilovic M, Bailey ST, Lee JT, Fishbein MC, Magyar C, Braas D, Graeber T, Jackson NJ, Czernin J, Emberley E, et al. 2017. Targeted inhibition of EGFR and glutaminase induces metabolic crisis in EGFR mutant lung cancer. *Cell Rep* **18:** 601–610.

Motz GT, Coukos G. 2011. The parallel lives of angiogenesis and immunosuppression: Cancer and other tales. *Nat Rev Immunol* **11:** 702–711.

Motz GT, Coukos G. 2013. Deciphering and reversing tumor immune suppression. *Immunity* **39:** 61–73.

Murdoch C, Giannoudis A, Lewis CE. 2004. Mechanisms regulating the recruitment of macrophages into hypoxic areas of tumors and other ischemic tissues. *Blood* **104:** 2224–2234.

Nakajima EC, Van Houten B. 2013. Metabolic symbiosis in cancer: Refocusing the Warburg lens. *Mol Carcinog* **52:** 329–337.

Paez-Ribes M, Allen E, Hudock J, Takeda T, Okuyama H, Vinals F, Inoue M, Bergers G, Hanahan D, Casanovas O. 2009. Anti-angiogenic therapy elicits malignant progression of tumors to increased local invasion and distant metastasis. *Cancer Cell* **15:** 220–231.

Palazon A, Goldrath AW, Nizet V, Johnson RS. 2014. HIF transcription factors, inflammation, and immunity. *Immunity* **41:** 518–528.

Pisarsky L, Bill R, Fagiani E, Dimeloe S, Goosen RW, Hagmann J, Hess C, Christofori G. 2016. Targeting metabolic symbiosis to overcome resistance to anti-angiogenic therapy. *Cell Rep* **15:** 1161–1174.

Polanski R, Hodgkinson CL, Fusi A, Nonaka D, Priest L, Kelly P, Trapani F, Bishop PW, White A, Critchlow SE, et al. 2014. Activity of the monocarboxylate transporter 1 inhibitor AZD3965 in small cell lung cancer. *Clin Cancer Res* **20:** 926–937.

Potente M, Gerhardt H, Carmeliet P. 2011. Basic and therapeutic aspects of angiogenesis. *Cell* **146:** 873–887.

Rigamonti N, Kadioglu E, Keklikoglou I, Wyser Rmili C, Leow CC, De Palma M. 2014. Role of angiopoietin-2 in adaptive tumor resistance to VEGF signaling blockade. *Cell Rep* **8:** 696–706.

Rivera LB, Bergers G. 2014. Targeting vascular sprouts. *Science* **344:** 1449–1450.

Rivera LB, Bergers G. 2015. Tumor angiogenesis, from foe to friend. *Science* **349:** 694–695.

Rivera LB, Meyronet D, Hervieu V, Frederick MJ, Bergsland E, Bergers G. 2015. Intratumoral myeloid cells regulate responsiveness and resistance to anti-angiogenic therapy. *Cell Rep* **11:** 577–591.

Rubenstein JL, Kim J, Ozawa T, Zhang M, Westphal M, Deen DF, Shuman MA. 2000. Anti-VEGF antibody treatment of glioblastoma prolongs survival but results in increased vascular cooption. *Neoplasia* **2:** 306–314.

Santoni M, Bracarda S, Nabissi M, Massari F, Conti A, Bria E, Tortora G, Santoni G, Cascinu S. 2014. CXC and CC chemokines as angiogenic modulators in nonhaematological tumors. *Biomed Res Int* **2014:** 768758.

Schioppa T, Uranchimeg B, Saccani A, Biswas SK, Doni A, Rapisarda A, Bernasconi S, Saccani S, Nebuloni M, Vago L, et al. 2003. Regulation of the chemokine receptor CXCR4 by hypoxia. *J Exp Med* **198:** 1391–1402.

Schoenfeld JD, Dranoff G. 2011. Anti-angiogenesis immunotherapy. *Hum Vaccin* **7:** 976–981.

Schoors S, Bruning U, Missiaen R, Queiroz KC, Borgers G, Elia I, Zecchin A, Cantelmo AR, Christen S, Goveia J, et al. 2015. Fatty acid carbon is essential for dNTP synthesis in endothelial cells. *Nature* **520:** 192–197.

Semenza GL. 2008. Tumor metabolism: Cancer cells give and take lactate. *J Clin Invest* **118:** 3835–3837.

Semenza GL. 2011. Oxygen sensing, homeostasis, and disease. *N Engl J Med* **365:** 537–547.

Sennino B, Ishiguro-Oonuma T, Wei Y, Naylor RM, Williamson CW, Bhagwandin V, Tabruyn SP, You WK, Chapman HA, Christensen JG, et al. 2012. Suppression of tumor invasion and metastasis by concurrent inhibition of c-Met and VEGF signaling in pancreatic neuroendocrine tumors. *Cancer Discov* **2:** 270–287.

Shojaei F, Ferrara N. 2007a. Antiangiogenesis to treat cancer and intraocular neovascular disorders. *Lab Invest* **87:** 227–230.

Shojaei F, Ferrara N. 2007b. Anti-angiogenic therapy for cancer: An update. *Cancer J* **13:** 345–348.

Shojaei F, Ferrara N. 2008. Role of the microenvironment in tumor growth and in refractoriness/resistance to anti-angiogenic therapies. *Drug Resist Updat* **11:** 219–230.

Sonveaux P, Vegran F, Schroeder T, Wergin MC, Verrax J, Rabbani ZN, De Saedeleer CJ, Kennedy KM, Diepart C, Jordan BF, et al. 2008. Targeting lactate-fueled respiration selectively kills hypoxic tumor cells in mice. *J Clin Invest* **118:** 3930–3942.

Sounni NE, Cimino J, Blacher S, Primac I, Truong A, Mazzucchelli G, Paye A, Calligaris D, Debois D, De Tullio P, et al. 2014. Blocking lipid synthesis overcomes tumor regrowth and metastasis after anti-angiogenic therapy withdrawal. *Cell Metab* **20:** 280–294.

Staller P, Sulitkova J, Lisztwan J, Moch H, Oakeley EJ, Krek W. 2003. Chemokine receptor CXCR4 downregulated by von Hippel-Lindau tumour suppressor pVHL. *Nature* **425:** 307–311.

Vander Heiden MG, Cantley LC, Thompson CB. 2009. Understanding the Warburg effect: The metabolic requirements of cell proliferation. *Science* **324:** 1029–1033.

Vasudev NS, Reynolds AR. 2014. Anti-angiogenic therapy for cancer: Current progress, unresolved questions and future directions. *Angiogenesis* **17:** 471–494.

Wang YY, Chen SM, Li H. 2010. Hydrogen peroxide stress stimulates phosphorylation of FoxO1 in rat aortic endothelial cells. *Acta Pharmacol Sin* **31:** 160–164.

Wang Q, Liang B, Shirwany NA, Zou MH. 2011. 2-Deoxy-D-glucose treatment of endothelial cells induces autophagy by reactive oxygen species-mediated activation of the AMP-activated protein kinase. *PLoS One* **6:** e17234.

Xu Y, An X, Guo X, Habtetsion TG, Wang Y, Xu X, Kandala S, Li Q, Li H, Zhang C, et al. 2014. Endothelial PFKFB3 plays a critical role in angiogenesis. *Arterioscler Thromb Vasc Biol* **34:** 1231–1239.

Yeh WL, Lin CJ, Fu WM. 2008. Enhancement of glucose transporter expression of brain endothelial cells by vascular endothelial growth factor derived from glioma exposed to hypoxia. *Mol Pharmacol* **73:** 170–177.

Cell of Origin and Cancer Stem Cells in Tumor Suppressor Mouse Models of Glioblastoma

Sheila R. Alcantara Llaguno, Xuanhua Xie, and Luis F. Parada
Brain Tumor Center and Cancer Biology and Genetics Program, Memorial Sloan Kettering Cancer Center, New York, New York 10065
Correspondence: paradal@mskcc.org

The cellular origins and the mechanisms of progression, maintenance of tumorigenicity, and therapeutic resistance are central questions in the glioblastoma multiforme (GBM) field. Using tumor suppressor mouse models, our group recently reported two independent populations of adult GBM-initiating central nervous system progenitors. We found different functional and molecular subtypes depending on the tumor-initiating cell lineage, indicating that the cell of origin is a driver of GBM subtype diversity. Using an in vivo model, we also showed that GBM cancer stem cells (CSCs) or glioma stem cells (GSCs) contribute to resistance to chemotherapeutic agents and that genetic ablation of GSCs leads to a delay in tumor progression. These studies are consistent with the cell of origin and CSCs as critical regulators of the pathogenesis of GBM.

Glioblastoma is the most aggressive and prevalent primary malignancy in the central nervous system (CNS). However, despite numerous advances in cancer therapy, these tumors remain incurable (Stupp et al. 2005). In recent years, large-scale genomic analyses have been used to stratify glioblastoma multiforme (GBM) based on gene expression signatures, driver mutations, copy number changes, and other molecular alterations that could reflect differing pathologic characteristics (TCGA 2008). The Cancer Genome Atlas (TCGA) analyses based on the above-mentioned criteria resolve GBM into four subtypes denoted mesenchymal, proneural, classical, and neural (Verhaak et al. 2010). Several groups have since reported various glioma subtyping schema based on newer molecular data and other criteria (Eckel-Passow et al. 2015; Ceccarelli et al. 2016). The clinical implications for therapeutic response of these genomic subclassifications, however, remain unclear.

GBM-INITIATING LINEAGES

The extensive cellular heterogeneity found within the adult mammalian brain affords a variety of cell types that could, in principle, give rise to malignant tumors. Our studies have progressively focused on stem and progenitor cells in the adult brain as attractive candidates for tumor origin based on self-renewal and proliferative potential that would provide an opportunity for the stochastic accumulation of the essential mutations required for malignant transformation. In the mouse, adult neural stem cells reside in the subventricular zone (SVZ) of the lateral ventricles and the subgranular zone of the hippocampal dentate gyrus, where they produce neuronal progenitor cells (NPCs) that primarily give rise to new neurons (Ming and Song 2011). In contrast, the widely dispersed oligodendrocyte progenitor cells (OPCs) represent a significant proportion of dividing cells in the adult brain and are the primary source of oligodendrocytes in both the developing and mature CNS (Nishiyama et al. 2009).

A TALE OF TWO TUMORS

Various genetically engineered mouse models (GEMMs) of brain tumors have been generated by oncogene overexpression and/or deletion of tumor suppressors in various cell types of the brain. Our laboratory has applied the GEMM technology to drive GBM formation through ablation of three of the most commonly mutated genes in human GBM: *NF1* (neurofibromin 1), *TP53* (tumor protein P53), and *PTEN (*phosphatase and tensin homolog). Initially, we used a human glial fibrillary acidic protein (GFAP) promoter transgene construct (*hGfap-cre*) to drive Cre recombinase expression in the embryonic and adult stem cell niche to generate compound heterozygotes at the tumor suppressor loci. Fully penetrant GBM formation was accompanied by loss of heterozygosity for the tumor suppressors in the malignant tissue (Zhu et al. 2005). These studies were further refined by targeting the adult neural stem cell compartment using a *Nestin-creER^{T2}* transgene to ablate tumor suppressors and also generated GBM with 100% efficiency (Alcantara Llaguno et al. 2009). More recently, we used an *Ascl1-creER^{TM}* transgene that targets two different populations of adult lineage-restricted CNS progenitors, the NPCs and OPCs. Adult bipotential progenitor-directed loss of *Nf1*, *Trp53*, and *Pten* in mice led to the development of two histologically and molecularly distinct GBM subtypes (Alcantara Llaguno et al. 2015). The type 1 GBM

Published by Cold Spring Harbor Laboratory Press; doi: 10.1101/sqb.2016.81.030973

is a more infiltrative, Gfap[hi] tumor and is preferentially found in the dorsal brain. The type 2 GBM has a Gfap[lo] signature and better defined tumor borders and is most frequently found in the ventral brain (Fig. 1). Histologic and genomic analyses showed that *Ascl1-creER*^{TM}-driven type 1 GBMs resemble *Nestin-creER*^{T2}-driven tumors. In contrast, the type 2 GBMs resembled tumors that were generated through mutation of the three tumor suppressors with an OPC-specific Cre driver (*NG2-creER*^{TM}).

Thus, at least two independent populations of GBM-initiating adult progenitors exist that, in the setting of identical driver mutations, give rise to GBM as determined by classic histopathological criteria. However, additional pathological, molecular, and morbidity assays reveal differing features for the two GBM types, indicating that the cell of origin contributes to GBM genomic profiles and resultant phenotypes. This was further illustrated when GBM derived from different tumor suppressor mouse models were found to be molecularly separable based on the lineage of the tumor-initiating cell, indicating that tumor cells have cell-of-origin memory and thus supporting the idea that the cell of origin is a determinant of GBM subtype (Alcantara Llaguno et al. 2015). Given these results, the variety of cell lineages and progenitor populations in the adult mammalian brain (Ming and Song 2011; Merkle et al. 2014) present the plausible scenario that yet additional to-be-defined GBM subtypes may exist (Fig. 1). Thus, the various populations of tumor-initiating cells may underlie a diversity of functional subtypes found in human GBM.

It has been reported that more differentiated cell types can give rise to GBM in mouse models (Bachoo et al. 2002; Friedmann-Morvinski et al. 2012). Although in the experimental setting this may be possible, it is unclear how fully differentiated, nondividing neurons or astrocytes might acquire successive transforming mutations in the natural setting. To date, the occurrence of dedifferentiation in adult postmitotic CNS cells and its physiologic relevance remains controversial. Additional work on this topic will be essential to understand these issues.

DEVELOPMENTAL MEMORY OF GBM CELLS

How does a divergent lineage, such as the OPC lineage, give rise to a CNS tumor that exhibits predominantly histologic astrocytic features? Recent work suggests a common developmental origin for stem cells and progenitors. For example, embryonic OPCs have been shown to generate not only oligodendrocytes but also protoplasmic astrocytes in the gray matter of the ventral forebrain (Nishiyama et al. 2009). This raises the possibility that embryonic identity can be reactivated during adult glioma development, which may allow previously bipotential OPCs to give rise to type 2–like astrocytomas. Moreover, recent data show that adult SVZ neural stem cells share a common lineage with embryonic stem cells that give rise to the cells of the cortex, striatum, and septum (Fuentealba et al. 2015), suggesting that transformed stem cells or progenitors may exhibit a developmental memory that preferentially leads them to migrate to these dorsal forebrain regions, where type 1 tumors are prevalent.

HUMAN GLIOMAS AND THE CELL OF ORIGIN

When we transplanted type 1 and type 2 GBMs as intracranial allografts, the engrafted tumors retained their original primary histological features, and type 1 tumor-transplanted mice showed significantly shorter median survival compared to type 2 tumor-transplanted mice (Alcantara Llaguno et al. 2015). This raises the possibility that GBM prognosis may be defined not only by the sum of pathological characterization and molecular driver genes but equally or more so by the cell of origin in which the mutations arise.

Our results show that GBM can be stratified into functional subtypes in mouse models through initiation in different progenitor cells. Expansion of such approaches to include the larger spectrum of possible initiating driver mutations and additional progenitor lineages integrated

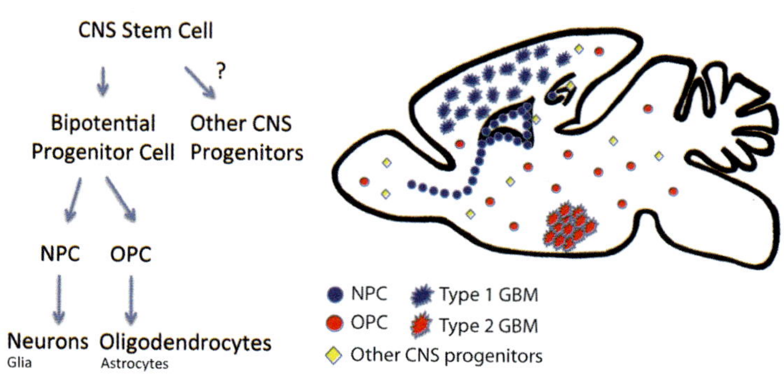

Figure 1. Two independent populations of glioblastoma multiforme (GBM)-initiating central nervous system (CNS) progenitors exist in the adult mouse brain. (*Right*) GBMs can be initiated by tumor suppressor mutations in adult neuronal progenitor cells (NPCs) and adult oligodendrocyte progenitor cells (OPCs), which give rise to histologically and molecularly distinct tumors. NPC-driven type 1 tumors are infiltrative gliomas mostly found in the dorsal brain, whereas OPC-driven type 2 tumors have more defined tumor borders prevalent in ventral brain regions. (*Left*) Lineage hierarchy of CNS cells: Multipotent CNS stem cells undergo asymmetric cell division to produce bipotential progenitor cells that give rise to lineage-restricted progenitors that differentiate into mature CNS cell types. Other potential unique subsets of adult progenitor lineages may give rise to additional GBM subtypes.

with existing genomic and clinical data should provide a more robust and complete picture of GBM. Should this lead to discovery of lineage or subtype-specific pathways that can be therapeutically targeted, it will allow functional stratification of patients into subgroups that may respond to more tailored therapies.

CANCER STEM CELLS IN GBM

The concept of the cancer stem cells (CSCs) relates to the hypothesis that some solid tumors may propagate in a hierarchical fashion in which a relatively quiescent tumor cell lies at the apex of the hierarchy. In such a model, the highly proliferating derivative tumor cells are limited in their self-renewal capacity and are not responsible for therapeutic resistance or tumor relapse. The CSC model was led by research in the hematopoietic system, where in-depth characterization has resulted in a toolkit of cell surface markers that distinguish between hematopoietic stem cells, progenitors, and differentiated cells, as well as their malignant derivatives (Bonnet and Dick 1997). In solid tumors, CSCs have been described in different organ cancers, including breast, prostate, pancreas, colon, skin, and brain (Hermann et al. 2007; Wang et al. 2009a; Pece et al. 2010; Chen et al. 2012a; Driessens et al. 2012; Schepers et al. 2012). However, faithful markers for CSCs in solid tumors are in infancy and most CSC assays for solid tumors rely on indirect surrogate approaches such as the limited repertoire of enrichment antibodies or transplantation assays that involve tissue dissociation, removal from microenvironment, and, in some cases, incubation with growth factor cocktails in vitro.

For GBM patients, survival has not improved significantly in the last several decades, suggesting that current treatment regimens fail to target the critical cells responsible for tumor growth. One explanation for this would be that GBM development and growth is hierarchical and that CSCs are impervious to current therapies. We have examined glioma stem cells (GSCs) in the context of a possible relationship with wild-type neural stem and progenitor cells, and their possible roles not only in tumor initiation and maintenance but also in therapy resistance and tumor recurrence.

Through genomic landscape studies, GBM core signaling pathways that may contribute to gliomagenesis have been proposed. Among them, the ARF/P53 pathway, the INK4A/RB pathway, the RTK pathway, and the NF-κB pathway are frequently dysregulated (Chen et al. 2012b). These same pathways are known to play essential roles in wild-type neural stem cell self-renewal, proliferation, and survival (Groszer et al. 2006; Meletis et al. 2006; Hegedus et al. 2007). A variety of mouse models manipulating these pathways in neural stem/progenitor cells have been shown to efficiently form high-grade gliomas (Zhu et al. 2005, 2014; Ligon et al. 2007; Alcantara Llaguno et al. 2009; Wang et al. 2009b; Zheng et al. 2013). These studies pointed to a critical role for neural stem/progenitor cells in glioma formation. However, whether GSCs exist in most or all forms of GBMs remains unknown.

GSCs were first isolated from human GBM samples by fluorescence-activated cell sorting (FACS) using the CD133/Prominin1 antibody (Singh et al. 2003). However, subsequent studies reported contradictory results showing that CD133-negative cells also possess similar properties as CD133-positive cells, such as neurosphere formation in defined culture media and the ability to form transplanted tumors (Nishide et al. 2009; Holmberg Olausson et al. 2014). Other markers have also been reported to enrich for GSCs, including integrin α6, CD44, CD15, and EphA3 (Son et al. 2009; Anido et al. 2010; Lathia et al. 2010; Day et al. 2013). Nevertheless, a thorough characterization of these potential GSC markers is still lacking. There are also a host of unanswered questions. Are these different GSC markers expressed in the same cells? How are they functionally related to GSC features? How do these markers relate to the traditional (primary versus secondary) and genomic (neural, proneural, mesenchymal, classical) classifications of GBM? Furthermore, different assays have been used to investigate the GSC features of primary or cultured cells. The gold standard in the field is to perform tumor cell transplantation assays in immunodeficient mice, requiring tumor cell dissociation and often with manipulation in culture outside the tumor microenvironment before transplantation. The study of GSCs studied in their native microenvironment may better reflect their biological properties. The use of spontaneous mouse GBM models afford a reproducible and reliable system to genetically identify and label GSCs, elucidate their unique features, and examine the critical mechanisms involved in tumor recurrence.

IN VIVO GBM CSC MODEL

To directly test the CSC hypothesis in GBM, we used the human GFAP-Cre (*hGfap-cre*)-driven mouse GBM model together with a *Nestin-ΔTK-GFP* transgene, validated to label adult neural stem cells in the SVZ. This same transgene also labels a subset of quiescent tumor cells in mouse model GBMs. The presence of thymidine kinase (TK) allows the ablation of dividing TK transgene-positive cells with gancyclovir (GCV). Administration of GCV into mutant mice carrying the *Nestin-ΔTK-GFP* transgene was shown to significantly prolong the life span of tumor-bearing mice (Chen et al. 2012a). In vivo tracing with BrdU analogs IdU and CldU indicated that they were resistant to temozolomide treatment and responsible for tumor recurrence. Combination treatment of GCV and temozolomide also efficiently eradicated the type 1 dorsal tumors. However, mutant mice still died from secondary type 2–like ventral tumors that developed concurrently with the type 1 dorsal tumors. This may have been caused by the widespread expression of the *hGfap-cre* transgene used to initiate the tumors. Hence, a more restricted and preferably inducible cre expression would be required to circumvent this problem. On the other hand, the development of a GSC reporter provided a powerful means by which we can extensively determine the properties of GSCs.

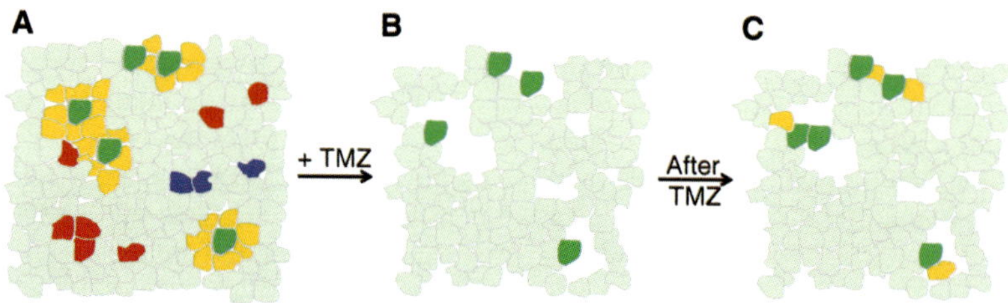

Figure 2. A restricted cell population in a spontaneous glioblastoma multiforme (GBM) mouse model is responsible for tumor recurrence after chemotherapy. (*A*) A cartoon depicts the heterogeneity of glioblastoma. Dark green cells represent the quiescent stem cells; yellow, red, and blue cells indicate the actively dividing cells with different proliferative potential; light green cells, which represent the bulk tumor cells, have exited the cell cycle. (*B*) Upon temozolomide (TMZ) treatment, most of the dividing cells are eradicated. (*C*) Quiescent stem-like cancer cells re-enter the cell cycle to repropagate the tumor. They can divide symmetrically to form more stem cells (self-renewal) or asymmetrically to generate more proliferative cells.

USING THE NESTIN-GFP TRANSGENE TO CHARACTERIZE GSCS

The ability to label a GSC population in spontaneous GBM in vivo with the *Nestin-ΔTK-GFP* transgene opens the possibility of performing a relatively unbiased analysis for potential markers. FACS sorting of the Nestin-GFP^+ cells enriches a unique population of tumor cells derived from the mouse model. Whole transcriptome analysis between the GFP^+ and GFP^- cells will potentially lead to a list of candidate genes for marker identification. It can also provide insight into the molecular characteristics of quiescent GFP^+ GSCs that can be used to design potential therapies. One potential issue in the CSC field is the resolution of the cellular and molecular data collected in different studies. The quiescent status of CSCs can affect the overall behavior of tumor cells significantly. For example, it might change the overall transcription level of GSCs. If this is the case, analysis of cells in the group level could be misleading. Advances in sequencing technologies and computational analysis have greatly reduced the cost and effort of studying biological events at the single-cell level.

Together with other single-cell technologies in DNA analysis, proteomics, and metabolomics, we will be able to draw a higher-resolution map for the unique features of GSCs to elucidate their various roles in tumor progression, maintenance, and resistance to therapies.

GSCS AND THERAPEUTIC RESISTANCE

How do GSCs contribute to resistance to current therapies? There could be several mechanisms at play. For example, like many normal stem cell populations, GSCs may have enhanced capacity to either block entry or efficiently export drugs. Certain membrane transporters, such as ABCG2, have been shown to be up-regulated in both normal stem cells and GSCs (Bleau et al. 2009), which may allow GSCs to pump out chemotherapeutic drugs like temozolomide. Another possibility is that GSCs harbor distinct features that current radiation and chemotherapy treatments are incapable of targeting. Most of these therapies were designed to act on dividing tumor cells. Hence, the less proliferative, the less likely these tumor cells will be effectively targeted by agents designed to target cell proliferation (cell cycle, mitosis, or DNA replication components). In addition, the relatively slow progression to cell cycle for GSCs would minimize the degree of mutation suffered compared to highly proliferative cells that would incur greater DNA damage during multiple rounds of treatment as they progress through cell division. Our studies with the *Nestin-ΔTK-GFP* transgene indicate that the GSCs are primarily in a quiescent state and exhibit relative resistance to the effects of chemotherapy (Fig. 2A,B). After drug treatment, quiescent GFP^+ GSCs re-enter the cell cycle to produce new cells that eventually repopulate the tumor (Fig. 2C). This leads to tumor recurrence after aggressive multimodal treatment that has been the norm in GBM patients. The challenge therefore is to identify CSC vulnerabilities. Such approaches can complement traditional therapies in order to effect more lasting tumor inhibition.

CONCLUSION

Our studies on tumor suppressor mouse models suggest important roles for the cell of origin and CSCs in the pathogenesis of glioblastoma. However, despite numerous advances, much remains to be accomplished to elucidate the various mechanisms involved. The direct link between cells of origin and CSCs is also not clear. Do cells of origin directly give rise to CSCs or is there a common pathway to CSCs, as in leukemias? How do these transformations occur? Do similar pathways govern tumor initiation, progression, and maintenance? What aspects of the cellular origin are maintained or acquired in CSCs? Do different functional GBM subtypes possess distinct GSC subsets? How can these be targeted? Continued research using animal and xenograft models as well as human data analysis will provide a clearer picture of their contributions in GBM formation, which will hopefully lead to better diagnostics and therapeutics for glioblastoma. The hope is that better functional stratification of GBM and direct targeting of GSCs will add to the direly needed armamentarium to prolong the lives of patients with GBM.

ACKNOWLEDGMENTS

This work was supported in part by Children's Tumor Foundation Young Investigator Award and National Institutes of Health (NIH) T32 Postdoctoral Trainee support (2T32CA124334-06; PI: Jerry Shay) to S.R.A.L. X.X. is supported by the American Brain Tumor Association Fellowship. L.F.P. is an American Cancer Society Research Professor and recipient of NIH R01 grant CA131313-01A1 and Cancer Prevention and Research Institute of Texas Grant RP100782.

Disclosure of conflicts of interest: The authors report no potential conflicts of interest.

REFERENCES

Alcantara Llaguno S, Chen J, Kwon CH, Jackson EL, Li Y, Burns DK, Alvarez-Buylla A, Parada LF. 2009. Malignant astrocytomas originate from neural stem/progenitor cells in a somatic tumor suppressor mouse model. *Cancer Cell* **15:** 45–56.

Alcantara Llaguno SR, Wang Z, Sun D, Chen J, Xu J, Kim E, Hatanpaa KJ, Raisanen JM, Burns DK, Johnson JE, et al. 2015. Adult lineage-restricted CNS progenitors specify distinct glioblastoma subtypes. *Cancer Cell* **28:** 429–440.

Anido J, Saez-Borderias A, Gonzalez-Junca A, Rodon L, Folch G, Carmona MA, Prieto-Sanchez RM, Barba I, Martinez-Saez E, Prudkin L, et al. 2010. TGF-β receptor inhibitors target the $CD44^{high}/Id1^{high}$ glioma-initiating cell population in human glioblastoma. *Cancer Cell* **18:** 655–668.

Bachoo RM, Maher EA, Ligon KL, Sharpless NE, Chan SS, You MJ, Tang Y, DeFrances J, Stover E, Weissleder R, et al. 2002. Epidermal growth factor receptor and Ink4a/Arf: Convergent mechanisms governing terminal differentiation and transformation along the neural stem cell to astrocyte axis. *Cancer Cell* **1:** 269–277.

Bleau AM, Huse JT, Holland EC. 2009. The ABCG2 resistance network of glioblastoma. *Cell Cycle* **8:** 2936–2944.

Bonnet D, Dick JE. 1997. Human acute myeloid leukemia is organized as a hierarchy that originates from a primitive hematopoietic cell. *Nat Med* **3:** 730–737.

Ceccarelli M, Barthel FP, Malta TM, Sabedot TS, Salama SR, Murray BA, Morozova O, Newton Y, Radenbaugh A, Pagnotta SM, et al. 2016. Molecular profiling reveals biologically discrete subsets and pathways of progression in diffuse glioma. *Cell* **164:** 550–563.

Chen J, Li Y, Yu TS, McKay RM, Burns DK, Kernie SG, Parada LF. 2012a. A restricted cell population propagates glioblastoma growth after chemotherapy. *Nature* **488:** 522–526.

Chen J, McKay RM, Parada LF. 2012b. Malignant glioma: Lessons from genomics, mouse models, and stem cells. *Cell* **149:** 36–47.

Day BW, Stringer BW, Al-Ejeh F, Ting MJ, Wilson J, Ensbey KS, Jamieson PR, Bruce ZC, Lim YC, Offenhauser C, et al. 2013. EphA3 maintains tumorigenicity and is a therapeutic target in glioblastoma multiforme. *Cancer Cell* **23:** 238–248.

Driessens G, Beck B, Caauwe A, Simons BD, Blanpain C. 2012. Defining the mode of tumour growth by clonal analysis. *Nature* **488:** 527–530.

Eckel-Passow JE, Lachance DH, Molinaro AM, Walsh KM, Decker PA, Sicotte H, Pekmezci M, Rice T, Kosel ML, Smirnov IV, et al. 2015. Glioma groups based on 1p/19q, IDH, and TERT promoter mutations in tumors. *N Engl J Med* **372:** 2499–2508.

Friedmann-Morvinski D, Bushong EA, Ke E, Soda Y, Marumoto T, Singer O, Ellisman MH, Verma IM. 2012. Dedifferentiation of neurons and astrocytes by oncogenes can induce gliomas in mice. *Science* **338:** 1080–1084.

Fuentealba LC, Rompani SB, Parraguez JI, Obernier K, Romero R, Cepko CL, Alvarez- Buylla A. 2015. Embryonic origin of postnatal neural stem cells. *Cell* **161:** 1644–1655.

Groszer M, Erickson R, Scripture-Adams DD, Dougherty JD, Le Belle J, Zack JA, Geschwind DH, Liu X, Kornblum HI, Wu H. 2006. PTEN negatively regulates neural stem cell self-renewal by modulating G0-G1 cell cycle entry. *Proc Natl Acad Sci* **103:** 111–116.

Hegedus B, Dasgupta B, Shin JE, Emnett RJ, Hart-Mahon EK, Elghazi L, Bernal-Mizrachi E, Gutmann DH. 2007. Neurofibromatosis-1 regulates neuronal and glial cell differentiation from neuroglial progenitors in vivo by both cAMP- and Ras-dependent mechanisms. *Cell Stem Cell* **1:** 443–457.

Hermann PC, Huber SL, Herrler T, Aicher A, Ellwart JW, Guba M, Bruns CJ, Heeschen C. 2007. Distinct populations of cancer stem cells determine tumor growth and metastatic activity in human pancreatic cancer. *Cell Stem Cell* **1:** 313–323.

Holmberg Olausson K, Maire CL, Haidar S, Ling J, Learner E, Nister M, Ligon KL. 2014. Prominin-1 (CD133) defines both stem and non-stem cell populations in CNS development and gliomas. *PLoS One* **9:** e106694.

Lathia JD, Gallagher J, Heddleston JM, Wang J, Eyler CE, Macswords J, Wu Q, Vasanji A, McLendon RE, Hjelmeland AB, et al. 2010. Integrin alpha 6 regulates glioblastoma stem cells. *Cell Stem Cell* **6:** 421–432.

Ligon KL, Huillard E, Mehta S, Kesari S, Liu H, Alberta JA, Bachoo RM, Kane M, Louis DN, Depinho RA, et al. 2007. Olig2-regulated lineage-restricted pathway controls replication competence in neural stem cells and malignant glioma. *Neuron* **53:** 503–517.

Meletis K, Wirta V, Hede SM, Nister M, Lundeberg J, Frisen J. 2006. p53 suppresses the self-renewal of adult neural stem cells. *Development* **133:** 363–369.

Merkle FT, Fuentealba LC, Sanders TA, Magno L, Kessaris N, Alvarez-Buylla A. 2014. Adult neural stem cells in distinct microdomains generate previously unknown interneuron types. *Nat Neurosci* **17:** 207–214.

Ming GL, Song H. 2011. Adult neurogenesis in the mammalian brain: Significant answers and significant questions. *Neuron* **70:** 687–702.

Nishide K, Nakatani Y, Kiyonari H, Kondo T. 2009. Glioblastoma formation from cell population depleted of Prominin1-expressing cells. *PLoS One* **4:** e6869.

Nishiyama A, Komitova M, Suzuki R, Zhu X. 2009. Polydendrocytes (NG2 cells): Multifunctional cells with lineage plasticity. *Nat Rev Neurosci* **10:** 9–22.

Pece S, Tosoni D, Confalonieri S, Mazzarol G, Vecchi M, Ronzoni S, Bernard L, Viale G, Pelicci PG, Di Fiore PP. 2010. Biological and molecular heterogeneity of breast cancers correlates with their cancer stem cell content. *Cell* **140:** 62–73.

Schepers AG, Snippert HJ, Stange DE, van den Born M, van Es JH, van de Wetering M, Clevers H. 2012. Lineage tracing reveals $Lgr5^{+}$ stem cell activity in mouse intestinal adenomas. *Science* **337:** 730–735.

Singh SK, Clarke ID, Terasaki M, Bonn VE, Hawkins C, Squire J, Dirks PB. 2003. Identification of a cancer stem cell in human brain tumors. *Cancer Res* **63:** 5821–5828.

Son MJ, Woolard K, Nam DH, Lee J, Fine HA. 2009. SSEA-1 is an enrichment marker for tumor-initiating cells in human glioblastoma. *Cell Stem Cell* **4:** 440–452.

Stupp R, Mason WP, van den Bent MJ, Weller M, Fisher B, Taphoorn MJ, Belanger K, Brandes AA, Marosi C, Bogdahn U, et al. 2005. Radiotherapy plus concomitant and adjuvant temozolomide for glioblastoma. *N Engl J Med* **352:** 987–996.

TCGA. 2008. Comprehensive genomic characterization defines human glioblastoma genes and core pathways. *Nature* **455:** 1061–1068.

Verhaak RG, Hoadley KA, Purdom E, Wang V, Qi Y, Wilkerson MD, Miller CR, Ding L, Golub T, Mesirov JP, et al. 2010. Integrated genomic analysis identifies clinically relevant subtypes of glioblastoma characterized by abnormalities in PDGFRA, IDH1, EGFR, and NF1. *Cancer Cell* **17:** 98–110.

Wang X, Kruithof-de Julio M, Economides KD, Walker D, Yu H, Halili MV, Hu YP, Price SM, Abate-Shen C,

Shen MM. 2009a. A luminal epithelial stem cell that is a cell of origin for prostate cancer. *Nature* **461:** 495–500.

Wang Y, Yang J, Zheng H, Tomasek GJ, Zhang P, McKeever PE, Lee EY, Zhu Y. 2009b. Expression of mutant p53 proteins implicates a lineage relationship between neural stem cells and malignant astrocytic glioma in a murine model. *Cancer Cell* **15:** 514–526.

Zheng S, Fu J, Vegesna R, Mao Y, Heathcock LE, Torres-Garcia W, Ezhilarasan R, Wang S, McKenna A, Chin L, et al. 2013. A survey of intragenic breakpoints in glioblastoma identifies a distinct subset associated with poor survival. *Genes Dev* **27:** 1462–1472.

Zhu Y, Guignard F, Zhao D, Liu L, Burns DK, Mason RP, Messing A, Parada LF. 2005. Early inactivation of p53 tumor suppressor gene cooperating with NF1 loss induces malignant astrocytoma. *Cancer Cell* **8:** 119–130.

Zhu Z, Khan MA, Weiler M, Blaes J, Jestaedt L, Geibert M, Zou P, Gronych J, Bernhardt O, Korshunov A, et al. 2014. Targeting self-renewal in high-grade brain tumors leads to loss of brain tumor stem cells and prolonged survival. *Cell Stem Cell* **15:** 185–198.

The Enigma of p53

Guillermina Lozano
The University of Texas MD Anderson Cancer Center, Houston, Texas 77030
Correspondence: gglozano@mdanderson.org

This perspective will focus on the physiological impact of wild-type and mutant p53 activities. In particular, the tissue-specific nature of activation of p53 targets and their subsequent effects on cell behavior will be discussed. Because mutations in p53 are common in human cancers, the regulation and physiological consequences of mutant p53 proteins will also be discussed.

p53, A SHORT-LIVED PROTEIN TRANSCRIPTION FACTOR

p53 is a potent transcription factor that binds two repeats of a specific DNA sequence (5′-RRRCWWGYYY-3′) as a tetramer and activates hundreds of genes—many in a tissue- and cell-specific manner (El-Deiry et al. 1992; Hoh et al. 2002). Most often this sequence is within 10 kb of the promoter but sometimes occurs further away at enhancers. p53 is stabilized and activated in response to numerous cellular signals that often sense DNA damage, stress, and inappropriate oncogene activation (which will be termed the stress response in this review) (Bieging et al. 2014). These p53-regulated genes have numerous functions from killing cells (*Puma*, *Noxa*, *Bax*) to arresting the cell cycle (*p21*, *Btg2*, *Ptprv*) and initiating senescence (*p21*, *Pml*, *Pai-1*) to altering the metabolic state (*TIGAR*, *Soc-2*, *DRAM*) for cell survival (Aubrey et al. 2016). Why does p53 activate so many genes? And how does it decide which genes to activate when? Answers to these questions remain elusive.

As with most transcription factors, p53 is normally a very short-lived protein (it has a 10- to 20-min half-life). In normal cells, p53 is kept at very low levels by two critical inhibitors Mdm2 and Mdm4 (Pant and Lozano 2014). Mdm2 encodes an E3 ubiquitin ligase that targets p53 for polyubiquitination and proteasomal degradation. Mdm4 on the other hand has a RING domain that lacks E3 ligase function but forms a heterodimer with Mdm2 to inhibit p53 activity. Mdm4 likely functions as an E4 ligase to enhance processivity of Mdm2 (Pant and Lozano 2014). Loss of either *Mdm2* or *Mdm4* in mice leads to embryo lethal phenotypes that are completely dependent on p53 as deletion of *p53* rescues these phenotypes (Jones et al. 1995; Montes de Oca Luna et al. 1995; Parant et al. 2001). In the central nervous system (CNS), a combination of *Mdm2* and *Mdm4* loss leads to more detrimental phenotypes, suggesting that both cooperate to inhibit p53 function at least in some cell types (Francoz et al. 2006; Xiong et al. 2006). Furthermore, disruption of the Mdm2–Mdm4 interaction is also an embryo lethal event, indicating that the Mdm2/4 heterodimer is the most potent inhibitor of p53 (Huang et al. 2011; Pant et al. 2011). That being said, an Mdm2 that lacks E3 ligase activity but retains binding to Mdm4 is not embryonic lethal, indicating that the ability to bind the transactivation domain is sufficient to keep p53 levels low under physiological conditions (Tollini et al. 2014). However, these mice are extremely sensitive to irradiation and die because p53 levels cannot return to baseline quickly enough. Thus, the Mdm proteins are essential inhibitors of p53 activity, but the levels required to maintain p53 at low physiological levels are less that those required to inhibit p53 levels after a stress response.

NUMEROUS *Mdm2* ALLELES DEFINE p53 PHYSIOLOGICAL OUTCOMES

Mice heterozygous for *Mdm2* or *Mdm4* are normal. However, a low dose of ionizing radiation, which causes DNA damage, kills $Mdm2^{+/-}$ and $Mdm4^{+/-}$ mice but not controls in a p53-dependent manner (Terzian et al. 2007). *Mdm2 Mdm4* double heterozygous mice, which express slightly more p53 (because of haploinsufficiency of these two inhibitors), suffer lethal developmental defects 30% of the time or hematopoietic failure within 2–3 wk after birth (70%). None survive beyond this age. The observation that $Mdm2^{+/-}$ $Mdm4^{+/-}$ $p53^{+/-}$ mice are alive and well indicates how exquisitely Mdm2 and Mdm4 regulate p53 levels (Terzian et al. 2007).

Studies of hypomorphic *Mdm2* alleles have also defined important tissue specificities. Mice inheriting a hypomorphic and a null allele of *Mdm2* express ~30% of Mdm2 levels and show increased p53 activity. These mice are small, lymphopenic, and radiosensitive (Mendrysa et al. 2003). Another hypomorphic *Mdm2* allele (which contains a 184-bp deletion and retains the *Neomycin* gene

Published by Cold Spring Harbor Laboratory Press; doi: 10.1101/sqb.2016.81.031062

in intron 1 in feedback loop defective mice) with decreased Mdm2 levels also exhibits unique p53-dependent pathologies: increased skin pigmentation of the epidermis, reduction of seminiferous tubules and decreased spermatocytes in testis, and decreased number of ovarian follicles (Pant et al. 2016). None of the defects is rescued by deletion of the p53 target *p21*, but the reproductive phenotypes are rescued by *Puma* loss. Thus, deletion of a single downstream target of p53 (*Puma*) could revert a p53-dependent phenotype. The skin phenotype appears to be driven by KitL signaling. *KitL* is another p53 target expressed in melanocytes and is linked to their migration and accumulation (Zeron-Medina et al. 2013). These studies show for the first time that depletion of a single p53 target can bypass the effects of p53 and further implicate tissue-specific nature of the p53 response in vivo.

Deletion of *Mdm2* using an *Mdm2* floxed allele and a tamoxifen-inducible universally expressed CreER transgene resulted in numerous pathologies in adult mice, which included extensive absence of white pulp in the spleen, erythroid depletion in the bone marrow, villus atrophy in the intestine, and tubular proteinuria and fibrosis in the kidney (Table 1; Zhang et al. 2014). Importantly, all phenotypes were obviated by *p53* loss highlighting the near-universal importance of Mdm2 in maintaining appropriate levels of p53 in vivo. An analysis of p53 transcriptional activity showed a varied response depending on tissue type (Table 1). Surprisingly, *p21* was elevated almost 200-fold in the kidney compared with seven- and 12-fold in liver and spleen, respectively. As an aside, *p21* is always up-regulated by p53 in response to numerous signals and serves as a great marker for p53 transcriptional activation. *Puma*, another p53 transcriptional target, was elevated 18-fold in the kidney but was not elevated in the liver. Global loss of *Mdm2* also resulted in apoptosis and senescence phenotypes that varied by tissue type (Table 1).

Besides affecting normal developmental/physiological functions, minor changes in the levels of Mdm2 and Mdm4 also affect tumor phenotypes. Haploinsufficiency at *Mdm2* or *Mdm4* loci delays tumor onset in mice carrying an *Eμ-myc* transgene (Alt et al. 2003; Terzian et al. 2007), and decreased *Mdm2* levels delay intestinal tumors in $Apc^{min/+}$ mice (Mendrysa et al. 2006). In addition, a single-nucleotide polymorphism in the *Mdm2* promoter that increases the levels of *Mdm2* also contributes to cancer risk (Bond et al. 2004; Post et al. 2010). Last, transgenic mice that have high levels of Mdm2 or Mdm4 are also prone to develop multiple cancers, indicating that Mdm2 and Mdm4 directly impact tumor development (Jones et al. 1998; Xiong et al. 2010).

p53 RESPONDS TO DNA DAMAGE/STRESS

An abnormal physiological state (that likely differs by cell type) is sensed by p53 and results in p53 activation and cellular outcomes (Bieging et al. 2014). Numerous upstream signals activate p53 and include DNA damage, telomere erosion, oxidative stress, hypoxia, and hyperproliferative signals. For these reasons, p53 was labeled the guardian of the genome (Lane 1992). As indicated above, the Mdm proteins keep p53 at low levels normally. However, in response to DNA damage and a myriad of other signals that sense abnormal cell behavior, p53 is phosphorylated, acetylated, and sumolyated and these posttranslational modifications unleash a tissue-specific transcriptional program.

THE p53 TUMOR SUPPRESSOR PATHWAY

In 1990, the wild-type sequence of p53 was finally determined, and it was shown to inhibit transformation in vitro (Finlay et al. 1989). At the same time, mutations in *Tp53* with concomitant loss of heterozygosity were identified in colorectal cancers (Baker et al. 1989). p53 mutations were subsequently identified in the germline of individuals with a rare cancer syndrome, the Li–Fraumeni syndrome (Malkin et al. 1990). And last, mice with deletion of the *p53* gene developed a variety of cancers that vary depending on the background of the animal (Donehower et al. 1992; Jacks et al. 1994; Donehower and Lozano 2009).

DNA sequencing of thousands of human tumors has revealed that *Tp53* is the most often mutated tumor suppressor (Hainaut and Pfeifer 2016). Decreased p53 levels by either deletion or mutation or increased levels of the p53 inhibitors Mdm2 and Mdm4 lead to tumor development in a mutually exclusive relationship (Wasylishen and Lozano 2016).

Table 1. Phenotypes of $Mdm2^{FM/-}$ *CAG-CreER* mice upon activation of Cre

Tissue	Phenotype[a]	Cdkn1a (p21)	Bbc3 (Puma)	Apoptosis	Senescence[b]
Spleen	Absence of white pulp	11.9 ± 4.20	2.11 ± 0.96	pos	neg
Bone marrow	Erythroid depletion[c]	NA	NA	NA	NA
Intestine	Villus atrophy, apoptotic crypts	NA	NA	pos	NA
Kidney	Protein casts, tubule dilation	184.56 ± 56.3	18.21 ± 5.32	pos	pos
Liver	Increased AST, fibrosis	6.80 ± 1.37	0.91 ± 0.27	pos	pos
Heart	Fibrosis	37.07 ± 12.76	7.37 ± 2.31	neg	pos
Cerebrum		16.21 ± 2.68	3.13 ± 0.48	pos	pos
Cerebellum		21.59 ± 4.58	2.46 ± 0.41	neg	NA
Retina		20.58 ± 7.08	6.97 ± 1.58	pos	pos

NA, not available; pos, positive; neg, negative; AST, aspartate aminotransferase.
[a]As observed by morphology for two different treatments.
[b]At least two cytokines/chemokines were elevated.
[c]Erythroid depletion occurs in Diamond–Blackfan anemia and is due to elevated p53 activity.

However, the most common alterations at the p53 locus are missense mutations (~75%) that occur in the DNA binding domain and disrupt interactions with DNA and thus transcriptional activity. These missense mutations are of two kinds: structural and conformational. Importantly for this discussion, most p53 missense mutations do not disrupt the amino and carboxyl domains of p53. These are the domains through which cells signal to p53. For example, phosphorylation at the amino terminus disrupts interactions with Mdm2/4 (Loughery and Meek 2013). Tetramerization of the carboxyl terminus forces conformational changes in the DNA binding domain that facilitate DNA binding (D'Abramo et al. 2016).

Modeling of the hotspot p53 mutations in mice have yielded several observations, the most important of which is that mutant p53 has a gain-of-function phenotype (Lang et al. 2004; Olive et al. 2004; Brosh and Rotter 2009; Freed-Pastor and Prives 2012). Tumor cells expressing a stable mutant p53 have a more aggressive and highly metastatic phenotype. Another important observation from the mouse models is that, similar to the half-life of wild-type p53, mutant p53 is an unstable protein under normal physiological conditions. Some tumors expressing only mutant p53 did not have a detectable p53 protein by immunohistochemistry, leading to the hypothesis that tumor-specific events stabilized p53 (Lang et al. 2004). As proof of concept, homozygosity of the *p53R172H* allele rescued the *Mdm2*-null lethal phenotype and resulted in increased levels of p53 mutant protein, increased tumor incidence, and increased metastasis (Terzian et al. 2008). Several p53-activating stress responses such as DNA damage and reactive oxygen species (ROS) also stabilize the p53 mutant protein in vivo (Suh et al. 2011). Thus, it appears that signals that regulate p53 via the amino and carboxyl termini also regulate mutant p53. An important question that needs to be addressed is what signals stabilize the p53 mutant protein in tumors.

These signals may be constitutive and result in a stable p53 protein in all cells of the tumors (Fig. 1). But these signals may also be intermittent, resulting in p53 stability only in some tumors cells at any one time, leading to tumor heterogeneity. The timing of p53 mutation and stabilization is critical. In ovarian cancer, for example, p53 mutation appears to be an early and probably the first event resulting in a stable mutant protein in all cells (Perets et al. 2013). In colorectal cancers, on the other hand, p53 mutations are late and correlate to a change in the behavior of tumor cells to a more aggressive and invasive one (Fearon and Vogelstein 1990). So what stabilizes mutant p53 in tumors? Many of these signals are likely the same mechanisms that stabilize wild-type p53. Inflammatory signals, DNA damage, hypoxia, and ROS are all likely to contribute to stabilization of mutant p53 proteins. Nonetheless, some p53 mutant mice develop tumors that do not have a stable p53 protein and these likely resemble tumors with deletions of *p53* (Lang et al. 2004).

Further studies are needed to understand how mutant p53 is stabilized, as a growing number of studies suggest tumor cells are addicted to mutant p53 (Alexandrova et al. 2015). Undermining the signals that stabilize mutant p53 (e.g., hsp90) may have important therapeutic consequences.

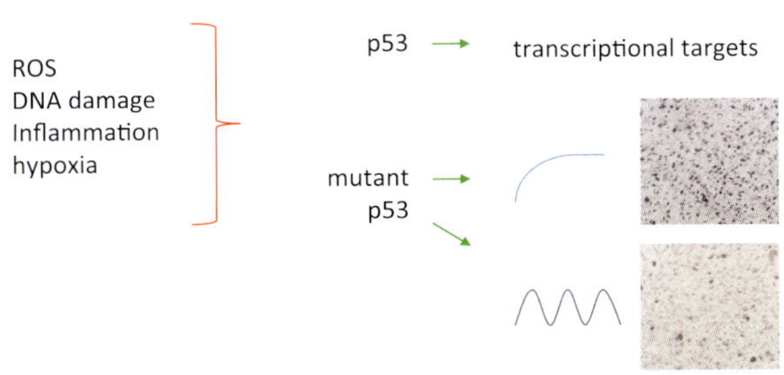

Figure 1. Numerous stress signals stabilize both wild-type and mutant p53. Wild-type p53 activates numerous transcriptional targets many in a cell-specific manner. Mutant p53, on the other hand, is transcriptionally inactive but binds other transcription factors to alter gene expression. When stabilized, mutant p53 shows gain-of-function activities. Some signals are constitutive (e.g., genomic instability) and lead to p53 stability in all tumor cells. Other signals (e.g., hypoxia) may result in only some cells having a mutant p53. ROS, reactive oxygen species.

CONCLUSION

p53 is a master regulator of a tissue-specific transcriptional program that alters the behavior of cells killing some and stopping others from proliferation. Because of these tumor suppressive activities, p53 is the most often altered gene and pathway. I will venture to say that all tumor cells have undermined the p53 pathway by some mechanism. Studies to understand how p53 is regulated will lead to important therapeutic options, as has already happened with the development of Mdm2 inhibitors that are being tested in clinical trials (Cheok and Lane 2016; Wang et al. 2016). A deeper understanding of how mutant p53 leads to more aggressive highly invasive phenotypes in vivo is important, as is an understanding of the signals that stabilize mutant p53 proteins.

REFERENCES

Alexandrova EM, Yallowitz AR, Li D, Xu S, Schulz R, Proia DA, Lozano G, Dobbelstein M, Moll UM. 2015. Improving survival by exploiting tumour dependence on stabilized mutant p53 for treatment. *Nature* **523:** 352–356.

Alt JR, Greiner TC, Cleveland JL, Eischen CM. 2003. Mdm2 haplo-insufficiency profoundly inhibits Myc-induced lymphomagenesis. *EMBO J* **22:** 1442–1450.

Aubrey BJ, Strasser A, Kelly GL. 2016. Tumor-suppressor functions of the TP53 pathway. *Cold Spring Harb Perspect Med* **6:** a026062.

Baker SJ, Fearon ER, Nigro JM, Hamilton SR, Preisinger AC, Jessup JM, van Tuinen P, Ledbetter DH, Barker DF, Nakamura Y, et al. 1989. Chromosome 17 deletions and p53 gene mutations in colorectal carcinomas. *Science* **244:** 217–221.

Bieging KT, Mello SS, Attardi LD. 2014. Unravelling mechanisms of p53-mediated tumour suppression. *Nat Rev Cancer* **14:** 359–370.

Bond GL, Hu W, Bond EE, Robins H, Lutzker SG, Arva NC, Bargonetti J, Bartel F, Taubert H, Wuerl P, et al. 2004. A single nucleotide polymorphism in the MDM2 promoter attenuates the p53 tumor suppressor pathway and accelerates tumor formation in humans. *Cell* **119:** 591–602.

Brosh R, Rotter V. 2009. When mutants gain new powers: News from the mutant p53 field. *Nat Rev Cancer* **9:** 701–713.

Cheok CF, Lane DP. 2016. Exploiting the p53 pathway for therapy. *Cold Spring Harb Perspect Med* **6:** a026310.

D'Abramo M, Besker N, Desideri A, Levine AJ, Melino G, Chillemi G. 2016. The p53 tetramer shows an induced-fit interaction of the C-terminal domain with the DNA-binding domain. *Oncogene* **35:** 3272–3281.

Donehower LA, Lozano G. 2009. 20 years studying p53 functions in genetically engineered mice. *Nat Rev Cancer* **9:** 831–841.

Donehower LA, Harvey M, Slagle BL, McArthur MJ, Montgomery CA Jr, Butel JS, Bradley A. 1992. Mice deficient for p53 are developmentally normal but susceptible to spontaneous tumours. *Nature* **356:** 215–221.

El-Deiry WS, Kern SE, Pietenpol JA, Kinzler KW, Vogelstein B. 1992. Definition of a consensus binding site for p53. *Nat Genet* **1:** 45–49.

Fearon ER, Vogelstein B. 1990. A genetic model for colorectal tumorigenesis. *Cell* **61:** 759–767.

Finlay CA, Hinds PW, Levine AJ. 1989. The p53 proto-oncogene can act as a suppressor of transformation. *Cell* **57:** 1083–1093.

Francoz S, Froment P, Bogaerts S, De Clercq S, Maetens M, Doumont G, Bellefroid E, Marine JC. 2006. Mdm4 and Mdm2 cooperate to inhibit p53 activity in proliferating and quiescent cells in vivo. *Proc Natl Acad Sci* **103:** 3232–3237.

Freed-Pastor WA, Prives C. 2012. Mutant p53: One name, many proteins. *Genes Dev* **26:** 1268–1286.

Hainaut P, Pfeifer GP. 2016. Somatic TP53 mutations in the era of genome sequencing. *Cold Spring Harb Perspect Med* **6:** a026179.

Hoh J, Jin S, Parrado T, Edington J, Levine AJ, Ott J. 2002. The p53MH algorithm and its application in detecting p53-responsive genes. *Proc Natl Acad Sci* **99:** 8467–8472.

Huang L, Yan Z, Liao X, Li Y, Yang J, Wang ZG, Zuo Y, Kawai H, Shadfan M, Ganapathy S, et al. 2011. The p53 inhibitors MDM2/MDMX complex is required for control of p53 activity in vivo. *Proc Natl Acad Sci* **108:** 12001–12006.

Jacks T, Remington L, Williams BO, Schmitt EM, Halachmi S, Bronson RT, Weinberg RA. 1994. Tumor spectrum analysis in p53-mutant mice. *Curr Biol* **4:** 1–7.

Jones SN, Roe AE, Donehower LA, Bradley A. 1995. Rescue of embryonic lethality in Mdm2-deficient mice by absence of p53. *Nature* **378:** 206–208.

Jones SN, Hancock AR, Vogel H, Donehower LA, Bradley A. 1998. Overexpression of Mdm2 in mice reveals a p53-independent role for Mdm2 in tumorigenesis. *Proc Natl Acad Sci* **95:** 15608–15612.

Lane DP. 1992. Cancer. p53, guardian of the genome. *Nature* **358:** 15–16.

Lang GA, Iwakuma T, Suh YA, Liu G, Rao VA, Parant JM, Valentin-Vega YA, Terzian T, Caldwell LC, Strong LC, et al. 2004. Gain of function of a p53 hot spot mutation in a mouse model of Li–Fraumeni syndrome. *Cell* **119:** 861–872.

Loughery J, Meek D. 2013. Switching on p53: An essential role for protein phosphorylation? *BioDiscovery* **8:** 1.

Malkin D, Li FP, Strong LC, Fraumeni JF Jr, Nelson CE, Kim DH, Kassel J, Gryka MA, Bischoff FZ, Tainsky MA, et al. 1990. Germ line p53 mutations in a familial syndrome of breast cancer, sarcomas, and other neoplasms. *Science* **250:** 1233–1238.

Mendrysa SM, McElwee MK, Michalowski J, O'Leary KA, Young KM, Perry ME. 2003. *mdm2* is critical for inhibition of p53 during lymphopoiesis and the response to ionizing irradiation. *Mol Cell Biol* **23:** 462–472.

Mendrysa SM, O'Leary KA, McElwee MK, Michalowski J, Eisenman RN, Powell DA, Perry ME. 2006. Tumor suppression and normal aging in mice with constitutively high p53 activity. *Genes Dev* **20:** 16–21.

Montes de Oca Luna R, Wagner DS, Lozano G. 1995. Rescue of early embryonic lethality in *mdm2*-deficient mice by deletion of p53. *Nature* **378:** 203–206.

Olive KP, Tuveson DA, Ruhe ZC, Yin B, Willis NA, Bronson RT, Crowley D, Jacks T. 2004. Mutant p53 gain of function in two mouse models of Li–Fraumeni syndrome. *Cell* **119:** 847–860.

Pant V, Lozano G. 2014. Limiting the power of p53 through the ubiquitin proteasome pathway. *Genes Dev* **28:** 1739–1751.

Pant V, Xiong S, Iwakuma T, Quintas-Cardama A, Lozano G. 2011. Heterodimerization of Mdm2 and Mdm4 is critical for regulating p53 activity during embryogenesis but dispensable for p53 and Mdm2 stability. *Proc Natl Acad Sci* **108:** 11995–12000.

Pant V, Xiong S, Chau G, Tsai K, Shetty G, Lozano G. 2016. Distinct downstream targets manifest p53-dependent pathologies in mice. *Oncogene* doi: 10.1038/onc.2016.111.

Parant J, Chavez-Reyes A, Little NA, Yan W, Reinke V, Jochemsen AG, Lozano G. 2001. Rescue of embryonic lethality in Mdm4-null mice by loss of Trp53 suggests a nonoverlapping pathway with MDM2 to regulate p53. *Nat Genet* **29:** 92–95.

Perets R, Wyant GA, Muto KW, Bijron JG, Poole BB, Chin KT, Chen JY, Ohman AW, Stepule CD, Kwak S, et al. 2013. Transformation of the fallopian tube secretory epithelium leads to high-grade serous ovarian cancer in Brca;Tp53;Pten models. *Cancer Cell* **24:** 751–765.

Post SM, Quintas-Cardama A, Pant V, Iwakuma T, Hamir A, Jackson JG, Maccio DR, Bond GL, Johnson DG, Levine AJ, et al. 2010. A high-frequency regulatory polymorphism in the p53 pathway accelerates tumor development. *Cancer Cell* **18:** 220–230.

Suh YA, Post SM, Elizondo-Fraire AC, Maccio DR, Jackson JG, El-Naggar AK, Van Pelt C, Terzian T, Lozano G. 2011. Multiple stress signals activate mutant p53 in vivo. *Cancer Res* **71:** 7168–7175.

Terzian T, Wang Y, Van Pelt CS, Box NF, Travis EL, Lozano G. 2007. Haploinsufficiency of Mdm2 and Mdm4 in tumorigenesis and development. *Mol Cell Biol* **27:** 5479–5485.

Terzian T, Suh YA, Iwakuma T, Post SM, Neumann M, Lang GA, Van Pelt CS, Lozano G. 2008. The inherent instability of mutant p53 is alleviated by Mdm2 or p16INK4a loss. *Genes Dev* **22:** 1337–1344.

Tollini LA, Jin A, Park J, Zhang Y. 2014. Regulation of p53 by Mdm2 E3 ligase function is dispensable in embryogenesis and development, but essential in response to DNA damage. *Cancer Cell* **26:** 235–247.

Wang S, Zhao Y, Aguilar A, Bernard D, Yang C. 2016. Targeting the MDM2-p53 protein–protein interaction for new cancer therapy: Progress and challenges. *Cold Spring Harb Perspect Med* **6:** a026245.

Wasylishen AR, Lozano G. 2016. Attenuating the p53 pathway in human cancers: Many means to the same end. *Cold Spring Harb Perspect Med* **6:** a026211.

Xiong S, Van Pelt CS, Elizondo-Fraire AC, Liu G, Lozano G. 2006. Synergistic roles of Mdm2 and Mdm4 for p53 inhibition in central nervous system development. *Proc Natl Acad Sci* **103:** 3226–3231.

Xiong S, Pant V, Suh YA, Van Pelt CS, Wang Y, Valentin-Vega YA, Post SM, Lozano G. 2010. Spontaneous tumorigenesis in mice overexpressing the p53-negative regulator Mdm4. *Cancer Res* **70:** 7148–7154.

Zeron-Medina J, Wang X, Repapi E, Campbell MR, Su D, Castro-Giner F, Davies B, Peterse EF, Sacilotto N, Walker GJ, et al. 2013. A polymorphic p53 response element in KIT ligand influences cancer risk and has undergone natural selection. *Cell* **155:** 410–422.

Zhang Y, Xiong S, Li Q, Hu S, Tashakori M, Van Pelt C, You MJ, Pageon L, Lozano G. 2014. Tissue-specific and age-dependent effects of global Mdm2 loss. *J Pathol* **233:** 380–391.

Alterations in Three-Dimensional Organization of the Cancer Genome and Epigenome

JOANNA ACHINGER-KAWECKA,[1,2] PHILLIPPA C. TABERLAY,[1,2,3] AND SUSAN J. CLARK[1,2]
[1]*Epigenetics Research Laboratory, Genomics and Epigenetics Division, Garvan Institute of Medical Research, Darlinghurst, New South Wales 2010, Australia*
[2]*St. Vincent's Clinical School, Faculty of Medicine, University of New South Wales, Darlinghurst, New South Wales 2010, Australia*
[3]*School of Medicine, Faculty of Health, University of Tasmania, Hobart, Tasmania 7000, Australia*
Correspondence: s.clark@garvan.org.au

The structural and functional basis of the genome is provided by the three-dimensional (3D) chromatin state. To enable accurate gene regulation, enhancer elements and promoter regions are brought into close spatial proximity to ensure proper, cell type–specific gene expression. In cancer, genetic and epigenetic processes can deregulate the transcriptional program. To investigate whether the 3D chromatin state is also disrupted in cancer we performed Hi-C chromosome conformation sequencing in normal and prostate cancer cells and compared the chromatin interaction maps with changes to the genome and epigenome. Notably, we find that additional topologically associated domain (TAD) boundaries are formed in cancer cells resulting in smaller TADs and altered gene expression profiles. The new TAD boundaries are commonly associated with copy-number changes observed in the cancer genome. We also identified new cancer-specific chromatin loops within TADs that are enriched for enhancers and promoters. Finally, we find that many of the long-range epigenetically silenced (LRES) and long-range epigenetically active (LREA) regions in cancer are characterized by differential chromatin interactions. Together our data provide a new insight into charting alterations in higher-order structure and the relationship with genetic, epigenetic, and transcriptional changes across the cancer genome.

The genome is not simply linear, but is exquisitely folded into a three-dimensional (3D) chromatin structure to facilitate appropriate cell type–specific gene expression patterns. The epigenome partitions the genome into active and inactive domains. Active regions are marked by open chromatin and active histone marks, whereas inactive regions are marked by repressive histone modifications and closed chromatin (Fig. 1). For a gene to be active there needs to be a physical interaction or chromatin interaction that enables genomic regions that are distant in linear space on chromosomes to come in close proximity in 3D space and relay signals—for example, via enhancer–promoter looping interactions (Fig. 1). Chromatin loops are maintained by architectural proteins—for example, insulator protein CCCTC-binding factor (CTCF) and SMC-family complex, cohesion (RAD21) (Dixon et al. 2012; Nora et al. 2012; Rao et al. 2014). Rapid advances of chromosome conformation capture (3C)-based (Dekker et al. 2002) technologies such as 4C (Simonis et al. 2006; Zhao et al. 2006), 5C (Dostie et al. 2006), Hi-C (Lieberman-Aiden et al. 2009), and ChIA-PET (Fullwood et al. 2009) have provided increasing insights into the 3D architecture of the genomes. It is now clear that the genome is organized into topologically associated domains (TADs) or contact domains that are hundreds of kilobases in size and encompass multiple genes and regulatory elements (Dixon et al. 2012; Nora et al. 2012). At a higher level, TADs are organized into functionally distinct megabase compartments, which comprise large blocks of chromatin that are either A-type (open, gene-rich) or B-type (closed, gene-poor) (Lieberman-Aiden et al. 2009; Zhang et al. 2012). TADs are characterized by preferential interactions within them and lack of interactions between different TADs, suggesting that TAD boundaries act as insulators by preventing communication between elements on either side of the boundary. TADs are conserved and reported to be largely invariant between different cell types (Dixon et al. 2012; Nora et al. 2012), whereas the chromatin interactions or loops within TADs are more tissue-specific (Dixon et al. 2012, 2015; Phillips-Cremins et al. 2013; Smith et al. 2016). Because the genome and the epigenome are both highly deregulated in many cancer types, we were interested to determine whether the 3D chromatin organization in cancer was also disrupted and if this was associated with the aberrant gene expression, such as activation of oncogenes or inactivation of tumor-suppressor genes.

TAD ORGANIZATION IS DISRUPTED IN PROSTATE CANCER CELLS

To elucidate the role of 3D genome architecture in prostate cancer biology, we performed Hi-C and RNA-seq experiments in normal prostate epithelial cells (PrECs)

Published by Cold Spring Harbor Laboratory Press; doi: 10.1101/sqb.2016.81.031013

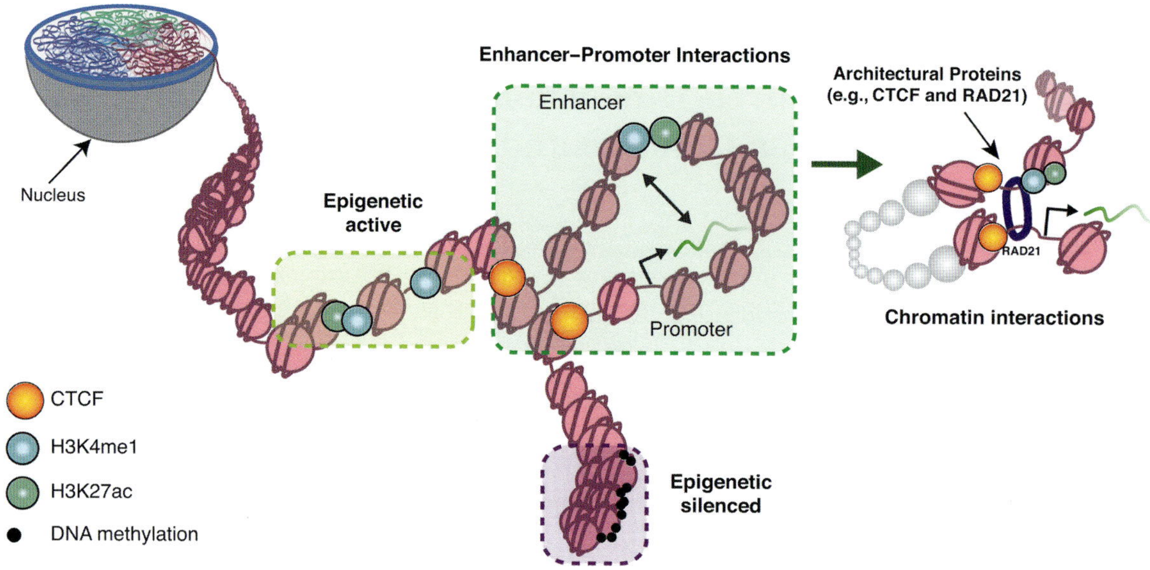

Figure 1. The epigenome partitions the genome into inactive or active domains. Inactive regions are marked by closed chromatin and repressive modifications (e.g., DNA methylation) and active regions are marked by open chromatin and active histone marks (e.g., H3K4me1 and H3K27ac). For a gene to be active there needs to be a physical chromatin interaction that enable genomic regions that are distant in a linear space on chromosomes to come in close proximity in three-dimensional space and relay signals (e.g., via enhancer–promoter interactions). These have been shown to be maintained by architectural proteins (e.g., CTCF and RAD21).

and two prostate cancer cell lines (LNCaP, androgen-sensitive, and PC3, androgen-resistant) (Taberlay et al. 2016). To further understand the relationship between cancer genome organization and genetic and epigenetic remodeling, we used ChIP-seq to probe binding patterns of transcription factors CTCF and RAD21 and regulatory elements, including enhancers (H3K4me1 and H3K27ac), promoters (H3K4me3), and Polycomb (H3K27me3).

Considering the importance of the topological organization of the genome, we next explored the segmentation of the normal and cancer genomes into TADs (Dixon et al. 2012; Nora et al. 2012). Changes within TADs (sub-TADs) have been reported to occur during cell differentiation (Dixon et al. 2015) and reprogramming (Beagan et al. 2016; Krijger et al. 2016) in embryonic stem cells, which suggests that within the hierarchical genome structure there is a discrete level of 3D organization that displays more plasticity to facilitate cell type–specific expression. We hypothesized that TAD organization would be disrupted in cancer cells and potentially resemble more closely the organization observed in undifferentiated cells. Comparison of the TAD organization in normal prostate cells and prostate cancer cells revealed that cancer cells retain the ability to segment their genomes into domains; however, we observed large differences in both their number (Fig. 2A) and in the position of the TAD boundaries across the genome (Fig. 2B). The size of TADs was significantly smaller in the two prostate cancer cell lines compared with normal cells. In PC3 cells, TADs were on average half the size of TADs in normal PrECs and the number of TADs increased by approximately twofold compared with PrECs (Fig. 2A). Interestingly, the large increase in the number of TADs and the decrease in their average size were even more prominent in LNCaP cells. In LNCaP cells, TADs were on average 0.3 times the size of those in PrECs and their number increased by 3.5-fold (Fig. 2A). The variability between the two prostate cancer cell lines could be potentially explained by the differences in the biology of these cells.

By visualizing topological domains as 2D interaction matrices, we observed that large domains in normal cells were often subdivided into smaller domains ("subdomains") in cancer cells (Fig. 2B). We further validated this observation, by comparing the number of TADs identified in cancer cells that are located entirely within the large TADs observed in PrECs to a random overlap. Indeed, the number of cancer-associated domains (identified in PC3 and LNCaP cells) found to be subdomains of normal cell (PrEC) TADs was significantly higher than expected by chance ($P < 0.01$). Similar observations were recently reported by Barutcu et al. (2015, 2017). Here the authors studied genome-wide chromatin interaction patterns in epithelial and breast cancer cells and discovered that TADs were frequently "broken" into multiple sub-TADs in cancer cells. Taken together, the findings show that although the segmentation of the genome into well-defined TADs is still maintained in the cancer genome, the characteristics of TADs in cancer cells are significantly different to TADs in normal cells.

To further delineate the differences we observed in the TAD organization between normal and cancer genomes, we intersected the TAD boundaries identified in each of the cell types and categorized the boundaries as either "constitutive" (present in all three cell types), "cancer-specific" (present in both cancer cell lines, but not in normal cells), or "facultative" (maintained in only one of the cell lines studied). Topological domain boundaries

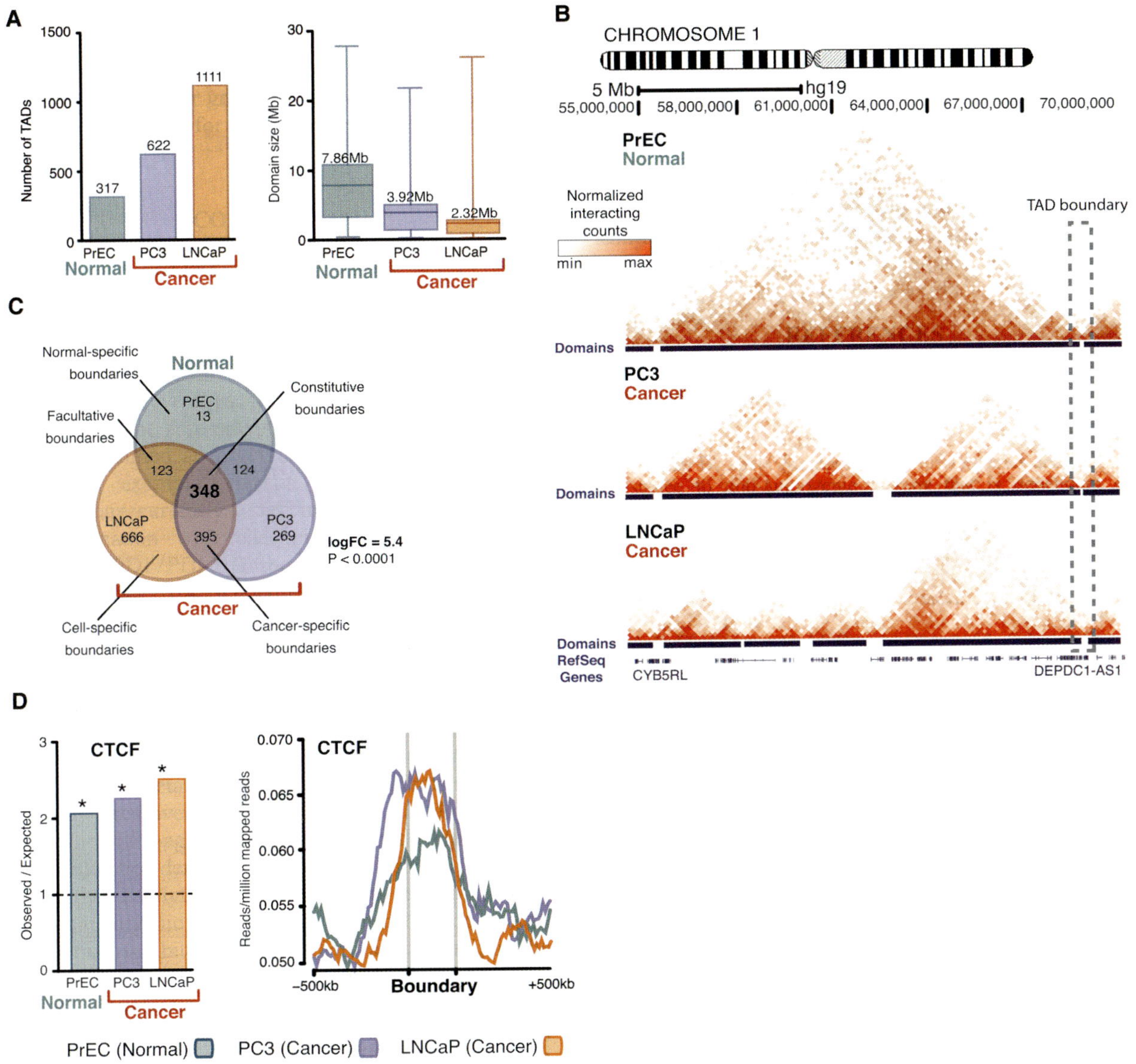

Figure 2. Topologically associated domains (TADs) in the cancer cell genome. (*A*) Number and average size (in Mb) of domains identified in normal prostate cells (PrEC) and prostate cancer cells (PC3 and LNCaP). (*B*) An example from chromosome 1 of a two-dimensional chromatin interaction matrix from PrEC, PC3, and LNCaP cells visualized in the WashU Epigenome Browser. Large TADs found in normal cells are frequently subdivided into two or more domains in cancer cells. (*C*) The majority of TAD boundaries present in normal cells (PrEC) are also present in cancer cells (constitutive boundaries), whereas ~20% of boundaries were maintained in only one of the two cancer cells (cell type–specific boundaries). Cancer cells (LNCaP and PC3) share a large proportion of their TAD boundaries (cancer-specific boundaries; logFC = 5.4, $P < 0.0001$). (*D*) Observed/expected fold enrichment of CTCF binding at the domain boundaries in PrEC, PC3, and LNCaP cells. Genome-wide average distribution of CTCF binding around the TAD boundaries in PrEC, PC3, and LNCaP cells. (Adapted, with permission, from Taberlay et al. 2016, © Cold Spring Harbor Laboratory Press.)

represent the region of strict division between two TADs and are characterized by high stability among normal cell types (Dixon et al. 2012, 2015). Epigenetic marks, gene expression, and replication timing were previously shown to be regulated by TAD architecture (Le Dily et al. 2014; Pope et al. 2014). Recent Hi-C studies revealed that architectural proteins play a critical role in determining the 3D organization of the genome. The insulator proteins, CTCF and cohesin, are the most well-known architectural proteins that are known to be involved in TAD formation and maintenance (Misteli 2007; Phillips and Corces 2009). However, the exact mechanisms by which architectural proteins function to form and maintain TAD boundaries are still not well-understood. In multiple cell types, TAD boundaries have been shown to work as insulators by restricting interaction with genes in the neighboring TADs (Nora et al. 2012; Andrey et al. 2013; Symmons et al. 2014; Giorgio et al. 2015; Lupianez et al. 2015; Tsujimura et al. 2015; Franke et al. 2016).

Thus, we were prompted to investigate what factors might be contributing to the loss of TAD boundary integrity in cancer. Comparing the locations of boundaries between the two prostate cancer cell lines, we found >90% of boundaries present in normal cells were also present in at least one cancer cell type (Fig. 2C). Of these, ~80% were "constitutive" (present in all three cell types), whereas ~20% were "facultative," as they were maintained in only one of the two cancer cell lines studied. Importantly, out of all of the nonconstitutive boundaries, 30% were cancer-specific TAD boundaries. This suggests that the acquisition of new sub-TAD boundaries at similar genomic locations is a cancer-specific process of 3D reorganization. In support of this concept, we observed high fold change (observed over expected) overlap between the two sets of new TAD boundaries in each cancer cell type (Fig. 2C). An intriguing explanation for the formation of new TAD boundaries in cancer is that these may be related to widespread epigenome remodeling and genetic alterations that are frequently observed in cancer. To test that possibility, we analyzed ChIP-seq data of architectural proteins and histone modifications. We found that TAD boundaries in all three cell types were highly enriched for CTCF (Fig. 2D), with ~7% of all CTCF binding sites being located at TAD boundaries. However, we did not observe major differences in CTCF enrichment between normal and cancer cells. This result may be expected because CTCF is known to be a conserved insulator element (Li et al. 2013).

Recently, there has been a growing interest into elucidating the role of genetic changes on the organization of the 3D chromatin in cancer. The vast majority of genetic alterations found in the cancer genome occur outside of protein coding regions, and therefore the functional consequences can be difficult to interpret. We hypothesized that alterations frequently found in the cancer genomes may result in the disruption of TADs and therefore provide an explanation for the large increase in sub-TADs observed in the cancer cells. It is well established that the genomes of prostate cancer cells can have recurrent copy-number alterations (Williams et al. 2014; Boutros et al. 2015). We therefore next asked whether copy-number variants (CNVs) could provide a compelling explanation for the formation of new TAD boundaries. We surmised that copy-number changes located within a TAD could result in change in local chromatin interactions and thus lead to aberrant contacts between two distant regions and formation of a new TAD boundary. To test this hypothesis, we identified 60 chromosomal variants, up to 10 Mb in size, that were present in both LNCaP and PC3 cells, identified using single-nucleotide polymorphism (SNP) arrays. We intersected the locations of cancer-specific TAD boundaries with the identified cancer CNVs. Interestingly, we observed that ~70% of the CNVs that were shared between LNCaP and PC3 cells were also located at or in close proximity to the newly formed cancer-specific domain boundaries. For example, a 400- to 600-kb cancer-specific deletion on Chromosome 17p13.1 is associated with the establishment of a new domain boundary in both prostate cancer cell lines and a resulting change in local interactions across this region (Fig. 3). Interestingly, this locus harbors the TP53 gene and is commonly deleted (14.8%) in all prostate cancers (Kluth et al. 2014). Consistent with this finding, previous reports suggested that genomic deletions at TAD borders in normal cells result in a change in local chromatin organization and in a formation of new domain boundaries (Nora et al. 2012; Ibn-Salem et al. 2014). These findings suggest an important role of genetic alterations in defining the cancer-specific disruption to the 3D genome organization. As it is known that TAD boundaries are crucial for controlling the action of enhancers on genes in adjacent TADs (Nora et al. 2012; Dowen et al. 2014; Symmons et al. 2014), our findings may help explain the altered gene expression profiles in cancer cells.

CANCER CELLS ARE CHARACTERIZED BY FREQUENT LOCAL DIFFERENTIAL CHROMATIN INTERACTIONS

Alterations in local chromatin interactions have been a focus of many studies into long-range gene regulation and oncogene activation in cancer. To further understand the role of atypical chromatin interactions in prostate cancer, we used a diffHiC package (Lun and Smyth 2015) to identify and characterize differential interactions between normal prostate and prostate cancer cells. diffHiC permits analysis of interactions with significant differences in intensity between two biological conditions. Then we used the diffHiC data to define anchor points of chromatin interactions (Fig. 4A). By comparing normal prostate (PrEC) interaction data independently to each of the prostate cancer data sets (PC3 and LNCaP), we obtained more than 2000 differential high-confidence interactions for each of the comparisons (Fig. 4B). By intersecting the two differential data sets, we found that majority of all differential interactions identified in a comparison of PrEC with PC3 were also found in a comparison of PrEC with LNCaP, suggesting that the majority are common, cancer-specific interactions (Fig. 4B). Interestingly, differential interactions were enriched in cancer cells, suggesting that cancer cells gain interactions more easily than lose interactions (Fig. 4C). This observation is in agreement with previous reports showing that cancer cells have atypical enhancer–promoter interactions that lead to disruption of gene regulation (for review, see Achinger-Kawecka and Clark 2017). Dixon et al. (2015) recently showed that majority of changes in local chromatin interactions during cell differentiation occur within the borders of the TADs, with only a very small number of interactions crossing the TAD boundaries. This suggests that atypical enhancer–promoter interactions in cancer cells may be restricted by the presence of conserved TAD boundaries in these cells. Notably, we observed that the most significant differential interactions are predominantly located within topological domains, further supporting that hypothesis.

To delineate the role of differential interactions in prostate cancer biology, it is crucial to understand their potential context in relation to epigenome remodeling. To determine whether regulatory elements may be associated with differential interactions, we used ChromHMM

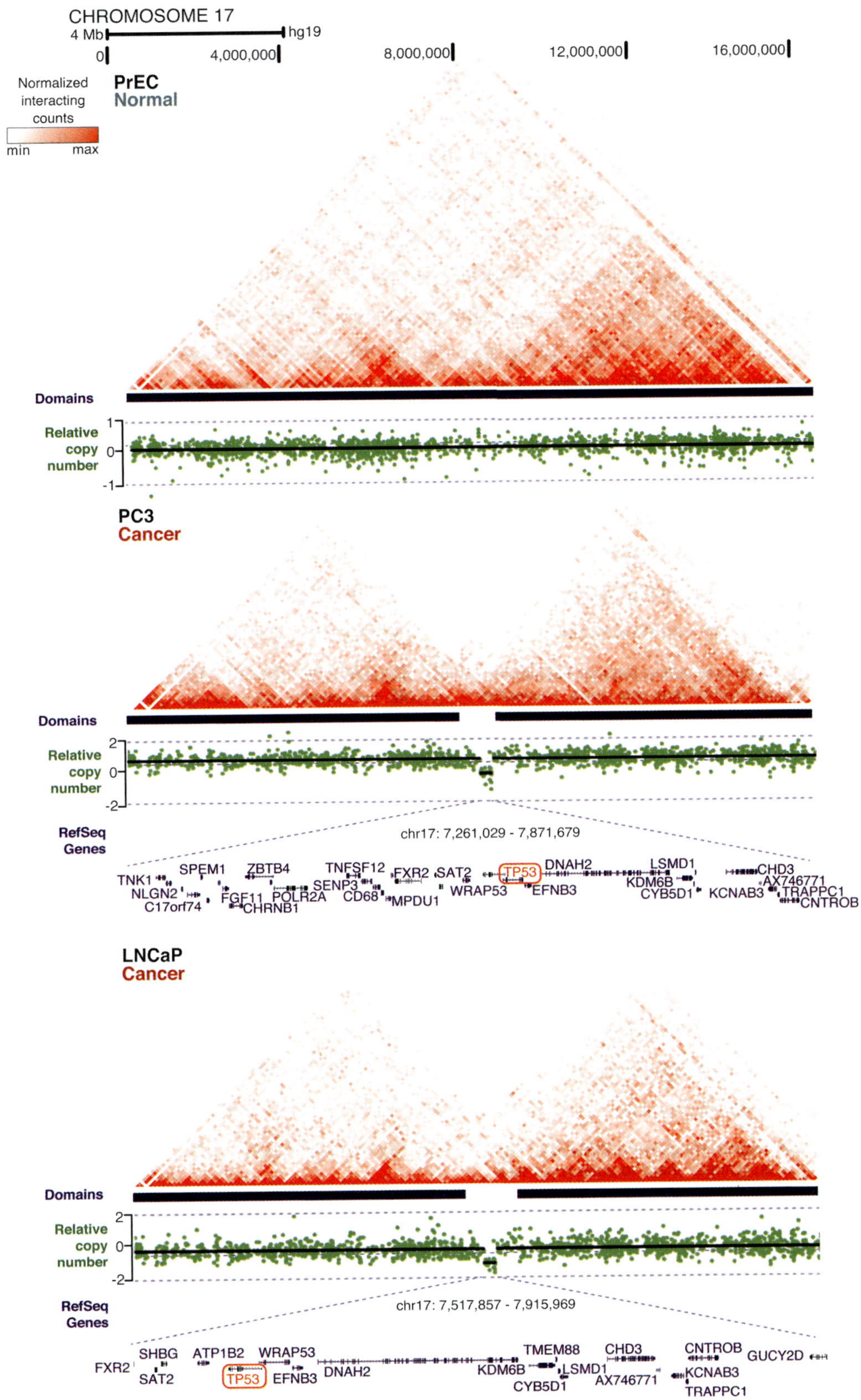

Figure 3. Copy-number variants (CNVs) are located at cancer-specific topologically associated domain (TAD) boundaries. A representative example from Chromosome 17 showing a deletion encompassing the TP53 locus (shown in green) that is present in both cancer cell lines is associated with a formation of a new, cancer-specific domain boundary and a change in local chromatin interactions in both cancer cell lines. (Reprinted, with permission, from Taberlay et al. 2016, © Cold Spring Harbor Laboratory Press.)

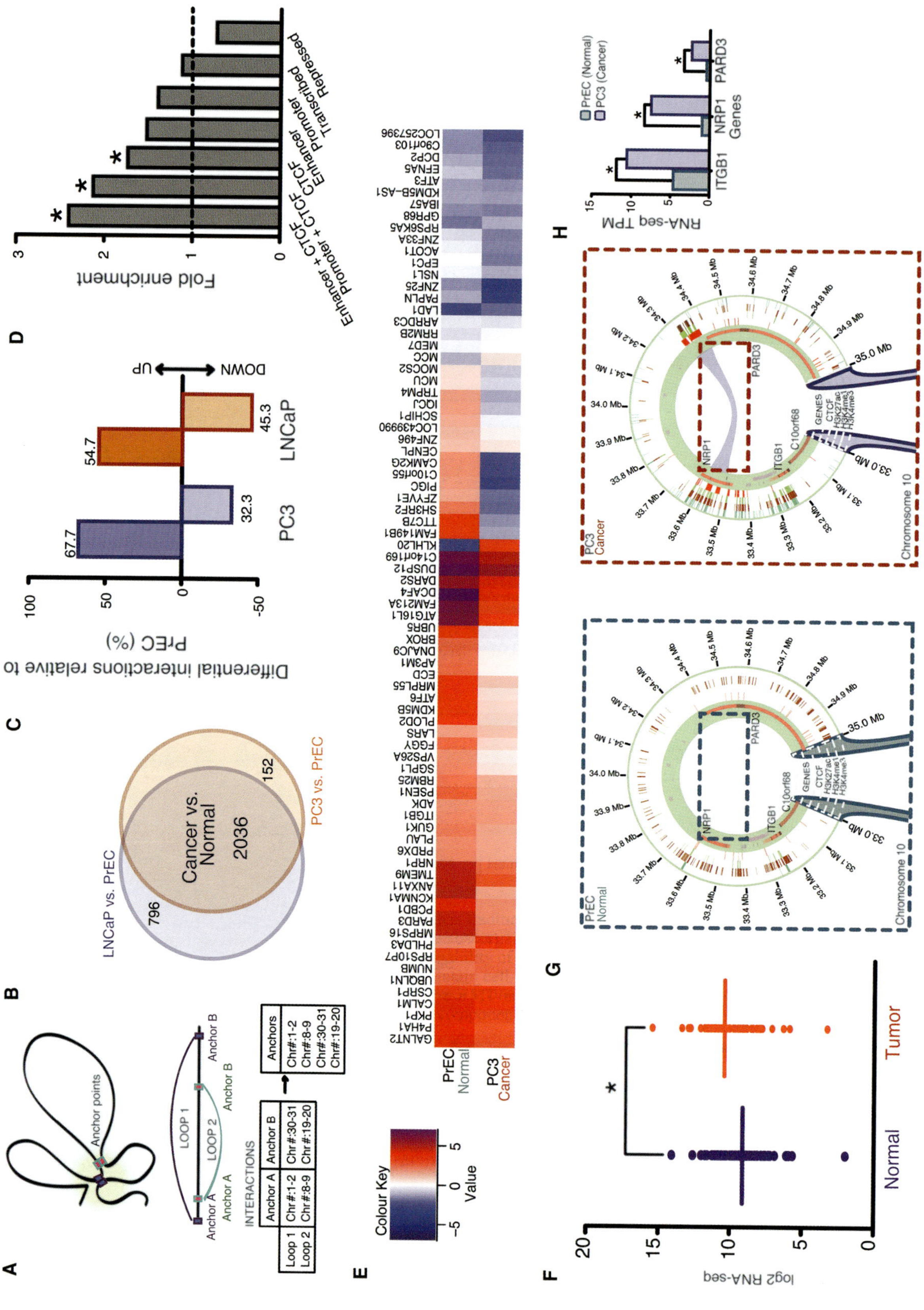

Figure 4. *(Legend on following page.)*

(Ernst and Kellis 2012) and annotated the epigenomes of each normal and cancer cell type into distinct chromatin states (obtained from Taberlay et al. 2014). When we intersected the identified chromatin states with the anchor points of cancer-specific interactions, we observed high enrichment for enhancers that were marked by CTCF, promoters marked by CTCF as well as distal CTCF sites, suggesting that functional elements occupied by CTCF may facilitate new interactions in cancer cells (Fig. 4D). The high overrepresentation of enhancers and promoters at differentially interacting sites prompt us to investigate whether changes in local chromatin interactions could affect expression of nearby genes. First, we compared the expression of genes located at anchor points of differential interactions between normal and cancer cells using RNA-seq and we discovered that novel cancer-specific chromatin interactions were associated with altered gene expression (Fig. 4E). Altogether, ~80% of all differential interactions in cancer cells correlated with differential expression of genes located at anchor points (Fig. 4E), which is consistent with their enrichment for enhancers and promoters. Second, we hypothesized that genes located at cancer-specific interactions may be involved in the prostate cancer pathology, and therefore their expression would be significantly disrupted in tumor samples from prostate cancer patients. To validate, we tested whether expression of genes located within differential interactions was also altered in the TCGA prostate cohort (Cancer Genome Atlas Research Network 2015). In support of this concept, we determined that expression of more than half of all genes located at differential interactions was significantly increased in tumor samples compared with normal prostate samples (Fig. 4F). It is interesting to speculate that functional elements, especially ones occupied by CTCF, may facilitate new chromatin interactions in cancer cells. Enhancers are known to engage in interactions that exist in a cell type–specific manner and our results support the hypothesis that enhancer dynamics play a role in regulating local interaction differences between normal and cancer cells. Furthermore, altered chromatin interactions may lead to deregulation of gene expression and contribute to increased expression of oncogenes in cancer cells and repression of tumor-suppressor genes.

Our findings highlight that differential chromatin interactions are enriched at enhancers and promoters; however, these potential new 3D loops remain difficult to visualize in a linear genome browser. With this in mind we developed *Rondo*, a web-based interface for visualizing Hi-C data (Taberlay et al. 2016). *Rondo* displays gene tracks and epigenome data sets, such as ChIP-seq and RNA-seq, allowing biological interpretation of Hi-C data in context of other "omic" data sets. Using *Rondo*, we visualized differential interactions alongside histone modifications, and then intersected these with gene expression data to determine the functional output of atypical interactions in the prostate cancer cells. A representative example of a cancer-specific interaction, which involves a change in enhancer and promoter marks, as well as in gene expression, is presented in Figure 4G. In cancer cells, a novel loop between the enhancer and promoter of *NRP1* results in regional activation of genes as shown by RNA-seq (Fig. 4H). In normal prostate cells *NRP1* is not active, and the putative enhancer is marked by residual H3K4me1 but is devoid of H3K27ac, H3K4me3, and CTCF binding in normal cells. In cancer cells, a marked increase in H3K4me1, H3K4me3, and H3K27ac signal is observed at both the putative enhancer and promoter regions of the gene, together with increase in CTCF binding. Remarkably, this epigenetic remodeling was not restricted to the *NRP1* gene boundaries and also encompassed neighboring enhancers and cancer-associated genes, such as *ITGB1* and *PARD3*. Expression of *ITGB1*, *NRP1*, and *PARD3* genes were all significantly increased in the cancer cells (Fig. 4H), suggesting that functional elements and long-range epigenetically regulated domains may be altered concomitant with the interactome. Interestingly, *NRP1* has been shown to participate in several different types of signaling pathways that control cell migration and overexpression of *NRP1* has been reported to predict distant relapse after radical prostatectomy in clinically localized prostate cancer (Talagas et al. 2013).

RELATIONSHIP BETWEEN LONG-RANGE EPIGENETIC REMODELING AND 3D CHROMATIN ORGANIZATION

The concordant inactivation of adjacent genes due to long-range epigenetic silencing (LRES) has been reported in various cancer types, including colorectal, bladder, non–small cell lung cancer, breast, prostate, and Wilms' tumor (Frigola et al. 2006; Stransky et al. 2006; Hitchins

Figure 4. *(Figure on previous page.)* Differential interactions observed in cancer cells are enriched for enhancers and promoters and are coincident with altered cancer epigenetic patterns. (*A*) Anchor points of chromatin interactions were defined as the genomic locations where an interaction is present. (*B*) Overlap between anchor points of differential interactions between PrEC and PC3 and between PrEC and LNCaP cell lines. (*C*) Differential interactions are enriched in cancer cells. (*D*) Enrichment of differential interactions for enhancers that were marked by CTCF (enhancer + CTCF), promoters marked by CTCF (promoter + CTCF), as well as distal CTCF sites (CTCF) using ChromHMM. (*E*) Differential interactions are associated with significantly altered gene expression in cancer cells. (*F*) Genes located at differential interactions are associated with altered expression in primary prostate cohort (TCGA PRAD) (log_2 RNA-seq, *P* value <0.001). (*G*) A representative example from Chromosome 10 showing epigenetic remodeling (activation) of genes located at the differentially interacting region in cancer cells. The promoter and a putative enhancer of the NRP1 gene are both marked by residual H3K4me1 in normal cells. In cancer cells there is a marked increase in H3K4me1, H3K4me3, and H3K27ac in cancer cells, indicative of increased gene activity that is association with a formation of a new cancer-specific interaction. (*H*) Genes located at differential interaction are significantly overexpressed in cancer cells (*q value < 0.0001). (Adapted, with permission, from Taberlay et al. 2016, © Cold Spring Harbor Laboratory Press.)

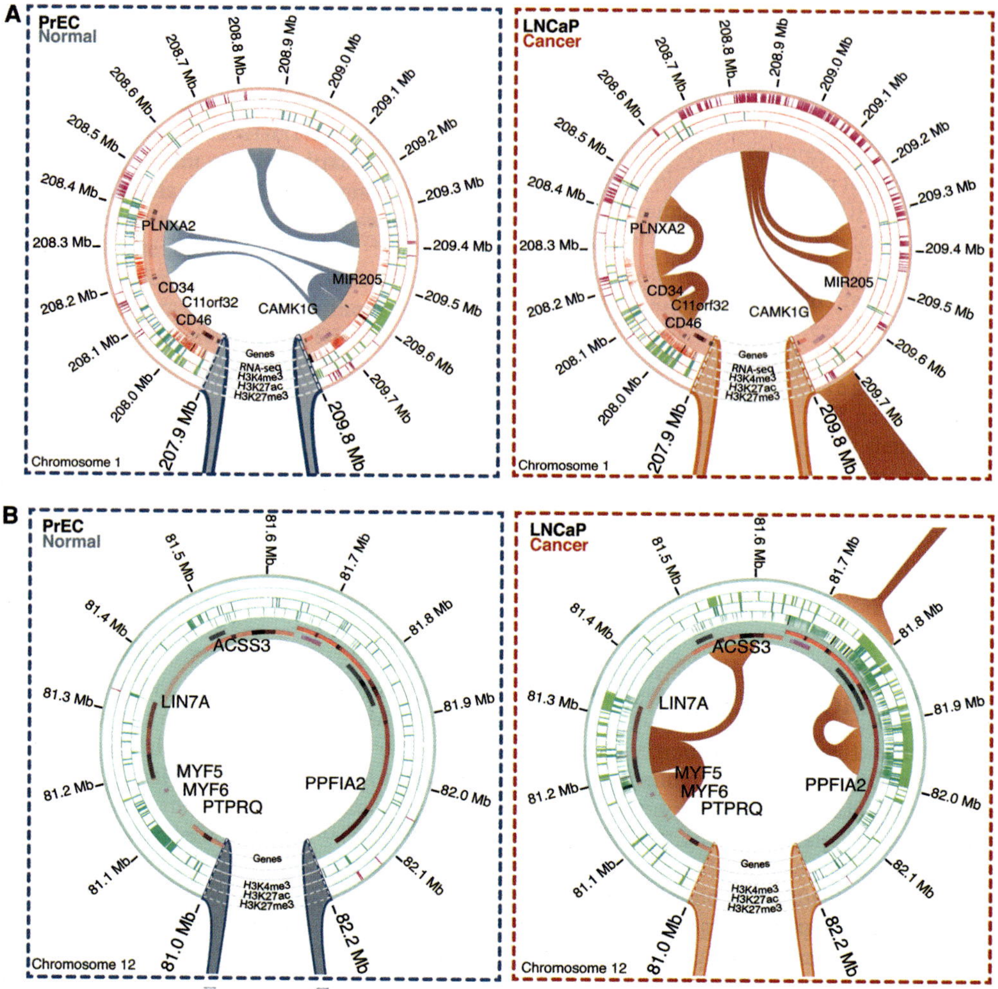

Figure 5. Long-range epigenetically silenced (LRES) and active (LREA) domains occur at differential interactions in cancer cells. (*A*) A representative example of LRES domain on Chromosome 1, showing three chromatin interactions enriched for highly expressed genes (RNA-seq) and active histone marks in normal cells (H3K4me3 and H3K27ac). In cancer cells, new cancer-specific interactions are present in that region and are associated with loss of gene expression (RNA-seq), decrease in active histone marks, and increase in repressive marks (H3K27me3). (*B*) A representative example of LREA domain on Chromosome 12, showing the absence of chromatin interactions in normal cells that is associated with low levels of active histone marks (H3K4me3 and H3K27ac) and gene inactivity (RNA-seq). Aberrant, cancer-specific interactions in cancer cells result in high expression (RNA-seq) of genes and associated increase in active histone marks (H3K4me3 and H3K27ac). (Adapted, with permission, from Taberlay et al. 2016, © Cold Spring Harbor Laboratory Press.)

et al. 2007; Novak et al. 2008; Seng et al. 2008; Dallosso et al. 2009; Rafique et al. 2015). In prostate cancer, we have shown that LRES is primarily characterized by regional gains of repressive histone marks and loss of active histone marks encompassing tumor-suppressor or cancer-associated genes (Coolen et al. 2010). Regional activation of adjacent genes also occurs in cancer, and this is due to long-range epigenetic activation (LREA) (Bert et al. 2013). We found that LREA domains were associated with simultaneous gains in active histone marks and loss of repressive histone marks, and these domains commonly harbor oncogenes (Bert et al. 2013). Although these findings highlight the important role of long-range epigenetic remodeling in cancer, the potential impact on higher-order chromatin interactions has not been characterized.

To study the relationship between long-range epigenetic deregulation and 3D chromatin organization, we tested the significance of the intersection between LRES and LREA regions and cancer-specific differential interactions. We confirmed changes in gene expression between normal and cancer cell lines with RNA-seq. Using *Rondo*, we visualized gene expression, histone modifications, and differential interactions overlapping known LRES and LREA domains. Figure 5A shows an example of differential interactions and epigenetic marks between normal PrECs and LNCaP cancer cells within a ~1-Mb region LRES region on chromosome 1. In normal cells, the LRES region that contains the genes *PLXNA2*, *MIR205*, and *CAMK1G* is actively expressed, as indicated by RNA-seq, whereas in cancer this region is silenced (Fig. 5A). Gene silencing across the LRES region is correlates with a gain of the repressive H3K27me3 mark and loss of the active H3K4me3 and H3K27ac chromatin marks (Fig. 5A). In *Rondo* we could observe a strong interaction between the active *PLXNA2* and *MIR205* loci in normal prostate cells. Notably, the chromatin interactions, spanning nearly 1 Mb, were located between putative active enhancers, marked by H3K27ac (Fig. 5A). A second interaction in the loci in the normal cells was observed in the intergenic region bounded by H3K27me3 marks. In contrast, in cancer cells, the chromatin interactions are notably altered. In fact the higher-order interactions between *PLXNA2* with a putative enhancer element in normal cells are no longer observed. Instead, we find new local interactions and these are associated with loci that gain H3K27me3. The data are consistent with the LRES region harboring inactive genes in a more condensed chromatin structure, and as such the epigenetically silenced MIR205 locus no longer interacts with the *PLXNA2* locus in the cancer cells (Fig. 5A).

We also observed alterations in the 3D chromatin interactions in LREA domains and these too were associated with gene expression and epigenetic changes. Figure 5B shows an example of differential interactions and epigenetic marks between normal PrECs and LNCaP cancer cells within a ~1.2-Mb region LREA region on chromosome 12. These loci contain the genes *PTPRQ*, *MYF6*, *MYF5*, *LIN7A*, *ACSS3*, and *PPFIA2*, which in normal prostate cells are not expressed but in prostate cancer are concordantly activated (Fig. 5B). Regional gene activation across the LREA region is associated with gain of the active H3K4me3 and H3K27ac chromatin marks and new chromatin interactions between the *MYF* loci and *LIN7A* and *LIN7A* and *ACSS3* (Fig. 5B). Notably, a new chromatin interaction at the *PPFIA2* gene loci is also formed which intersects two potentially discrete enhancer regions enriched for H3K27ac (Fig. 5B). The data are consistent with the LREA region forming a more open chromatin structure and as such the epigenetically activated genes form new chromatin contacts with promoter and enhancer elements that are associated with concordant gene activation of the region. Together, these data indicate that long-range epigenetic dysregulation in cancer is also associated with an alteration in higher-order chromatin interactions commonly at regulatory loci.

CONCLUSION

Cancer is associated with widespread alterations to the genome and the epigenome. We propose that these changes are instrumental in the creation of a new 3D architecture that results in both localized and regional changes in gene expression. In contrast, the overall higher-order spatial organization of chromatin in cancer cells is highly similar to normal cells suggesting that on a large scale, chromosomes are folded similarly and retain their physical positions within the nucleus. Where the spatial organization of chromatin was altered in the cancer cells was restricted to the size of the TADs, which are smaller in cancer. Notably, these smaller TADs reside within the normal TAD architecture, rather than forming new TADs spanning domains. In fact, we found that the majority of domain boundaries that were present in normal cells were also present in cancer cells. We were surprised to find that more than half of the new smaller subdomain TADs that were acquired in both prostate cancer cell lines shared similar new boundaries. Clearly more chromatin conformation studies are required in many more cancer samples to determine whether there is indeed a cancer-specific process in 3D reorganization. Second, we identified new cancer-specific interactions within the smaller TADs. We found that these cancer differential interactions were enriched for regulatory elements (enhancers, promoters, and CTCF binding sites) suggesting that new chromatin interactions in cancer cells are involved in reprogramming the unique epigenetic programs of cancer cells. Enhancers are known to engage in interactions that exist in a cell type–specific manner and our results support the hypothesis that enhancer dynamics play a role in regulating local interaction differences between normal and cancer cells. Finally, we found that differential interactions also occur within long-range epigenetically deregulated domains in cancer cells. The regulatory regions within the long-range epigenetically deregulated domains show local changes in chromatin organization and abnormal chromatin interactions in both LRES and LREA regions. Of interest, we found new interactions at putative enhancers associated with gene activation in the cancer cells in LREA regions and loss of interactions at putative enhancer elements in cancer cells that are silenced in cancer in LRES regions.

In summary the implementation of a novel visualization tool has allowed us to explore Hi–C interaction data in relation to the transcription and epigenetic changes between normal and prostate cancer cells. Our data support the hypothesis that alterations in the 3D chromatin state underpin the functional basis of the cancer genome by changing the spatial architecture of regulatory elements and proximity to genes resulting in aberrant cancer gene expression. Together our findings provide a new insight into long-range epigenetic dysregulation and higher-order chromatin interactions in cancer and suggests that long-range epigenetic dysregulation in cancer is associated with an alteration in higher-order chromatin interactions commonly at regulatory loci that potentially results in atypical promoter–enhancer and enhancer–enhancer interactions in cancer cells.

ACKNOWLEDGMENTS

S.J.C. is a National Health and Medical Research Council (NHMRC) Senior Principal Research Fellow #1063559; P.C.T. is a NHMRC Career Development Fellow. This work was supported by grant #1070613 and NHMRC Project grants #1011447 and #1051757.

REFERENCES

Achinger-Kawecka J, Clark SJ. 2017. Disruption of the 3D cancer genome blueprint. *Epigenomics* **9:** 47–55.

Andrey G, Montavon T, Mascrez B, Gonzalez F, Noordermeer D, Leleu M, Trono D, Spitz F, Duboule D. 2013. A switch between topological domains underlies HoxD genes collinearity in mouse limbs. *Science* **340:** 1234167.

Barutcu AR, Lajoie BR, McCord RP, Tye CE, Hong D, Messier TL, Browne G, van Wijnen AJ, Lian JB, Stein JL, et al. 2015. Chromatin interaction analysis reveals changes in small chromosome and telomere clustering between epithelial and breast cancer cells. *Genome Biol* **16:** 214.

Barutcu AR, Lian JB, Stein JL, Stein GS, Imbalzano AN. 2017. The connection between BRG1, CTCF and topoisomerases at TAD boundaries. *Nucleus*: 1–6.

Beagan JA, Gilgenast TG, Kim J, Plona Z, Norton HK, Hu G, Hsu SC, Shields EJ, Lyu X, Apostolou E, et al. 2016. Local genome topology can exhibit an incompletely rewired 3D-folding state during somatic cell reprogramming. *Cell Stem Cell* **18:** 611–624.

Bert SA, Robinson MD, Strbenac D, Statham AL, Song JZ, Hulf T, Sutherland RL, Coolen MW, Stirzaker C, Clark SJ. 2013. Regional activation of the cancer genome by long-range epigenetic remodeling. *Cancer Cell* **23:** 9–22.

Boutros PC, Fraser M, Harding NJ, de Borja R, Trudel D, Lalonde E, Meng A, Hennings-Yeomans PH, McPherson A, Sabelnykova VY, et al. 2015. Spatial genomic heterogeneity within localized, multifocal prostate cancer. *Nat Genet* **47:** 736–745.

Cancer Genome Atlas Research Network. 2015. The molecular taxonomy of primary prostate cancer. *Cell* **163:** 1011–1025.

Coolen MW, Stirzaker C, Song JZ, Statham AL, Kassir Z, Moreno CS, Young AN, Varma V, Speed TP, Cowley M, et al. 2010. Consolidation of the cancer genome into domains of repressive chromatin by long-range epigenetic silencing (LRES) reduces transcriptional plasticity. *Nat Cell Biol* **12:** 235–246.

Dallosso AR, Hancock AL, Szemes M, Moorwood K, Chilukamarri L, Tsai HH, Sarkar A, Barasch J, Vuononvirta R, Jones C, et al. 2009. Frequent long-range epigenetic silencing of protocadherin gene clusters on chromosome 5q31 in Wilms' tumor. *PLoS Genet* **5:** e1000745.

Dekker J, Rippe K, Dekker M, Kleckner N. 2002. Capturing chromosome conformation. *Science* **295:** 1306–1311.

Dixon JR, Selvaraj S, Yue F, Kim A, Li Y, Shen Y, Hu M, Liu JS, Ren B. 2012. Topological domains in mammalian genomes identified by analysis of chromatin interactions. *Nature* **485:** 376–380.

Dixon JR, Jung I, Selvaraj S, Shen Y, Antosiewicz-Bourget JE, Lee AY, Ye Z, Kim A, Rajagopal N, Xie W, et al. 2015. Chromatin architecture reorganization during stem cell differentiation. *Nature* **518:** 331–336.

Dostie J, Richmond TA, Arnaout RA, Selzer RR, Lee WL, Honan TA, Rubio ED, Krumm A, Lamb J, Nusbaum C, et al. 2006. Chromosome Conformation Capture Carbon Copy (5C): A massively parallel solution for mapping interactions between genomic elements. *Genome Res* **16:** 1299–1309.

Dowen JM, Fan ZP, Hnisz D, Ren G, Abraham BJ, Zhang LN, Weintraub AS, Schuijers J, Lee TI, Zhao K, et al. 2014. Control of cell identity genes occurs in insulated neighborhoods in mammalian chromosomes. *Cell* **159:** 374–387.

Ernst J, Kellis M. 2012. ChromHMM: Automating chromatin-state discovery and characterization. *Nat Methods* **9:** 215–216.

Franke M, Ibrahim DM, Andrey G, Schwarzer W, Heinrich V, Schopflin R, Kraft K, Kempfer R, Jerkovic I, Chan WL, et al. 2016. Formation of new chromatin domains determines pathogenicity of genomic duplications. *Nature* **538:** 265–269.

Frigola J, Song J, Stirzaker C, Hinshelwood RA, Peinado MA, Clark SJ. 2006. Epigenetic remodeling in colorectal cancer results in coordinate gene suppression across an entire chromosome band. *Nat Genet* **38:** 540–549.

Fullwood MJ, Liu MH, Pan YF, Liu J, Xu H, Mohamed YB, Orlov YL, Velkov S, Ho A, Mei PH, et al. 2009. An oestrogen-receptor-α-bound human chromatin interactome. *Nature* **462:** 58–64.

Giorgio E, Robyr D, Spielmann M, Ferrero E, Di Gregorio E, Imperiale D, Vaula G, Stamoulis G, Santoni F, Atzori C, et al. 2015. A large genomic deletion leads to enhancer adoption by the lamin B1 gene: A second path to autosomal dominant adult-onset demyelinating leukodystrophy (ADLD). *Hum Mol Genet* **24:** 3143–3154.

Hitchins MP, Lin VA, Buckle A, Cheong K, Halani N, Ku S, Kwok CT, Packham D, Suter CM, Meagher A, et al. 2007. Epigenetic inactivation of a cluster of genes flanking MLH1 in microsatellite-unstable colorectal cancer. *Cancer Res* **67:** 9107–9116.

Ibn-Salem J, Kohler S, Love MI, Chung HR, Huang N, Hurles ME, Haendel M, Washington NL, Smedley D, Mungall CJ, et al. 2014. Deletions of chromosomal regulatory boundaries are associated with congenital disease. *Genome Biol* **15:** 423.

Kluth M, Harasimowicz S, Burkhardt L, Grupp K, Krohn A, Prien K, Gjoni J, Hass T, Galal R, Graefen M, et al. 2014. Clinical significance of different types of p53 gene alteration in surgically treated prostate cancer. *Int J Cancer* **135:** 1369–1380.

Krijger PH, Di Stefano B, de Wit E, Limone F, van Oevelen C, de Laat W, Graf T. 2016. Cell-of-origin-specific 3D genome structure acquired during somatic cell reprogramming. *Cell Stem Cell* **18:** 597–610.

Le Dily F, Bau D, Pohl A, Vicent GP, Serra F, Soronellas D, Castellano G, Wright RH, Ballare C, Filion G, et al. 2014. Distinct structural transitions of chromatin topological domains correlate with coordinated hormone-induced gene regulation. *Genes Dev* **28:** 2151–2162.

Li Y, Huang W, Niu L, Umbach DM, Covo S, Li L. 2013. Characterization of constitutive CTCF/cohesin loci: A possible role in establishing topological domains in mammalian genomes. *BMC Genomics* **14:** 553.

Lieberman-Aiden E, van Berkum NL, Williams L, Imakaev M, Ragoczy T, Telling A, Amit I, Lajoie BR, Sabo PJ, Dorschner MO, et al. 2009. Comprehensive mapping of long-range interactions reveals folding principles of the human genome. *Science* **326:** 289–293.

Lun AT, Smyth GK. 2015. diffHic: A Bioconductor package to detect differential genomic interactions in Hi-C data. *BMC Bioinformatics* **16:** 258.

Lupianez DG, Kraft K, Heinrich V, Krawitz P, Brancati F, Klopocki E, Horn D, Kayserili H, Opitz JM, Laxova R, et al. 2015. Disruptions of topological chromatin domains cause pathogenic rewiring of gene-enhancer interactions. *Cell* **161:** 1012–1025.

Misteli T. 2007. Beyond the sequence: Cellular organization of genome function. *Cell* **128:** 787–800.

Nora EP, Lajoie BR, Schulz EG, Giorgetti L, Okamoto I, Servant N, Piolot T, van Berkum NL, Meisig J, Sedat J, et al. 2012. Spatial partitioning of the regulatory landscape of the X-inactivation centre. *Nature* **485:** 381–385.

Novak P, Jensen T, Oshiro MM, Watts GS, Kim CJ, Futscher BW. 2008. Agglomerative epigenetic aberrations are a common event in human breast cancer. *Cancer Res* **68:** 8616–8625.

Phillips JE, Corces VG. 2009. CTCF: Master weaver of the genome. *Cell* **137:** 1194–1211.

Phillips-Cremins JE, Sauria ME, Sanyal A, Gerasimova TI, Lajoie BR, Bell JS, Ong CT, Hookway TA, Guo C, Sun Y, et al. 2013. Architectural protein subclasses shape 3D organization of genomes during lineage commitment. *Cell* **153:** 1281–1295.

Pope BD, Ryba T, Dileep V, Yue F, Wu W, Denas O, Vera DL, Wang Y, Hansen RS, Canfield TK, et al. 2014. Topologically associating domains are stable units of replication-timing regulation. *Nature* **515:** 402–405.

Rafique S, Thomas JS, Sproul D, Bickmore WA. 2015. Estrogen-induced chromatin decondensation and nuclear re-organization linked to regional epigenetic regulation in breast cancer. *Genome Biol* **16:** 145.

Rao SS, Huntley MH, Durand NC, Stamenova EK, Bochkov ID, Robinson JT, Sanborn AL, Machol I, Omer AD, Lander ES, et al. 2014. A 3D map of the human genome at kilobase resolution reveals principles of chromatin looping. *Cell* **159:** 1665–1680.

Seng TJ, Currey N, Cooper WA, Lee CS, Chan C, Horvath L, Sutherland RL, Kennedy C, McCaughan B, Kohonen-Corish MR. 2008. DLEC1 and MLH1 promoter methylation are associated with poor prognosis in non-small cell lung carcinoma. *Br J Cancer* **99:** 375–382.

Simonis M, Klous P, Splinter E, Moshkin Y, Willemsen R, de Wit E, van Steensel B, de Laat W. 2006. Nuclear organization of active and inactive chromatin domains uncovered by chromosome conformation capture-on-chip (4C). *Nat Genet* **38:** 1348–1354.

Smith EM, Lajoie BR, Jain G, Dekker J. 2016. Invariant TAD Boundaries Constrain Cell-Type-Specific Looping Interactions between Promoters and Distal Elements around the CFTR Locus. *Am J Hum Genet* **98:** 185–201.

Stransky N, Vallot C, Reyal F, Bernard-Pierrot I, de Medina SG, Segraves R, de Rycke Y, Elvin P, Cassidy A, Spraggon C, et al. 2006. Regional copy number-independent deregulation of transcription in cancer. *Nat Genet* **38:** 1386–1396.

Symmons O, Uslu VV, Tsujimura T, Ruf S, Nassari S, Schwarzer W, Ettwiller L, Spitz F. 2014. Functional and topological characteristics of mammalian regulatory domains. *Genome Res* **24:** 390–400.

Taberlay PC, Statham AL, Kelly TK, Clark SJ, Jones PA. 2014. Reconfiguration of nucleosome-depleted regions at distal regulatory elements accompanies DNA methylation of enhancers and insulators in cancer. *Genome Res* **24:** 1421–1432.

Taberlay PC, Achinger-Kawecka J, Lun AT, Buske FA, Sabir K, Gould CM, Zotenko E, Bert SA, Giles KA, Bauer DC, et al. 2016. Three-dimensional disorganization of the cancer genome occurs coincident with long-range genetic and epigenetic alterations. *Genome Res* **26:** 719–731.

Talagas M, Uguen A, Garlantezec R, Fournier G, Doucet L, Gobin E, Marcorelles P, Volant A, DE Braekeleer M. 2013. VEGFR1 and NRP1 endothelial expressions predict distant relapse after radical prostatectomy in clinically localized prostate cancer. *Anticancer Res* **33:** 2065–2075.

Tsujimura T, Klein FA, Langenfeld K, Glaser J, Huber W, Spitz F. 2015. A discrete transition zone organizes the topological and regulatory autonomy of the adjacent tfap2c and bmp7 genes. *PLoS Genet* **11:** e1004897.

Williams JL, Greer PA, Squire JA. 2014. Recurrent copy number alterations in prostate cancer: An in silico meta-analysis of publicly available genomic data. *Cancer Genet* **207:** 474–488.

Zhang Y, McCord RP, Ho YJ, Lajoie BR, Hildebrand DG, Simon AC, Becker MS, Alt FW, Dekker J. 2012. Spatial organization of the mouse genome and its role in recurrent chromosomal translocations. *Cell* **148:** 908–921.

Zhao Z, Tavoosidana G, Sjolinder M, Gondor A, Mariano P, Wang S, Kanduri C, Lezcano M, Sandhu KS, Singh U, et al. 2006. Circular chromosome conformation capture (4C) uncovers extensive networks of epigenetically regulated intra- and interchromosomal interactions. *Nat Genet* **38:** 1341–1347.

Composition and Function of Mammalian SWI/SNF Chromatin Remodeling Complexes in Human Disease

JOHN L. PULICE[1,2] AND CIGALL KADOCH[1,2]
[1]Department of Pediatric Oncology, Dana-Farber Cancer Institute and Harvard Medical School, Boston, Massachusetts 02215
[2]Broad Institute of MIT and Harvard, Cambridge, Massachusetts 02142
Correspondence: cigall_kadoch@dfci.harvard.edu

Mammalian SWI/SNF (BAF) chromatin remodeling complexes play critical roles in maintaining chromatin architecture and gene expression. Genomic sequencing efforts over the past several years have unveiled a major role for these complexes in the development of human cancer as well as neurologic disease, prompting the need to interrogate underlying mechanisms and to develop new methods to comprehensively understand mSWI/SNF complex function. Here we discuss the emerging insights from genetic, biochemical, and functional genomic studies in the field and suggest approaches toward further basic investigations, as well as therapeutic targeting of chromatin remodeling machinery.

Epigenetic regulation plays a critical role in the maintenance of proper gene expression and is achieved via several mechanisms including DNA methylation, histone modification, and ATP-dependent chromatin remodeling. Chromatin remodeling complexes alter the positioning of nucleosomes in an ATP-dependent manner, enabling dynamic control of chromatin accessibility (Clapier and Cairns 2009). First discovered in yeast through screens for genes that regulate mating type switching (switch; SWI) and nutrient switching (sucrose nonfermenting; SNF), these genes were then identified as subunits of a transcriptional activating complex, thereby termed SWI/SNF (Neigeborn and Carlson 1984; Stern et al. 1984; Peterson and Herskowitz 1992). Evolutionary homologs of the yeast SWI/SNF complex were then identified in *Drosophila* (BAP complex), and ultimately in mammals (BAF complex) (Tamkun et al. 1992; Kwon et al. 1994).

Mammalian SWI/SNF (mSWI/SNF) complexes are 12- to 15-subunit macromolecular assemblies combinatorially assembled from the products of 29 genes. In vitro reconstitution studies have showed that the BRG1 (SMARCA4) ATPase subunit can alone remodel nucleosome positioning on nucleosomal templates, and that a four-subunit minimal complex composed of BRG1, BAF155 (SMARCC1), BAF170 (SMARCC2), and BAF47 (SMARCB1) achieves efficient chromatin remodeling in vitro (Phelan et al. 1999). mSWI/SNF complex subunits have been shown to play critical roles in many tissues and developmental processes, including vertebrate nervous system development, cardiac development, and induced pluripotency, as well as several other developmental contexts (Lessard et al. 2007; Ho et al. 2009; Ho and Crabtree 2010; Singhal et al. 2010). Homozygous inactivation of the large majority of core mSWI/SNF complex subunits results in embryonic lethality, underscoring a critical requirement for mSWI/SNF complex genes in early development (Bultman et al. 2000; Klochendler-Yeivin et al. 2000; Kim et al. 2001; Gao et al. 2008). Additional roles for mSWI/SNF complexes in decatenation of DNA in mitosis through association with TOP2A, as well as in DNA damage repair, have also more recently been demonstrated (Dykhuizen et al. 2013; Kakarougkas et al. 2014). Although the majority of subunits are ubiquitously expressed, a set of mSWI/SNF complex subunits have been shown to be cell/tissue type–specific, with switching of paralog subunit gene expression during stem cell and neural differentiation (Lessard et al. 2007; Ho et al. 2009; Yoo et al. 2009). For example, during the transition from neural progenitors to mature, postmitotic neurons, a set of three subunit families, BAF53A/B, BAF45(A-D)/(B-C), and SS18/SS18L1, exhibit changes in gene expression, resulting in swapping of subunits at these positions within the complex. Taken together, these data suggest the requirements for distinct complex assemblies in specific lineages. Importantly, these mSWI/SNF subunit switches are instructive in nature; for example, the BAF53A to BAF53B switch is required for the conversion of human fibroblasts to neurons (Yoo et al. 2011).

mSWI/SNF complexes exist in two distinct assemblies, BAF and PBAF complexes, which are marked by the incorporation of a set of mutually exclusive subunits (Fig. 1). This includes subunits such as BAF200 (ARID2), BAF180 (PBRM1), BAF45A (PHF10), and BRD7, which exclusively incorporate into PBAF but not BAF complexes, and BAF250A/B (ARID1A/B), BAF45D (DPF2), and SS18, which exclusively incorporate into BAF but not PBAF complexes. Both BAF and PBAF complexes contain homologs that have been conserved throughout eukaryotic evolution, with BAF com-

Published by Cold Spring Harbor Laboratory Press; doi: 10.1101/sqb.2016.81.031021

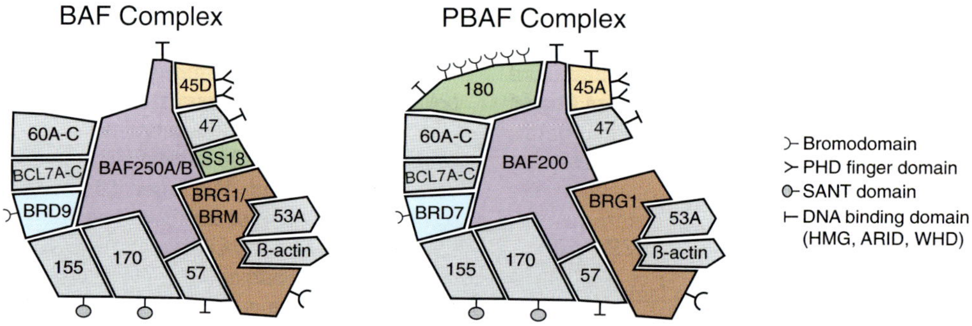

Figure 1. Mammalian SWI/SNF (BAF) complexes: two distinct assemblies. Subunits of BAF complexes and PBAF complexes are annotated with DNA binding domains as well as histone recognition domains such as bromodomains, PHD finger domains, and SANT domains. Subunits with paralog exclusivity to BAF or PBAF complexes are highlighted in corresponding colors, and subunits entirely exclusive to BAF or PBAF complexes are labeled in green.

plexes sharing homology with yeast SWI/SNF complexes and *Drosophila* BAP complexes, whereas PBAF complexes share homology with yeast RSC complexes as well as *Drosophila* PBAP complexes (Xue et al. 2000). Recent in vitro yeast work has showed a critical role for RSC but not SWI/SNF in creating nucleosome free regions at promoters, suggesting differential requirements for the two complexes in promoter regulation (Krietenstein et al. 2016). Despite this, several core subunits are shared between BAF and PBAF complexes, which implicates a branched assembly pathway leading to the formation of these two distinct complexes.

THE SEQUENCING ERA UNVEILS EXTENSIVE ROLES FOR mSWI/SNF COMPLEXES IN CANCER

Evidence supporting a driving role for mSWI/SNF complex perturbations in cancer was first documented in malignant rhabdoid tumor (MRT), an aggressive pediatric cancer characterized by biallelic inactivation of *SMARCB1*, which encodes the BAF47 subunit (Versteege et al. 1998). Studies also documented loss of BAF47 as a hallmark feature of atypical teratoid/rhabdoid tumor (AT/RT) and in ~90% of epithelioid sarcomas (EpS), suggesting that the BAF47 subunit acts as a classical tumor suppressor (via loss of heterozygosity) in multiple contexts (Biegel et al. 1999; Modena et al. 2005). In addition, early studies documented mutations to *BRG1*, as well as loss of both BRG1 and BRM ATPase subunits, in human cancer cell lines (Dunaief et al. 1994; Wong et al. 2000; Reisman et al. 2002). These early studies provided genetic foundational evidence implicating mSWI/SNF complexes as tumor suppressors that can be altered through loss of specific subunits.

Beginning in 2011, whole-exome sequencing studies began to vastly substantiate these early observations, identifying a surprisingly high frequency of mutations to the genes encoding mSWI/SNF complex subunits across a wide range of human cancers. Among the first reports to emerge, sequencing of ovarian clear cell carcinomas showed that *BAF250A (ARID1A)* is mutated in 46% of these cancers (Wiegand et al. 2010), and soon after, *BAF180 (PBRM1)* was found to be mutated in 41% of clear cell renal cell carcinomas (Varela et al. 2011). Early exome-sequencing studies also identified frequent mutations to mSWI/SNF complex subunits in numerous other cancers, including bladder cancer, gastric cancer, and lung cancer (Gui et al. 2011; Wang et al. 2011; Cancer Genome Atlas Research Network 2012). From these, several themes emerged: (1) Mutations in specific subunits were found in a restricted set of human malignancies, suggesting tissue-specific protective functions hallmark to each subunit; (2) these mutations can occur in the genes encoding subunits of either mSWI/SNF assembly, BAF or PBAF complexes, as well as those shared between both; (3) mutations were found to be disproportionately early frameshift (truncating) or gene loss mutations, resulting in absent protein expression in human tumors and cell lines as assessed by immunohistochemistry, similar to the previously documented loss of BAF47 in MRT; (4) the subunits most frequently mutated in cancer, such as BAF250A and BAF180, are dispensable for in vitro chromatin remodeling (Phelan et al. 1999), suggesting additional, undiscovered functions for mSWI/SNF complexes that can be perturbed in cancer. These surprising results have since catalyzed a surge of interest in both understanding and targeting these complexes among investigators across a range of fields, including cancer biology, epigenetics and transcription, and drug discovery.

More recent efforts spearheaded by The Cancer Genome Atlas (TCGA) and others have further uncovered recurrent mutations in key mSWI/SNF subunit genes, such as BAF250A (*ARID1A*) in endometrioid carcinoma and bladder cancer, BRG1 in lung carcinomas, and BAF180 in clear cell renal cell carcinoma (Lawrence et al. 2013). Ranking of tumor-suppressor genes in a pan-cancer study highlights BAF250A (fourth), BAF180 (sixth), BAF200 (14th), BAF250B (42nd), BRG1 (51st), and BAF47 (75th) among all candidates (Davoli et al. 2013). Pan-cancer analyses aided by the identification of the full constellation of mSWI/SNF complex subunits via biochemical approaches have shown that mSWI/SNF complexes (all subunits in total) are mutated in >20% of human cancers (Fig. 2; Kadoch et al. 2013).

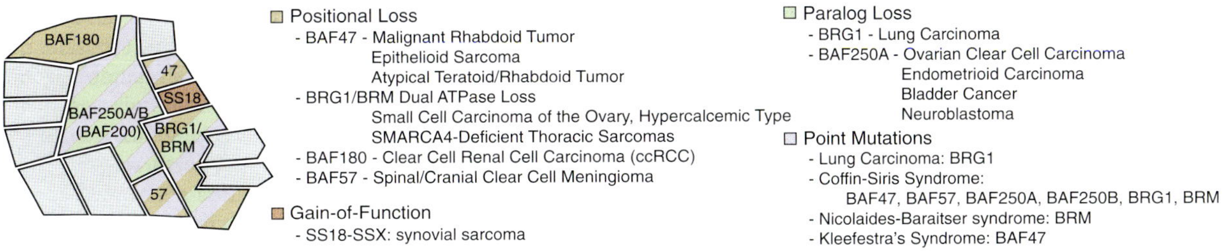

Figure 2. Classes of mSWI/SNF complex subunit perturbation in cancer and neurodevelopmental disorders. Perturbations are categorized by positional loss (yellow), paralog loss (green), point mutations (purple), and gain of function perturbations (red). mSWI/SNF complex subunits can be perturbed via mutation to yield several distinct oncogenic mechanisms.

Given the high frequency of inactivating mutations to the mSWI/SNF complex subunit genes and their global functions, it is likely that exome sequencing alone vastly underestimates the full spectrum of perturbations to mSWI/SNF complex structure and activity in cancer. Other routes to perturbation of mSWI/SNF genes may include enhancer mutations, promoter hypermethylation, or other mechanisms of epigenetic mSWI/SNF subunit gene silencing, all of which may serve to alter subunit gene expression, and hence stoichiometric availability for proper assembly into complexes. Current whole-genome sequencing efforts coupled with gene expression studies across cancer subtypes as well as proteomic studies assessing cancer-specific mSWI/SNF composition will continue to unmask novel routes to mSWI/SNF complex perturbation.

TUMOR SUPPRESSION—MECHANISMS AND EMERGING CONCEPTS

The majority of mSWI/SNF complex mutations are suggested to result in loss of tumor suppressive functions, such as with the well-studied biallelic inactivation of *SMARCB1* (BAF47) in MRT. Because BAF47 has no coexpressed and interchangeable paralog, loss of this subunit yields loss of a positional subunit of mSWI/SNF complexes that is not compensated by another. Similarly, loss of BAF57 (SMARCE1) in clear cell meningioma results in a similar mechanism of oncogenesis through complete loss of a subunit position, as BAF57 has no paralog in mSWI/SNF complexes (Smith et al. 2013). Surprisingly, positional subunit loss does not only occur in subunits containing no known mammalian paralogs, as more recently it was identified that coinactivation of both BRG1 and BRM ATPase subunits is the hallmark feature of small cell carcinoma of the ovary, hypercalcemic type (SCCOHT) as well as SMARCA4-deficient thoracic sarcoma (SA4-DTS) (Jelinic et al. 2014; Ramos et al. 2014; Witkowski et al. 2014; Le Loarer et al. 2015). A critical question regarding positional subunit loss remains how loss of these subunits affects the structural integrity and function of residual mSWI/SNF complexes. Indeed, reports to date highlight contradicting findings regarding loss of the BAF47 subunit, with some studies suggesting that BAF47 loss has a destabilizing effect on mSWI/SNF complexes (Wei et al. 2014; Wang et al. 2017) and others, including our own, demonstrating that BAF47-deficient mSWI/SNF complexes assemble nearly completely in the absence of BAF47 (Doan et al. 2004; Wang et al. 2009a; Jamshidi et al. 2016; Kadoch et al. 2017; RT Nakayama, JL Pulice, AM Valencia, et al., unpubl.). Understanding how loss of a positional subunit affects mSWI/SNF complexes biochemically and genomically is critical to understanding the mechanism of oncogenesis and to the identification of potential therapeutic approaches (Fig. 3).

Several of the most common mSWI/SNF complex mutations in human cancer occur in genes encoding only one paralog subunit of a multiparalog family that incorporate into complexes in a mutually exclusive manner. For example, in cancers bearing either ARID1A (BAF250A) or SMARCA4 (BRG1) loss, the paralogs of these subunits ARID1B (BAF250B) and SMARCA2 (BRM), respectively, are still expressed and residual mSWI/SNF complexes assemble normally in the paralog's absence. Indeed, studies have shown that both BAF250B and BRM are synthetic lethal genes in the absence of BAF250A and BRG1, respectively (Helming et al. 2014; Hoffman et al. 2014). This unique synthetic lethal dependency distinguishes BRG1-mutant cancers from those lacking both BRG1 and BRM, and suggests distinct mechanisms of action for cancers driven by paralog subunit loss versus positional subunit loss. Of note, BRG1 and BAF250A are consistently more highly expressed in the tissues of origin for these cancers, which raises the question as to whether these cancers are driven by dosage insufficiency upon loss of the dominant paralog, or by a unique, paralog-specific regulatory property that cannot be compensated by the remaining paralog. Ongoing studies aim to define the specific mechanistic underpinnings of these robust intracomplex synthetic lethal interactions.

Interestingly, emerging evidence suggests that the roles for mSWI/SNF complexes in tumor suppression may not be limited to mutations in the genes encoding the stable, dedicated subunits themselves. Recent work has showed that expression of GATA3, a transcription factor frequently mutated in breast cancer, results in altered targeting of mSWI/SNF complexes (Takaku et al. 2016). This suggests that mSWI/SNF complexes can be tumor suppressors whose regulation is lost not through mutation of a subunit, but rather through loss of a required recruitment to specific loci genome-wide. Growing evidence from our laboratory and others suggests that transcription factors can also render complexes oncogenic (GJ Sandoval, JL Pulice, MA Schenone, et al., unpubl.).

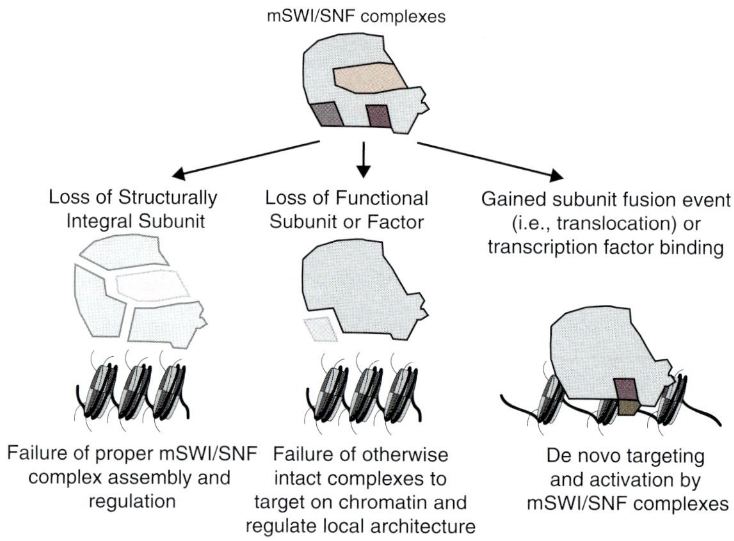

Figure 3. Functional outcomes of mSWI/SNF alterations. mSWI/SNF complexes can be perturbed in two distinct yet overlapping manners: loss of structural integrity, in which mSWI/SNF complexes cannot assemble normally in the absence of a subunit, and loss of functional regulation, in which mSWI/SNF complexes can fully assemble in the absence of the lost subunit but cannot be properly targeted on chromatin or properly regulate target genes. In addition, gain-of-function alterations through translocations or tethered transcription factors can also occur, in which functional regulation is gained to support and/or drive oncogenesis.

mSWI/SNF complex perturbations can be categorized between two groups: structural aberration and functional alteration. Structural aberrations constitute settings in which subunit loss alters the assembly or integrity of mSWI/SNF complexes, such that residual complexes are unable to form and therefore properly regulate the genome in the subunit's absence. In contrast, functional alterations result in either a change to the targeting or the activity of the complex absent significant changes to mSWI/SNF complex assembly. Functional alterations can result via several routes, including dosage insufficiency, loss of a subunit required to facilitate mSWI/SNF complex regulation, or loss of genomic targeting by a subunit domain or bound transcription factor. Indeed, gain-of-function mechanisms such as mSWI/SNF subunit fusion proteins resulting from translocations or transcription factor binding also fit the definition of a functional perturbation, harnessing the DNA accessibility and gene activation properties of mSWI/SNF complexes to support oncogenesis.

MECHANISMS OF GAIN-OF-FUNCTION PERTURBATIONS TO mSWI/SNF COMPLEXES

Although the majority of mSWI/SNF complex perturbations are genetic loss-of-function mutations, one genetic event in particular, the SS18-SSX translocation in human synovial sarcoma, has been shown to drive gain-of-function complex activity. Our group discovered and characterized SS18 as a stable, dedicated member of mSWI/SNF complexes; the hallmark genetic event in human synovial sarcoma is the t(X;18) chromosomal translocation which results in the fusion of 78 amino acids of SSX1,2, or 4 to the carboxyl terminus of the SS18 subunit. We found that the SS18-SSX oncogenic fusion dominantly incorporates into mSWI/SNF complexes to drive oncogenesis in synovial sarcoma (Kadoch and Crabtree 2013). Interestingly, SS18-SSX also displaces the BAF47 canonical tumor suppressor subunit from mSWI/SNF complexes. Our data suggest that SS18-SSX mediates retargeting of mSWI/SNF complexes to new loci, resulting in activation of oncogenic target genes such as *SOX2* via opposition of Polycomb-mediated repression. However, the relative contributions of the gain of the SS18-SSX fusion and the concomitant loss of BAF47 subunit are not yet fully understood. Decoupling these two concomitant events will be critical to further understanding the mechanism of this mSWI/SNF perturbation, and in informing therapeutic approaches, particularly those involving inhibitors of Polycomb-repressor complex function (i.e., EZH2 and EED inhibitors). The original identification of this gain-of-function role of the SS18-SSX oncogenic fusion on mSWI/SNF complexes suggests that transcription factors or other oncogenic fusions involving mSWI/SNF complex subunits may alter complex targeting in a similar manner. Taken together, these results indicate that the activating role of mSWI/SNF complexes can be harnessed as a driver event, not only a tumor suppressor.

INTELLECTUAL DISABILITY SYNDROMES: POINT MUTATIONS LEND INSIGHTS INTO DOMAIN FUNCTION

In addition to human cancers, mutations to genes encoding mammalian SWI/SNF components have been identified in intellectual disability (ID) syndromes. In particular, exome sequencing has uncovered mutations in Coffin–Siris syndrome (CSS), Nicolaides–Baraitser syndrome, and others (Kleefstra et al. 2012; Santen et al.

2012; Tsurusaki et al. 2012; Van Houdt et al. 2012). Intriguingly, these mutations are nearly uniformly heterozygous, de novo, and often missense. Missense point mutations, in particular, present a unique opportunity to gain insight into the function of specialized domains on subunits. For example, in CSS, mutations fall within the HMG domain of SMARCE1 (BAF57), a putative DNA binding domain. Additionally, the most frequently recurrent mutation in CSS across several cohorts of patients results in deletion of one amino acid in the highly conserved coiled-coil region of the BAF47 subunit (K364del).

Interestingly, the point mutations uncovered in these neurological disorders can also function similarly to mutations in cancer, but perhaps in a more dosage-sensitive manner. For instance, mutations to BRM and BRG1 largely occur within the ATPase and helicase domains of these proteins, but their heterozygous nature suggests a dosage requirement in the nervous system as opposed to the complete subunit protein loss more often observed in cancer. Mutations to BAF250A and BAF250B are largely truncating or frameshift events, but the expression of the second allele suggests these syndromes result from haploinsufficiency due to inadequate gene dosage (Santen et al. 2012). It is noteworthy that although BAF250A and BRG1 are more frequently mutated in cancer, their paralogs, BAF250B and BRM are more frequently mutated in developmental disorders, suggesting differential expression dominance or functions of paralogs between the nervous system and other cell lineages. The high frequency of domain-centered point mutations in disease settings such as CSS presents an opportunity to mechanistically determine domain function, since heterozygous mutations resulting in both subunit loss and domain perturbations produce disease.

mSWI/SNF COMPLEXES PLAY CRITICAL ROLES IN PIONEERING AND MAINTENANCE OF PROMOTER AND ENHANCER STATE

The opposition of mSWI/SNF complexes and Polycomb repressive complexes plays a critical role in development and disease. Members of mSWI/SNF complexes have been identified as trithorax genes due to their ability to oppose developmental aberrations caused by Polycomb mutations in *Drosophila* (Kennison and Tamkun 1988). We developed an assay to study opposition of Polycomb-mediated repression using a rapamycin-inducible recruitment system in mouse embryonic fibroblasts (MEFs) (Kadoch et al. 2017). Using this system, we showed that opposition of Polycomb-mediated repression by mSWI/SNF complexes is a rapid, step-wise process, resulting in eviction of PRC1 and PRC2 complexes, followed by removal of H2AK119ub1 followed by H3K27me3 marks from the recruitment site, yielding the formation of open, accessible chromatin within 40 min of recruitment, independent of replication or transcription (Fig. 4). Interestingly, we show this opposition is ATP-dependent, and that loss of the BAF47 subunit results in a similar failure to evict Polycomb complexes as recruitment of an ATPase-dead complex. In contrast, recruitment of SS18-SSX-containing mSWI/SNF complexes results in robust Polycomb opposition at the recruitment site and throughout the gene body, suggesting differential oncogenic processes rendered by the functional differences in the ability of these oncogenic complexes to oppose Polycomb complexes on heterochromatin. Interestingly, our data suggest that mSWI/SNF–Polycomb complex opposition may be critical in developmental processes at defined sets of developmental gene loci, as mSWI/SNF complexes can activate bivalent genes through opposition of Polycomb-mediated repression. Taken together, these data suggest broad-spanning gene regulatory mechanisms for mSWI/SNF complexes in both development and disease.

Recent genome-wide studies have shown that mSWI/SNF complexes are targeted to active promoters and enhancers, and that this targeting is concomitant with other chromatin remodelers and regulators (Hnisz et al. 2013; de Dieuleveult et al. 2016). However, the relative roles and contributions of these different regulators and complexes are not well understood. Our studies and others have suggested a widespread role for mSWI/SNF complexes in pioneering enhancer activation, such that loss of BAF47 or BAF250A subunits, for example, can

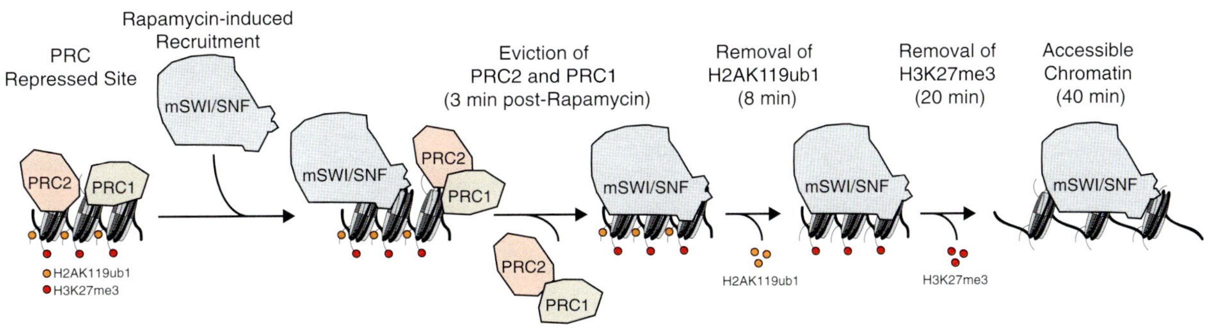

Figure 4. mSWI/SNF complexes oppose Polycomb complex–mediated repression on chromatin. Recruitment of mSWI/SNF complexes to a Polycomb complex–repressed locus via a rapamycin-inducible system (Kadoch et al. 2017) results in rapid, ATP-dependent eviction of PRC1 and PRC2 complexes, followed by removal of H2AK119ub1 and H3K27me3 marks and DNA accessibility. This opposition is inhibited or potentiated by cancer-associated mSWI/SNF complex perturbations such as BAF47 loss (i.e., in malignant rhabdoid tumor [MRT]) or incorporation of the SS18-SSX (in synovial sarcoma), respectively.

impair enhancer activation (Mathur et al. 2017; RT Nakayama, JL Pulice, AM Valencia, et al., unpubl.). Interestingly, our data do resolve additional enhancer sites that are active and occupied by mSWI/SNF complexes in the absence of BAF47, suggesting these enhancers may not require fully intact mSWI/SNF complexes to pioneer activation, but require some remodeling or maintenance function in concert with other regulators. The distinction between pioneering and maintenance of enhancer activation may explain the tissue specificity of certain oncogenic mutations, as multiple chromatin remodeling complexes can likely pioneer enhancer activation in different contexts.

At promoters, we find a similar distinction between the pioneering and maintenance roles for mSWI/SNF complexes. Interestingly, our data suggest that opposition of Polycomb complexes may function as an analog to the enhancer activation observed. We demonstrate mSWI/SNF complexes retain occupancy at active promoters in the absence of BAF47 (RT Nakayama, JL Pulice, AM Valencia, et al., unpubl.), suggesting that dependency on the BRG1 ATPase in MRT cell lines is due to a requirement for a retained or aberrant regulation by residual BAF47-deficient mSWI/SNF complexes at promoters (Wang et al. 2009a) Indeed, we suggest that there are distinct roles for pioneering and maintaining active chromatin that can be differentially perturbed in different human cancers. It is likely that loss of particular subunits can affect some of these functions to drive oncogenesis while leaving others intact, yielding unique, cancer-specific dependencies on the remaining functions of the perturbed mSWI/SNF complexes.

CONCLUSION

The mechanistic basis underlying the overrepresentation of mSWI/SNF complex disruption in human cancer and developmental disorders remains to be comprehensively understood and will require a battery of new biochemical insights and functional assays to completely resolve. The demonstrated functions for this complex in enhancer activation and Polycomb complex opposition suggest multiple critical functions that can be singly or dually perturbed in different contexts. However, the fundamental question regarding the unique role for mSWI/SNF complexes, which represent only one family of ATP-dependent chromatin remodeling complexes among 28 others, in disease will require a more developed mechanistic understanding of other chromatin remodeling complexes. For instance, albeit less frequent than mSWI/SNF, other chromatin remodeler complex genes are mutated in cancer, such as *ATRX* in gliomas and *CHD1* in prostate cancer, suggesting that mSWI/SNF complexes are not the only chromatin remodelers implicated in oncogenesis. However, it is important to note that our emerging understanding of the contribution of mSWI/SNF complexes to cancer has strongly relied on our understanding of the subunit composition of these protein complexes; without this, beyond BRG1 and BRM ATPase mutations in lung and other cancers, we would not have determined the outsized contribution of mSWI/SNF complexes to human cancer.

Our studies have shown that BAF47 is required for BAF complex-mediated activation of enhancers, suggesting a pioneering role for BAF complexes that can be affected by subunit loss to alter enhancer-mediated cell state. It is likely that regulation of activation by mSWI/SNF complexes and other chromatin regulators occurs in a manner similar to the opposition observed between HATs and HDACs at active promoters (Wang et al. 2009b). This regulatory dynamic has been shown with the opposition between mSWI/SNF and NuRD complexes in embryonic stem cells (Yildirim et al. 2011) as one example of the types of dynamics at work at promoters and enhancers. In contrast, our studies have showed that mSWI/SNF–Polycomb complex opposition results in durable eviction of PRC2 complexes, such that although this can be reversed by removal of mSWI/SNF complex recruitment, the two complexes do not show significant genome-wide co-occupancy in cells. Understanding the precise dynamics and regulation of these complexes on chromatin in normal states and in the context of different classes of mSWI/SNF complex aberrancies will prove critical for therapeutic targeting.

ACKNOWLEDGMENTS

We thank members of the Kadoch laboratory for their assistance in reviewing this work and for their emerging mechanistic insights, especially Brittany C. Michel, Alfredo M. Valencia, and Joshua Pan. We thank our funding sources, including the National Institutes of Health (NIH) DP2 New Innovator Award 1DP2 OD019696-01 (C.K.), the American Cancer Society Research Scholar Award RSG-14-051-01-DMC (C.K.), the NIH SARC Sarcoma SPORE (5U54 CA168512-04), and the Pew-Stewart Scholars in Cancer Research Grant (C.K.) for their generous support of our work.

REFERENCES

Biegel JA, Zhou JY, Rorke LB, Stenstrom C, Wainwright LM, Fogelgren B. 1999. Germ-line and acquired mutations of INI1 in atypical teratoid and rhabdoid tumors. *Cancer Res* **59:** 74–79.

Bultman S, Gebuhr T, Yee D, La Mantia C, Nicholson J, Gilliam A, Randazzo F, Metzger D, Chambon P, Crabtree G, et al. 2000. A Brg1 null mutation in the mouse reveals functional differences among mammalian SWI/SNF complexes. *Mol Cell* **6:** 1287–1295.

Cancer Genome Atlas Research Network. 2012. Comprehensive genomic characterization of squamous cell lung cancers. *Nature* **489:** 519–525.

Clapier CR, Cairns BR. 2009. The biology of chromatin remodeling complexes. *Annu Rev Biochem* **78:** 273–304.

Davoli T, Xu AW, Mengwasser KE, Sack LM, Yoon JC, Park PJ, Elledge SJ. 2013. Cumulative haploinsufficiency and triplosensitivity drive aneuploidy patterns and shape the cancer genome. *Cell* **155:** 948–962.

de Dieuleveult M, Yen K, Hmitou I, Depaux A, Boussouar F, Bou Dargham D, Jounier S, Humbertclaude H, Ribierre F, Baulard C, et al. 2016. Genome-wide nucleosome specificity and function of chromatin remodellers in ES cells. *Nature* **530:** 113–116.

Doan DN, Veal TM, Yan Z, Wang W, Jones SN, Imbalzano AN. 2004. Loss of the INI1 tumor suppressor does not impair the expression of multiple BRG1-dependent genes or the assembly of SWI/SNF enzymes. *Oncogene* **23:** 3462–3473.

Dunaief JL, Strober BE, Guha S, Khavari PA, Alin K, Luban J, Begemann M, Crabtree GR, Goff SP. 1994. The retinoblastoma protein and BRG1 form a complex and cooperate to induce cell cycle arrest. *Cell* **79:** 119–130.

Dykhuizen EC, Hargreaves DC, Miller EL, Cui K, Korshunov A, Kool M, Pfister S, Cho YJ, Zhao K, Crabtree GR. 2013. BAF complexes facilitate decatenation of DNA by topoisomerase IIα. *Nature* **497:** 624–627.

Gao X, Tate P, Hu P, Tjian R, Skarnes WC, Wang Z. 2008. ES cell pluripotency and germ-layer formation require the SWI/SNF chromatin remodeling component BAF250a. *Proc Natl Acad Sci* **105:** 6656–6661.

Gui Y, Guo G, Huang Y, Hu X, Tang A, Gao S, Wu R, Chen C, Li X, Zhou L, et al. 2011. Frequent mutations of chromatin remodeling genes in transitional cell carcinoma of the bladder. *Nat Genet* **43:** 875–878.

Helming KC, Wang X, Wilson BG, Vazquez F, Haswell JR, Manchester HE, Kim Y, Kryukov GV, Ghandi M, Aguirre AJ, et al. 2014. ARID1B is a specific vulnerability in ARID1A-mutant cancers. *Nat Med* **20:** 251–254.

Hnisz D, Abraham BJ, Lee TI, Lau A, Saint-Andre V, Sigova AA, Hoke HA, Young RA. 2013. Super-enhancers in the control of cell identity and disease. *Cell* **155:** 934–947.

Ho L, Crabtree GR. 2010. Chromatin remodelling during development. *Nature* **463:** 474–484.

Ho L, Ronan JL, Wu J, Staahl BT, Chen L, Kuo A, Lessard J, Nesvizhskii AI, Ranish J, Crabtree GR. 2009. An embryonic stem cell chromatin remodeling complex, esBAF, is essential for embryonic stem cell self-renewal and pluripotency. *Proc Natl Acad Sci* **106:** 5181–5186.

Hoffman GR, Rahal R, Buxton F, Xiang K, McAllister G, Frias E, Bagdasarian L, Huber J, Lindeman A, Chen D, et al. 2014. Functional epigenetics approach identifies BRM/SMARCA2 as a critical synthetic lethal target in BRG1-deficient cancers. *Proc Natl Acad Sci* **111:** 3128–3133.

Jamshidi F, Bashashati A, Shumansky K, Dickson B, Gokgoz N, Wunder JS, Andrulis IL, Lazar AJ, Shah SP, Huntsman DG, et al. 2016. The genomic landscape of epithelioid sarcoma cell lines and tumours. *J Pathol* **238:** 63–73.

Jelinic P, Mueller JJ, Olvera N, Dao F, Scott SN, Shah R, Gao J, Schultz N, Gonen M, Soslow RA, et al. 2014. Recurrent SMARCA4 mutations in small cell carcinoma of the ovary. *Nat Genet* **46:** 424–426.

Kadoch C, Crabtree GR. 2013. Reversible disruption of mSWI/SNF (BAF) complexes by the SS18-SSX oncogenic fusion in synovial sarcoma. *Cell* **153:** 71–85.

Kadoch C, Hargreaves DC, Hodges C, Elias L, Ho L, Ranish J, Crabtree GR. 2013. Proteomic and bioinformatic analysis of mammalian SWI/SNF complexes identifies extensive roles in human malignancy. *Nat Genet* **45:** 592–601.

Kadoch C, Williams RT, Calarco JP, Miller EL, Weber CM, Braun SM, Pulice JL, Chory EJ, Crabtree GR. 2017. Dynamics of BAF-Polycomb complex opposition on heterochromatin in normal and oncogenic states. *Nat Genet* **49:** 213–222.

Kakarougkas A, Ismail A, Chambers AL, Riballo E, Herbert AD, Kunzel J, Lobrich M, Jeggo PA, Downs JA. 2014. Requirement for PBAF in transcriptional repression and repair at DNA breaks in actively transcribed regions of chromatin. *Mol Cell* **55:** 723–732.

Kennison JA, Tamkun JW. 1988. Dosage-dependent modifiers of Polycomb and antennapedia mutations in *Drosophila*. *Proc Natl Acad Sci* **85:** 8136–8140.

Kim JK, Huh SO, Choi H, Lee KS, Shin D, Lee C, Nam JS, Kim H, Chung H, Lee HW, et al. 2001. Srg3, a mouse homolog of yeast SWI3, is essential for early embryogenesis and involved in brain development. *Mol Cell Biol* **21:** 7787–7795.

Kleefstra T, Kramer JM, Neveling K, Willemsen MH, Koemans TS, Vissers LE, Wissink-Lindhout W, Fenckova M, van den Akker WM, Kasri NN, et al. 2012. Disruption of an EHMT1-associated chromatin-modification module causes intellectual disability. *Am J Hum Genet* **91:** 73–82.

Klochendler-Yeivin A, Fiette L, Barra J, Muchardt C, Babinet C, Yaniv M. 2000. The murine SNF5/INI1 chromatin remodeling factor is essential for embryonic development and tumor suppression. *EMBO Rep* **1:** 500–506.

Krietenstein N, Wal M, Watanabe S, Park B, Peterson CL, Pugh BF, Korber P. 2016. Genomic nucleosome organization reconstituted with pure proteins. *Cell* **167:** 709–721 e712.

Kwon H, Imbalzano AN, Khavari PA, Kingston RE, Green MR. 1994. Nucleosome disruption and enhancement of activator binding by a human SW1/SNF complex. *Nature* **370:** 477–481.

Lawrence MS, Stojanov P, Polak P, Kryukov GV, Cibulskis K, Sivachenko A, Carter SL, Stewart C, Mermel CH, Roberts SA, et al. 2013. Mutational heterogeneity in cancer and the search for new cancer-associated genes. *Nature* **499:** 214–218.

Le Loarer F, Watson S, Pierron G, de Montpreville VT, Ballet S, Firmin N, Auguste A, Pissaloux D, Boyault S, Paindavoine S, et al. 2015. SMARCA4 inactivation defines a group of undifferentiated thoracic malignancies transcriptionally related to BAF-deficient sarcomas. *Nat Genet* **47:** 1200–1205.

Lessard J, Wu JI, Ranish JA, Wan M, Winslow MM, Staahl BT, Wu H, Aebersold R, Graef IA, Crabtree GR. 2007. An essential switch in subunit composition of a chromatin remodeling complex during neural development. *Neuron* **55:** 201–215.

Mathur R, Alver BH, San Roman AK, Wilson BG, Wang X, Agoston AT, Park PJ, Shivdasani RA, Roberts CW. 2017. ARID1A loss impairs enhancer-mediated gene regulation and drives colon cancer in mice. *Nat Genet* **49:** 296–302.

Modena P, Lualdi E, Facchinetti F, Galli L, Teixeira MR, Pilotti S, Sozzi G. 2005. SMARCB1/INI1 tumor suppressor gene is frequently inactivated in epithelioid sarcomas. *Cancer Res* **65:** 4012–4019.

Neigeborn L, Carlson M. 1984. Genes affecting the regulation of SUC2 gene expression by glucose repression in *Saccharomyces cerevisiae*. *Genetics* **108:** 845–858.

Peterson CL, Herskowitz I. 1992. Characterization of the yeast SWI1, SWI2, and SWI3 genes, which encode a global activator of transcription. *Cell* **68:** 573–583.

Phelan ML, Sif S, Narlikar GJ, Kingston RE. 1999. Reconstitution of a core chromatin remodeling complex from SWI/SNF subunits. *Mol Cell* **3:** 247–253.

Ramos P, Karnezis AN, Hendricks WP, Wang Y, Tembe W, Zismann VL, Legendre C, Liang WS, Russell ML, Craig DW, et al. 2014. Loss of the tumor suppressor SMARCA4 in small cell carcinoma of the ovary, hypercalcemic type (SCCOHT). *Rare Dis* **2:** e967148.

Reisman DN, Strobeck MW, Betz BL, Sciariotta J, Funkhouser WJr, Murchardt C, Yaniv M, Sherman LS, Knudsen ES, Weissman BE. 2002. Concomitant down-regulation of BRM and BRG1 in human tumor cell lines: Differential effects on RB-mediated growth arrest vs CD44 expression. *Oncogene* **21:** 1196–1207.

Santen GW, Aten E, Sun Y, Almomani R, Gilissen C, Nielsen M, Kant SG, Snoeck IN, Peeters EA, Hilhorst-Hofstee Y, et al. 2012. Mutations in SWI/SNF chromatin remodeling complex gene ARID1B cause Coffin–Siris syndrome. *Nat Genet* **44:** 379–380.

Singhal N, Graumann J, Wu G, Arauzo-Bravo MJ, Han DW, Greber B, Gentile L, Mann M, Scholer HR. 2010. Chromatin-remodeling components of the BAF complex facilitate reprogramming. *Cell* **141:** 943–955.

Smith MJ, O'Sullivan J, Bhaskar SS, Hadfield KD, Poke G, Caird J, Sharif S, Eccles D, Fitzpatrick D, Rawluk D, et al. 2013. Loss-of-function mutations in SMARCE1 cause an inherited disorder of multiple spinal meningiomas. *Nat Genet* **45:** 295–298.

Stern M, Jensen R, Herskowitz I. 1984. Five SWI genes are required for expression of the HO gene in yeast. *J Mol Biol* **178:** 853–868.

Takaku M, Grimm SA, Shimbo T, Perera L, Menafra R, Stunnenberg HG, Archer TK, Machida S, Kurumizaka H, Wade PA. 2016. GATA3-dependent cellular reprogramming requires activation-domain dependent recruitment of a chromatin remodeler. *Genome Biol* **17:** 36.

Tamkun JW, Deuring R, Scott MP, Kissinger M, Pattatucci AM, Kaufman TC, Kennison JA. 1992. brahma: A regulator of *Drosophila* homeotic genes structurally related to the yeast transcriptional activator SNF2/SWI2. *Cell* **68:** 561–572.

Tsurusaki Y, Okamoto N, Ohashi H, Kosho T, Imai Y, Hibi-Ko Y, Kaname T, Naritomi K, Kawame H, Wakui K, et al. 2012. Mutations affecting components of the SWI/SNF complex cause Coffin–Siris syndrome. *Nat Genet* **44:** 376–378.

Van Houdt JK, Nowakowska BA, Sousa SB, van Schaik BD, Seuntjens E, Avonce N, Sifrim A, Abdul-Rahman OA, van den Boogaard MJ, Bottani A, et al. 2012. Heterozygous missense mutations in SMARCA2 cause Nicolaides–Baraitser syndrome. *Nat Genet* **44:** 445–449, S441.

Varela I, Tarpey P, Raine K, Huang D, Ong CK, Stephens P, Davies H, Jones D, Lin ML, Teague J, et al. 2011. Exome sequencing identifies frequent mutation of the SWI/SNF complex gene PBRM1 in renal carcinoma. *Nature* **469:** 539–542.

Versteege I, Sevenet N, Lange J, Rousseau-Merck MF, Ambros P, Handgretinger R, Aurias A, Delattre O. 1998. Truncating mutations of hSNF5/INI1 in aggressive paediatric cancer. *Nature* **394:** 203–206.

Wang X, Sansam CG, Thom CS, Metzger D, Evans JA, Nguyen PT, Roberts CW. 2009a. Oncogenesis caused by loss of the SNF5 tumor suppressor is dependent on activity of BRG1, the ATPase of the SWI/SNF chromatin remodeling complex. *Cancer Res* **69:** 8094–8101.

Wang Z, Zang C, Cui K, Schones DE, Barski A, Peng W, Zhao K. 2009b. Genome-wide mapping of HATs and HDACs reveals distinct functions in active and inactive genes. *Cell* **138:** 1019–1031.

Wang K, Kan J, Yuen ST, Shi ST, Chu KM, Law S, Chan TL, Kan Z, Chan AS, Tsui WY, et al. 2011. Exome sequencing identifies frequent mutation of ARID1A in molecular subtypes of gastric cancer. *Nat Genet* **43:** 1219–1223.

Wang X, Lee RS, Alver BH, Haswell JR, Wang S, Mieczkowski J, Drier Y, Gillespie SM, Archer TC, Wu JN, et al. 2017. SMARCB1-mediated SWI/SNF complex function is essential for enhancer regulation. *Nat Genet* **49:** 289–295.

Wei D, Goldfarb D, Song S, Cannon C, Yan F, Sakellariou-Thompson D, Emanuele M, Major MB, Weissman BE, Kuwahara Y. 2014. SNF5/INI1 deficiency redefines chromatin remodeling complex composition during tumor development. *Mol Cancer Res* **12:** 1574–1585.

Wiegand KC, Shah SP, Al-Agha OM, Zhao Y, Tse K, Zeng T, Senz J, McConechy MK, Anglesio MS, Kalloger SE, et al. 2010. ARID1A mutations in endometriosis-associated ovarian carcinomas. *N Engl J Med* **363:** 1532–1543.

Witkowski L, Carrot-Zhang J, Albrecht S, Fahiminiya S, Hamel N, Tomiak E, Grynspan D, Saloustros E, Nadaf J, Rivera B, et al. 2014. Germline and somatic SMARCA4 mutations characterize small cell carcinoma of the ovary, hypercalcemic type. *Nat Genet* **46:** 438–443.

Wong AK, Shanahan F, Chen Y, Lian L, Ha P, Hendricks K, Ghaffari S, Iliev D, Penn B, Woodland AM, et al. 2000. BRG1, a component of the SWI-SNF complex, is mutated in multiple human tumor cell lines. *Cancer Res* **60:** 6171–6177.

Xue Y, Canman JC, Lee CS, Nie Z, Yang D, Moreno GT, Young MK, Salmon ED, Wang W. 2000. The human SWI/SNF-B chromatin-remodeling complex is related to yeast rsc and localizes at kinetochores of mitotic chromosomes. *Proc Natl Acad Sci* **97:** 13015–13020.

Yildirim O, Li R, Hung JH, Chen PB, Dong X, Ee LS, Weng Z, Rando OJ, Fazzio TG. 2011. Mbd3/NURD complex regulates expression of 5-hydroxymethylcytosine marked genes in embryonic stem cells. *Cell* **147:** 1498–1510.

Yoo AS, Staahl BT, Chen L, Crabtree GR. 2009. MicroRNA-mediated switching of chromatin-remodelling complexes in neural development. *Nature* **460:** 642–646.

Yoo AS, Sun AX, Li L, Shcheglovitov A, Portmann T, Li Y, Lee-Messer C, Dolmetsch RE, Tsien RW, Crabtree GR. 2011. MicroRNA-mediated conversion of human fibroblasts to neurons. *Nature* **476:** 228–231.

The Essential Transcriptional Function of BRD4 in Acute Myeloid Leukemia

JAE-SEOK ROE AND CHRISTOPHER R. VAKOC
Cold Spring Harbor Laboratory, Cold Spring Harbor, New York 11724
Correspondence: vakoc@cshl.edu

Acute myeloid leukemia (AML) is often initiated by genetic alterations of machineries that regulate chromatin and transcription, thereby blocking cell differentiation. Such mechanisms may also render leukemia cells vulnerable to perturbations of transcriptional regulators, which includes small molecules targeting the coactivator protein BRD4. Numerous studies have validated BRD4 as a therapeutic target in diverse subtypes of AML; however, the vital function of BRD4 in this disease is only beginning to be understood. Here we discuss the recent progress in elucidating the transcriptional function of BRD4 in AML cells, with an emphasis on the desirable attributes, but also the inherent limitations, of targeting general coactivator proteins as cancer therapy.

Chromatin alterations are intimately linked with the control of transcription, and such processes are commonly deregulated in the pathogenesis of cancer (Dawson and Kouzarides 2012). This is highlighted by many large-scale cancer genome studies, which have identified frequent mutations in the genes encoding chromatin regulators during tumor evolution (Garraway and Lander 2013). Although the majority of mutations in chromatin regulators found in cancer are predicted to be loss-of-function, a smaller subset of these mutations result in a gain of function that drives oncogenic gene expression. In addition, unmutated chromatin regulators can support the function of other oncoproteins to drive tumorigenesis, thereby acting as nononcogene dependencies in cancer. Thus, cancer cells are often addicted to certain chromatin regulators, which represent viable targets for direct therapeutic intervention in this disease (Shen and Vakoc 2015).

Acute myeloid leukemia (AML) was one of the first malignancies in which genes encoding chromatin regulators were identified as drivers of cell transformation, and therefore AML serves as a paradigm for studying chromatin deregulation in cancer (Greenblatt and Nimer 2014; Wouters and Delwel 2016). As an example, the histone H3 lysine 4 methyltransferase protein MLL (also known as KMT2A) is often altered in AML via chromosomal translocations or mutations, thereby deregulating chromatin to cause a blockade in myeloid maturation (Krivtsov and Armstrong 2007). Other commonly mutated genes in AML include *DNMT3A* or *TET2*, which encode enzymes involved in DNA methylation and demethylation, respectively (Delhommeau et al. 2009; Ley et al. 2010). Genetic alterations of transcription factors (TFs) are also common in AML and lead to indirect alterations of chromatin structure and function to initiate disease progression (Rosenbauer and Tenen 2007).

The widespread involvement of aberrant chromatin states in AML provides a rationale to explore the use of chromatin modulating therapeutics in this disease. One of the first examples of such an agent is 5-azacytidine, which inhibits DNA methyltransferases. 5-Azacytidine and other related molecules are approved therapies in AML management, despite uncertainty regarding the mechanisms underlying their therapeutic efficacy (Silverman et al. 2002; Kantarjian et al. 2006; Issa et al. 2015). A major effort in the field in recent years has been to uncover other epigenetic vulnerabilities in AML, which has culminated in recently initiated clinical trials evaluating inhibitors of histone modifying enzymes (DOT1L and LSD1) and inhibitors of the chromatin reader protein BRD4 (Bernt et al. 2011; Nguyen et al. 2011; Harris et al. 2012). Here, we will review our current understanding of BRD4 as therapeutic target in AML.

BRD4 IS A DEPENDENCY AND A THERAPEUTIC TARGET IN AML

BRD4 can be classified as a chromatin reader protein, as it contains two tandem bromodomain modules that engage in acetyllysine recognition to tether BRD4 to chromatin (Wu and Chiang 2007). BRD4 can also be classified as a transcriptional coactivator, as it can bind directly (or indirectly) to TFs to promote downstream effects on RNA polymerase II (Pol II) activation (Shi and Vakoc 2014). Most of the available evidence suggests that BRD4 functions as a scaffold protein that links acetylated histones and TFs with the local assembly of the transcriptional apparatus at specific *cis*-regulatory elements. However, it was recently proposed that BRD4 employs intrinsic kinase and acetyltransferase activities,

Published by Cold Spring Harbor Laboratory Press; doi: 10.1101/sqb.2016.81.031039

although the functional relevance of these activities to the transcriptional function of BRD4 remains to be validated (Devaiah et al. 2012, 2016). Studies over nearly the past 20 years have implicated BRD4 in numerous cellular processes, including the regulation of transcription, DNA replication, and DNA repair (Wu and Chiang 2007; Floyd et al. 2013).

Although it might have been assumed that BRD4 performs a general role in regulating genomic processes, several groups identified small-molecule inhibitors of BRD4 bromodomains (and its closest homologs BRD2/BRD3/BRDT) in cell-based screens evaluating for discrete biological effects reviewed in Filippakopoulos et al. 2010; Nicodeme et al. (2010). These small molecules were found to bind to the bromodomains in a competitive manner with acetyllysine, and hence result in the release of BRD4 and its homologs (which are often referred to as the bromodomain and extraterminal domain–containing proteins, or BET protein family) from cellular chromatin. Since the description of these molecules in 2010, numerous studies have shown that chemical inhibition of BET proteins results in gene-specific effects on transcription, which are distinct from global transcription inhibition (Shi and Vakoc 2014). The specificity of transcriptional and cellular effects of BET inhibition has also allowed an investigation of disease-specific roles of BRD4, particularly in cancer.

In 2011, several independent groups showed that BET inhibitors exhibit anti-AML activity in animal models of this disease (Dawson et al. 2011; Mertz et al. 2011; Zuber et al. 2011b). One of these studies was from our own laboratory, in which we identified BRD4 using a short-hairpin RNA (shRNA) screen that evaluated the essentiality of chromatin regulators for disease progression in a mouse model of AML initiated by the MLL-AF9 and $Nras^{G12D}$ oncogenes (Zuber et al. 2011b). This screen identified BRD4 as a top vulnerability in this disease, based on the overall potency of the growth-arrest phenotype observed following BRD4 knockdown. Notably, a modest fourfold knockdown of BRD4 is sufficient to trigger terminal differentiation and apoptosis in AML cells, whereas BRD4 knockdown only elicited cytostatic effects in other immortalized cell types. Through a collaboration with James Bradner's laboratory, we obtained the BET inhibitor JQ1 and showed that the phenotypic consequences observed upon BRD4 knockdown were faithfully recapitulated following exposure to nanomolar concentrations of this compound. The close similarity in the effects of BRD4 knockdown and JQ1 exposure at the phenotypic and global transcriptional level provided strong support that BRD4 was the relevant target underlying the effects of JQ1 in this cell type (Zuber et al. 2011b). Most notably, intraperitoneal injection of JQ1 into AML-bearing animals was shown to delay disease progression in vivo at doses that showed minimal effects on normal hematopoiesis. Although our mechanistic understanding of these effects was limited during the early stages of this research, two observations were made at this time that would come to shape our mechanistic evaluation of BRD4 in AML. First, the sensitivity to JQ1 was identified in MLL-rearranged subtype of AML but was equally apparent in the other diverse genetic subtypes of this disease. Second, JQ1 caused a rapid, and pronounced, suppression of c-Myc transcription in AML cells, whereas Myc expression was unaffected by JQ1 in immortalized fibroblasts. Collectively, these experiments led us to propose that BRD4 was a novel epigenetic target in AML and highlighted the utility of genetic screening to nominate therapeutic opportunities in cancer.

An independent study from the laboratory of Tony Kouzarides published at the same time also showed the efficacy of BET inhibitors in the MLL-rearranged subtype of AML (Dawson et al. 2011). In this study, a series of proteomic screens performed in AML nuclear lysates had found BET proteins in association with the transcription elongation regulators PAF and P-TEFb, which are known interacting partners of MLL-fusion oncoproteins (Muntean et al. 2010; Yokoyama et al. 2010). These findings motivated an evaluation of the BET inhibitor I-BET, which inhibits BET bromodomains in a similar manner to JQ1, in various preclinical models of the MLL-fusion subtype of AML. I-BET was shown to suppress AML progression in these models, in association with transcriptional suppression of MYC, BCL2, and CDK6 (Dawson et al. 2011). Although it is unlikely that BET proteins act directly to support the function of MLL-fusion proteins, this work provided strong evidence supporting BET proteins as therapeutic targets in AML. The laboratory of Robert Sims at Constellation Pharmaceuticals also showed independently that human AML cell lines are sensitive to JQ1 in association with transcriptional suppression of *MYC* (Mertz et al. 2011). Collectively, these convergent studies showed that diverse subtypes of AML are vulnerable to BRD4 inhibition, and set the stage for evaluation of BET inhibitors as AML therapy in early stage clinical trials.

Clinical studies were initiated in 2013 to evaluate the effect of BET inhibitors in human cancer patients. The first Phase I study of a BET bromodomain inhibitor (called OTX-015) in leukemia patients was published in 2016 (Berthon et al. 2016). This dose escalation study evaluated OTX-015 in relapsed-refractory acute leukemia patients. This study provided the first glimpse into the toxicities of BET inhibition in humans, which includes reduced platelet counts, gastrointestinal symptoms, fatigue, and altered levels of coagulation factors, and hence led to the identification of suitable dosing regimens for future clinical investigation. Clinical responses were identified in a subset of patients treated on this trial at doses that were near or below the maximum tolerated dose, with a subset of patients achieving a complete remission. However, all of these patients showing a complete remission eventually relapsed while receiving therapy (Berthon et al. 2016). Although this study provides encouraging support for BET inhibition as a therapeutic strategy in human AML, numerous challenges remain in identifying predictive correlates of BET inhibitor sensitivity and in determining optimal combination drug regimens to make responses more durable. It should be emphasized that sensitivity to BET inhibition is not limited to AML but has also been established in lymphoid neoplasms, including lymphoma, B- and T-cell leukemias, multiple myeloma, and various solid tumors, which are indications under investigation in clinical trials (Cai et al. 2015).

One of the clear challenges to BET inhibitor therapy in AML will be the rapid emergence of drug-resistant cells. Recent studies have dissected the molecular mechanisms of resistance BET inhibition in AML mouse models, which have implicated nongenetic mechanisms (Fong et al. 2015; Rathert et al. 2015). Instead, it appears that cells with reprogrammed transcriptional states can be selected for in the presence of BET inhibition that are able to bypass the essential function of BRD4 to sustain growth-promoting genes, such as Myc. One of these BRD4-bypass pathways is via the WNT signaling pathway and suggests that dual targeting BRD4 and WNT could lead to durable therapeutic responses (Rathert et al. 2015). Numerous studies have identified effective BET inhibitor drug synergies, including combinations of BET inhibitors with FLT3 kinase inhibitors in cells harboring activating mutations of FLT3 (Fiskus et al. 2014). This study also suggests that resistance to FLT3 inhibition might be overcome by targeting BET proteins. In the MLL-fusion subtype of AML, co-inhibition of BRD4 and DOT1L can also lead to synergistic antileukemia responses (Gilan et al. 2016). These studies highlight how an expanding body of basic and translational research strategies can inform how BET inhibitors proceed in future clinical trials.

Despite the speed in which BET inhibitors have moved into clinical evaluation, it remains unknown why chemical inhibition of a general coactivator protein like BRD4 leads to therapeutic effects in AML. At the heart of this issue is the deficiency in our understanding of the leukemia maintenance function of BRD4, which has been the subject of intense investigation over the past 5 years. Below, we summarize our efforts, and those of other laboratories, in characterizing the essential transcriptional function of BRD4 in AML.

HEMATOPOIETIC LINEAGE TRANSCRIPTION FACTORS USE BRD4 AS A COFACTOR

The central hypothesis in the field regarding the therapeutic index of BET inhibitors in cancer is that targeting BRD4 disproportionately suppresses cancer-promoting genes versus other gene classes. The mechanisms that underlie this gene specificity are not well understood but may relate to the manner in which BRD4 is recruited to chromatin. ChIP-seq analysis in murine MLL-AF9/Nras^{G12D}-transformed AML cells revealed that BRD4 tends to occupy acetylated nucleosomes, as well as nucleosome-free regions across genome (Roe et al. 2015). The presence of BRD4 at nucleosome depleted sites implicated the possibility that transcription factors were involved in this process. Indeed, prior work had implicated BET proteins in recognizing acetylated transcription factors GATA-1, NF-κB, and Twist in other cellular contexts (Lamonica et al. 2011; Shi et al. 2014). Using an analysis of overrepresented motifs at BRD4-occupied sites in leukemia cells, it was found that the TFs ERG, FLI1, PU.1, C/EBPα, C/EBPβ, and MYB are highly correlated with BRD4 in this cell type (Roe et al. 2015). Notably, each of these TFs are lineage determinants that are vital for normal hematopoietic stem and progenitor cells (Zhang et al. 1997; Fisher and Scott 1998; Spyropoulos et al. 2000; Hirai et al. 2006; Greig et al. 2008; Loughran et al. 2008). Ectopic expression of these TFs in heterologous cell types (fibroblasts) will reprogram BRD4 to the unique binding sites found in leukemia cells, which suggests that TFs, either directly or indirectly, promote BRD4 recruitment to the genome in AML (Roe et al. 2015).

The mechanism by which these hematopoietic TFs stimulate BRD4 recruitment is via the use of the lysine acetyltransferase p300 as a coactivator, which prior work had shown can bind to each of these TFs (Dai et al. 1996; Parker et al. 1999; Yamamoto et al. 1999; Tomita et al. 2000; Erickson et al. 2001; Sano and Ishii 2001; Bai et al. 2005; Sandberg et al. 2005; Cesena et al. 2007; Li et al. 2011; Schlottmann et al. 2012; Roe et al. 2015). Genetic targeting and chemical inhibition of p300 in AML closely phenocopies the transcriptional effects of BRD4 knockdown and leads to the release of BRD4 from chromatin (Roe et al. 2015). Not only is p300 recruited to chromatin by these TFs, but it can also acetylate each of these TFs at functionally important sites (Tomita et al. 2000; Erickson et al. 2001; Bai et al. 2005; Cesena et al. 2007; Schlottmann et al. 2012; Blee et al. 2016). We characterized one of these sites on ERG in detail, which is a KGGK motif that can be acetylated by p300 one both lysines. The diacetylated ERG peptide can bind directly to BRD4 and is likely to contribute to BRD4 tethering to the AML genome. However, it is likely that multiple TFs and histones associate with BRD4 via low-affinity interactions to account for its recruitment. Collectively, our study showed the critical role of hematopoietic lineage TFs in using BRD4 as a cofactor to promote transcriptional activation in AML.

NSD3, CHD8, MEDIATOR, AND P-TEFb ARE EFFECTORS OF BRD4 IN AML CELLS

Another key aspect of BRD4 function in AML concerns the downstream mechanisms by which this regulator activates RNA polymerase II. Proteomic studies have repeatedly shown that BRD4 engages in numerous protein–protein interactions and thus acts as a scaffold for other regulators. One BRD4-interacting partner is the kinase complex P-TEFb, which consists of Cdk9 and a cyclin subunit (Jang et al. 2005; Yang et al. 2005). P-TEFb phosphorylates the pausing factors DSIF and NELF, as well as RNA polymerase II itself to promote transcription elongation (Peterlin and Price 2006). Studies in AML and in various other cell types have shown that JQ1 promotes the rapid eviction of P-TEFb from chromatin in parallel with an accumulation of paused RNA polymerase at nearby genes (Loven et al. 2013; Bhagwat et al. 2016). Moreover, CRISPR exon scanning experiments have revealed that the carboxy-terminal motif of BRD4, which engages in a direct interaction with

P-TEFb, is also vital for AML cell proliferation (Bisgrove et al. 2007; Shi et al. 2015).

Multiple subunits of the Mediator complex can also bind to BRD4 (Jiang et al. 1998; Donner et al. 2010). Although the MED23 subunit of Mediator appears to be required for the BRD4 interaction, the precise binding surface that links these two factors has yet to be mapped (Wang et al. 2013). In AML and in other cell types, BRD4 occupancy across the genome is tightly correlated with the Mediator complex, with JQ1 releasing Mediator complexes from chromatin (Donner et al. 2010; Loven et al. 2013; Bhagwat et al. 2016). In addition, genetic targeting of certain Mediator subunits in AML cells, such as MED12, MED13, MED23, and MED24 will lead to a similar differentiation phenotype and transcriptional alterations as observed upon targeting BRD4 (Bhagwat et al. 2016). Mediator is also known to regulate P-TEFb activity via its head module (Conaway and Conaway 2013), which leads to the coordination, and perhaps synergy, between these two positive regulators of RNA polymerase II activity (Allen and Taatjes 2015).

One of the defining domains of all four BET proteins is the extraterminal (ET) domain, which has been shown to activate transcription when artificially tethered to a minimal promoter (Shen et al. 2015). In AML cells, the ET domain physically associates with a short isoform of NSD3, which acts as an adaptor protein that couples BRD4 with the CHD8 chromatin remodeler (Shen et al. 2015). Notably, genetic targeting of NSD3 or CHD8 leads to similar phenotypes and transcriptional effects as observed upon BRD4 inhibition (Shen et al. 2015). Collectively, these studies highlight how BRD4 uses several interacting partners to facilitate chromatin remodeling and transcriptional activation in AML cells.

Although BRD4-containing protein complexes are broadly associated with active promoters and enhancers across the AML genome, JQ1 leads to the disproportionate eviction of these complexes from specific *cis* elements (Loven et al. 2013; Shi et al. 2013). This includes a distal cluster of enhancers (or super-enhancer) at the *Myc* locus, thus accounting for the hypersensitivity of *Myc* transcription to BRD4 inhibition in AML (Shi et al. 2013; Bhagwat et al. 2016). Remarkably, the locations of severe BRD4 eviction following JQ1 exposure are highly correlated with genes having pro-leukemic functions, as revealed through gene ontology analysis (Bhagwat et al. 2016). Moreover, these *cis* elements are skewed in having more Myb occupancy than the other hematopoietic TFs described above (Bhagwat et al. 2016). This result implies that Myb is more suppressed by JQ1 than other hematopoietic TFs, which might be a contributor to the overall sensitivity of leukemia cells to JQ1, because prior studies have shown that AML cells are more sensitive to Myb suppression than normal myeloid cells (Anfossi et al. 1989; Calabretta et al. 1991; Zuber et al. 2011a). These findings suggest that biased release of BRD4 complexes for specific *cis* elements is a key determinant of the gene-specific transcriptional effects of BET bromodomain inhibition.

One key question yet to be addressed is why some *cis* elements are more sensitive than others to JQ1. One possibility is that clustered enhancers recruit BRD4 in a cooperative fashion, which is more vulnerable to chemical inhibition than sites lacking cooperative BRD4 recruitment (Loven et al. 2013). Another possibility is that many sites in the genome recruit BRD4 in a bromodomain-independent manner and hence resist JQ1-mediated displacement. A third possibility is that specific *cis* elements show highly dynamic ON–OFF rates of BRD4 recruitment, which consequently are preferentially evicted by JQ1 when compared with more static *cis* elements. Regardless of the underlying mechanism, the skewed pattern of BRD4 release from the genome from leukemia-relevant genes is the clearest mechanism to date that accounts for the therapeutic index of BET inhibitors in AML.

CONCLUSION

Mechanistic studies of BRD4 in the MLL-fusion subtype of AML have unexpectedly revealed that this regulator is almost entirely engaged as a cofactor for hematopoietic lineage transcription factors, and not via an association with the MLL-fusion oncoprotein (Roe et al. 2015). Thus, the antileukemia activity of JQ1 occurs despite the output of MLL-fusion oncoproteins being unperturbed. Thus, targeting BRD4 in AML may share more in common conceptually with targeting lineage-specific TFs in prostate (androgen receptor) and breast (estrogen receptor) cancers (Bhagwat and Vakoc 2015). However, it must be recognized that, unlike AR and ER examples, BRD4 is likely to function as a key coactivator in all human tissues, and hence the largest concern for BET inhibitors as therapeutics lies in their potential for undesirable toxicities, which is just beginning to be exposed in clinical trials (Amorim et al. 2016; Berthon et al. 2016). A key priority for future investigation will be to expose coactivator perturbations that are capable of achieving potent anticancer activity while minimizing the broad suppression of multiple TFs. In AML, this would best be achieved by targeting coactivators in a manner that suppresses Myb function while preserving other vital TFs for normal hematopoiesis. Continued efforts aimed at defining the fundamental interaction between TFs and coactivators might expose improved modalities to reprogram transcription in a manner suitable for therapeutic efficacy in cancer.

ACKNOWLEDGMENTS

C.R.V. is supported by the Leukemia and Lymphoma Society, the Burroughs-Wellcome Fund, the Pershing Square Sohn Cancer Research Alliance, the Starr Cancer Consortium, and the National Institutes of Health (NIH)/National Cancer Institute (NCI) grants RO1 CA174793 and 5P01CA013106.

REFERENCES

Allen BL, Taatjes DJ. 2015. The Mediator complex: A central integrator of transcription. *Nat Rev Mol Cell Biol* **16:** 155–166.

Amorim S, Stathis A, Gleeson M, Iyengar S, Magarotto V, Leleu X, Morschhauser F, Karlin L, Broussais F, Rezai K, et al. 2016. Bromodomain inhibitor OTX015 in patients with lymphoma or multiple myeloma: A dose-escalation, open-label, pharmacokinetic, phase 1 study. *Lancet Haematol* **3:** e196–e204.

Anfossi G, Gewirtz AM, Calabretta B. 1989. An oligomer complementary to c-myb-encoded mRNA inhibits proliferation of human myeloid leukemia cell lines. *Proc Natl Acad Sci* **86:** 3379–3383.

Bai Y, Srinivasan L, Perkins L, Atchison ML. 2005. Protein acetylation regulates both PU.1 transactivation and Ig κ3′ enhancer activity. *J Immunol* **175:** 5160–5169.

Bernt KM, Zhu N, Sinha AU, Vempati S, Faber J, Krivtsov AV, Feng Z, Punt N, Daigle A, Bullinger L, et al. 2011. MLL-rearranged leukemia is dependent on aberrant H3K79 methylation by DOT1L. *Cancer Cell* **20:** 66–78.

Berthon C, Raffoux E, Thomas X, Vey N, Gomez-Roca C, Yee K, Taussig DC, Rezai K, Roumier C, Herait P, et al. 2016. Bromodomain inhibitor OTX015 in patients with acute leukaemia: A dose-escalation, phase 1 study. *Lancet Haematol* **3:** e186–e195.

Bhagwat AS, Vakoc CR. 2015. Targeting transcription factors in cancer. *Trends Cancer* **1:** 53–65.

Bhagwat AS, Roe JS, Mok BY, Hohmann AF, Shi J, Vakoc CR. 2016. BET bromodomain inhibition releases the Mediator complex from select *cis*-regulatory elements. *Cell Rep* **15:** 519–530.

Bisgrove DA, Mahmoudi T, Henklein P, Verdin E. 2007. Conserved P-TEFb-interacting domain of BRD4 inhibits HIV transcription. *Proc Natl Acad Sci* **104:** 13690–13695.

Blee AM, Liu S, Wang L, Huang H. 2016. BET bromodomain-mediated interaction between ERG and BRD4 promotes prostate cancer cell invasion. *Oncotarget* **7:** 38319–38332.

Cai SF, Chen CW, Armstrong SA. 2015. Drugging chromatin in cancer: Recent advances and novel approaches. *Mol Cell* **60:** 561–570.

Calabretta B, Sims RB, Valtieri M, Caracciolo D, Szczylik C, Venturelli D, Ratajczak M, Beran M, Gewirtz AM. 1991. Normal and leukemic hematopoietic cells manifest differential sensitivity to inhibitory effects of c-myb antisense oligodeoxynucleotides: An in vitro study relevant to bone marrow purging. *Proc Natl Acad Sci* **88:** 2351–2355.

Cesena TI, Cardinaux JR, Kwok R, Schwartz J. 2007. CCAAT/enhancer-binding protein (C/EBP) β is acetylated at multiple lysines: Acetylation of C/EBPβ at lysine 39 modulates its ability to activate transcription. *J Biol Chem* **282:** 956–967.

Conaway RC, Conaway JW. 2013. The Mediator complex and transcription elongation. *Biochim Biophys Acta* **1829:** 69–75.

Dai P, Akimaru H, Tanaka Y, Hou DX, Yasukawa T, Kanei-Ishii C, Takahashi T, Ishii S. 1996. CBP as a transcriptional coactivator of c-Myb. *Genes Dev* **10:** 528–540.

Dawson MA, Kouzarides T. 2012. Cancer epigenetics: From mechanism to therapy. *Cell* **150:** 12–27.

Dawson MA, Prinjha RK, Dittmann A, Giotopoulos G, Bantscheff M, Chan WI, Robson SC, Chung CW, Hopf C, Savitski MM, et al. 2011. Inhibition of BET recruitment to chromatin as an effective treatment for MLL-fusion leukaemia. *Nature* **478:** 529–533.

Delhommeau F, Dupont S, Della Valle V, James C, Trannoy S, Masse A, Kosmider O, Le Couedic JP, Robert F, Alberdi A, et al. 2009. Mutation in TET2 in myeloid cancers. *N Engl J Med* **360:** 2289–2301.

Devaiah BN, Lewis BA, Cherman N, Hewitt MC, Albrecht BK, Robey PG, Ozato K, Sims RJ III, Singer DS. 2012. BRD4 is an atypical kinase that phosphorylates serine2 of the RNA polymerase II carboxy-terminal domain. *Proc Natl Acad Sci* **109:** 6927–6932.

Devaiah BN, Case-Borden C, Gegonne A, Hsu CH, Chen Q, Meerzaman D, Dey A, Ozato K, Singer DS. 2016. BRD4 is a histone acetyltransferase that evicts nucleosomes from chromatin. *Nat Struct Mol Biol* **23:** 540–548.

Donner AJ, Ebmeier CC, Taatjes DJ, Espinosa JM. 2010. CDK8 is a positive regulator of transcriptional elongation within the serum response network. *Nat Struct Mol Biol* **17:** 194–201.

Erickson RL, Hemati N, Ross SE, MacDougald OA. 2001. p300 coactivates the adipogenic transcription factor CCAAT/enhancer-binding protein α. *J Biol Chem* **276:** 16348–16355.

Filippakopoulos P, Qi J, Picaud S, Shen Y, Smith WB, Fedorov O, Morse EM, Keates T, Hickman TT, Felletar I, et al. 2010. Selective inhibition of BET bromodomains. *Nature* **468:** 1067–1073.

Fisher RC, Scott EW. 1998. Role of PU.1 in hematopoiesis. *Stem Cells* **16:** 25–37.

Fiskus W, Sharma S, Qi J, Shah B, Devaraj SG, Leveque C, Portier BP, Iyer S, Bradner JE, Bhalla KN. 2014. BET protein antagonist JQ1 is synergistically lethal with FLT3 tyrosine kinase inhibitor (TKI) and overcomes resistance to FLT3-TKI in AML cells expressing FLT-ITD. *Mol Cancer Ther* **13:** 2315–2327.

Floyd SR, Pacold ME, Huang Q, Clarke SM, Lam FC, Cannell IG, Bryson BD, Rameseder J, Lee MJ, Blake EJ, et al. 2013. The bromodomain protein Brd4 insulates chromatin from DNA damage signalling. *Nature* **498:** 246–250.

Fong CY, Gilan O, Lam EY, Rubin AF, Ftouni S, Tyler D, Stanley K, Sinha D, Yeh P, Morison J, et al. 2015. BET inhibitor resistance emerges from leukaemia stem cells. *Nature* **525:** 538–542.

Garraway LA, Lander ES. 2013. Lessons from the cancer genome. *Cell* **153:** 17–37.

Gilan O, Lam EY, Becher I, Lugo D, Cannizzaro E, Joberty G, Ward A, Wiese M, Fong CY, Ftouni S, et al. 2016. Functional interdependence of BRD4 and DOT1L in MLL leukemia. *Nat Struct Mol Biol* **23:** 673–681.

Greenblatt SM, Nimer SD. 2014. Chromatin modifiers and the promise of epigenetic therapy in acute leukemia. *Leukemia* **28:** 1396–1406.

Greig KT, Carotta S, Nutt SL. 2008. Critical roles for c-Myb in hematopoietic progenitor cells. *Semin Immunol* **20:** 247–256.

Harris WJ, Huang X, Lynch JT, Spencer GJ, Hitchin JR, Li Y, Ciceri F, Blaser JG, Greystoke BF, Jordan AM, et al. 2012. The histone demethylase KDM1A sustains the oncogenic potential of MLL-AF9 leukemia stem cells. *Cancer Cell* **21:** 473–487.

Hirai H, Zhang P, Dayaram T, Hetherington CJ, Mizuno S, Imanishi J, Akashi K, Tenen DG. 2006. C/EBPβ is required for 'emergency' granulopoiesis. *Nat Immunol* **7:** 732–739.

Issa JP, Roboz G, Rizzieri D, Jabbour E, Stock W, O'Connell C, Yee K, Tibes R, Griffiths EA, Walsh K, et al. 2015. Safety and tolerability of guadecitabine (SGI-110) in patients with myelodysplastic syndrome and acute myeloid leukaemia: A multicentre, randomised, dose-escalation phase 1 study. *Lancet Oncol* **16:** 1099–1110.

Jang MK, Mochizuki K, Zhou M, Jeong HS, Brady JN, Ozato K. 2005. The bromodomain protein Brd4 is a positive regulatory component of P-TEFb and stimulates RNA polymerase II-dependent transcription. *Mol Cell* **19:** 523–534.

Jiang YW, Veschambre P, Erdjument-Bromage H, Tempst P, Conaway JW, Conaway RC, Kornberg RD. 1998. Mammalian mediator of transcriptional regulation and its possible role as an end-point of signal transduction pathways. *Proc Natl Acad Sci* **95:** 8538–8543.

Kantarjian H, Issa JP, Rosenfeld CS, Bennett JM, Albitar M, DiPersio J, Klimek V, Slack J, de Castro C, Ravandi F, et al. 2006. Decitabine improves patient outcomes in myelodysplastic syndromes: Results of a phase III randomized study. *Cancer* **106:** 1794–1803.

Krivtsov AV, Armstrong SA. 2007. MLL translocations, histone modifications and leukaemia stem-cell development. *Nat Rev Cancer* **7:** 823–833.

Lamonica JM, Deng W, Kadauke S, Campbell AE, Gamsjaeger R, Wang H, Cheng Y, Billin AN, Hardison RC, Mackay JP, et al. 2011. Bromodomain protein Brd3 associates with acetylated GATA1 to promote its chromatin occupancy at erythroid target genes. *Proc Natl Acad Sci* **108:** E159–E168.

Ley TJ, Ding L, Walter MJ, McLellan MD, Lamprecht T, Larson DE, Kandoth C, Payton JE, Baty J, Welch J, et al. 2010. DNMT3A mutations in acute myeloid leukemia. *N Engl J Med* **363:** 2424–2433.
Li Y, Kong D, Wang Z, Ahmad A, Bao B, Padhye S, Sarkar FH. 2011. Inactivation of AR/TMPRSS2-ERG/Wnt signaling networks attenuates the aggressive behavior of prostate cancer cells. *Cancer Prev Res (Phila)* **4:** 1495–1506.
Loughran SJ, Kruse EA, Hacking DF, de Graaf CA, Hyland CD, Willson TA, Henley KJ, Ellis S, Voss AK, Metcalf D, et al. 2008. The transcription factor Erg is essential for definitive hematopoiesis and the function of adult hematopoietic stem cells. *Nat Immunol* **9:** 810–819.
Loven J, Hoke HA, Lin CY, Lau A, Orlando DA, Vakoc CR, Bradner JE, Lee TI, Young RA. 2013. Selective inhibition of tumor oncogenes by disruption of super-enhancers. *Cell* **153:** 320–334.
Mertz JA, Conery AR, Bryant BM, Sandy P, Balasubramanian S, Mele DA, Bergeron L, Sims RJ III. 2011. Targeting MYC dependence in cancer by inhibiting BET bromodomains. *Proc Natl Acad Sci* **108:** 16669–16674.
Muntean AG, Tan J, Sitwala K, Huang Y, Bronstein J, Connelly JA, Basrur V, Elenitoba-Johnson KS, Hess JL. 2010. The PAF complex synergizes with MLL fusion proteins at HOX loci to promote leukemogenesis. *Cancer Cell* **17:** 609–621.
Nguyen AT, Taranova O, He J, Zhang Y. 2011. DOT1L, the H3K79 methyltransferase, is required for MLL-AF9-mediated leukemogenesis. *Blood* **117:** 6912–6922.
Nicodeme E, Jeffrey KL, Schaefer U, Beinke S, Dewell S, Chung CW, Chandwani R, Marazzi I, Wilson P, Coste H, et al. 2010. Suppression of inflammation by a synthetic histone mimic. *Nature* **468:** 1119–1123.
Parker D, Rivera M, Zor T, Henrion-Caude A, Radhakrishnan I, Kumar A, Shapiro LH, Wright PE, Montminy M, Brindle PK. 1999. Role of secondary structure in discrimination between constitutive and inducible activators. *Mol Cell Biol* **19:** 5601–5607.
Peterlin BM, Price DH. 2006. Controlling the elongation phase of transcription with P-TEFb. *Mol Cell* **23:** 297–305.
Rathert P, Roth M, Neumann T, Muerdter F, Roe JS, Muhar M, Deswal S, Cerny-Reiterer S, Peter B, Jude J, et al. 2015. Transcriptional plasticity promotes primary and acquired resistance to BET inhibition. *Nature* **525:** 543–547.
Roe JS, Mercan F, Rivera K, Pappin DJ, Vakoc CR. 2015. BET bromodomain inhibition suppresses the function of hematopoietic transcription factors in acute myeloid leukemia. *Mol Cell* **58:** 1028–1039.
Rosenbauer F, Tenen DG. 2007. Transcription factors in myeloid development: Balancing differentiation with transformation. *Nat Rev Immunol* **7:** 105–117.
Sandberg ML, Sutton SE, Pletcher MT, Wiltshire T, Tarantino LM, Hogenesch JB, Cooke MP. 2005. c-Myb and p300 regulate hematopoietic stem cell proliferation and differentiation. *Dev Cell* **8:** 153–166.
Sano Y, Ishii S. 2001. Increased affinity of c-Myb for CREB-binding protein (CBP) after CBP-induced acetylation. *J Biol Chem* **276:** 3674–3682.
Schlottmann S, Erkizan HV, Barber-Rotenberg JS, Knights C, Cheema A, Uren A, Avantaggiati ML, Toretsky JA. 2012. Acetylation increases EWS-FLI1 DNA binding and transcriptional activity. *Front Oncol* **2:** 107.
Shen C, Vakoc CR. 2015. Gain-of-function mutation of chromatin regulators as a tumorigenic mechanism and an opportunity for therapeutic intervention. *Curr Opin Oncol* **27:** 57–63.
Shen C, Ipsaro JJ, Shi J, Milazzo JP, Wang E, Roe JS, Suzuki Y, Pappin DJ, Joshua-Tor L, Vakoc CR. 2015. NSD3-short is an adaptor protein that couples BRD4 to the CHD8 chromatin remodeler. *Mol Cell* **60:** 847–859.
Shi J, Vakoc CR. 2014. The mechanisms behind the therapeutic activity of BET bromodomain inhibition. *Mol Cell* **54:** 728–736.
Shi J, Whyte WA, Zepeda-Mendoza CJ, Milazzo JP, Shen C, Roe JS, Minder JL, Mercan F, Wang E, Eckersley-Maslin MA, et al. 2013. Role of SWI/SNF in acute leukemia maintenance and enhancer-mediated Myc regulation. *Genes Dev* **27:** 2648–2662.
Shi J, Wang Y, Zeng L, Wu Y, Deng J, Zhang Q, Lin Y, Li J, Kang T, Tao M, et al. 2014. Disrupting the interaction of BRD4 with diacetylated Twist suppresses tumorigenesis in basal-like breast cancer. *Cancer Cell* **25:** 210–225.
Shi J, Wang E, Milazzo JP, Wang Z, Kinney JB, Vakoc CR. 2015. Discovery of cancer drug targets by CRISPR-Cas9 screening of protein domains. *Nat Biotechnol* **33:** 661–667.
Silverman LR, Demakos EP, Peterson BL, Kornblith AB, Holland JC, Odchimar-Reissig R, Stone RM, Nelson D, Powell BL, DeCastro CM, et al. 2002. Randomized controlled trial of azacitidine in patients with the myelodysplastic syndrome: A study of the cancer and leukemia group B. *J Clin Oncol* **20:** 2429–2440.
Spyropoulos DD, Pharr PN, Lavenburg KR, Jackers P, Papas TS, Ogawa M, Watson DK. 2000. Hemorrhage, impaired hematopoiesis, and lethality in mouse embryos carrying a targeted disruption of the Fli1 transcription factor. *Mol Cell Biol* **20:** 5643–5652.
Tomita A, Towatari M, Tsuzuki S, Hayakawa F, Kosugi H, Tamai K, Miyazaki T, Kinoshita T, Saito H. 2000. c-Myb acetylation at the carboxyl-terminal conserved domain by transcriptional co-activator p300. *Oncogene* **19:** 444–451.
Wang W, Yao X, Huang Y, Hu X, Liu R, Hou D, Chen R, Wang G. 2013. Mediator MED23 regulates basal transcription in vivo via an interaction with P-TEFb. *Transcription* **4:** 39–51.
Wouters BJ, Delwel R. 2016. Epigenetics and approaches to targeted epigenetic therapy in acute myeloid leukemia. *Blood* **127:** 42–52.
Wu SY, Chiang CM. 2007. The double bromodomain-containing chromatin adaptor Brd4 and transcriptional regulation. *J Biol Chem* **282:** 13141–13145.
Yamamoto H, Kihara-Negishi F, Yamada T, Hashimoto Y, Oikawa T. 1999. Physical and functional interactions between the transcription factor PU.1 and the coactivator CBP. *Oncogene* **18:** 1495–1501.
Yang Z, Yik JH, Chen R, He N, Jang MK, Ozato K, Zhou Q. 2005. Recruitment of P-TEFb for stimulation of transcriptional elongation by the bromodomain protein Brd4. *Mol Cell* **19:** 535–545.
Yokoyama A, Lin M, Naresh A, Kitabayashi I, Cleary ML. 2010. A higher-order complex containing AF4 and ENL family proteins with P-TEFb facilitates oncogenic and physiologic MLL-dependent transcription. *Cancer Cell* **17:** 198–212.
Zhang DE, Zhang P, Wang ND, Hetherington CJ, Darlington GJ, Tenen DG. 1997. Absence of granulocyte colony-stimulating factor signaling and neutrophil development in CCAAT enhancer binding protein α-deficient mice. *Proc Natl Acad Sci* **94:** 569–574.
Zuber J, Rappaport AR, Luo W, Wang E, Chen C, Vaseva AV, Shi J, Weissmueller S, Fellmann C, Taylor MJ, et al. 2011a. An integrated approach to dissecting oncogene addiction implicates a Myb-coordinated self-renewal program as essential for leukemia maintenance. *Genes Dev* **25:** 1628–1640.
Zuber J, Shi J, Wang E, Rappaport AR, Herrmann H, Sison EA, Magoon D, Qi J, Blatt K, Wunderlich M, et al. 2011b. RNAi screen identifies Brd4 as a therapeutic target in acute myeloid leukaemia. *Nature* **478:** 524–528.

Reexamining How Cancer Cells Exploit the Body's Metabolic Resources

CRAIG B. THOMPSON AND WILHELM PALM
Department of Cancer Biology and Genetics, Memorial Sloan Kettering Cancer Center, New York, New York 10065
Correspondence: thompsonc@mskcc.org

A key feature of multicellular life is the sharing of nutritional resources by all cells of the organism. In animals, the allocation of nutrients to individual cells is not determined in a cell-autonomous fashion. Instead, growth factor and hormonal signaling pathways have evolved to regulate cellular nutrient uptake, which prevents individual cells from parasitizing the organism's metabolic reserves. Cancer is a condition where the highly ordered regulation of nutrient distribution is disrupted. During carcinogenesis, transformed cells acquire mutations in signaling pathways that render nutrient uptake cell-autonomous. A deeper understanding of how the acquisition of potential nutrients is regulated may help develop improved approaches to cancer therapy.

Unlike their unicellular counterparts, mammalian cells do not have to forage for nutrients to maintain their bioenergetics. The extracellular fluid that bathes cells in the mammalian body has an abundant supply of reduced carbon in the form of glucose that can be used to support ATP production and macromolecular synthesis (Vander Heiden et al. 2009). In addition, extracellular fluid has an ample supply of reduced nitrogen in the form of glutamine. Glutamine-derived nitrogen is used for the production of nucleic acids, nonessential amino acids, phospholipids, and hexosamines (DeBerardinis et al. 2007). A robust supply of dietary and endogenously produced lipids circulates in lipoprotein particles or bound to albumin. Essential amino acids derived from our diet are also present in micromolar quantities (Hosios et al. 2016).

NUTRIENT UPTAKE IS REGULATED BY SIGNAL TRANSDUCTION

Over the last 15 years, we have come to appreciate that the cellular uptake of nutrients is regulated by growth factor or hormonal signal transduction. Unlike single-cell eukaryotes, mammalian cells lack the cell-autonomous ability to acquire the shared resources in the extracellular environment (Rathmell et al. 2000; Lindsten et al. 2003; DeBerardinis et al. 2008). Instead, the control of nutrient uptake is regulated in mammals as a result of growth factor engagement of receptor tyrosine kinases, which control the expression and localization of nutrient transporters necessary to take up glucose, amino acids, and lipids (DeBerardinis et al. 2008). The strength of growth factor signals received by a cell through its extracellular receptors determines the intracellular supply of nutrients available as substrates for cellular bioenergetics and macromolecular synthesis to sustain cell survival. When growth factor signal transduction is of sufficient magnitude, cells are forced to take up extracellular nutrients in quantities that exceed their bioenergetic needs. When this occurs, cellular metabolism is reprogrammed, resulting in the production of anabolic metabolites that can fuel cell growth or support effector function (Fig. 1A).

The dependency of mammalian cells on growth factors to direct nutrient uptake provides a molecular explanation for why cells can live only at sites within the body where they receive sufficient levels of lineage-specific growth factors. Signaling from such growth factors is required to maintain sufficient nutrient uptake to support the cell's metabolic needs. In the absence of growth factor signaling, cells cannot take up adequate nutrients to maintain themselves (Rathmell et al. 2000; Lindsten et al. 2003). Under these circumstances, mammalian cells support their bioenergetic needs through autophagy, using the lysosomal degradation of intracellular macromolecules to sustain ATP production (Lum et al. 2005). Cell survival can only be maintained transiently in the absence of growth factor–induced nutrient uptake. Ultimately, the loss of sufficient nutrient supply to maintain bioenergetics and macromolecular integrity leads to cellular atrophy and/or apoptosis (Fig. 1B). The induction of apoptosis as a result of insufficient growth factor stimulation of nutrient uptake is referred to as death by neglect.

LIVING ON GLUCOSE AND GLUTAMINE

Over the last 10 years, a great deal has been learned about the signaling pathways that control the regulation of glucose and glutamine uptake (Vander Heiden et al. 2009). The presentation of bulk nutrient transporters at the cell surface is controlled by effector molecules down-

Published by Cold Spring Harbor Laboratory Press; doi: 10.1101/sqb.2016.81.030734

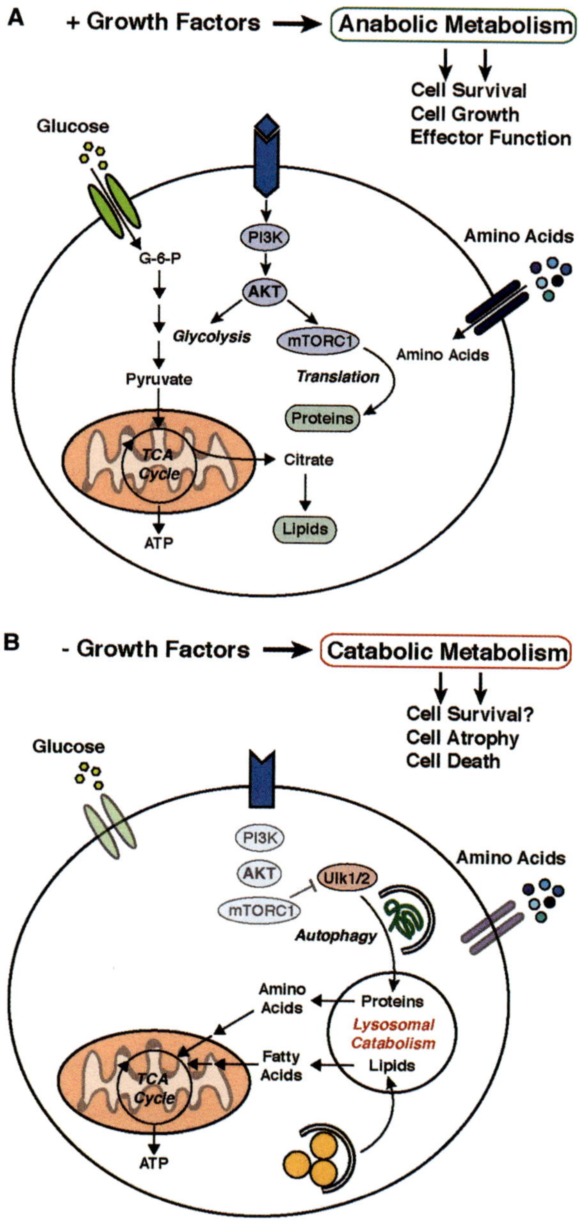

Figure 1. Growth factor–induced nutrient uptake fuels anabolic metabolism. (*A*) Growth factor signaling supports anabolic metabolism to enhance cell survival and growth. The PI3K/AKT/mTORC1 signaling axis stimulates glucose and amino acid uptake. If nutrient uptake exceeds a cell's bioenergetic needs, the excess carbon and nitrogen are directed into anabolic pathways. (*B*) In the absence of growth factor signals, cells lose the ability to take up extracellular nutrients. The ensuing derepression of autophagy initiates degradation of intracellular contents to generate bioenergetic substrates to support ATP needs. The autophagic degradation of intracellular contents leads to cellular atrophy and ultimately cell death. PI3K, phosphoinositide 3-kinase; mTORC1, mammalian target of rapamycin complex 1; TCA, tricarboxylic acid.

stream from receptor tyrosine kinase–initiated activation of phosphatidylinositide 3-kinase (PI3K) (Fig. 2; Edinger and Thompson 2002; Frauwirth et al. 2002). The membrane-associated production of phosphatidylinositol 3-phosphate leads to the recruitment and activation of the serine/threonine kinase AKT. As a PI3K effector, AKT controls surface translocation of glucose transporters, the intracellular capture of glucose by phosphorylation, and the commitment of glucose to glycolytic metabolism (Rathmell et al. 2003; Elstrom et al. 2004). AKT also activates the growth-promoting mammalian target of rapamycin complex 1 (mTORC1) pathway through phosphorylation and degradation of the negative mTORC1 regulator, tuberous sclerosis complex 2 (TSC2) (Plas and Thompson 2003). This results in increased surface expression of bulk amino acid transporters to capture available essential amino acids (Edinger and Thompson 2004).

Cellular uptake and use of glutamine is regulated by the transcription factor Myc (Wise et al. 2008). For example, glutamine uptake in hematopoietic cells is regulated through cytokine receptors that activate JAK/STAT signaling (Fig. 2). The resulting induction of the transcription factor Myc up-regulates/promotes the expression of a battery of proteins that allow cells to take up glutamine and initiate its metabolism by glutaminase-dependent deamination to glutamate. The resulting glutamate is then used to maintain tricarboxylic acid (TCA) cycle integrity and nonessential amino acid synthesis. Glutamine also plays a key role in de novo nucleotide biosynthesis. Studies of cancer cells have shown that cells overexpressing Myc consume glutamine in excess of their ability to use glutamine in biosynthetic reactions and excrete the excess carbon in the form of lactate, alanine, and/or glutamate.

Cancer cells with mutations in either the PI3K or JAK/STAT signaling pathways take up nutrients in excess of their synthetic and bioenergetic needs (Hammerman et al. 2005). Cells can deal with this imbalance by secreting excess carbon in the form of lactate. Recently, studies have implicated lactate production as an inducer of vasculogenesis and as a modulator of innate immune responses (Colegio et al. 2014; Carmona-Fontaine et al. 2017). Both of these properties have been implicated in enhancing tissue regeneration and wound repair. Thus, the induction of excess nutrient metabolism in cancer cells has consequences beyond cell intrinsic effects. These new insights into the regulation of nutrient uptake and waste disposal have spurred a renewed interest in the role of nutrients and the intermediate metabolites produced by their catabolism in regulating processes as diverse as differentiation, tissue regeneration, and immune defense (Pavlova and Thompson 2016).

SURVIVING NUTRIENT DEPLETION

Most of the above insights have come from experiments performed with cells cultured in media, in which extracellular nutrients are present in essentially unlimited quantities. Increasingly, the field of cellular metabolism is investigating how cells adapt to changes in the availability of extracellular nutrients. This problem is particularly relevant for the understanding of cancer biology. As tumor cells grow at their site of origin, they frequently outstrip vasculature's ability to supply sufficient nutrients and oxygen to maintain the anabolic metabolism required for cell

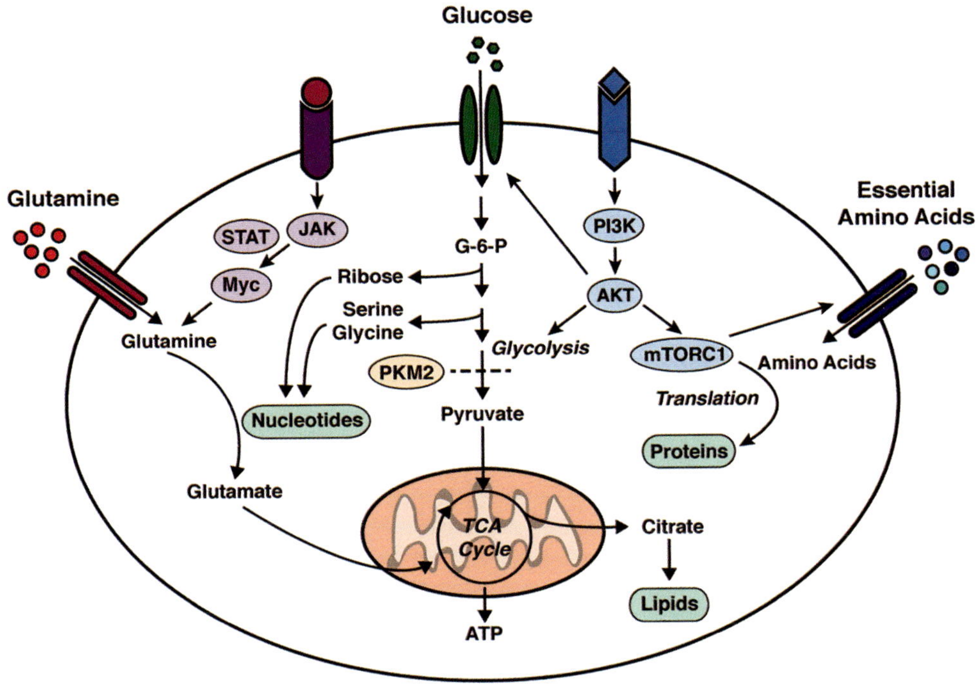

Figure 2. Growth factor–directed signals enhance the uptake of both essential amino acids and glutamine. In some cells, such as hematopoietic cells, the uptake of glutamine is regulated by induction of the transcription factor Myc through JAK/STAT signaling. In contrast, essential amino acid and glucose uptake is promoted by the PI3K/AKT/mTORC1 signaling axis. PI3K, phosphoinositide 3-kinase; mTORC1, mammalian target of rapamycin complex 1; TCA, tricarboxylic acid.

proliferation. Under these conditions, cells must adapt to nutrient limitation (Pavlova and Thompson 2016).

Over the last several years, a number of groups have studied cellular adaptation to hypoxia or low levels of glucose and/or amino acids (Commisso et al. 2013; Kamphorst et al. 2013; Maddocks et al. 2013; Ye et al. 2014; Mayers et al. 2016; Sousa et al. 2016). These studies have identified that a common consequence of limitation in any of these nutrients is redox stress. This has led to renewed interest in the regulation of redox defense. Cell survival under redox stress depends on increasing production of nicotinamide adenine dinucleotide phosphate (NADPH) and glutathione for the detoxification of reactive oxygen species (Schafer et al. 2009; Fan et al. 2014; Piskounova et al. 2015). Studies of how cells survive nutrient limitation have led to the discovery of a critical role for oxygen and α-ketoglutarate-dependent dioxygenases in regulating adaptation to the redox stress brought on by nutrient limitation including metabolic reprogramming, transcriptional/translational adaptation, and changes in chromatin accessibility (Kaelin and McKnight 2013).

CELL GROWTH IN VASCULARLY COMPROMISED TISSUES AND TUMORS

In vascularly compromised tissues, the delivery of all nutrients required to support cell growth is impaired—glucose, glutamine, and oxygen levels are all reduced. Moreover, the levels of other nutrients can also be severely compromised. This is a critical issue because cell growth depends on a number of nutrients that cannot be synthesized by mammalian cells. Among these are essential amino acids and polyunsaturated lipids (Kamphorst et al. 2013; Hosios et al. 2016). How cells maintain sufficient quantities of these vital nutrients to support viability and growth in a vascularly compromised tissue remains poorly understood. In the last several years, it has been discovered that tumor cells residing in vascularly compromised areas can derive these essential nutrients through the bulk uptake and degradation of extracellular macromolecules in their environment (Commisso et al. 2013; Kamphorst et al. 2013, 2015; Palm et al. 2015).

VASCULARLY COMPROMISED CELLS USE MACROMOLECULES TO SUPPORT GROWTH

Based on tissue culture studies, it had been believed that mammalian cells did not rely on extracellular macromolecules as a source of building blocks with which to generate cell mass. Pioneering studies by Eagle and his colleagues in the 1950s had shown that HeLa cells did not use extracellular proteins as a meaningful source of amino acids (Eagle and Piez 1960). However, several recent studies have shown that Ras-transformed cells can maintain growth in nutrient-deprived environments by taking up extracellular macromolecules to support cell growth (Commisso et al. 2013; Kamphorst et al. 2013; Palm et al. 2015). For example, when lipid desaturation is compromised by the lack of oxygen availability, Ras-transformed cells maintain a supply of unsaturated fatty acids for membrane synthesis by acquiring albumin-bound fatty acids (Kamphorst et al. 2013). Similarly, under conditions of glutamine compromise, it was shown that growing cancer cells could use macropinocytosis of albumin as an alternative source of nitrogen and carbon for production of nonessential amino acids (Commisso et al. 2013).

These initial observations have led to the discovery that mammalian cells retain the ability to take up and degrade extracellular macromolecules through macropinocytosis, a nonselective endocytic pathway, in a manner similar to amoeboid single-cell eukaryotes. During macropinocytosis, extracellular macromolecules are taken up in bulk and delivered to the lysosome where pH-sensitive lipases and proteases release the constituent amino acids and acyl chains as an intracellular nutrient source. In cancer cells, oncogenic forms of Ras have been shown to regulate the rate at which macropinocytosis occurs. Since many human cancers have activated Ras alleles, Ras-induced macropinocytosis can potentially allow these cancers to use macromolecules as alternative nutrients. This would be especially important in a vascularly compromised environment.

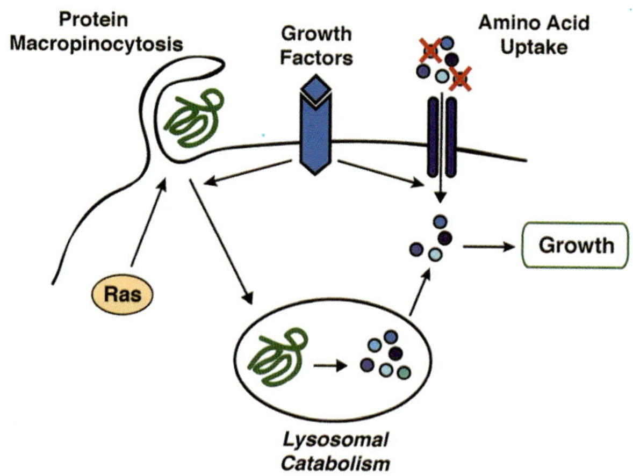

Figure 3. Ras-transformed cells can use protein macropinocytosis to acquire amino acids. Oncogenic Ras GTPases increase macropinocytosis of extracellular proteins. Degradation of ingested proteins in the lysosome can supplement the uptake of free amino acids to support protein synthesis and cell growth.

REGULATION OF MACROPINOCYTOSIS

Recent studies have been able to document that macropinocytosis can provide not only a supplemental source of nutrients but can also be a sufficient source of essential amino acids to support growth even when free essential amino acids are eliminated from the extracellular environment (Palm et al. 2015). For example, when tissue culture medium is supplemented with the levels of protein, in the form of albumin, normally present in extracellular fluid, Ras-transformed cells are able to proliferate in medium completely deficient of the essential amino acid leucine. The ability of cells to take up sufficient quantities of proteins to support their essential amino acid needs through lysosomal degradation is not a property shared by all cancer cell lines. Cell lines with activating mutations in the oncogene AKT are unable to use proteins as a source of essential amino acids to support their cell growth. Based on these findings, there appear to be two distinct mechanisms of essential amino acid acquisition by mammalian cells (Fig. 3). When present in the extracellular space, essential amino acids are taken up by specific transporters (Edinger and Thompson 2002). However, when the supply of free essential amino acids is compromised in the extracellular space, cells can engage in macropinocytosis to take up extracellular proteins and recover their amino acid content through lysosomal catabolism (Palm et al. 2015).

mTORC1 AS A REGULATOR OF LYSOSOMAL PROTEIN DEGRADATION

The potent ability of Ras to stimulate macropinocytosis may be one of the molecular mechanisms by which Ras mutations are selected for in human carcinogenesis. Surprisingly, although Ras accelerates the rate of macropinocytosis, the rate at which extracellular fluid is taken up into the endosomal compartment does not determine the rate at which ingested proteins are catabolized. Endosomes produced by macropinocytosis can either re-circulate to the cell surface and disgorge their contents or can be trafficked to the lysosome for nutrient capture.

Recently, we have discovered that the fate of the macropinosome within the endocytic pathway is primarily controlled by the mTORC1 complex (Palm et al. 2015). When mTORC1 is activated, it appears to prevent endosomes from fusing with the lysosome, thus restricting the use of extracellular proteins as a source of amino acids. When extracellular free amino acid levels are sufficient, their uptake through transporters leads to lysosomal recruitment and activation of mTORC1. This activation then suppresses endosomal trafficking and lysosomal degradation of extracellular proteins. This provides an explanation for why cells growing in standard tissue culture medium do not use extracellular proteins as a source of nutrients, as mTORC1 activation suppresses it. However, when one or more amino acids becomes depleted in the extracellular space, transfer RNA (tRNA) charging becomes compromised, the accumulation of uncharged tRNAs leads to the activation of GCN2 (Ye et al. 2010). The activation of GCN2 results in eIF2α/ATF4-dependent transcription of Sestrin2, a negative regulator of mTORC1 (Ye et al. 2015). Sestrin induction suppresses mTORC1, allowing cells to support continued growth through the uptake and degradation of extracellular proteins (Fig. 4). The resulting ability of cells to grow and divide using extracellular protein as a source of amino acids for translation is paradoxically dependent on suppressing mTORC1, which is widely believed to be essential for eukaryotic cell growth (Howell et al. 2013).

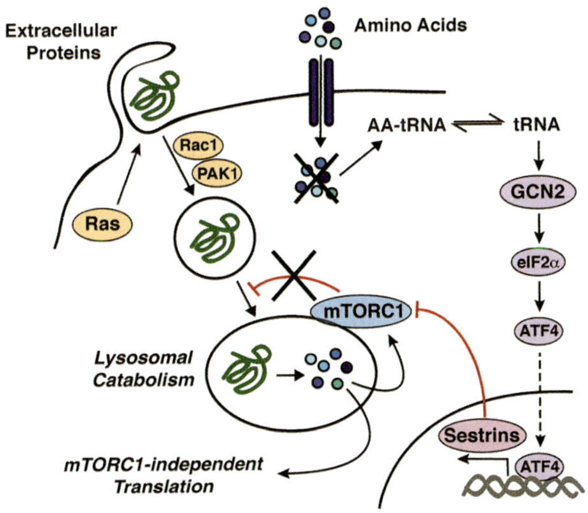

Figure 4. mTORC1 regulates cellular preference for proteins or free amino acids as nutrients. The intracellular amino acid sensor, mTORC1, can suppress lysosomal degradation of extracellular proteins taken up by macropinocytosis. This results in coupling of cell growth to the uptake of extracellular amino acids. In contrast, when uncharged tRNAs build up, GCN2-dependent transcriptional induction of Sestrin proteins suppresses mTORC1 to stimulate amino acid recovery from the degradation of extracellular proteins in the lysosome. AA-tRNA, aminoacyl transfer RNA; mTORC1, mammalian target of rapamycin complex 1

IS mTORC1 A REQUIRED EFFECTOR OF CELL GROWTH?

The above results suggest that multicellular animals have evolved a system that allows cells to switch between free amino acids and extracellular proteins as a source of amino acids to support protein translation and cell growth. This work also suggests that mTORC1 has evolved as a critical modulator of cell growth allowing cells to couple growth tightly to free amino acid availability. These results challenge the current view that mTORC1 is a required mediator of cell growth (Howell et al. 2013). The idea that cells can grow using extracellular proteins as an amino acid source in an mTORC1-independent mechanism is heretical. However, if true, these findings have the potential to explain some of the recent paradoxical results concerning mTORC1 inhibitors in clinical trials of cancer treatment. Based on the hypothesis that mTORC1 is a required effector of anabolic growth, it has been believed that mTORC1 inhibitors would be potent suppressors of cancer growth. Despite this, mTOR inhibitors have been disappointing in the clinic. Most recently, some tumors have been suggested to experience accelerated growth when treated with mTOR inhibitors. To determine whether this resulted from de-repression of the use of extracellular proteins to support cell growth, we have performed studies to determine the effects of mTORC1 inhibitors in an animal model of Ras-induced pancreatic cancer (Palm et al. 2015). Under this circumstance, mTORC1 inhibition enhanced the rate of growth of established tumors. Cell growth was accelerated specifically in the vascularly compromised regions of the tumor. These results suggest that the ability of cells to grow in the presence of mTORC1 inhibitors is not an in vitro artifact. Thus, mTORC1 inhibitors can accelerate the growth of tumors in which the majority of tumor cells exist in vascularly compromised environments.

CONSEQUENCES OF USING PROTEINS AS NUTRIENTS TO SUPPORT GROWTH

The idea that metazoan cells can use the degradation of extracellular proteins as a source of amino acids to support cell growth has been viewed as physiologically unlikely. Cellular degradation of extracellular proteins for use as nutrients is energetically costly because it requires continuous synthesis of plasma protein by the liver. However, there appears to be increasing evidence that extracellular proteins can be used as a source of amino acids to support cell survival and proliferation in vascularly compromised tissues in adult organisms (Kamphorst et al. 2015). Over the last 70 years, it has been documented that in a wide variety of vascularly compromised tumors nonessential amino acid levels becomes severely compromised while essential amino acids become significantly elevated relative to surrounding normal tissue. The explanation of this has been that under vascularly compromised conditions, the resident cells do not grow and therefore do not use essential amino acids for protein translation. At the same time, it was believed they continue to consume nonessential amino acids to support their bioenergetics. An extension of this hypothesis has been generated based on the ability of cells to grow and repair vascularly compromised tissues by using extracellular proteins to maintain cell growth and energy production. Protein synthesis is maintained but there is still a local depletion of nonessential amino acids and an elevation of essential amino acids (Kamphorst et al. 2015). This imbalance is created because the nonessential amino acids released by protein degradation are used to support bioenergetics as well as nucleotide, fatty acid, and protein synthesis. In contrast, essential amino acids will be used primarily to support protein synthesis. As a result, essential amino acids will accumulate relative to nonessential amino acids.

CONCLUSION

In conclusion, the finding that extracellular macromolecules function as important nutrients for the support of cell growth has the potential to open up a new way of thinking about how tumor cells regulate the acquisition and utilization of extracellular nutrients. These findings, if validated in other model systems, have the potential to provide important new insights that may be therapeutically exploited in the treatment of cancers based on specific oncogenic mutations. The validation of this hypothesis is still a work in progress. However, at the very least, these recent studies suggest that we have much more to learn concerning the molecular mechanisms by which tumor cells exploit their environment for metabolic resources to maintain cell growth and proliferation.

REFERENCES

Carmona-Fontaine C, Deforet M, Akkari L, Thompson CB, Joyce JA, Xavier JB. 2017. Metabolic origins of spatial organization in the tumor microenvironment. *Proc Natl Acad Sci*. doi:10.1073/pnas.1700600114.

Colegio OR, Chu NQ, Szabo AL, Chu T, Rhebergen AM, Jairam V, Cyrus N, Brokowski CE, Eisenbarth SC, Phillips GM, et al. 2014. Functional polarization of tumour-associated macrophages by tumour-derived lactic acid. *Nature* **513:** 559–563.

Commisso C, Davidson SM, Soydaner-Azeloglu RG, Parker SJ, Kamphorst JJ, Hackett S, Grabocka E, Nofal M, Drebin JA, Thompson CB, et al. 2013. Macropinocytosis of protein is an amino acid supply route in Ras-transformed cells. *Nature* **497:** 633–637.

DeBerardinis RJ, Mancuso A, Daikhin E, Nissim I, Yudkoff M, Wehrli S, Thompson CB. 2007. Beyond aerobic glycolysis: Transformed cells can engage in glutamine metabolism that exceeds the requirement for protein and nucleotide synthesis. *Proc Natl Acad Sci* **104:** 19345–19350.

DeBerardinis RJ, Lum JJ, Hatzivassiliou G, Thompson CB. 2008. The biology of cancer: Metabolic reprogramming fuels cell growth and proliferation. *Cell Metab* **7:** 11–20.

Eagle H, Piez KA. 1960. The utilization of proteins by cultured human cells. *J Biol Chem* **235:** 1095–1097.

Edinger AL, Thompson CB. 2002. Akt maintains cell size and survival by increasing mTOR-dependent nutrient uptake. *Mol Biol Cell* **13:** 2276–2288.

Edinger AL, Thompson CB. 2004. An activated mTOR mutant supports growth factor-independent, nutrient-dependent cell survival. *Oncogene* **23:** 5654–5663.

Elstrom RL, Bauer DE, Buzzai M, Karnauskas R, Harris MH, Plas DR, Zhuang H, Cinalli RM, Alavi A, Rudin CM, et al. 2004. Akt stimulates aerobic glycolysis in cancer cells. *Cancer Res* **64:** 3892–3899.

Fan J, Ye J, Kamphorst JJ, Shlomi T, Thompson CB, Rabinowitz JD. 2014. Quantitative flux analysis reveals folate-dependent NADPH production. *Nature* **510:** 298–302.

Frauwirth KA, Riley JL, Harris MH, Parry RV, Rathmell JC, Plas DR, Elstrom RL, June CH, Thompson CB. 2002. The CD28 signaling pathway regulates glucose metabolism. *Immunity* **16:** 769–777.

Hammerman PS, Fox CJ, Birnbaum MJ, Thompson CB. 2005. Pim and Akt oncogenes are independent regulators of hematopoietic cell growth and survival. *Blood* **105:** 4477–4483.

Hosios AM, Hecht VC, Danai LV, Johnson MO, Rathmell JC, Steinhauser ML, Manalis SR, Vander Heiden MG. 2016. Amino acids rather than glucose account for the majority of cell mass in proliferating mammalian cells. *Dev Cell* **36:** 540–549.

Howell JJ, Ricoult SJ, Ben-Sahra I, Manning BD. 2013. A growing role for mTOR in promoting anabolic metabolism. *Biochem Soc Trans* **41:** 906–912.

Kaelin WG Jr, McKnight SL. 2013. Influence of metabolism on epigenetics and disease. *Cell* **153:** 56–69.

Kamphorst JJ, Cross JR, Fan J, de Stanchina E, Mathew R, White EP, Thompson CB, Rabinowitz JD. 2013. Hypoxic and Ras-transformed cells support growth by scavenging unsaturated fatty acids from lysophospholipids. *Proc Natl Acad Sci* **110:** 8882–8887.

Kamphorst JJ, Nofal M, Commisso C, Hackett SR, Lu W, Grabocka E, Vander Heiden MG, Miller G, Drebin JA, Bar-Sagi D, et al. 2015. Human pancreatic cancer tumors are nutrient poor and tumor cells actively scavenge extracellular protein. *Cancer Res* **75:** 544–553.

Lindsten T, Golden JA, Zong WX, Minarcik J, Harris MH, Thompson CB. 2003. The proapoptotic activities of Bax and Bak limit the size of the neural stem cell pool. *J Neurosci* **23:** 11112–11119.

Lum JJ, Bauer DE, Kong M, Harris MH, Li C, Lindsten T, Thompson CB. 2005. Growth factor regulation of autophagy and cell survival in the absence of apoptosis. *Cell* **120:** 237–248.

Maddocks OD, Berkers CR, Mason SM, Zheng L, Blyth K, Gottlieb E, Vousden KH. 2013. Serine starvation induces stress and p53-dependent metabolic remodelling in cancer cells. *Nature* **493:** 542–546.

Mayers JR, Torrence ME, Danai LV, Papagiannakopoulos T, Davidson SM, Bauer MR, Lau AN, Ji BW, Dixit PD, Hosios AM, et al. 2016. Tissue of origin dictates branched-chain amino acid metabolism in mutant Kras-driven cancers. *Science* **353:** 1161–1165.

Palm W, Park Y, Wright K, Pavlova NN, Tuveson DA, Thompson CB. 2015. The utilization of extracellular proteins as nutrients is suppressed by mTORC1. *Cell* **162:** 259–270.

Pavlova NN, Thompson CB. 2016. The emerging hallmarks of cancer metabolism. *Cell Metab* **23:** 27–47.

Piskounova E, Agathocleous M, Murphy MM, Hu Z, Huddlestun SE, Zhao Z, Leitch AM, Johnson TM, DeBerardinis RJ, Morrison SJ. 2015. Oxidative stress inhibits distant metastasis by human melanoma cells. *Nature* **527:** 186–191.

Plas DR, Thompson CB. 2003. Akt activation promotes degradation of tuberin and FOXO3a via the proteasome. *J Biol Chem* **278:** 12361–12366.

Rathmell JC, Vander Heiden MG, Harris MH, Frauwirth KA, Thompson CB. 2000. In the absence of extrinsic signals, nutrient utilization by lymphocytes is insufficient to maintain either cell size or viability. *Mol Cell* **6:** 683–692.

Rathmell JC, Fox CJ, Plas DR, Hammerman PS, Cinalli RM, Thompson CB. 2003. Akt-directed glucose metabolism can prevent Bax conformation change and promote growth factor-independent survival. *Mol Cell Biol* **23:** 7315–7328.

Schafer ZT, Grassian AR, Song L, Jiang Z, Gerhart-Hines Z, Irie HY, Gao S, Puigserver P, Brugge JS. 2009. Antioxidant and oncogene rescue of metabolic defects caused by loss of matrix attachment. *Nature* **461:** 109–113.

Sousa CM, Biancur DE, Wang X, Halbrook CJ, Sherman MH, Zhang L, Kremer D, Hwang RF, Witkiewicz AK, Ying H, et al. 2016. Pancreatic stellate cells support tumour metabolism through autophagic alanine secretion. *Nature* **536:** 479–483.

Vander Heiden MG, Cantley LC, Thompson CB. 2009. Understanding the Warburg effect: The metabolic requirements of cell proliferation. *Science* **324:** 1029–1033.

Wise DR, DeBerardinis RJ, Mancuso A, Sayed N, Zhang XY, Pfeiffer HK, Nissim I, Daikhin E, Yudkoff M, McMahon SB, et al. 2008. Myc regulates a transcriptional program that stimulates mitochondrial glutaminolysis and leads to glutamine addiction. *Proc Natl Acad Sci* **105:** 18782–18787.

Ye J, Kumanova M, Hart LS, Sloane K, Zhang H, De Panis DN, Bobrovnikova-Marjon E, Diehl JA, Ron D, Koumenis C. 2010. The GCN2-ATF4 pathway is critical for tumour cell survival and proliferation in response to nutrient deprivation. *EMBO J* **29:** 2082–2096.

Ye J, Fan J, Venneti S, Wan YW, Pawel B, Zhang J, Finley L, Lu C, Lindsten T, Cross J, et al. 2014. Serine catabolism regulates mitochondrial redox control during hypoxia. *Cancer Disc* **4:** 1406–1417.

Ye J, Palm WP, Peng M, King B, Lindsten T, Li MO, Koumenis C, Thompson CB. 2015. GCN2 sustains mTORC1 suppression by inducing Sestrin2. *Genes Dev* **29:** 2331–2336.

Autophagy, Metabolism, and Cancer

Jessie Yanxiang Guo[1,2,3] and Eileen White[1,4]
[1]*Rutgers Cancer Institute of New Jersey, New Brunswick, New Jersey 08903*
[2]*Department of Medicine, Rutgers Robert Wood Johnson Medical School, New Brunswick, New Jersey 08901*
[3]*Department of Chemical Biology, Rutgers Ernest Mario School of Pharmacy, Piscataway, New Jersey 08854*
[4]*Department of Molecular Biology and Biochemistry, Rutgers University, Piscataway, New Jersey 08854*
Correspondence: epwhite@cinj.rutgers.edu

Macroautophagy (autophagy hereafter) is a process that collects cytoplasmic components, particularly mitochondria, and degrades them in lysosomes. In mammalian systems, basal autophagy levels are normally low but are profoundly stimulated by starvation and essential for survival. Cancer cells up-regulate autophagy and can be more autophagy-dependent than most normal tissues. Genetic deficiency in essential autophagy genes in tumors in many autochthonous mouse models for cancer reduces tumor growth. In K-ras^{G12D}-driven non–small cell lung cancer (NSCLC) and other models, autophagy sustains metabolism and survival. The mechanism by which autophagy promotes tumorigenesis varies in different contexts, but evidence points to a critical role for autophagy in sustaining metabolism, thereby preventing p53 activation, energy crisis, growth arrest, apoptosis, senescence, and activation of the immune response. Autophagy in NSCLC preserves mitochondrial quality and regulates their abundance. By degrading macromolecules in lysosomes, autophagy provides mitochondria with substrates to prevent energy crisis and fatal nucleotide pool depletion in starvation. We review here how autophagy supports mammalian survival and how cancer cells usurp this survival mechanism to maintain mitochondrial metabolism for their own benefit. Insights from these studies provide the rationale and approach to target the autophagy survival pathway for cancer therapy.

Autophagy captures and degrades intracellular proteins and organelles in lysosomes (Mizushima and Komatsu 2011). The autophagy pathway was first identified by yeast genetic approaches to confer survival to nitrogen starvation. Later it became apparent that newborn mice also require autophagy to survive starvation (Kuma et al. 2004; Komatsu et al. 2005). Thus, autophagy is an intracellular nutrient scavenging pathway whereby cells eat parts of themselves and presumably recycle the breakdown products, to sustain survival in starvation.

Basal levels of autophagy are normally low but are profoundly induced by stress and starvation, indicating that rapid autophagy mobilization is part of the integrated stress response (Mizushima et al. 2004). Genetic ablation of autophagy in various settings causes the accumulation of autophagy substrates, particularly damaged proteins and organelles such as abnormal mitochondria (Mizushima and Komatsu 2011). Thus, autophagy is needed for metabolic homeostasis and for protein and organelle quality and numerical control. Without autophagy, survival to stress is compromised, and cells and tissues are subjected to the toxic effects of vast accumulation of autophagy substrates. Because the levels of half the cellular proteins are affected by the functional status of autophagy, this pathway represents a major mechanism for protein turnover and homeostasis (Mathew et al. 2009, 2014).

Unlike normal cells, tumor cells can have high levels of basal autophagy and can be constitutively dependent on autophagy for survival (Degenhardt et al. 2002; Guo et al. 2011; Yang et al. 2011). Autophagy is also induced in hypoxic tumor regions and confers a survival advantage (Degenhardt et al. 2002). These findings suggest that autophagy inhibition may compromise tumor survival and may be a new approach ripe to target for cancer therapy (Amaravadi et al. 2011, 2016). Indeed, genetically engineered mouse models (GEMMs) for cancer where essential autophagy genes are ablated show the functional importance of autophagy for various aspects of tumorigenesis (Amaravadi et al. 2016). Autophagy in these GEMMs for cancer promotes the growth, survival, and malignancy of a broad spectrum of cancers that was reviewed in detail recently (Amaravadi et al. 2011). Herein we provide an overview and discuss the settings and underlying mechanisms by which autophagy promotes tumorigenesis.

AUTOPHAGY IS REQUIRED FOR MAMMALIAN SURVIVAL

First insight into the role of autophagy in sustaining mammalian survival came from studies where an essential autophagy gene was deleted in mice. These mice are

Published by Cold Spring Harbor Laboratory Press; doi: 10.1101/sqb.2016.81.030981

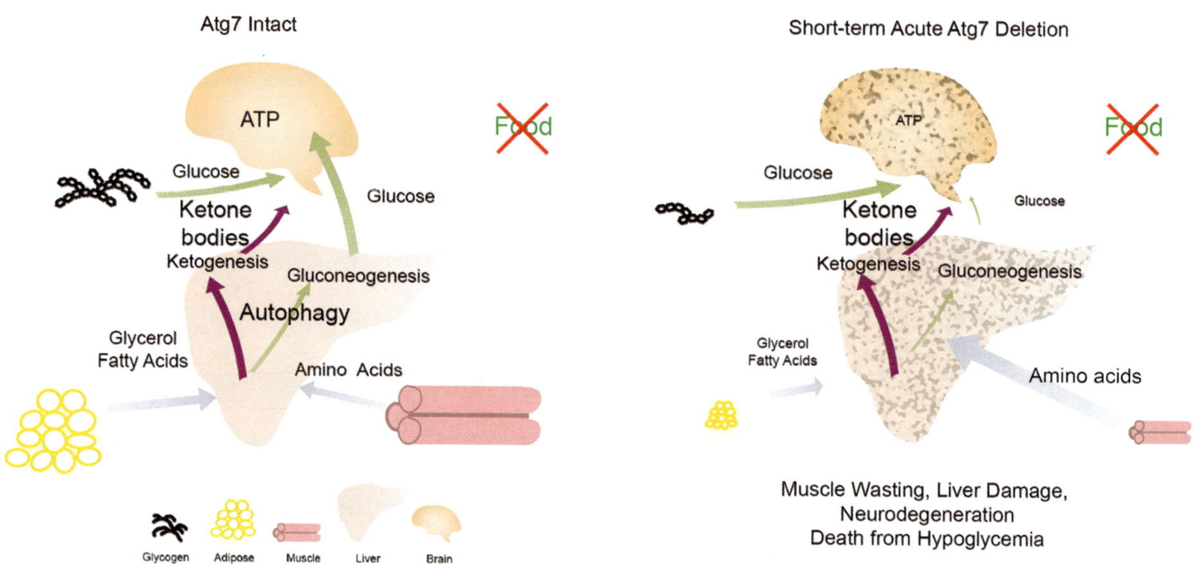

Figure 1. Mechanism by which autophagy promotes the survival of adult mice during fasting. During fasting, healthy adult mice activate autophagy and degrade glycogen to maintain circulating glucose and mobilize lipid stores to produce glycerol and fatty acids for ketogenesis and gluconeogenesis. As a last resort, muscle degradation by atrophy-associated E3 ligases may supply amino acids for ketogenesis and gluconeogenesis (Karsli-Uzunbas et al. 2014). As a result, glucose homeostasis is maintained by autophagy in mice to surviving fasting. However, fasting adult mice with acute autophagy ablation causes excessive depletion of nutrient stores including lipid and glycogen, accelerated muscle catabolism, and liver damage, which causes severe systemic hypoglycemia, brain damage, and mouse death. (Modified, with permission, from Karsli-Uzunbas et al. 2014, © AACR.)

born but fail to survive the neonatal starvation period (Kuma et al. 2004; Komatsu et al. 2005). Feeding autophagy-deficient neonatal mice extends survival only briefly indicating that autophagy is essential for mammalian survival to neonatal starvation but also generally. In contrast, adult mice subjected to conditional whole-body deletion of an essential autophagy gene live for 2–3 months, indicating that adult mice are less autophagy-dependent than neonates (Karsli-Uzunbas et al. 2014). Adult mice are physiologically and metabolically different from neonates and possess dedicated nutrient stores that may reduce autophagy dependence. Fasting these adult mice without autophagy, however, is lethal (Karsli-Uzunbas et al. 2014). Thus, the main function of autophagy conserved from yeast to mammals is to confer survival to starvation.

How does autophagy promote survival to starvation at the cellular and organismal level? During fasting, normal adult mice activate autophagy at the cellular level to sustain cell survival and they mobilize nutrient stores at the organismal level to sustain circulating nutrients essential for survival (Karsli-Uzunbas et al. 2014). For example, glycogen is degraded to generate circulating glucose (Fig. 1). Lipid in white adipose tissue (WAT) is also degraded to produce glycerol and fatty acids that are used as substrates for ketogenesis and gluconeogenesis in the liver (Fig. 1). With prolonged starvation, muscle protein is degraded to produce amino acids for use by the liver to synthesize ketone bodies and glucose. This prioritized mobilization of circulating nutrients is aimed at maintaining circulating substrates for brain metabolism, the absence of which is fatal (Fig. 1).

In contrast, adult mice without autophagy owing to conditional deletion of the essential autophagy gene *Atg7* have a systemic metabolic defect that is fatal during fasting (Karsli-Uzunbas et al. 2014). The inability to activate autophagy at the cellular level causes rapid depletion of glycogen and WAT lipid stores, and extensive muscle catabolism (cachexia) (Fig. 1). The accelerated depletion of nutrient stores causes autophagy-deficient mice to fail to maintain circulating glucose levels, resulting in death within 16 h from hypoglycemia (Karsli-Uzunbas et al. 2014). Autophagy in the liver appears necessary to sustain circulating glucose levels and brain function (Ezaki et al. 2011; Karsli-Uzunbas et al. 2014). Thus, autophagy in mammals is essential to supply circulating nutrients for systemic metabolism during fasting. The inability of cells to use autophagy may create a systemic metabolic deficit, accelerating catabolism of nutrient stores (Karsli-Uzunbas et al. 2014). Alternatively, autophagy may be directly involved in the secretion of essential metabolic substrates into the circulation, the absence of which increases dependency on nutrient stores. These findings also raise the interesting possibility that suppression of autophagy may promote cachexia, a significant medical problem for advanced cancer patients (Karsli-Uzunbas et al. 2014).

As autophagy is essential for glucose homeostasis in fasted adult mice, is there a role also in the fed state? A small fraction of the adult mice conditionally deleted for *Atg7* die soon after deletion from streptococcus infection (Karsli-Uzunbas et al. 2014). In in vitro experiments autophagy plays a role in host defense by eliminating intracellular bacteria including streptococcus, and this finding indicates that this may also be relevant in vivo. Most adult mice conditionally deleted for *Atg7* die 2–3 months postdeletion (Karsli-Uzunbas et al. 2014). This contrasts neonatal mice that fail to tolerate *Atg7* deletion (Kuma et al. 2004; Komatsu et al. 2005). Neurodegeneration is the main cause of death of *Atg7* conditionally ablated adult mice (Karsli-Uzunbas et al. 2014), which is not surprising given that autophagy deficiency targeted to the brain also causes neurodegeneration (Hara et al. 2006; Komatsu et al. 2006). What is surprising is that neurodegeneration upon conditional deletion of *Atg7* in the whole body is greatly accelerated in comparison to tissue-specific deletion (Karsli-Uzunbas et al. 2014). This suggests that autophagy deficiency in non-neuronal tissues also contributes to neurodegeneration, perhaps through defective systemic metabolism. These *Atg7*-ablated adult mice have a deficit in circulating arginine levels, and it will be of great interest to explore if this contributes to the neurodegeneration phenotype (Karsli-Uzunbas et al. 2014). Another important unanswered question is why tissues like the brain are more autophagy-dependent than other tissues, such as lung. In summary, switching off autophagy in adult mice is tolerated for weeks in the fed state.

AUTOPHAGY PROMOTES TUMORIGENESIS IN GEMMs FOR CANCER

The first clue that autophagy is important for the survival of tumor cells came from the observation that, in stark contrast to normal cells, some cancer cells have high levels of basal autophagy (Guo et al. 2011; Yang et al. 2011). Up-regulation of autophagy in cancer cells is caused by direct activation of the transcription factors of the microphthalmia-associated transcription factor (MiTF)/TFE family that control autophagy and lysosomal biogenesis or by removal of a repressive phosphorylation on the autophagy initiation machinery (Perera et al. 2015; Wong et al. 2015). Moreover, deletion of an essential autophagy gene compromises the survival of those tumor cells in vitro and in tumors in vivo (Guo et al. 2011; Yang et al. 2011). This prompted many laboratories to examine the role of autophagy in GEMMs for cancer where essential autophagy genes are deleted in tumor cells arising spontaneously in the context of a functional immune system. In most of these models, loss of autophagy reduces tumor growth, survival, and proliferation (Amaravadi et al. 2016). In models of K-ras-driven lung and pancreatic cancer, loss of autophagy limits tumor progression to benign disease, which is attributed in part to amplified activation of p53. The requirement for autophagy for tumor growth and survival is also apparent with p53 deletion. In lung cancer GEMMs, deletion of *Atg7* dramatically alters tumor pathology from carcinomas to that of benign oncocytomas, which are human tumors that massively accumulate defective mitochondria (Guo et al. 2013; Strohecker et al. 2013). As defective mitochondria are a major autophagy substrate, this indicates that benign human tumors manifest a phenotype of defective autophagy, perhaps explaining their benign status. Indeed, genomic and biological characterization of human renal oncocytomas revealed a metabolic defect stemming from pathogenic mitochondrial genome mutations and defective intracellular trafficking, lysosome, and autophagy function (Joshi et al. 2015). These benign tumors represent a natural setting where tumor progression is halted due to deficient metabolism, which can be caused by loss of mitochondrial energy homeostasis and quality control.

TUMORS ARE MORE AUTOPHAGY-DEPENDENT THAN MOST NORMAL TISSUES

These collective findings suggest that autophagy inhibition may be a therapeutic approach for some cancers (Amaravadi et al. 2016). As autophagy is important for some normal tissues, this raised the question as to whether the selective dependency of tumors on autophagy is sufficient to produce a therapeutic window. To address this, the essential autophagy gene *Atg7* was deleted conditionally in the whole body of mice with established K-ras-driven lung cancer. Conditional, systemic autophagy loss has remarkable antitumor activity, more so than tumor-specific loss (Karsli-Uzunbas et al. 2014). This suggests that host autophagy in addition to tumor cell autonomous autophagy contributes to tumor growth. Indeed, stromal cell autophagy in models of pancreatic cancer leads to secretion of alanine that is used by tumor cells as metabolic fuel (Sousa et al. 2016). Blockade of this tumor metabolism-promoting metabolic cross talk between stroma and tumor is an additional anticancer benefit of autophagy inhibition (Amaravadi et al. 2016). Whether systemic autophagy beyond stroma also contributes to tumor growth remains to be explored.

AUTOPHAGY RECYCLES MACROMOLECULES

Although the experiments above begin to address the role of autophagy in sustaining whole-body mammalian metabolism, what happens at the cellular level is still an open question. How does autophagy promote survival of cells in starvation? Is this through recycling of cellular components? Do cancer cells do this differently from normal cells and tissues? To begin to address these points we developed an assay to directly measure autophagy-mediated recycling of macromolecules into cancer cell metabolism to find out whether recycling occurs, and if so, what it is used for.

Cancer cells derived directly from the tumors arising in the GEMMs for lung cancer are autophagy dependent as isogenic *Atg7*-deficient tumor cells are more sensitive to starvation than *Atg7* wild-type tumor cells (Guo et al. 2013; Strohecker et al. 2013). To address why autophagy deficiency causes sensitivity to starvation, individual nutrients were added back to determine which were essential for survival. We found that glutamine supple-

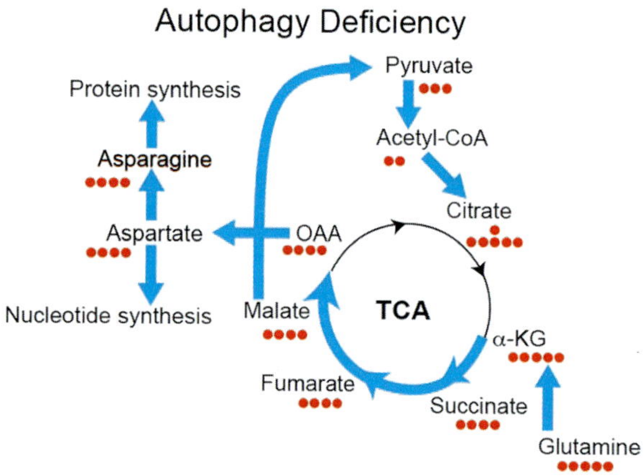

Figure 2. Defective autophagy in tumor cells alters metabolic flux from glutamine to aspartate and to de novo nucleotide synthesis. K-ras-driven tumor cells depend on autophagy to supply the substrate glutamine and its derivatives to sustain mitochondria metabolism for starvation survival. To compensate, *Atg7*-deficient tumor cells significantly increase glutamine flux to the tricarboxylic acid (TCA) cycle, de novo nucleotide synthesis, and nicotinamide adenine dinucleotide phosphate (NADPH) generation by up-regulating flux from malate, pyruvate, acetyl-CoA, and citrate for antioxidant defense (Guo et al. 2016). OAA, oxaloacetate; α-KG, α-ketogluturate.

mentation in starvation conditions is most effective at rescuing the survival of autophagy-deficient tumor cells (Guo et al. 2013; Strohecker et al. 2013). Because glutamine is a major anapleurotic substrate for the tricarboxylic acid (TCA) cycle (Zong et al. 2016), this suggests that autophagy may recycle intracellular macromolecules to provide amino acid substrates such as glutamine to maintain mitochondrial function (Fig. 2). To test this hypothesis, macromolecules of *Atg7* wild-type and deficient tumor cells were stably labeled with heavy isotopes of carbon and nitrogen. The label was then briefly chased out of metabolic pathways with cold media, and then the cells were challenged with starvation to force autophagy-mediated recycling specifically in the *Atg7* competent cells and not in the *Atg7*-deficient cells. Reappearance of heavy isotope labeled metabolites into central carbon metabolism was then assessed by liquid chromatography and mass spectrometry (Guo et al. 2016).

Upon challenge with starvation, labeled metabolites were enriched in the *Atg7* competent compared with deficient tumor cells, particularly in TCA cycle intermediates and amino acids synthesized in mitochondria (Guo et al. 2016). What surprised us was that labeling of metabolism was also apparent in the *Atg7*-deficient tumor cells, although it was significantly less than *Atg7* wild-type cells. This indicates that autophagy does recycle macromolecules in starvation, that this is specific to mitochondrial metabolism, but that there also exists autophagy-independent mechanisms for recycling (Guo et al. 2016). What those autophagy-independent mechanisms of recycling are is not known but proteasome-mediated or other forms of lysosome-mediated degradation may be responsible.

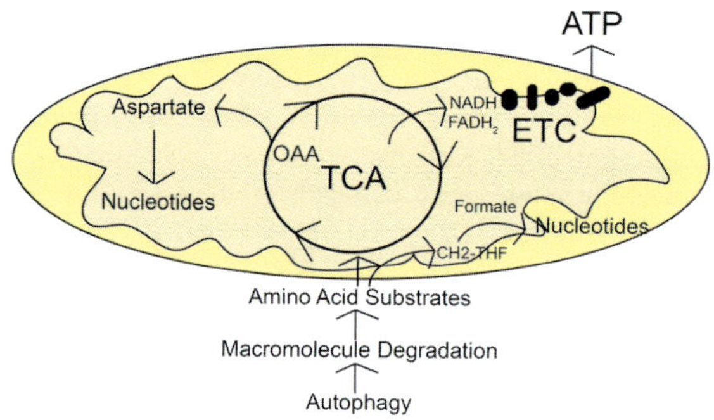

Figure 3. Autophagy recycles macromolecules to provide mitochondrial substrates for nucleotide synthesis and energy homeostasis. Tumor cells rely on mitochondria metabolism for their proliferation and tumorigenesis. In starvation, autophagy is essential to mediate substrate recycling to supply tricarboxylic acid (TCA)-associated amino acids, which are further metabolized through the TCA cycle for energy production and de novo nucleotide biosynthesis. OAA, oxaloacetate; ETC, electron transport chain.

AUTOPHAGY-MEDIATED RECYCLING PROMOTES CANCER METABOLISM

If autophagy is degrading macromolecules in starvation to supply substrates for mitochondrial metabolism, what are the substrates used for? Isotope tracing helped answer this question. Carbon flux from glutamine is accelerated in *Atg7*-deficient tumor cells in starvation where it flows from the TCA cycle to pyruvate and to aspartate (Fig. 3). Using glutamine carbon to synthesize pyruvate can produce nicotinamide adenine dinucleotide phosphate (NADPH) via malic enzyme for use in generating reduced glutathione for antioxidant defense (Guo et al. 2016). Aspartate is a major biosynthetic product of oxaloacetate (OAA) from the TCA cycle and is a precursor for de novo nucleotide biosynthesis (Guo et al. 2016). Finally, the general supply of substrates to mitochondria through autophagy can drive energy production (Guo et al. 2016). These findings pointed to a role for autophagy-supplied mitochondrial substrates in redox homeostasis, nucleotide biosynthesis, and energy homeostasis (Fig. 4).

Atg7 deficiency increases reactive oxygen species (ROS), which is extinguished by glutamine supplementation in starvation (Guo et al. 2016). Scavenging ROS, however, fails to rescue survival of starved *Atg7*-deficient

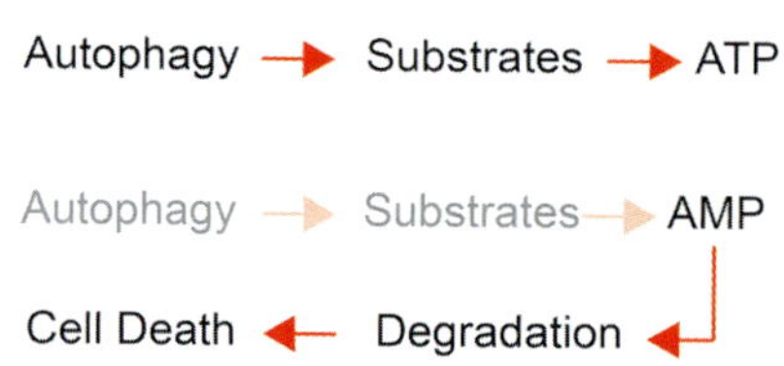

Figure 4. Autophagy-supplied substrates prevent energy crisis and fatal nucleotide depletion. Substrates derived from autophagy are required to generate ATP via tricarboxylic acid (TCA) cycle–driven oxidative phosphorylation in starvation. In the absence of autophagy, defective de novo nucleotide synthesis and increased nucleotide degradation cause nucleotide pool depletion and cell death.

tumor cells. Thus, autophagy-supplied substrates function in maintaining redox homeostasis, although restoring redox balance alone is insufficient to rescue survival in starvation.

Starving *Atg7*-deficient tumor cells severely depletes nucleotide pools, particularly adenosine nucleotides, consistent with a requirement for autophagy to supply substrates for aspartate and de novo nucleotide synthesis (Guo et al. 2016). Supplementation of autophagy-deficient tumor cells with nucleosides during starvation rescues survival, indicating that autophagy prevents fatal nucleotide pool depletion. In starvation, autophagy also prevents loss of mitochondrial respiration, ATP loss and AMP accumulation, and AMP kinase activation (Guo et al. 2016). Thus, a major function of autophagy is to supply substrates for mitochondrial metabolism to maintain energy homeostasis.

The depletion of nucleotide pools and energy crisis in starved autophagy-deficient tumor cells may be related. In yeast, autophagy loss also causes AMP accumulation and energy crisis in starvation, triggering AMP degradation to escape energy crisis in the short term (Walther et al. 2010). In tumor cells, as in yeast, the requirement for autophagy to supply substrates for de novo nucleotide synthesis in starvation may also be caused by energy crisis and AMP degradation as a means to escape energy crisis (Fig. 4). Indeed, nucleotide degradation products accumulate to high levels in starved *Atg7*-deficient but not *Atg7* wild-type tumor cells (Guo et al. 2016). Autophagy therefore becomes essential to provide metabolic substrates to prevent energy crisis upon nutrient deprivation and thereby the subsequent nucleotide pool depletion.

CONCLUSION

From the study of genetically engineered mice, it is clear that autophagy is important for systemic adult mammalian metabolism. During fed conditions, autophagy is less critical, conferring loss of WAT, select tissue damage, susceptibility to infection, and neurodegeneration after 2–3 months. Autophagy is, however, critical to prevent fatal hypoglycemia during fasting. We discovered that autophagy is a component of nutrient sharing among tissues, as rapid depletion of WAT and glycogen in mice with conditional loss of *Atg7* is also associated with excessive muscle catabolism. The metabolic defect at the cellular level creates a systemic metabolic deficit, accelerating exhaustion of nutrient stores and compensatory destruction of muscle, which is ultimately fatal.

Many tumors up-regulate autophagy transcriptionally and posttranscriptionally and depend on it for macromolecular recycling and survival, a requirement that is particularly acute in stress in the tumor microenvironment and during starvation. Indeed, the antitumor activity of caloric restriction is enhanced by deficiency in autophagy (Lashinger et al. 2016). In K-ras-driven non–small cell lung cancer, macromolecular recycling by autophagy suppresses oxidative stress, but preventing energy crisis and nucleotide pool depletion are essential for survival to stress. As some tumors are more autophagy-dependent than most normal tissues, it will be important to determine the therapeutic utility of autophagy inhibition in cancer therapy (Amaravadi et al. 2016).

DISCLOSURE OF POTENTIAL CONFLICTS OF INTEREST

E.W. is on the Scientific Advisory Board of Forma Therapeutics.

ACKNOWLEDGMENTS

J.Y.G. is supported by the National Institutes of Health (NIH) award K22 CA190521. E.W. is supported by the NIH under award numbers R01CA163591, R01CA130893, R01CA188096, and R01CA193970, the Robert Wood Johnson Foundation, and P30CA72720 to the Rutgers Cancer Institute of New Jersey.

REFERENCES

Amaravadi RK, Lippincott-Schwartz J, Yin XM, Weiss WA, Takebe N, Timmer W, DiPaola RS, Lotze MT, White E. 2011. Principles and current strategies for targeting autophagy for cancer treatment. *Clin Cancer Res* **17:** 654–666.

Amaravadi R, Kimmelman AC, White E. 2016. Recent insights into the function of autophagy in cancer. *Genes Dev* **30:** 1913–1930.

Degenhardt K, Chen G, Lindsten T, White E. 2002. BAX and BAK mediate p53-independent suppression of tumorigenesis. *Cancer Cell* **2:** 193–203.

Ezaki J, Matsumoto N, Takeda-Ezaki M, Komatsu M, Takahashi K, Hiraoka Y, Taka H, Fujimura T, Takehana K, Yoshida M, et al. 2011. Liver autophagy contributes to the maintenance of blood glucose and amino acid levels. *Autophagy* **7:** 727–736.

Guo JY, Chen HY, Mathew R, Fan J, Strohecker AM, Karsli-Uzunbas G, Kamphorst JJ, Chen G, Lemons JM, Karantza V, et al. 2011. Activated Ras requires autophagy to maintain oxidative metabolism and tumorigenesis. *Genes Dev* **25:** 460–470.

Guo JY, Karsli-Uzunbas G, Mathew R, Aisner SC, Kamphorst JJ, Strohecker AM, Chen G, Price S, Lu W, Teng X, et al. 2013. Autophagy suppresses progression of K-ras-induced lung tumors to oncocytomas and maintains lipid homeostasis. *Genes Dev* **27:** 1447–1461.

Guo JY, Teng X, Laddha SV, Ma S, Van Nostrand SC, Yang Y, Khor S, Chan CS, Rabinowitz JD, White E. 2016. Autophagy provides metabolic substrates to maintain energy charge and nucleotide pools in Ras-driven lung cancer cells. *Genes Dev* **30:** 1704–1717.

Hara T, Nakamura K, Matsui M, Yamamoto A, Nakahara Y, Suzuki-Migishima R, Yokoyama M, Mishima K, Saito I, Okano H, et al. 2006. Suppression of basal autophagy in neural cells causes neurodegenerative disease in mice. *Nature* **441:** 885–889.

Joshi S, Tolkunov D, Aviv H, Hakimi AA, Yao M, Hsieh JJ, Ganesan S, Chan CS, White E. 2015. The genomic landscape of renal oncocytoma identifies a metabolic barrier to tumorigenesis. *Cell Rep* **13:** 1895–1908.

Karsli-Uzunbas G, Guo JY, Price S, Teng X, Laddha SV, Khor S, Kalaany NY, Jacks T, Chan CS, Rabinowitz JD, et al. 2014. Autophagy is required for glucose homeostasis and lung tumor maintenance. *Cancer Discov* **4:** 914–927.

Komatsu M, Waguri S, Ueno T, Iwata J, Murata S, Tanida I, Ezaki J, Mizushima N, Ohsumi Y, Uchiyama Y, et al. 2005. Impairment of starvation-induced and constitutive autophagy in Atg7-deficient mice. *J Cell Biol* **169:** 425–434.

Komatsu M, Waguri S, Chiba T, Murata S, Iwata J, Tanida I, Ueno T, Koike M, Uchiyama Y, Kominami E, et al. 2006. Loss of autophagy in the central nervous system causes neurodegeneration in mice. *Nature* **441:** 880–884.

Kuma A, Hatano M, Matsui M, Yamamoto A, Nakaya H, Yoshimori T, Ohsumi Y, Tokuhisa T, Mizushima N. 2004. The role of autophagy during the early neonatal starvation period. *Nature* **432:** 1032–1036.

Lashinger LM, O'Flanagan CH, Dunlap SM, Rasmussen AJ, Sweeney S, Guo JY, Lodi A, Tiziani S, White E, Hursting SD. 2016. Starving cancer from the outside and inside: Separate and combined effects of calorie restriction and autophagy inhibition on Ras-driven tumors. *Cancer Metab* **4:** 18.

Mathew R, Karp CM, Beaudoin B, Vuong N, Chen G, Chen HY, Bray K, Reddy A, Bhanot G, Gelinas C, et al. 2009. Autophagy suppresses tumorigenesis through elimination of p62. *Cell* **137:** 1062–1075.

Mathew R, Khor S, Hackett SR, Rabinowitz JD, Perlman DH, White E. 2014. Functional role of autophagy-mediated proteome remodeling in cell survival signaling and innate immunity. *Mol Cell* **55:** 916–930.

Mizushima N, Komatsu M. 2011. Autophagy: Renovation of cells and tissues. *Cell* **147:** 728–741.

Mizushima N, Yamamoto A, Matsui M, Yoshimori T, Ohsumi Y. 2004. In vivo analysis of autophagy in response to nutrient starvation using transgenic mice expressing a fluorescent autophagosome marker. *Mol Biol Cell* **15:** 1101–1111.

Perera RM, Stoykova S, Nicolay BN, Ross KN, Fitamant J, Boukhali M, Lengrand J, Deshpande V, Selig MK, Ferrone CR, et al. 2015. Transcriptional control of autophagy-lysosome function drives pancreatic cancer metabolism. *Nature* **524:** 361–365.

Sousa CM, Biancur DE, Wang X, Halbrook CJ, Sherman MH, Zhang L, Kremer D, Hwang RF, Witkiewicz AK, Ying H, et al. 2016. Pancreatic stellate cells support tumour metabolism through autophagic alanine secretion. *Nature* **536:** 479–483.

Strohecker AM, Guo JY, Karsli-Uzunbas G, Price SM, Chen GJ, Mathew R, McMahon M, White E. 2013. Autophagy sustains mitochondrial glutamine metabolism and growth of BrafV600E-driven lung tumors. *Cancer Discov* **3:** 1272–1285.

Walther T, Novo M, Rossger K, Letisse F, Loret MO, Portais JC, Francois JM. 2010. Control of ATP homeostasis during the respiro-fermentative transition in yeast. *Mol Syst Biol* **6:** 344.

Wong PM, Feng Y, Wang J, Shi R, Jiang X. 2015. Regulation of autophagy by coordinated action of mTORC1 and protein phosphatase 2A. *Nat Commun* **6:** 8048.

Yang S, Wang X, Contino G, Liesa M, Sahin E, Ying H, Bause A, Li Y, Stommel JM, Dell'Antonio G, et al. 2011. Pancreatic cancers require autophagy for tumor growth. *Genes Dev* **25:** 717–729.

Zong WX, Rabinowitz JD, White E. 2016. Mitochondria and cancer. *Mol Cell* **61:** 667–676.

A Time for MYC: Metabolism and Therapy

CHI V. DANG
Abramson Cancer Center, Perelman School of Medicine, University of Pennsylvania, Philadelphia, Pennsylvania 19104
Correspondence: dangvchi@upenn.edu

The MYC oncogene is frequently deregulated in human cancers, whereas the proto-oncogene is exquisitely, tightly regulated in normal cells. Deregulated MYC drives transcriptional imbalance, thereby altering metabolism and disrupting the circadian Bmal1-Clock E-box-dependent transcriptional circuitry. Sustained oncogenic MYC expression drives a constitutive growth program with mammalian target of rapamycin (mTOR) activation that renders cells dependent on nutrients, such that glucose or glutamine deprivation could trigger cell death and key enzymes such as lactate dehydrogenase A (LDHA) and glutaminase (GLS) amenable for targeting in cancers. Further, MYC-mediated suppression of the circadian clock is surmised to suspend the inhibitory effect of Bmal1-Clock on metabolism, allowing for MYC-driven cancer cells to reach a higher state of anabolic metabolism. Hence, metabolic therapy could be deployed, particularly at specific times of the day, to diminish side effects to normal tissues while maintaining antitumor efficacy.

Signals, which stimulate cell growth leading to cell division, emanate from growth factor receptor signaling and nutrient sensors that together amplify growth signals. Growth factor and nutrient sensing signaling stimulate mammalian target of rapamycin (mTOR) and MYC expression (Laplante and Sabatini 2012; Stine et al. 2015), which together transmit growth signals by amplifying translation and transcription to generate transcripts that are in turn translated to build components of the growing cell (Fig. 1). MYC stimulates transcription by binding to open chromatin and largely through releasing paused RNA polymerases (Kress et al. 2015; Wolf et al. 2015). As such, "housekeeping" metabolic genes and genes involved in mitochondrial function and ribosome biogenesis are accessible by MYC, which readily activates these genes through E-boxes in open chromatin found in virtually all nucleated cells that universally use core metabolic pathways such as glycolysis and glutaminolysis (Fig. 2; Dang 2012, 2013). Distinct from physiologic levels of MYC, a high oncogenic level of MYC is surmised to bind lower-affinity DNA binding sites, such as those bound by other E-box binding transcription factors (Clock-Bmal1, Srebp, or hypoxia-inducible factors [HIFs]), and bind enhancer sequences, resulting in nonlinear amplification of transcripts (Fig. 2; Dang 2012). The imbalanced transcriptional program is likely to result in nonstoichiometric production of RNAs or proteins in multicomponent complexes and protein overload and hence is responsible for integrated stress responses in oncogenic MYC-driven cells (Hart et al. 2012; Cunningham et al. 2014). This stress condition also provides a therapeutic window.

Hence, the MYC transcriptional response program is characterized by metabolic reprogramming that is shown experimentally through isotopic tracer studies using labeled glucose or glutamine. In the model of human Burkitt lymphoma with an inducible MYC gene, both glycolysis and glutaminolysis were stimulated by MYC that also sustains glutamine metabolism under hypoxic conditions (Fig. 3; Le et al. 2012; Murphy et al. 2013). These observations suggest that pathways involved in glucose and glutamine metabolism could be targeted for cancer therapy (Fig. 4). Indeed, proof-of-concept studies provided evidence that inhibition of lactate dehydrogenase, the key enzyme involved in the Warburg effect, and glutaminase could delay tumorigenesis (Le et al. 2010; Xie et al. 2014; Xiang et al. 2015). However, use of a single metabolic therapeutic agent can result in metabolic rewiring and subsequent progression of resistant disease (Yuneva et al. 2012; Davidson et al. 2016). As such, deeper understanding of metabolic rewiring driven by specific oncogenic drivers (MYC, MET, p53, PTEN, PI3 K, Ras, and Akt) in specific tissues is essential for the future use of metabolic therapy, which is likely most effective in strategic combination therapies.

We surmise that high levels of oncogenic MYC could perturb the circadian clock by MYC's invasion of clock driven E-boxes (Fig. 5; Altman et al. 2015). The molecular clock is driven by the central clock Clock-Bmal1 transcriptional heterodimer, which regulates clock-controlled genes as well as induces two negative feedback loops that provide the oscillatory activity of Clock-Bmal1. One feedback loop involves posttranscriptional regulation by the Clock-Bmal1 target genes CRYs and PERs. The other involves the induction of REV-ERBs, which transcriptionally repress the expression of Bmal1. Using an inducible MYC system in a U2OS cell line that bears the Bmal1 promoter-luciferase reporter, we document the ability of MYC to disrupt and suppress Bmal1 expression (Fig. 5).

Published by Cold Spring Harbor Laboratory Press; doi: 10.1101/sqb.2016.81.031153

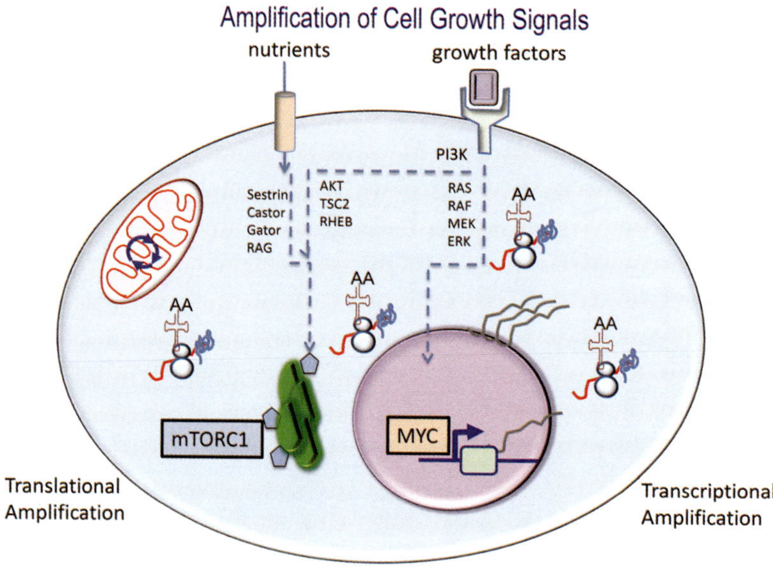

Figure 1. Amplification of cell growth signals. The diagram depicts the dependency of growing mammalian cells on nutrients, which are imported by transporters, and growth factors, which are sensed by growth factor receptors. Nutrients, particularly amino acids, are sensed along with growth factor signaling through tuberous sclerosis complex 2 (TSC2) and AKT resulting in the activation of mammalian target of rapamycin complex 1 (mTORC1), which increases the cell's protein synthetic capacity resulting in translational amplification of growth and nutrient signaling. On the other hand, growth factor signaling through phosphoinositide 3-kinase (PI3K), RAS–RAF–MEK–ERK pathways together with nutrient sensing triggers transcriptional amplification of growth through activation of MYC, a key transcriptional factor for cell growth.

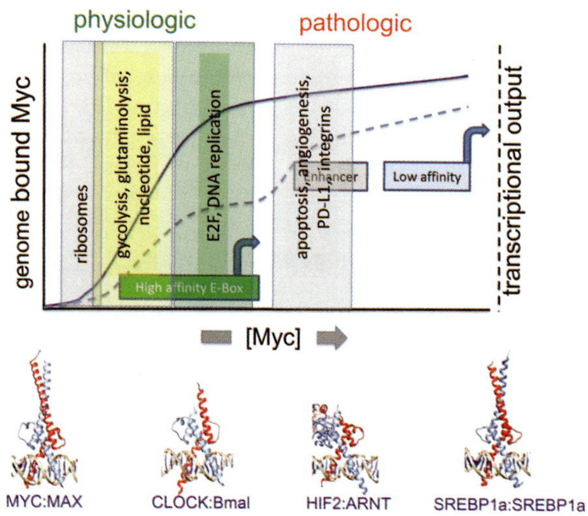

Figure 2. A graph illustrating that increased MYC levels result in the binding of high-affinity E-box sites and activation of "housekeeping genes" such as those involved in ribosome biogenesis or central metabolic pathways (glycolysis, glutaminolysis). With cell growth reaching a critical size, MYC collaborates with E2F transcription factors for cell entry in S phase for DNA replication. When MYC is deregulated, it could ectopically bind enhancer sequences and low-affinity sites, which result in imbalanced transcriptional amplification triggering apoptosis, angiogenesis, or activation of immune checkpoints through specific "ectopic" target gene expression. The low-affinity sites may consist of E-boxes that are targets of other E-box binding transcription factors shown at the *bottom* of the figure.

These studies show that E-box-driven negative regulators (Crys, Pers, and Rev-erbs) were all activated by ectopic MYC, whereas Bmal1 expression was diminished in a fashion that could be partially rescued by knocking down Rev-erb expression. Subsequent to our findings, a report corroborated that MYC could suppress Bmal1 expression and the clock, but it implicates a role for direct suppression of Clock and Bmal1 expression via Miz1 (Shostak et al. 2016). Hence, MYC consistently suspends the clock, but this is likely to involve a direct Miz1-dependent suppression and an indirect suppression of Bmal1 via MYC induction of Rev-erbs.

Based on the observation that MYC could suspend the clock and drive anabolic metabolism, we hypothesize that effective anticancer metabolic therapy may depend on the time of drug administration during the day—particularly for short-lived drugs (Fig. 6). We hypothesize that toxicity to normal tissues, which have oscillatory metabolic gene expression and circadian metabolism, would be reduced depending on the time of day of drug administration, whereas efficacy against MYC-driven tumors would be retained because the tumors have constitutive high expression of the target gene in question such as NAMPT, which is a target gene of both Clock-Bmal1 in normal tissues and oncogenic MYC in MYC-driven cancers. As such, we examine the levels on NAD^+ and expression of NAMPT, a rate-limiting enzyme for NAD^+ synthesis, and found that ectopic MYC induced a sustained expression of

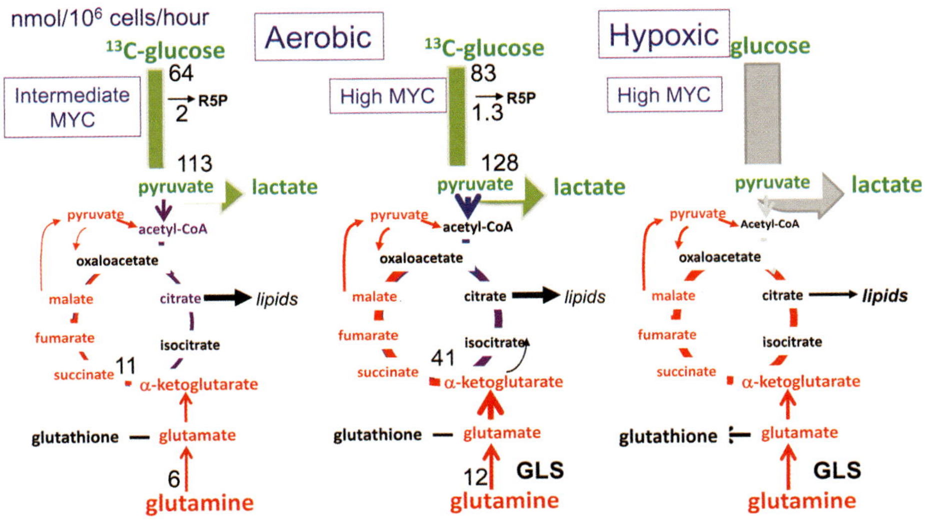

Figure 3. The cartoon summarizes the flux of metabolites through glycolysis and the TCA cycle as determined by isotopic labeling of a MYC-inducible human B-cell lymphoma model. The two metabolic diagrams on the *left* illustrate the findings of Murphy et al. (2013) comparing intermediate MYC expression under a nontumorigenic state with ectopic high MYC expression under a tumorigenic state. Note that the thickness of the arrows approximates flux, which is shown as numbers with values of nmol/10^6 cells/h. Although ectopic high MYC increased glycolytic flux by 20%, the flux through the TCA cycle increased by 400%. The metabolic diagram on the *right* summarizes the findings of Le et al. (2012) indicating that hypoxic high MYC expressing B lymphocytes had increased production of lactate from glucose and continued glutamine flux into the TCA cycle as well as increased glutathione production under hypoxia.

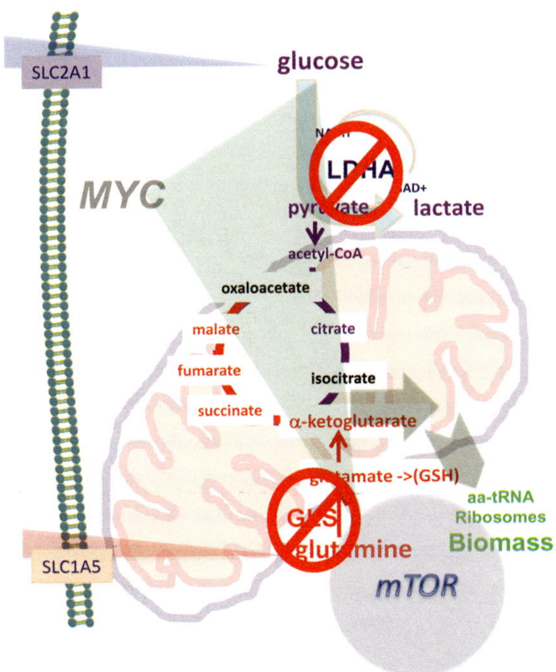

Figure 4. The diagram depicts the ability of ectopic MYC expression to drive the expression of genes involved in glucose and glutamine transport, glycolysis, glutaminolysis, and protein synthesis (aa-tRNA, aminoacyl-tRNAs). Together with mammalian target of rapamycin (mTOR), MYC drives deregulated cell growth rendering cells dependent on key metabolic enzymes such as lactate dehydrogenase A (LDHA) and glutaminase (GLS), which are marked as being targeted for proof of concept that metabolic inhibitors could curb tumor growth.

NAMPT accompanied by high levels of intracellular NAD^+. To assess the potential importance of NAD^+ dysregulation by MYC, we studied a potent inhibitor of NAMPT, FK866, that is effective against in a xenograft model of human Burkitt lymphoma (Le et al. 2010). Because NAMPT is expressed with remarkable circadian oscillation with nadirs in the morning and peaks in early evening in mouse livers, we empirically chose to treat animals at 10 a.m. or 6 p.m. FK866 has been under clinical development and clinical trials uncovered a dose-limiting toxicity of low platelet counts or thrombocytopenia in patients (Sampath et al. 2015). As such, we hypothesized that FK866 given at two different times of the day would result in similar efficacy but distinctly difference toxicities. Indeed, we found that administration of FK866 at 6 p.m., but not 10 a.m., was associated with thrombocytopenia, but antitumor efficacy was indistinguishable. These observations suggest that suspension of the clock by MYC could provide a new therapeutic window for timed metabolic therapies and those metabolic therapies could be affected by the circadian biology of normal tissues and cancer cells.

ACKNOWLEDGMENTS

The work described herein was all performed by many members of the Dang lab and supported by the National Cancer Institute, Leukemia Lymphoma Society, and Stand-Up-to-Cancer. This short overview is a summary of the presentation given at the 81st CSHL Quantitative Biology Meeting and is not meant to be an exhaustive

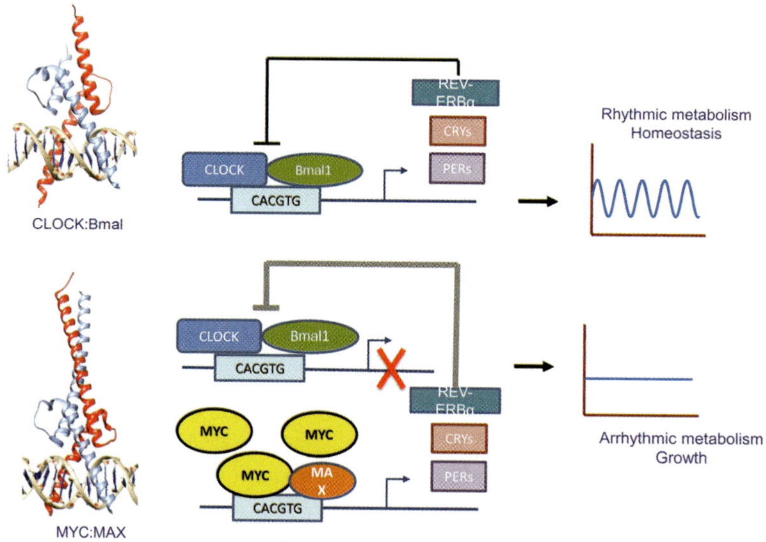

Figure 5. The cartoons illustrate the circadian transcriptional clock machinery (*top*) and its putative disruption by ectopic MYC expression (*bottom*). The central transcription factor Clock-Bmal1 (*top*) induces two negative feedback loops via the induction of CRYs and PERs (posttranscriptional negative feedback) and REV-REVs (transcriptional negative feedback), which create the oscillatory activity of Clock-Bmal1 that drives circadian expression of target metabolic genes. Ectopic MYC (*bottom*) is depicted to bind Clock-driven E-boxes to activate the negative regulators constitutively that resulted in continuous suppression of Bmal1 expression and inhibit circadian expression of Clock target genes. Crystal structures of Clock-Bmal1 and Myc-Max are shown on the *left*, illustrating the structural similarities of these helix–loop–helix transcription factors and their binding to E-boxes (5′-CACGTG-3′).

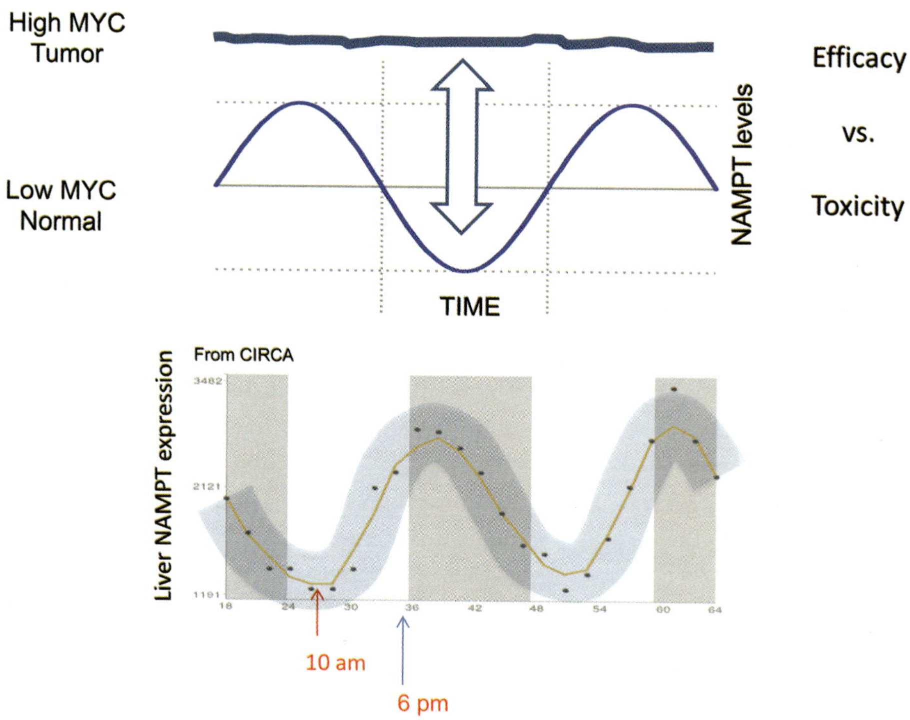

Figure 6. The figure illustrates the hypothesis that MYC deregulation of metabolic genes that are regulated normally by the Clock-Bmal1 transcription factor, such as nicotinamide phosphoribosyltransferase (NAMPT), could open up a therapy window for metabolic cancer therapy. It is surmised that toxicity to normal tissues, which have oscillatory metabolic gene expression and circadian metabolism, would be reduced depending on the time of day of drug administration, whereas efficacy against MYC-driven tumors would be retained because the tumors have constitutive high expression of the target gene in question such as NAMPT, which is a target gene of both Clock-Bmal1 (in normal tissues, such as liver shown at the *bottom*) and oncogenic MYC.

review of the literature, which is enriched by many papers not cited here.

REFERENCES

Altman BJ, Hsieh AL, Sengupta A, Krishnanaiah SY, Stine ZE, Walton ZE, Gouw AM, Venkataraman A, Li B, Goraksha-Hicks P, et al. 2015. MYC disrupts the circadian clock and metabolism in cancer cells. *Cell Metab* **22:** 1009–1019.

Cunningham JT, Moreno MV, Lodi A, Ronen SM, Ruggero D. 2014. Protein and nucleotide biosynthesis are coupled by a single rate-limiting enzyme, PRPS2, to drive cancer. *Cell* **157:** 1088–1103.

Dang CV. 2012. MYC on the path to cancer. *Cell* **149:** 22–35.

Dang CV. 2013. MYC, metabolism, cell growth, and tumorigenesis. *Cold Spring Harb Perspect Med* **3:** a014217.

Davidson SM, Papagiannakopoulos T, Olenchock BA, Heyman JE, Keibler MA, Luengo A, Bauer MR, Jha AK, O'Brien JP, Pierce KA, et al. 2016. Environment impacts the metabolic dependencies of Ras-driven non-small cell lung cancer. *Cell Metab* **23:** 517–528.

Hart LS, Cunningham JT, Datta T, Dey S, Tameire F, Lehman SL, Qiu B, Zhang H, Cerniglia G, Bi M, et al. 2012. ER stress-mediated autophagy promotes Myc-dependent transformation and tumor growth. *J Clin Invest* **122:** 4621–4634.

Kress TR, Sabò A, Amati B. 2015. MYC: Connecting selective transcriptional control to global RNA production. *Nat Rev Cancer* **15:** 593–607.

Laplante M, Sabatini DM. 2012. mTOR signaling in growth control and disease. *Cell* **149:** 274–293.

Le A, Cooper CR, Gouw AM, Dinavahi R, Maitra A, Deck LM, Royer RE, Vander Jagt DL, Semenza GL, Dang CV. 2010. Inhibition of lactate dehydrogenase A induces oxidative stress and inhibits tumor progression. *Proc Natl Acad Sci* **107:** 2037–2042.

Le A, Lane AN, Hamaker M, Bose S, Gouw A, Barbi J, Tsukamoto T, Rojas CJ, Slusher BS, Zhang H, et al. 2012. Glucose-independent glutamine metabolism via TCA cycling for proliferation and survival in B cells. *Cell Metab* **15:** 110–121.

Murphy TA, Dang CV, Young JD. 2013. Isotopically nonstationary ^{13}C flux analysis of Myc-induced metabolic reprogramming in B-cells. *Metab Eng* **15:** 206–217.

Sampath D, Zabka TS, Misner DL, O'Brien T, Dragovich PS. 2015. Inhibition of nicotinamide phosphoribosyltransferase (NAMPT) as a therapeutic strategy in cancer. *Pharmacol Ther* **151:** 16–31.

Shostak A, Ruppert B, Ha N, Bruns P, Toprak UH; ICGC MMML-Seq Project, Eils R, Schlesner M, Diernfellner A, Brunner M. 2016. MYC/MIZ1-dependent gene repression inversely coordinates the circadian clock with cell cycle and proliferation. *Nat Commun* **7:** 11807.

Stine ZE, Walton ZE, Altman BJ, Hsieh AL, Dang CV. 2015. MYC, metabolism, and cancer. *Cancer Discov* **5:** 1024–1039.

Wolf E, Lin CY, Eilers M, Levens DL. 2015. Taming of the beast: Shaping Myc-dependent amplification. *Trends Cell Biol* **25:** 241–248.

Xiang Y, Stine ZE, Xia J, Lu Y, O'Connor RS, Altman BJ, Hsieh AL, Gouw AM, Thomas AG, Gao P, et al. 2015. Targeted inhibition of tumor-specific glutaminase diminishes cell-autonomous tumorigenesis. *J Clin Invest* **125:** 2293–2306.

Xie H, Hanai J, Ren JG, Kats L, Burgess K, Bhargava P, Signoretti S, Billiard J, Duffy KJ, Grant A, et al. 2014. Targeting lactate dehydrogenase A inhibits tumorigenesis and tumor progression in mouse models of lung cancer and impacts tumor-initiating cells. *Cell Metab* **19:** 795–809.

Beyond the Oncogene Revolution: Four New Ways to Combat Cancer

THORSTEN BERGER, MARY E. SAUNDERS, AND TAK W. MAK
The Campbell Family Institute for Breast Cancer Research and Ontario Cancer Institute, University Health Network, Toronto, Ontario M5G 2M9, Canada
Correspondence: tmak@uhnres.utoronto.ca

It has become clear that tumorigenesis results from much more than just the activation of an oncogene and/or the inactivation of a tumor-suppressor gene, and that the cancer cell genome contains many more alterations than can be specifically targeted at once. This observation has led our group to a search for alternative ways to kill cancer cells (while sparing normal cells) by focusing on properties unique to the former. We have identified four approaches with the potential to generate new anticancer therapies: combatting the tactics by which cancers evade antitumor immune responses, targeting metabolic adaptations that tumor cells use to survive conditions that would kill normal cells, manipulating a cancer cell's response to excessive oxidative stress, and exploiting aneuploidy. This review describes our progress to date on these fronts.

Work over the past several years has made it clear that numerous alterations are involved in the malignant transformation of a normal cell into a cancer cell, much more than the mere activation of oncogenes and inactivation of tumor-suppressor genes. It is therefore too difficult to treat cancer by applying one or more agents each designed to target a specific mutational change. Instead, our group (and others) has investigated several promising novel approaches that target an aspect of the overall transformed state of a cancer cell. These aspects of transformation include the ability of tumors to evade immune responses mounted against them, the altered metabolic pathways tumors use to thrive under conditions that would kill normal cells, and the unusual mechanisms cancer cells use to maintain genomic stability in the face of hurdles such as excessive oxidative stress and aneuploidy. Manipulation of these altered cellular states may open up exciting new avenues for cancer therapy that do not depend on the targeting of an individual oncogene or tumor-suppressor gene. This review summarizes some of our work on the exploration of these intriguing new possibilities for cancer therapies.

COMBATING TUMOR EVASION TACTICS

Immunotherapy has become one of the most promising new strategies for fighting cancer. Although the immune system works every day to prevent incipient tumor cells from establishing, clearly the system fails often enough that cancers develop. Moreover, tumor cells deploy numerous tactics to evade and neutralize the immune responses they provoke. Thus, cancer immunologists have devised ways to stimulate specific components of the immune system to increase cancer cell killing and to reverse the evasive actions of cancer cells, rendering them once more vulnerable to immune attack.

Unleashing Antitumor T-Cell Responses

Cancers frequently express non-self-antigens that are recognized by a host's immune system. The binding of a complex containing a major histocompatibility complex (MHC) molecule bound to antigenic peptide (in this context, derived from the tumor) to a T cell's T-cell receptors (TCRs) is the first step in activating that T cell. The second step is co-stimulation in the form of binding between the T cell's CD28 molecules and B7 molecules expressed by the antigen-presenting cell (APC) bearing the peptide–MHC complex. Only upon successful co-stimulation is full T-cell activation achieved and an immune response mounted. However, to prevent excessive damage to normal tissues, any T-cell response is naturally limited in duration and magnitude by negative regulation exerted by other molecules up-regulated on the surface of activated T cells, including PD-1 and CTLA-4. Both PD-1 and CTLA-4 bind to B7 molecules with much higher affinity than does CD28, interrupting the co-stimulation mediated by this molecule and consequently shutting down the proliferation and IL-2 secretion of activated T cells. Although helpful in normal tissues, in the case of cancer, this PD-1- or CTLA-4-mediated inhibition may turn off an antitumor immune response before it has completed its job of eliminating cancer cells. Thus, researchers have sought a means of neutralizing this negative regulation so as to sustain anticancer immune responses and maintain the attack on the tumor.

Published by Cold Spring Harbor Laboratory Press; doi: 10.1101/sqb.2016.81.031161

Several groups have independently developed monoclonal antibodies that block PD-1 or CTLA-4 from binding to CD28, creating an "immune blockade" that prevents the shutting down of the antitumor T-cell response and allows T cells to continue their assault on cancer cells (Page et al. 2014). These landmark experiments have led to the now-established anti-CTLA-4 and anti-PD-1 therapies that are currently used for the treatment of end-stage melanoma patients. Clinical trials are under way to evaluate these therapies in several other cancer types, including ovarian cancer, non–small cell lung carcinoma (NSCLC), colorectal cancer, renal cell carcinoma (RCC), metastatic hormone-refractory prostate cancer, and gastric cancer (Page et al. 2014).

Overcoming Anergy

So-called tumor-specific antigens (TSAs) expressed by cancer cells are often present either because of modification of a host's self-antigen or atypical expression of an unmodified self-antigen. These antigens will not usually trigger an immune response because the central and peripheral immune tolerance mechanisms designed to prevent autoimmunity will have resulted in the elimination of T cells responding to the original self-antigens (Ohashi 2003). Surviving T cells that do recognize such TSAs are usually rendered inactive by the induction of anergy (a state of nonresponsiveness) (Willimsky et al. 2008). There is some evidence in cancer patients that anergy of antitumor T cells can also be induced by the tumor itself, and that the transcription factor Egr2 may be involved in controlling such T-cell anergy (Safford et al. 2005; Zheng et al. 2012; Pardoll 2015). However, the downstream molecular mechanisms involved in establishing and maintaining the anergic state are incompletely understood. Moreover, the lack of a specific surface marker identifying an anergic T cell makes this line of research a difficult challenge. Nevertheless, if anergy can be reversed, as by the appropriate delivery of a second stimulatory signal, any TSA-specific T cells present might become activated and mount an effective antitumor response (Pellegrini et al. 2010).

Restoring Priming and Innate Responses

The molecules delivering co-stimulatory signals to T cells are up-regulated on the surfaces of APCs such as dendritic cells (DCs) after these cells use their Toll-like receptors (TLRs) to sense microbial products and become activated. This engagement of TLRs and initiation of DC maturation also contributes to activation of the innate immune system. In the case of a tumor that expresses no non-self-antigens that could trigger TLR binding and DC maturation, the innate immune response is limited and few cytokines and chemokines are produced. As a result, the recruitment of innate and adaptive immune cells able to attack tumor cells does not occur, and the cancer is allowed to progress. Several groups have experimented with the application of endogenous factors known to boost DC maturation in vitro, including heat shock proteins, uric acid, and damage-associated molecular patterns (DAMPs) such as high-mobility group box 1 protein (HMGB1) (Tesniere et al. 2008). In vivo chemotherapy increases the abundance of these factors due to increased cell death, leading to a more robust immune response (Pellegrini et al. 2010).

Overcoming the Negative Influence of the Tumor Microenvironment

Mechanisms designed to prevent autoimmunity in healthy individuals can compromise anticancer immune responses when they take effect in the tumor microenvironment (TME). These mechanisms include the actions of T regulatory (Treg) cells, myeloid-derived suppressor cells (MDSCs), and other inhibitory cell types (Rabinovich et al. 2007; Murdoch et al. 2008; Pittet 2009). Furthermore, cancer cells actively influence the TME to prevent or dampen antitumor immune responses by expressing negative regulatory ligands such as PDL1 or B7H4. The binding of these regulators to their cognate receptors on T cells inhibits the activity of these cells and impairs the adaptive antitumor response (Krambeck et al. 2006; Blank and Mackensen 2007). In addition, tumor cells actively secrete immunosuppressive signaling molecules and cytokines such as interleukin-10, transforming growth factor-β, and/or indoleamine 2,3-dioxygenase (IDO) (Katz et al. 2008; Mantovani et al. 2008). Treatment with monoclonal antibodies that can block these regulatory cells and mechanisms might therefore significantly enhance antitumor responses (Pellegrini et al. 2010). Indeed, newly devised immunotherapy drugs such as ipilimumab release these brakes, enabling the immune system to keep up the attack (Crouch 2016).

Preventing Immune Escape

Although the killing of tumor cells by cytolytic $CD8^+$ T cells may initially reduce a cancerous mass, it may eventually promote the selective outgrowth of tumor cells that have lost or mutated the antigens targeted by those T cells, blunting the response. This selective pressure may also cause the tumor cell to alter its MHC expression or antigen processing, again abetting immune escape. Therefore, strategies such as local radiotherapy, inhibition of histone deactylase and/or DNA methyltransferase, or stimulation with type I interferons can enhance levels of MHC expression and antigen presentation on tumor cells (Pellegrini et al. 2010). In addition, natural killer cells that target tumor cells lacking surface MHC can be harvested from a cancer patient, expanded and activated ex vivo, and then given back to the patient to boost antitumor activity (Chan et al. 2008).

Penetrating the Wall

All of the above tumor properties can counter responses mediated by T cells and/or permit the cancer cells to act like stealth bombers and hide from the immune system. However, recent work has revealed that nefarious co-opting of immune cells in the service of the cancer also exists. For example, analysis of human papillomavirus (HPV)-associated head and neck squamous cell carcinomas revealed the colocalization of tumor-associated macrophages, PDL-1-expressing tumor cells, and CD8$^+$-infiltrating lymphocytes expressing high levels of PD-1 in a "wall" blocking access to the tumor's interior (Lyford-Pike et al. 2013). This cell collection creates an immune-privileged site that allows the tumor cells to shut down the responses of any T cells already in the cancerous mass and to resist the penetration of any new T cells (Lyford-Pike et al. 2013). Some have likened the outer layer of macrophages in this wall to the fortifications that prevented the Greek soldiers from marching into Troy to rescue Helen (Apple 2016). The search for a microscopic Trojan horse that can break through or slither over the wall and expose the tumor cells to fresh immune attack is under way.

ALTERED METABOLIC PATHWAYS

Warburg Was Right

In 1924, Otto Warburg postulated that, "The cause of cancer is the replacement of the respiration of oxygen in normal body cells by a fermentation of sugar" (Warburg et al. 1924; Koppenol et al. 2011). This hypothesis was subsequently dubbed the "Warburg effect." In 1966, Warburg repeated his theory that cancer is caused mainly by a switch from normal cell respiration to fermentation, and that this switch might be induced by damage to the enzymes needed for respiration (http://www.mediatheque.lindau-nobel.org/videos/31517/on-the-primary-causes-and-on-the-secondary-causes-of-cancer-german-presentation-1966/laureate-warburg). Many scientists are now coming around to the view that Warburg was indeed right, and that cancer cells undergo metabolic changes that might make them vulnerable to new types of anticancer treatment.

Altered Metabolism and Reactive Oxygen Species Handling

Cells undergoing malignant transformation show certain metabolic adaptations (in addition to their genetic and epigenetic changes) that may be induced by their altered TME. The signaling pathways that are up-regulated in the course of these adaptations are not intrinsically tumorigenic but permit incipient tumor cells to thrive under conditions killing normal cells (Galluzzi et al. 2013). Tumor cell metabolism is altered to maximize the addressing of the fundamental needs of proliferating cells: increased macromolecule biosynthesis, rapid ATP production to generate energy, and heightened regulation of intracellular redox status. Alterations to carbohydrate, nucleic acid, lipid, and protein metabolism ensue in the cancer cell's fight to survive and divide (Cairns et al. 2011). These changes cause precancerous and cancerous cells to become metabolically "addicted," meaning that new avenues of therapeutic intervention may abound. Because normal cells do not endure the same degree of energy stress as frantically dividing malignant cells, an agent targeting a metabolic adaptation present only in tumor cells might spare normal cells, reducing deleterious side effects.

In addition to nimble metabolism, the regulation of reactive oxygen species (ROS) is crucial for the survival and functions of tumor cells. The increased growth of cancer cells produces increased ROS, so that tumor cells have to alter their signaling pathways mediating ROS regulation to cope. Elevated ROS in cancers are generated by endoplasmic reticulum (ER) stress, defective metabolism, hypoxia, and oncogene activity (Gorrini et al. 2013b). At the biochemical level, such ROS are routinely neutralized via glutathione or NADPH and by dietary antioxidants. Cells under stress conditions also activate transcription factors such as NRF2 that drive the expression of antioxidant genes. The activities of tumor suppressors such as PTEN, p53, BRCA1, and ATM are also increased in stressed cells (Gorrini et al. 2013b). Transformation itself is a form of stress, and so as ROS increase in a precancerous or cancerous cell, it up-regulates its antioxidant pathways to protect itself from ROS-induced death.

Although antioxidants have been proposed as anticancer agents, there is controversy over their role in tumorigenesis. In a recent study in our laboratory, we showed that synthesis of the antioxidant glutathione (GSH), which requires the activity of the modifier subunit of glutamate-cysteine ligase (GCLM), is in fact necessary for cancer initiation (Harris et al. 2015). Targeted therapeutics that block antioxidant pathways may thus paradoxically be helpful because they induce the apoptotic death of tumor cells. Alternatively, a drug that could increase ROS production above the level that the cancer cell's up-regulated antioxidant mechanisms could handle might kill the tumor cell. Once again, because these antioxidant pathways are not activated in resting normal cells, they would in theory be unaffected by the drug.

Mutant IDH1 and IDH2

Much recent research in the metabolic adaptation field has been focused on isocitrate dehydrogenases (IDHs) 1 and 2, which usually influence the cytoplasmic/mitochondrial NADP:NADPH ratio and thus a cell's reducing potential (Cairns and Mak 2013). Cancer-associated mutations in *IDH1* were identified through cancer cell genome sequencing efforts (Parsons et al. 2008; Mardis et al. 2009) and were followed by the discovery of similar mutations in *IDH2* (Yan et al. 2009). The normal biochemical function of IDH1/2 is to convert isocitrate to α-ketoglutarate and concomitantly reduce NADP to NADPH while producing CO_2 (Cairns and Mak 2013). IDH1 carries out this reaction in the cytoplasm, whereas IDH2 does so in the mitochondria (Fig. 1). Cancer-associated *IDH1* mutations predominantly alter arginine-132 (R132) in the enzyme's active site, whereas those altering

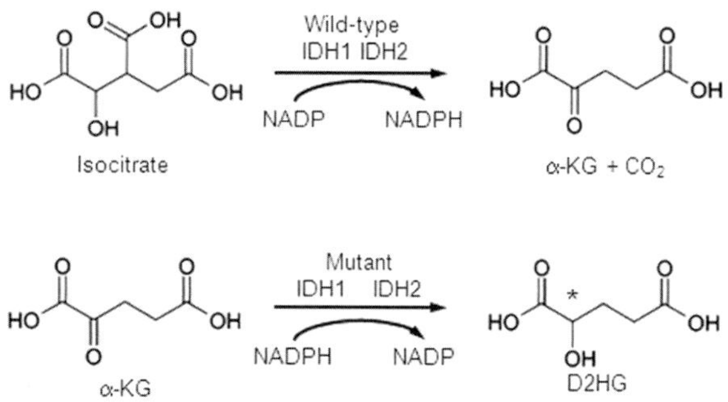

Figure 1. Enzymatic reactions catalyzed by wild-type and mutant IDH enzymes. (*Top*) Wild-type IDH1 and IDH2 convert isocitrate to α-KG and CO_2, with concomitant production of NADPH from NADP. (*Bottom*) Tumor-associated mutant IDH1 and IDH2 enzymes convert α-KG to D2HG, with concomitant production of NADP from NADPH. D2HG is a chiral molecule very similar in structure to α-KG. The chiral center in D2HG is denoted by *. (Adapted from Cairns and Mak 2013.)

IDH2 usually occur at R172 and R140 (Ward et al. 2010; Cairns and Mak 2013). However, rather than removing IDH activity, the R132, R172, and R140 mutations of IDH1/2 give these altered enzymes a new function. In 2009, researchers at Agios Pharmaceuticals used a metabolite profiling strategy to show that the normal IDH product α-ketoglutarate is converted by mutated IDH1/2 to 2-hydroxyglutarate (D2HG) in a reaction that consumes, rather than produces, NADPH (Fig. 1; Dang et al. 2009; Cairns and Mak 2013). D2HG is now viewed by many in the field to be an oncometabolite.

IDH1 mutations have been found at relatively high frequency in acute myeloid leukemia (AML) (Mardis et al. 2009), cholangiocarcinoma (Borger et al. 2012; Wang et al. 2013), glioblastoma multiforme (GBM) (Parsons et al. 2008), and chondrosarcoma (Amary et al. 2011). Some melanomas and colon, prostate, and NSCLC cancers also possess mutated *IDH1* genes (Kang et al. 2009). *IDH2* mutations occur in AML (Mardis et al. 2009), cholangiocarcinoma (Borger et al. 2012; Wang et al. 2013), myelodysplastic syndrome (MDS), and myeloproliferative disorder (MPD) (Patnaik et al. 2012; Rakheja et al. 2012; Shih et al. 2012), D2HG aciduria (Kranendijk et al. 2010), angioimmunoblastic T-cell lymphoma (AITL) (Cairns et al. 2012), and chondrosarcoma (Amary et al. 2011).

Mouse Models of Mutated IDH1/2

To generate accurate mouse models of oncogenesis related to IDH mutations, we used the lox–stop–lox (LSL) system to create conditional knock-in (KI) mice in which Cre recombinase drives the excision of the stop codon upstream of *IDH1* exon 4. The mutated IDH1 protein is then expressed from the endogenous locus (Sasaki et al. 2012a,b). Without Cre, neither the wild-type (WT) *IDH* allele nor the LSL IDH1 R132 mutated allele is expressed. It turns out that WT *IDH1* function is not necessary for mouse survival under laboratory conditions because complete IDH1 knockout mice are fertile and viable (Cairns and Mak 2013). Mice heterozygous for the *IDH1* R132 KI allele that also constitutively express Cre do not survive beyond the early embryonic stage, suggesting that the D2HG produced by the mutated IDH1 enzyme is lethal during development (Cairns and Mak 2013). When we crossed LysM-Cre mice with our IDH1 KI mice to examine the effect on the myeloid compartment, the mutant mice were born normally and grew as expected until 6 months of age, when they began to show evidence of extramedullary hematopoiesis, splenomegaly, and reduced bone marrow cellularity (Sasaki et al. 2012b). Because tumor cells in human AML and glioma patients with mutated IDH1 or IDH2 show altered DNA methylation (Figueroa et al. 2010; Noushmehr et al. 2010), we examined DNA methylation in our IDH1 KI mice. An increased proportion of hypermethylated CpG sites was present in LSK cells of LysM-Cre IDH1 KI mice, with a significant increase in CpG sites showing >80% methylation (Sasaki et al. 2012b). Moreover, LysM-Cre IDH1 KI cells showed alterations to DNA methylation and histone hypermethylation reminiscent of those in human IDH-mutant AML (Sasaki et al. 2012b).

Involvement of ATM and Notch1

Our most recent work has shown that the *Atm* gene is down-regulated in IDH1 KI mice as a consequence of increased histone methylation and the closing of the chromatin structure (Inoue et al. 2016). By using CyTOF mass cytometry to identify proteins that were differentially expressed between WT and IDH-KI mice, we showed that ATM was significantly underexpressed in hematopoietic stem cells (HSCs) of the mutant animals. The *Atm* promoter in these mutant cells showed an accumulation of methylated histone H3K9 and a closed chromatin structure (Inoue et al. 2016). Furthermore, the number of long-term hematopoietic stem cells (LT-HSCs) in these mice was reduced and the self-renewal capacity of these cells was impaired. In addition to their decreased ATM expression, IDH1-mutant LT-HSCs showed increased sensitivity to DNA damage as measured by the formation of 53BP1 and γH2AX foci, regardless of whether this damage was spontaneous or induced by irradiation. Importantly, all these effects were found to be independent of TET2 (Inoue et al. 2016). Our findings provide mechanistic insight into why IDH1 KI mice develop myeloid malignancies. Significantly, like our mutant mice, patients with IDH-mutated AML show low expression levels of ATM and other DNA damage repair-associated genes (Inoue et al. 2016; Penard-Lacronique and Bernard 2016).

We have also introduced the IDH1 R132 mutation into the entire hematopoietic system by crossing our IDH1 KI mice with Vav-Cre animals, and we have investigated the role of IDH1 mutations in T-cell malignancies (Hao et al. 2016). Vav-Cre IDH1-KI mutants spontaneously developed T-cell acute lymphoblastic leukemia (T-ALL) that was transplantable to fresh recipient mice and that maintained its mutant IDH1 expression (Hao et al. 2016).

When we performed whole-exome sequencing on isolated tumors, we identified a spontaneous activating mutation in Notch1, a gene commonly mutated in human T-ALL (Ferrando 2009). These results suggested that IDH1 mutations may have the capacity to cooperate with Notch1 mutation to drive T-ALL. Crossing our Vav-Cre IDH1 mutants with conditional Trp53 null mice accelerated the onset of T-cell lymphomagenesis (Hao et al. 2016). Interestingly, metabolomics analysis revealed that tumor cells derived from these double mutants showed increased dependence on both glucose and glutamine. Thus, mutant IDH1 may contribute to malignancy in the T-cell lineage and alter the metabolic profile of these cells (Hao et al. 2016).

Oncogenicity of D2HG

Our studies of IDH1/2-mutated and -related mutant mice have reinforced our hypothesis that it is the novel activity of these mutant IDH enzymes (resulting in massive D2HG production) that is the basis of their tumorigenicity. But why is D2HG oncogenic? D2HG competitively inhibits a family of 2-OG-dependent dioxygenases that all use α-ketoglutarate as a substrate (Cairns and Mak 2013). Mammalian cells contain more than 60 2-OG-dependent dioxygenases contributing to biologic processes as diverse as fatty acid metabolism, collagen biosynthesis, hypoxia sensing, modifications of chromatin and RNA, and DNA repair (Rose et al. 2011). In general, these enzymes convert α-ketoglutarate to succinate and CO_2 in a reaction requiring ascorbate, iron, and oxygen as cofactors (Rose et al. 2011). Competitive inhibition of 2-OG-dependent dioxygenases by D2HG has been shown in vitro (Chowdhury et al. 2011), and the increased levels of D2HG in IDH-mutant tumors imply that D2HG can also impair the activities of 2-OG-dependent dioxygenases in vivo (Cairns and Mak 2013). Potential targets of D2HG-mediated inhibition include the JumonjiC domain–containing histone demethylases, the TET proteins involved in DNA methylation, the prolyl hydroxylases (PHDs) that regulate hypoxia-inducible factor (HIF) signaling, and the PHD and lysyl hydroxylases (LHDs) required for collagen maturation (Cairns and Mak 2013). Work is ongoing to determine exactly how D2HG-mediated inhibition of any or all of these enzymes leads to malignancy.

TARGETING REACTIVE OXYGEN SPECIES

As noted above, ROS are key intracellular stress factors that are elevated in cancer cells. Our group has engaged in extensive study of why excess ROS are oncogenic. We have shown that BRCA1 deletion increases ROS in breast cancer cells because the lack of BRCA1 impairs the Nrf2-mediated antioxidant response (Gorrini et al. 2013a). BRCA1 normally interacts directly with Nrf2 to influence Keap1-mediated Nrf2 ubiquitination and thus Nrf2 activation and stability (Gorrini et al. 2013a). A critical finding has been that estrogen treatment partly restores Nrf2 activity in mammary tumor cells lacking BRCA1, protecting them from excessive oxidative stress and furthering their growth (Gorrini et al. 2013a, 2014). These results provide the long-sought explanation for why it is almost exclusively breast and ovarian tumors that arise in carriers of *BRCA1* mutations (Fig. 2). We hypothesize that somatic loss of BRCA1 function in heterozygous carriers of *BRCA1* mutations has tissue-specific effects. In tissues lacking high levels of estrogen, BRCA1 deficiency blocks Nrf2 antioxidant signaling such that ROS accumulate to levels that kill the BRCA1-deficient cells.

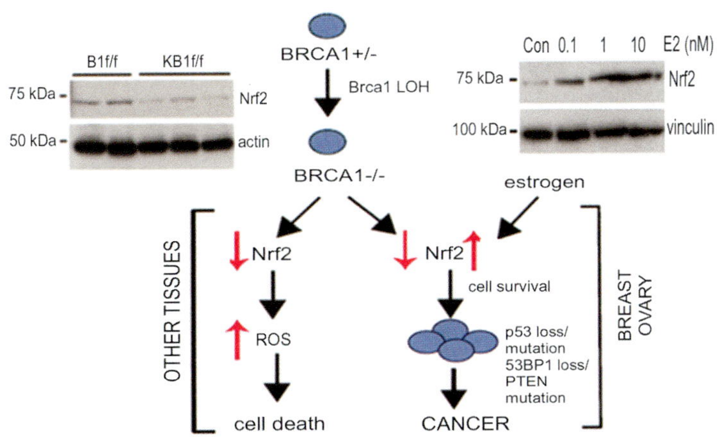

Figure 2. Model of the role of NRF2 regulation in BRCA1-associated tumorigenesis. (*Top left*) Immunoblot comparing levels of the antioxidant transcription factor Nrf2 in mammary epithelial cells of control mice (B1f/f) and double-mutant BRCA1- and P53-deficient mice (KB1f/f). Nrf2 is decreased in the absence of BRCA1. (*Top right*) Immunoblot demonstrating the induction of Nrf2 in wild-type (WT) mammary epithelial cells treated with increasing doses of estrogen (E2). (*Center*) When the function of the WT BRCA1 allele is lost in a heterozygous carrier of a BRCA1 mutation, the outcome depends on the tissue type. In breast and ovary, where estrogen is abundant, NRF2 is up-regulated so that incipient tumor cells survive to acquire additional mutations, such as loss of p53 or PTEN function, and eventually undergo full transformation. In other tissues lacking high levels of estrogen, NRF2 activity is decreased in the absence of BRCA1. The excessive ROS (reactive oxygen species) generated in the incipient cancer cell cannot be combated effectively and the cell dies, preventing tumor formation. LOH, loss of heterozygosity. (Adapted from Gorrini et al. 2014.)

Tumor cells do not get a chance to develop. However, in the estrogen-rich environment of mammary and ovarian tissues, Nrf2 can be activated by a PI3K–AKT-dependent mechanism even in the absence of BRCA1. The BRCA1-deficient incipient tumor cells are protected against oxidative stress-induced death and survive to acquire additional oncogenic mutations. If such a BRCA1-deficient cell sustains loss of PTEN function, the PI3K–AKT pathway may run uncontrolled and further stimulate estrogen-driven NRF2 signaling. Pathways activated downstream from AKT, including those with antioxidant or mitogenic effects, may combine with the genomic instability induced by the failure of BRCA1-mediated DNA repair to ultimately drive malignant transformation of the BRCA1-deficient cells (Gorrini et al. 2014).

TARGETING ANEUPLOIDY

A long-term goal of our group is to come up with new classes of anticancer therapies that may be more effective for certain cancers than existing agents. As noted above, no one approach is likely to eliminate all types of malignancies because of the extent of variation in cancer cell genomes. Our latest efforts have targeted aneuploidy, an alteration present in most advanced cancer cells but absent from normal cells.

We systematically combined gene expression analysis with RNAi screening of human breast tumor samples and cancer cell lines that show aneuploidy. Polo-like kinase-4 (PLK4), an enzyme crucial for the maintenance of aneuploidy, was identified as a promising target (Mason et al. 2014). Our academic drug discovery team isolated a selective and potent small molecule inhibitor of PLK4 called CFI-400945 (Mason et al. 2014). Before they died, cancer cells treated with CFI-400945 showed centriole duplication and mitotic defects similar to those resulting from PLK4 kinase inhibition in vitro (Mason et al. 2014). Mice that bore xenografted human ovarian or breast cancer tissues and were treated with CFI-400945 showed significant tumor growth inhibition, which was influenced by the PTEN status of the cancer cells (Mason et al. 2014). Xenografts lacking PTEN function were more responsive to CFI-400945 treatment than were those with WT PTEN (Mason et al. 2014). Thus, PTEN status may be a predictive biomarker for our new anticancer agent.

CFI400945 recently entered a first-in-human phase I trial to establish its safety, tolerability, and pharmacokinetics (PK) and to determine the recommended phase II dose (RP2D) (Bedard et al. 2016). For this trial, 31 patients with advanced solid tumors of any type were enrolled between April 2014 and December 2015. The trial followed a standard 3+3 dose escalation design, with a starting dose of 3 mg delivered orally once daily. The primary end point was the incidence of dose-limiting toxicities (DLTs). Over the course of the trial, dosing reached 72 mg without observation of any DLT events. The most frequent treatment-related adverse events (trAEs) were fatigue (24%), diarrhea (17.2%), nausea (17.2%), decreased appetite (13.8%), and vomiting (6.9%), all of grade 1 or grade 2 in severity (Bedard et al. 2016). To date, two patients enrolled at the 48 mg dose level have completed >6 cycles, including a patient with KRAS mutant colorectal cancer who achieved a 24% reduction in target lesions and >50% reduction in serum carcinoembryonic antigen (CEA) levels (Bedard et al. 2016). Our data so far indicate that CFI-400945 is well tolerated at doses up to 72 mg and has a favorable PK profile. Most importantly, preliminary evidence of effective antitumor activity has been observed. Exploration of 96 mg daily dosing of CFI400945 is ongoing (Bedard et al. 2016).

CONCLUSION

We believe that the future looks bright for the development of new types of anticancer therapies that can truly make a difference for cancer patients. We now recognize that cancer differs in its genetics from individual to individual, and that no one strategy focusing on tumor-associated mutations will cure all. Approaches that can boost the body's antitumor responses, shut down the tumor's escape mechanisms, or undermine its metabolic or replicative adaptations have the potential to help many sufferers of this dreaded disease.

ACKNOWLEDGMENTS

This work was supported by grants from the California Institute for Regenerative Medicine, the Leukemia & Lymphoma Society, and the Canadian Institutes of Health Research to T.W.M.

REFERENCES

Amary MF, Bacsi K, Maggiani F, Damato S, Halai D, Berisha F, Pollock R, O'Donnell P, Grigoriadis A, Diss T, et al. 2011. IDH1 and IDH2 mutations are frequent events in central chondrosarcoma and central and periosteal chondromas but not in other mesenchymal tumours. *J Pathol* **224:** 334–343.

Apple S. 2016. At the Vatican, a search for cancer's miracle cure. *The New Yorker*. www.newyorker.com/tech/elements/at-the-vatican-a-search-for-cancers-miracle-cure.

Bedard PL, Cescon DW, Fletcher G, Denny T, Brokx R, Sampson P, Bray MR, Slamon DJ, Mak TW, Wainberg ZA. 2016. Abstract CT066: First-in-human phase I trial of the oral PLK4 inhibitor CFI-400945 in patients with advanced solid tumors. *Cancer Res* **76:** CT066.

Blank C, Mackensen A. 2007. Contribution of the PD-L1/PD-1 pathway to T cell exhaustion: An update on implications for chronic infections and tumor evasion. *Cancer Immunol Immunother* **56:** 739–745.

Borger DR, Tanabe KK, Fan KC, Lopez HU, Fantin VR, Straley KS, Schenkein DP, Hezel AF, Ancukiewicz M, Liebman HM, et al. 2012. Frequent mutation of isocitrate dehydrogenase *(IDH)1* and *IDH2* in cholangiocarcinoma identified through broad-based tumor genotyping. *Oncologist* **17:** 72–79.

Cairns RA, Mak TW. 2013. Oncogenic isocitrate dehydrogenase mutations: Mechanisms, models, and clinical opportunities. *Cancer Discov* **3:** 730–741.

Cairns RA, Harris IS, Mak TW. 2011. Regulation of cancer cell metabolism. *Nat Rev Cancer* **11:** 85–95.

Cairns RA, Iqbal J, Lemonnier F, Kucuk C, de Leval L, Jais JP, Parrens M, Martin A, Xerri L, Brousset P, et al. 2012. *IDH2* mutations are frequent in angioimmunoblastic T-cell lymphoma. *Blood* **119:** 1901–1903.

Chan CJ, Andrews DM, Smyth MJ. 2008. Can NK cells be a therapeutic target in human cancer? *Eur J Immunol* **38:** 2964–2968.

Chowdhury R, Yeoh KK, Tian YM, Hillringhaus L, Bagg EA, Rose NR, Leung IK, Li XS, Woon EC, Yang M, et al. 2011. The oncometabolite 2-hydroxyglutarate inhibits histone lysine demethylases. *EMBO Rep* **12:** 463–469.

Crouch M. 2016. 21 completely reassuring things scientists wish you knew about cancer. *Reader's Digest.* www.rd.com/health/healthcare/genius-cancer-breakthroughs.

Dang L, White DW, Gross S, Bennett BD, Bittinger MA, Driggers EM, Fantin VR, Jang HG, Jin S, Keenan MC, et al. 2009. Cancer-associated IDH1 mutations produce 2-hydroxyglutarate. *Nature* **462:** 739–744.

Ferrando AA. 2009. The role of NOTCH1 signaling in T-ALL. *Hematol Am Soc Hematol Educ Program* **2009:** 353–361.

Figueroa ME, Abdel-Wahab O, Lu C, Ward PS, Patel J, Shih A, Li Y, Bhagwat N, Vasanthakumar A, Fernandez HF, et al. 2010. Leukemic IDH1 and IDH2 mutations result in a hypermethylation phenotype, disrupt TET2 function, and impair hematopoietic differentiation. *Cancer Cell* **18:** 553–567.

Galluzzi L, Kepp O, Vander Heiden MG, Kroemer G. 2013. Metabolic targets for cancer therapy. *Nat Rev Drug Discov* **12:** 829–846.

Gorrini C, Baniasadi PS, Harris IS, Silvester J, Inoue S, Snow B, Joshi PA, Wakeham A, Molyneux SD, Martin B, et al. 2013a. BRCA1 interacts with Nrf2 to regulate antioxidant signaling and cell survival. *J Exp Med* **210:** 1529–1544.

Gorrini C, Harris IS, Mak TW. 2013b. Modulation of oxidative stress as an anticancer strategy. *Nat Rev Drug Discov* **12:** 931–947.

Gorrini C, Gang BP, Bassi C, Wakeham A, Baniasadi SP, Hao Z, Li WY, Cescon DW, Li YT, Molyneux S, et al. 2014. Estrogen controls the survival of BRCA1-deficient cells via a PI3K-NRF2-regulated pathway. *Proc Natl Acad Sci* **111:** 4472–4477.

Hao Z, Cairns RA, Inoue S, Li WY, Sheng Y, Lemonnier F, Wakeham A, Snow BE, Dominguez-Brauer C, Ye J, et al. 2016. *Idh1* mutations contribute to the development of T-cell malignancies in genetically engineered mice. *Proc Natl Acad Sci* **113:** 1387–1392.

Harris IS, Treloar AE, Inoue S, Sasaki M, Gorrini C, Lee KC, Yung KY, Brenner D, Knobbe-Thomsen CB, Cox MA, et al. 2015. Glutathione and thioredoxin antioxidant pathways synergize to drive cancer initiation and progression. *Cancer Cell* **27:** 211–222.

Inoue S, Li WY, Tseng A, Beerman I, Elia AJ, Bendall SC, Lemonnier F, Kron KJ, Cescon DW, Hao Z, et al. 2016. Mutant IDH1 downregulates ATM and alters DNA repair and sensitivity to DNA damage independent of TET2. *Cancer Cell* **30:** 337–348.

Kang MR, Kim MS, Oh JE, Kim YR, Song SY, Seo SI, Lee JY, Yoo NJ, Lee SH. 2009. Mutational analysis of IDH1 codon 132 in glioblastomas and other common cancers. *Int J Cancer* **125:** 353–355.

Katz JB, Muller AJ, Prendergast GC. 2008. Indoleamine 2,3-dioxygenase in T-cell tolerance and tumoral immune escape. *Immunol Rev* **222:** 206–221.

Koppenol WH, Bounds PL, Dang CV. 2011. Otto Warburg's contributions to current concepts of cancer metabolism. *Nat Rev Cancer* **11:** 325–337.

Krambeck AE, Thompson RH, Dong H, Lohse CM, Park ES, Kuntz SM, Leibovich BC, Blute ML, Cheville JC, Kwon ED. 2006. B7-H4 expression in renal cell carcinoma and tumor vasculature: Associations with cancer progression and survival. *Proc Natl Acad Sci* **103:** 10391–10396.

Kranendijk M, Struys EA, van Schaftingen E, Gibson KM, Kanhai WA, van der Knaap MS, Amiel J, Buist NR, Das AM, de Klerk JB, et al. 2010. IDH2 mutations in patients with D-2-hydroxyglutaric aciduria. *Science* **330:** 336.

Lyford-Pike S, Peng S, Young GD, Taube JM, Westra WH, Akpeng B, Bruno TC, Richmon JD, Wang H, Bishop JA, et al. 2013. Evidence for a role of the PD-1:PD-L1 pathway in immune resistance of HPV-associated head and neck squamous cell carcinoma. *Cancer Res* **73:** 1733–1741.

Mantovani A, Romero P, Palucka AK, Marincola FM. 2008. Tumour immunity: Effector response to tumour and role of the microenvironment. *Lancet* **371:** 771–783.

Mardis ER, Ding L, Dooling DJ, Larson DE, McLellan MD, Chen K, Koboldt DC, Fulton RS, Delehaunty KD, McGrath SD, et al. 2009. Recurring mutations found by sequencing an acute myeloid leukemia genome. *N Engl J Med* **361:** 1058–1066.

Mason JM, Lin DC, Wei X, Che Y, Yao Y, Kiarash R, Cescon DW, Fletcher GC, Awrey DE, Bray MR, et al. 2014. Functional characterization of CFI-400945, a Polo-like kinase 4 inhibitor, as a potential anticancer agent. *Cancer Cell* **26:** 163–176.

Murdoch C, Muthana M, Coffelt SB, Lewis CE. 2008. The role of myeloid cells in the promotion of tumour angiogenesis. *Nat Rev Cancer* **8:** 618–631.

Noushmehr H, Weisenberger DJ, Diefes K, Phillips HS, Pujara K, Berman BP, Pan F, Pelloski CE, Sulman EP, Bhat KP, et al. 2010. Identification of a CpG island methylator phenotype that defines a distinct subgroup of glioma. *Cancer Cell* **17:** 510–522.

Ohashi PS. 2003. Negative selection and autoimmunity. *Curr Opin Immunol* **15:** 668–676.

Page DB, Postow MA, Callahan MK, Allison JP, Wolchok JD. 2014. Immune modulation in cancer with antibodies. *Annu Rev Med* **65:** 185–202.

Pardoll D. 2015. Cancer and the immune system: Basic concepts and targets for intervention. *Semin Oncol* **42:** 523–538.

Parsons DW, Jones S, Zhang X, Lin JC, Leary RJ, Angenendt P, Mankoo P, Carter H, Siu IM, Gallia GL, et al. 2008. An integrated genomic analysis of human glioblastoma multiforme. *Science* **321:** 1807–1812.

Patnaik MM, Hanson CA, Hodnefield JM, Lasho TL, Finke CM, Knudson RA, Ketterling RP, Pardanani A, Tefferi A. 2012. Differential prognostic effect of *IDH1* versus *IDH2* mutations in myelodysplastic syndromes: A Mayo Clinic Study of 277 patients. *Leukemia* **26:** 101–105.

Pellegrini M, Mak TW, Ohashi PS. 2010. Fighting cancers from within: Augmenting tumor immunity with cytokine therapy. *Trends Pharmacol Sci* **31:** 356–363.

Penard-Lacronique V, Bernard OA. 2016. IDH1, histone methylation, and so forth. *Cancer Cell* **30:** 501.

Pittet MJ. 2009. Behavior of immune players in the tumor microenvironment. *Curr Opin Oncol* **21:** 53–59.

Rabinovich GA, Gabrilovich D, Sotomayor EM. 2007. Immunosuppressive strategies that are mediated by tumor cells. *Annu Rev Immunol* **25:** 267–296.

Rakheja D, Konoplev S, Medeiros LJ, Chen W. 2012. IDH mutations in acute myeloid leukemia. *Hum Pathol* **43:** 1541–1551.

Rose NR, McDonough MA, King ON, Kawamura A, Schofield CJ. 2011. Inhibition of 2-oxoglutarate dependent oxygenases. *Chem Soc Rev* **40:** 4364–4397.

Safford M, Collins S, Lutz MA, Allen A, Huang CT, Kowalski J, Blackford A, Horton MR, Drake C, Schwartz RH, Powell JD. 2005. Egr-2 and Egr-3 are negative regulators of T cell activation. *Nat Immunol* **6:** 472–480.

Sasaki M, Knobbe CB, Itsumi M, Elia AJ, Harris IS, Chio II, Cairns RA, McCracken S, Wakeham A, Haight J, et al. 2012a. D-2-hydroxyglutarate produced by mutant IDH1 perturbs collagen maturation and basement membrane function. *Genes Dev* **26:** 2038–2049.

Sasaki M, Knobbe CB, Munger JC, Lind EF, Brenner D, Brustle A, Harris IS, Holmes R, Wakeham A, Haight J, et al. 2012b. IDH1(R132H) mutation increases murine haematopoietic progenitors and alters epigenetics. *Nature* **488:** 656–659.

Shih AH, Abdel-Wahab O, Patel JP, Levine RL. 2012. The role of mutations in epigenetic regulators in myeloid malignancies. *Nat Rev Cancer* **12:** 599–612.

Tesniere A, Apetoh L, Ghiringhelli F, Joza N, Panaretakis T, Kepp O, Schlemmer F, Zitvogel L, Kroemer G. 2008. Immu-

nogenic cancer cell death: A key-lock paradigm. *Curr Opin Immunol* **20:** 504–511.
Wang P, Dong Q, Zhang C, Kuan PF, Liu Y, Jeck WR, Andersen JB, Jiang W, Savich GL, Tan TX, et al. 2013. Mutations in isocitrate dehydrogenase 1 and 2 occur frequently in intrahepatic cholangiocarcinomas and share hypermethylation targets with glioblastomas. *Oncogene* **32:** 3091–3100.
Warburg O, Posener K, Negelein E. 1924. Über den Stoffwechsel der Tumoren. *Biochemische Zeitschrift* **152:** 309–344.
Ward PS, Patel J, Wise DR, Abdel-Wahab O, Bennett BD, Coller HA, Cross JR, Fantin VR, Hedvat CV, Perl AE, et al. 2010. The common feature of leukemia-associated IDH1 and IDH2 mutations is a neomorphic enzyme activity converting α-ketoglutarate to 2-hydroxyglutarate. *Cancer Cell* **17:** 225–234.
Willimsky G, Czeh M, Loddenkemper C, Gellermann J, Schmidt K, Wust P, Stein H, Blankenstein T. 2008. Immunogenicity of premalignant lesions is the primary cause of general cytotoxic T lymphocyte unresponsiveness. *J Exp Med* **205:** 1687–1700.
Yan H, Parsons DW, Jin G, McLendon R, Rasheed BA, Yuan W, Kos I, Batinic-Haberle I, Jones S, Riggins GJ, et al. 2009. *IDH1* and *IDH2* mutations in gliomas. *N Engl J Med* **360:** 765–773.
Zheng Y, Zha Y, Driessens G, Locke F, Gajewski TF. 2012. Transcriptional regulator early growth response gene 2 (Egr2) is required for T cell anergy in vitro and in vivo. *J Exp Med* **209:** 2157–2163.

Lipid Synthesis Is a Metabolic Liability of Non–Small Cell Lung Cancer

Robert U. Svensson and Reuben J. Shaw
Molecular and Cell Biology Laboratory, The Salk Institute for Biological Studies, La Jolla, California 92037
Correspondence: shaw@salk.edu

The renaissance in the study of cancer metabolism has refocused efforts to identify and target metabolic dependencies of tumors as an approach for cancer therapy. One of the unique metabolic requirements that cancer cells possess to sustain their biosynthetic growth demands is altered fatty acid metabolism, in particular the synthesis of de novo fatty acids that are required as cellular building blocks to support cell division. Enhanced fatty acid synthesis that is observed in many tumor types has been postulated to open a therapeutic window for cancer therapy and, correspondingly, efforts to pharmacologically inhibit key enzymes of fatty acid synthesis are being pursued. However, despite these efforts, whether inhibition of fatty acid synthesis stunts tumor growth in vivo has been poorly understood. In this review, we focus on the recent evidence that pharmacologic inhibition of acetyl-CoA carboxylase, the enzyme that regulates the rate-limiting step of de novo fatty acid synthesis, exposes a metabolic liability of non–small cell lung cancer and represses tumor growth in preclinical models.

ALTERED FATTY ACID SYNTHESIS IN CANCER

Although most normal cells and tissues satisfy their requirement for fatty acids by importing them from the circulation, the observation more than 60 years ago that tumors undergo high rates of de novo lipogenesis suggested that cancer cells satisfy their lipid demand differently compared with their normal counterparts (Medes et al. 1953). Surprisingly, the rate of lipogenesis measured in tumors was comparable to that of the liver, an organ that requires a high rate of fatty acid (FA) synthesis. Since then, enhanced FA synthesis has been documented in a number of tumors (reviewed in Swinnen et al. 2006; Abramson 2011). The elevated de novo FA synthesis observed in tumors is reflected by the increased expression of lipogenic enzymes that has been described in a number of cancers (Szutowicz et al. 1979; Witters et al. 1994; Milgraum et al. 1997). A seminal finding in this respect was the identification of the oncoprotein OA-519 as fatty acid synthase (FASN), which was overexpressed in aggressive breast cancer (Kuhajda et al. 1994) and now confirmed to be up-regulated in a wide number of other cancers such as those of the prostate, ovary, colon, lung, and bladder (Menendez and Lupu 2007).

Lipogenic enzymes in the FA synthesis pathway are regulated at various levels. Strikingly, many of the genes encoding these enzymes are controlled transcriptionally by the sterol regulatory element-binding proteins (SREBPs) (Fig. 1). SREBPs are basic helix–loop–helix transcription factors that are crucial for maintaining cellular lipid homeostasis (Horton 2002). There are three SREBP isoforms in mammals: SREBP1a and SREBP1c, which are products of the *SREBF1* gene, and SREBP2, a product of the *SREBPF2* gene. Although SREBP1c mainly regulates genes in the FA synthesis pathway and SREBP2 regulates genes involved in cholesterol synthesis, redundancy among the SREBPs allows each isoform to regulate genes within each pathway (Horton 2002). Inhibition of SREBP1 and SREBP2 in cancer cells led to altered cellular lipid composition, which triggered endoplasmic reticulum (ER) stress and the unfolded protein response (UPR) (Griffiths et al. 2013), demonstrating that SREBPs play a critical role in maintaining cellular lipid homeostasis in cancer. Emerging evidence also indicates that SREBPs provide a crucial link between oncogene signaling and altered fatty acid metabolism. SREBP1 activity is induced by AKT in a manner that is dependent on mTORC1 (mammalian target of rapamycin complex 1), demonstrating that lipogenesis and cell growth are intimately connected through the AKT-mTORC1 pathway (Fig. 1; Porstmann et al. 2008; Duvel et al. 2010).

In addition to transcriptional regulation, the activity of lipogenic enzymes is modulated posttranslationally by phosphorylation. For example, AKT directly phosphorylates and stimulates ATP citrate lyase (ACLY) (Bauer et al. 2005) contributing to the control of histone acetylation in response to growth factor stimulation (Lee et al. 2014). In contrast to this phosphorylation stimulating ACLY by oncogenic signaling, in response to low energy levels, the AMP activated protein kinase (AMPK) phosphorylates and inactivates key regulators of FA synthesis, including acetyl-CoA carboxylase 1 and 2 (ACC1 and ACC2) as well as SREBP1c itself (Fig. 1; Carling et al. 1989; Li et al. 2011). Thus, during periods of energy deprivation, AMPK dominantly inhibits lipogenesis. The acti-

Published by Cold Spring Harbor Laboratory Press; doi: 10.1101/sqb.2016.81.030874

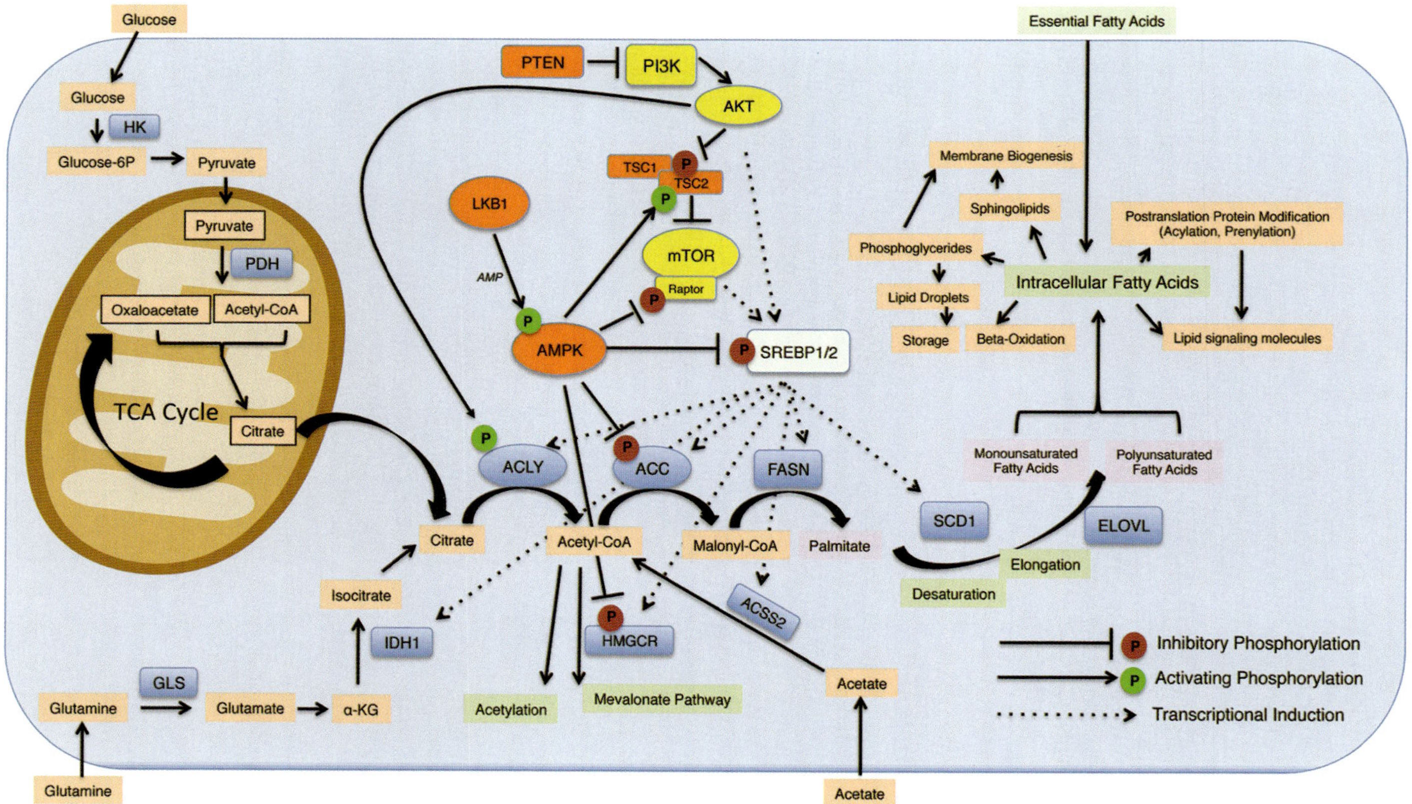

Figure 1. Overview of fatty acid biosynthesis. Glucose- or glutamine-derived citrate is converted into acetyl-CoA by ATP-citrate lyase (ACLY). Acetyl-CoA is also generated from exogenous uptake of acetate by acetyl-CoA synthetase short-chain family member 2 (ACSS2). Acetyl-CoA carboxylase (ACC) carboxylates acetyl-CoA to form malonyl-CoA, which is then used as a substrate for fatty acid synthase (FASN) to form the 16 carbon saturated fatty acid palmitate (C16:0). Palmitate is then elongated and desaturated to generate pools of monounsaturated and polyunsaturated fatty acids. Essential fatty acids that cannot be endogenously synthesized have to be taken up from the bloodstream. Cellular pools of fatty acids can then be synthesized into additional fatty acid and lipid species that are used by tumor cells for a number of essential processes. Lipogenic enzymes in the fatty acid synthesis pathway are regulated at multiple levels. The AMP-activated protein kinase (AMPK) phosphorylates and inactivates Raptor (mTOR), ACC, sterol regulatory binding protein 1c (SREBP1c), and 3-hydroxy-3-methyl-glutaryl-coenzyme A reductase (HMGCR) during periods of energy deprivation to eliminate ATP consumption by fatty acid synthesis. AKT directly phosphorylates ACLY and indirectly stimulates SREBP. SREBP activation promotes the transcriptional induction of several lipogenic enzymes in a manner that is also dependent on activation by mTOR. Proteins inactivated in cancer shown in orange and those hyperactivated in cancers shown in yellow. α-KG, α ketoglutarate; ELOVL, elongation of very-long-chain fatty acids; GLS, glutaminase; HK, hexokinase; IDH1, isocitrate dehydrogenase 1; LKB1, liver kinase B1; PDH, pyruvate dehydrogenase; PI3K, phosphoinositide 3-kinase; PTEN, phosphatase and tensin homolog; TCA, tricarboxylic acid; TSC1/2, tuberous sclerosis complex.

vation of AMPK is induced by phosphorylation by the tumor suppressor serine threonine kinase 11 (STK11), otherwise known as liver kinase B1 (LKB1), suggesting that ACC activity and lipogenesis itself may be hyperactivated in the broad variety of cancers lacking this tumor suppressor (Shackelford and Shaw 2009), paralleling the hyperactivation of ACLY predicted in the many cancers mutated for AKT, PIK3CA (phosphatidylinositol-4,5-biphosphate 3-kinase catalytic subunit alpha), or PTEN (phosphatase and tensin homolog).

Generally, the importance of FA synthesis in cancer cell proliferation and survival in vitro has now been described in a number of studies (Rohrig and Schulze 2016). However, whether autochthonous tumors in vivo possess a similar requirement is less understood. Because of the complex nature of tumors and the availability of lipids from both the circulation and the tumor microenvironment, the effects of de novo FA synthesis inhibition in vitro may not be translatable in vivo. Furthermore, the need to rigorously evaluate the anticancer effects of FA synthesis inhibitors is required to understand whether the enhanced synthesis observed in tumors is truly a metabolic dependency of cancer or rather simply a by-product of altered oncogene signaling and metabolism (Kim and DeBerardinis 2016).

FASN: THE FIRST TARGETED COMPONENT OF THE FA SYNTHESIS PATHWAY

Given elevated FASN was the first molecular discovery linking fat metabolism to cancer (Kuhajda et al. 1994), many studies have since shown increased FASN expression and enhanced activity in a wide variety of tumors (Menendez and Lupu 2007). Correspondingly, efforts focused on pharmacologically inhibiting FASN activity have been pursued for a number of years, and several FASN inhibitors have been described to date. The mycotoxin cerulenin inhibits the activity of FASN and has shown the ability to inhibit proliferation and induce apoptosis of cancer cells and xenograft tumors in vivo (Pizer et al. 1996; Menendez et al. 2004). Similarly, the more stable analog, C75, displayed antitumor activity in xenograft models of breast, prostate, and lung cancer (Pizer et al. 2000; Chen et al. 2012; Relat et al. 2012). C75 was also shown to significantly delay breast cancer progression in HER2 transgenic mice (Alli et al. 2005). TVB-3166, a potent and orally bioavailable inhibitor of FASN, induced apoptosis in breast and prostate cancer cell lines and displayed antitumor activity in several tumor xenograft models (Ventura et al. 2015). The thiophenopyrimidine-based FASN inhibitor Fasnall recently showed potent antitumor activity in the mouse mammary tumor virus (MMTV)-Neu model of $HER2^{+}$ breast cancer, where it also synergized with carboplatin (Alwarawrah et al. 2016). Collectively, the preclinical activity of FASN inhibitors has been encouraging, and recently the first FASN inhibitor was entered into clinical trials for cancer. However, the mechanism underlying the antitumor efficacy of various FASN inhibitors is poorly understood, as FASN inhibition was shown in some settings to increase malonyl-CoA levels (Thupari et al. 2001), which led to reduced food intake and body weight in mice (Loftus et al. 2000). Newer studies have also shed light on multiple mechanisms underlying sensitivity to FASN inhibitors (Benjamin et al. 2015; Ventura et al. 2015).

ACLY AND ACSS2: MORE THAN ONE WAY TO MAKE ACETYL-CoA

Other enzymes within the FA synthesis pathway have been hypothesized to be attractive therapeutic targets in cancer. In the cytoplasm, glucose- or glutamine-derived citrate is converted into acetyl-CoA by ACLY (Fig. 1). Acetyl-CoA is an essential substrate for FA synthesis; hence, ACLY inhibition should result in depleted acetyl-CoA pools and offer a way to shut down FA synthesis in cancer. Inhibition of ACLY by either RNAi or pharmacological inhibitors results in cell cycle arrest and induction of apoptosis in vitro and in vivo (Hatzivassiliou et al. 2005). However, whether these effects are specifically due to loss of FA synthesis in cancer cells is questioned, as fatty acid supplementation did not rescue the effects of ACLY inhibition (Migita et al. 2008). Moreover, given that acetyl-CoA contributes to the mevalonate pathway as well as to histone acetylation (Wellen et al. 2009), ACLY activity may modulate cancer growth on multiple levels.

Importantly, ACLY activity does not provide the only source of cellular acetyl-CoA. The enzyme acetyl-CoA synthetase short-chain family member 2 (ACSS2) produces acetyl-CoA using acetate as a substrate (Fig. 1; Yoshii et al. 2009; Comerford et al. 2014). Thus, ACSS2 may compensate for ACLY inhibition if cellular pools of acetate are not limiting. To this end, exogenous treatment of cancer cells with acetate was able to rescue the effects of ACLY inhibition on proliferation arrest and histone acetylation (Wellen et al. 2009; Hanai et al. 2012). Furthermore, ACSS2 expression is up-regulated in ACLY-deficient cells in vitro and in vivo, suggesting that acetate metabolism underlies a crucial mechanism for ACLY deficiency (Zhao et al. 2016). The contribution of acetate to acetyl-CoA production plays a fundamental role in intermediary metabolism and proliferation of cancer cells (Lyssiotis and Cantley 2014). Exogenous acetate is readily captured and metabolized by cancer cells to fuel cell growth and interestingly was found to contribute a greater proportion of carbon to the acetyl-CoA pool than glucose or glutamine (Mashimo et al. 2014). ACSS2 has been shown to be essential in this process, as ACSS2 deletion in mouse models of hepatocellular carcinoma led to a significant reduction in tumor burden (Comerford et al. 2014). Additionally, ACSS2 contributes to acetyl-CoA generation during hypoxia and is essential for cancer cell survival in tumor xenografts (Schug et al. 2015). Taken together, the pivotal roles of ACLY and ACSS2 in generating acetyl-CoA pools that are required for cancer cell growth make them attractive therapeutic targets for cancer. However, because of their relative redundancies in maintaining these pools and the ability of cancer cells to readily uptake exogenous sources of acetate in vivo, whether pharmacological inhibition of either ACLY or

ACSS2 will provide a maximal therapeutic benefit still needs to be fully analyzed.

ACC: A CANCER TARGET WITH A DISTINCT ADVANTAGE?

An alternative approach to inhibit FA synthesis in cancer is the pharmacological inhibition of ACC, which controls the enzymatic step in between ACLY and FASN; the first-committed step of the lipogenesis pathway. ACC inhibition may offer a favorable advantage over other lipogenic enzymes because of its critical role in controlling the rate-limiting reaction of FA synthesis, the production of malonyl-CoA, which serves as the substrate for the de novo synthesis of all endogenously made fatty acids (Fig. 1). ACC uses acetyl-CoA for malonyl-CoA production regardless of the source from which it was derived. Thus the generation of malonyl-CoA pools that are essential for FA production depends entirely on the activity of ACC and is not confounded by functional redundancies among other lipogenic enzymes such as ACLY and ACSS2. Furthermore, inhibition of ACCs should lead to greater lipid depletion than inhibition of any other lipogenic enzyme in the pathway, because of loss of FA synthesis and simultaneous induction of fatty acid oxidation (FAOxn) that occurs when ACC is inhibited (see below). Correspondingly, efforts to target ACC have been pursued in the bio-pharmaceutical industry for the past two decades; however, until recently the development of potent and specific ACC inhibitors with favorable drug-like properties has been difficult (reviewed in Tong and Harwood 2006; Abramson 2011). Here we focus on the recent development of a novel small-molecule allosteric ACC inhibitor that has shown promise as an anticancer compound in preclinical models of non–small cell lung cancer (NSCLC) (Svensson et al. 2016).

ACC REGULATION OF FATTY ACID METABOLISM

Eukaryotic ACCs are large, single-chain multidomain enzymes with a biotin carboxylase (BC) domain, a biotin carboxyl carrier protein (BCCP) domain, and a carboxyltransferase (CT) domain (Fig. 1). ACC catalyzes the ATP-dependent carboxylation of acetyl-CoA to form malonyl-CoA, which is the first committed step in fatty acid synthesis and the rate-limiting reaction in the pathway (Kim 1997). The carboxylation of acetyl-CoA to malonyl-CoA is catalyzed by the BC domain of ACC and involves the ATP-dependent carboxylation of biotin with bicarbonate serving as the donor of CO_2. CT then transfers the carboxyl group from biotin to acetyl-CoA to form malonyl-CoA. In mammals, ACC exists as two isoforms, a cytoplasmic 265-kDa ACC1 isoform that is expressed in lipogenic tissues such as the liver and a 280-kDa ACC2 isoform expressed in oxidative tissues such as heart and muscle, where it is localized exclusively to the outer mitochondrial membrane (Bianchi et al. 1990; Kim 1997).

Dimerization of ACC is required for enzymatic activation, and the active form of mammalian ACC is a large linear polymer. Elegant studies have shown that ACC forms multiple dimers within the BC domain of the enzyme, which promotes its activity as a carboxylase to enable the generation of malonyl-CoA (Fig. 1; Shen et al. 2004; Cho et al. 2008). The original model in the field was that the malonyl-CoA produced from ACC1 serves as a substrate for FASN and results in the formation of fatty acids, and the malonyl-CoA produced by ACC2 at the outer mitochondrial membrane functions to allosterically inhibit carnitine palmitoyltransferase 1 (CPT-1) to prevent fatty acid entry into mitochondria to undergo β-oxidation (McGarry et al. 1978). However, recent findings from genetic knockout and knock-in mice with point mutations resulting in mild gain-of-function alleles of ACC reveal that this strict division of ACC isoforms in regulation of FA synthesis and FAOxn is probably not accurate, as the evidence suggests that there are substantial redundancies between each ACC isoform (Savage et al. 2006; Harada et al. 2007; Fullerton et al. 2013). Thus, the ability of malonyl-CoA to promote fatty acid synthesis or to inhibit FAOxn independent of the ACC isoform may be an important feature for tumor cell adaptation. However, regardless of the acetyl-CoA and malonyl-CoA source, the net effect of ACC activation in a tumor results in enhanced fatty acid synthesis and reduced fatty acid degradation via FAOxn inhibition (Fig. 2A). The de novo synthesis of intracellular fatty acids results in the production of fatty acids and lipids that can be used by tumor cells for a number of purposes—including membrane biogenesis, posttranslational modification of oncoproteins, to function as signaling molecules, or stored for energy production—all of which favor the growth of a tumor (Fig. 1).

Physiological inhibition of ACC activity is controlled by phosphorylation by AMPK (Steinberg and Kemp 2009). AMPK is activated under conditions of low ATP, such as following nutrient deprivation or mitochondrial inhibitors, but also modulated by metabolic hormones and whole-body effects from exercise and dietary restriction. AMPK phosphorylates conserved serine residues Ser79 in mouse ACC1/Ser212 in ACC2 to inhibit its dimerization that is required for activation. In addition, ACC is activated allosterically by citrate, which provides the signal that acetyl-CoA pools are sufficient to enable FA synthesis. Citrate promotes dimerization of ACC and dephosphorylation of the AMPK phosphorylated serines, which serves to induce ACC activity and malonyl-CoA formation. The stimulatory effect of citrate on ACC is reversed by palmitoyl-CoA, which is abundant when there is an excess of fatty acids and therefore functions to promote the inactive conformation of ACC.

ACC AS AN ANTICANCER DRUG TARGET

Several lines of evidence support the rational for targeted ACC inhibition in oncology. Short interfering RNA (siRNA) knockdown of ACC1 expression induced apo-

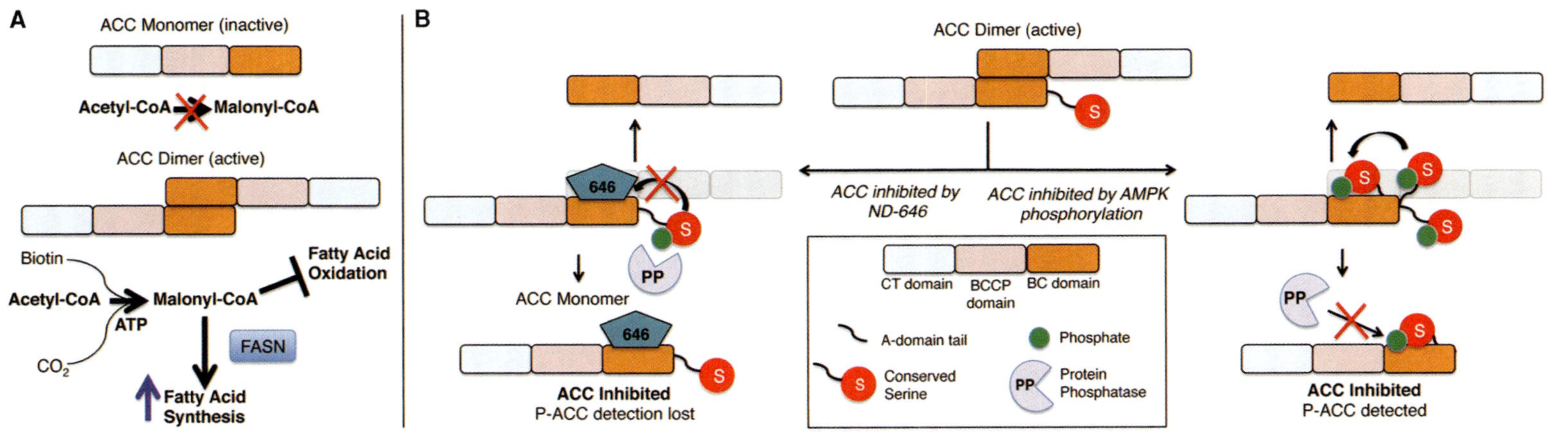

Figure 2. Model of acetyl-CoA carboxylase (ACC) regulation of fatty acid synthesis and inhibition by ND-646. (*A*) Eukaryotic ACC is a multidomain enzyme. ACC is inactive in its monomeric form and requires dimerization at the biotin carboxylase (BC) domain for activation. Activated dimeric ACC produces malonyl-CoA in a reaction that requires biotin, ATP, and CO_2. Malonyl-CoA promotes induction of fatty acid synthesis and inhibition of fatty acid oxidation. (*B*) ACC inhibition by ND-646 mimics physiologic inhibition of ACC by AMP-activated protein kinase (AMPK). (*Right*) AMPK phosphorylates conserved serine residues of the A-domain tail in the BC domain of ACC1 and ACC2 to disrupt enzyme dimerization and inhibit ACC activity. Upon phosphorylation the tail undergoes a conformation change and the phosphorylated serine binds to the dimer interface of the BC domain. Upon binding ACC dimers are disrupted into monomers and inactivated. Phosphorylated serine is protected from dephosphorylation while bound to the BC domain, and phosphorylated ACC (P-ACC) is detected. (*Left*) ND-646 binds to the dimer interface of the BC domain to disrupt dimerization and promote monomeric ACC. The AMPK phosphorylated serine of ACC is unable to bind into the BC domain because of the presence of ND-646 and subsequently becomes accessible to phosphatases that dephosphorylate ACC. Phosphorylated ACC is not detected when ND-646 is bound to ACC. CT, carboxyltransferase; BCCP, biotin carboxyl carrier protein.

ptosis in prostate (Brusselmans et al. 2005) and breast (Chajes et al. 2006) cancer cells and in cisplatin-resistant lung cancer cell lines (Wangpaichitr et al. 2012), and chemical inhibition of ACC1 and ACC2 by the macrocyclic myxobacterial natural product Soraphen A led to growth arrest in breast cancer cells (Beckers et al. 2007) and induction of apoptosis in prostate cancer cells (Corominas-Faja et al. 2014). Constitutively active ACC mutants also protected head and neck squamous cancer cells from cetuximab-induced growth inhibition (Luo et al. 2016). However, although these observations provide compelling data that ACC mediates cellular proliferation, these studies were exclusively in vitro and did not address whether ACC activation was required for tumor growth in vivo, as the genetic role of ACC in tumor growth had not been investigated. In an attempt to understand this and also to determine whether loss of cell proliferation and survival that had been observed with ACC inhibition was a direct result of FA synthesis inhibition and lipid depletion, our laboratory recently analyzed the role of ACC in NSCLC cells using CRISPR–Cas9 (Svensson et al. 2016). Consistent with the lipogenic phenotype that has been observed in tumors, we found that ACC1 expression levels were high in the panel of NSCLC cells that we analyzed and interestingly were significantly higher than ACC2, which had expression ranging from low to undetectable. CRISPR–Cas9 deletion of ACC1 led to pronounced proliferation defects and cell death, which was completely rescued by the exogenous addition of the fatty acid palmitate to the media, suggesting that the FA synthesis was impaired by ACC1 deletion. Metabolic tracing experiments to quantitate fatty acid production revealed that ACC1 deletion led to a complete loss of de novo FA synthesis, consistent with the critical role of ACC1 in lipogenesis. When we implanted ACC1-deleted NSCLC cells subcutaneously into mice, we found that tumor growth was severely impaired relative to the growth of cells bearing endogenous ACC1 expression. However, tumor growth in ACC1-deleted cells was completely rescued by stable re-expression of ACC1 into deleted cells. Taken together, our findings revealed that ACC1 mediates a metabolic liability of NSCLC, providing the rationale for targeted pharmacologic inhibition of ACC.

Consistent with the important role of fatty acid biosynthesis in cancer and consistent with our RNA-sequencing profiling, the expression of ACC1 has been shown to be frequently up-regulated in several tumor types (Witters et al. 1994; Milgraum et al. 1997; Swinnen et al. 2000; Yahagi et al. 2005). Conversely, and consistent with our studies describing low expression levels of ACC2 mRNA in NSCLC cells, the expression or activation of ACC2 has been found to be down-regulated in tumors (Corbet et al. 2016; German et al. 2016). This may be due to the important role of FAOxn in tumors to support tumor growth and survival (Carracedo et al. 2013). ACC2 inactivation in tumors can occur via phosphorylation by AMPK and recent evidence suggests that ACC2 activity is also modulated by hydroxylation (German et al. 2016). Hydroxylation of ACC2 by prolyl hydroxylase 3 (PHD3) was required for ACC2 activation to repress FAOxn and cancer cell proliferation, and it was found that cancer cells down-regulate expression of PHD3, which directly correlated with loss of ACC2 activity. Additionally, ACC2 expression may be regulated by the acetylome, providing an epigenetic mechanism for ACC2 down-regulation in tumors (Corbet et al. 2016). Ultimately, despite the fact that ACC1 and ACC2 function can be redundant, certain conditions may still allow for concomitant activation of FA synthesis and FAOxn to support tumor growth.

FAOxn inhibition itself is a focus of cancer therapy and inhibition of FAOxn has shown therapeutic efficacy in cancer models (Samudio et al. 2010; Camarda et al. 2016). This may present a conundrum when discussing the value of ACC inhibition as a strategy for cancer therapy as an ACC2 inhibitor would be expected to promote FAOxn, which in theory may enhance tumor growth rather than inhibit it. However, this conundrum can be solved by development of compounds that allow the simultaneous inhibition of ACC1 and ACC2. In principle, a dual ACC1/-2 inhibitor would result in FA synthesis inhibition and induction of FAOxn, which should result in severe lipid depletion in tumors by shutting off fatty acid production and enhancing fatty acid degradation. Thus, if tumors require the mass abundance of lipids to support cell growth, then the induction of FAOxn that may support tumor growth would be negated by the severe lipid depletion achieved by ACC1 and ACC2 inhibition, which ultimately should repress tumor growth.

ACC: OLD TARGET, NEW DRUG

Because of the pivotal role of ACC in regulating fatty acid metabolism, ACC has been under intense investigation as a clinical drug target in several metabolic diseases, including nonalcoholic fatty liver disease (NAFLD) and nonalcoholic steatohepatitis (NASH), and, correspondingly, ACC has been the focus of several clinical pipelines in the pharmaceutical industry. Inhibition of FA synthesis and stimulation of FAOxn via ACC1/-2 inhibition may favorably affect the morbidity and mortality associated with metabolic syndromes (Wakil and Abu-Elheiga 2009). To this end, several isoform nonselective ACC inhibitors have been described to date. However, because of the high degree of hydrophobicity in the regions of ACC that these inhibitors interact, they lack the optimal drug-like properties that are required for their clinical use. Recently, however, the development of small-molecule ACC inhibitors with more favorable drug-like properties was described by Harriman et al. (2016). In an effort to identify potent inhibitors of ACC1/-2 that display enhanced pharmacokinetics (PK) and pharmacodynamics (PD), Harriman et al. targeted the dimerization site in the BC domain of the enzyme in an attempt to identify novel allosteric protein–protein interaction inhibitors of ACC. Using state-of-the-art structure-based drug design, Harriman et al. identified a series of allosteric ACC1/-2 dimerization inhibitors and reported the ability of the liver selective compound ND-630 to reduce hepatic steatosis, improve insulin sensitivity, and reduce weight gain in experiment animal models of metabolic disease. Our own laboratory recently described the ability of another member of this series, ND-646, a more broadly distributed ACC inhibitor, to repress FA synthesis and tumor growth of NSCLC in genetically engineered preclinical models (Svensson et al. 2016).

MECHANISM OF ACTION OF ND-646

ND-646 and other members of the ACC inhibitor series developed by Harriman et al. function as allosteric inhibitors of ACC1 and ACC2, by preventing the dimerization that is required for activity (Harriman et al. 2016; Svensson et al. 2016). The site that ND-646 binds to in the BC domain of ACC is of importance, as it is the exact same site that the AMPK phosphorylated serine of ACC interacts to inhibit dimerization and activity of ACC. Thus, ACC inhibition by ND-646 mimics physiological inhibition of ACC by AMPK. Phosphorylation of conserved serine residues by AMPK in ACC1 (mouse serine 79, human serine 117) and ACC2 (mouse serine 212, human serine 222) in the A-domain tail of the BC domain promotes a conformational change of the tail and binding of the phospho-peptide to the BC dimer interface (Fig. 2). Binding of the phosphorylated serine to the BC domain disrupts dimerization of ACC and promotes the formation of the inactive monomer. Given that ND-646 binds to the same site in the BC domain of ACC to analogously disrupt enzyme dimerization, it also fortuitously facilitates analysis of the phosphorylation state of ACC as a biomarker to determine whether ND-646 is bound. When ND-646 binds to the BC domain, it prevents binding of the phosphorylated serine to the same site, which results in constitutive dephosphorylation of ACC as the phosphate is no longer protected from protein phosphatases (Fig. 2). Thus, phosphorylated ACC is not detected when ND-646 is bound and can be used as an efficient biomarker in cells and tissues (Svensson et al. 2016).

ACC REGULATION OF FATTY ACID SYNTHESIS AS A METABOLIC LIABILITY OF NSCLC

The development of ND-646 as a broadly distributed ACC inhibitor with good drug-like properties enabled us to characterize the in vivo PD properties of ND-646 and to determine whether ACC inhibition in tumors in vivo would be therapeutically beneficial. To determine whether ACC inhibition by ND-646 would lead to decreased de novo fatty acid synthesis in autochthonous tumors in vivo, our laboratory, in collaboration with the laboratory of Christian Metallo, performed metabolic labeling analyses in genetically engineered mouse models of NSCLC using deuterated water (^2H_2O). ^2H_2O is incorporated into a number of biosynthetic intermediates that are required for fatty acid production and only labels fatty acids that are newly synthesized (Lee et al. 2000). Thus, quantitation of deuterium-labeled fatty acids in harvested tumors provides a powerful method to measure the synthesis rates of newly made fatty acids. The ability to administer deuterated water directly into the drinking water of mice may offer advantages over alternative approaches to quantitate metabolic flux in vivo, as it is a facile procedure and well tolerated. Consistent with the studies of Medes et al. more than 60 years ago, we found that lung tumors from Kras$^{G12D/+}$; Trp53$^{-/-}$ (KP), and Kras^{G12D}; LKB1$^{-/-}$ (KL) mouse models were undergoing high rates of de novo fatty acid synthesis (Svensson et al. 2016). In our unpublished studies, we also directly compared the synthesis rates in tumors of KP mice to livers in the same animals and found that the rate of fatty acid synthesis of the most predominant saturated fatty acid palmitate was higher in the tumor tissue, supporting the original conclusions of Medes et al. and suggesting that this high rate of synthesis may be required to support the fast growth of these tumors. We found that treatment of lung tumors with ND-646 led to a significant decrease in the rate of fatty acid synthesis. Furthermore, plasma levels of free fatty acids were also significantly reduced in ND-646-treated mice, demonstrating the ability of ND-646 to inhibit FA synthesis in other tissues. To the best of our knowledge, our study describing these effects is the first published report of an ACC inhibitor with these pharmacodynamic properties in vivo.

Given that ND-646 was able to significantly inhibit fatty acid synthesis in lung tumors in these autochthonous models, it led us to investigate whether this inhibition would affect the growth of the tumors after chronic dosing. We set up large-scale preclinical trials in KP and KL mice to test the therapeutic efficacy of ND-646, as a single agent and also in combination with the standard of care chemotherapeutic agent carboplatin. We recently reported the results from these trials, where we found that KP mice were particularly sensitive to ND-646 monotherapy and that the combination of ND-646 and carboplatin led to striking suppression of lung tumor growth in both KP and KL mouse models. In summary, our study showed that ACC inhibition in biologically relevant preclinical mouse models of NSCLC led to suppression of fatty acid synthesis and tumor growth, revealing that ACC mediates a metabolic liability for this particular subset of cancers (Fig. 3).

CONCLUSION AND FUTURE PERSPECTIVES

Medes et al. concluded more than 60 years ago that the elevated rates of de novo lipogenesis they measured in tumors was probably required to support the rapid proliferation of cancer cells. Our work further supports this model and demonstrates the therapeutic viability of targeting this pathway (Svensson et al. 2016).

An open question that remains from our study is understanding whether decreased levels of plasma free fatty acids that is caused by ND-646 treatment contributes to the antitumor efficacy evoked by ACC inhibition (Fig. 3). We found that chronic 6 weeks of dosing of ND-646 led to significant reductions in the levels of plasma free fatty

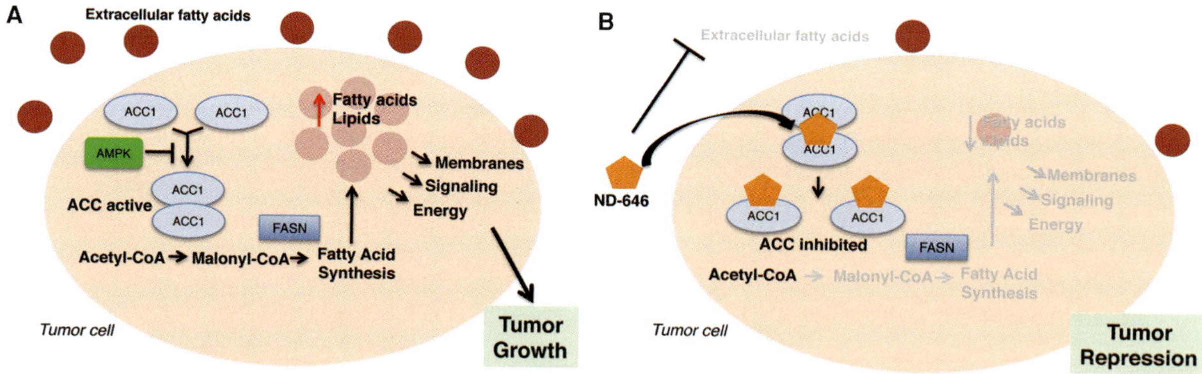

Figure 3. Model depicting the therapeutic mechanism of action of ND-646 in non–small cell lung cancer. (*A*) Lung tumors with active acetyl-CoA carboxylase (ACC) undergo high rates of de novo FA synthesis to provide fatty acids and lipids to support tumor growth. (*B*) ND-646 treatment inhibits ACC activity and suppresses fatty acid synthesis and tumor growth and also lowers plasma levels of fatty acids. AMPK, AMP-activated protein kinase; FASN, fatty acid synthase.

acids and given the ability of exogenous palmitate to rescue the effects of ACC inhibition in vitro suggests that the levels of fatty acids that are taken up by autochthonous tumors are limiting and unable to compensate for the loss of de novo synthesis. This leads to the possibility that the antitumor effects of ND-646 occur by at least a two-prong mechanism: (1) direct inhibition of FA synthesis in tumors, and (2) inhibition of FA synthesis in the tumor microenvironment and other tissues such as the liver, which is a major contributor to plasma fatty acid levels, thus reducing the ability of circulatory and microenvironmental fatty acids to rescue the effects of ND-646. The availability of the liver-directed ACC inhibitor, ND-630, may aid in determining whether the reduction in plasma free fatty acids levels alone has any effect on in vivo lung tumor growth. Additionally, whether diet modulates fatty acid bioavailability to tumors and the response to ND-646 will need to be determined.

Our study warrants further investigation of the utility of ACC inhibitors in oncology. Identification of specific tumor types and subsets of cancers that respond to ND-646 will be of great interest. Furthermore, whether the mutational status of the tumors dictates the therapeutic response to ACC inhibitors will need to be examined. Our work showed that LKB1 deletion in Kras^{G12D} lung tumors led to increased rates of de novo fatty acid synthesis compared with p53 deletion, consistent with hyperactivation of ACC activity due to LKB1 loss; however, KL tumors were less sensitive to ND-646 as a single agent. This may be related to the absolute levels of fatty acids in the tumors and the degree of suppression by ND-646, which may dictate the therapeutic efficacy of ACC inhibition. However, KL lung tumors were exquisitely sensitized to the combined effects of ND-646 and carboplatin. Because of the genetic integrity of *p53* in the KL tumors, we found that carboplatin induced nuclear p53 accumulation in these tumors, which might explain the striking synergy that we observed with this combination treatment, as p53 activation is known to promote FAOxn via its induction of mRNAs for *Cpt1c*, *Lipin1*, and *Acad11* (Assaily et al. 2011; Zaugg et al. 2011; Jiang et al. 2015).

Thus, the ability of ND-646 to inhibit FA synthesis, combined with carboplatin-induced p53-dependent FAOxn, would be expected to significantly deplete the cellular pools of fatty acids that are needed for cellular division. Given that ND-646 inhibits both ACC1 and ACC2, ND-646 by itself would be expected to inhibit FA synthesis and promote FAOxn; however, the majority of tumor cell lines that we have analyzed express little to no endogenous ACC2.

Although not directly assessed in our study with ND-646, ACC inhibition with ND-630 led to potent suppression of malonyl-CoA production (Harriman et al. 2016). However, whether the intracellular levels of acetyl-CoA are affected by decreased conversion into malonyl-CoA has not been analyzed. Although we showed that non-FA synthesis pathways that use acetyl-CoA, such as cholesterol synthesis, were unaffected, we did not determine whether acetylation of histones were affected, as has been described with genetic deletion of ACC in yeast and mouse liver (Galdieri and Vancura 2012; Chow et al. 2014). It will be of great interest to determine these effects and whether ACC provides an additional link between fatty acid metabolism and epigenetic control of transcriptional regulation.

Finally, given that many oncogenic proteins are modulated by lipidation, it raises the intriguing possibility that the antitumor activity of ND-646 might also be explained in part by decreased posttranslational modification of Kras, which is the oncogenic driver of NSCLC growth in our models. Kras undergoes extensive posttranslational modification, including farnesylation and palmitoylation (Ahearn et al. 2012); thus, whether loss of FA synthesis through ACC inhibition affects these modifications and the signaling ability of Kras should be investigated.

In conclusion, the development of ND-646 as a potent and specific ACC inhibitor has revealed that ACC mediates a metabolic liability of cancer. The continued discovery of compounds targeting other members of the FA synthesis pathway and the elucidation of new targets will shed light on whether inhibition of FA synthesis will translate from bench to bedside in oncology.

ACKNOWLEDGMENTS

R.J.S. holds the William R. Brody Chair and is a professor in the Molecular and Cell Biology Department at the Salk Institute of Biological Studies. The work from our laboratory described in this review is supported by grants from the National Institutes of Health (R01CA172229, P01CA120964), the Samuel Waxman Cancer Research Foundation, and The Leona M. and Harry B. Helmsley Charitable Trust (grant #2012-PG-MED002). R.U.S. was supported by a postdoctoral fellowship from the American Cancer Society (grant ACS#124183-PF-13-023-01-CSM).

REFERENCES

Abramson HN. 2011. The lipogenesis pathway as a cancer target. *J Med Chem* **54:** 5615–5638.

Ahearn IM, Haigis K, Bar-Sagi D, Philips MR. 2012. Regulating the regulator: Post-translational modification of RAS. *Nat Rev Mol Cell Biol* **13:** 39–51.

Alli PM, Pinn ML, Jaffee EM, McFadden JM, Kuhajda FP. 2005. Fatty acid synthase inhibitors are chemopreventive for mammary cancer in neu-N transgenic mice. *Oncogene* **24:** 39–46.

Alwarawrah Y, Hughes P, Loiselle D, Carlson DA, Darr DB, Jordan JL, Xiong J, Hunter LM, Dubois LG, Thompson JW, et al. 2016. Fasnall, a selective FASN inhibitor, shows potent anti-tumor activity in the MMTV-Neu model of $HER2^+$ breast cancer. *Cell Chem Biol* **23:** 678–688.

Assaily W, Rubinger DA, Wheaton K, Lin Y, Ma W, Xuan W, Brown-Endres L, Tsuchihara K, Mak TW, Benchimol S. 2011. ROS-mediated p53 induction of Lpin1 regulates fatty acid oxidation in response to nutritional stress. *Mol Cell* **44:** 491–501.

Bauer DE, Hatzivassiliou G, Zhao F, Andreadis C, Thompson CB. 2005. ATP citrate lyase is an important component of cell growth and transformation. *Oncogene* **24:** 6314–6322.

Beckers A, Organe S, Timmermans L, Scheys K, Peeters A, Brusselmans K, Verhoeven G, Swinnen JV. 2007. Chemical inhibition of acetyl-CoA carboxylase induces growth arrest and cytotoxicity selectively in cancer cells. *Cancer Res* **67:** 8180–8187.

Benjamin DI, Li DS, Lowe W, Heuer T, Kemble G, Nomura DK. 2015. Diacylglycerol metabolism and signaling is a driving force underlying FASN inhibitor sensitivity in cancer cells. *ACS Chem Biol* **10:** 1616–1623.

Bianchi A, Evans JL, Iverson AJ, Nordlund AC, Watts TD, Witters LA. 1990. Identification of an isozymic form of acetyl-CoA carboxylase. *J Biol Chem* **265:** 1502–1509.

Brusselmans K, De Schrijver E, Verhoeven G, Swinnen JV. 2005. RNA interference-mediated silencing of the acetyl-CoA-carboxylase-α gene induces growth inhibition and apoptosis of prostate cancer cells. *Cancer Res* **65:** 6719–6725.

Camarda R, Zhou AY, Kohnz RA, Balakrishnan S, Mahieu C, Anderton B, Eyob H, Kajimura S, Tward A, Krings G, et al. 2016. Inhibition of fatty acid oxidation as a therapy for MYC-overexpressing triple-negative breast cancer. *Nat Med* **22:** 427–432.

Carling D, Clarke PR, Zammit VA, Hardie DG. 1989. Purification and characterization of the AMP-activated protein kinase. Copurification of acetyl-CoA carboxylase kinase and 3-hydroxy-3-methylglutaryl-CoA reductase kinase activities. *Eur J Biochem* **186:** 129–136.

Carracedo A, Cantley LC, Pandolfi PP. 2013. Cancer metabolism: Fatty acid oxidation in the limelight. *Nat Rev Cancer* **13:** 227–232.

Chajes V, Cambot M, Moreau K, Lenoir GM, Joulin V. 2006. Acetyl-CoA carboxylase α is essential to breast cancer cell survival. *Cancer Res* **66:** 5287–5294.

Chen HW, Chang YF, Chuang HY, Tai WT, Hwang JJ. 2012. Targeted therapy with fatty acid synthase inhibitors in a human prostate carcinoma LNCaP/tk-luc-bearing animal model. *Prostate Cancer Prostatic Dis* **15:** 260–264.

Cho YS, Lee JI, Shin D, Kim HT, Cheon YH, Seo CI, Kim YE, Hyun YL, Lee YS, Sugiyama K, et al. 2008. Crystal structure of the biotin carboxylase domain of human acetyl-CoA carboxylase 2. *Proteins* **70:** 268–272.

Chow JDY, Lawrence RT, Healy ME, Dominy JE, Liao JA, Breen DS, Byrne FL, Kenwood BM, Lackner C, Okutsu S, et al. 2014. Genetic inhibition of hepatic acetyl-CoA carboxylase activity increases liver fat and alters global protein acetylation. *Mol Metab* **3:** 419–431.

Comerford SA, Huang Z, Du X, Wang Y, Cai L, Witkiewicz AK, Walters H, Tantawy MN, Fu A, Manning HC, et al. 2014. Acetate dependence of tumors. *Cell* **159:** 1591–1602.

Corbet C, Pinto A, Martherus R, Santiago de Jesus JP, Polet F, Feron O. 2016. Acidosis drives the reprogramming of fatty acid metabolism in cancer cells through changes in mitochondrial and histone acetylation. *Cell Metab* **24:** 311–323.

Corominas-Faja B, Cuyas E, Gumuzio J, Bosch-Barrera J, Leis O, Martin AG, Menendez JA. 2014. Chemical inhibition of acetyl-CoA carboxylase suppresses self-renewal growth of cancer stem cells. *Oncotarget* **5:** 8306–8316.

Duvel K, Yecies JL, Menon S, Raman P, Lipovsky AI, Souza AL, Triantafellow E, Ma QC, Gorski R, Cleaver S, et al. 2010. Activation of a metabolic gene regulatory network downstream of mTOR complex 1. *Mol Cell* **39:** 171–183.

Fullerton MD, Galic S, Marcinko K, Sikkema S, Pulinilkunnil T, Chen ZP, O'Neill HM, Ford RJ, Palanivel R, O'Brien M, et al. 2013. Single phosphorylation sites in Acc1 and Acc2 regulate lipid homeostasis and the insulin-sensitizing effects of metformin. *Nat Med* **19:** 1649–1654.

Galdieri L, Vancura A. 2012. Acetyl-CoA carboxylase regulates global histone acetylation. *J Biol Chem* **287:** 23865–23876.

German NJ, Yoon H, Yusuf RZ, Murphy JP, Finley LWS, Laurent G, Haas W, Satterstrom FK, Guarnerio J, Zaganjor E, et al. 2016. PHD3 loss in cancer enables metabolic reliance on fatty acid oxidation via deactivation of ACC2. *Mol Cell* **63:** 1006–1020.

Griffiths B, Lewis CA, Bensaad K, Ros S, Zhang Q, Ferber EC, Konisti S, Peck B, Miess H, East P, et al. 2013. Sterol regulatory element binding protein-dependent regulation of lipid synthesis supports cell survival and tumor growth. *Cancer Metab* **1:** 3.

Hanai J, Doro N, Sasaki AT, Kobayashi S, Cantley LC, Seth P, Sukhatme VP. 2012. Inhibition of lung cancer growth: ATP citrate lyase knockdown and statin treatment leads to dual blockade of mitogen-activated protein kinase (MAPK) and phosphatidylinositol-3-kinase (PI3K)/AKT pathways. *J Cell Physiol* **227:** 1709–1720.

Harada N, Oda Z, Hara Y, Fujinami K, Okawa M, Ohbuchi K, Yonemoto M, Ikeda Y, Ohwaki K, Aragane K, et al. 2007. Hepatic de novo lipogenesis is present in liver-specific ACC1-deficient mice. *Mol Cell Biol* **27:** 1881–1888.

Harriman G, Greenwood J, Bhat S, Huang X, Wang R, Paul D, Tong L, Saha AK, Westlin WF, Kapeller R, et al. 2016. Acetyl-CoA carboxylase inhibition by ND-630 reduces hepatic steatosis, improves insulin sensitivity, and modulates dyslipidemia in rats. *Proc Natl Acad Sci* **113:** E1796–E1805.

Hatzivassiliou G, Zhao FP, Bauer DE, Andreadis C, Shaw AN, Dhanak D, Hingorani SR, Tuveson DA, Thompson CB. 2005. ATP citrate lyase inhibition can suppress tumor cell growth. *Cancer Cell* **8:** 311–321.

Horton JD. 2002. Sterol regulatory element-binding proteins: Transcriptional activators of lipid synthesis. *Biochem Soc Trans* **30:** 1091–1095.

Jiang D, LaGory EL, Kenzelmann Broz D, Bieging KT, Brady CA, Link N, Abrams JM, Giaccia AJ, Attardi LD. 2015. Analysis of p53 transactivation domain mutants reveals Acad11 as a metabolic target important for p53 pro-survival function. *Cell Rep* **10:** 1096–1109.

Kim KH. 1997. Regulation of mammalian acetyl-coenzyme A carboxylase. *Annu Rev Nutr* **17:** 77–99.

Kim J, DeBerardinis RJ. 2016. Blocking fatty acid synthesis reduces lung tumor growth in mice. *Nat Med* **22:** 1077–1078.

Kuhajda FP, Jenner K, Wood FD, Hennigar RA, Jacobs LB, Dick JD, Pasternack GR. 1994. Fatty acid synthesis: A potential selective target for antineoplastic therapy. *Proc Natl Acad Sci* **91:** 6379–6383.

Lee WN, Bassilian S, Lim S, Boros LG. 2000. Loss of regulation of lipogenesis in the Zucker diabetic (ZDF) rat. *Am J Physiol Endocrinol Metab* **279:** E425–E432.

Lee JV, Carrer A, Shah S, Snyder NW, Wei S, Venneti S, Worth AJ, Yuan Z-F, Lim H-W, Liu S, et al. 2014. Akt-dependent metabolic reprogramming regulates tumor cell histone acetylation. *Cell Metab* **20:** 306–319.

Li Y, Xu S, Mihaylova MM, Zheng B, Hou X, Jiang B, Park O, Luo Z, Lefai E, Shyy JY, et al. 2011. AMPK phosphorylates and inhibits SREBP activity to attenuate hepatic steatosis and atherosclerosis in diet-induced insulin-resistant mice. *Cell Metab* **13:** 376–388.

Loftus TM, Jaworsky DE, Frehywot GL, Townsend CA, Ronnett GV, Lane MD, Kuhajda FP. 2000. Reduced food intake and body weight in mice treated with fatty acid synthase inhibitors. *Science* **288:** 2379–2381.

Luo J, Hong Y, Lu Y, Qiu S, Chaganty BK, Zhang L, Wang X, Li Q, Fan Z. 2016. Acetyl-CoA carboxylase rewires cancer metabolism to allow cancer cells to survive inhibition of the Warburg effect by cetuximab. *Cancer Lett* **384:** 39–49.

Lyssiotis CA, Cantley LC. 2014. Acetate fuels the cancer engine. *Cell* **159:** 1492–1494.

Mashimo T, Pichumani K, Vemireddy V, Hatanpaa KJ, Singh DK, Sirasanagandla S, Nannepaga S, Piccirillo SG, Kovacs Z, Foong C, et al. 2014. Acetate is a bioenergetic substrate for human glioblastoma and brain metastases. *Cell* **159:** 1603–1614.

McGarry JD, Leatherman GF, Foster DW. 1978. Carnitine palmitoyltransferase. I. The site of inhibition of hepatic fatty acid oxidation by malonyl-CoA. *J Biol Chem* **253:** 4128–4136.

Medes G, Thomas A, Weinhouse S. 1953. Metabolism of neoplastic tissue. IV. A study of lipid synthesis in neoplastic tissue slices in vitro. *Cancer Res* **13:** 27–29.

Menendez JA, Lupu R. 2007. Fatty acid synthase and the lipogenic phenotype in cancer pathogenesis. *Nat Rev Cancer* **7:** 763–777.

Menendez JA, Vellon L, Mehmi I, Oza BP, Ropero S, Colomer R, Lupu R. 2004. Inhibition of fatty acid synthase (FAS) suppresses HER2/neu (erbB-2) oncogene overexpression in cancer cells. *Proc Natl Acad Sci* **101:** 10715–10720.

Migita T, Narita T, Nomura K, Miyagi E, Inazuka F, Matsuura M, Ushijima M, Mashima T, Seimiya H, Satoh Y, et al. 2008. ATP citrate lyase: Activation and therapeutic implications in non–small cell lung cancer. *Cancer Res* **68:** 8547–8554.

Milgraum LZ, Witters LA, Pasternack GR, Kuhajda FP. 1997. Enzymes of the fatty acid synthesis pathway are highly expressed in in situ breast carcinoma. *Clin Cancer* **3:** 2115–2120.

Pizer ES, Wood FD, Heine HS, Romantsev FE, Pasternack GR, Kuhajda FP. 1996. Inhibition of fatty acid synthesis delays disease progression in a xenograft model of ovarian cancer. *Cancer Res* **56:** 1189–1193.

Pizer ES, Thupari J, Han WF, Pinn ML, Chrest FJ, Frehywot GL, Townsend CA, Kuhajda FP. 2000. Malonyl-coenzyme-A is a potential mediator of cytotoxicity induced by fatty-acid synthase inhibition in human breast cancer cells and xenograft. *Cancer Res* **60:** 213–218.

Porstmann T, Santos CR, Griffiths B, Cully M, Wu M, Leevers S, Griffiths JR, Chung YL, Schulze A. 2008. SREBP activity is regulated by mTORC1 and contributes to Akt-dependent cell growth. *Cell Metab* **8:** 224–236.

Relat J, Blancafort A, Oliveras G, Cufi S, Haro D, Marrero PF, Puig T. 2012. Different fatty acid metabolism effects of (−)-epigallocatechin-3-gallate and C75 in adenocarcinoma lung cancer. *BMC Cancer* **12:** 280.

Rohrig F, Schulze A. 2016. The multifaceted roles of fatty acid synthesis in cancer. *Nat Rev Cancer* **16:** 732–749.

Samudio I, Harmancey R, Fiegl M, Kantarjian H, Konopleva M, Korchin B, Kaluarachchi K, Bornmann W, Duvvuri S, Taegtmeyer H, et al. 2010. Pharmacologic inhibition of fatty acid oxidation sensitizes human leukemia cells to apoptosis induction. *J Clin Invest* **120:** 142–156.

Savage DB, Choi CS, Samuel VT, Liu ZX, Zhang D, Wang A, Zhang XM, Cline GW, Yu XX, Geisler JG, et al. 2006. Reversal of diet-induced hepatic steatosis and hepatic insulin resistance by antisense oligonucleotide inhibitors of acetyl-CoA carboxylases 1 and 2. *J Clin Invest* **116:** 817–824.

Schug ZT, Peck B, Jones DT, Zhang QF, Grosskurth S, Alam IS, Goodwin LM, Smethurst E, Mason S, Blyth K, et al. 2015. Acetyl-CoA synthetase 2 promotes acetate utilization and maintains cancer cell growth under metabolic stress. *Cancer Cell* **27:** 57–71.

Shackelford DB, Shaw RJ. 2009. The LKB1-AMPK pathway: Metabolism and growth control in tumour suppression. *Nat Rev Cancer* **9:** 563–575.

Shen Y, Volrath SL, Weatherly SC, Elich TD, Tong L. 2004. A mechanism for the potent inhibition of eukaryotic acetyl-coenzyme A carboxylase by soraphen A, a macrocyclic polyketide natural product. *Mol Cell* **16:** 881–891.

Steinberg GR, Kemp BE. 2009. AMPK in health and disease. *Physiol Rev* **89:** 1025–1078.

Svensson RU, Parker SJ, Eichner LJ, Kolar MJ, Wallace M, Brun SN, Lombardo PS, Van Nostrand JL, Hutchins A, Vera L, et al. 2016. Inhibition of acetyl-CoA carboxylase suppresses fatty acid synthesis and tumor growth of non-small-cell lung cancer in preclinical models. *Nat Med* **22:** 1108–1119.

Swinnen JV, Vanderhoydonc F, Elgamal AA, Eelen M, Vercaeren I, Joniau S, Van Poppel H, Baert L, Goossens K, Heyns W, et al. 2000. Selective activation of the fatty acid synthesis pathway in human prostate cancer. *Int J Cancer* **88:** 176–179.

Swinnen JV, Brusselmans K, Verhoeven G. 2006. Increased lipogenesis in cancer cells: New players, novel targets. *Curr Opin Clin Nutr Metab Care* **9:** 358–365.

Szutowicz A, Kwiatkowski J, Angielski S. 1979. Lipogenetic and glycolytic enzyme activities in carcinoma and nonmalignant diseases of the human breast. *Br J Cancer* **39:** 681–687.

Thupari JN, Pinn ML, Kuhajda FP. 2001. Fatty acid synthase inhibition in human breast cancer cells leads to malonyl-CoA-induced inhibition of fatty acid oxidation and cytotoxicity. *Biochem Biophys Res Commun* **285:** 217–223.

Tong L, Harwood HJ Jr. 2006. Acetyl-coenzyme A carboxylases: Versatile targets for drug discovery. *J Cell Biochem* **99:** 1476–1488.

Ventura R, Mordec K, Waszczuk J, Wang ZT, Lai JL, Fridlib M, Buckley D, Kemble G, Heuer TS. 2015. Inhibition of de novo palmitate synthesis by fatty acid synthase induces apoptosis in tumor cells by remodeling cell membranes, inhibiting signaling pathways, and reprogramming gene expression. *EBiomedicine* **2:** 808–824.

Wakil SJ, Abu-Elheiga LA. 2009. Fatty acid metabolism: Target for metabolic syndrome. *J Lipid Res* **50:** S138–S143.

Wangpaichitr M, Sullivan EJ, Theodoropoulos G, Wu C, You M, Feun LG, Lampidis TJ, Kuo MT, Savaraj N. 2012. The relationship of thioredoxin-1 and cisplatin resistance: Its impact on ROS and oxidative metabolism in lung cancer cells. *Mol Cancer Ther* **11:** 604–615.

Wellen KE, Hatzivassiliou G, Sachdeva UM, Bui TV, Cross JR, Thompson CB. 2009. ATP-citrate lyase links cellular metabolism to histone acetylation. *Science* **324:** 1076–1080.

Witters LA, Widmer J, King AN, Fassihi K, Kuhajda F. 1994. Identification of human acetyl-CoA carboxylase isozymes in tissue and in breast cancer cells. *Int J Biochem* **26:** 589–594.

Yahagi N, Shimano H, Hasegawa K, Ohashi K, Matsuzaka T, Najima Y, Sekiya M, Tomita S, Okazaki H, Tamura Y, et al. 2005. Co-ordinate activation of lipogenic enzymes in hepatocellular carcinoma. *Eur J Cancer* **41:** 1316–1322.

Yoshii Y, Furukawa T, Yoshii H, Mori T, Kiyono Y, Waki A, Kobayashi M, Tsujikawa T, Kudo T, Okazawa H, et al. 2009. Cytosolic acetyl-CoA synthetase affected tumor cell survival under hypoxia: The possible function in tumor acetyl-CoA/acetate metabolism. *Cancer Sci* **100:** 821–827.

Zaugg K, Yao Y, Reilly PT, Kannan K, Kiarash R, Mason J, Huang P, Sawyer SK, Fuerth B, Faubert B, et al. 2011. Carnitine palmitoyltransferase 1C promotes cell survival and tumor growth under conditions of metabolic stress. *Genes Dev* **25:** 1041–1051.

Zhao S, Torres A, Henry RA, Trefely S, Wallace M, Lee JV, Carrer A, Sengupta A, Campbell SL, Kuo YM, et al. 2016. ATP-citrate lyase controls a glucose-to-acetate metabolic switch. *Cell Rep* **17:** 1037–1052.

Cancer Immunogenomics: Computational Neoantigen Identification and Vaccine Design

Jasreet Hundal,[1] Christopher A. Miller,[1,2] Malachi Griffith,[1,2] Obi L. Griffith,[1,2] Jason Walker,[1] Susanna Kiwala,[1] Aaron Graubert,[1] Joshua McMichael,[1] Adam Coffman,[1] and Elaine R. Mardis[3]

[1]*McDonnell Genome Institute at Washington University School of Medicine, St. Louis, Missouri 63108*

[2]*Department of Medicine, Division of Oncology, Washington University School of Medicine, St. Louis, Missouri 63110*

[3]*Nationwide Children's Hospital, Institute for Genomic Medicine, Columbus, Ohio 43205*

Correspondence: elaine.mardis@nationwidechildrens.org

The application of modern high-throughput genomics to the study of cancer genomes has exploded in the past few years, yielding unanticipated insights into the myriad and complex combinations of genomic alterations that lead to the development of cancers. Coincident with these genomic approaches have been computational analyses that are capable of multiplex evaluations of genomic data toward specific therapeutic end points. One such approach is called "immunogenomics" and is now being developed to interpret protein-altering changes in cancer cells in the context of predicted preferential binding of these altered peptides by the patient's immune molecules, specifically human leukocyte antigen (HLA) class I and II proteins. One goal of immunogenomics is to identify those cancer-specific alterations that are likely to elicit an immune response that is highly specific to the patient's cancer cells following stimulation by a personalized vaccine. The elements of such an approach are outlined herein and constitute an emerging therapeutic option for cancer patients.

Cancer immunologists developed experimentally testable hypotheses around the idea that mutated proteins in cancer cells provided favored targets for immune response in the mid-1980s (De Plaen et al. 1988; Monach et al. 1995). Their hypotheses were inspired by earlier observations of immune-capable mice that developed spontaneous cancers and, after removal and rechallenge, exhibited resistance to their original cancers (Foley 1953; Prehn and Main 1957; Old 1982). This implied that cancer-specific immunity was a plausible resistance mechanism, but the mechanism for immunity was unclear. Experimental approaches to test for the presence of tumor-specific mutant antigens or "neoantigens" were aided by discoveries in immunology that elucidated the role and functions of the human leukocyte antigen (HLA) proteins (Bjorkman et al. 1987; Babbitt et al. 2005), and by the advent of molecular cloning techniques. This combination led to the cloning and identification of the first tumor neoantigen by Boon and colleagues, a peptide called MAGE-A1 that was derived from a melanoma patient sample (van der Bruggen et al. 1991). As exciting as this proof of hypothesis was at the time, the laborious and lengthy procedures required to identify a single tumor-specific neoantigen that was highly unique to the patient's disease meant there was no clear trajectory for this approach in a clinical setting.

IMMUNOGENOMICS AND NEOANTIGEN PREDICTION

Several recent developments have made it more plausible to revisit the notion of identifying tumor-unique neoantigenic peptides, however. In particular, the development and widespread use of next-generation sequencing (NGS) platforms has figured prominently in this trend (Mardis 2017). Introduced in the mid-2000s, these instruments and their accompanying techniques led to the first cancer whole-genome sequencing study that compared a tumor with normal genome for a single patient, establishing that somatic mutations could be discovered in the comparison (Ley et al. 2008). Selective hybridization-based techniques known as "hybrid capture" emerged in 2009 and permitted isolation and sequencing of only the known protein-coding exons from a whole-genome library (the exome). This "whole-exome sequencing" approach significantly simplified data analysis for the identification of DNA-level changes that were somatic (i.e., specific to the tumor cells) and led to changes in the amino acid sequences of the resulting proteins. The foundational underpinning for mutation discovery lies in the reference human genome sequence, which serves as a template for NGS read alignment. Over time, the increasing length of NGS reads, the advent of read-pair data (both ends of each library fragment are sequenced), and

Published by Cold Spring Harbor Laboratory Press; doi: 10.1101/sqb.2016.81.030726

the improved mapping algorithms have led to overall improvements in the quality of NGS read alignment on the highly repetitive and complex human genome reference.

Beyond these advances, the most important step in this process—namely, the proper identification of variants in aligned NGS read data—has also improved over time. There are various algorithmic approaches to somatic variant identification from NGS data that typically are highly tuned to the type of variant being detected. The basic workflow involves detecting variants in the aligned read data from the tumor sample, and separately in the normal sample, and then comparing the variants at each locus to eliminate those shared between tumor and normal from further consideration because they are constitutional or "germline" in origin (Ding et al. 2010). In practice, each variant identification algorithm has strengths and weaknesses (Griffith et al. 2015), and therefore most data analysis approaches to somatic variant identification use more than one variant caller and consider most strongly the shared or consensus variant lists called by each algorithm as likely true positives. Furthermore, variant detection accuracy is a function of data coverage, so most variant detection approaches utilize a pre- or post-variant call filtering scheme to remove all variants identified at loci with suboptimal coverage depth. Generally speaking, detection of somatic single-nucleotide variants (SNVs) or point mutations is the most straightforward variant type to identify. Detecting the insertion or deletion (indel) of one or more nucleotides is more difficult, as gapped alignment or assembly of the sequence read data that define the indel is required. Here, some proportion of the indel-containing reads will not map without special alignment algorithms, because of the need to produce a gapped alignment with sufficient length of sequence similarity to anchor the read accurately to the surrounding sequence. As a result, coverage is almost always lower at indel variant sites because of read mapping difficulties. Finally, structural variants are the most difficult to detect with high accuracy. These variants include chromosomal-scale events such as inversions, translocations, and deletions and are detected based on mismapped read-pair data following NGS read alignment. Both indels and structural variants, when they occur in protein-coding genes, can provide important and potentially immunogenic changes to the amino acid sequences of expressed proteins in a cancer cell.

The next step is annotating the variants that have been identified, by interpreting their altered nucleotide sequences in the context of their protein-coding impact. This is straightforward in SNVs, but indels that result in a shift in the open reading frame (frameshift) and gene-fusing structural variants are more difficult to interpret properly. Automated annotation software such as Variant Effect Predictor (VEP) (McLaren et al. 2016) or ANNOVAR (Wang et al. 2010) typically is used for variant interpretation. For many genes in the human genome, multiple transcripts have been identified, which complicates the annotation process because one must specify which transcript is to be annotated. In the absence of adequate RNA-seq data to infer the correct transcript, one typically annotates from the transcript with the highest evidence for expression. Tissue-specific expression data sets, such as GTEx (www.gtexportal.org/), also can be consulted for transcript information, if desired. Once the variants and their proper notation are determined along with their flanking sequences, the results are parsed for use in the neoantigen detection schema described below. Typically for major histocompatibility complex (MHC) class I restricted neoantigen discovery, we tile through the variant-containing region of each protein with peptides of eight to 11 amino acids to generate the test sequences for neoantigen evaluation. For MHC class II restricted neoantigen discovery, the tiled peptides are 15 amino acids in length.

Stringent alignment of NGS reads from the normal tissue comparator onto the HLA coding genes permits the haplotype of each patient to be determined from NGS exome data. There are several approaches to HLA typing from NGS data, including HLAminer (Warren et al. 2012), OptiType (Szolek et al. 2014), and ATHLATES (Liu et al. 2013). HLAminer and OptiType also can call HLA haplotypes from RNA-seq data. In high-mutation-load tumor types, it also is advised to evaluate and annotate the HLA gene sequences from the tumor tissue NGS data, because there may be inactivating (nonsense or frameshift) mutations in one or more of these genes that would preclude further consideration of these mutation-inactivated HLA proteins in neoantigen prediction.

The final step in neoantigen discovery is the prediction of binding affinities for each novel somatic mutation-translated set of peptides to the patient's HLA proteins. The development of modeling algorithms that calculate HLA binding affinities has been an active area of research that has resulted in several approaches such as analysis by neural network-based algorithms that are trained on measured binding affinities (e.g., NetMHC), scattering-matrix method (SMM)-based approaches, and others (Peters and Sette 2005; Lundegaard et al. 2008; Srivastava et al. 2013). The large number of algorithms indicates the attendant difficulty of identifying neoantigens, although precision has improved over time, especially for class I HLA proteins, for which more experimental data exist, even for the more rare haplotypes. Class II neoantigen binding is more difficult to predict because of the nature of the peptide binding site, which is open at both ends, permitting the flanking amino acids around the nine-amino-acid-core binding motif to influence binding affinity in a context-dependent manner (Paul et al. 2015).

Because of the complexities of the aforementioned set of steps required to generate predicted neoantigens, our group published a computational pipeline called pVAC-Seq (personalized Variant Antigens by Cancer Sequencing) that is designed to perform these steps in sequence (Hundal et al. 2016). pVAC-Seq requires several input data, including an annotated list of somatic variants, comparison sequences from the native peptide correlates, predicted HLA class I haplotypes for the patient, and RNA-seq-derived data regarding mutant expression levels for each of the input peptides. The overall workflow for neoantigen prediction is represented in Figure 1.

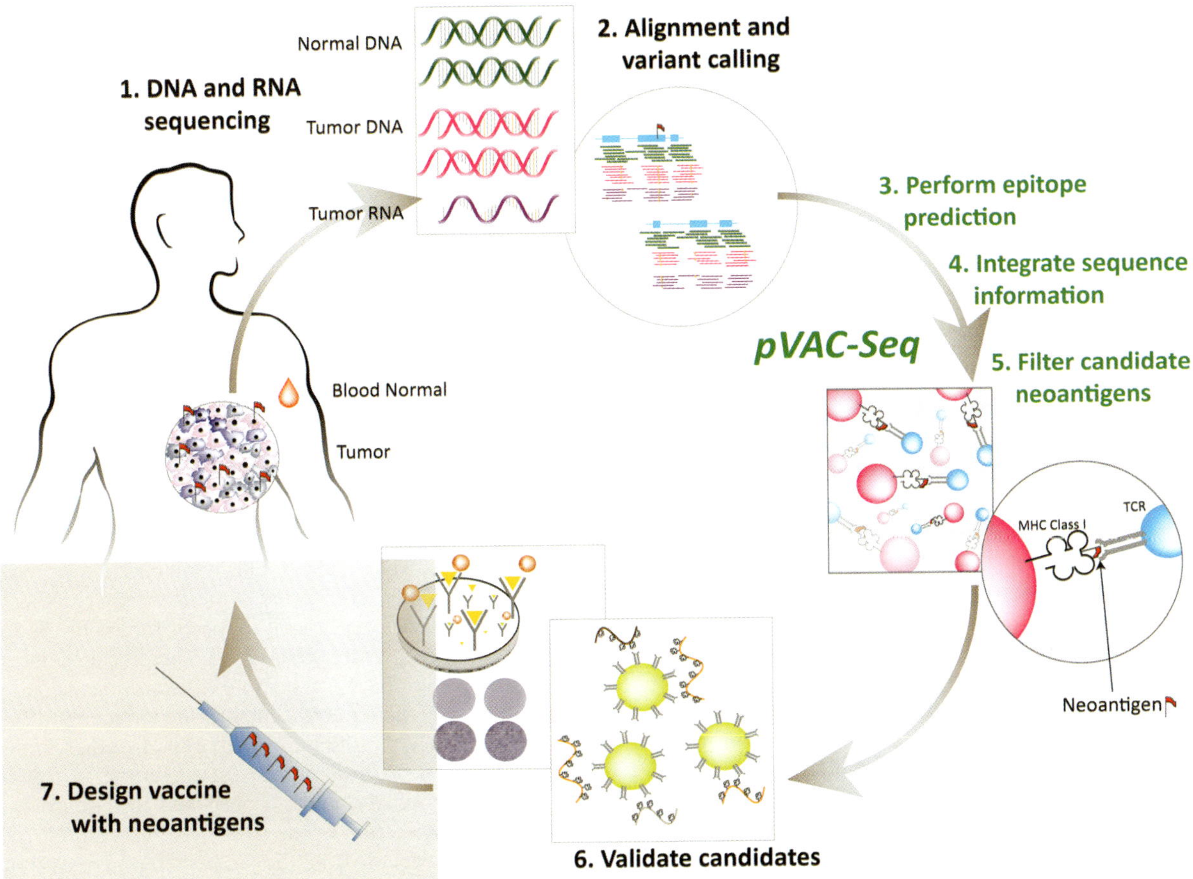

Figure 1. Generalized workflow diagram for identification of tumor-specific mutant antigens ("neoantigens") from next-generation sequencing (NGS) data. MHC, major histocompatibility complex; TCR, T-cell receptor.

Neoantigen prediction can be used clinically to achieve several therapeutic goals, depending on the nature of the study. Two case reports are presented to illustrate the clinical utility of neoantigen prediction to (1) predict neoantigen load as a consequence of an identified mismatch repair gene defect and as an indicator of likely response to checkpoint blockade therapy, (2) identify changes in a temporal series of tumors from the same individual patient, and (3) design a personalized vaccine for a patient with recurrent cancer.

CASE REPORTS

We recently reported (Johanns et al. 2016) on an individual case of a male patient (early 30s) with a history of colon polyps who presented with a brain tumor that was removed by craniotomy and diagnosed as a glioblastoma multiforme (GBM), World Health Organization (WHO) grade IV with primitive neuroectodermal tumor (PNET) features. Following resection, the patient received radiation and temozolomide (TMZ) treatment per standard of care, but new symptoms subsequent to this treatment during high-dose maintenance TMZ led to the laminectomy and diagnosis by histopathology of a drop spinal metastasis. Intermediate to the primary and secondary diagnoses, a panel gene test (Foundation One) identified an abnormally elevated mutation rate, and a reflex assay of blood DNA for germline susceptibility identified a known pathogenic mutation in polymerase epsilon (POLE L424 V). Based on this result, the patient received a PD-1 checkpoint blockade therapy (pembrolizumab) for about 3 wk following the removal of the drop metastasis. Upon experiencing further complications after 3 wk of pembrolizumab therapy, the patient was subsequently diagnosed with a second drop metastasis lower on the spinal cord. This tumor also was resected by laminectomy and diagnosed as a GBM with PNET features by histopathology. Since all three occurrences of this patient's cancer were banked, we isolated DNA and RNA from each sample. An exome capture-based sequencing comparison of all three tumors to a matched blood normal from the patient coupled with RNA-seq from each of the three tumors were used to characterize the patient's changes in tumor heterogeneity and neoantigen load over time. We also performed immunohistochemistry (IHC) on slides from each tumor resection sample to identify whether the use of checkpoint blockade therapy had influenced the tumor infiltration of different immune molecules behind the blood–brain barrier.

As demonstrated in our retrospective sequencing and analysis, genomic characterization of a tumor series can

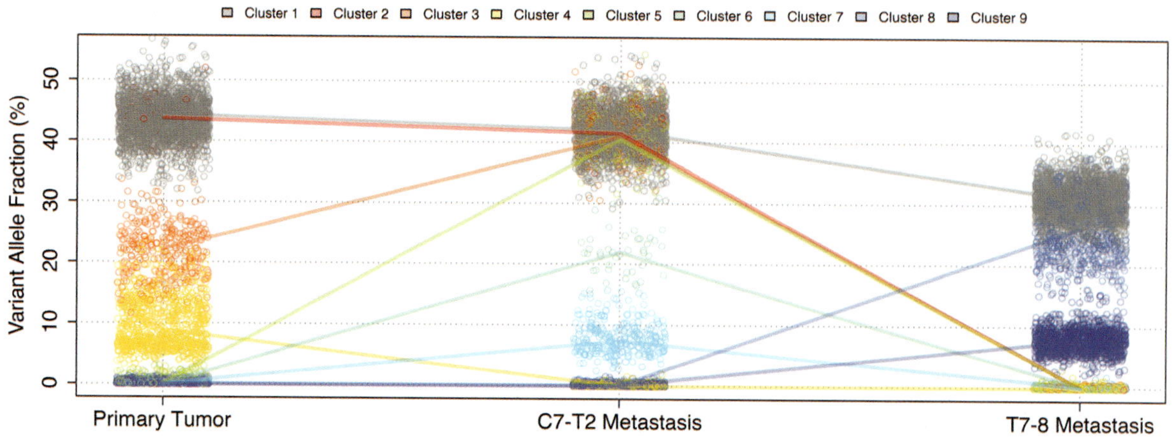

Figure 2. Comparison of tumor heterogeneity and subclonal population variation in sequential samples from a single patient. (Reprinted, with permission, from Johanns et al. 2016, © AACR.)

verify an anticipated "hyper" mutation rate for patients with a POLE pathogenic mutation and can illustrate response to treatment in the form of dramatic changes in tumor heterogeneity and subclonal tumor cell populations. This was especially evident in the comparison of the first drop metastasis to the second, with an anti PD-1 therapy intermediate (Fig. 2). Similarly, we utilized our genomic data to identify significant flux in the neoantigens identified in all three tumors. Our comparative analysis of RNA expression data obtained from each tumor illustrated the significant changes in immune cell infiltrates following anti-PD1 checkpoint blockade therapy, and these were supported by IHC studies that further illustrated the significant influx of immune cells following anti-PD1 therapy (Fig. 3). Importantly, this study provided the first combined genomic and molecular evidence that checkpoint blockade could influence immune activity behind the blood–brain barrier in a central nervous system (CNS) tumor.

We also have used immunogenomic methods to identify neoantigens in individual patient cancers for the purposes of designing a personalized vaccine-based therapeutic (Carreno et al. 2015). This work built upon proof-of-concept studies completed with Robert Schreiber's laboratory using carcinogen-induced sarcoma mouse models with a high mutational load. These studies investigated whether the combination of exome sequencing and RNA sequencing could identify neoantigens in this tumor model (Matsushita et al. 2012) and further evaluated whether a peptide vaccine based on the identified neoantigens could be used to vaccinate the mouse and lead to the eradication of the tumor (Gubin et al. 2014). In the human vaccine trial, the pVAC-Seq pipeline (Hundal et al. 2016) identified neoantigens from melanoma samples in three patients with BRAF-mutated melanomas that had received prior treatment with a BRAF small-molecule-inhibitor therapy and with an anti-CTLA4 checkpoint blockade therapy (ipilimumab). After identifying neoantigens, the vaccines were produced by isolating dendritic cells from each patient and culturing them to maturity based on established in vitro methods, and then coupling the mature dendritic cells with synthetic peptides for each unique neoantigen plus two control peptides (based on gp-100 shared antigens). These den-

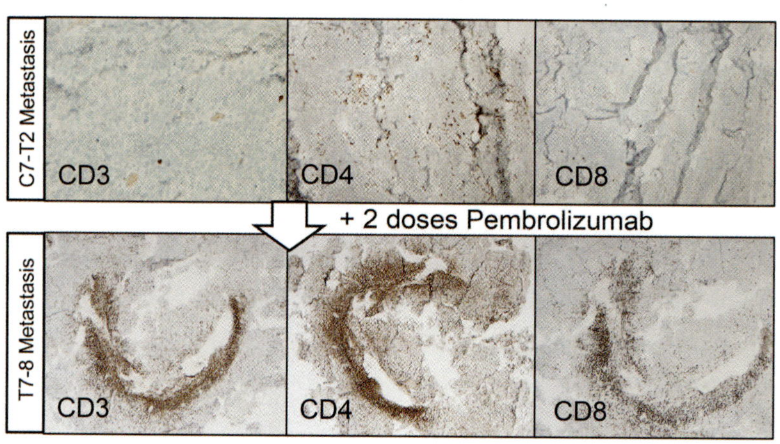

Figure 3. Immunohistochemistry-based evaluation of sequential glioblastoma multiforme (GBM) samples for immune infiltration before and after anti PD-1 therapy. (Reprinted, with permission, from Johanns et al. 2016, © AACR.)

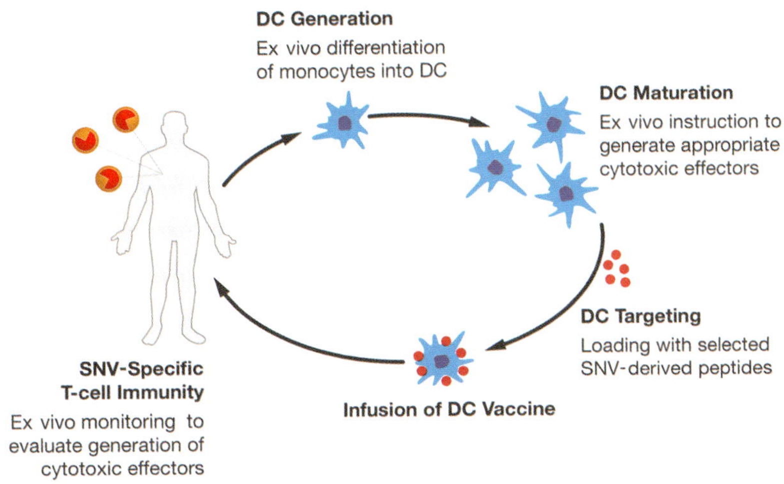

Figure 4. Overall workflow for creating a dendritic cell (DC) vaccine based on mis-sense somatic point mutations (single-nucleotide variants [SNVs]) that generate putative neoantigens.

dritic cell cultures were infused into the patients in a series of three vaccine infusions at 2-wk intervals. The overall workflow is shown in Figure 4.

Postvaccination, we then monitored each patient's response using flow cytometry–based dextramer assays designed to specifically identify which of the neoantigens were eliciting a specific $CD8^+$ T-cell response. In each case, we saw strong responses from three of the seven neoantigens utilized per patient (each set of neoantigens is completely unique to each patient), and we saw strong responses from our control gp-100 peptides. To further investigate the T-cell response, we evaluated the complexity of the T-cell receptor (TCR) repertoire using the Adaptive Biosciences immunoSEQ assay (http://www.adaptivebiotech.com/immunoseq), which also uses next-generation sequencing to provide the TCR data for analysis. Our results demonstrated that, for neoantigens with a discernable response by dextramer-flow cytometry assay, the corresponding T cells showed increased diversity in the TCR repertoire post-vaccine compared with pre-vaccine.

These initial studies have led to new clinical trials of personalized vaccines in tumor types with a lower average mutational load than UV-associated melanomas, such as triple-negative breast cancer. Here, the challenge is to identify a sufficient number of neoantigens using the pVAC-Seq pipeline, for the purposes of designing the vaccine. To address this, we have been expanding the types of variants considered in our analysis beyond SNVs (point mutations). For example, although they are less frequent and more difficult to identify precisely, insertion or deletion (indel) variants can shift the open reading frame of a protein, thereby creating a novel tumor-unique peptide that may be a neoantigen. Similarly, the identification of structural variants that generate fusion proteins can also potentially encode a neoantigenic peptide at the junction sequence of the two proteins. We recently used an approach to mine fusion transcripts from cancer RNA-seq data for purposes of evaluating neoantigenicity of predicted fusions. The implementation of this approach, called INTEGRATE-Neo, was tested in the context of TCGA prostate cancer data (exome and RNA-seq) for a common fusion peptide created by the translocation that fuses TMPRSS2 to ERG in ~50% of prostate cancers. As shown in Figure 5, the HLA binding energy calculated for this novel fusion peptide varies across patients according to HLA haplotype. Like class I neoantigens, we also wish to evaluate class II restricted neoantigens and add these to our vaccines. We are actively pursuing these and other changes to our neoantigen discovery pipeline (https://github.com/griffithlab/pVAC-Seq).

Finally, we would like to expand our personalized neoantigen vaccine approaches to tumor types with significantly lower mutational burden, such as pediatric cancers. In particular, the opportunity to pursue these studies lies in patients emerging with recurrent disease, post-standard-of-care therapies such as radiation and alkylating chemotherapies like TMZ. In a study published by Costello's laboratory, ~30% of pediatric glioma patients with recurrent tumors post-radiation and TMZ demonstrated a significant elevation in their tumor mutational burden posttherapy (van Thuijl et al. 2015). These children might represent reasonable subjects for a clinical trial to evaluate a personalized neoantigen vaccine in the recurrent setting. In this regard, since the TMZ-induced mutations are subclonal (i.e., not in the founder clone), we would need to include a combination of founder clone variants from the primary tumor and newer, TMZ-induced subclonal variants to increase the likelihood of clinical benefit.

REMAINING QUESTIONS IN IMMUNOGENOMICS

Although these new applications of genomics to clinical cancer diagnostics and therapeutics are exciting and

TCGA ID	HLA allele	Epitope	Affinity (nM)	Spanning reads
TCGA-EJ-5530-01	HLA-A02:01	ALNSEALSV	38.93	38
TCGA-2A-A8VV-01	HLA-A02:01	ALNSEALSV	38.93	18
TCGA-J4-A6M7-01	HLA-A02:01	ALNSEALSV	38.93	15
TCGA-CH-5769-01	HLA-A02:05	ALNSEALSV	370.79	11
TCGA-HC-7230-01	HLA-A02:01	ALNSEALSV	38.93	10
TCGA-HI-7171-01	HLA-A02:01	ALNSEALSV	38.93	9
TCGA-KK-A8IH-01	HLA-A02:01	ALNSEALSV	38.93	3
TCGA-KK-A7B1-01	HLA-C14:02	KMALNSEAL	142.47	20
TCGA-VN-A88L-01	HLA-C03:03	MALNSEAL	285.75	58
TCGA-FC-7708-01	HLA-C03:03	MALNSEAL	285.75	8

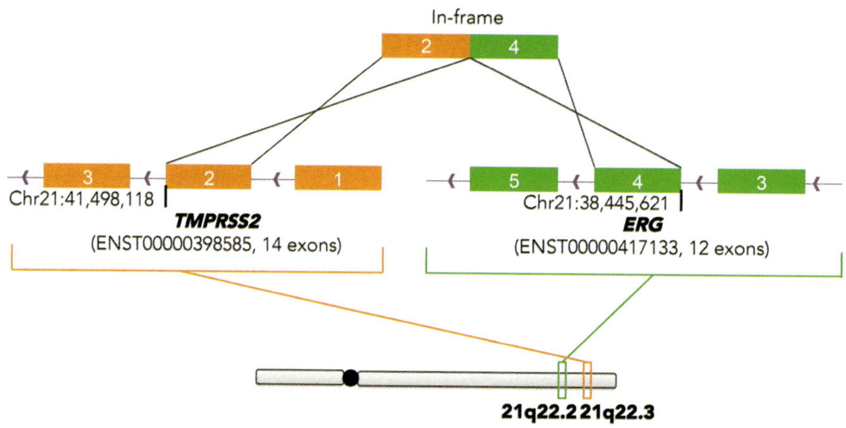

Figure 5. Evaluation of human leukocyte antigen (HLA) binding affinity for a common fusion peptide in prostate cancer (TMPRSS2-ERG) as evaluated by INTEGRATE-Neo. (Reprinted from Zhang et al. 2016, by permission of Oxford University Press.)

potentially transformational, we have many remaining questions to answer. Examples include questions about the optimal number and types of neoantigens that are ideal for vaccine efficacy, the type of vaccine platform (dendritic cell–based, peptide-based, DNA-based, and so on) that is scalable, rapid to produce, and affordable so that many patients can benefit (if indeed, benefit is demonstrated). Better information about how patients respond to a vaccine across the spectrum of tumor burden is needed, so we can more effectively design our clinical trials. For example, are patients with advanced tumor burden less likely to have a clinical response or benefit because the immune system, although activated by the vaccine, cannot respond fast enough to eradicate their cancers? Conversely, do patients require some minimal level of tumor burden to achieve a benefit by giving the T cells elicited by the vaccine a target to attack? And finally, with a tumor burden eliminated by surgery, will the immune surveillance remain? Or will surveillance potentially be eluded by the emergence of new tumors that have suppressed the expression of the neoantigens targeted by the vaccine? Since this phenomenon of neoepitope editing already has been demonstrated in acquired resistance to checkpoint blockade therapies (Anagnostou et al. 2016), it is likely also in the therapeutic vaccine setting.

CONCLUSION

The evolving applications of modern genomics and computational analysis to the immunology-based study of cancer have transformed our studies since these central concepts of cancer immunology were hypothesized in the 1980s. Certainly, the facility with which somatic mutations can be identified in individual tumor samples has encouraged progress, but vital contributions were required toward the algorithmic evaluation of mutant peptides in the context of each patient's HLA haplotypes and in devising novel immunotherapeutics. Immunogenomics has been demonstrated to be clinically informative in several early studies, outlined here, and in research that builds the case for clinical translation. As our level of sophistication increases in identifying neoantigens across the spectrum of mutation types and HLA classes, foundational knowledge about cancer immunology will result as will our understanding of how best to apply this information to clinical care.

ACKNOWLEDGMENTS

We acknowledge our clinical collaborators on the case studies reported, including Beatriz Carreno, Gerald Linette, Gavin Dunn, and Tanner Johanns, and we thank

Robert Schreiber for his incredible guidance and encouragement to pursue these studies. We acknowledge the patients who were studied herein and thank them for their courage and willingness to contribute to clinical research.

REFERENCES

Anagnostou V, Smith KN, Forde PM, Niknafs N, Bhattacharya R, White J, Zhang T, Adleff V, Phallen J, Wali N, et al. 2016. Evolution of neoantigen landscape during immune checkpoint blockade in non-small cell lung cancer. *Cancer Discov* **7:** 264–276.

Babbitt BP, Allen PM, Matsueda G, Haber E, Unanue ER. 2005. Binding of immunogenic peptides to Ia histocompatibility molecules. 1985. *J Immunol* **175:** 4163–4165.

Bjorkman PJ, Saper MA, Samraoui B, Bennett WS, Strominger JL, Wiley DC. 1987. Structure of the human class I histocompatibility antigen, HLA-A2. *Nature* **329:** 506–512.

Carreno BM, Magrini V, Becker-Hapak M, Kaabinejadian S, Hundal J, Petti AA, Ly A, Lie WR, Hildebrand WH, Mardis ER, et al. 2015. Cancer immunotherapy. A dendritic cell vaccine increases the breadth and diversity of melanoma neoantigen-specific T cells. *Science* **348:** 803–808.

De Plaen E, Lurquin C, Van Pel A, Mariamé B, Szikora JP, Wölfel T, Sibille C, Chomez P, Boon T. 1988. Immunogenic (tum^-) variants of mouse tumor P815: Cloning of the gene of tum^- antigen P91A and identification of the tum^- mutation. *Proc Natl Acad Sci* **85:** 2274–2278.

Ding L, Wendl MC, Koboldt DC, Mardis ER. 2010. Analysis of next-generation genomic data in cancer: Accomplishments and challenges. *Hum Mol Genet* **19:** R188–R196.

Foley EJ. 1953. Antigenic properties of methylcholanthrene-induced tumors in mice of the strain of origin. *Cancer Res* **13:** 835–837.

Griffith M, Miller CA, Griffith OL, Krysiak K, Skidmore ZL, Ramu A, Walker JR, Dang HX, Trani L, Larson DE, et al. 2015. Optimizing cancer genome sequencing and analysis. *Cell Syst* **1:** 210–223.

Gubin MM, Zhang X, Schuster H, Caron E, Ward JP, Noguchi T, Ivanova Y, Hundal J, Arthur CD, Krebber WJ, et al. 2014. Checkpoint blockade cancer immunotherapy targets tumour-specific mutant antigens. *Nature* **515:** 577–581.

Hundal J, Carreno BM, Petti AA, Linette GP, Griffith OL, Mardis ER, Griffith M. 2016. pVAC-Seq: A genome-guided in silico approach to identifying tumor neoantigens. *Genome Med* **8:** 11.

Johanns Tanner M, Miller CA, Dorward IG, Tsien C, Chang E, Perry A, Uppaluri R, Ferguon C, Schmidt RE, Dahiya S, et al. 2016. Immunogenomics of hypermutated glioblastoma: A patient with germline POLE deficiency treated with checkpoint blockade immunotherapy. *Cancer Discov* **6:** 1230–1236.

Ley TJ, Mardis ER, Ding L, Fulton B, McLellan MD, Chen K, Dooling D, Dunford-Shore BH, McGrath S, Hickenbotham M, et al. 2008. DNA sequencing of a cytogenetically normal acute myeloid leukaemia genome. *Nature* **456:** 66–72.

Liu C, Yang X, Duffy B, Mohanakumar T, Mitra RD, Zody MC, Pfeifer JD. 2013. ATHLATES: Accurate typing of human leukocyte antigen through exome sequencing. *Nucleic Acids Res* **41:** e142.

Lundegaard C, Lamberth K, Harndahl M, Buus S, Lund O, Nielsen M. 2008. NetMHC-3.0: Accurate web accessible predictions of human, mouse and monkey MHC Class I affinities for peptides of length 8-11. *Nucleic Acids Res* **36**(Web Server): W509–W512.

Mardis ER. 2017. DNA sequencing technologies: 2006–2016. *Nat Protoc* **12:** 213–218.

Matsushita H, Vesely MD, Koboldt DC, Rickert CG, Uppaluri R, Magrini VJ, Arthur CD, White JM, Chen YS, Shea LK, et al. 2012. Cancer exome analysis reveals a T-cell-dependent mechanism of cancer immunoediting. *Nature* **482:** 400–404.

McLaren W, Gil L, Hunt SE, Riat HS, Ritchie GR, Thormann A, Flicek P, Cunningham F. 2016. The Ensembl Variant Effect Predictor. *Genome Biol* **17:** 122.

Monach PA, Meredith SC, Siegel CT, Schreiber H. 1995. A unique tumor antigen produced by a single amino acid substitution. *Immunity* **2:** 45–59.

Old LJ. 1982. Cancer immunology: The search for specificity. *Natl Cancer Inst Monogr* **60:** 193–209.

Paul S, Lindestam Arlehamn CS, Scriba TJ, Dillon MB, Oseroff C, Hinz D, McKinney DM, Carrasco Pro S, Sidney J, Peters B, et al. 2015. Development and validation of a broad scheme for prediction of HLA class II restricted T cell epitopes. *J Methods Immunol* **422:** 28–34.

Peters B, Sette A. 2005. Generating quantitative models describing the sequence specificity of biological processes with the stabilized matrix method. *BMC Bioinformatics* **6:** 132.

Prehn RT, Main JM. 1957. Immunity to methylcholanthrene-induced sarcomas. *J Natl Cancer Inst* **18:** 769–778.

Srivastava A, Ghosh S, Anantharaman N, Jayaraman VK. 2013. Hybrid biogeography based simultaneous feature selection and MHC class I peptide binding prediction using support vector machines and random forests. *J Immunol Methods* **387:** 284–292.

Szolek A, Schubert B, Mohr C, Sturm M, Feldhahn M, Kohlbacher O. 2014. OptiType: Precision HLA typing from next-generation sequencing data. *Bioinformatics* **30:** 3310–3316.

van der Bruggen P, Traversari C, Chomez P, Lurquin C, De Plaen E, Van den Eynde B, Knuth A, Boon T. 1991. A gene encoding an antigen recognized by cytolytic T lymphocytes on a human melanoma. *Science* **254:** 1643–1647.

van Thuijl HF, Mazor T, Johnson BE, Fouse SD, Aihara K, Hong C, Malmström A, Hallbeck M, Heimans JJ, Kloezeman JJ, et al. 2015. Evolution of DNA repair defects during malignant progression of low-grade gliomas after temozolomide treatment. *Acta Neuropathol* **129:** 597–607.

Wang K, Li M, Hakonarson H. 2010. ANNOVAR: Functional annotation of genetic variants from high-throughput sequencing data. *Nucleic Acids Res* **38:** e164.

Warren RL, Choe G, Freeman DJ, Castellarin M, Munro S, Moore R, Holt RA. 2012. Derivation of HLA types from Shotgun sequence datasets. *Genome Med* **4:** 95.

Zhang J, Mardis ER, Maher CA. 2016. INTEGRATE-neo: A pipeline for personalized gene fusion neoantigen discovery. *Bioinformatics* **2016:** pii: btw674.

Targeting HIF2 in Clear Cell Renal Cell Carcinoma

HYEJIN CHO AND WILLIAM G. KAELIN
Howard Hughes Medical Institute, Dana-Farber Cancer Institute and Brigham and Women's Hospital, Harvard Medical School, Boston, Massachusetts 022145
Correspondence: william_kaelin@dfci.harvard.edu

Inactivation of the von Hippel–Lindau tumor-suppressor protein (pVHL) is the signature "truncal" event in clear cell renal cell carcinoma, which is the most common form of kidney cancer. pVHL is part of a ubiquitin ligase the targets the α subunit of the hypoxia-inducible factor (HIF) transcription factor for destruction when oxygen is available. Preclinical studies strongly suggest that deregulation of HIF, and particularly HIF2, drives pVHL-defective renal carcinogenesis. Although HIF2α was classically considered undruggable, structural and chemical work by Rick Bruick and Kevin Gardner at University of Texas Southwestern laid the foundation for the development of small molecule direct HIF2α antagonists (PT2385 and the related tool compound PT2399) by Peloton Therapeutics that block the dimerization of HIF2α with its partner protein ARNT1. These compounds inhibit clear cell renal cell carcinoma growth in preclinical models, and PT2385 has now entered the clinic. Nonetheless, the availability of such compounds, together with clustered regularly interspaced short palindromic repeat (CRISPR)-based gene editing approaches, has revealed a previously unappreciated heterogeneity among clear cell renal carcinomas and patient-derived xenografts with respect to HIF2 dependence, suggesting that predictive biomarkers will be needed to optimize the use of such agents in the clinic.

Epithelial cancers, such as breast, lung, and colon cancer, account for most cancer deaths in the developed world. Accordingly, it is imperative that we make strides toward preventing and treating such cancers if we are going to significantly decrease cancer mortality in countries such as the United States. Kidney cancer is an epithelial cancer and in the United States is one of the 10 most common cancers in men and women and one of the 10 most common causes of cancer deaths in men (Siegel et al. 2016). Kidney cancer can be subdivided based on its histological appearance into clear cell renal cell carcinoma, which is by far the most common form of kidney cancer, papillary renal cell carcinoma, chromophobe renal cell carcinoma, collecting duct, and a few other variants. This review will discuss the role of the von Hippel–Lindau (VHL) tumor-suppressor protein (pVHL) and the HIF2 transcription factor in clear cell renal cell carcinoma and recent progress toward targeting HIF2 with drug-like small molecules.

CLEAR CELL RENAL CELL CARCINOMA GENETICS

Rare hereditary forms of cancer have often been informative with respect to the genetics of sporadic cancers. VHL disease presents as a hereditary cancer syndrome characterized by an increased risk of clear cell renal cell carcinoma, blood vessel tumors called hemangioblastomas, and adrenal gland tumors called pheochromocytomas (Maher and Kaelin 1997; Kaelin 2002). Individuals with VHL disease have inherited a defective version of the *VHL* gene, which resides at chromosome 3p25, from one of their parents or acquired a de novo mutation, resulting in heterozygosity or mosaicism, respectively, at the *VHL* locus. Tumors develop when the remaining wild-type *VHL* alleles are mutated, deleted, or otherwise silenced in a susceptible cell. Importantly, biallelic *VHL* inactivation is also extremely common (>50%) in sporadic clear cell renal cell carcinomas. The actual prevalence might be significantly higher if one excludes clear cell renal carcinomas with atypical histological features and uses state of the art DNA sequencing methods. In sporadic clear cell renal carcinoma, inactivation of the maternal and paternal *VHL* alleles is caused by sporadic mutations (in contrast to classical VHL disease, where the first mutation or "hit" has been inherited), including point mutations and deletions or, less commonly, hypermethylation.

Although biallelic *VHL* inactivation is a common, and possibly nearly universal, first step in clear cell renal carcinogenesis, it is not sufficient to cause this disease. This has been most clearly shown by careful studies of kidneys from VHL patients, which can contain hundreds of small dysplastic lesions and cysts. Genotyping such lesions reveals biallelic *VHL* inactivation (Walther et al.

Published by Cold Spring Harbor Laboratory Press; doi: 10.1101/sqb.2016.81.030833

1995; Zhuang et al. 1995; Lubensky et al. 1996; Mandriota et al. 2002; Montani et al. 2010). It therefore appears that *VHL* loss is sufficient to cause renal dysplasia but not cancer. The same appears to be true in genetically engineered mouse models (Gnarra et al. 1997; Frew et al. 2008; Schietke et al. 2012; Albers et al. 2013). Another line of evidence has come from genomic analyses of hereditary (VHL disease) and sporadic clear cell renal carcinomas to identify recurrent mutations, including intragenic mutations and copy-number changes (Beroukhim et al. 2009; Guo et al. 2011; Varela et al. 2011; Dondeti et al. 2012; Duns et al. 2012; Pena-Llopis et al. 2012; Cancer Genome Atlas Research Network 2013). Clear cell renal carcinomas typically include specific copy-number changes, most commonly gain of chromosome 5q, loss of chromosomes 3p and 14q, and specific intragenic mutations affecting genes linked to chromatin regulation, phosphoinositide 3-kinase (PI3K) signaling, and response to redox stress or DNA damage. Examples of such genes include *PBRM1*, *BAP1*, *SETD2*, *KDM5C*, *TSC1*, *PIK3CA*, *PTEN*, *MTOR*, and *TP53*. Importantly, inactivation of *VHL* has preceded the acquisition of these other mutations in every tumor examined to date in which deep sequencing of spatially distinct sites was used to infer its evolutionary history (Gerlinger et al. 2012, 2014; Xu et al. 2012; Fisher et al. 2014; Sankin et al. 2014). These studies have confirmed that *VHL* inactivation is an early, or "truncal," lesion in clear cell renal cell carcinomas and that such tumor often display significant heterogeneity, or "branching," with respect to the subsequent genetic events that conspired with *VHL* loss to cause cancer. Interestingly, *VHL*, *PBRM1*, *BAP1*, and *SETD2* all reside on chromosome 3p. Therefore three "hits," including two intragenic mutations and loss of chromosome 3p, can inactivate both *VHL* and another "Knudson 2-hit" renal cancer suppressor. For reasons that are still not well understood, *VHL* mutations are exceedingly rare in other neoplasms with the exceptions of pheochromocytomas and hemangioblastomas.

THE VHL TUMOR-SUPPRESSOR PROTEIN

The *VHL* gene encodes two different proteins, with apparent molecular mass of ~28 kDa and 19 kDa after sodium dodecyl sulfate-polyacrylamide gel electrophoresis (SDS-PAGE), because of alternative in-frame translational start sites (Kaelin 2007). For simplicity let us refer to both of these as pVHL in this review, because their functions appear to be highly similar, and disease-associated *VHL* mutations almost invariably affect both isoforms. pVHL can be found in both the cytoplasm and the nucleus, shuttling dynamically between the two compartments (Kaelin 2007). Some pVHL can also be detected associated with membranes, the endoplasmic reticulum, and mitochondria (Kaelin 2007).

Reintroducing wild-type pVHL into pVHL-defective clear cell renal carcinoma lines suppresses their ability to form tumors in nude mice xenograft assays but does not affect their viability or proliferation under standard cell culture conditions (Iliopoulos et al. 1995). pVHL does inhibit pVHL-defective clear cell renal carcinoma cells ex vivo, however, under specific conditions such as when such cells are grown as three-dimensional spheroids or at confluence under growth factor–poor conditions (Lieubeau-Teillet et al. 1998; Pause et al. 1998; Davidowitz et al. 2001).

pVHL is the substrate adapter of a ubiquitin ligase that contains elongin B, elongin C, Rbx1, and Cullin-2 (Kaelin 2007). pVHL has two domains that are hotspots for missense mutations in VHL disease: the α domain, which recruits the ubiquitination machinery, and the β domain, which serves as a substrate docking site. Although many potential pVHL substrates have been identified, the best documented substrate, and the one believed to be most tightly linked to clear cell renal carcinogenesis, is the HIF (hypoxia-inducible factor) transcription factor.

THE HIF TRANSCRIPTION FACTOR

Active HIF is a heterodimer of two basic helix–loop–helix PAS domain–containing DNA-binding proteins, an unstable α subunit and a stable β subunit, and binds to specific DNA sequences called hypoxia-response elements (HREs) (Kaelin and Ratcliffe 2008; Semenza 2011). The human genome contains three HIFα genes (HIF1α, HIF2α, HIF3α) and two HIFβ genes (HIF1β, also called ARNT1, and HIF2β, also called ARNT2). HIF1α is widely expressed and, together with HIF1β, forms the well-studied, canonical, HIF transcription factor. The expression of HIF2α is more restricted. HIF1α and HIF2α each contain two transactivation domains, an amino-terminal transactivation domain (NTAD) and carboxy-terminal transactivation domain (CTAD), that enable them, upon DNA-binding, to recruit coactivators such as p300 and CREB-binding protein (CBP) and activate transcription. HIF3α has been far less studied, but encodes multiple splice variants that do not activate transcription but instead act as dominant-negative inhibitors of HIF-dependent transcription.

HIF regulates hundreds of genes, many of which play roles in acute or chronic adaptation to hypoxia. HIF regulates genes linked to processes such as anaerobic glycolysis, mitochondrial biosynthesis and function, autophagy, angiogenesis, cell proliferation, invasion, and migration (Fig. 1). The ability of HIF to activate some HRE-containing HIF target genes is highly cell type– and context-dependent, presumably because of the influence of other *cis*-acting transcription factors as well as changes in chromatin accessibility. For example, vascular endothelial growth factor (VEGF) responds to HIF in many tissues, whereas erythropoietin (EPO) in adults responds to HIF largely in dedicated cells in the kidney. Likewise, the sets of genes regulated by HIF1 and HIF2, although overlapping, are not identical. For example, some glycolytic and autophagy genes are primarily regulated by HIF1, whereas EPO is primarily regulated by HIF2 (Hu et al. 2003; Rankin et al. 2007; Kapitsinou et al. 2010).

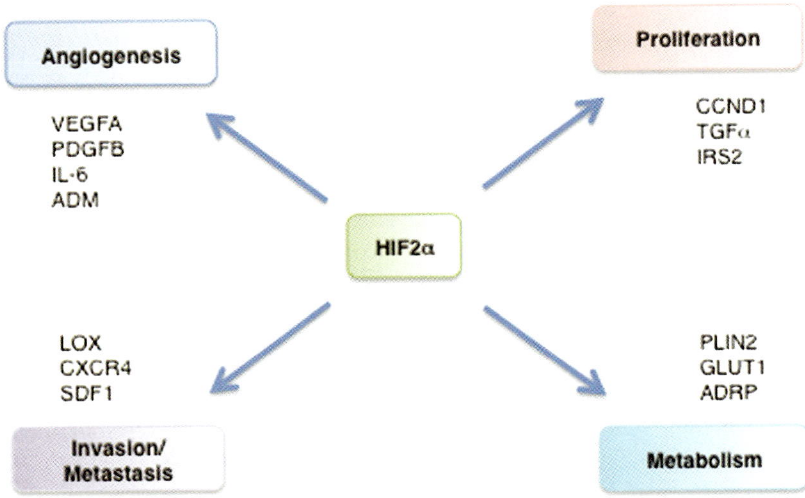

Figure 1. Hypoxia-inducible factor 2 (HIF2)-regulated biological processes germane to kidney cancer. Examples of HIF2-responsive mRNAs linked to biological processes implicated in renal carcinogenesis are shown.

REGULATION OF HIF BY pVHL AND OXYGEN

Under well-oxygenated conditions HIFα is prolyl-hydroxylated on one (or both) of two potential prolyl residues by members of the EglN (also called PHD) prolyl hydroxylase family, which serve as oxygen sensors (for review, see Kaelin and Ratcliffe 2008). Once prolyl-hydroxylated, HIFα binds directly to the pVHL ubiquitin ligase complex and is targeted for proteasomal degradation. Under low-oxygen conditions EglN activity is impaired, which allows HIFα to accumulate, bind to HIFβ, and activate transcription. Another layer of oxygen-dependent HIF regulation is provided by the FIH1 asparaginyl hydroxylase. In the presence of oxygen FIH1 hydroxylates a conserved asparaginyl residue located within the HIFα CTAD, preventing it from binding to coactivators. Importantly, FIH1 has a higher oxygen affinity than the EglNs, and thus can remain active at intermediate levels of hypoxia that are sufficient to stabilize HIFα. Some HIF-responsive genes depend primarily on the NTAD and others on the CTAD. Moreover, the HIF2α is relatively resistant to FIH1. Therefore, FIH1 can potentially tune the HIF response at intermediate levels of hypoxia by altering the balance of NTAD and CTAD function and possibly the balance of HIF1 and HIF2 activity.

ROLE OF HIF2 IN CLEAR CELL RENAL CARCINOGENESIS

Multiple lines of evidence suggest that HIF, and particularly HIF2, is a critical target in clear cell renal cell carcinoma. Every *VHL* mutation linked to clear cell renal carcinoma to date results in a protein that is defective with respect to HIF regulation, and genotype–phenotype correlations suggest that the risk of developing clear cell renal carcinoma in VHL disease families is linked to the degree to which their *VHL* alleles compromise pVHL's ability to suppress HIF (Li et al. 2007; Kaelin 2008). In particular, VHL families with a low risk of developing clear cell renal carcinoma, even if at high risk of hemangioblastomas and pheochromocytomas, have *VHL* alleles that encode proteins that retain significant HIF ubiquitin ligase activity, whereas families at high risk of clear cell renal carcinoma are grossly defective in this regard (Kaelin 2008).

In the laboratory, activating HIF2-responsive genes, such as by expressing a HIF2α variant that cannot be prolyl-hydroxylated, can override pVHL's tumor-suppressor activity in nude mice xenograft assays (Kondo et al. 2002; Raval et al. 2005; Biswas et al. 2010). Conversely, down-regulating HIF2α with short-hairpin RNA (shRNA) technology suppresses tumor formation by pVHL-defective clear cell renal carcinomas in mice (Kondo et al. 2003; Zimmer et al. 2004; Gordan et al. 2008).

Currently there are no faithful, highly penetrant, genetically engineered mouse models for $VHL^{-/-}$ clear cell renal carcinomas. Nonetheless, the pathological changes induced after inactivating VHL in various mouse tissues has, when tested, been prevented by concurrent inactivation of HIF2α (Rankin et al. 2007, 2008, 2009) and can, when tested, be mimicked by expressing a version of HIF2α that cannot be prolyl-hydroxylated in wild-type mice (Kim et al. 2006). Notably, a human genetic polymorphism linked to HIF2α is associated with the risk of developing clear cell renal cell carcinoma (Purdue et al. 2011).

In stark contrast HIF1α appears capable of suppressing clear cell renal carcinomas. Forced expression of HIF1α, unlike HIF2α, does not antagonize pVHL's tumor-suppressor activity in nude mouse experiments (Maranchie et al. 2002; Raval et al. 2005; Biswas et al. 2010; Shen et al. 2011). Interestingly, HIF1α resides on chromosome 14q, which is often deleted in clear cell renal cell carcinoma, and many clear cell renal carcinoma lines have sustained homozygous deletions that specifically inactivate HIF1α such that they solely produce HIF2α (Shen et al. 2011). Restoring HIF1α expression in such cell lines suppresses tumor growth, whereas eliminating HIF1α in lines that retain a wild-type *HIF1α* allele promotes tumor growth (Shen et al. 2011). Intragenic *HIF1α* mutations have rarely been identified in clear cell renal cancers and, when tested, have uniformly been loss of function (Morris et al. 2009; Dalgliesh et al. 2010; Shen et al. 2011). Moreover, the appearance of HIF2α, and apparent loss of HIF1α, in preneoplastic renal lesions in VHL patients correlates with worsening cellular atypia and signs of impending transformation (Mandriota et al. 2002; Schietke et al. 2012). Interestingly, genetic ablation of *HIF1α* worsens the renal pathology associated with papillary renal cancers driven by fumarate hydratase loss in the mouse, providing a precedent for HIF1α as a renal cancer suppressor (Adam et al. 2011).

Most 14q deleted clear cell renal cancers, in contrast to clear cell renal carcinoma cell lines, retain one *HIF1α* allele and the remaining *HIF1α* allele is usually wild-type. It is not yet clear whether the homozygous *HIF1α* deletions detected in cell lines occurred in vivo, possibly associated with disease progression, or were selected for ex vivo. Nor is it clear whether *HIF1α* can act as a haploinsufficient tumor suppressor in vivo.

Nonetheless, these findings suggest that HIF1α and HIF2α have nonidentical, and probably opposing, roles in clear cell renal cancer (Keith et al. 2012). In further support of this conclusion, HIF2α has been shown in the renal carcinoma setting to cooperate with c-Myc and to suppress p53, whereas HIF1α has the opposite effects (Keith et al. 2012). These different roles presumably reflect differences in their ability to activate different HIF target genes, for the reasons cited above, as well as differences with respect to other noncanonical functions.

TARGETING HIF2-RESPONSIVE GENE PRODUCTS WITH SMALL MOLECULES

Drugs exist for a number of HIF2-responsive gene products suspected of playing a role in clear cell renal carcinogenesis including VEGF, PDGF B, c-Met, TGFα (ligand for EGFR), cyclin D1, and both SDF1 and its receptor, CXCR4. Among epithelial cancers, clear cell renal carcinomas have the highest VEGF levels, presumably driven by pVHL loss and HIF2 deregulation. Moreover, the induction of VEGF early during the evolution of clear cell renal carcinomas probably minimizes the selection pressure to activate alternative angiogenic factors. Perhaps for this reason, clear cell carcinomas have proven to be particularly sensitive to drugs that inhibit VEGF or its receptor, kinase insert domain receptor (KDR). Six such agents have now been approved by the U.S. Federal Drug Administration (FDA) for this indication (Table 1).

Platelet-derived growth factor subunit B (PDGFB) acts to support vascular pericytes, and immature vessels without pericyte coverage are more sensitive to VEGF withdrawal than are mature vessels (Benjamin and Keshet 1997; Benjamin et al. 1998, 1999). Many of the currently available KDR inhibitors also inhibit the PDGF receptor, which could theoretically enhance their activity. On the other hand, PDGF receptor inhibitors have not proven to be active in clear cell renal carcinoma, either alone or added to VEGF inhibitors (Polite et al. 2006; Vuky et al. 2006; Hainsworth et al. 2007; Ryan et al. 2011).

HIF2, in ways that are still not completely understood, enhances signaling by c-Met and its ligand hepatocyte growth factor (HGF) (Koochekpour et al. 1999; Pennacchietti et al. 2003; Nakaigawa et al. 2006), and pVHL-defective renal carcinoma cells are hypersensitive to c-Met depletion with small interfering RNAs (siRNAs) (Bommi-Reddy et al. 2008). Cabozantinib, which inhibits KDR and c-Met, is highly active against clear cell renal carcinoma, including in patients who have failed multiple VEGF inhibitors (Choueiri et al. 2015, 2016). It is unclear, however, whether this is truly because of its ability to inhibit c-Met. It is formally possible, for example, that cabozantinib is simply a better KDR inhibitor in vivo than other drugs in this class.

Clear cell renal carcinomas express high level of the HIF2-responsive growth factor transforming growth factor α (TGF-α) as well as its receptor, epidermal growth factor receptor (EGFR) (Uhlman et al. 1995; Ramp et al. 1997; Knebelmann et al. 1998; de Paulsen et al. 2001; Merseburger et al. 2005; Smith et al. 2005). HIF2 promotes the translation of EGFR (Franovic et al. 2007; Uniacke et al. 2012), whereas pVHL has been reported to promote the endocytosis and degradation of EGFR (Zhou and Yang 2011; Uniacke et al. 2012). Although clear cell renal carcinoma lines are sensitive to EGFR inactivation in mouse models, EGFR inhibitors have thus far been essentially inactive against clear cell renal carcinoma patients. The reason for this conundrum is unclear, but it could be due, at least partly, to selection for EGFR dependence ex vivo during cell line selection and propagation.

The cells giving rise to clear cell renal carcinoma are unusual in that HIF2α can induce cyclin D1 in this

Table 1. List of genes associated with methylation imprint, which are confirmed by knockout mouse studies

Agent	U.S. Brand Name	Company	Description	Approval Year for Kidney Cancer
Sorafenib	Nexavar	Onyx/Bayer	TKI	2005
Sunitinib	Sutent	Pfizer	TKI	2006
Bevacizumab	Avastin	Genentech	Anti-VEGF Ab	2009
Pazopanib	Votrient	GlaxoSmithKline	TKI	2009
Axitinib	Inlyta	Pfizer	TKI	2012
Cabozantinib	Cabometyx, Cometriq	Exelixis	TKI	2016

context, whereas HIF and hypoxia usually down-regulate cyclin D1 (Bindra et al. 2002; Zatyka et al. 2002; Baba et al. 2003). Moreover, pVHL-defective clear cell renal carcinoma cells appear to be more dependent on cdk6 than isogenic cells in which pVHL has been restored and are sensitive to shRNA-mediated depletion of cyclin D1 (Bommi-Reddy et al. 2008; Zhang et al. 2013). Finally, a human polymorphism linked to the risk of clear cell renal cell carcinoma alters the binding of HIF2α to a *cyclin D* transcriptional enhancer element (Schodel et al. 2012). It will be of interest to see if newer, more specific, cdk4/6 inhibitors, such as palbociclib and abemaciclib, are active in this disease.

CXCR4 has been implicated in the maintenance of clear cell renal carcinoma cancer–initiating cells, in tumor cell invasion and metastasis, and in mobilization of myeloid-derived suppressor cells that have been linked to resistance to VEGF blockade (Staller et al. 2003; Zagzag et al. 2005; Pan et al. 2006; Struckmann et al. 2008; Gassenmaier et al. 2013; Panka et al. 2013; Vanharanta et al. 2013; Micucci et al. 2015). A number of CXCR4 inhibitors have been proven safe in man, and clear cell renal carcinoma trials are contemplated.

DEVELOPMENT OF DIRECT HIF2 ANTAGONISTS

In theory targeting HIF2 itself would be more efficacious than targeting any one HIF2-responsive gene product. Unfortunately, however, DNA-binding transcription factors have classically been viewed as undruggable by the pharmaceutical industry. The one exception to this rule has been the steroid hormone receptors, which have ligand-binding hydrophobic pockets that can serve as an entry point for drug design.

Fortunately, pioneering work by Rick Bruick and Kevin Gardner identified a potentially druggable hydrophobic pocket in the HIF2α PAS B domain (Scheuermann et al. 2009, 2013; Rogers et al. 2013). Moreover, these investigators identified chemical matter that can bind to this pocket and allosterically prevent dimer formation between HIF2α and aryl hydrocarbon receptor nuclear translocator (ARNT) (Scheuermann et al. 2009, 2013; Rogers et al. 2013). These discoveries formed the basis for medicinal chemistry efforts at Peloton Therapeutics that led to the HIF2α inhibitor PT2385 and the highly related tool compound PT2399 (Chen et al. 2016; Cho et al. 2016; Wallace et al. 2016).

PT2385 and PT2399 inhibit HIF2α-ARNT1 dimerization and HIF2α-dependent transcription in cells at high-nanomolar/low-micromolar concentrations (Chen et al. 2016; Cho et al. 2016; Wallace et al. 2016) (Fig. 2). Importantly, these effects are highly specific because these two compounds do not affect HIF1 (Chen et al. 2016; Cho et al. 2016; Wallace et al. 2016). PT2399 also inhibits the growth of pVHL-defective clear cell renal carcinoma cells in soft agar assays, in subcutaneous and orthotopic xenograft assays, and in lung colonization assays aimed at modeling established pulmonary metastases (Chen et al. 2016; Cho et al. 2016; Wallace et al. 2016). These effects of PT2399 on transcription and tumor growth are on-target because they can be reversed with HIF2α variants that contain single missense mutations that prevent them from binding to PT2399 (Chen et al. 2016; Cho et al. 2016). PT2385 and PT2399 are also active in patient-derived xenograft models, including some models that are relatively resistant to the KDR inhibitor sunitinib (Chen et al. 2016; Cho et al. 2016; Wallace et al. 2016). Based on these data PT2385 has entered human clinical trials, with promising early signs of activity (Chen et al. 2016).

Nonetheless, it is already clear that some pVHL-defective clear cell renal carcinoma lines and patient-derived xenografts are resistant to PT2385/PT2399, as are some patients (Chen et al. 2016; Cho et al. 2016). This is because of, at least in part, previously unappreciated variability with respect to HIF2α dependence among

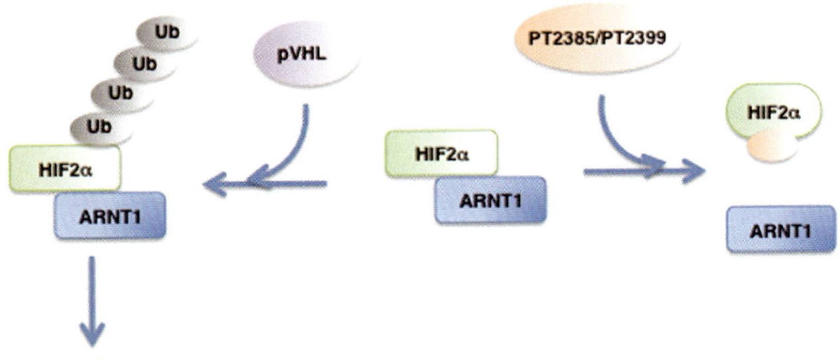

Figure 2. Control of hypoxia-inducible factor 2 (HIF2) by von Hippel–Lindau tumor-suppressor protein (pVHL) and PT2385/PT2399. In the presence of oxygen pVHL binds directly to the HIF2α subunit and polyubiquitylates it, thereby targeting HIF2α for proteasomal degradation. In pVHL-defective cells HIF2 accumulates inappropriately. PT2385 and the related compound PT2399 bind directly to HIF2α and induce an allosteric change that prevents HIF2α from binding to its obligate partner, aryl hydrocarbon receptor nuclear translocator 1 (ARNT1).

pVHL-defective clear cell renal carcinoma lines and tumors and suggests that predictive biomarkers will be needed to optimize the use of such agents in the clinic (Chen et al. 2016; Cho et al. 2016). In the preclinical models the most sensitive renal carcinoma lines and patient-derived xenografts had the highest HIF2α levels, suggesting these two properties are linked (Chen et al. 2016; Cho et al. 2016). Conversely, the presence of p53 mutations predicts for resistance (Chen et al. 2016; Cho et al. 2016). Although p53 mutations are relatively rare in primary clear cell renal carcinomas, their true prevalence in metastatic disease is not known. Moreover, it is possible that p53 mutations are selected for by prior therapy, including therapy with VEGF inhibitors.

CONCLUSION

Biallelic inactivation of the *VHL* tumor-suppressor gene is the most frequent initiating, or truncal, event in clear cell renal carcinoma. The VHL gene product, pVHL, is the substrate recognition subunit of a ubiquitin ligase that targets HIFα subunits for proteasomal degradation when oxygen is present. Loss of pVHL leads the inappropriate accumulation of active HIF1 and HIF2. HIF2 drives clear cell renal carcinogenesis, and drugs that inhibit the HIF2-responsive gene product VEGF are now cornerstones of kidney cancer therapy. HIF2α was recently found to have a druggable hydrophobic pocket, and a first-generation HIF2α inhibitor that blocks its dimerization with ARNT, and hence inactivates HIF2, has now entered the clinic based on promising preclinical data. Nonetheless, HIF2α dependence among clear cell renal carcinomas appears to be heterogeneous. Moreover, p53 pathway mutations confer resistance to HIF2α antagonists in preclinical models. These two observations imply the need for predictive biomarkers to optimize the use of HIF2α antagonists in the clinic.

ACKNOWLEDGMENTS

W.G.K. is supported by grants from the National Institutes of Health and is a Howard Hughes Medical Institute (HHMI) investigator. He is a paid consultant for, and has equity in, Peloton Therapeutics.

REFERENCES

Adam J, Hatipoglu E, O'Flaherty L, Ternette N, Sahgal N, Lockstone H, Baban D, Nye E, Stamp GW, Wolhuter K, et al. 2011. Renal cyst formation in Fh1-deficient mice is independent of the Hif/Phd pathway: Roles for fumarate in KEAP1 succination and Nrf2 signaling. *Cancer Cell* **20:** 524–537.

Albers J, Rajski M, Schonenberger D, Harlander S, Schraml P, von Teichman A, Georgiev S, Wild PJ, Moch H, Krek W, et al. 2013. Combined mutation of Vhl and Trp53 causes renal cysts and tumours in mice. *EMBO Mol Med* **5:** 949–964.

Baba M, Hirai S, Yamada-Okabe H, Hamada K, Tabuchi H, Kobayashi K, Kondo K, Yoshida M, Yamashita A, Kishida T, et al. 2003. Loss of von Hippel–Lindau protein causes cell density dependent deregulation of CyclinD1 expression through hypoxia-inducible factor. *Oncogene* **22:** 2728–2738.

Benjamin LE, Keshet E. 1997. Conditional switching of vascular endothelial growth factor (VEGF) expression in tumors: Induction of endothelial cell shedding and regression of hemangioblastoma-like vessels by VEGF withdrawal. *Proc Natl Acad Sci* **94:** 8761–8766.

Benjamin LE, Hemo I, Keshet E. 1998. A plasticity window for blood vessel remodelling is defined by pericyte coverage of the preformed endothelial network and is regulated by PDGF-B and VEGF. *Development* **125:** 1591–1598.

Benjamin LE, Golijanin D, Itin A, Pode D, Keshet E. 1999. Selective ablation of immature blood vessels in established human tumors follows vascular endothelial growth factor withdrawal. *J Clin Invest* **103:** 159–165.

Beroukhim R, Brunet JP, Di Napoli A, Mertz KD, Seeley A, Pires MM, Linhart D, Worrell RA, Moch H, Rubin MA, et al. 2009. Patterns of gene expression and copy-number alterations in von Hippel–Lindau disease—Associated and sporadic clear cell carcinoma of the kidney. *Cancer Res* **69:** 4674–4681.

Bindra RS, Vasselli JR, Stearman R, Linehan WM, Klausner RD. 2002. VHL-mediated hypoxia regulation of cyclin D1 in renal carcinoma cells. *Cancer Res* **62:** 3014–3019.

Biswas S, Troy H, Leek R, Chung YL, Li JL, Raval RR, Turley H, Gatter K, Pezzella F, Griffiths JR, et al. 2010. Effects of HIF-1α and HIF2α on growth and metabolism of clear-cell renal cell carcinoma 786-0 xenografts. *J Oncol* **2010:** 757908.

Bommi-Reddy A, Almeciga I, Sawyer J, Geisen C, Li W, Harlow E, Kaelin WG Jr, Grueneberg DA. 2008. Kinase requirements in human cells: III. Altered kinase requirements in $VHL^{-/-}$ cancer cells detected in a pilot synthetic lethal screen. *Proc Natl Acad Sci* **105:** 16484–16489.

Cancer Genome Atlas Research Network. 2013. Comprehensive molecular characterization of clear cell renal cell carcinoma. *Nature* **499:** 43–49.

Chen W, Hill H, Christie A, Kim MS, Holloman E, Pavia-Jimenez A, Homayoun F, Ma Y, Patel N, Yell P, et al. 2016. Targeting renal cell carcinoma with a HIF-2 antagonist. *Nature* **539:** 112–117.

Cho H, Du X, Rizzi JP, Liberzon E, Chakraborty AA, Gao W, Carvo I, Signoretti S, Bruick R, Josey JA, et al. 2016. On-target efficacy of a HIF2α antagonist in preclinical kidney cancer models. *Nature* **539:** 107–111.

Choueiri TK, Escudier B, Powles T, Mainwaring PN, Rini BI, Donskov F, Hammers H, Hutson TE, Lee JL, Peltola K, et al. 2015. Cabozantinib versus everolimus in advanced renal-cell carcinoma. *N Engl J Med* **373:** 1814–1823.

Choueiri TK, Escudier B, Powles T, Tannir NM, Mainwaring PN, Rini BI, Hammers HJ, Donskov F, Roth BJ, Peltola K, et al. 2016. Cabozantinib versus everolimus in advanced renal cell carcinoma (METEOR): Final results from a randomised, open-label, phase 3 trial. *Lancet Oncol* **17:** 917–927.

Dalgliesh GL, Furge K, Greenman C, Chen L, Bignell G, Butler A, Davies H, Edkins S, Hardy C, Latimer C, et al. 2010. Systematic sequencing of renal carcinoma reveals inactivation of histone modifying genes. *Nature* **463:** 360–363.

Davidowitz E, Schoenfeld A, Burk R. 2001. VHL induces renal cell differentiation and growth arrest through integration of cell-cell and cell-extracellular matrix signaling. *Mol Cell Biol* **21:** 865–874.

de Paulsen N, Brychzy A, Fournier M-C, Klausner RD, Gnarra JR, Pause A, Lee S. 2001. Role of transforming growth factor-α in $VHL^{-/-}$ clear cell renal carcinoma cell proliferation: A possible mechanism coupling von Hippel–Lindau tumor suppressor inactivation and tumorigenesis. *Proc Natl Acad Sci* **13:** 1387–1392.

Dondeti VR, Wubbenhorst B, Lal P, Gordan JD, D'Andrea K, Attiyeh EF, Simon MC, Nathanson KL. 2012. Integrative genomic analyses of sporadic clear cell renal cell carcinoma define disease subtypes and potential new therapeutic targets. *Cancer Res* **72:** 112–121.

Duns G, Hofstra RM, Sietzema JG, Hollema H, van Duivenbode I, Kuik A, Giezen C, Jan O, Bergsma JJ, Bijnen H, et al. 2012. Targeted exome sequencing in clear cell renal cell carcinoma

tumors suggests aberrant chromatin regulation as a crucial step in ccRCC development. *Hum Mutat* **33:** 1059–1062.
Fisher R, Horswell S, Rowan A, Salm MP, de Bruin EC, Gulati S, McGranahan N, Stares M, Gerlinger M, Varela I, et al. 2014. Development of synchronous VHL syndrome tumors reveals contingencies and constraints to tumor evolution. *Genome Biol* **15:** 433.
Franovic A, Gunaratnam L, Smith K, Robert I, Patten D, Lee S. 2007. Translational up-regulation of the EGFR by tumor hypoxia provides a nonmutational explanation for its overexpression in human cancer. *Proc Natl Acad Sci* **104:** 13092–13097.
Frew IJ, Thoma CR, Georgiev S, Minola A, Hitz M, Montani M, Moch H, Krek W. 2008. pVHL and PTEN tumour suppressor proteins cooperatively suppress kidney cyst formation. *EMBO J* **27:** 1747–1757.
Gassenmaier M, Chen D, Buchner A, Henkel L, Schiemann M, Mack B, Schendel DJ, Zimmermann W, Pohla H. 2013. CXC chemokine receptor 4 is essential for maintenance of renal cell carcinoma-initiating cells and predicts metastasis. *Stem Cells* **31:** 1467–1476.
Gerlinger M, Rowan AJ, Horswell S, Larkin J, Endesfelder D, Gronroos E, Martinez P, Matthews N, Stewart A, Tarpey P, et al. 2012. Intratumor heterogeneity and branched evolution revealed by multiregion sequencing. *N Engl J Med* **366:** 883–892.
Gerlinger M, Horswell S, Larkin J, Rowan AJ, Salm MP, Varela I, Fisher R, McGranahan N, Matthews N, Santos CR, et al. 2014. Genomic architecture and evolution of clear cell renal cell carcinomas defined by multiregion sequencing. *Nat Genet* **46:** 225–233.
Gnarra J, Ward J, Porter F, Wagne J, Devor D, Grinberg A, Emmert-Buck M, Westphal H, Klausner R, Linehan W. 1997. Defective placental vasculogenesis causes embryonic lethality in VHL-deficient mice. *Proc Natl Acad Sci* **94:** 9102–9107.
Gordan JD, Lal P, Dondeti VR, Letrero R, Parekh KN, Oquendo CE, Greenberg RA, Flaherty KT, Rathmell WK, Keith B, et al. 2008. HIF-α effects on c-Myc distinguish two subtypes of sporadic VHL-deficient clear cell renal carcinoma. *Cancer Cell* **14:** 435–446.
Guo G, Gui Y, Gao S, Tang A, Hu X, Huang Y, Jia W, Li Z, He M, Sun L, et al. 2011. Frequent mutations of genes encoding ubiquitin-mediated proteolysis pathway components in clear cell renal cell carcinoma. *Nat Genet* **44:** 17–19.
Hainsworth JD, Spigel DR, Sosman JA, Burris HA III, Farley C, Cucullu H, Yost K, Hart LL, Sylvester L, Waterhouse DM, et al. 2007. Treatment of advanced renal cell carcinoma with the combination bevacizumab/erlotinib/imatinib: A phase I/II trial. *Clin Genitourin Cancer* **5:** 427–432.
Hu C-J, Wang L-Y, Chodosh LA, Keith B, Simon MC. 2003. Differential roles of hypoxia-inducible factor 1a (HIF-1a) and HIF-2a in hypoxic gene regulation. *Mol Cell Biol* **23:** 9361–9374.
Iliopoulos O, Kibel A, Gray S, Kaelin WG. 1995. Tumor suppression by the human von Hippel–Lindau gene product. *Nat Med* **1:** 822–826.
Kaelin WG. 2002. Molecular basis of the VHL hereditary cancer syndrome. *Nat Rev Cancer* **2:** 673–682.
Kaelin WG Jr. 2007. von Hippel–Lindau disease. *Annu Rev Pathol* **2:** 145–173.
Kaelin WG Jr. 2008. The von Hippel–Lindau tumour suppressor protein: O_2 sensing and cancer. *Nat Rev Cancer* **8:** 865–873.
Kaelin WG Jr, Ratcliffe PJ. 2008. Oxygen sensing by metazoans: The central role of the HIF hydroxylase pathway. *Mol Cell* **30:** 393–402.
Kapitsinou PP, Liu Q, Unger TL, Rha J, Davidoff O, Keith B, Epstein JA, Moores SL, Erickson-Miller CL, Haase VH. 2010. Hepatic HIF-2 regulates erythropoietic responses to hypoxia in renal anemia. *Blood* **116:** 3039–3048.
Keith B, Johnson RS, Simon MC. 2012. HIF1α and HIF2α: Sibling rivalry in hypoxic tumour growth and progression. *Nat Rev Cancer* **12:** 9–22.
Kim WY, Safran M, Buckley MR, Ebert BL, Glickman J, Bosenberg M, Regan M, Kaelin WG Jr. 2006. Failure to prolyl hydroxylate hypoxia-inducible factor α phenocopies VHL inactivation in vivo. *EMBO J* **25:** 4650–4662.
Knebelmann B, Ananth S, Cohen H, Sukhatme V. 1998. Transforming growth factor α is a target for the von Hippel–Lindau tumor suppressor. *Cancer Res* **58:** 226–231.
Kondo K, Klco J, Nakamura E, Lechpammer M, Kaelin WG. 2002. Inhibition of HIF is necessary for tumor suppression by the von Hippel–Lindau protein. *Cancer Cell* **1:** 237–246.
Kondo K, Kim WY, Lechpammer M, Kaelin WG Jr. 2003. Inhibition of HIF2α is sufficient to suppress pVHL-defective tumor growth. *PLoS Biol* **1:** 439–444.
Koochekpour S, Jeffers M, Wang P, Gong C, Taylor G, Roessler L, Stearman R, Vasselli J, Stetler-Stevenson W, Kaelin WJ, et al. 1999. The von Hippel–Lindau tumor suppressor gene inhibits hepatocyte growth factor/scatter factor-induced invasion and branching morphogenesis in renal carcinoma cells. *Mol Cell Biol* **19:** 5902–5912.
Li L, Zhang L, Zhang X, Yan Q, Minamishima YA, Olumi AF, Mao M, Bartz S, Kaelin WG Jr. 2007. Hypoxia-inducible factor linked to differential kidney cancer risk seen with type 2A and type 2B VHL mutations. *Mol Cell Biol* **27:** 5381–5392.
Lieubeau-Teillet B, Rak J, Jothy S, Iliopoulos O, Kaelin W, Kerbel R. 1998. von Hippel–Lindau gene-mediated growth suppression and induction of differentiation in renal cell carcinoma cells grown as multicellular tumor spheroids. *Cancer Res* **58:** 4957–4962.
Lubensky IA, Gnarra JR, Bertheau P, Walther MM, Linehan WM, Zhuang Z. 1996. Allelic deletions of the VHL gene detected in multiple microscopic clear cell renal lesions in von Hippel–Lindau disease patients. *Am J Pathol* **149:** 2089–2094.
Maher E, Kaelin WG. 1997. von Hippel–Lindau disease. *Medicine* **76:** 381–391.
Mandriota SJ, Turner KJ, Davies DR, Murray PG, Morgan NV, Sowter HM, Wykoff CC, Maher ER, Harris AL, Ratcliffe PJ, et al. 2002. HIF activation identifies early lesions in VHL kidneys: Evidence for site-specific tumor suppressor function in the nephron. *Cancer Cell* **1:** 459–468.
Maranchie JK, Vasselli JR, Riss J, Bonifacino JS, Linehan WM, Klausner RD. 2002. The contribution of VHL substrate binding and HIF1-α to the phenotype of VHL loss in renal cell carcinoma. *Cancer Cell* **1:** 247–255.
Merseburger AS, Hennenlotter J, Simon P, Kruck S, Koch E, Horstmann M, Kuehs U, Kufer R, Stenzl A, Kuczyk MA. 2005. Membranous expression and prognostic implications of epidermal growth factor receptor protein in human renal cell cancer. *Anticancer Res* **25:** 1901–1907.
Micucci C, Matacchione G, Valli D, Orciari S, Catalano A. 2015. HIF2α is involved in the expansion of CXCR4-positive cancer stem-like cells in renal cell carcinoma. *Br J Cancer* **113:** 1178–1185.
Montani M, Heinimann K, von Teichman A, Rudolph T, Perren A, Moch H. 2010. VHL-gene deletion in single renal tubular epithelial cells and renal tubular cysts: Further evidence for a cyst-dependent progression pathway of clear cell renal carcinoma in von Hippel–Lindau disease. *Am J Surg Pathol* **34:** 806–815.
Morris MR, Hughes DJ, Tian YM, Ricketts CJ, Lau KW, Gentle D, Shuib S, Serrano-Fernandez P, Lubinski J, Wiesener MS, et al. 2009. Mutation analysis of hypoxia-inducible factors HIF1A and HIF2A in renal cell carcinoma. *Anticancer Res* **29:** 4337–4343.
Nakaigawa N, Yao M, Baba M, Kato S, Kishida T, Hattori K, Nagashima Y, Kubota Y. 2006. Inactivation of von Hippel–Lindau gene induces constitutive phosphorylation of MET protein in clear cell renal carcinoma. *Cancer Res* **66:** 3699–3705.
Pan J, Mestas J, Burdick MD, Phillips RJ, Thomas GV, Reckamp K, Belperio JA, Strieter RM. 2006. Stromal derived factor-1 (SDF-1/CXCL12) and CXCR4 in renal cell carcinoma metastasis. *Mol Cancer* **5:** 56.

Panka DJ, Liu Q, Geissler AK, Mier JW. 2013. Effects of HDM2 antagonism on sunitinib resistance, p53 activation, SDF-1 induction, and tumor infiltration by CD11b^{+}/Gr-1^{+} myeloid derived suppressor cells. *Mol Cancer* **12:** 17.

Pause A, Lee S, Lonergan KM, Klausner RD. 1998. The von Hippel–Lindau tumor suppressor gene is required for cell cycle exit upon serum withdrawal. *Proc Natl Acad Sci* **95:** 993–998.

Pena-Llopis S, Vega-Rubin-de-Celis S, Liao A, Leng N, Pavia-Jimenez A, Wang S, Yamasaki T, Zhrebker L, Sivanand S, Spence P, et al. 2012. BAP1 loss defines a new class of renal cell carcinoma. *Nat Genet* **44:** 751–759.

Pennacchietti S, Michieli P, Galluzzo M, Mazzone M, Giordano S, Comoglio PM. 2003. Hypoxia promotes invasive growth by transcriptional activation of the met protooncogene. *Cancer Cell* **3:** 347–361.

Polite BN, Desai AA, Manchen B, Stadler WM. 2006. Combination therapy of imatinib mesylate and interferon-α demonstrates minimal activity and significant toxicity in metastatic renal cell carcinoma: Results of a single-institution phase II trial. *Clin Genitourin Cancer* **4:** 275–280.

Purdue MP, Johansson M, Zelenika D, Toro JR, Scelo G, Moore LE, Prokhortchouk E, Wu X, Kiemeney LA, Gaborieau V, et al. 2011. Genome-wide association study of renal cell carcinoma identifies two susceptibility loci on 2p21 and 11q13.3. *Nat Genet* **43:** 60–65.

Ramp U, Jaquet K, Reinecke P, Schardt C, Friebe U, Nitsch T, Marx N, Gabbert HE, Gerharz CD. 1997. Functional intactness of stimulatory and inhibitory autocrine loops in human renal carcinoma cell lines of the clear cell type. *J Urol* **157:** 2345–2350.

Rankin EB, Biju MP, Liu Q, Unger TL, Rha J, Johnson RS, Simon MC, Keith B, Haase VH. 2007. Hypoxia-inducible factor-2 (HIF-2) regulates hepatic erythropoietin in vivo. *J Clin Invest* **117:** 1068–1077.

Rankin EB, Rha J, Unger TL, Wu CH, Shutt HP, Johnson RS, Simon MC, Keith B, Haase VH. 2008. Hypoxia-inducible factor-2 regulates vascular tumorigenesis in mice. *Oncogene* **27:** 5354–5358.

Rankin EB, Rha J, Selak MA, Unger TL, Keith B, Liu Q, Haase VH. 2009. HIF-2 regulates hepatic lipid metabolism. *Mol Cell Biol* **29:** 4527–4538.

Raval RR, Lau KW, Tran MG, Sowter HM, Mandriota SJ, Li JL, Pugh CW, Maxwell PH, Harris AL, Ratcliffe PJ. 2005. Contrasting properties of hypoxia-inducible factor 1 (HIF-1) and HIF-2 in von Hippel–Lindau-associated renal cell carcinoma. *Mol Cell Biol* **25:** 5675–5686.

Rogers JL, Bayeh L, Scheuermann TH, Longgood J, Key J, Naidoo J, Melito L, Shokri C, Frantz DE, Bruick RK, et al. 2013. Development of inhibitors of the PAS-B domain of the HIF-2α transcription factor. *J Med Chem* **56:** 1739–1747.

Ryan CW, Vuky J, Chan JS, Chen Z, Beer TM, Nauman D. 2011. A phase II study of everolimus in combination with imatinib for previously treated advanced renal carcinoma. *Invest New Drugs* **29:** 374–379.

Sankin A, Hakimi AA, Mikkilineni N, Ostrovnaya I, Silk MT, Liang Y, Mano R, Chevinsky M, Motzer RJ, Solomon SB, et al. 2014. The impact of genetic heterogeneity on biomarker development in kidney cancer assessed by multiregional sampling. *Cancer Med* **3:** 1485–1492.

Scheuermann TH, Tomchick DR, Machius M, Guo Y, Bruick RK, Gardner KH. 2009. Artificial ligand binding within the HIF2α PAS-B domain of the HIF2 transcription factor. *Proc Natl Acad Sci* **106:** 450–455.

Scheuermann TH, Li Q, Ma HW, Key J, Zhang L, Chen R, Garcia JA, Naidoo J, Longgood J, Frantz DE, et al. 2013. Allosteric inhibition of hypoxia inducible factor-2 with small molecules. *Nat Chem Biol* **9:** 271–276.

Schietke RE, Hackenbeck T, Tran M, Gunther R, Klanke B, Warnecke CL, Knaup KX, Shukla D, Rosenberger C, Koesters R, et al. 2012. Renal tubular HIF-2α expression requires VHL inactivation and causes fibrosis and cysts. *PLoS One* **7:** e31034.

Schodel J, Bardella C, Sciesielski LK, Brown JM, Pugh CW, Buckle V, Tomlinson IP, Ratcliffe PJ, Mole DR. 2012. Common genetic variants at the 11q13.3 renal cancer susceptibility locus influence binding of HIF to an enhancer of cyclin D1 expression. *Nat Genet* **44:** 420–425, S421–S422.

Semenza GL. 2011. Oxygen sensing, homeostasis, and disease. *N Engl J Med* **365:** 537–547.

Shen C, Beroukhim R, Schumacher SE, Zhou J, Chang M, Signoretti S, Kaelin WG Jr. 2011. Genetic and functional studies implicate HIF1α as a 14q kidney cancer suppressor gene. *Cancer Discov* **1:** 222–235.

Siegel RL, Miller KD, Jemal A. 2016. Cancer statistics, 2016. *CA Cancer J Clin* **66:** 7–30.

Smith K, Gunaratnam L, Morley M, Franovic A, Mekhail K, Lee S. 2005. Silencing of epidermal growth factor receptor suppresses hypoxia-inducible factor-2–driven VHL$^{-/-}$ renal cancer. *Cancer Res* **65:** 5221–5230.

Staller P, Sulitkova J, Lisztwan J, Moch H, Oakeley EJ, Krek W. 2003. Chemokine receptor CXCR4 downregulated by von Hippel–Lindau tumour suppressor pVHL. *Nature* **425:** 307–311.

Struckmann K, Mertz K, Steu S, Storz M, Staller P, Krek W, Schraml P, Moch H. 2008. pVHL co-ordinately regulates CXCR4/CXCL12 and MMP2/MMP9 expression in human clear-cell renal cell carcinoma. *J Pathol* **214:** 464–471.

Uhlman DL, Nguyen P, Manivel JC, Zhang G, Hagen K, Fraley E, Aeppli D, Niehans GA. 1995. Epidermal growth factor receptor and transforming growth factor α expression in papillary and nonpapillary renal cell carcinoma: Correlation with metastatic behavior and prognosis. *Clin Cancer Res* **1:** 913–920.

Uniacke J, Holterman CE, Lachance G, Franovic A, Jacob MD, Fabian MR, Payette J, Holcik M, Pause A, Lee S. 2012. An oxygen-regulated switch in the protein synthesis machinery. *Nature* **486:** 126–129.

Vanharanta S, Shu W, Brenet F, Hakimi AA, Heguy A, Viale A, Reuter VE, Hsieh JJ, Scandura JM, Massague J. 2013. Epigenetic expansion of VHL-HIF signal output drives multiorgan metastasis in renal cancer. *Nat Med* **19:** 50–56.

Varela I, Tarpey P, Raine K, Huang D, Ong CK, Stephens P, Davies H, Jones D, Lin ML, Teague J, et al. 2011. Exome sequencing identifies frequent mutation of the SWI/SNF complex gene *PBRM1* in renal carcinoma. *Nature* **469:** 539–542.

Vuky J, Isacson C, Fotoohi M, dela Cruz J, Otero H, Picozzi V, Malpass T, Aboulafia D, Jacobs A. 2006. Phase II trial of imatinib (Gleevec) in patients with metastatic renal cell carcinoma. *Invest New Drugs* **24:** 85–88.

Wallace EM, Rizzi JP, Han G, Wehn PM, Cao Z, Du X, Cheng T, Czerwinski RM, Dixon DD, Goggin BS, et al. 2016. A small-molecule antagonist of HIF2α is efficacious in preclinical models of renal cell carcinoma. *Cancer Res* **76:** 5491–5500.

Walther M, Lubensky I, Venzon D, Zbar B, Linehan W. 1995. Prevalence of microscopic lesions in grossly normal renal parenchyma from patients with von Hippel–Lindau disease, sporadic renal cell carcinoma and no renal disease: Clinical implications. *J Urol* **154:** 2010–2014.

Xu X, Hou Y, Yin X, Bao L, Tang A, Song L, Li F, Tsang S, Wu K, Wu H, et al. 2012. Single-cell exome sequencing reveals single-nucleotide mutation characteristics of a kidney tumor. *Cell* **148:** 886–895.

Zagzag D, Krishnamachary B, Yee H, Okuyama H, Chiriboga L, Ali MA, Melamed J, Semenza GL. 2005. Stromal cell-derived factor-1α and CXCR4 expression in hemangioblastoma and clear cell-renal cell carcinoma: von Hippel–Lindau loss-of-function induces expression of a ligand and its receptor. *Cancer Res* **65:** 6178–6188.

Zatyka M, da Silva NF, Clifford SC, Morris MR, Wiesener MS, Eckardt KU, Houlston RS, Richards FM, Latif F, Maher ER. 2002. Identification of cyclin D1 and other novel targets for the von Hippel–Lindau tumor suppressor gene by expression array analysis and investigation of cyclin D1 genotype as a modifier in von Hippel–Lindau disease. *Cancer Res* **62:** 3803–3811.

Zhang T, Niu X, Liao L, Cho EA, Yang H. 2013. The contributions of HIF-target genes to tumor growth in RCC. *PLoS One* **8:** e80544.

Zhou L, Yang H. 2011. The von Hippel–Lindau tumor suppressor protein promotes c-Cbl-independent poly-ubiquitylation and degradation of the activated EGFR. *PLoS One* **6:** e23936.

Zhuang Z, Bertheau P, Emmert-Buck MR, Liotta L, Gnarra J, Linehan WM, Lubensky IA. 1995. A microscopic dissection technique for archival DNA analysis of specific cell populations in lesions <1 mm in size. *Am J Pathol* **146:** 620–625.

Zimmer M, Doucette D, Siddiqui N, Iliopoulos O. 2004. Inhibition of hypoxia-inducible factor is sufficient for growth suppression of $VHL^{-/-}$ tumors. *Mol Cancer Res* **2:** 89–95.

BET Bromodomain Proteins as Cancer Therapeutic Targets

Shaokun Shu and Kornelia Polyak

Department of Medical Oncology, Dana-Farber Cancer Institute; Department of Medicine, Brigham and Women's Hospital; and Department of Medicine, Harvard Medical School, Boston, Massachusetts 02215

Correspondence: Kornelia_polyak@dfci.harvard.edu

Epigenetic regulators are emerging therapeutic targets in a wide variety of human cancers. BET bromodomain proteins have been identified as key regulators of oncogenic transcription factors including MYC; therefore, their inhibition might provide a way to block these "undruggable" targets. Several BET bromodomain inhibitors are in clinical development with promising preliminary findings. However, tumors acquire resistance to these agents in several different ways. In this review, we summarize the role that BET bromodomain proteins play in tumorigenesis as well as the molecular mechanisms underlying therapeutic responses and resistance to their inhibition with emphasis on BRD4 and breast cancer.

Epigenetic regulation of gene expression is a dynamic and reversible process that establishes normal cellular phenotypes but also contributes to human diseases including cancer. Epigenetic changes that occur during tumor progression are increasingly recognized as potential therapeutic targets in various human cancer types. Histone modifications (e.g., methylation and acetylation) regulate chromatin structure and transcription, and abnormalities in this process lead to perturbed development and differentiation. The demonstration of frequent somatic mutations in histones and chromatin-modifying factors in a wide range of human tumor types accentuates the importance of epigenetic regulation in tumorigenesis (van Haaften et al. 2009; Dalgliesh et al. 2010; Morin et al. 2011). These regulators are broadly categorized into (1) writers—enzymes that catalyze histone or DNA modifications such as histone and DNA methyl transferases, (2) erasers—enzymes that remove histone or DNA modifications including histone and DNA demethylases, and (3) readers—proteins that recognize a specific histone or DNA modification and modulate transcription by binding to these modifications. The bromodomain-containing family of proteins represents an important class of histone modification readers that recognize acetylated lysine residues and play critical roles in maintaining epigenetic memory and gene transcription (Dey et al. 2009; Zhao et al. 2011).

In this review, we briefly summarize the role of BET bromodomain proteins in various cancer types, the mechanisms of response and resistance to BET bromodomain inhibitors (BETis), and their clinical development.

BET BROMODOMAIN PROTEINS

The bromodomain was first described in 1992 as an evolutionarily conserved domain of ~110 amino acids present in several proteins in humans, fruit flies, and yeast with important roles in transcription (Haynes et al. 1992). Its structure consists of four α helices separated by variable loop regions, which together form a hydrophobic cavity that recognizes acetyllysine. The human genome encodes for 42 bromodomain-containing proteins with 61 unique bromodomains, where differences in the amino acid residues around the acetyllysine binding site impart binding specificity (Filippakopoulos and Knapp 2012). The BET (bromodomain and extraterminal domain) subfamily of bromodomain proteins, composed of BRD2, BRD3, BRD4, and BRDT, perform diverse roles in regulating transcription by RNA polymerase II (POLII) (Dhalluin et al. 1999). Common to all four BET proteins are two conserved amino-terminal bromodomains (BD1 and BD2) that are critical for their chromatin binding via recognition of acetylated lysine residues on histone tails and other nuclear proteins. A well-studied member of the BET family is BRD4, which was first identified as a MED1-interacting protein and a component of the mammalian Mediator complex (Jiang et al. 1998). Consistent with this interaction, BRD4 and Mediator co-occupy similar sites across the genome and can stabilize one another's occupancy at certain regions, including thousands of enhancers and promoters associated with active genes (Loven et al. 2013). Exceptionally high levels of these factors occur at a small set of large enhancer regions defined as super-enhancers (SEs). SEs drive genes that are essential for establishing and maintaining cellular identity and also play key roles in driving the expression of oncogenes during cancer development and progression (Chapuy et al. 2013; Hnisz et al. 2013; Loven et al. 2013). BRD4 is also a crucial regulator of transcriptional elongation that recruits the positive transcription elongation factor b (P-TEFb) complex to chromatin and mediates the

Published by Cold Spring Harbor Laboratory Press; doi: 10.1101/sqb.2016.81.030908

formation of the active form of P-TEFb, which phosphorylates and activates RNA polymerase II. The BRD4/P–TEFb interaction is important for rapid transcriptional reinitiation after mitosis (Jang et al. 2005; Yang et al. 2005).

A number of selective and potent small-molecule inhibitors have been developed that can compete for the acetyl-binding pockets present in the bromodomains of BET proteins and block gene expression (Filippakopoulos et al. 2010; Nicodeme et al. 2010). One of the most notable inhibitors is JQ1, a thienodiazepine-based small molecule that shows excellent inhibition against the BET subfamily in the low-nanomolar range and is especially effective against BRD4 (Filippakopoulos et al. 2010). The crystal structure of JQ1 and the first bromodomain of BRD4 showed excellent structural complementarity between the small molecule and the protein. JQ1 engages the bromodomain pocket in a manner that is competitive with acetylated peptide binding, thereby causing the displacement of BET proteins from chromatin with high affinity and specificity and excellent cell permeability. The development of these inhibitors has revealed novel insights into the physiological role and therapeutic potential of inhibiting BET function, especially in various human cancers and systemic inflammatory conditions (Nicodeme et al. 2010; Delmore et al. 2011; Zuber et al. 2011). Treatment of tumor cells with the BETi JQ1 causes a preferential loss of BRD4, Mediator, and P-TEFb at SEs, leading to the relatively selective decrease of transcription at SE-associated genes, including the c-MYC oncogene in c-MYC-driven malignancies (Delmore et al. 2011; Loven et al. 2013). Thus, key oncogenic drivers can be regulated by SEs, which can confer disproportionate sensitivity to loss of BRD4 leading to rather selective inhibition of transcription by BETis. Further investigations have now shown that BETis can repress the expression of oncogenic transcription factors beyond c-MYC, thus expanding their potential clinical utility to several solid tumors, including colorectal cancer, castration-resistant prostate cancer, glioblastoma, non–small cell lung cancer, N-MYC-amplified neuroblastoma, ovarian carcinomas, and breast cancer (Lockwood et al. 2012; Puissant et al. 2013; Sahai et al. 2014, 2016; Shu et al. 2016). BRD4 was also identified as a component of a recurrent t(15;19) chromosomal translocation in NUT midline carcinoma, an aggressive form of human squamous carcinoma (French et al. 2008). The resulting BRD4-NUT oncoprotein is an aberrant transcriptional regulator that relies on the bromodomains of BRD4 for its oncogenic function. JQ1 treatment induces differentiation in patient-derived NUT cancer cell lines and also showed excellent efficacy in xenograft models without obvious toxicity (Filippakopoulos et al. 2010).

To date, 10 BETis have entered clinical trials (Galdeano and Ciulli 2016). One of them, RVX-208 (Resverlogic) has reached Phase III clinical trials in a total of seven clinical trials for the treatment of atherosclerosis and associated cardiovascular disease (Galdeano and Ciulli 2016). Three additional BETis have also reached Phase II clinical trials (OTX015/MK-8628 [Merck], BMS-9861158 [BMS], and GSK525762 [GSK]), while six other BETis have completed or are in ongoing Phase I clinical trials (TEN-010 [Roche], CPI-0610 [Constellation], GS-5829 [Gilead Sciences], BAY1238097 [Bayer], ABBV-075 [AbbVie], and INCB054329 [Incyte]), for both solid tumors and hematological malignancies. Clinical responses have been seen in NUT midline carcinoma patients and in leukemia, lymphoma, and multiple myeloma, primarily as a result of repressing c-MYC expression (Stathis et al. 2016). Additional trials are planned testing BETis in combination with chemotherapy, other targeted therapies, and immunotherapy.

BET INHIBITORS IN BREAST CANCER

Breast cancer is currently the second leading cause of cancer-related deaths in women worldwide. More than 200,000 women in the United States and more than one million across the world are diagnosed with breast cancer each year. Despite improvements in adjuvant therapy, many women still experience recurrences of their cancer sometimes even decades after initial diagnosis and treatment. Once metastatic, breast cancer is generally incurable and is responsible for approximately 40,000 deaths in the United States and more than 400,000 deaths worldwide annually (Siegel et al. 2016). Molecular interrogation of human breast cancers has led to the identification of at least four distinct tumor subtypes distinguished by their gene expression profiles and clinical behavior: luminal A, luminal B, $HER2^+$, and basal-like (Perou et al. 2000; Sorlie et al. 2001). These molecular subtypes complement, but do not fully overlap with pathologic classification based on the expression of estrogen receptor (ER), progesterone receptor (PR), and HER2. Luminal tumors are typically ER^+PR^+, and basal-like tumors are commonly "triple-negative" (ER^-, PR^-, and $HER2^-$), whereas $HER2^+$ tumors have amplification of the *ERBB2* oncogene.

ER^+ breast cancers represent ~70% of all breast cancers and depend on estrogen-induced activation of ER for their growth. Estrogen (E2)-induced transcription requires coordinated recruitment of estrogen receptor α (ERα) and multiple factors at the promoter of activated genes. However, the precise mechanism by which this complex stimulates the activity of RNA polymerase II required to execute transcription is largely unresolved. Genome-wide studies revealed an enrichment of BRD4 at transcriptional start sites of active genes and a requirement of BRD4 for H2B mono-ubiquitination in the transcribed regions of estrogen-responsive genes (Nagarajan et al. 2014). Importantly, BRD4 occupies distal EREs enriched for the histone H3 lysine 27acetyl (H3K27ac) mark and regulates enhancer RNA synthesis by affecting RNAPII recruitment and elongation. Consistently, BRD4 activity is required for the proliferation of ER^+ breast and endometrial cancer cells and for uterine growth in mice (Nagarajan et al. 2014). JQ1 suppresses E2-induced growth and transcription in MCF7 and T47D ER^+ breast cancer cell lines (Nagarajan et al. 2014). JQ1 treatment does not disrupt E2-induced recruitment of ER, SRC3, or histone acetylation status at the promoters of genes but blocks the E2-induced transition of RNA polymerase II from initiation to elongation by stalling it at the promoter region of the responsive genes upstream of the transcription start site.

Treatment options for ER^+ breast cancer patients have relied heavily on antihormonal strategies with varying anti-estrogen modalities including SERMs (selective estrogen receptor modulators), such as tamoxifen and raloxifen, and aromatase inhibitors (Early Breast Cancer Trialists' Collaborative Group 2005). Although most patients respond to endocrine agents, a significant fraction will eventually relapse (Lupien et al. 2010). A number of studies have investigated the potential use of BETis in endocrine-resistant breast tumors. BET bromodomain proteins contribute to tamoxifen resistance by recruiting WHSC1, a histone H3K6 methyltransferase, to ERα for facilitating ERα gene expression (Feng et al. 2014). JQ1 potently suppresses the classic ERα signaling pathway and the growth of tamoxifen-resistant breast cancer cells in vitro and in vivo (Feng et al. 2014). Importantly, tamoxifen-resistant cell lines are more sensitive to the BETi JQ1 than are parental lines, implying its potential use as a combination agent to overcome endocrine resistance.

The PI3K/mTOR pathway has also been shown to mediate resistance to anti-estrogens in ER^+ breast cancer patients (Miller et al. 2010). Everolimus, an allosteric inhibitor of mTOR complex 1, has been approved in combination with exemestane (an aromatase inhibitor) for the treatment of breast cancer patients with ER^+HER2^- tumors who progressed on anastrozole or letrozole aromatase inhibitor treatment alone (Baselga et al. 2012). However, despite initial efficacy, breast tumors develop acquired resistance to everolimus. Acquired resistance to PI3K/mTOR/Akt pathway inhibitors is often associated with compensatory feedback loops involving the activation of oncogenes (Piccart et al. 2014). c-MYC was identified as a driver of mTOR inhibitor resistance in long-term estrogen deprivation models (Bihani et al. 2015). Both everolimus resistance and acute treatment of everolimus can lead to the up-regulation of MYC expression and the enrichment of MYC signatures. Importantly, depletion of c-MYC using siRNA or the BET inhibitor JQ1 restored everolimus sensitivity, whereas a combination of everolimus and JQ1 led to synergistic growth inhibition in 3D Matrigel cultures and in xenograft models.

BETis were also shown to overcome resistance to class I PI3K inhibitors in a murine model of metastatic breast cancer driven by mutant PI3K and MYC (Stratikopoulos et al. 2015). BETis decreased PI3K signaling and dissociated BRD4 from chromatin at genomic regions that regulate the expression of insulin receptor and epidermal growth factor receptor (EGFR) family receptor tyrosine kinases (RTKs). Combined PI3K and BET inhibition induced cell death, tumor regression, and clamped inhibition of PI3K signaling.

The *ERBB2* oncogene is amplified or overexpressed in ~25% of breast cancers and serves as the primary driver of $HER2^+$ tumors. Therapeutics that target HER2, such as trastuzumab (Herceptin) and lapatinib, often provide initial clinical benefit, but resistance frequently develops. Resistance to lapatinib can result from kinome signaling reprograming in tumor cells including up-regulation of HER3 and activation of multiple tyrosine kinases. Genetic and chemical inhibition of BET bromodomain proteins suppresses transcription of many lapatinib-induced kinases involved in resistance, including ERBB3, IGF1R, DDR1, MET, and fibroblast growth factor receptors (FGFRs), preventing downstream SRC/FAK signaling and AKT reactivation (Stuhlmiller et al. 2015). In line with this, combination treatment with lapatinib and JQ1 blocked the growth of lapatinib-resistant cells. Thus, BETis may also be useful in decreasing resistance to $HER2^+$ targeted therapies in $HER2^+$ breast cancer.

BET INHIBITORS IN TRIPLE-NEGATIVE BREAST CANCER

Triple-negative breast cancers (TNBCs) that are negative for ER, PR, and HER2 account for ~20% of breast cancers in the United States. TNBCs have a high rate of recurrence to distant metastatic sites and are currently the only major breast tumor subtype that lacks targeted therapies (Lehmann et al. 2011). Hence, new therapeutic options are clearly needed. We hypothesized that because of the central role for c-MYC in TNBCs (Daemen et al. 2013), these tumors may be dependent on BET bromodomain proteins. To test this hypothesis, we screened a large panel of breast cancer cell lines with BETis and found that TNBC cell lines showed significant growth inhibition in cell culture when treated with low-nanomolar concentrations of BETis, compared with luminal cell lines, the majority of which were relatively resistant (Shu et al. 2016). The growth of established xenografts derived from TNBC cell lines and from primary patient TNBC samples was also efficiently inhibited by JQ1 treatment. The expression of BRD4 is higher in TNBCs compared with luminal lines, and its down-regulation by TET-inducible short-hairpin RNAs (shRNAs) arrested TNBC cell growth in vitro whereas in vivo it led to complete tumor regression without any regrowth even after stopping doxycycline treatment. c-MYC is more abundantly expressed in TNBC lines compared with luminal ones, but its expression did not correlate with JQ1 response and did not change consistently after JQ1 treatment. In line with the role of BRD4 in transcription restart after G_2/M, JQ1 treatment prevented cell cycle reentry, arrested TNBC cells in early G_1, and induced apoptosis, senescence, and luminal features in basal-like breast cancer cell lines. Using integrated epigenomic analysis, JQ1 was found to efficiently displace chromatin-bound BRD4 in sensitive TNBC cells. This effect was more pronounced at genomic regions showing elevated Bio-JQ1 binding based on Chem-seq, implying strong association between BRD4 and JQ1 genomic binding patterns. Genome-wide studies with ChIP-seq and RNA-seq have also revealed preferential disruption of SE-associated genes by JQ1, leading to deregulation of coordinated transcriptional pathways involved in cell proliferation,

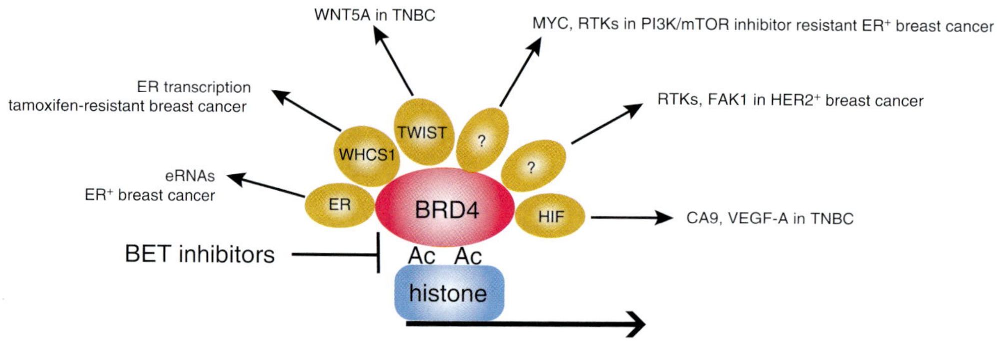

Figure 1. Schematic model of BRD4 mechanism of action in breast cancer. BRD4 binds to the indicated proteins leading to activation of downstream genes. ER, estrogen receptor; TNBC, triple-negative breast cancer; RTKs, receptor tyrosine kinases; HER2, human epidermal growth factor receptor 2; VEGF-A, vascular endothelial growth factor A; HIF, hypoxia-inducible factor.

invasion, and survival. These results provide strong preclinical data indicating that BETis may represent novel therapeutic agents for the treatment of TNBCs.

Hypoxia enables progression of TNBCs partly by driving metabolic adaptation, angiogenesis, and metastasis through up-regulation of hypoxia-regulated genes (da Motta et al. 2017). JQ1 treatment significantly down-regulates the hypoxic transcriptome response including CA9 and VEGF-A by preventing HIF binding to the hypoxia response element in CA9 promoter but does not alter HIF expression or activity (da Motta et al. 2017). JQ1 reduced TNBC growth in vitro and in vivo and inhibited xenograft vascularization. Zhou and his colleagues showed that the binding of BRD4 with diacetylated TWIST directs WNT5A expression in basal-like breast cancer (BLBC) (Shi et al. 2014). TIP60 diacetylates TWIST at its ''histone H4-mimic'' GK-X-GK motif, which binds to the second bromodomain of BRD4, whose first bromodomain interacts with acetylated H4, thereby constructing an activated TWIST/BRD4/P-TEFb/RNA-Pol II complex at the WNT5A promoter and enhancer. JQ1 treatment disrupted the TWIST–BRD4 interaction and resulted in significant WNT5A reduction, leading to inhibition of invasion, cancer stem cell (CSC)-like properties and tumorigenicity of BLBC. TWIST–BRD4 interaction is required for the recruitment of BRD4 at the WNT5A SE to activate WNT5A transcription. Histone H4 and TWIST synergistically interact with BRD4. Overall, they concluded that the TWIST–BRD4–WNT5A axis is critical for the growth of BLBC in vitro and in vivo.

Based on these studies BETis appear to be promising novel therapeutic agents in all types of breast tumors, especially in combination with other therapies, with underlying mechanisms varying depending on tumor subtype (Fig. 1).

MECHANISMS OF BETi RESISTANCE

Targeted therapies inevitably select for resistance during prolonged treatment. Furthermore, studying mechanisms of acquired resistance can also help delineate the detailed mechanism of action of a particular agent. Therefore, we and others have investigated potential resistance mechanisms to BETis in various cancer types and identified various mechanisms of resistance in different cancer types (Fig. 2). In TNBC, we identified a unique mechanism of epigenomic resistance to BETi in paired cell lines selected for acquired resistance to BET inhibition from previously sensitive TNBCs by long-term culture in escalating doses of JQ1 (Shu et al. 2016). BETi resistance was not attributable to drug export or alterations of BET bromodomain proteins and other driver genes, and resistant cells maintained their dependency on BRD4. By comparing Bio-JQ1 Chem-seq profiles of sensitive and resistant cells, we found a significant gain in the number of SEs in resistant cells, and a less pronounced loss of fewer SEs. The gain of Bio-JQ1 SEs was associated with gain of BRD4 binding to these genomic loci and also with increased transcription of the associated genes. Among the top gained super-enhancers in JQ1-resistant SUM159 cells was an upstream and intragenic region of H3K27ac enrichment at the *BCL-xL* locus, consistent with higher BCL-XL expression in resistant cells. BRD4 is not displaced from chromatin following JQ1 treatment in resistant cells because of the recruitment of BRD4 to SE loci in a bromodomain-independent manner. By performing an unbiased proteomic analysis using RIME (rapid immunoprecipitation mass spectrometry of endogenous protein) (Mohammed et al. 2013), we identified strong association of BRD4 with MED1 in resistant cells. Cheng-Ming Chiang's group has previously reported that BRD4 is phosphorylated at several residues by casein kinase II (CK2) and this increases the stability and nuclear localization of BRD4 (Wu et al. 2013). We found a marked increase of phospho-BRD4 (pBRD4) in resistant cells and also identified PP2A as a BRD4 phosphatase. We determined that BRD4 is hyperphosphorylated in JQ1-resistant cells and this increases the recruitment of BRD4 to MED1 and contributes to BETi resistance. We confirmed this by expressing BRD4 mutants that cannot be phosphorylated (7A) or that mimic constitutive phosphorylation (7D) and showing that these increase and decrease BETi sensitivity, respectively. Last, based on our genomic and proteomic profiling of resistant cells, we identified potential drug combinations that can overcome BETi resistance. Specifically, inhibitors of BCL-XL (e.g., ABT263) and CK2 (e.g., CX-4945) and activators of PP2A (e.g., perphena-

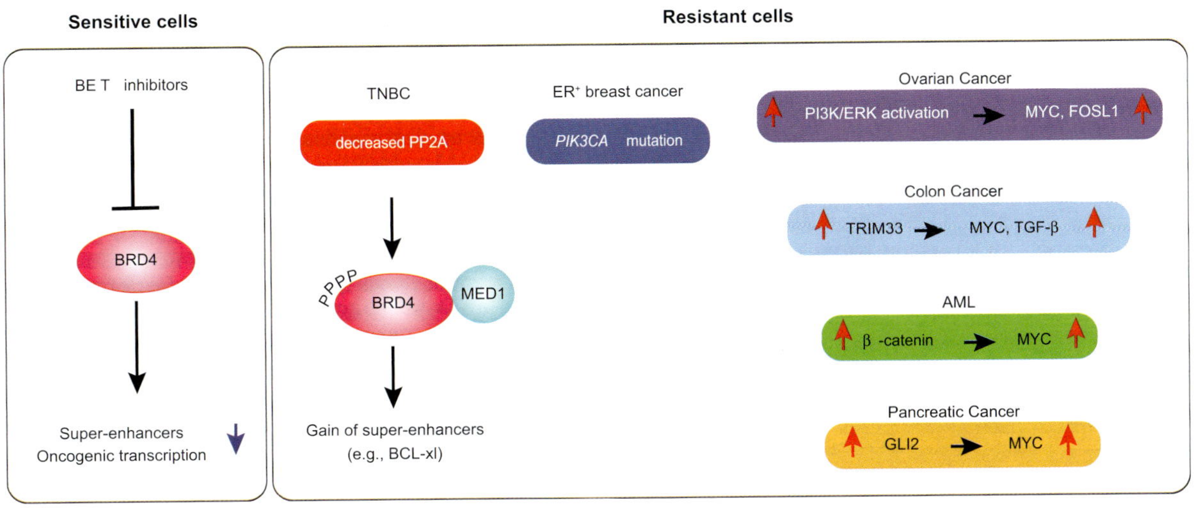

Figure 2. BET bromodomain inhibitor (BETi) resistance mechanisms. In sensitive cells BETis inhibit BRD4 function through disrupting its binding to super-enhancers, thereby turning off downstream oncogenic transcription. The model depicts mechanisms of BETi resistance in the indicated cancer types. TNBC, triple-negative breast cancer; ER, estrogen receptor; PI3K, phosphoinositide 3-kinase; TRIM33, tripartite motif-containing protein 33.

zine) synergize with JQ1 and can overcome BETi resistance in TNBCs.

Ben Neel's group has also identified BRD4 as a potential target in luminal breast cancers by performing an shRNA screen and integrated molecular analyses in a large panel of breast cancer cell lines (Marcotte et al. 2016). They tested the effect of JQ1 in a subset of the cell lines used in the shRNA screen and found that lines with high JQ1 sensitivity underwent apoptosis, whereas resistant lines had slower cell cycle progression. Many luminal HER2^{+} breast cancer cell lines sensitive to BRD4 down-regulation were JQ1-resistant, but most basal-like lines that were sensitive to BRD4 loss were JQ1-sensitive. JQ1 sensitivity did not reflect impaired MYC expression, and exogenous MYC did not convert JQ1-sensitive lines to JQ1-resistant ones. Integrative analysis revealed a strong correlation between JQ1 resistance and *PIK3CA* mutation. Overexpression of wild-type or mutant *PIK3CA* conferred JQ1 resistance in JQ1-sensitive cells. PI3K-specific kinase inhibitors increased the JQ1 sensitivity of resistant cells in vitro and in vivo.

In pancreatic cancer cells, acquired resistance to BETis was found to be associated with resistance to BRD4 down-regulation, increased expression of JQ1 targets including c-MYC, FOSL1, and HMGA2, and epithelial–mesenchymal transition (EMT) (Kumar et al. 2015). JQ1-resistant cells remain dependent on c-MYC through increased binding of GLI2 to the *c-MYC* promoter. Targeting GLI2 resensitized pancreatic cancer cells to BET inhibitors and BRD4 down-regulation.

To identify potential mediators of BETi resistance in MLL-AF9-driven AML (acute myeloid leukemia), Johannes Zuber's group performed a chromatin-focused RNAi screen in a BETi-sensitive MLL–AF9;Nras^{G12D}-driven AML mouse cell line (Rathert et al. 2015). They found that resistance to BET bromodomain inhibitors is associated with profound changes in the enhancer landscape. The suppression of the PRC2 complex promotes BET inhibitor resistance in AML by facilitating the remodeling of regulatory pathways that restore the transcription of key targets such as MYC. This process involves the activation of a focal MYC enhancer that activates WNT genes in response to BET inhibition. WNT signaling promotes primary and acquired BETi resistance in leukemia, establishing it as a major driver and candidate biomarker of BETi resistance in leukemia.

Mark A. Dawson's group derived BETi-resistant mouse AML cells by continuously exposing the cells to BET inhibitor eventually yielding drug-resistant clonal populations in vitro and in vivo (Fong et al. 2015). BETi resistance in this model is not mediated by increased drug efflux or metabolism but was shown to emerge from leukemia stem cells, suggesting that the AML cancer-stem-cell population, or a subset thereof, does not respond to BET inhibitors. Chromatin-bound BRD4 is globally reduced in resistant cells, whereas the expression of key target genes such as Myc remains unaltered. They also showed that Wnt/β-catenin activation has a role in BETi resistance and inhibition of WNT/β-catenin signaling restores BETi sensitivity in vitro and in vivo.

In colon cancer, loss of tripartite motif-containing protein 33 (TRIM33), a chromatin-associated E3 ubiquitin ligase, was identified to confer resistance to BETi based on shRNA-based genetic screen (Shi et al. 2016). TRIM33 silencing reprograms cancer cells to a more resistant state through at least two mechanisms. TRIM33 loss maintains MYC expression following BETi treatment. Moreover, TRIM33 silencing enhances TGF-β receptor expression and signaling. The authors also showed that inhibition of TGF-β signaling increased sensitivity to BET bromodomain inhibitors.

The James S. Duncan group reported that BETi resistance in ovarian cancers is associated with adaptive kinome reprogramming (Kurimchak et al. 2016). Despite initial inhibition by JQ1, ovarian carcinoma (OC) cells acquired resistance following sustained treatment with BETi. Using a mass spectrometry approach that globally measures kinase signaling at the proteomic level, they investigated changes following JQ1 treatment in a panel of BRD4-dependent OC cell lines. Ovarian cancer cells chronically exposed to JQ1 acquire dependency on RTK, PI3K, and/or ERK signaling and show exquisite sensitivity to kinase inhibitors targeting these pathways. Elevated PI3K/ERK activity stabilizes MYC/FOSL1 proteins in JQ1-resistant cells. Cotargeting BET proteins and RTK, PI3K or ERK signaling enhances BET inhibitor therapy.

In summary, it appears that resistance to BETis might be due to nongenetic mechanisms and involves the activation of different pathways or downstream genes depending on tumor type.

CONCLUSION

BET bromodomain inhibitors are being explored as promising therapeutic agents in numerous hematopoietic and solid tumor types including NUT midline carcinoma, leukemia, and breast cancer with several ongoing clinical trials actively enrolling patients. BET inhibitors may also have a role in preventing or overcoming acquired resistance to U.S. Federal Drug Administration (FDA)-approved targeted therapies such as endocrine therapy and PI3K/mTOR inhibitors in ER^+ breast cancer patients, and HER2-targeted therapy in $HER2^+$ breast cancers. Furthermore, as mechanisms of BETi resistance have been investigated in several cancer types, combination treatment with inhibitors of pathways mediating resistance may significantly improve the clinical efficacy of BETis. Finally, several studies have shown that BRD4 regulates the expression of PD-L1 and CD47 immune regulatory proteins in ovarian cancer, leukemia, and lymphomas, and BETi treatment increases cytotoxic T-cell activity (Casey et al. 2016; Kagoya et al. 2016; Zhu et al. 2016). Therefore, the antitumor effects of BET inhibitors may include the targeting of both tumor cells as well as the tumor-promoting immune environment. Thus, combination therapy of BETi with immune checkpoint blockade may provide new opportunities to increase the efficacy of immunotherapies.

ACKNOWLEDGMENTS

We thank members of our laboratory for their critical reading of our manuscript and helpful discussions. S.S. is funded by a Susan G. Komen Postdoctoral Fellowship. This work was supported by the National Institutes of Health (P50CA168504) SPORE in Breast Cancer and the Susan G. Komen Foundation.

Conflict of interest: The authors have no conflicts of interest to declare.

REFERENCES

Baselga J, Campone M, Piccart M, Burris HA III, Rugo HS, Sahmoud T, Noguchi S, Gnant M, Pritchard KI, Lebrun F, et al. 2012. Everolimus in postmenopausal hormone-receptor-positive advanced breast cancer. *N Engl J Med* **366:** 520–529.

Bihani T, Ezell SA, Ladd B, Grosskurth SE, Mazzola AM, Pietras M, Reimer C, Zinda M, Fawell S, D'Cruz CM. 2015. Resistance to everolimus driven by epigenetic regulation of MYC in ER^+ breast cancers. *Oncotarget* **6:** 2407–2420.

Casey SC, Tong L, Li Y, Do R, Walz S, Fitzgerald KN, Gouw AM, Baylot V, Gutgemann I, Eilers M, et al. 2016. MYC regulates the antitumor immune response through CD47 and PD-L1. *Science* **352:** 227–231.

Chapuy B, McKeown MR, Lin CY, Monti S, Roemer MG, Qi J, Rahl PB, Sun HH, Yeda KT, Doench JG, et al. 2013. Discovery and characterization of super-enhancer-associated dependencies in diffuse large B cell lymphoma. *Cancer Cell* **24:** 777–790.

Daemen A, Griffith OL, Heiser LM, Wang NJ, Enache OM, Sanborn Z, Pepin F, Durinck S, Korkola JE, Griffith M, et al. 2013. Modeling precision treatment of breast cancer. *Genome Biol* **14:** R110.

Dalgliesh GL, Furge K, Greenman C, Chen L, Bignell G, Butler A, Davies H, Edkins S, Hardy C, Latimer C, et al. 2010. Systematic sequencing of renal carcinoma reveals inactivation of histone modifying genes. *Nature* **463:** 360–363.

da Motta LL, Ledaki I, Purshouse K, Haider S, De Bastiani MA, Baban D, Morotti M, Steers G, Wigfield S, Bridges E, et al. 2017. The BET inhibitor JQ1 selectively impairs tumour response to hypoxia and downregulates CA9 and angiogenesis in triple negative breast cancer. *Oncogene* **36:** 122–132.

Delmore JE, Issa GC, Lemieux ME, Rahl PB, Shi J, Jacobs HM, Kastritis E, Gilpatrick T, Paranal RM, Qi J, et al. 2011. BET bromodomain inhibition as a therapeutic strategy to target c-Myc. *Cell* **146:** 904–917.

Dey A, Nishiyama A, Karpova T, McNally J, Ozato K. 2009. Brd4 marks select genes on mitotic chromatin and directs postmitotic transcription. *Mol Biol Cell* **20:** 4899–4909.

Dhalluin C, Carlson JE, Zeng L, He C, Aggarwal AK, Zhou MM. 1999. Structure and ligand of a histone acetyltransferase bromodomain. *Nature* **399:** 491–496.

Early Breast Cancer Trialists' Collaborative Group. 2005. Effects of chemotherapy and hormonal therapy for early breast cancer on recurrence and 15-year survival: An overview of the randomised trials. *Lancet* **365:** 1687–1717.

Feng Q, Zhang Z, Shea MJ, Creighton CJ, Coarfa C, Hilsenbeck SG, Lanz R, He B, Wang L, Fu X, et al. 2014. An epigenomic approach to therapy for tamoxifen-resistant breast cancer. *Cell Res* **24:** 809–819.

Filippakopoulos P, Knapp S. 2012. The bromodomain interaction module. *FEBS Lett* **586:** 2692–2704.

Filippakopoulos P, Qi J, Picaud S, Shen Y, Smith WB, Fedorov O, Morse EM, Keates T, Hickman TT, Felletar I, et al. 2010. Selective inhibition of BET bromodomains. *Nature* **468:** 1067–1073.

Fong CY, Gilan O, Lam EY, Rubin AF, Ftouni S, Tyler D, Stanley K, Sinha D, Yeh P, Morison J, et al. 2015. BET inhibitor resistance emerges from leukaemia stem cells. *Nature* **525:** 538–542.

French CA, Ramirez CL, Kolmakova J, Hickman TT, Cameron MJ, Thyne ME, Kutok JL, Toretsky JA, Tadavarthy AK, Kees UR, et al. 2008. BRD-NUT oncoproteins: A family of closely related nuclear proteins that block epithelial differentiation and maintain the growth of carcinoma cells. *Oncogene* **27:** 2237–2242.

Galdeano C, Ciulli A. 2016. Selectivity on-target of bromodomain chemical probes by structure-guided medicinal chemistry and chemical biology. *Future Med Chem* **8:** 1655–1680.

Haynes SR, Dollard C, Winston F, Beck S, Trowsdale J, Dawid IB. 1992. The bromodomain: A conserved sequence found in human, *Drosophila* and yeast proteins. *Nucleic Acids Res* **20:** 2603.

Hnisz D, Abraham BJ, Lee TI, Lau A, Saint-Andre V, Sigova AA, Hoke HA, Young RA. 2013. Super-enhancers in the control of cell identity and disease. *Cell* **155:** 934–947.

Jang MK, Mochizuki K, Zhou M, Jeong HS, Brady JN, Ozato K. 2005. The bromodomain protein Brd4 is a positive regulatory component of P-TEFb and stimulates RNA polymerase II-dependent transcription. *Mol Cell* **19:** 523–534.

Jiang YW, Veschambre P, Erdjument-Bromage H, Tempst P, Conaway JW, Conaway RC, Kornberg RD. 1998. Mammalian mediator of transcriptional regulation and its possible role as an end-point of signal transduction pathways. *Proc Natl Acad Sci* **95:** 8538–8543.

Kagoya Y, Nakatsugawa M, Yamashita Y, Ochi T, Guo T, Anczurowski M, Saso K, Butler MO, Arrowsmith CH, Hirano N. 2016. BET bromodomain inhibition enhances T cell persistence and function in adoptive immunotherapy models. *J Clin Invest* **126:** 3479–3494.

Kumar K, Raza SS, Knab LM, Chow CR, Kwok B, Bentrem DJ, Popovic R, Ebine K, Licht JD, Munshi HG. 2015. GLI2-dependent c-MYC upregulation mediates resistance of pancreatic cancer cells to the BET bromodomain inhibitor JQ1. *Sci Rep* **5:** 9489.

Kurimchak AM, Shelton C, Duncan KE, Johnson KJ, Brown J, O'Brien S, Gabbasov R, Fink LS, Li Y, Lounsbury N, et al. 2016. Resistance to BET bromodomain inhibitors is mediated by kinome reprogramming in ovarian cancer. *Cell Rep* **16:** 1273–1286.

Lehmann BD, Bauer JA, Chen X, Sanders ME, Chakravarthy AB, Shyr Y, Pietenpol JA. 2011. Identification of human triple-negative breast cancer subtypes and preclinical models for selection of targeted therapies. *J Clin Invest* **121:** 2750–2767.

Lockwood WW, Zejnullahu K, Bradner JE, Varmus H. 2012. Sensitivity of human lung adenocarcinoma cell lines to targeted inhibition of BET epigenetic signaling proteins. *Proc Natl Acad Sci* **109:** 19408–19413.

Loven J, Hoke HA, Lin CY, Lau A, Orlando DA, Vakoc CR, Bradner JE, Lee TI, Young RA. 2013. Selective inhibition of tumor oncogenes by disruption of super-enhancers. *Cell* **153:** 320–334.

Lupien M, Meyer CA, Bailey ST, Eeckhoute J, Cook J, Westerling T, Zhang X, Carroll JS, Rhodes DR, Liu XS, et al. 2010. Growth factor stimulation induces a distinct ERα cistrome underlying breast cancer endocrine resistance. *Genes Dev* **24:** 2219–2227.

Marcotte R, Sayad A, Brown KR, Sanchez-Garcia F, Reimand J, Haider M, Virtanen C, Bradner JE, Bader GD, Mills GB, et al. 2016. Functional genomic landscape of human breast cancer drivers, vulnerabilities, and resistance. *Cell* **164:** 293–309.

Miller TW, Hennessy BT, Gonzalez-Angulo AM, Fox EM, Mills GB, Chen H, Higham C, Garcia-Echeverria C, Shyr Y, Arteaga CL. 2010. Hyperactivation of phosphatidylinositol-3 kinase promotes escape from hormone dependence in estrogen receptor-positive human breast cancer. *J Clin Invest* **120:** 2406–2413.

Mohammed H, D'Santos C, Serandour AA, Ali HR, Brown GD, Atkins A, Rueda OM, Holmes KA, Theodorou V, Robinson JL, et al. 2013. Endogenous purification reveals GREB1 as a key estrogen receptor regulatory factor. *Cell Rep* **3:** 342–349.

Morin RD, Mendez-Lago M, Mungall AJ, Goya R, Mungall KL, Corbett RD, Johnson NA, Severson TM, Chiu R, Field M, et al. 2011. Frequent mutation of histone-modifying genes in non-Hodgkin lymphoma. *Nature* **476:** 298–303.

Nagarajan S, Hossan T, Alawi M, Najafova Z, Indenbirken D, Bedi U, Taipaleenmaki H, Ben-Batalla I, Scheller M, Loges S, et al. 2014. Bromodomain protein BRD4 is required for estrogen receptor–dependent enhancer activation and gene transcription. *Cell Rep* **8:** 460–469.

Nicodeme E, Jeffrey KL, Schaefer U, Beinke S, Dewell S, Chung CW, Chandwani R, Marazzi I, Wilson P, Coste H, et al. 2010. Suppression of inflammation by a synthetic histone mimic. *Nature* **468:** 1119–1123.

Perou CM, Sorlie T, Eisen MB, van de Rijn M, Jeffrey SS, Rees CA, Pollack JR, Ross DT, Johnsen H, Akslen LA, et al. 2000. Molecular portraits of human breast tumours. *Nature* **406:** 747–752.

Piccart M, Hortobagyi GN, Campone M, Pritchard KI, Lebrun F, Ito Y, Noguchi S, Perez A, Rugo HS, Deleu I, et al. 2014. Everolimus plus exemestane for hormone-receptor-positive, human epidermal growth factor receptor-2-negative advanced breast cancer: Overall survival results from BOLERO-2. *Ann Oncol* **25:** 2357–2362.

Puissant A, Frumm SM, Alexe G, Bassil CF, Qi J, Chanthery YH, Nekritz EA, Zeid R, Gustafson WC, Greninger P, et al. 2013. Targeting MYCN in neuroblastoma by BET bromodomain inhibition. *Cancer Discov* **3:** 308–323.

Rathert P, Roth M, Neumann T, Muerdter F, Roe JS, Muhar M, Deswal S, Cerny-Reiterer S, Peter B, Jude J, et al. 2015. Transcriptional plasticity promotes primary and acquired resistance to BET inhibition. *Nature* **525:** 543–547.

Sahai V, Kumar K, Knab LM, Chow CR, Raza SS, Bentrem DJ, Ebine K, Munshi HG. 2014. BET bromodomain inhibitors block growth of pancreatic cancer cells in three-dimensional collagen. *Mol Cancer Ther* **13:** 1907–1917.

Sahai V, Redig AJ, Collier KA, Eckerdt FD, Munshi HG. 2016. Targeting bet bromodomain proteins in solid tumors. *Oncotarget* doi: 10.18632/oncotarget.9804.

Shi J, Wang Y, Zeng L, Wu Y, Deng J, Zhang Q, Lin Y, Li J, Kang T, Tao M, et al. 2014. Disrupting the interaction of BRD4 with diacetylated Twist suppresses tumorigenesis in basal-like breast cancer. *Cancer Cell* **25:** 210–225.

Shi X, Mihaylova VT, Kuruvilla L, Chen F, Viviano S, Baldassarre M, Sperandio D, Martinez R, Yue P, Bates JG, et al. 2016. Loss of TRIM33 causes resistance to BET bromodomain inhibitors through MYC- and TGF-β-dependent mechanisms. *Proc Natl Acad Sci* **113:** E4558–E4566.

Shu S, Lin CY, He HH, Witwicki RM, Tabassum DP, Roberts JM, Janiszewska M, Huh SJ, Liang Y, Ryan J, et al. 2016. Response and resistance to BET bromodomain inhibitors in triple-negative breast cancer. *Nature* **529:** 413–417.

Siegel RL, Miller KD, Jemal A. 2016. Cancer statistics, 2016. *CA Cancer J Clin* **66:** 7–30.

Sorlie T, Perou CM, Tibshirani R, Aas T, Geisler S, Johnsen H, Hastie T, Eisen MB, van de Rijn M, Jeffrey SS, et al. 2001. Gene expression patterns of breast carcinomas distinguish tumor subclasses with clinical implications. *Proc Natl Acad Sci* **98:** 10869–10874.

Stathis A, Zucca E, Bekradda M, Gomez-Roca C, Delord JP, de La Motte Rouge T, Uro-Coste E, de Braud F, Pelosi G, French CA. 2016. Clinical response of carcinomas harboring the BRD4-NUT oncoprotein to the targeted bromodomain inhibitor OTX015/MK-8628. *Cancer Discov* **6:** 492–500.

Stratikopoulos EE, Dendy M, Szabolcs M, Khaykin AJ, Lefebvre C, Zhou MM, Parsons R. 2015. Kinase and BET inhibitors together clamp inhibition of PI3 K signaling and overcome resistance to therapy. *Cancer Cell* **27:** 837–851.

Stuhlmiller TJ, Miller SM, Zawistowski JS, Nakamura K, Beltran AS, Duncan JS, Angus SP, Collins KA, Granger DA, Reuther RA, et al. 2015. Inhibition of Lapatinib-induced kinome reprogramming in ERBB2-positive breast cancer by targeting BET family bromodomains. *Cell Rep* **11:** 390–404.

van Haaften G, Dalgliesh GL, Davies H, Chen L, Bignell G, Greenman C, Edkins S, Hardy C, O'Meara S, Teague J, et al. 2009. Somatic mutations of the histone H3K27 demethylase gene UTX in human cancer. *Nat Genet* **41:** 521–523.

Wu SY, Lee AY, Lai HT, Zhang H, Chiang CM. 2013. Phospho switch triggers Brd4 chromatin binding and activator recruitment for gene-specific targeting. *Mol Cell* **49:** 843–857.

Yang Z, Yik JH, Chen R, He N, Jang MK, Ozato K, Zhou Q. 2005. Recruitment of P-TEFb for stimulation of transcriptional elongation by the bromodomain protein Brd4. *Mol Cell* **19:** 535–545.

Zhao R, Nakamura T, Fu Y, Lazar Z, Spector DL. 2011. Gene bookmarking accelerates the kinetics of post-mitotic transcriptional re-activation. *Nat Cell Biol* **13:** 1295–1304.

Zhu H, Bengsch F, Svoronos N, Rutkowski MR, Bitler BG, Allegrezza MJ, Yokoyama Y, Kossenkov AV, Bradner JE, Conejo-Garcia JR, et al. 2016. BET bromodomain inhibition promotes anti-tumor immunity by suppressing PD-L1 expression. *Cell Rep* **16:** 2829–2837.

Zuber J, Shi J, Wang E, Rappaport AR, Herrmann H, Sison EA, Magoon D, Qi J, Blatt K, Wunderlich M, et al. 2011. RNAi screen identifies Brd4 as a therapeutic target in acute myeloid leukaemia. *Nature* **478:** 524–528.

To Prime, or Not to Prime: That Is the Question

DANIELLE S. POTTER AND ANTHONY LETAI
Dana-Farber Cancer Institute, and Harvard Medical School, Boston, Massachusetts 02215
Correspondence: Anthony_Letai@dfci.harvard.edu

Mitochondrial priming is regulated by the B-cell lymphoma 2 (BCL-2) family of proteins and determines a cell's "readiness" for apoptosis. A highly primed cell will undergo apoptosis more easily than an unprimed cell in response to apoptotic stimuli via the intrinsic apoptotic pathway. Priming can be measured via BH3 profiling, which uses BH3 peptides derived from the BH3 domain of pro-apoptotic BH3-only BCL-2 family members to provoke a response from viable mitochondria. BH3 profiling can be performed on tumor cells and can identify mechanisms a cell uses to evade apoptosis and anti-apoptotic dependency to the anti-apoptotic BCL-2 family members. Priming correlates with chemosensitivity of patients in multiple cancers. Therapeutics that enhances priming of patient tumor cells ex vivo could be used to aid therapeutic decisions for patients in the future.

Apoptosis is a form of physiological programmed cell death in multicellular organisms, and because of its functionality, it is tightly regulated (Wyllie 1997). Apoptosis acts as a barrier against tumorigenesis and, consequently, tumors select for mechanisms to resist apoptosis. Such mechanisms are hallmarks of cancer and are recognized to result in resistance to anticancer therapies (Hanahan and Weinberg 2000, 2011). There is a real need for new and improved therapies to overcome apoptotic resistance in cancer. The two major apoptotic pathways are the intrinsic (the focus of this review) and the extrinsic pathways. Both ultimately result in the activation of initiator caspases (cysteinyl aspartate specific protease; caspase-8 and -9) that in turn activate downstream effector caspases (caspase-3 and -7), triggering a caspase cascade (Li et al. 1997; Fadok et al. 2000; Savill and Fadok 2000).

INTRINSIC APOPTOSIS AND THE BCL-2 FAMILY

The intrinsic apoptotic pathway, once activated, results in mitochondrial outer membrane permeabilization (MOMP), usually the point of no return. MOMP causes the release of pro-apoptotic factors such as cytochrome *c* (Fig. 1) and SMAC (second mitochondrial derived activator of caspases) from the mitochondrial intermembrane space (Wang 2001). Once cytochrome *c* enters the cytosol, it interacts with APAF1 (apoptotic protease activating factor-1), the initiator procaspase-9, and ATP to form a holoenzyme known as the apoptosome. The formation of the apoptosome results in activation of caspase-9. Caspase-9 then cleaves and activates caspase-3 (Li et al. 1997). The activated caspase cascade causes cell-wide specific proteolysis, dysfunction, decrease in ATP production, and remodeling of the cell surface with "molecular flags" such as externalization of phosphatidylserine. This allows the cell to be recognized by phagocytic cells like macrophages that then engulf the apoptotic cell (Fadok et al. 2000, 2001; Toda et al. 2012; Segawa et al. 2014). This efficient removal of dying cells allows biological "building blocks" to be recycled, an immune response to dead cells is contained, and damage to healthy cells in the vicinity of apoptotic cells is minimized, in contrast to necrosis and necroptosis (Li et al. 1997; Savill and Fadok 2000; Guerriero et al. 2011; Meng et al. 2016).

MOMP is regulated by the B-cell lymphoma 2 (BCL-2) family of proteins and can be activated by various mechanisms such as developmental cues, DNA damage, growth factor deprivation, nutrient deprivation, and treatment with a cytotoxic agent (Lovell et al. 2008; Youle and Strasser 2008; Fu et al. 2009; Spencer et al. 2009). BCL-2 family members can be grouped into three subfamilies, the pro-apoptotic effectors, the anti-apoptotic effector, and the pro-apoptotic BH3-only proteins (Fig. 1). The effector subfamily are multidomain proteins, containing BH (BCL-2 homology) 1, 2, 3, and 4 domains (Suzuki et al. 2000; Moldoveanu et al. 2006; Kvansakul et al. 2008). The members of this subfamily are BAK (BCL2-antagonist/killer), BAX (BCL-2-associated protein), and BOK (BCL2-related ovarian killer) (Youle and Strasser 2008). BAK is an integral mitochondrial protein found on the cytosolic side of the outer mitochondrial membrane (OMM). A prelude to MOMP after apoptotic stimuli is BAK undergoing a conformational change in the amino terminus revealing a newly exposed epitope (Griffiths et al. 1999). This leads to a second step activation in which BAK's BH1 domain is revealed and parallels the release of anti-apoptotic BCL-2 family members from BAK (Griffiths et al. 2001). In

Published by Cold Spring Harbor Laboratory Press; doi: 10.1101/sqb.2016.81.030841

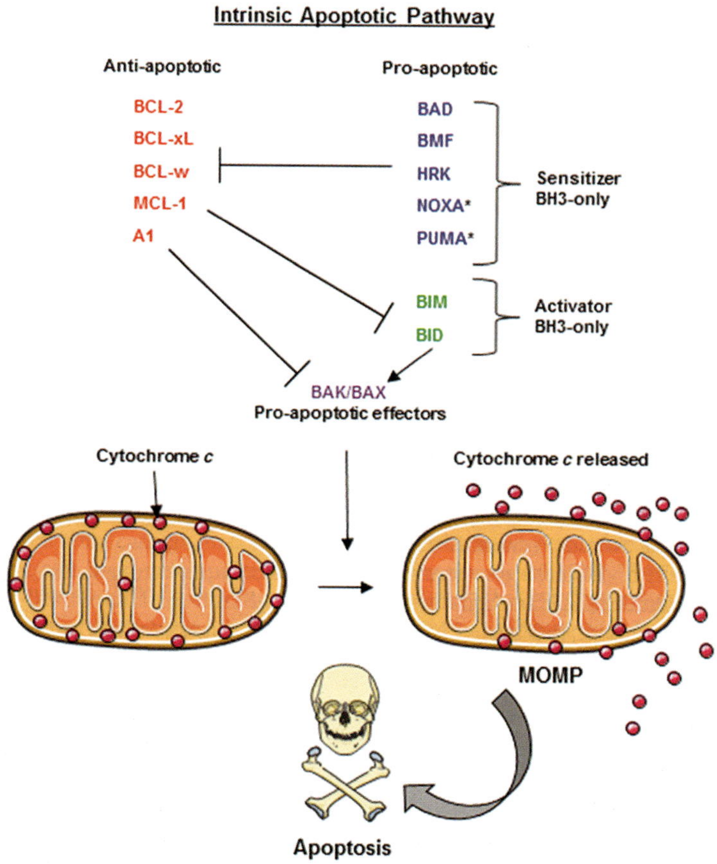

Figure 1. Intrinsic apoptotic pathway. The intrinsic apoptotic pathway can be activated by various factors such as growth factor deprivation and cellular stress. Apoptotic signals result in an increase in active/available pro-apoptotic B-cell lymphoma 2 (BCL-2) family members (green, blue, and purple) and/or decrease in active/available anti-apoptotic BCL-2 family members (red) shifting the cell survival equilibrium toward apoptosis. This allows the pro-apoptotic effectors BCL-2-antagonist/killer (BAK) and BCL-2-associated protein (BAX) (purple) to be released and activated by pro-apoptotic BH3-only proteins BCL-2-interacting mediator of cell death (BIM) or BH3-interacting domain death agonist (BID) (green) resulting in BAX/BAK homo-oligomerization in the outer mitochondrial membrane, mitochondrial outer membrane permeabilization (MOMP), and apoptosis. *NOXA and PUMA (p53 up-regulated modulator of apoptosis) can act as activators. (Illustration made using Servier Medical Art.)

healthy cells, BAX is cytosolic (Hsu and Youle 1997; Wolter et al. 1997). In response to apoptotic stimuli, BAX undergoes a conformational change in the amino terminus resulting in BAX translocation to the OMM (Makin et al. 2001). On the OMM, activated BAK and BAX form homo-oligomer, creating pores and therefore MOMP (Fig. 1; Eskes et al. 2000; Korsmeyer et al. 2000; Wei et al. 2001). BAK and BAX are functionally redundant; however, loss of both prevents intrinsic apoptosis, showing that BAX and BAK are essential for MOMP (Lindsten et al. 2000). BOK is the least studied of the effector proteins and is primarily expressed in reproductive tissue (Hsu et al. 1997) but also in brain and myeloid cells (Ke et al. 2012). BOK has been shown to have redundant function with BAK and BAX in the hematopoietic system (Ke et al. 2015). In contrast to BAX and BAK, it may be primarily controlled by protein stability (Llambi et al. 2016).

The anti-apoptotic subfamily consists of the multidomain proteins BCL-2, BCL-xL (BCL-2-like protein 1, isoform long), BCL-w (BCL-2-like protein 2), MCL-1 (myeloid cell leukemia 1), and A1 (BCL-2-related protein A1, also known as BFL-1) (Fig. 1). These conserved proteins share homology in BH1-4 domains and are predominantly located in the OMM, but they are also present in the cytosol and endoplasmic reticulum (Lithgow et al. 1994; Pinton and Rizzuto 2006). The BCL-2 anti-apoptotic members prevent MOMP via antagonizing interactions with the pro-apoptotic BCL-2 family members (Chipuk et al. 2010). The BH1, -2, and -3 domains of the anti-apoptotic family members fold into a globular tertiary structure forming a hydrophobic groove on its surface. The hydrophobic groove provides a platform for the pro-apoptotic BCL-2 family members to interact via their BH3 domains (Sattler et al. 1997; Liu et al. 2003; Chonghaile and Letai 2008).

The pro-apoptotic BH3-only subfamily includes BAD (BCL-2-associated agonist of cell death), BID (BH3-interacting domain death agonist), BIK (BCL-2-interacting killer), BIM (BCL-2-interacting mediator of cell death), BMF (BCL-2-modifying factor), HRK (Harakiri), NOXA, and PUMA (p53 up-regulated modulator of apoptosis) (Chipuk et al. 2010). The BH3-only proteins are so-called because of their shared homology with BH3

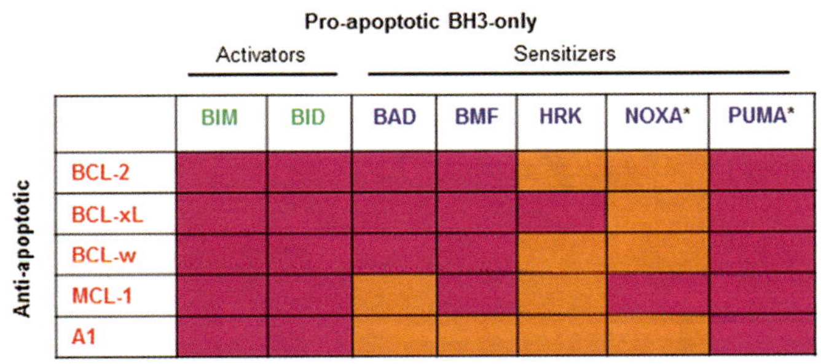

Figure 2. Specificity in binding of the pro-apoptotic BH3-only proteins to the anti-apoptotic BCL-2 family members. BH3-only proteins can be further divided into activator and/or sensitizers depending on their apoptotic properties. BH3-only peptides bind and antagonize specific anti-apoptotic family members. Pink indicates high-affinity binding and orange indicates poor/no binding (Letai et al. 2002; Certo et al. 2006; Deng et al. 2007). *NOXA and PUMA can act as activators. BIM, BCL-2-interacting mediator of cell death; BID, BH3-interacting domain death agonist; BAD, BCL-2-associated agonist of cell death; BMF, BCL-2-modifying factor; HRK, Harakiri; PUMA, p53 up-regulated modulator of apoptosis.

domains but little shared homology with the other BCL-2 family members (Aouacheria et al. 2005). BH3-only proteins show distinct specificities in their binding to the anti-apoptotic BCL-2 family members (Fig. 2) and can act as activators of BAK/BAX and/or sensitizer proteins (anti-apoptotic antagonists). BIM and BID are potent activators (Letai et al. 2002; Certo et al. 2006) that bind promiscuously to all the anti-apoptotic BCL-2 family members. BAD is a sensitizer and a BCL-2, BCL-w, and BCL-xL antagonist. HRK is a sensitizer that is a specific BCL-xL antagonist. BIK and BMF are sensitizers that antagonize BCL-2, BCL-w, BCL-xL, and MCL-1. Studies on the BH3 domain of PUMA (antagonist to all anti-apoptotic BCL-2 family members) showed no activator function (Letai et al. 2002; Kuwana et al. 2005; Certo et al. 2006). However, studies using full protein show that PUMA has activator properties (Kim et al. 2006). Double-knockout (DKO) BIM/BID mice do not recapitulate the BAK/BAX DKO mice phenotype, but BIM/BID/PUMA triple-knockout mice phenocopy BAK/BAX DKO mice (Ren et al. 2010; Du et al. 2011), suggesting that PUMA acts as an activator. NOXA is a sensitizer, a specific MCL-1 antagonist, that has been shown more recently to have an activator function (Du et al. 2011).

PRIMING

MOMP is a switch-like event that is regulated by the balance of pro- and anti-apoptotic proteins. One can consider MOMP like a cliff and imagine that some cells are closer to the cliff's edge than others. The cells proximity to this cliff's edge would carry with it significant physiologic consequences, including increased likelihood of commitment to cell death following toxic perturbations. We consider cells that are close to the cliff's edge to be relatively highly "primed" for apoptosis, whereas those far from the cliff's edge to be relatively "unprimed." It is important to discuss how priming can be measured so that primed and unprimed cells can be distinguished and what the molecular basis of this priming is. Priming (also referred to as mitochondrial priming or apoptotic priming) is also regulated by the BCL-2 family of proteins. The balance between the BCL-2 anti-apoptotic and pro-apoptotic family members and the respective availability of their interaction sites determine how primed a cell is and in turn whether MOMP will occur. Priming is relative; the more primed a cell becomes, the lower the apoptotic threshold and tolerance for apoptotic stimuli (Fig. 3). Like apoptosis, priming is affected by a number of physiological conditions such as growth factor deprivation, changes in metabolism, or genetic aberrations (Certo et al. 2006). Drug treatment can enhance priming, proving advantageous in treating cancer (discussed in detail later).

BH3 profiling was developed as a functional tool to measure priming experimentally (Letai et al. 2002; Certo et al. 2006; Ryan et al. 2010). The assay uses synthetic BH3 peptides derived from the BH3 domain of the pro-apoptotic BH3-only proteins to provoke a response from the mitochondria. Priming is measured in terms of MOMP induced by a BH3 peptide. The different platforms of BH3 profiling measure a surrogate for MOMP, such as loss of JC-1 red fluorescence or cytochrome *c* released from the mitochondria (Ryan and Letai 2013; Ryan et al. 2016).

As discussed previously BH3-only proteins can either be promiscuous or selective in their affinity for anti-apoptotic BCL-2 family members. Therefore, the specificity of BH3 peptides to the anti-apoptotic BCL-2 family members allows BH3 profiling to differentiate between promiscuous overall priming (determined by increased MOMP after BIM, BID, or PUMA exposure) or a more selective measure of anti-apoptotic dependency. For example, if a BH3 profile revealed that the sensitizer BH3 peptide HRK caused MOMP, this would suggest that these cells are likely dependent on BCL-xL because HRK is a selective antagonist for BCL-xL (Letai et al. 2002; Certo et al. 2006).

When a carefully selected panel of BH3 peptides are used, BH3 profiling can also reveal what apoptotic block cancer cells are likely using to evade apoptosis and thus provides a measure of how primed that cell is (Fig. 4; Deng et al. 2007). For instance, a cell that evades apoptosis via down-regulation of pro-apoptotic BH3-only activators will only respond to activator peptides in the

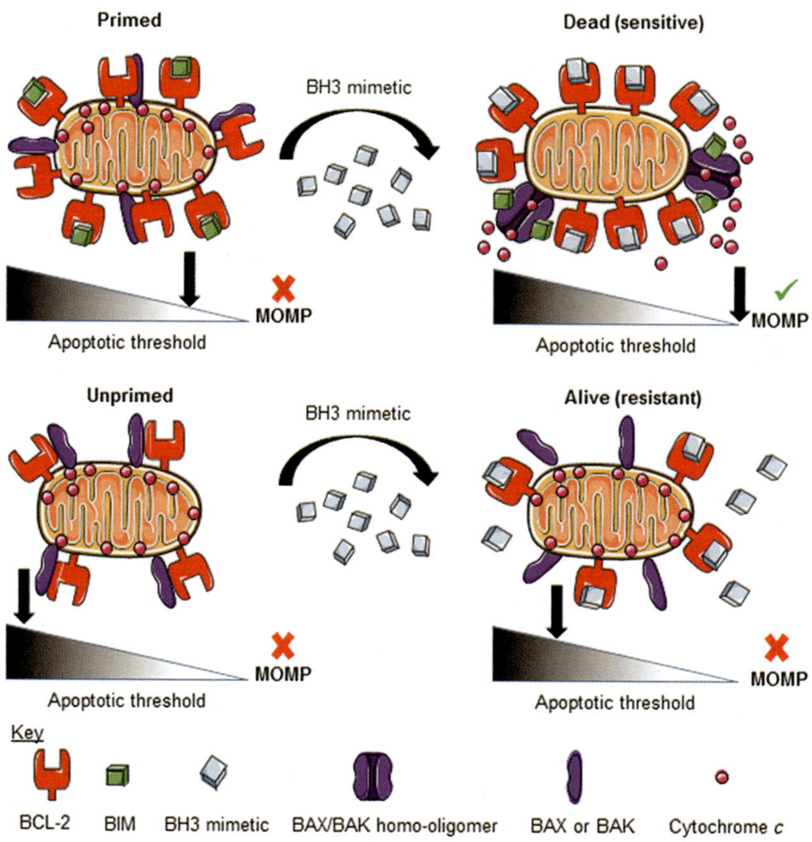

Figure 3. Primed versus unprimed cellular response to apoptotic stimuli. Priming is relative; therefore, a primed cell has a relatively lower apoptotic threshold than an unprimed cell. An apoptotic assault such as treatment with a BH3 mimetic is likely to cause mitochondrial outer membrane permeabilization (MOMP) on a primed cell but not an unprimed cell. MOMP with a red × indicates no MOMP occurred. MOMP with a green ✓ indicates MOMP occurred.

BH3 profiling assay (Fig. 4A [green bars]). The inability of either activator or sensitizer BH3 peptides to induce cytochrome *c* release by BH3 profiling would suggest a lack of functional BAX/BAK in those cells (Fig. 4B). If both activator and sensitizer BH3 peptides could induce cytochrome *c* release, then a mechanism of evading apoptosis via up-regulation of anti-apoptotic BCL-2 family members would be suggested (Fig. 4C [green and blue bars]). This type of cell would be classed as primed because all the components to induce apoptosis are present, but they are constantly neutralized by the anti-apoptotic BCL-2 family members (Deng et al. 2007).

PRINCIPLES OF BH3 PROFILING

The BH3 domain was first described in BIK more than 20 years ago (Boyd et al. 1995) and is essential for interactions among BCL-2 members (Zha et al. 1996; Sattler et al. 1997). Short synthetic BH3 peptides were used to investigate BCL-2 family member interactions and showed that BIM and BID BH3 peptides cause BAX and BAK oligomerization (Letai et al. 2002; Certo et al. 2006). At this point, BH3 profiling was performed on isolated mitochondria using an enzyme-linked immunoabsorbent assay (ELISA)-based method to measure cytochrome *c* released from the mitochondria, after exposure to a fixed concentration of BH3 peptide. Released cytochrome *c* is used as a surrogate of MOMP (Letai et al. 2002; Certo et al. 2006). The early method of BH3 profiling had limitations such as lack of incorporation cell surface markers, low throughput, and requirements for a relatively large number of cells as starting material (10^8–10^9 cells). Limitations were addressed and BH3 profiling on whole cells was developed, reducing the number of cells needed by 100-fold. The current BH3 profiling techniques most commonly used in the Letai laboratory are plate-based JC-1 BH3 profiling (Ryan and Letai 2013) and iBH3 (intracellular BH3) (Ryan et al. 2016).

For BH3 peptides to engage with the mitochondria of viable cells, the plasma membrane must be permeabilized to allow the diffusion of the BH3 peptides into the cell. Digitonin is a glycoside that precipitates cholesterol, which is relatively more abundant in the plasma membrane than organelles. A low concentration of digitonin (5–25 µg/mL), lightly permeabilizes the plasma membrane but keeps organelles intact, allowing BH3 peptides to interact with intact mitochondria. Once MOMP occurs the cell is committed to intrinsic apoptosis and therefore MOMP is the end point for BH3 profiling (Ryan and Letai 2013; Ryan et al. 2016). Activation of MOMP causes depolarization of the transmembrane mitochondrial

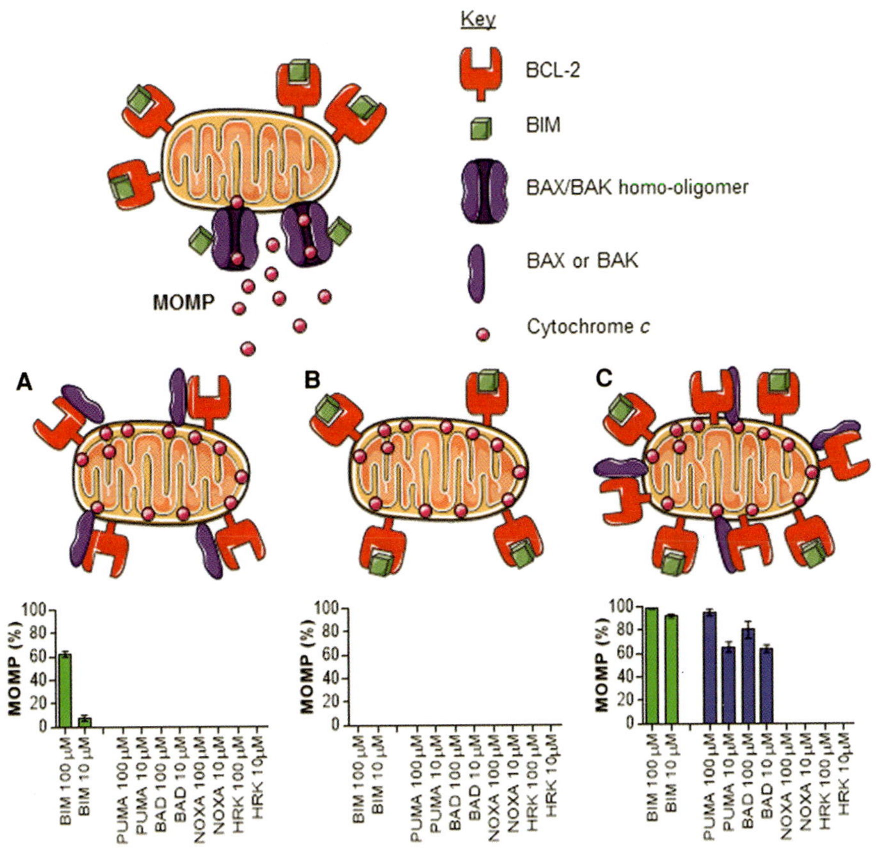

Figure 4. Anti-apoptotic mechanisms in cancer cells. Cancer cells can evade apoptosis via three main mechanisms involving the BCL-2 family, and BH3 profiling can distinguish these mechanisms via induction of mitochondrial outer membrane permeabilization (MOMP) after exposure to individual activator or sensitizer BH3 peptides. Down-regulation of pro-apoptotic BH3-only activators would evade apoptosis because pro-apoptotic effectors BCL-2-associated protein (BAX)/ BCL-2-antagonist/killer (BAK) cannot be activated in the absence of activators, and therefore no MOMP. BH3 profiling on this type of cell would be determined because activator peptides (green bars) would induce MOMP, but sensitizer peptides (blue bars) would not (*A*). The second mechanism of evading apoptosis is through inactivation or down-regulation of BAK/BAX, because without BAX/BAK, there is no MOMP. BH3 profiling on a cell that uses this block would be determined when none of the BH3 peptides (activators or sensitizers) cause MOMP (*B*). The final mechanism to evade apoptosis is via up-regulation of anti-apoptotic BCL-2 family members, as this would neutralize any pro-apoptotic effectors or BH3-only proteins preventing MOMP. BH3 profiling on cells that use this block would be determined when both activator and sensitizer BH3 peptides cause MOMP (*C*) (Deng et al. 2007). (Illustration made using Servier Medical Art.)

potential ($\Delta\Psi m$). JC-1 is a fluorescent dual-emission mitochondrial intermembrane probe that fluoresces red (JC-1 aggregates) when concentrated in the matrix of a healthy, polarized mitochondria or green when $\Delta\Psi m$ has depolarized because of MOMP (JC-1 monomers). Once cells have been permeabilized and treated with a fixed concentration of BH3 peptide, loss of JC-1 red fluorescence caused by $\Delta\Psi m$ depolarization, is used as a surrogate for the end point MOMP in plate-based JC-1 BH3 profiling (Fig. 5). This BH3 profiling platform using JC-1 allows real-time kinetic measurements of loss of mitochondrial integrity (Ryan and Letai 2013).

If the cell population is homogenous, such as in cell lines or a cell-sorted population, then a plate-based approach can provide high throughput for the lowest cost. However, if BH3 profiling is performed on a mixed population of cells, such as a primary sample, the iBH3 approach would be more useful. Intercellular BH3 is a flow cytometry–based assay that measures retained cytochrome *c* as a surrogate to MOMP. Primary samples are dissociated using enzymatic and mechanical forces to break down the extracellular matrix, leaving a single-cell suspension. Cells can be labeled for cell surface markers to identify cell subpopulation of interest. Labeled cells are then permeabilized with digitonin, exposed to BH3 peptides of known concentration for a given time, and fixed with formaldehyde. Cells are then incubated with a cytochrome *c* antibody to detect retained cytochrome *c* (Fig. 5). The cell fixation step provides an advantage for iBH3 because infectious primary samples including HIV can be handled; also, samples can be analyzed weeks after fixation (Ryan et al. 2016).

BH3 MIMETICS TO TARGET THE BCL-2 FAMILY

Alterations in the function and expression of the BCL-2 family of proteins contribute to cancer progression, highlighting their potential as rational drug targets. Compared

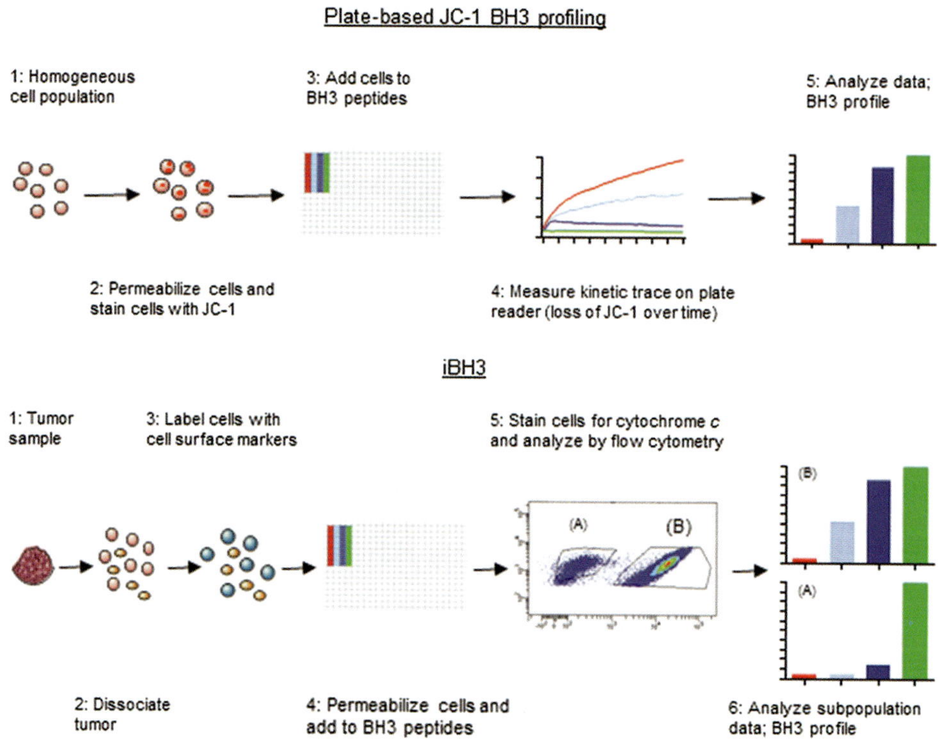

Figure 5. BH3 profiling platforms. Plate-based JC-1 BH3 profiling versus iBH3 (intracellular BH3).

to healthy cells, cancer cells have increased apoptotic signaling caused by factors such as oncogene activation and changes in the microenvironment (Hanahan and Weinberg 2000, 2011). Cancer cells become reliant on anti-apoptotic signaling to survive, known as "oncogene addiction," and drugs have been developed that target and exploit this. This led to the development of small-molecule antagonists that can block interactions between anti-apoptotic and pro-apoptotic BCL-2 family members, known as BH3 mimetics (Oltersdorf et al. 2005; Souers et al. 2013).

BH3 mimetics compete with pro-apoptotic BCL-2 family proteins for binding in the BH3 binding site of anti-apoptotic proteins. Their binding to anti-apoptotic proteins releases the pro-apoptotic BCL-2 family members to induce MOMP (Fig. 3). The first BH3 mimetic, ABT-737, was developed in 2005 (AbbVie) and is functionally analogous to the BH3 domain of BAD, binding BCL-2, BCL-xL, and BCL-w (Oltersdorf et al. 2005). Preclinical studies showed that ABT-737 has single-agent activity in some small cell lung cancer (SCLC) and lymphoma cell lines as well as in primary chronic lymphocytic leukemia (CLL) and lymphoma cells (Oltersdorf et al. 2005; Deng et al. 2007; Tahir et al. 2007; Del Gaizo Moore et al. 2008; Mason et al. 2008). ABT-737 was also found to act synergistically with conventional chemotherapeutic agents and targeted therapies in a variety of cancer types (Premkumar et al. 2012; Gardner et al. 2014; Potter et al. 2014, 2016). Cells that rely on BCL-2, BCL-w, or BCL-xL for survival are particularly sensitive to ABT-737. ABT-737 is unable to interact with MCL-1 and, thus, expression of this BCL-2 family member is a resistance factor for ABT-737 efficacy (Vogler et al. 2009). Indeed, ABT-737 sensitization caused by other drugs or the tumor microenvironment is often associated with MCL-1 down-regulation (Harrison et al. 2011; Gardner et al. 2014).

A clinical derivative of ABT-737 is navitoclax (ABT-263), an orally bioavailable BH3 mimetic (Tse et al. 2008). Navitoclax has shown promise in the phase I setting, in which biomarkers of apoptotic cell death were detected in patients treated with navitoclax (Gandhi et al. 2011). It has been tested in phase I/II clinical trials in SCLC, ovarian, acute lymphocytic leukemia (ALL), and CLL patients as a single agent and in combination with chemotherapy (clinicaltrials.gov). Navitoclax showed promising single-agent efficacy in phase II clinical trials in CLL (Roberts et al. 2012) but not in in SCLC (Gandhi et al. 2011). Combing navitoclax with targeted agents such as MEK (MAPK/ERK kinase), mTOR (mammalian target of rapamycin), or PI3K (phosphoinositide 3-kinase) inhibitors appears to increase the potency of navitoclax in preclinical models in NSCLC (non–small cell lung cancer), pancreatic cancer, and SCLC (Tan et al. 2013; Faber et al. 2014, 2015; Potter et al. 2016), although the clinical impact of these findings remains to be determined. Thrombocytopenia is a dose-limiting toxicity of navitoclax, caused by the on-target inhibition of BCL-xL, as platelets are BCL-xL-dependent (Tse et al. 2008; Schoenwaelder et al. 2011). This led to the development of a more specific BCL-2 antagonist (Souers et al. 2013).

Venetoclax (ABT-199) is a BCL-2-selective BH3 mimetic that does not cause thrombocytopenia. Eleven years after the first BH3 mimetic was developed, venetoclax (co-developed by AbbVie and Genentech), was approved by the FDA in CLL patients with 17p deletion after receiving at least one prior treatment (Anderson et al. 2016; Roberts et al. 2016). Venetoclax is currently in clinical trials in other hematological malignancies (acute myeloid leukemia [AML] and lymphoma) (clinicaltrials.gov). It was shown that the Tec family kinase inhibitor, ibrutinib, sensitized to venetoclax in CLL, in which ibrutinib has also been approved for second-line treatment. Sensitization is via inhibition of the Tec family kinase, BTK, and was associated with a decrease in MCL-1 and BCL-xL levels (Cervantes-Gomez et al. 2015). In another study in CLL, venetoclax-resistant cells were sensitized to venetoclax when venetoclax was combined with the BEZ235 (dual PI3K/mTOR inhibitor), associated with MCL-1 down-regulation due to PI3K pathway inhibition (Choudhary et al. 2015). As of this writing, the first clinical trial of an MCL-1 inhibitor (AMG 176), made by Amgen, is under way in relapsed or refractory myeloma (clinicaltrials.gov).

BH3 PROFILING AS A PREDICTOR FOR CHEMOSENSITIVITY

For the past 50 years, conventional chemotherapy has been the mainstay of cancer therapy, curing millions of people worldwide. In recent years, a more targeted approach has further improved therapeutic outcome in some cancers. However, not all cancer patients respond to chemotherapy. BH3 profiling has been used in many studies, in different cancer types, to successfully predict chemosensitivity and clinical outcome.

In multiple myeloma, serum levels of the monoclonal protein (M protein) are measured as an indication of tumor burden (Durie et al. 2006). BH3 profiling using BMF and PUMA peptides to measure priming was performed on 51 multiple myeloma patient before treatment. This study showed that priming correlated with a decrease in M protein and therefore tumor burden after chemotherapy. Priming also correlated with better clinical outcome for patients after chemotherapy. Highly primed patients had a longer progression-free survival than the poorly primed patients (Ni Chonghaile et al. 2011).

Pediatric ALL has a long-term survival rate of 80%–90% compared to adult ALL, with a survival rate of 40%, suggesting that age is a factor in prognosis. BH3 profiling on both pediatric and adult ALL samples revealed that pediatric ALL is more primed than adult. This increased priming observed in pediatric ALL could explain the superior response rate to chemotherapy. A further look into patient's clinical response showed that patients who responded and then relapsed were less primed compared to patients that responded but did not relapse (Ni Chonghaile et al. 2011).

Two studies in AML showed that priming correlated with clinical response to chemotherapy. In one study BH3 profiling on malignant myeloblasts from patients' bone marrow before any therapy revealed that patients who had complete remission were more primed than those that did not (Ni Chonghaile et al. 2011). In a second study, priming of malignant myeloblasts from AML patients (measured by BH3 profiling with BIM peptide) correlated with response to chemotherapy. Patients with highly primed myeloblasts had a better clinical outcome to chemotherapy compared to patients with poorly primed myeloblasts (Vo et al. 2012).

BH3 profiling was performed on CLL cells from patient blood before any chemotherapy. The promiscuous BH3 peptide BIM was used to measure priming. Patients were then treated and followed for clinical response. Patients who had either a partial or complete response to therapy were significantly more primed than patients that had stable or progressive disease. In CLL, priming successfully predicts chemosensitivity and could be used as a predictive biomarker for response (Davids et al. 2012).

BH3 profiling as a predictor of chemosensitivity has also been shown in solid tumors. In ovarian cancer, BIM, PUMA, and BMF peptides revealed that highly primed tumors correlated with superior progression-free survival (Ni Chonghaile et al. 2011).

To determine whether priming is responsible and not just correlative for chemosensitivity, a study investigated mechanisms to enhance priming to assess whether this increased sensitivity to chemotherapy agents. As previously mentioned, BH3 mimetics directly target the antiapoptotic BCL-2 family members and therefore may increase priming. BH3 mimetics treatment reduces anti-apoptotic binding site availability and free activators to activate BAK and/or BAX, lowering the apoptotic threshold (Fig. 3). The myeloid leukemia cell line K562 was treated with ABT-737 and the effect on chemosensitivity was assessed. ABT-737 caused increased sensitivity to doxorubicin, vincristine, and etoposide, suggesting that priming is a determinant of chemosensitivity (Ni Chonghaile et al. 2011).

PRIMING OVER TIME

The difference in pediatric and adult priming observed in ALL suggested possible age-related differences in priming of normal tissues as well. Priming in young and adult mice was investigated using BH3 profiling. This revealed that in healthy young mice, brain, heart, and kidney cells were extremely sensitive to BH3 peptides and were highly primed. In contrast, mitochondria in the tissue of healthy adult mice were completely insensitive to BH3 peptides. Intrinsic apoptosis could only be triggered in these cells with the addition of exogenous BAX protein, suggesting that healthy, adult brain, heart, and kidney tissue are incompetent for apoptosis. This was confirmed with immunoblotting levels of BCL-2 family members in a series of developmental time points from prenatal to postnatal into adulthood. BAX and BAK levels were dramatically down-regulated during postnatal development to nearly undetectable levels by adulthood.

Without BAX and BAK, cells cannot undergo MOMP and are therefore incompetent for intrinsic apoptosis. BH3 profiling in healthy, human brain tissue confirmed that adult brain cells are less primed than child brain cells, strengthening the findings in mice. BH3 profiling in young and adult spleen and bone marrow revealed that high levels of priming continue into adulthood, emphasizing the organ-specific regulation of apoptotic machinery (Sarosiek et al. 2015).

MECHANISM TO ENHANCE PRIMING

Priming correlates with clinical response to therapy; therefore, mechanisms to enhance priming would be beneficial to the patient. Mechanisms to enhance priming led to the development of dynamic BH3 profiling, in which tumor cells are pretreated with a drug and the effect on priming is measured using BH3 profiling. Dynamic BH3 profiling uses a functional approach to precision medicine; an increase in priming compared to untreated control is referred to as delta priming. Dynamic BH3 profiling integrates net changes in BCL-2 family members after treatment with a therapy, measuring total proapoptotic signaling. Delta priming distinguishes cells to be killed by a therapy and therefore predicts therapies that will have a cytotoxic effect before the event in various models and on patient samples. Delta priming can be used to predict a therapy's cytotoxic response in vitro, in vivo, and importantly in the clinic (Montero et al. 2015).

CONCLUSION

Predicting the most efficacious drug for a patient is ideal. Unless there is specific knowledge of that patient's tumor available, such as actionable driver mutations identified through tumor genotyping, dynamic BH3 profiling could be used to overcome imperfect knowledge and predict therapies that enhance priming. As well as identifying individual compounds that enhance priming, dynamic BH3 profiling can also reveal combination regimens of said compound with BH3 mimetics, improving therapeutic outcome for patients. The goal of precision medicine is to match a patient to the best treatment available. Genomics approaches dominate precision medicine although they currently benefit only a minority of cancer patients. We anticipate that functional precision medicine approaches, including dynamic BH3 profiling, by measuring what happens when the drug of interest contacts the tumor cell of interest, will play a major role in the future of precision medicine in cancer.

REFERENCES

Anderson MA, Deng J, Seymour JF, Tam C, Kim SY, Fein J, Yu L, Brown JR, Westerman D, Si EG, et al. 2016. The BCL2 selective inhibitor venetoclax induces rapid onset apoptosis of CLL cells in patients via a TP53-independent mechanism. *Blood* **127:** 3215–3224.

Aouacheria A, Brunet F, Gouy M. 2005. Phylogenomics of life-or-death switches in multicellular animals: Bcl-2, BH3-Only, and BNip families of apoptotic regulators. *Mol Biol Evol* **22:** 2395–2416.

Boyd JM, Gallo GJ, Elangovan B, Houghton AB, Malstrom S, Avery BJ, Ebb RG, Subramanian T, Chittenden T, Lutz RJ, et al. 1995. Bik, a novel death-inducing protein, shares a distinct sequence motif with Bcl-2 family proteins and interacts with viral and cellular survival-promoting proteins. *Oncogene* **11:** 1921–1928.

Certo M, Del Gaizo Moore V, Nishino M, Wei G, Korsmeyer S, Armstrong SA, Letai A. 2006. Mitochondria primed by death signals determine cellular addiction to antiapoptotic BCL-2 family members. *Cancer Cell* **9:** 351–365.

Cervantes-Gomez F, Lamothe B, Woyach JA, Wierda WG, Keating MJ, Balakrishnan K, Gandhi V. 2015. Pharmacological and protein profiling suggests venetoclax (ABT-199) as optimal partner with ibrutinib in chronic lymphocytic leukemia. *Clin Cancer Res* **21:** 3705–3715.

Chipuk JE, Moldoveanu T, Llambi F, Parsons MJ, Green DR. 2010. The BCL-2 family reunion. *Mol Cell* **37:** 299–310.

Chonghaile TN, Letai A. 2008. Mimicking the BH3 domain to kill cancer cells. *Oncogene* **27**(Suppl 1): S149–S157.

Choudhary GS, Al-Harbi S, Mazumder S, Hill BT, Smith MR, Bodo J, Hsi ED, Almasan A. 2015. MCL-1 and BCL-xL-dependent resistance to the BCL-2 inhibitor ABT-199 can be overcome by preventing PI3 K/AKT/mTOR activation in lymphoid malignancies. *Cell Death Dis* **6:** e1593.

Davids MS, Deng J, Wiestner A, Lannutti BJ, Wang L, Wu CJ, Wilson WH, Brown JR, Letai A. 2012. Decreased mitochondrial apoptotic priming underlies stroma-mediated treatment resistance in chronic lymphocytic leukemia. *Blood* **120:** 3501–3509.

Del Gaizo Moore V, Schlis KD, Sallan SE, Armstrong SA, Letai A. 2008. BCL-2 dependence and ABT-737 sensitivity in acute lymphoblastic leukemia. *Blood* **111:** 2300–2309.

Deng J, Carlson N, Takeyama K, Dal Cin P, Shipp M, Letai A. 2007. BH3 profiling identifies three distinct classes of apoptotic blocks to predict response to ABT-737 and conventional chemotherapeutic agents. *Cancer Cell* **12:** 171–185.

Du H, Wolf J, Schafer B, Moldoveanu T, Chipuk JE, Kuwana T. 2011. BH3 domains other than Bim and Bid can directly activate Bax/Bak. *J Biol Chem* **286:** 491–501.

Durie BG, Harousseau JL, Miguel JS, Blade J, Barlogie B, Anderson K, Gertz M, Dimopoulos M, Westin J, Sonneveld P, et al. 2006. International uniform response criteria for multiple myeloma. *Leukemia* **20:** 1467–1473.

Eskes R, Desagher S, Antonsson B, Martinou JC. 2000. Bid induces the oligomerization and insertion of Bax into the outer mitochondrial membrane. *Mol Cell Biol* **20:** 929–935.

Faber AC, Coffee EM, Costa C, Dastur A, Ebi H, Hata AN, Yeo AT, Edelman EJ, Song Y, Tam AT, et al. 2014. mTOR inhibition specifically sensitizes colorectal cancers with KRAS or BRAF mutations to BCL-2/BCL-XL inhibition by suppressing MCL-1. *Cancer Discov* **4:** 42–52.

Faber AC, Farago AF, Costa C, Dastur A, Gomez-Caraballo M, Robbins R, Wagner BL, Rideout WM III, Jakubik CT, Ham J, et al. 2015. Assessment of ABT-263 activity across a cancer cell line collection leads to a potent combination therapy for small-cell lung cancer. *Proc Natl Acad Sci* **112:** E1288–E1296.

Fadok VA, Bratton DL, Rose DM, Pearson A, Ezekewitz RA, Henson PM. 2000. A receptor for phosphatidylserine-specific clearance of apoptotic cells. *Nature* **405:** 85–90.

Fadok VA, Xue D, Henson P. 2001. If phosphatidylserine is the death knell, a new phosphatidylserine-specific receptor is the bellringer. *Cell Death Differ* **8:** 582–587.

Fu NY, Sukumaran SK, Kerk SY, Yu VC. 2009. Baxbeta: A constitutively active human Bax isoform that is under tight regulatory control by the proteasomal degradation mechanism. *Mol Cell* **33:** 15–29.

Gandhi L, Camidge DR, Ribeiro de Oliveira M, Bonomi P, Gandara D, Khaira D, Hann CL, McKeegan EM, Litvinovich E, Hemken PM, et al. 2011. Phase I study of Navitoclax

(ABT-263), a novel Bcl-2 family inhibitor, in patients with small-cell lung cancer and other solid tumors. *J Clin Oncol* **29:** 909–916.

Gardner EE, Connis N, Poirier JT, Cope L, Dobromilskaya I, Gallia GL, Rudin CM, Hann CL. 2014. Rapamycin rescues ABT-737 efficacy in small cell lung cancer. *Cancer Res* **74:** 2846–2856.

Griffiths GJ, Dubrez L, Morgan CP, Jones NA, Whitehouse J, Corfe BM, Dive C, Hickman JA. 1999. Cell damage–induced conformational changes of the pro-apoptotic protein Bak in vivo precede the onset of apoptosis. *J Cell Biol* **144:** 903–914.

Griffiths GJ, Corfe BM, Savory P, Leech S, Esposti MD, Hickman JA, Dive C. 2001. Cellular damage signals promote sequential changes at the N-terminus and BH-1 domain of the pro-apoptotic protein Bak. *Oncogene* **20:** 7668–7676.

Guerriero JL, Ditsworth D, Catanzaro JM, Sabino G, Furie MB, Kew RR, Crawford HC, Zong WX. 2011. DNA alkylating therapy induces tumor regression through an HMGB1-mediated activation of innate immunity. *J Immunol* **186:** 3517–3526.

Hanahan D, Weinberg RA. 2000. The hallmarks of cancer. *Cell* **100:** 57–70.

Hanahan D, Weinberg RA. 2011. Hallmarks of cancer: The next generation. *Cell* **144:** 646–674.

Harrison LR, Micha D, Brandenburg M, Simpson KL, Morrow CJ, Denneny O, Hodgkinson C, Yunus Z, Dempsey C, Roberts D, et al. 2011. Hypoxic human cancer cells are sensitized to BH-3 mimetic-induced apoptosis via downregulation of the Bcl-2 protein Mcl-1. *J Clin Invest* **121:** 1075–1087.

Hsu YT, Youle RJ. 1997. Nonionic detergents induce dimerization among members of the Bcl-2 family. *J Biol Chem* **272:** 13829–13834.

Hsu SY, Kaipia A, McGee E, Lomeli M, Hsueh AJ. 1997. Bok is a pro-apoptotic Bcl-2 protein with restricted expression in reproductive tissues and heterodimerizes with selective anti-apoptotic Bcl-2 family members. *Proc Natl Acad Sci* **94:** 12401–12406.

Ke F, Voss A, Kerr JB, O'Reilly LA, Tai L, Echeverry N, Bouillet P, Strasser A, Kaufmann T. 2012. BCL-2 family member BOK is widely expressed but its loss has only minimal impact in mice. *Cell Death Differ* **19:** 915–925.

Ke F, Grabow S, Kelly GL, Lin A, O'Reilly LA, Strasser A. 2015. Impact of the combined loss of BOK, BAX and BAK on the hematopoietic system is slightly more severe than compound loss of BAX and BAK. *Cell Death Dis* **6:** e1938.

Kim H, Rafiuddin-Shah M, Tu HC, Jeffers JR, Zambetti GP, Hsieh JJ, Cheng EH. 2006. Hierarchical regulation of mitochondrion-dependent apoptosis by BCL-2 subfamilies. *Nat Cell Biol* **8:** 1348–1358.

Korsmeyer SJ, Wei MC, Saito M, Weiler S, Oh KJ, Schlesinger PH. 2000. Pro-apoptotic cascade activates BID, which oligomerizes BAK or BAX into pores that result in the release of cytochrome *c*. *Cell Death Differ* **7:** 1166–1173.

Kuwana T, Bouchier-Hayes L, Chipuk JE, Bonzon C, Sullivan BA, Green DR, Newmeyer DD. 2005. BH3 domains of BH3-only proteins differentially regulate Bax-mediated mitochondrial membrane permeabilization both directly and indirectly. *Mol Cell* **17:** 525–535.

Kvansakul M, Yang H, Fairlie WD, Czabotar PE, Fischer SF, Perugini MA, Huang DC, Colman PM. 2008. Vaccinia virus anti-apoptotic F1L is a novel Bcl-2-like domain-swapped dimer that binds a highly selective subset of BH3-containing death ligands. *Cell Death Differ* **15:** 1564–1571.

Letai A, Bassik MC, Walensky LD, Sorcinelli MD, Weiler S, Korsmeyer SJ. 2002. Distinct BH3 domains either sensitize or activate mitochondrial apoptosis, serving as prototype cancer therapeutics. *Cancer Cell* **2:** 183–192.

Li P, Nijhawan D, Budihardjo I, Srinivasula SM, Ahmad M, Alnemri ES, Wang X. 1997. Cytochrome *c* and dATP-dependent formation of Apaf-1/caspase-9 complex initiates an apoptotic protease cascade. *Cell* **91:** 479–489.

Lindsten T, Ross AJ, King A, Zong WX, Rathmell JC, Shiels HA, Ulrich E, Waymire KG, Mahar P, Frauwirth K, et al. 2000. The combined functions of proapoptotic Bcl-2 family members bak and bax are essential for normal development of multiple tissues. *Mol Cell* **6:** 1389–1399.

Lithgow T, van Driel R, Bertram JF, Strasser A. 1994. The protein product of the oncogene bcl-2 is a component of the nuclear envelope, the endoplasmic reticulum, and the outer mitochondrial membrane. *Cell Growth Differ* **5:** 411–417.

Liu X, Dai S, Zhu Y, Marrack P, Kappler JW. 2003. The structure of a Bcl-xL/Bim fragment complex: Implications for Bim function. *Immunity* **19:** 341–352.

Llambi F, Wang YM, Victor B, Yang M, Schneider DM, Gingras S, Parsons MJ, Zheng JH, Brown SA, Pelletier S, et al. 2016. BOK is a non-canonical BCL-2 family effector of apoptosis regulated by ER-associated degradation. *Cell* **165:** 421–433.

Lovell JF, Billen LP, Bindner S, Shamas-Din A, Fradin C, Leber B, Andrews DW. 2008. Membrane binding by tBid initiates an ordered series of events culminating in membrane permeabilization by Bax. *Cell* **135:** 1074–1084.

Makin GW, Corfe BM, Griffiths GJ, Thistlethwaite A, Hickman JA, Dive C. 2001. Damage-induced Bax N-terminal change, translocation to mitochondria and formation of Bax dimers/complexes occur regardless of cell fate. *EMBO J* **20:** 6306–6315.

Mason KD, Vandenberg CJ, Scott CL, Wei AH, Cory S, Huang DC, Roberts AW. 2008. In vivo efficacy of the Bcl-2 antagonist ABT-737 against aggressive Myc-driven lymphomas. *Proc Natl Acad Sci* **105:** 17961–17966.

Meng MB, Wang HH, Cui YL, Wu ZQ, Shi YY, Zaorsky NG, Deng L, Yuan ZY, Lu Y, Wang P. 2016. Necroptosis in tumorigenesis, activation of anti-tumor immunity, and cancer therapy. *Oncotarget*. doi: 10.18632/oncotarget.10548.

Moldoveanu T, Liu Q, Tocilj A, Watson M, Shore G, Gehring K. 2006. The X-ray structure of a BAK homodimer reveals an inhibitory zinc binding site. *Mol Cell* **24:** 677–688.

Montero J, Sarosiek KA, DeAngelo JD, Maertens O, Ryan J, Ercan D, Piao H, Horowitz NS, Berkowitz RS, Matulonis U, et al. 2015. Drug-induced death signaling strategy rapidly predicts cancer response to chemotherapy. *Cell* **160:** 977–989.

Ni Chonghaile T, Sarosiek KA, Vo TT, Ryan JA, Tammareddi A, Moore Vdel G, Deng J, Anderson KC, Richardson P, Tai YT, et al. 2011. Pretreatment mitochondrial priming correlates with clinical response to cytotoxic chemotherapy. *Science* **334:** 1129–1133.

Oltersdorf T, Elmore SW, Shoemaker AR, Armstrong RC, Augeri DJ, Belli BA, Bruncko M, Deckwerth TL, Dinges J, Hajduk PJ, et al. 2005. An inhibitor of Bcl-2 family proteins induces regression of solid tumours. *Nature* **435:** 677–681.

Pinton P, Rizzuto R. 2006. Bcl-2 and Ca^{2+} homeostasis in the endoplasmic reticulum. *Cell Death Differ* **13:** 1409–1418.

Potter DS, Kelly P, Denneny O, Juvin V, Stephens LR, Dive C, Morrow CJ. 2014. BMX acts downstream of PI3 K to promote colorectal cancer cell survival and pathway inhibition sensitizes to the BH3 mimetic ABT-737. *Neoplasia* **16:** 147–157.

Potter DS, Galvin M, Brown S, Lallo A, Hodgkinson CL, Blackhall F, Morrow CJ, Dive C. 2016. Inhibition of PI3 K/BMX Cell Survival Pathway Sensitizes to BH3 Mimetics in SCLC. *Mol Cancer Ther* **15:** 1248–1260.

Premkumar DR, Jane EP, DiDomenico JD, Vukmer NA, Agostino NR, Pollack IF. 2012. ABT-737 synergizes with bortezomib to induce apoptosis, mediated by Bid cleavage, Bax activation, and mitochondrial dysfunction in an Akt-dependent context in malignant human glioma cell lines. *J Pharmacol Exp Ther* **341:** 859–872.

Ren D, Tu HC, Kim H, Wang GX, Bean GR, Takeuchi O, Jeffers JR, Zambetti GP, Hsieh JJ, Cheng EH. 2010. BID, BIM, and PUMA are essential for activation of the BAX- and BAK-dependent cell death program. *Science* **330:** 1390–1393.

Roberts AW, Seymour JF, Brown JR, Wierda WG, Kipps TJ, Khaw SL, Carney DA, He SZ, Huang DC, Xiong H, et al.

2012. Substantial susceptibility of chronic lymphocytic leukemia to BCL2 inhibition: Results of a phase I study of navitoclax in patients with relapsed or refractory disease. *J Clin Oncol* **30:** 488–496.

Roberts AW, Davids MS, Pagel JM, Kahl BS, Puvvada SD, Gerecitano JF, Kipps TJ, Anderson MA, Brown JR, Gressick L, et al. 2016. Targeting BCL2 with venetoclax in relapsed chronic lymphocytic leukemia. *N Engl J Med* **374:** 311–322.

Ryan J, Letai A. 2013. BH3 profiling in whole cells by fluorimeter or FACS. *Methods* **61:** 156–164.

Ryan JA, Brunelle JK, Letai A. 2010. Heightened mitochondrial priming is the basis for apoptotic hypersensitivity of $CD4^+$ $CD8^+$ thymocytes. *Proc Natl Acad Sci* **107:** 12895–12900.

Ryan J, Montero J, Rocco J, Letai A. 2016. iBH3: Simple, fixable BH3 profiling to determine apoptotic priming in primary tissue by flow cytometry. *Biol Chem* **397:** 671–678.

Sarosiek KA, Ziller M, Fraser C, Bhola P, Ryan J, Deng J, Jian B, Goldenberg M, Madsen J, Carrasco R, et al. 2015. Abstract 4728: Apoptotic priming is regulated by a developmental program and predisposes children to therapy-induced toxicity. *Cancer Res* **75:** 4728–4728.

Sattler M, Liang H, Nettesheim D, Meadows RP, Harlan JE, Eberstadt M, Yoon HS, Shuker SB, Chang BS, Minn AJ, et al. 1997. Structure of Bcl-xL-Bak peptide complex: Recognition between regulators of apoptosis. *Science* **275:** 983–986.

Savill J, Fadok V. 2000. Corpse clearance defines the meaning of cell death. *Nature* **407:** 784–788.

Schoenwaelder SM, Jarman KE, Gardiner EE, Hua M, Qiao J, White MJ, Josefsson EC, Alwis I, Ono A, Willcox A, et al. 2011. Bcl-xL-inhibitory BH3 mimetics can induce a transient thrombocytopathy that undermines the hemostatic function of platelets. *Blood* **118:** 1663–1674.

Segawa K, Kurata S, Yanagihashi Y, Brummelkamp TR, Matsuda F, Nagata S. 2014. Caspase-mediated cleavage of phospholipid flippase for apoptotic phosphatidylserine exposure. *Science* **344:** 1164–1168.

Souers AJ, Leverson JD, Boghaert ER, Ackler SL, Catron ND, Chen J, Dayton BD, Ding H, Enschede SH, Fairbrother WJ, et al. 2013. ABT-199, a potent and selective BCL-2 inhibitor, achieves antitumor activity while sparing platelets. *Nat Med* **19:** 202–208.

Spencer SL, Gaudet S, Albeck JG, Burke JM, Sorger PK. 2009. Non-genetic origins of cell-to-cell variability in TRAIL-induced apoptosis. *Nature* **459:** 428–432.

Suzuki M, Youle RJ, Tjandra N. 2000. Structure of Bax: Coregulation of dimer formation and intracellular localization. *Cell* **103:** 645–654.

Tahir SK, Yang X, Anderson MG, Morgan-Lappe SE, Sarthy AV, Chen J, Warner RB, Ng SC, Fesik SW, Elmore SW, et al. 2007. Influence of Bcl-2 family members on the cellular response of small-cell lung cancer cell lines to ABT-737. *Cancer Res* **67:** 1176–1183.

Tan N, Wong M, Nannini MA, Hong R, Lee LB, Price S, Williams K, Savy PP, Yue P, Sampath D, et al. 2013. Bcl-2/Bcl-xL inhibition increases the efficacy of MEK inhibition alone and in combination with PI3 kinase inhibition in lung and pancreatic tumor models. *Mol Cancer Ther* **12:** 853–864.

Toda S, Hanayama R, Nagata S. 2012. Two-step engulfment of apoptotic cells. *Mol Cell Biol* **32:** 118–125.

Tse C, Shoemaker AR, Adickes J, Anderson MG, Chen J, Jin S, Johnson EF, Marsh KC, Mitten MJ, Nimmer P, et al. 2008. ABT-263: A potent and orally bioavailable Bcl-2 family inhibitor. *Cancer Res* **68:** 3421–3428.

Vo TT, Ryan J, Carrasco R, Neuberg D, Rossi DJ, Stone RM, Deangelo DJ, Frattini MG, Letai A. 2012. Relative mitochondrial priming of myeloblasts and normal HSCs determines chemotherapeutic success in AML. *Cell* **151:** 344–355.

Vogler M, Dinsdale D, Dyer MJ, Cohen GM. 2009. Bcl-2 inhibitors: Small molecules with a big impact on cancer therapy. *Cell Death Differ* **16:** 360–367.

Wang X. 2001. The expanding role of mitochondria in apoptosis. *Genes Dev* **15:** 2922–2933.

Wei MC, Zong WX, Cheng EH, Lindsten T, Panoutsakopoulou V, Ross AJ, Roth KA, MacGregor GR, Thompson CB, Korsmeyer SJ. 2001. Proapoptotic BAX and BAK: A requisite gateway to mitochondrial dysfunction and death. *Science* **292:** 727–730.

Wolter KG, Hsu YT, Smith CL, Nechushtan A, Xi XG, Youle RJ. 1997. Movement of Bax from the cytosol to mitochondria during apoptosis. *J Cell Biol* **139:** 1281–1292.

Wyllie AH. 1997. Apoptosis: An overview. *Br Med Bull* **53:** 451–465.

Youle RJ, Strasser A. 2008. The BCL-2 protein family: Opposing activities that mediate cell death. *Nat Rev Mol Cell Biol* **9:** 47–59.

Zha H, Aime-Sempe C, Sato T, Reed JC. 1996. Proapoptotic protein Bax heterodimerizes with Bcl-2 and homodimerizes with Bax via a novel domain (BH3) distinct from BH1 and BH2. *J Biol Chem* **271:** 7440–7444.

Genetic Dissection of Cancer Development, Therapy Response, and Resistance in Mouse Models of Breast Cancer

STEFANO ANNUNZIATO,[1] MARCO BARAZAS,[1] SVEN ROTTENBERG,[1,2] AND JOS JONKERS[1,3]
[1]*Division of Molecular Pathology, The Netherlands Cancer Institute, 1066 CX Amsterdam, The Netherlands*
[2]*Institute of Animal Pathology, Vetsuisse Faculty, University of Bern, 3012 Bern, Switzerland*
[3]*Cancer Genomics Netherlands, The Netherlands Cancer Institute, 1066 CX Amsterdam, The Netherlands*
Correspondence: j.jonkers@nki.nl; sven.rottenberg@vetsuisse.unibe.ch

The cancer genomics revolution has rapidly expanded the inventory of somatic mutations characterizing human malignancies, highlighting a previously underappreciated extent of molecular variability between and within patients. Also in breast cancer, the most commonly diagnosed malignancy in women, this heterogeneity complicates the understanding of the stepwise sequence of pathogenic events and the design of effective and long-lasting target therapies. To disentangle this complexity and pinpoint which molecular perturbations are crucial to hijack the cellular machinery and lead to tumorigenesis and drug resistance, functional studies are needed in model systems that faithfully and comprehensively recapitulate all the salient aspects of their cognate human counterparts. Mouse models of breast cancer have been instrumental for the study of tumor initiation and drug response but also involve cost and time limitations that represent serious bottlenecks in translational research. To keep pace with the overwhelming amount of hypotheses that warrant in vivo testing, continuous refinement of current breast cancer models and implementation of new technologies is crucial. In this review, we summarize the current state of the art in modeling human breast cancer in mice, and we put forward our vision for future developments.

Breast cancer is the most commonly diagnosed invasive cancer worldwide, with more than 1.6 million new cases each year. Rather than a single disease, it represents a spectrum of malignancies, encompassing several distinct biological entities and subtypes, each associated with specific histopathological and molecular characteristics, responses to therapy, and clinical outcomes. Multiple taxonomies have been developed to divide breast cancer cases into different categories. Histopathological classification comprises several morphological and immunohistochemical phenotypes that can be further divided into different grades. Among the various morphologies, advanced mammary tumors mostly fall into the class of invasive ductal carcinomas (IDCs), followed by invasive lobular carcinomas (ILCs). Molecular classification based on gene expression patterns distinguishes five major subtypes of breast cancer: luminal A and B, $ErbB2^+$, basal-like, and claudin-low (Perou et al. 2000). Although these distinctions have proven useful for clinical decision-making, there are limitations in predicting disease prognosis and response to therapy. For example, a recent prospective, randomized phase III study showed that nearly half of the women with early breast cancer who are at high risk based on standard clinicopathological parameters might not require adjuvant chemotherapy (Cardoso et al. 2016). The additional use of a 70-gene expression signature may help to identify breast cancer patients who do not require adjuvant chemotherapy, but the identification of molecular signatures that reliably predict chemotherapy response remains elusive. Moreover, sequencing studies have shown that even within the same molecular subtype an extreme heterogeneity in the mutational landscape exists, which may account for discrepancies in prognosis and therapy response between different patients (Cancer Genome Atlas Network 2012; Stephens et al. 2012; Nik-Zainal et al. 2016). Another complicating factor is intratumoral heterogeneity. Individual tumors are mosaics of multiple clones of neoplastic cells, each characterized by a distinct genetic makeup and differential responses to the selective pressures to which they are exposed, making the tumor mass not static but continuously shaped by a branching evolutionary process resembling Darwinian evolution. Distinguishing causal disease variants (driver mutations) from background alterations (passenger mutations) is a major goal in breast cancer research, as it can pinpoint evolutionary conserved processes that mammary tumor cells apply during stepwise transformation and to which they might be addicted. To exploit these potential Achilles' heels, we require a comprehensive knowledge of how these signaling networks physiologically function, how they become aberrant, and how they can be directly or indirectly disrupted.

Published by Cold Spring Harbor Laboratory Press; doi: 10.1101/sqb.2016.81.030924

Given this complexity, genetically engineered mouse models (GEMMs) of breast cancer, together with patient-derived tumor xenografts (PDXs) and GEMM-derived tumor allografts, have proven valuable resources for deepening our understanding of how mammary tumors initiate, progress, metastasize, and respond to therapy in a physiologically relevant in vivo setting (Vargo-Gogola and Rosen 2007). These mouse models are increasingly being used in longitudinal preclinical studies for translation of novel therapies to clinical testing. Moreover, GEMMs provide unique opportunities to infer cause–effect relationships on de novo induced malignancies growing in intact organisms, rather than correlative observations on end-stage patient tumor samples.

Over the past 15 years, our research has been focused on the generation and characterization of mouse models for two breast cancer subtypes: ILCs and basal-like IDCs. To achieve this, we engineered a number of tumor-specific driver mutations in the relevant target cells of mouse models, recapitulating the key dependencies of the resulting lesions to the corresponding deranged signaling pathways. In this review, we will discuss how these models can be used for functional dissection of tumorigenic cascades, unraveling new therapeutic vulnerabilities and mechanisms of therapy resistance—in particular, in light of the advent of new technologies such as clustered regularly interspersed short palindromic repeat (CRISPR)–Cas9 gene editing, which are opening new avenues in breast cancer modeling in mice.

Table 1. Characteristics of human ILC and BRCA1-associated breast cancer and the corresponding GEMMs developed in our laboratory

Invasive lobular carcinoma	Human	*WAPcre;Cdh1$^{F/F}$;Pten$^{F/F}$* mouse model
Morphology	Lobular	Lobular
Invasive	Yes	Yes
Grade	Low	Low
Mitotic index	Low	Low
ER expression	Yes	Yes
Molecular subtype	Luminal	Luminal
Collagen deposition	Yes	Yes
Stroma-rich	Yes	Yes
BRCA1-associated breast cancer	**Human**	***K14cre;Brca1$^{F/F}$;p53$^{F/F}$* mouse model**
Morphology	Ductal	Ductal
Invasive	Yes	Yes
Grade	High	High
Mitotic index	High	High
ER/PR expression	No	No
Molecular subtype	Basal-like	Basal-like
Genomically instable	Yes	Yes
HR-deficient	Yes	Yes

ILC, invasive lobular carcinoma; BRCA1, breast cancer gene 1; GEMMs, genetically engineered mouse models; ER, estrogen receptor; PR, progesterone receptor; HR, homologous recombination.

INVASIVE LOBULAR BREAST CARCINOMA (ILC) MODELS

ILC accounts for 8%–14% of all breast cancer cases and is hallmarked at the morphological level by tumor cells growing in single "Indian files" within a dense fibrous stroma. This phenotype can be explained at the molecular level by loss of integrity of cell adherens junctions because of mutations or methylation of the *CDH1* gene, which encodes the transmembrane protein E-cadherin (Martinez and Azzopardi 1979; Borst and Ingold 1993; Moll et al. 1993; Vos et al. 1997; Droufakou et al. 2001). To our surprise, we found that mammary gland–specific Cre-mediated inactivation of *Cdh1* alleles in mice was insufficient to induce mammary tumors, probably because normal cells undergo apoptosis and are counterselected when E-cadherin is lost (Boussadia et al. 2002; Derksen et al. 2006, 2011). This prompted us to investigate which cooperating oncogenic events are required for malignant transformation of E-cadherin-deficient mammary epithelial cells. We have found that multifocal ILC formation is promoted by dual mammary-specific loss of E-cadherin and p53 (Derksen et al. 2006, 2011) or E-cadherin and PTEN (phosphatase and tensin homolog) (Boelens et al. 2016), with tumor architecture and molecular profiles closely resembling their human ILC counterparts (Table 1). However, it remains elusive which biological processes are rescued by codepletion of E-cadherin with one of these factors. To identify novel candidate cancer genes and networks that collaborate with E-cadherin loss in mammary tumorigenesis, we used the *Sleeping Beauty* (SB) transposon system (Collier et al. 2005; Dupuy et al. 2005) to perform an insertional mutagenesis screen in *WAPcre;Cdh1$^{F/F}$* mice (SM Kas, J de Ruiter, K Schipper, et al., in prep.). Retrieval of recurrent integrations in SB-induced *WAPcre;Cdh1$^{F/F}$* mammary tumors identified common insertion sites in several genes, some known to be mutated in human ILC, suggesting that mutagenesis of these genes leads to malignant transformation. Moreover, analysis of enriched targeted pathways and mutually exclusive insertions revealed the regulation of the actin cytoskeleton as a completely novel oncogenic pathway in both mouse and human ILC.

Furthermore, recent genomic studies on collections of human ILCs have unveiled that, in addition to somatic inactivation of E-cadherin, activation of phosphoinositide 3-kinase (PI3K)/AKT signaling appears to be a common event in this breast cancer subtype (Ciriello et al. 2015; Desmedt et al. 2016; Michaut et al. 2016). To validate these findings, we developed GEMMs of ILC that combine mammary gland–specific ablation of E-cadherin and activation of different oncogenic *Pik3ca* or *Akt* mutants (MHAM van Miltenburg, et al., in prep.). To rapidly generate breast cancer models carrying these allelic variants, we used a novel strategy for fast-track production of GEMMs, called GEMM-ESC, which is based on Flp-recombinase-mediated introduction of additional mutant alleles into the *Col1a1* locus of embryonic stem cells (ESCs) derived from existing GEMMs (Huijbers et al. 2014). Interestingly, the resulting mice showed rapid development of tumors with strong resemblance to human ILC in terms of morphology, gene expression, and invasiveness, on which we are now testing a panel of anticancer therapeutics to identify promising genotype-specific drug sensitivities.

BASAL-LIKE BREAST CANCER MODELS

Basal-like breast cancers represent a heterogeneous class of malignancies with poor clinical outcome that accounts in total for 10%–15% of all breast cancer cases (Perou et al. 2000; Badve et al. 2011). The majority of basal-like tumors lack expression of estrogen receptor (ER), progesterone receptor (PR), and human epidermal growth factor receptor 2 (HER2) and are therefore referred to as triple-negative breast cancers. These tumors are not targetable with hormonal therapy or HER2 inhibitors, which leaves clinicians with only few effective options for therapeutic intervention.

Approximately 50% of basal-like breast cancers display a dysfunctional BRCA pathway because of germline or somatic mutations in *BRCA1/2* or *BRCA1* promoter hypermethylation (Cancer Genome Atlas Network 2012; Nik-Zainal et al. 2016). Also a fraction of non-basal-like tumors are BRCA-deficient, mostly because of germline mutations in *BRCA2*. As these genes are crucial in the error-free repair of DNA double-strand breaks (DSBs) by homologous recombination (HR), BRCA defects are associated with chromosomal instability and hypersensitivity to DNA DSB-inducing drugs such as alkylating agents, poly ADP ribose polymerase inhibitors (PARPi), and radiotherapy (Jaspers et al. 2009; Bouwman and Jonkers 2012; M Barazas, et al., in prep.). However, drug resistance mechanisms have been described in both clinical and preclinical studies of BRCA-associated tumors, posing serious concerns, as no other therapies are currently available for relapsing patients.

To study tumorigenesis and drug resistance mechanisms, we developed several conditional mouse models for BRCA1- and BRCA2-associated breast cancer (Evers and Jonkers 2006; Bouwman and Jonkers 2008). In our *K14cre;Brca1*$^{F/F}$*;p53*$^{F/F}$ (KB1P) and *K14cre;Brca2*$^{F/F}$*;p53*$^{F/F}$ (KB2P) models, mammary inactivation of *Brca1/2* is accompanied by loss of p53, as mutations in this tumor suppressor frequently co-occur with *BRCA1/2* mutations in breast cancer (Jonkers et al. 2001; Liu et al. 2007). These mice develop mammary tumors after a latency period of 6–8 mo, suggesting that additional mutations are required for tumorigenesis (Table 1). However, in contrast to ILC, in which point mutations are the most common somatic alterations, *BRCA*-mutated breast cancers are characterized by complex patterns of DNA copy number aberrations (CNAs), including translocations and gains/losses of entire chromosome arms (Vollebergh et al. 2012). Using cross-species oncogenomics, we identified *MYC* amplification and *RB1* loss as recurrent CNAs in both mouse and human *BRCA1/2*-mutated breast cancers (Holstege et al. 2010). Exploiting the GEMM-ESC strategy, we could model conditional overexpression of MYC in our *WAPcre;Brca1*$^{F/F}$*;p53*$^{F/F}$ (WB1P) mouse model and found that mammary tumor development was indeed strongly accelerated compared to the original line (L Henneman, et al., in prep.). Moreover, we observed that the number of CNAs in WB1P-MYC tumors was markedly reduced compared to WB1P tumors, showing only few recurrent CNAs that most likely harbor additional cancer drivers that collaborate with MYC overexpression and loss of BRCA1/p53 in breast tumorigenesis. We are currently performing cross-species comparisons of the recurrent CNAs in WB1P-MYC tumors with CNA profiles from human breast cancers to identify candidate cancer genes, which will be validated in the WB1P-MYC model. We believe that this iterative CNA profiling approach in progressively complex GEMMs will be instrumental for deciphering the key driver events in BRCA1-associated breast cancer and for uncovering novel therapeutic vulnerabilities.

PRECLINICAL TRIALS IN BRCA-ASSOCIATED BREAST CANCER MODELS

Although phase-I and -II clinical trials are mostly carried out in heavily pretreated volunteer patients who suffer from end-stage metastatic cancer, mouse models provide the opportunity to initiate treatment on naïve tumors in a clinically relevant in vivo setting. Treatment of mammary tumor-bearing KB1P mice with a panel of DSB-inducing agents showed heterogeneous responses between individual tumors but also marked differences in tumors treated with doxorubicin or docetaxel and those treated with cisplatin (Rottenberg et al. 2007). Although KB1P tumors eventually developed resistance to doxorubicin and docetaxel, no acquired resistance was observed for cisplatin. Even though these tumors could never be completely eradicated by maximum tolerated dose concentrations of cisplatin, the relapsing tumors remained responsive to subsequent treatments, resulting in a typical sawtooth tumor response. A major breakthrough came when it was found that spontaneous KB1P and KB2P tumors could be orthotopically allografted in syngeneic mice while maintaining their genetic characteristics and drug sensitivity profile. This approach reduced the time to produce cohorts of tumor-bearing mice from 7–9 mo to 4–6 wk, and enabled large-scale intervention studies in which the response of a single donor to different chemotherapeutic strategies could be compared, ruling out any intertumor heterogeneity (Fig. 1). Intervention studies with the PARP inhibitor olaparib in KB1P tumor allografts led to the development of carboplatin and olaparib switch-maintenance therapy for *BRCA1*-mutated breast cancer (Rottenberg et al. 2008). This preclinical concept was confirmed in a clinical trial with olaparib maintenance therapy in *BRCA*-mutation carriers with platinum-sensitive ovarian cancer (Ledermann et al. 2012, 2014) and eventually led to clinical approval of olaparib (Deeks 2015). Similarly, intervention studies in KB2P tumor allografts showed that alkylators such as nimustine could induce complete tumor eradication (Evers et al. 2010). Eradication of *BRCA*-mutated and BRCA-like cancer by high-dose alkylating chemotherapy was subsequently confirmed by

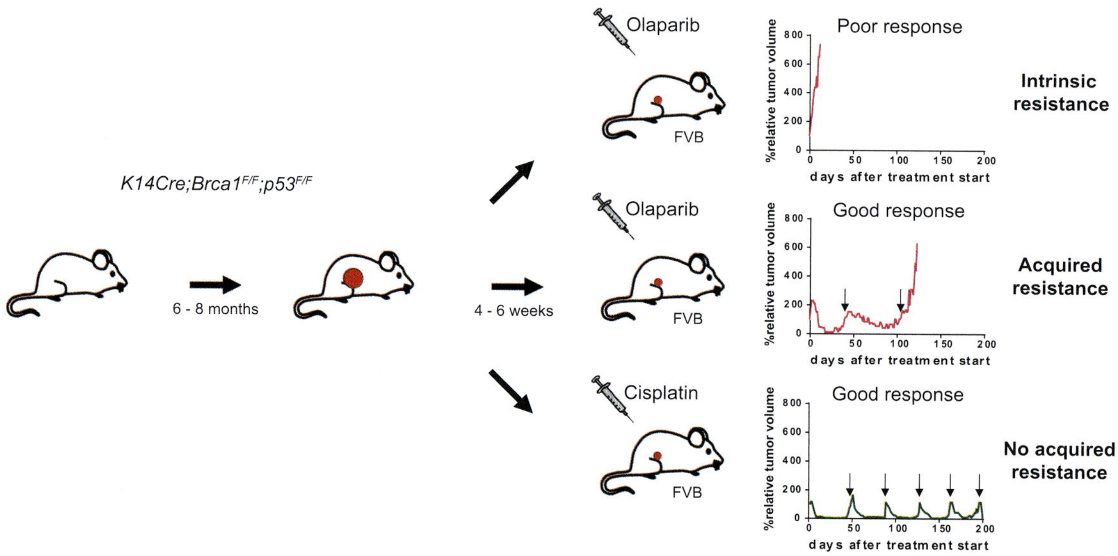

Figure 1. Large-scale intervention studies using breast cancer gene (BRCA)-deficient orthotopic allografts. Spontaneous tumors develop with a latency of 6–8 mo in *K14cre;Brca1$^{F/F}$;p53$^{F/F}$* (KB1P) or *K14cre;Brca2$^{F/F}$;p53$^{F/F}$* (KB2P) genetically engineered mouse models (GEMMs). Retransplantation of spontaneous tumors in syngeneic FVB wild-type mice highlighted intertumor heterogeneity in treatment response. In the case of olaparib, some tumors displayed intrinsic resistance, whereas others displayed initial good response followed by the emergence of acquired resistance. In contrast, resistance never developed in cisplatin-treated animals, despite multiple consecutive rounds of treatment (indicated by black arrows).

retrospective analysis of data from clinical trials (Vollebergh et al. 2011, 2014; Schouten et al. 2015). These and other studies illustrate the utility of GEMMs of human cancer in translational cancer medicine.

PARPi RESISTANCE MECHANISMS IN BRCA-ASSOCIATED BREAST CANCER MODELS

In addition to accelerating preclinical trials, the KB1P and KB2P allograft platforms also enabled large-scale induction of acquired resistance to a drug of choice and subsequent identification of the underlying resistance mechanisms. The power of this approach was demonstrated with the PARP inhibitor olaparib, which was described to display selective toxicity against BRCA1/2-deficient cells (Bryant et al. 2005; Farmer et al. 2005). Indeed, KB1P tumor allografts initially responded well to treatment but eventually relapsed and developed stable resistance (Rottenberg et al. 2008). This has provided a valuable collection of matched treatment-naïve and treatment-resistant tumors, which could be analyzed using next-generation sequencing or (phospho)-proteomics, thereby taking advantage of the clean genetic background of inbred mice and the known genetic profile of treatment-naïve tumors. We found that *Abcb1a* and *Abcb1b*, encoding P-glycoprotein (P-gp) efflux pumps, were up-regulated in resistant tumors and we confirmed that P-gp played an important role in mediating export of olaparib from tumor cells (Fig. 2A; Rottenberg et al. 2008). Resistance could be reversed when P-gp-mediated drug efflux was inhibited by coadministration of tariquidar. Although the clinical relevance of P-gp up-regulation as cause of drug resistance remains controversial (Amiri-Kordestani et al. 2012), expression of *MDR1*, the human counterpart of *Abcb1*, was recently found to be inversely correlated to olaparib response in human ovarian cancer cells (Vaidyanathan et al. 2016). Such increased expression may result from complex genomic rearrangements that fuse a distant promoter to the *MDR1* gene and thereby bypass the *MDR1* promoter methylation (Patch et al. 2015). The case of P-gp shows that a thorough mechanistic understanding is instrumental to combat resistant tumors—for example, by coadministration of tariquidar or by switching treatment to chemotherapeutics that are poor substrates for P-gp (Jaspers et al. 2013).

To dissect P-gp-independent mechanisms of PARPi resistance, the KB1P mouse model was refined through germline genetic deletion of *Mdr1* resulting in the *K14cre;Brca1$^{F/F}$;p53$^{F/F}$;Mdr1a/b$^{-/-}$* (KB1PM) model (Jaspers et al. 2013). Alternatively, KB1P tumors were treated with the PARP inhibitor AZD2461, which is a poor substrate for P-gp (Oplustil O'Connor et al. 2016). PARPi resistance developed in these models despite the exclusion of P-gp-related mechanisms. To identify the underlying resistance mechanisms, next-generation sequencing data from treatment-naïve and PARPi-resistant tumors were combined with data from unbiased functional genetic screens in vitro. Through an insertional mutagenesis screen in conditional BRCA1-knockout mouse ESCs, we found that loss of 53BP1 rescues the proliferation defect, HR deficiency, and PARPi hypersensitivity of BRCA1-deficient cells by enhancing DSB end resection (Fig. 2B; Bouwman et al. 2010). This work from our laboratory and similar studies from the Nussenzweig laboratory (Bunting et al. 2010) have led to novel mechanistic insights in DSB repair and to date several downstream effector proteins of 53BP1 have been shown to suppress

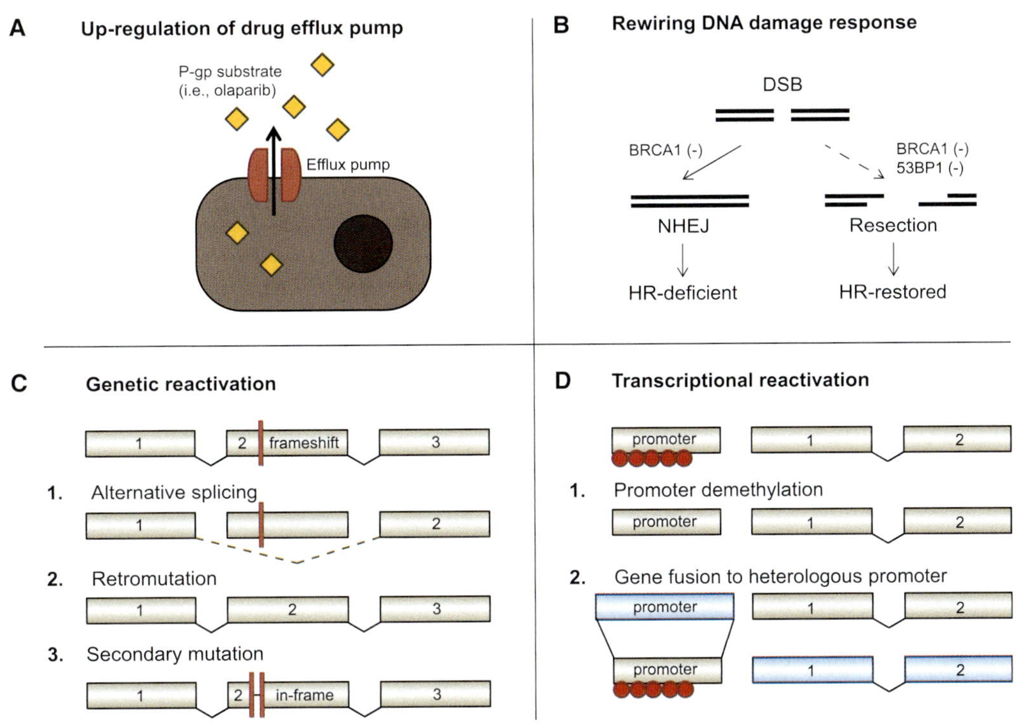

Figure 2. Overview of resistance mechanisms identified using mouse models of breast cancer gene (BRCA)-deficient breast cancer. (*A*) Up-regulation of drug efflux pumps (i.e., MDR1) reduces intracellular drug concentration. (*B*) The DNA damage response pathway can be rewired to restore homologous recombination (HR), in this case by loss of 53BP1. (*C*) Genetic reactivation of *BRCA1*-mutated alleles can occur because of alternative splicing, retromutations, or secondary mutations restoring the *BRCA1* reading frame. (*D*) Transcription of silenced *BRCA1* alleles can be restored upon promoter demethylation or gene fusions to distant promoters. P-gp, P-glycoprotein; DSB, double-strand break; NHEJ, nonhomologous end joining; HR, homologous recombination.

HR in BRCA1-deficient cells, including RIF1 (Chapman et al. 2013; Di Virgilio et al. 2013; Escribano-Diaz et al. 2013; Zimmermann et al. 2013), PTIP (Callen et al. 2013), Artemis (Wang et al. 2014), and REV7/MAD2L2 (Boersma et al. 2015; Xu et al. 2015). Thorough analysis of mutational status and expression levels of 53BP1 and REV7 in PARPi-resistant KB1P(M) tumors confirmed that loss of 53BP1 or REV7 causes in vivo resistance to PARPi (Jaspers et al. 2013; Xu et al. 2015). Interestingly, although KB1P(M) tumors with 53BP1 loss are cross-resistant to topotecan and doxorubicin, they are still responsive to cisplatin, suggesting that platinum drugs may be a useful salvage therapy for this class of PARPi-resistant tumors (Jaspers et al. 2013).

Although the majority of KB1P(M) tumors acquired PARPi resistance through restoration of HR, a substantial fraction of PARPi-resistant tumors remained defective in the formation of ionizing radiation-induced nuclear RAD51 foci (RAD51-IRIFs), which are a hallmark of HR. Moreover, when we analyzed the BRCA2-deficient KB2P tumors with acquired PARPi resistance, none of these showed restoration of HR as measured by RAD51-IRIF assays (E Gogola, et al., in prep.). This suggests the existence of alternative resistance mechanisms. It was recently shown that chemoresistance in BRCA2-deficient cells might be mediated through protection of replication forks (RFs)—for instance, by depletion of PAX transcription activation domain interacting protein (PTIP) (Chaudhuri et al. 2016). It will be important to investigate whether RF protection is a common feature of PARPi-resistant KB1P(M) and KB2P tumors.

In patients, mutated BRCA1 or BRCA2 proteins are often still expressed in tumors. Therefore, the large intragenic *Brca1/2* deletions present in KB1P(M) and KB2P tumors—although instrumental in genetic studies—might not fully recapitulate the biology of BRCA-associated tumors in mutation carriers. To this end, we generated several mouse models mimicking pathogenic *BRCA1* variants that are often encountered in the clinic (Drost et al. 2011, 2016). These models provided evidence that the type and location of the *BRCA1* mutation can have significant implications for the response of these tumors to DSB-inducing agents and PARPi. It was found that tumor cells harboring the $BRCA1^{185delAG}$ allelic variant, which was modeled in mice by a $Brca1^{185stop}$ allele, can use a downstream alternative start site leading to the expression of a RING-less BRCA1 protein (Drost et al. 2016). This RING-less BRCA1 protein maintains hypomorphic HR activity, which is sufficient to induce a poor response to platinum drugs or olaparib. These results illustrate the importance of testing *BRCA1* allelic variants not only for genetic counseling but also for providing adequate treatment.

PDX models provide a solution to narrow the gap between mouse and human cancer biology and as such represent a novel in vivo platform for studying therapy response and resistance. Although PDX models have been relatively difficult to generate in the past, recent advances have made it possible to generate PDX biobanks covering a heterogeneous population of tumors (Hidalgo et al. 2014). Once the (epi)genetic landscape of a PDX model is characterized, this provides an effective tool to study the drug response of a specific tumor and to predict which resistance mechanisms might evolve during treatment. We recently demonstrated the feasibility of such an approach by treatment of PDX models of BRCA1-deficient breast cancer with alkylating agents or olaparib (Ter Brugge et al. 2016). Similar to the GEMM tumors, these PDX tumors generally responded well to treatment, but eventually developed resistance. The underlying mechanism was dependent on the type of *BRCA1* inactivation: Whereas therapy-resistant *BRCA1*-methylated PDX tumors frequently showed *BRCA1* promoter demethylation, *BRCA1*-mutated tumors acquired resistance via genetic reversion through secondary mutations that restored the *BRCA1* reading frame (Fig. 2C,D). These events have also been known to mediate resistance in ovarian cancer patients (Swisher et al. 2008; Patch et al. 2015), showing the predictive potential of PDX models. The PDX models also revealed a novel resistance mechanism involving gene fusions that placed *BRCA1* under transcriptional control of a heterologous promoter. It is intriguing that resistance mechanisms in PDX tumors are mainly centered on re-expression of functional BRCA1 protein rather than inactivation of 53BP1 or related factors, highlighting the strong selective pressure on complete restoration of BRCA1 function when BRCA1-deficient tumor cells are exposed to DSB-inducing therapy. However, a fraction of tumors acquired resistance in the absence of BRCA1 reexpression, showing that alternative resistance mechanisms also occur in PDX models.

Taken together, these studies illustrate the power of mouse models in unraveling resistance mechanisms before their emergence in patients. It will be important to investigate to what extent these play a role in the clinic. This is not trivial, as they likely occur in a limited group of *BRCA* patients and thus require careful patient selection. It is noteworthy that resistance caused by mutations in additional DNA repair genes such as 53BP1 or REV7 might expose new treatment vulnerabilities (e.g., sensitivity to combined PARP and ataxia telangiectasia mutated [ATM] inhibition [Bunting et al. 2010]). It will therefore be important to determine if and how each resistance mechanism can be exploited therapeutically. Ultimately, this may provide a framework for oncologists to combat resistance in the clinic.

NONGERMLINE GEMMs OF BREAST CANCER

Large-scale cancer genome sequencing studies and forward genetic screens have jointly boosted the discrimination between passenger and driver mutations and the identification of genetic determinants of drug sensitivity and resistance in breast cancer. The systematic translation of these long catalogs of structural aberrations into functional information requires the assessment of the pathophysiological impact of candidate gene perturbations in reliable preclinical models. This inevitably poses a practical challenge for in vivo validation experiments, because of the considerable costs and time requirements associated with establishing new breast cancer GEMMs. Novel technologies, especially CRISPR–Cas9-based methods, are revolutionizing the genetic engineering field by providing fast ways for precise and efficient ESC manipulation and GEMM development (Wang et al. 2013). However, as sequencing expenses of human tumors keep decreasing, research will shift from testing oncogenicity of single driver alleles to investigating the impact of multiple allelic variants on tumor development and therapy response. At the same time, forward genetics strategies will evolve from genome-wide approaches based on simple gene (in)-activation to more refined chemical mutagenesis and gene-based CRISPR screens capable of identifying novel hypomorphic, dominant-negative, and separation-of-function mutants at the base pair level. We foresee that the number of testable hypotheses will far exceed the capacity of transgenic facilities, warranting the development of new in vivo platforms for systematic, multiplexed interrogation of putative cancer drivers. Ideally, such models should sort out current temporal and economical limitations of GEMM establishment and bypass extensive mouse husbandry but also allow a high degree of manipulability and flexibility by enabling spatiotemporal control of tumor initiation and progression.

To develop such a platform for breast cancer, we explored the possibility of nongermline modeling of mammary tumors by exploiting intraductal injection in the nipple of adult female mice as a way to deliver high-titer lentiviral or adenoviral preparations to mammary epithelium and achieve somatic genome engineering. We have shown that intraductally injected lentiviruses can target tumor-initiating cells of both the basal and the luminal compartment, allowing modeling of both ILC and basal-like tumors in mice with the corresponding set of relevant predisposing alleles. For example, intraductal injection of Cre-encoding lentiviruses in $Cdh1^{F/F}$;$Pten^{F/F}$ mice induced the formation of ILCs that were undistinguishable from the ILCs arising in the original *WAPcre*;$Cdh1^{F/F}$;$Pten^{F/F}$ model (Annunziato et al. 2016). Somatic Cre delivery may more accurately recapitulate sporadic tumor initiation by allowing titratable and spatiotemporally controlled delivery of viruses to mammary tissue. Moreover, targeting specificity can be modulated by using viruses with cell type–specific promoters and/or posttranscriptional control elements (Tao et al. 2014).

Importantly, the potential of nongermline modeling extends far beyond simple exogenous administration of Cre to established GEMMs. A diverse array of viral and nonviral constructs can be employed to achieve desired permutations of specific candidate genes even in the absence of germline conditional alleles: (a) vectors for overexpression of wild-type, truncated, or mutated cDNAs; (b) vectors for shRNA-mediated down-regulation or CRISPR-mediated (epi)genetic manipulation of single or multiple

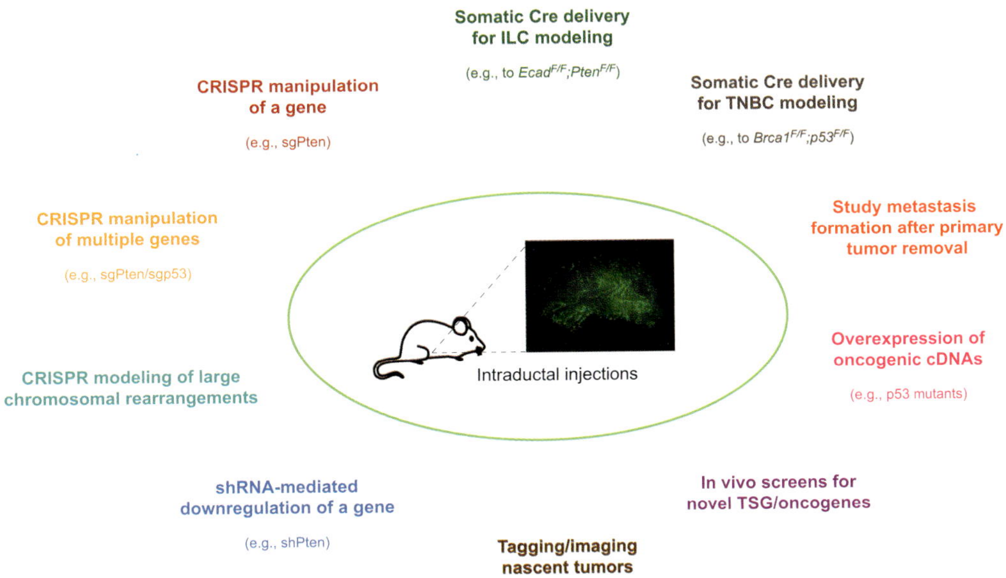

Figure 3. Multiple applications for somatic genome engineering of the mammary tissue via intraductal injection to study tumor biology. ILC, invasive lobular carcinoma; TNBC, triple-negative breast cancer; sgPten, single-guide phosphatase and tensin homolog; shPTEN, short-hairpin phosphatase and tensin homolog; TSG, tumor suppressor gene.

endogenous genes (Sander and Joung 2014); (c) CRISPR vectors for modeling large chromosomal rearrangements (Maddalo et al. 2014); and (d) vectors for tagging and imaging of tumors (Fig. 3). Regarding CRISPR-based in vivo editing approaches, we and others have shown that somatic delivery of the bacterial Cas9 protein has the considerable drawback of eliciting strong and specific immune responses in immunocompetent animals (Wang et al. 2015; Annunziato et al. 2016). This problem can be overcome by employing knock-in models that are tolerant to Cas9 because of constitutive or conditional expression of Cas9 or catalytically inactive dCas9-effector fusions (which allow for transcriptional silencing/activation of endogenous alleles) (Platt et al. 2014; Sánchez-Rivera and Jacks 2015). We have recently reported somatic induction of oncogenic loss-of-function mutations in mice with mammary-specific expression of Cas9 by intraductal injection of single-guide RNA (sgRNA)-encoding lentiviruses, which eventually led to ILC formation (Annunziato et al. 2016).

MAMMARY TUMOR ORGANOIDS

Another exciting technological breakthrough came from the possibility to derive organotypic 3D culture models of normal and malignant mammary tissue. Human and murine tumor organoid cultures retain key features of donor tumors, including cellular heterogeneity and molecular characteristics (Clevers 2016; Fatehullah et al. 2016). Compared to the laborious and time-consuming establishment of 2D cell lines, which requires adaptation to monolayer growth on plastic surfaces, tumor organoid cultures are much easier to derive, can be expanded indefinitely ex vivo, and upon xenografting/allografting undergo polyclonal expansion and efficiently produce tumors that preserve the cellular heterogeneity and drug response profiles of the original tumors (AA Duarte, E Gogola, N Sachs, et al., in prep.). For example, we found the differential olaparib sensitivity of isogenic treatment-naïve and PARPi-resistant KB1P mammary tumors to be stable upon organoid derivation and subsequent retransplantation. Using CRISPR–Cas9 technology, we were able to introduce *Trp53bp1* frameshift mutations in the treatment-naïve KB1P organoid line and demonstrate that this permutation rendered the organoid-derived tumors refractory to olaparib. We are exploiting the KB1P tumor organoid platform to test additional candidate drug resistance genes for their in vivo relevance, including candidates retrieved from forward genetic screens and from sequencing of drug-resistant tumors (Fig. 4). Moreover,

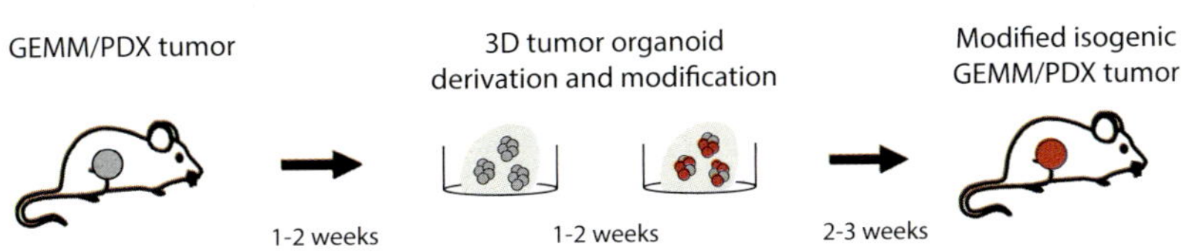

Figure 4. 3D tumor organoid cultures can be rapidly derived from established genetically engineered mouse model (GEMM)/patient-derived xenograft (PDX) tumors, modified ex vivo with desired genetic permutations and retransplanted in mice to evaluate in vivo treatment responses.

given the short latency period and polyclonal tumor outgrowth, GEMM and PDX tumor organoids are particularly amenable for in vivo genetic screens using short-hairpin RNA (shRNA), CRISPR, CRISPRi, and CRISPRa libraries.

CONCLUSION

A number of known and unknown biological discrepancies inevitably exist between mouse models and humans. Moreover, refinements in mouse modeling should be compliant with practical and ethical issues associated with model establishment. Nevertheless, the systematic and synergistic deployment of complementary in vitro and in vivo platforms (GEMMs, PDX models, organoids, nongermline models) is envisioned to provide a quantum leap in the oncology arena and in breast cancer research in particular. Cutting-edge mouse cancer clinics will enable so-called coclinical trials, in which clinical studies will be paralleled by preclinical intervention studies in mouse avatars. This will allow clinicians to infer in real-time genotype-specific drug response profiles from mouse models and design more effective and long-lasting patient-tailored treatment schemes. The emergence of drug resistance is an invariable and intrinsic consequence of Darwinian tumor growth dynamics, but instead of "whack-a-mole" treatment schedules, coclinical trials could assist in the design of more sophisticated and personalized regimens in which tumors are forced through evolutionary bottlenecks that render them exquisitely sensitive to secondary therapies. Reiteration of this adaptive process is possible only by the use of ever-smarter mouse models, which will ultimately lead to improved long-term management of this devastating disease.

ACKNOWLEDGMENTS

We are grateful to Peter Bouwman for critical reading of the manuscript. Research in the Jonkers and Rottenberg laboratories described in this review has been supported by grants from the Dutch Cancer Society (NKI 2007-3772, NKI 2008-4116, NKI 2009-4303, NKI 2011-5197, NKI 2011-5220, NKI 2012-5220, NKI 2014-6532, NKI 2015-7877, EMCR 2014-7048), the Netherlands Organization for Scientific Research (NWO), Cancer Genomics Netherlands (CGCNL), Cancer Systems Biology Center (CSBC), NGI Zenith 93512009, VIDI 91711302, VICI 91814643, the Swiss National Science Foundation (310030-156869), the Swiss Cancer Research Foundation (MD-PhD-3446-01-2014), the European Union Seventh Framework Programme (EurocanPlatform project 260791, Infrafrontier-I3 project 312325, CHEMORES project 037665), and the European Research Council (ERC SyG CombatCancer, ERC CoG-681572). Additional support is provided by the Mouse Clinic for Cancer and Aging (MCCA), financed by NWO as part of the National Roadmap for Large-Scale Research Facilities.

REFERENCES

Amiri-Kordestani L, Basseville A, Kurdziel K, Fojo AT, Bates SE. 2012. Targeting MDR in breast and lung cancer: Discriminating its potential importance from the failure of drug resistance reversal studies. *Drug Resist Updat* **15:** 50–61.

Annunziato S, Kas SM, Nethe M, Yucel H, Del Bravo J, Pritchard C, Bin Ali R, van Gerwen B, Siteur B, Drenth AP, et al. 2016. Modeling invasive lobular breast carcinoma by CRISPR/Cas9-mediated somatic genome editing of the mammary gland. *Genes Dev* **30:** 1470–1480.

Badve S, Dabbs DJ, Schnitt SJ, Baehner FL, Decker T, Eusebi V, Fox SB, Ichihara S, Jacquemier J, Lakhani SR, et al. 2011. Basal-like and triple-negative breast cancers: A critical review with an emphasis on the implications for pathologists and oncologists. *Mod Pathol* **24:** 157–167.

Boelens MC, Nethe M, Klarenbeek S, de Ruiter JR, Schut E, Bonzanni N, Zeeman AL, Wientjens E, van der Burg E, Wessels L, et al. 2016. PTEN loss in E-cadherin-deficient mouse mammary epithelial cells rescues apoptosis and results in development of classical invasive lobular carcinoma. *Cell Rep* **16:** 2087–2101.

Boersma V, Moatti N, Segura-Bayona S, Peuscher MH, van der Torre J, Wevers BA, Orthwein A, Durocher D, Jacobs JJ. 2015. MAD2L2 controls DNA repair at telomeres and DNA breaks by inhibiting 5′ end resection. *Nature* **521:** 537–540.

Borst MJ, Ingold JA. 1993. Metastatic patterns of invasive lobular versus invasive ductal carcinoma of the breast. *Surgery* **114:** 637–641; discussion 641–642.

Boussadia O, Kutsch S, Hierholzer A, Delmas V, Kemler R. 2002. E-cadherin is a survival factor for the lactating mouse mammary gland. *Mech Dev* **115:** 53–62.

Bouwman P, Jonkers J. 2008. Mouse models for BRCA1 associated tumorigenesis: From fundamental insights to preclinical utility. *Cell Cycle* **7:** 2647–2653.

Bouwman P, Jonkers J. 2012. The effects of deregulated DNA damage signalling on cancer chemotherapy response and resistance. *Nat Rev Cancer* **12:** 587–598.

Bouwman P, Aly A, Escandell JM, Pieterse M, Bartkova J, van der Gulden H, Hiddingh S, Thanasoula M, Kulkarni A, Yang Q, et al. 2010. 53BP1 loss rescues BRCA1 deficiency and is associated with triple-negative and BRCA-mutated breast cancers. *Nat Struct Mol Biol* **17:** 688–695.

Bryant HE, Schultz N, Thomas HD, Parker KM, Flower D, Lopez E, Kyle S, Meuth M, Curtin NJ, Helleday T. 2005. Specific killing of BRCA2-deficient tumours with inhibitors of poly(ADP-ribose) polymerase. *Nature* **434:** 913–917.

Bunting SF, Callen E, Wong N, Chen HT, Polato F, Gunn A, Bothmer A, Feldhahn N, Fernandez-Capetillo O, Cao L, et al. 2010. 53BP1 inhibits homologous recombination in *Brca1*-deficient cells by blocking resection of DNA breaks. *Cell* **141:** 243–254.

Callen E, Di Virgilio M, Kruhlak MJ, Nieto-Soler M, Wong N, Chen HT, Faryabi RB, Polato F, Santos M, Starnes LM, et al. 2013. 53BP1 mediates productive and mutagenic DNA repair through distinct phosphoprotein interactions. *Cell* **153:** 1266–1280.

Cancer Genome Atlas Network. 2012. Comprehensive molecular portraits of human breast tumours. *Nature* **490:** 61–70.

Cardoso F, van't Veer LJ, Bogaerts J, Slaets L, Viale G, Delaloge S, Pierga JY, Brain E, Causeret S, DeLorenzi M, et al. 2016. 70-gene signature as an aid to treatment decisions in early-stage breast cancer. *N Engl J Med* **375:** 717–729.

Chapman JR, Barral P, Vannier JB, Borel V, Steger M, Tomas-Loba A, Sartori AA, Adams IR, Batista FD, Boulton SJ. 2013. RIF1 is essential for 53BP1-dependent nonhomologous end joining and suppression of DNA double-strand break resection. *Mol Cell* **49:** 858–871.

Chaudhuri AR, Callen E, Ding X, Gogola E, Duarte AA, Lee JE, Wong N, Lafarga V, Calvo JA, Panzarino NJ, et al. 2016. Replication fork stability confers chemoresistance in BRCA-deficient cells. *Nature* **535:** 382–387.

Ciriello G, Gatza ML, Beck AH, Wilkerson MD, Rhie SK, Pastore A, Zhang H, McLellan M, Yau C, Kandoth C, et al. 2015. Comprehensive molecular portraits of invasive lobular breast cancer. *Cell* **163:** 506–519.

Clevers H. 2016. Modeling development and disease with organoids. *Cell* **165:** 1586–1597.

Collier LS, Carlson CM, Ravimohan S, Dupuy AJ, Largaespada DA. 2005. Cancer gene discovery in solid tumours using transposon-based somatic mutagenesis in the mouse. *Nature* **436:** 272–276.

Deeks ED. 2015. Olaparib: First global approval. *Drugs* **75:** 231–240.

Derksen PW, Liu X, Saridin F, van der Gulden H, Zevenhoven J, Evers B, van Beijnum JR, Griffioen AW, Vink J, Krimpenfort P, et al. 2006. Somatic inactivation of E-cadherin and p53 in mice leads to metastatic lobular mammary carcinoma through induction of anoikis resistance and angiogenesis. *Cancer Cell* **10:** 437–449.

Derksen PW, Braumuller TM, van der Burg E, Hornsveld M, Mesman E, Wesseling J, Krimpenfort P, Jonkers J. 2011. Mammary-specific inactivation of E-cadherin and p53 impairs functional gland development and leads to pleomorphic invasive lobular carcinoma in mice. *Dis Model Mech* **4:** 347–358.

Desmedt C, Zoppoli G, Gundem G, Pruneri G, Larsimont D, Fornili M, Fumagalli D, Brown D, Rothe F, Vincent D, et al. 2016. Genomic characterization of primary invasive lobular breast cancer. *J Clin Oncol* **34:** 1872–1881.

Di Virgilio M, Callen E, Yamane A, Zhang W, Jankovic M, Gitlin AD, Feldhahn N, Resch W, Oliveira TY, Chait BT, et al. 2013. Rif1 prevents resection of DNA breaks and promotes immunoglobulin class switching. *Science* **339:** 711–715.

Drost R, Bouwman P, Rottenberg S, Boon U, Schut E, Klarenbeek S, Klijn C, van der Heijden I, van der Gulden H, Wientjens E, et al. 2011. BRCA1 RING function is essential for tumor suppression but dispensable for therapy resistance. *Cancer Cell* **20:** 797–809.

Drost R, Dhillon KK, van der Gulden H, van der Heijden I, Brandsma I, Cruz C, Chondronasiou D, Castroviejo-Bermejo M, Boon U, Schut E, et al. 2016. *BRCA1*^{185delAG} tumors may acquire therapy resistance through expression of RING-less BRCA1. *J Clin Invest* **126:** 2903–2918.

Droufakou S, Deshmane V, Roylance R, Hanby A, Tomlinson I, Hart IR. 2001. Multiple ways of silencing E-cadherin gene expression in lobular carcinoma of the breast. *Int J Cancer* **92:** 404–408.

Dupuy AJ, Akagi K, Largaespada DA, Copeland NG, Jenkins NA. 2005. Mammalian mutagenesis using a highly mobile somatic *Sleeping Beauty* transposon system. *Nature* **436:** 221–226.

Escribano-Diaz C, Orthwein A, Fradet-Turcotte A, Xing M, Young JT, Tkac J, Cook MA, Rosebrock AP, Munro M, Canny MD, et al. 2013. A cell cycle–dependent regulatory circuit composed of 53BP1-RIF1 and BRCA1-CtIP controls DNA repair pathway choice. *Mol Cell* **49:** 872–883.

Evers B, Jonkers J. 2006. Mouse models of BRCA1 and BRCA2 deficiency: Past lessons, current understanding and future prospects. *Oncogene* **25:** 5885–5897.

Evers B, Schut E, van der Burg E, Braumuller TM, Egan DA, Holstege H, Edser P, Adams DJ, Wade-Martins R, Bouwman P, et al. 2010. A high-throughput pharmaceutical screen identifies compounds with specific toxicity against BRCA2-deficient tumors. *Clin Cancer Res* **16:** 99–108.

Farmer H, McCabe N, Lord CJ, Tutt AN, Johnson DA, Richardson TB, Santarosa M, Dillon KJ, Hickson I, Knights C, et al. 2005. Targeting the DNA repair defect in BRCA mutant cells as a therapeutic strategy. *Nature* **434:** 917–921.

Fatehullah A, Tan SH, Barker N. 2016. Organoids as an in vitro model of human development and disease. *Nat Cell Biol* **18:** 246–254.

Hidalgo M, Amant F, Biankin AV, Budinska E, Byrne AT, Caldas C, Clarke RB, de Jong S, Jonkers J, Maelandsmo GM, et al. 2014. Patient-derived xenograft models: An emerging platform for translational cancer research. *Cancer Discov* **4:** 998–1013.

Holstege H, van Beers E, Velds A, Liu X, Joosse SA, Klarenbeek S, Schut E, Kerkhoven R, Klijn CN, Wessels LF, et al. 2010. Cross-species comparison of aCGH data from mouse and human BRCA1- and BRCA2-mutated breast cancers. *BMC Cancer* **10:** 455.

Huijbers IJ, Bin Ali R, Pritchard C, Cozijnsen M, Kwon MC, Proost N, Song JY, de Vries H, Badhai J, Sutherland K, et al. 2014. Rapid target gene validation in complex cancer mouse models using re-derived embryonic stem cells. *EMBO Mol Med* **6:** 212–225.

Jaspers JE, Rottenberg S, Jonkers J. 2009. Therapeutic options for triple-negative breast cancers with defective homologous recombination. *Biochim Biophys Acta* **1796:** 266–280.

Jaspers JE, Kersbergen A, Boon U, Sol W, van Deemter L, Zander SA, Drost R, Wientjens E, Ji J, Aly A, et al. 2013. Loss of 53BP1 causes PARP inhibitor resistance in *Brca1*-mutated mouse mammary tumors. *Cancer Discov* **3:** 68–81.

Jonkers J, Meuwissen R, van der Gulden H, Peterse H, van der Valk M, Berns A. 2001. Synergistic tumor suppressor activity of BRCA2 and p53 in a conditional mouse model for breast cancer. *Nat Genet* **29:** 418–425.

Ledermann J, Harter P, Gourley C, Friedlander M, Vergote I, Rustin G, Scott C, Meier W, Shapira-Frommer R, Safra T, et al. 2012. Olaparib maintenance therapy in platinum-sensitive relapsed ovarian cancer. *N Engl J Med* **366:** 1382–1392.

Ledermann J, Harter P, Gourley C, Friedlander M, Vergote I, Rustin G, Scott CL, Meier W, Shapira-Frommer R, Safra T, et al. 2014. Olaparib maintenance therapy in patients with platinum-sensitive relapsed serous ovarian cancer: A preplanned retrospective analysis of outcomes by BRCA status in a randomised phase 2 trial. *Lancet Oncol* **15:** 852–861.

Liu X, Holstege H, van der Gulden H, Treur-Mulder M, Zevenhoven J, Velds A, Kerkhoven RM, van Vliet MH, Wessels LF, Peterse JL, et al. 2007. Somatic loss of BRCA1 and p53 in mice induces mammary tumors with features of human *BRCA1*-mutated basal-like breast cancer. *Proc Natl Acad Sci* **104:** 12111–12116.

Maddalo D, Manchado E, Concepcion CP, Bonetti C, Vidigal JA, Han YC, Ogrodowski P, Crippa A, Rekhtman N, de Stanchina E, et al. 2014. In vivo engineering of oncogenic chromosomal rearrangements with the CRISPR/Cas9 system. *Nature* **516:** 423–427.

Martinez V, Azzopardi JG. 1979. Invasive lobular carcinoma of the breast: Incidence and variants. *Histopathology* **3:** 467–488.

Michaut M, Chin SF, Majewski I, Severson TM, Bismeijer T, de Koning L, Peeters JK, Schouten PC, Rueda OM, Bosma AJ, et al. 2016. Integration of genomic, transcriptomic and proteomic data identifies two biologically distinct subtypes of invasive lobular breast cancer. *Sci Rep* **6:** 18517.

Moll R, Mitze M, Frixen UH, Birchmeier W. 1993. Differential loss of E-cadherin expression in infiltrating ductal and lobular breast carcinomas. *Am J Pathol* **143:** 1731–1742.

Nik-Zainal S, Davies H, Staaf J, Ramakrishna M, Glodzik D, Zou X, Martincorena I, Alexandrov LB, Martin S, Wedge DC, et al. 2016. Landscape of somatic mutations in 560 breast cancer whole-genome sequences. *Nature* **534:** 47–54.

Oplustil O'Connor L, Rulten SL, Cranston AN, Odedra R, Brown H, Jaspers JE, Jones L, Knights C, Evers B, Ting A, et al. 2016. The PARP inhibitor AZD2461 provides insights into the role of PARP3 inhibition for both synthetic lethality and tolerability with chemotherapy in preclinical models. *Cancer Res* doi: 10.1158/0008-5472.CAN-15-3240.

Patch AM, Christie EL, Etemadmoghadam D, Garsed DW, George J, Fereday S, Nones K, Cowin P, Alsop K, Bailey PJ, et al. 2015. Whole-genome characterization of chemoresistant ovarian cancer. *Nature* **521:** 489–494.

Perou CM, Sorlie T, Eisen MB, van de Rijn M, Jeffrey SS, Rees CA, Pollack JR, Ross DT, Johnsen H, Akslen LA, et al. 2000.

Molecular portraits of human breast tumours. *Nature* **406:** 747–752.

Platt RJ, Chen S, Zhou Y, Yim MJ, Swiech L, Kempton HR, Dahlman JE, Parnas O, Eisenhaure TM, Jovanovic M, et al. 2014. CRISPR–Cas9 knockin mice for genome editing and cancer modeling. *Cell* **159:** 440–455.

Rottenberg S, Nygren AO, Pajic M, van Leeuwen FW, van der Heijden I, van de Wetering K, Liu X, de Visser KE, Gilhuijs KG, van Tellingen O, et al. 2007. Selective induction of chemotherapy resistance of mammary tumors in a conditional mouse model for hereditary breast cancer. *Proc Natl Acad Sci* **104:** 12117–12122.

Rottenberg S, Jaspers JE, Kersbergen A, van der Burg E, Nygren AO, Zander SA, Derksen PW, de Bruin M, Zevenhoven J, Lau A, et al. 2008. High sensitivity of BRCA1-deficient mammary tumors to the PARP inhibitor AZD2281 alone and in combination with platinum drugs. *Proc Natl Acad Sci* **105:** 17079–17084.

Sánchez-Rivera FJ, Jacks T. 2015. Applications of the CRISPR–Cas9 system in cancer biology. *Nat Rev Cancer* **15:** 387–395.

Sander JD, Joung JK. 2014. CRISPR–Cas systems for editing, regulating and targeting genomes. *Nat Biotechnol* **32:** 347–355.

Schouten PC, Marme F, Aulmann S, Sinn HP, van Essen HF, Ylstra B, Hauptmann M, Schneeweiss A, Linn SC. 2015. Breast cancers with a BRCA1-like DNA copy number profile recur less often than expected after high-dose alkylating chemotherapy. *Clin Cancer Res* **21:** 763–770.

Stephens PJ, Tarpey PS, Davies H, Van Loo P, Greenman C, Wedge DC, Nik-Zainal S, Martin S, Varela I, Bignell GR, et al. 2012. The landscape of cancer genes and mutational processes in breast cancer. *Nature* **486:** 400–404.

Swisher EM, Sakai W, Karlan BY, Wurz K, Urban N, Taniguchi T. 2008. Secondary BRCA1 mutations in BRCA1-mutated ovarian carcinomas with platinum resistance. *Cancer Res* **68:** 2581–2586.

Tao L, van Bragt MP, Laudadio E, Li Z. 2014. Lineage tracing of mammary epithelial cells using cell-type-specific cre-expressing adenoviruses. *Stem Cell Reports* **2:** 770–779.

Ter Brugge P, Kristel P, van der Burg E, Boon U, de Maaker M, Lips E, Mulder L, de Ruiter J, Moutinho C, Gevensleben H, et al. 2016. Mechanisms of therapy resistance in patient-derived xenograft models of BRCA1-deficient breast cancer. *J Natl Cancer Inst* **108:** djw148.

Vaidyanathan A, Sawers L, Gannon AL, Chakravarty P, Scott AL, Bray SE, Ferguson MJ, Smith G. 2016. ABCB1 (MDR1) induction defines a common resistance mechanism in paclitaxel- and olaparib-resistant ovarian cancer cells. *Br J Cancer* **115:** 431–441.

Vargo-Gogola T, Rosen JM. 2007. Modelling breast cancer: One size does not fit all. *Nat Rev Cancer* **7:** 659–672.

Vollebergh MA, Lips EH, Nederlof PM, Wessels LF, Schmidt MK, van Beers EH, Cornelissen S, Holtkamp M, Froklage FE, de Vries EG, et al. 2011. An aCGH classifier derived from BRCA1-mutated breast cancer and benefit of high-dose platinum-based chemotherapy in HER2-negative breast cancer patients. *Ann Oncol* **22:** 1561–1570.

Vollebergh MA, Jonkers J, Linn SC. 2012. Genomic instability in breast and ovarian cancers: Translation into clinical predictive biomarkers. *Cell Mol Life Sci* **69:** 223–245.

Vollebergh MA, Lips EH, Nederlof PM, Wessels LF, Wesseling J, Vd Vijver MJ, de Vries EG, van Tinteren H, Jonkers J, Hauptmann M, et al. 2014. Genomic patterns resembling *BRCA1*- and *BRCA2*-mutated breast cancers predict benefit of intensified carboplatin-based chemotherapy. *Breast Cancer Res* **16:** R47.

Vos CB, Cleton-Jansen AM, Berx G, de Leeuw WJ, ter Haar NT, van Roy F, Cornelisse CJ, Peterse JL, van de Vijver MJ. 1997. E-cadherin inactivation in lobular carcinoma in situ of the breast: An early event in tumorigenesis. *Br J Cancer* **76:** 1131–1133.

Wang H, Yang H, Shivalila CS, Dawlaty MM, Cheng AW, Zhang F, Jaenisch R. 2013. One-step generation of mice carrying mutations in multiple genes by CRISPR/Cas-mediated genome engineering. *Cell* **153:** 910–918.

Wang J, Aroumougame A, Lobrich M, Li Y, Chen D, Chen J, Gong Z. 2014. PTIP associates with Artemis to dictate DNA repair pathway choice. *Genes Dev* **28:** 2693–2698.

Wang D, Mou H, Li S, Li Y, Hough S, Tran K, Li J, Yin H, Anderson DG, Sontheimer EJ, et al. 2015. Adenovirus-mediated somatic genome editing of Pten by CRISPR/Cas9 in mouse liver in spite of Cas9-specific immune responses. *Hum Gene Ther* **26:** 432–442.

Xu G, Chapman JR, Brandsma I, Yuan J, Mistrik M, Bouwman P, Bartkova J, Gogola E, Warmerdam D, Barazas M, et al. 2015. REV7 counteracts DNA double-strand break resection and affects PARP inhibition. *Nature* **521:** 541–544.

Zimmermann M, Lottersberger F, Buonomo SB, Sfeir A, de Lange T. 2013. 53BP1 regulates DSB repair using Rif1 to control 5′ end resection. *Science* **339:** 700–704.

Tumor–Stroma Interactions in Bone Metastasis: Molecular Mechanisms and Therapeutic Implications

Hanqiu Zheng,[1,2] Wenyang Li,[1] and Yibin Kang[1]
[1]Department of Molecular Biology, Princeton University, Princeton, New Jersey 08544
Correspondence: ykang@princeton.edu

Metastasis and associated complications are the major cause of death for cancer patients. The incidence of bone metastasis is among the highest in cancers arising from breast, prostate, and lung. Common skeletal-related events caused by bone metastasis include aberrant bone remodeling (osteolytic, osteoblastic, and mixed), bone pain, fracture, spinal cord compression, and life-threatening hypercalcemia. It is now known that interactions between tumor cells and bone stroma lie at the core of major steps of bone-metastasis progression. Approved pharmaceutical drugs for the treatment of bone metastasis, including bisphosphonate and denosumab, were designed to target bone stromal cell components. In recent years, research in our laboratory and others has revealed intricate tumor–stroma interactions as the driving force behind osteolytic bone-metastasis development, providing a set of new candidates for future drug development. Moreover, recent studies also indicate existence of distinct bone niches in supporting hematopoietic stem cell renewal and differentiation. These niche components are likely utilized by metastatic cancer cells for seeding, progression, and therapy resistance of bone metastasis. Future studies in this direction may discover additional therapeutic targets for bone-metastasis treatment.

Metastasis is responsible for most of cancer-related morbidity and mortality (Wan et al. 2013). Bone is one of the prominent organs that suffered from the metastatic spread of cancer. In breast and prostate cancers, it is estimated that >70% of late-stage patients develop skeletal metastasis (Mundy 2002; Weilbaecher et al. 2011). Bone is also extensively involved in many malignancies originating from hematopoietic cells, such as multiple myeloma and lymphomas (Melton et al. 2005; Wu et al. 2014). The common skeletal-related events (SREs) for bone metastasis include aberrant bone remodeling (osteolytic, osteoblastic, and mixed bone metastasis), bone pain, fracture, spinal cord compression, and life-threatening hypercalcemia (Mundy 2002; Weilbaecher et al. 2011; Ren et al. 2015). The initial clinical presentation and diagnosis for bone metastasis, including increased calcium and alkaline phosphatase levels and pain, could be varying for patients. The diagnosis and confirmation for bone metastasis involve regular X-ray imaging, bone scintillation, biopsy, and advanced medical imaging such as computerized tomography (CT), magnetic resonance imaging (MRI), and positron emission tomography (PET).

Once bone metastasis is confirmed, the goals of therapeutic treatments and management are to alleviate bone compression and bone pain, maintain intact bone structure and function, and slow down the metastatic tumor progression, although it is unlikely that the tumor cells can be completely eliminated from the bone. To achieve these goals, a set of treatment options can be personalized to meet the specific needs of each patient. For example, pain-relieving medicine/analgesia is commonly administrated to patients along with other systematic or localized treatments, including osteoclast inhibitors (bisphosphonate and denosumab), hormone therapy, chemotherapy, localized radiation, bone-targeting radioactive pharmaceuticals, and surgeries for patients with high risks of developing fractures. Most of these options are generic treatments for killing tumor cells or maintaining the quality of life for patients. Only bisphosphonate and denosumab are specifically designed for treating bone metastasis based on the molecular mechanisms of bone-metastasis development—by blocking osteolytic activity of the bone.

The exact molecular mechanisms underlying bone-metastasis progression could be varying in different cancer types. However, we now know that tumor–stroma interaction in the bone is critical for bone metastasis. In normal bone homeostasis, bone-building osteoblasts and bone-degrading osteoclasts keep their delicate balance to maintain the normal healthy bone structure. This balance is in part controlled by the secretion of an appropriate level of RANKL (receptor activator of nuclear factor κB ligand) by osteoblast, which signals to RANK receptor in preosteoclasts to promote their maturation (Fig. 1). However, when bone metastasis develops, tumor cells often break the balance between the two bone cell types by signaling interaction with osteoblasts and osteoclasts. For example, in osteolytic bone metastasis of breast cancer, tumor cells can increase RANKL level in the microenvironment either by directly secreting RANKL or by inducing RANKL production by osteoblasts. Bone

[2]Present address: Department of Basic Medical Sciences, School of Medicine, Tsinghua University, Beijing, China

Published by Cold Spring Harbor Laboratory Press; doi: 10.1101/sqb.2016.81.030775

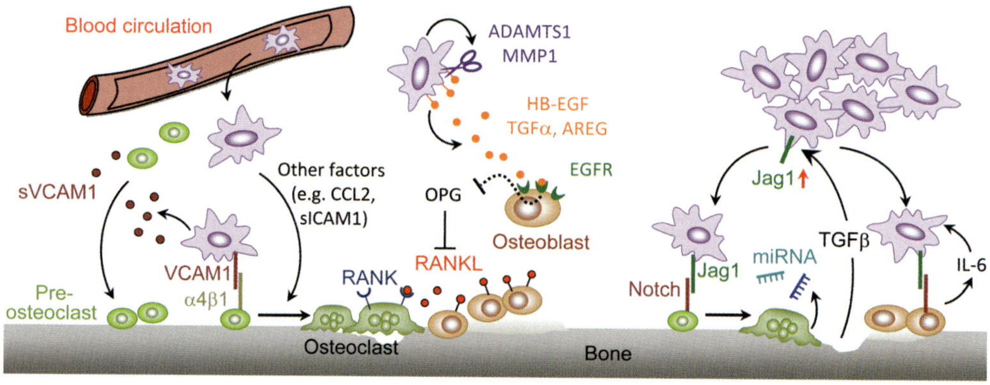

Figure 1. Tumor–stroma interactions in bone metastasis. Key pathways uncovered by our laboratory are highlighted here, including (1) VCAM1-activated osteolytic expansion of indolent bone micrometastasis (*left*); (2) osteolytic paracrine signaling cascade initiated by matrix metalloproteinases (MMPs, *middle*); (3) a positive-feedback loop in bone metastasis mediated by Jagged1/Notch and transforming growth factor β (TGF-β) signaling pathways (*right*). See text for details. sVCAM1, soluble vascular cell adhesion molecule 1; CCL2, chemokine (C–C motif) ligand 2; sICAM1, soluble intercellular adhesion molecule 1; IL-11, interleukin 1; CTGF, connective tissue growth factor; ADAMTS1, ADAM metallopeptidase with thrombospondin type 1 motif 1; HB-EGF, heparin-binding epidermal growth factor-like growth factor; AREG, amphiregulin; EGFR, epidermal growth factor receptor; OPG, osteoprotegerin; RANK, receptor activator of nuclear factor κB;, RANKL, receptor activator of nuclear factor κB ligand; Jag1, Jagged1; miRNA, microRNA; IL-6, interleukin 6. (Reprinted, with permission, from Kang 2016.)

resorption by osteoclasts also leads to the release of many bone-derived growth factors and cytokines that feed back to tumor cells to promote tumor cell survival and growth in the bone. This extremely simplified model exemplified a powerful feed-forward loop of bone degradation and tumor growth process termed the "vicious cycle of bone metastasis" (Mundy 2002; Weilbaecher et al. 2011; Ell and Kang 2012). Bisphosphonate, by binding to bone surface and inhibiting osteoclast activity, protects bone from degradation. Similarly, denosumab, a RANKL-neutralizing antibody, blocks the binding of RANKL to its receptor on preosteoclasts, thus reducing the number of mature osteoclasts and inhibiting bone resorption (Baron et al. 2011).

Recent studies in the field, including work done in our laboratory, revealed multiple layers of tumor–stroma interactions in bone metastasis of breast cancer. Our study revealed multiple molecular and signaling pathways promoting bone-metastasis progression, including matrix metalloproteinase 1 (MMP1) and a disintregin and metalloproteinase with thrombospondin motifs 1 (ADAMTS1), vascular cell adhesion molecule 1 (VCAM1), transforming growth factor β (TGF-β) pathway and Jagged1–NOTCH pathway, intercellular adhesion molecule 1 (ICAM1), and microRNAs (Fig. 1). These results provide important candidate molecules as targets for treating bone metastasis.

METASTASIS TO BONE IS DRIVEN BY A UNIQUE REPERTOIRE OF BONE-METASTASIS GENES

The development and widespread application of the genomic profiling method at the turn of the century opened up the possibility of search for organotropic metastasis genes in a comprehensive and unbiased way. However, there was a very limited number of breast cancer cell lines capable of generating bone metastasis reliably in mouse models. The Massagué group applied the in vivo selection strategy to derive highly bone-metastatic variants from the heterogeneous MDA-MB-231 human breast cancer cell line that was established from the pleural effusion of a metastatic breast cancer patient (Fig. 2A; Cailleau et al. 1978). Tumor cells were injected into the left ventricle of nude mice to generate bone metastasis. When bone metastasis could be detected by X-ray imaging, tumor cells were isolated from bone lesions for in vitro expansion (Kang et al. 2003). The process can be reiterated for further enrichment of bone-metastatic variants and to quantify the metastatic efficiency of each subline. Using the parental MDA-MB-231 and 12 subpopulations with differential bone-metastatic potentials, a gene signature was generated that was linked to bone-metastatic potential, including 43 up-regulated genes and 59 down-regulated genes. Interestingly, the majority of the mostly highly enriched genes encode membrane or secretory proteins, including chemokine receptor CXCR4 for bone-homing, angiogenic factor fibroblast growth factor-5 (FGF5), osteoclast stimulating factor interleukin-11 (IL-11), and osteopontin (OPN), as well as MMP1 and ADAMTS1 (Fig. 2B). This specific pattern of gene enrichment associated with bone-metastatic ability indicates that tumor–stroma interactions mediated by secreted factors are likely to be critical for tumor cells to survive and thrive in the bone microenvironment. A few of these candidate molecules have indeed been functionally validated to be important for bone metastasis.

An interesting question is how do these sublines with various bone-metastatic abilities emerge from in vivo selection? Do they preexist in the original cancer cell line because of tumor heterogeneity, or do they evolve and acquire the bone tropism de novo during in vivo

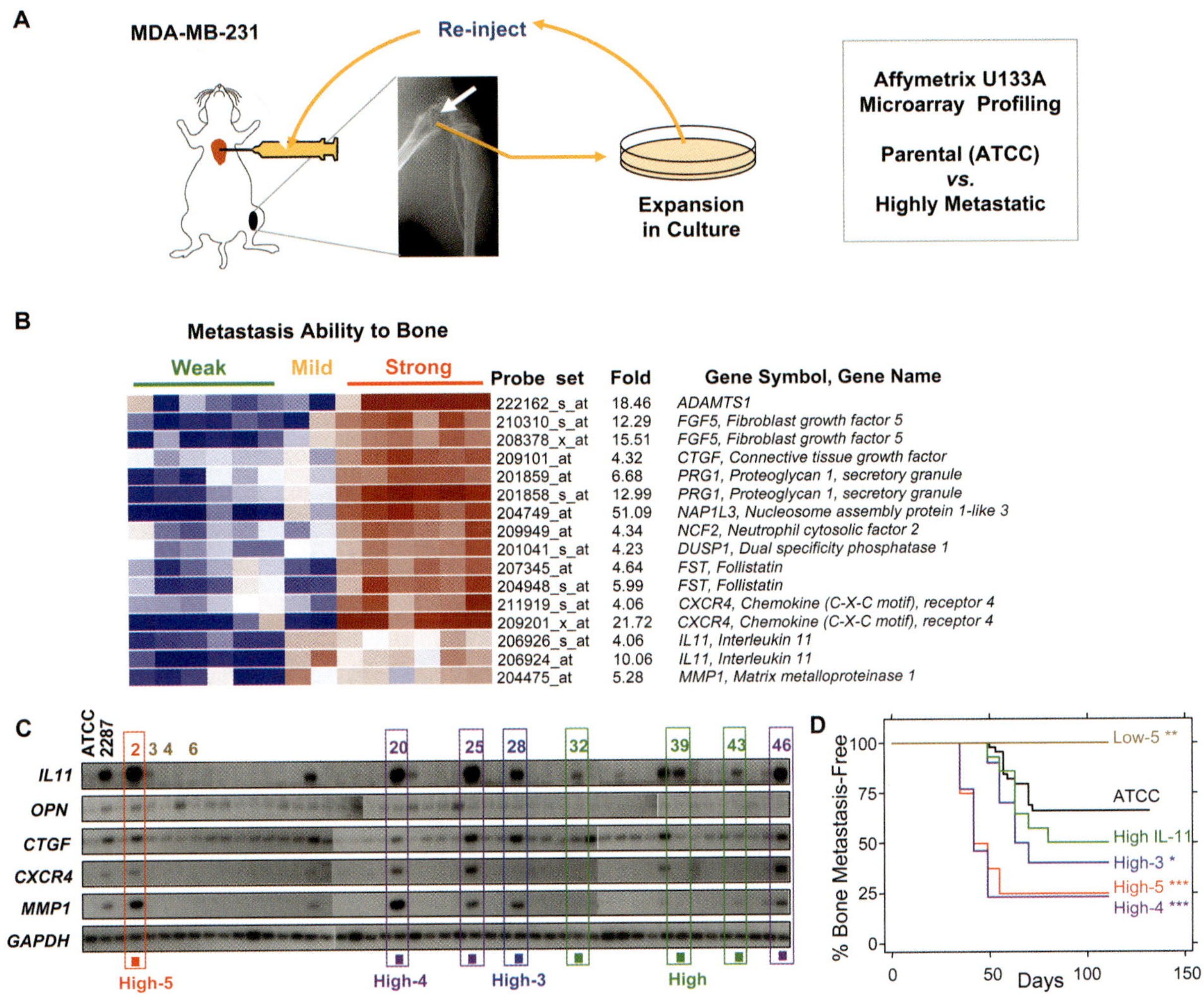

Figure 2. Discover bone-metastatic genes. (*A*) Procedures of in vivo selection for bone-metastatic cells. Parental MDA-MB-231 cells were inoculated into the left cardiac ventricle of nude mice. Tumor cells were isolated and expanded from X-ray imaging–confirmed bone lesions and reselected for another round. Reisolated cells were expanded and injected into nude mice to gauge their metastatic potentials. (*B*) RNAs from these generated cell lines were subjected to gene expression profiling analysis to identify candidate metastasis-enhancing genes. A subset of hierarchical cluster of 11 genes (with 16 probes) from the original 127 probe sets, which differed by more than fourfold in expression level between weak and strong metastatic groups. (*C*) Randomly picked single clones from the parental MDA-MB-231 population were expanded in culture, and subjected to northern blot analysis for five bone-metastasis genes. (*D*) Kaplan–Meier plot displays the bone-metastatic incidence of the parental MDA-MB-231 population and the indicated single-cell progeny clones. *, $P < 0.05$; ***, $P < 0.001$, compared with parental cells. (Adapted from Kang et al. 2003, with permission from Elsevier.)

selection in mice? The preexisting model seems to be more realistic as the selection time in the mouse is relatively short (6–10 wk). To confirm this, 46 single-cell-derived clonal populations (termed SCP for "single-cell progenies") were isolated in vitro from parental MDA-MB-231 cells and the expression level of five major bone-metastasis signature genes were measured. Although most SCPs do not express any of the bone-metastasis genes, a few SCPs express various numbers of the five genes at high levels (Fig. 2C). Remarkably, the bone-metastatic potential of these SCPs correlated well with their different degree of metastasis gene expression (Fig. 2D). This result not only confirmed that high-metastatic variants preexist in the heterogeneous population of the parental MDA-MB-231 cell line but also underscored the cooperative nature of metastasis genes in promoting the metastatic behavior of cancer cells.

ADAMTS1 AND MMP1 SYNERGISTICALLY PROMOTE METASTASIS BY ALTERING CYTOKINE PRODUCTION IN THE MICROENVIRONMENT

With the identification of these secreted proteins and cell surface proteins in highly bone-metastatic cell lines, we started to systematically evaluate the functional importance of these individual genes in the bone metastasis signature and explore the therapeutic options for targeting these genes. We first focused our analysis on two metalloproteinases, MMP1 and ADAMTS1 (Lu et al. 2009), which are among the two highest expression genes in the bone-metastasis gene signature (Fig. 2B). The expression of MMP1 and ADAMTS1 were knocked down separately or in combination by short-hairpin RNAs (shRNAs) in SCP20 cells, a highly bone-metastatic variant of MDA-MB-231 (Fig. 2C,D). These cells were then intracardially (IC) injected into nude mice to generate bone metastasis. Interestingly, single knockdown of either MMP1 or ADAMTS1 did not show any bone-metastasis reduction. However, combined knockdown of both genes led to a drastic reduction in bone-metastasis progression (Fig. 3A), as showed by in vivo bioluminescence imaging (BLI) (Fig. 3B). X-ray imaging and tartrate-resistant acid phosphatase (TRAP) staining for osteoclast cells revealed that reduced bone metastasis is accompanied by smaller osteolytic area and decreased TRAP-positive osteoclasts (Fig. 3B). Using in vitro assays for osteoclast differentiation from primary bone marrow cells, we concluded that MMP1 and ADAMTS1 altered the production of secreted growth factors from tumor cells that signal to osteoblasts to reduce the expression of osteoprotegerin

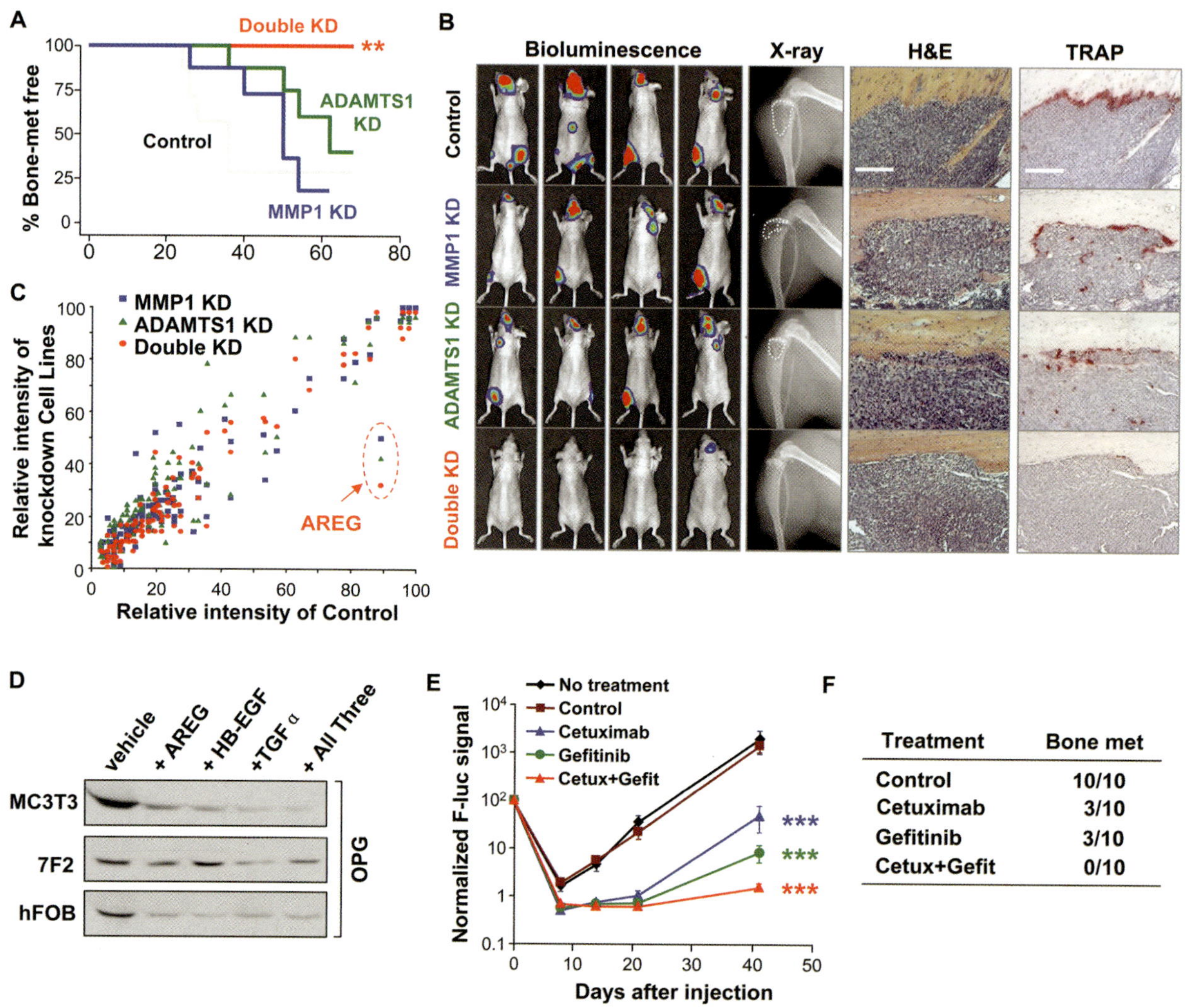

Figure 3. Combined knockdown (KD) of ADAMTS1 and MMP1 inhibits bone metastasis. (*A*) Kaplan–Meier plot indicates the incidence of bone metastases after intracardiac injection of various cell lines in immune compromised nude mice, including SCP20 stably transduced with control short-hairpin RNA (shRNA) or shRNAs targeting *MMP1*, *ADAMTS1*, or both. $n = 10$. **, $P < 0.01$, based on log-rank test. (*B*) Representative bioluminescence, radiographic, and histological (H&E and TRAP staining) images from each group at day 50 postinjection. (*C*) A human cytokine array kit was used to determine the differentially expressed cytokines in the conditional media from either MMP1 or/and ADAMTS1 knockdown or control SCP20 cells. AREG was the most down-regulated protein in knockdown cells. (*D*) Treatment of recombinant AREG, HB-EGF, and TGF-α individually or in combination in three different osteoblast cells repressed OPG expression as measured by immunoblotting. (*E*) In the SCP20 bone-metastasis model, single or combined treatment of cetuximab and gefitinib decreased bone-metastasis progression as measured by bioluminescence. (*F*) Single or combined treatment of cetuximab and gefitinib decreased the incidence of bone metastasis in nude mice. ADAMST1, a disintegrin and metalloproteinase with thrombospondin motifs 1; MMP1, matrix metalloproteinase 1; H&E, hematoxylin and eosin; TRAP, tartrate-resistant acid phosphatase; HB-EGF, heparin-binding epidermal growth factor; TGF-α, transforming growth factor α; hFOB, human fetal osteoblastic; OPG, osteoprotegerin. (Adapted, with permission, from Lu et al. 2009, Cold Spring Harbor Laboratory Press.)

(OPG), a natural decoy receptor and inhibitor of RANKL. Through cytokine array profiling, we observed a significant reduction of Amphiregulin (AREG) in the conditional media from MMP1 and/or ADAMTS1 knockdown cells, with the most profound reduction in double knockdown cells (Fig. 3C). AREG belongs to the epidermal growth factor (EGF) family ligands, and its membrane form can be proteolytically released by MMP1 and ADAMTS1 into the microenvironment for paracrine signaling. Recombinant AREG and its related family members were able to reduce OPG expression in osteoblasts (Fig. 3D) and subsequently boost osteoclast differentiation. These results thus underscored the potential application of EGF receptor (EGFR) inhibitors in bone-metastasis treatment. To this end, we performed preclinical testing of EGFR inhibitors, cetuximab and gefitinib, in bone-metastasis models. Indeed, treatment with either cetuximab or gefitinib or both in combination significantly reduced bone metastasis incidence and metastatic tumor burden. (Fig. 3E,F). These studies presented proof of concept that inhibiting EGFR signaling in tumor–stroma cross talk using a clinically approved EGFR inhibitor may reduce bone-metastasis progression.

TARGETING TGF-β AND Jagged1–Notch SIGNALING PATHWAYS IN BONE METASTASIS

TGF-β is one of the most abundant growth factors stored in the bone and has been suspected to be released during bone destruction to provide a feedback signal to cancer cells to promote their metastatic behavior. To directly determine whether TGF-β signaling is activated during bone metastasis, we developed a xenograft system in which we could detect TGF-β–SMAD pathway activation status while simultaneously monitoring tumor burden in the bone. In this system, tumor cells were transduced with a constitutive Renilla luciferase (R-Luc) expression cassette to monitor tumor burden and served as an internal control for normalizing tumor burden, whereas a TGF-β-responsive element was placed upstream of the firefly luciferase (F-Luc) coding sequence to provide a direct readout of TGF-β signaling intensity in cancer cells (Fig. 4A; Korpal et al. 2009). When the tumor cells were injected to nude mice via the left cardiac ventricle to form bone metastasis, the relative level of F-Luc/R-Luc gradually increased over the course of bone-metastasis progression with the development of osteolytic lesions, whereas no such increase was observed in nonosteolytic lesions (Fig. 4B,C). Furthermore, the increase of F-Luc activity could be blocked by treating the mice with bisphosphonate, which inhibits osteoclastogenesis and prevents osteolytic degradation of the bone. The result suggests that TGF-β pathway activation in tumor cells is enhanced by TGF-β released from osteolytic bone degradation. Blocking this feedback loop using TGF-β receptor kinase inhibitors thus could potentially be used to slow down bone-metastasis development. Interestingly, although early treatment of mice with TGF-β inhibitors can significantly reduce bone-metastasis burden, late treatment of established lesions can reduce TGF-β signaling activity but did not result in a reduction of tumor burden. These findings highlight the notion that the therapeutic window is critical to achieve the optimal chemotherapeutic outcome of bone-metastasis-targeting agents.

We noticed the Notch ligand, Jagged1, is among the genes that is regulated by the TGF-β–Smad pathway and was also highly expressed in bone-metastatic breast cancer cell lines (Fig. 4D,E; Sethi et al. 2011). Furthermore, elevated expression of Jagged1 in primary breast tumors is associated with higher risk of bone metastasis in patients. The Notch pathway has been extensively studied for its roles in bone development (Engin et al. 2008; Hilton et al. 2008) and developmentally conserved pathways are well-known to be often hijacked by cancer cells to promote their malignant behaviors (Sethi and Kang 2011). Therefore, we explored whether the Jagged1–Notch pathway is functionally involved in breast cancer bone metastasis. shRNA-mediated stable knockdown of Jagged1 expression in highly metastatic breast cancer cell lines drastically reduced bone-metastasis burden (Fig. 4F,G). On the other hand, overexpression of Jagged1 in weakly bone-metastatic cells promoted bone-metastasis progression. A close examination of the bone tissues showed that Jagged1-overexpressing cells generated much larger bone lytic areas with increased mature osteoclast numbers in the bone (Fig. 4F,G).

To understand the molecular mechanism by which Jagged1 promotes bone metastasis, we decided to coculture tumor cells with either osteoblasts or preosteolcasts in vitro. We discovered that tumor-derived Jagged1 interacts with osteoblast cells to induce Notch activation and IL-6 production, which in turn feeds back to tumor cells to promote their proliferation (Fig. 4H,I). This feedback loop could be completely blocked by either knockdown of Hey1, a Notch downstream effector gene in osteoblasts, or by administration of γ-secretase inhibitor (GSI, MRK-003 from Merck) to block Notch activation in the coculture system (Fig. 4H). Jagged1 also promotes osteolytic bone metastasis by engaging Notch signaling in osteoclasts. Tumor-derived Jagged1 induces the osteoclastogenesis through which monocytic preosteoclasts fused to form large multinucleated, TRAP-positive osteoclasts within several days in cell culture. This osteoclast maturation process could be completely diminished by administration of the GSI (Fig. 4J). In vivo administration of GSI inhibitor also dramatically decreased bone-metastasis burden, osteolytic lesion area, and osteoclast cell number in the bone (Fig. 4K). Although it is now known that GSIs have severe adverse effects on the gastrointestinal track of the patients, future modification of the inhibitor could potentially alleviate some of these side effects. Moreover, the development of pharmaceutical agents specifically targeting Jagged1 while sparing other Notch ligand may achieve similar therapeutic benefit without the adverse side effects incurred by pan-Notch inhibition by GSI.

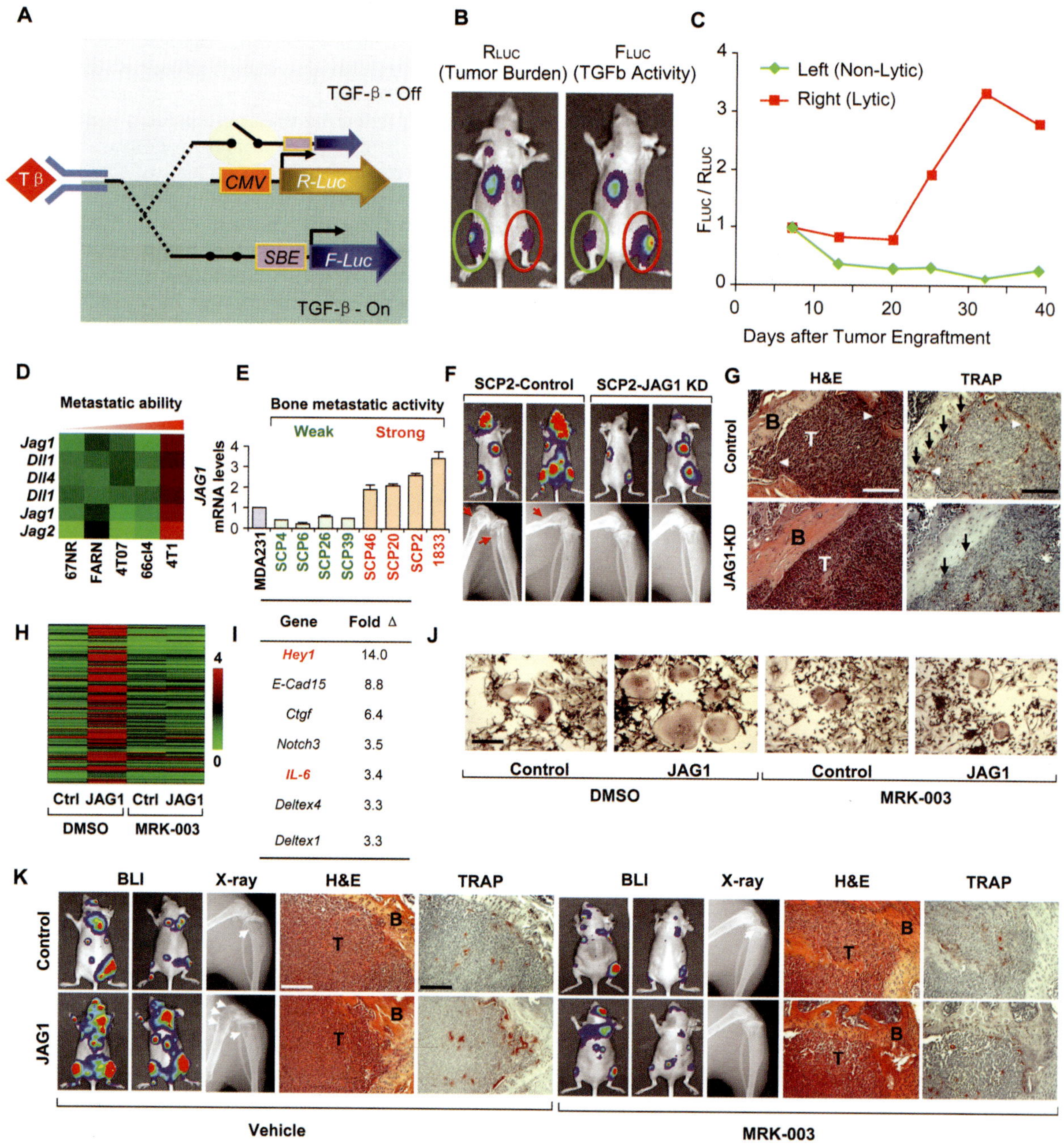

Figure 4. TGF-β and Jagged1–Notch pathway in bone-metastasis progression. (*A*) A schematic representation of the dual-luciferase reporter system to track tumor burden and TGF-β pathway activation and to conditional control of TGF-β–Smad signaling activity. (*B*) Osteolytic bone metastasis (*right* leg) displayed enhanced TGF-β pathway activation compared with the nonosteolytic lesion (*left* leg). (*C*) Quantification of normalized F-Luc/R-Luc intensity from mouse in *B*. (*D*) Heatmap of gene-expression profiling of Notch ligands in the 4T1 series with differential bone-metastatic potentials. (*E*) The mRNA levels of *Jagged1* in parental MDA231 cell and its derivative sub-cell lines with differential bone-metastatic potentials. (*F*) Control or Jagged1 knockdown SCP2 cells were intracardially injected into nude mice for bone metastasis. Metastasis progression was monitored by BLI and X-ray imaging. Red arrows indicate osteolytic bone lesions. (*G*) H&E and TRAP staining from experimental groups. Arrows indicate TRAP$^+$ osteoclasts. Arrowheads indicate areas of overt bone destruction. (*H*) Heatmap of microarray gene expression profiles of MC3T3-E1 osteoblasts that were sorted from cocultures of each experimental group (with coculture with control of Jagged1-overexpressing cancer cells, with or without α-secretase inhibitor [GSI] treatment). (*I*) A list of genes selected from *H* with expression levels greater than threefold in osteoblasts cocultured with Jagged1-overexpressing tumor cells versus with control tumor cells. (*J*) Control or Jagged1 overexpressing tumor cells were cocultured with Raw264.7 preosteoclast cells. Dimethylsulfoxide (DMSO) or GSI were immediately added into coculture. Representative TRAP staining images were shown from the experiment. (*K*) Control or Jagged1-overexpressing SCP28 cells were inoculated into nude mice by intracardiac injection and then treated with either control or GSI (MRK-003). BLI, X-ray, H&E, and TRAP staining images of bone lesions from representative mice in each experimental group were displayed on day 42. TGF-β, transforming growth factor β; R-Luc (R_{LUC}), Renilla luciferase; F-Luc (F_{LUC}), firefly luciferase; KD, knockdown; BLI, bioluminescence; H&E, hematoxylin and eosin; TRAP, tartrate-resistant acid phosphatase. (Adapted, with permission, from Korpal et al. 2009; Sethi et al. 2011.)

OSTEOCLAST microRNAs AS BIOMARKER AND REGULATOR OF BONE METASTASIS

Our previous studies mostly focused on protein-coding genes in bone metastasis. However, with better understanding of the human and mouse genomes, microRNAs, long noncoding RNAs, and other regulatory RNAs have been recognized for their important functional roles in many biological processes, including cancer progression and metastasis. Different from regular protein-coding mRNAs, microRNAs are very stable in the circulation, making microRNA-based biomarker and therapeutic development a promising direction in metastasis. Based on the central importance of osteoclasts in osteolytic bone metastasis, we started by investigating how tumor-derived factors affect miRNA expression during osteoclast maturation. We first used conditional media (CM) from highly bone-metastatic cancer cell lines 4T1.2 and TSu-Pr1-B2 to treat RAW264.7 preosteoclast cells at the limited level of RANKL at a concentration that is not sufficient to induce full osteoclastogenesis. CM treatment significantly promoted osteoclast differentiation when combined with a low level of RANKL (Ell et al. 2013). In contrast, CM from weakly bone-metastatic counterpart cell lines (4T1 and TSU-Pr1 cells) had no effect on osteoclast differentiation. Similar results were also observed in tumor-preosteoclast coculture assay. We performed miRNA microarray analysis of Raw264.7 cells treated with CM from either highly or weakly bone-metastatic cells. Among the detected microRNAs, 42 miRNAs were up-regulated and 45 miRNAs were down-regulated with a >2.2-fold change across treatment groups. Eventually, five down-regulated miRNAs with multiple osteoclast regulatory gene targets, including miR-33a-5p, miR-133a, miR-141-3p, miR-190, and miR-219-5p (hereafter miR-33a, miR-141, and miR-219), were confirmed to block osteoclastogenesis when transfected into preosteoclasts, and were further validated in the in vitro bone resorption assays. The strongest repressing effect was seen in miR-133a, miR-141, and miR-219-transfected cells, whereas the other two miRNAs showed mild inhibition. Nevertheless, we proceeded to evaluate all five miRNAs' effect on bone development and bone metastasis in vivo. First, we injected pre-miRNA via tail-vein injection to Balb/c mice weekly for 4 wk. The bone density of these mice was evaluated by weekly X-ray imaging and by micro-CT at 4 wk. These examinations revealed increased trabecular bone density, bone volume, and thickness in hind limb bones in miRNA treated mice, especially those injected with miR-141, mir-190, or miR-219 (Fig. 5A). These changes were associated with decreased TRAP-positive osteoclasts in the bone, whereas no obvious difference in osteoblast activity was observed. The same five miRNAs were subsequently analyzed for their possible effect on reducing bone metastasis. Treatment with MiR-141 or MiR-219 significantly decreased bone-metastasis burden in hind limbs and drastic reduction of TRAP-positive osteoclast number in the bone (Fig. 5B–E). Importantly, among miRNAs up-regulated during osteoclastogenesis, miR-16 and miR-378 can be used as serum biomarkers to detect bone metastasis in breast cancer patients. Our result supports the further development of miRNAs as biomarkers and therapeutic agents of bone metastasis.

BONE-METASTATIC NICHES—LOOKING BEYOND OSTEOCLASTOGENESIS

The new candidate therapeutic targets of bone metastasis described above were mainly derived from a better understanding of the molecular mechanisms for the "vicious cycle" of bone metastasis. Bone is a nurturing yet extremely complicated microenvironment for metastasis progression and contains many other stromal cells. In normal physiological conditions, bone is not only the structural framework to support the human body but also harbors most of the hematopoietic stem cells (HSCs) for the renewal and lineage differentiation to constitute the whole hematopoietic system (Morrison and Scadden 2014). Studies suggest there are multiple bone stromal cell types occupying specific regions, termed "niches," to support HSC formation. Accumulating evidence in recent research indicated that there are two major niches—the perivascular niche and the osteoblast niche (Calvi et al. 2003; Zhang et al. 2003). The osteoblast niche is localized at the inner surface of the bone cavity and potentially at the trabecular bone surfaces. Quiescent HSCs are initially proposed to reside in the osteoblast niche. However, recent studies suggest these HSCs are localized to the perivascular niche, which is composed of an endothelial cell, a mesenchymal stromal cell (MSC), and some neuronal cells. Many stem cell factors, such as CXCL12 and stem cell factor (SCF), were produced by a combination of these perivascular niche cells and to some lower extent by osteoblast niche cells (Ding et al. 2012; Greenbaum et al. 2013; Morrison and Scadden 2014). Notably, it has been reported that CXCL12 is a bone-derived chemokine that promotes the bone homing of metastatic tumor cells expressing its receptors with CXCR4 and CXCR7. (Muller et al. 2001; Wang et al. 2006).

As the seeds for distant metastasis, circulating and disseminated tumor cells face a hostile environment during the metastatic cascade, such as the sheer stress of the blood flow and the attacks from immune cells (Gay and Felding-Habermann 2011). When tumor cells eventually arrive at the bone microenvironment, how these cells interact with bone niches is an essential question for understanding the microenvironment for bone metastasis, especially at the early stages of bone metastasis and in therapy resistance. It seems that the interaction with either the perivascular niche or the osteoblast niche could be important in different models (Fig. 6). In a prostate-metastasis model, bone-metastatic tumor cells interact with osteoblast niche cells and prevent the engraftment of the introduced HSC cells (Shiozawa et al. 2011). Interestingly, when HSCs were inoculated into the bone of lethally irradiated mice to rescue the mice from mortality, its protection effect could be reversed by simultaneously inoculating with bone-metastatic cells. Direct imaging

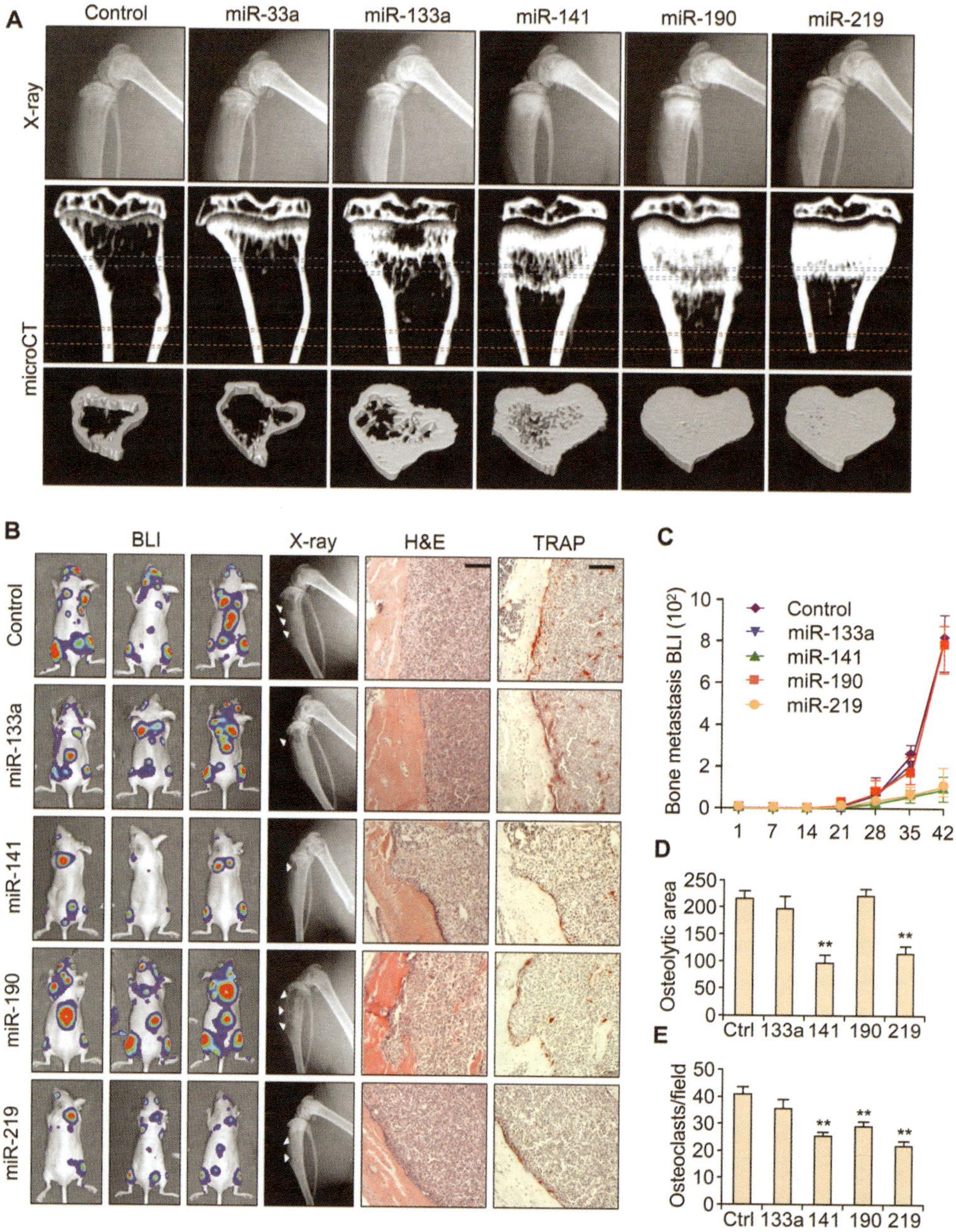

Figure 5. MicroRNAs as therapeutic agents for blocking bone-metastasis progression. (*A*) Representative X-ray and micro-computed tomography (μCT) images of bones from mice treated with indicated miRNAs or a nontargeting control miRNA. (*B*) SCP28 cells were injected into nude mice to generate bone metastasis and subsequently treated with indicated miRNAs. BLI, X-ray, H&E, and TRAP images from each group were presented. White arrows indicate osteolytic areas in the X-ray images. (*C*) Normalized bioluminescence (BLI) signals from mice in *A*. (*D*) Quantification of osteolytic lesion areas. **, $P < 0.01$. (*E*) Quantification of TRAP-positive osteoclasts from TRAP staining experiment in *A*. **, $P < 0.01$. H&E, hematoxylin and eosin; TRAP, tartrate-resistant acid phosphatase. (Adapted, with permission, from Ell et al. 2013.)

of these cells in the bone niche suggests that they prefer to lodge to similar regions of the bone, most likely to the osteoblast cells. In breast cancer, disseminated tumor cells engage the osteogenic niches in bone through heterotypic adherin junctions (E-cadherin from tumor cells and N-cadherin from osteoblast niche cells). The activation of the mTOR pathway in tumor cells by this interaction fuels the initial metastatic expansion (Wang et al. 2015). When N–E heterotypic cadherin junctions were disengaged by administration of E-cadherin neutralizing antibody, bone-metastasis seeding and colonization were significantly prohibited.

Bone-metastatic tumor cells also engage perivascular niche cells for multiple functions. For example, leukemia cells in the bone marrow localize to and disrupt the normal HSC perivascular niche. In mice bearing leukemia cells, the number of $CD34^+$ cells drastically decreased over time and were not able to enter the peripheral circulation upon stimulation (Colmone et al. 2008). In the cancer dormancy model, dormant disseminated tumor

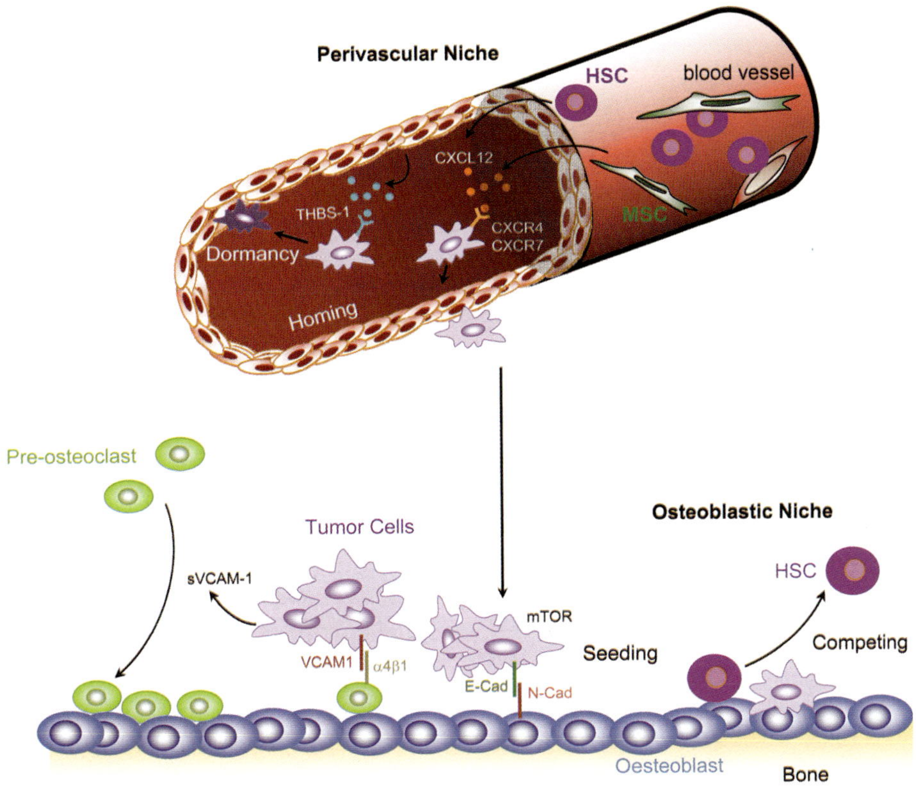

Figure 6. Various bone niches and their role in bone metastasis. In the perivascular niche, CXCL12 and CXCR4/CXCR7 interaction is critical for tumor cell homing to the bone. Tumor dormancy is mediated by endothelial-derived thrombospondin-1. The osteoblastic niche is critical for initial tumor seeding and micrometastatic growth in prostate and breast cancer. The N/E-cadherin junction between tumor and osteoblasts promotes micrometastatic tumor growth through activating the AKT-mTOR pathway. Tumor-derived VCAM1 attracts preosteoclasts and promotes their differentiation to facilitate osteolytic outgrowth. HSC, hematopoietic stem cell; VCAM1, vascular cell adhesion molecule 1; mTOR, mammalian target of rapamycin.

cells (DTCs) reside on the vasculature of metastatic sites, including the bone (Ghajar et al. 2013). The interaction of tumor cells with microvascular niche cells, specifically mediated by endothelial-derived thrombospondin-1, leads to tumor cell quiescence. However, this repressive signal is lost upon the sprouting of neovasculature and the switch to a growth-permissive microenvironment. Our recent analysis of metastatic dormancy and activation in bone also revealed an important role of VCAM-1 in this process. Tumor-derived VCAM-1 attracts monocytic osteoclast progenitor cells to promote osteoclast activity so that the tumor cells can outgrow from indolent micrometastasis to form overt metastasis (Lu et al. 2011). Despite these new insights regarding different niche environments in bone metastasis, the temporospatial dynamics of tumor cell occupancy and utilization of these niches during different stages of metastatic progression require further investigations.

Systematic cancer therapy, like hormone therapy, targeted therapy, and chemotherapy, as well as localized radiation therapy, could reduce bone-metastasis incidence, slow down metastatic growth, and most importantly alleviate some of the symptoms caused by metastasis. However, these treatments also have other unavoidable effects on bone stromal components. Whether the changes in the bone niche cells have a negative or positive effect on metastasis progression is an important yet poorly explored question. A few recent studies showed that there are fundamental changes in bone niche cells during chemotherapy of hematopoietic malignancies, which promote chemoresistance. Treatment of bone engrafted acute lymphoblastic leukemia (ALL) with cytarabine or daunorubicin for a few days induced a striking bone niche switch (Duan et al. 2014). At the early stage of the treatment, ALL cells were mainly localized at the normal bone niches of vascular type. However, 12 d after treatment, most residual cells were present in locations full of mesenchymal cell types. Molecular characterization of these cells revealed they are Nestin^{+}NG2^{-} mesenchymal cells, which express high level of GDF15 to activate the TGF-β pathway in the leukemia cells and protect the cells from chemotherapy-induced apoptosis. Other studies also suggest that systematic therapy could induce chemokines and growth factors from bone stromal cells, including CXCL-12, CCL2, VEGF, and hepatocyte growth factor (HGF), to inhibit apoptosis (Jankowski et al. 2003; Park et al. 2012). A deeper understanding is needed to reveal the complete tumor–bone niche interaction and modifications during every major step of metastasis and in the development of therapy resistance. This knowledge will serve as the foundation to improve targeted therapies for bone metastasis.

CONCLUSION

With the development of bone-metastatic cell lines, in vivo BLI imaging, and μCT imaging of the mice, we are able to study bone metastasis in preclinical mouse models qualitatively and quantitatively. Genomic profiling of in vivo selected bone-metastatic variants combined with functional analysis in mouse models and clinical prognosis studies have led to the identification and validation of multiple novel bone-metastasis genes with diverse functional mechanisms. MMP1 and ADMATS1 proteolytically release membrane-bound EGF ligands to promote osteoclast differentiation. Tumor-derived Jagged1, which is induced by bone-derived TGF-β, promotes osteoclastogenesis and survival cytokine production from osteoblast cells. VCAM-1 expression in disseminated tumor cells and bone micrometastases can be activated in inflammatory cytokines and functions to recruit and activate preosteoclasts to promote the aggressive outgrowth of bone metastasis. miRNAs involved in regulating osteoclast maturation can also be used as biomarkers and therapeutic agents of bone metastasis. These basic understandings of bone metastasis have revealed multiple new druggable targets for reducing or blocking bone metastasis, including TGF-β, Jagged1–Notch, and VCAM-1.

Recent studies also revealed two critical niches in HSC development—the perivascular niche and the osteoblastic niche. We have started to appreciate the importance of these two niches in attracting tumor cells to the bone, initializing bone seeding, maintaining and breaking the dormancy, and promoting bone-metastasis outgrowth and therapy resistance. Further understanding of the molecular mechanisms in each of these steps will undoubtedly provide additional new targets for future drug development.

ACKNOWLEDGMENTS

We thank members of the Kang Laboratory for a critical reading of the manuscript. Research in our laboratory is supported by grants from the National Institutes of Health, Department of Defense, Susan G. Komen Foundation, and the Brewster Foundation. H.Z. was supported by a Komen Postdoctoral Fellowship (KG111164).

REFERENCES

Baron R, Ferrari S, Russell RG. 2011. Denosumab and bisphosphonates: Different mechanisms of action and effects. *Bone* **48:** 677–692.

Cailleau R, Olive M, Cruciger QV. 1978. Long-term human breast carcinoma cell lines of metastatic origin: Preliminary characterization. *In vitro* **14:** 911–915.

Calvi LM, Adams GB, Weibrecht KW, Weber JM, Olson DP, Knight MC, Martin RP, Schipani E, Divieti P, Bringhurst FR, et al. 2003. Osteoblastic cells regulate the haematopoietic stem cell niche. *Nature* **425:** 841–846.

Colmone A, Amorim M, Pontier AL, Wang S, Jablonski E, Sipkins DA. 2008. Leukemic cells create bone marrow niches that disrupt the behavior of normal hematopoietic progenitor cells. *Science* **322:** 1861–1865.

Ding L, Saunders TL, Enikolopov G, Morrison SJ. 2012. Endothelial and perivascular cells maintain haematopoietic stem cells. *Nature* **481:** 457–462.

Duan CW, Shi J, Chen J, Wang B, Yu YH, Qin X, Zhou XC, Cai YJ, Li ZQ, Zhang F, et al. 2014. Leukemia propagating cells rebuild an evolving niche in response to therapy. *Cancer Cell* **25:** 778–793.

Ell B, Kang Y. 2012. SnapShot: Bone metastasis. *Cell* **151:** 690–690.e691.

Ell B, Mercatali L, Ibrahim T, Campbell N, Schwarzenbach H, Pantel K, Amadori D, Kang Y. 2013. Tumor-induced osteoclast miRNA changes as regulators and biomarkers of osteolytic bone metastasis. *Cancer Cell* **24:** 542–556.

Engin F, Yao Z, Yang T, Zhou G, Bertin T, Jiang MM, Chen Y, Wang L, Zheng H, Sutton RE, et al. 2008. Dimorphic effects of Notch signaling in bone homeostasis. *Nat Med* **14:** 299–305.

Gay LJ, Felding-Habermann B. 2011. Contribution of platelets to tumour metastasis. *Nat Rev Cancer* **11:** 123–134.

Ghajar CM, Peinado H, Mori H, Matei IR, Evason KJ, Brazier H, Almeida D, Koller A, Hajjar KA, Stainier DY, et al. 2013. The perivascular niche regulates breast tumour dormancy. *Nat Cell Biol* **15:** 807–817.

Greenbaum A, Hsu YM, Day RB, Schuettpelz LG, Christopher MJ, Borgerding JN, Nagasawa T, Link DC. 2013. CXCL12 in early mesenchymal progenitors is required for haematopoietic stem-cell maintenance. *Nature* **495:** 227–230.

Hilton MJ, Tu X, Wu X, Bai S, Zhao H, Kobayashi T, Kronenberg HM, Teitelbaum SL, Ross FP, Kopan R, et al. 2008. Notch signaling maintains bone marrow mesenchymal progenitors by suppressing osteoblast differentiation. *Nat Med* **14:** 306–314.

Jankowski K, Kucia M, Wysoczynski M, Reca R, Zhao D, Trzyna E, Trent J, Peiper S, Zembala M, Ratajczak J, et al. 2003. Both hepatocyte growth factor (HGF) and stromal-derived factor-1 regulate the metastatic behavior of human rhabdomyosarcoma cells, but only HGF enhances their resistance to radiochemotherapy. *Cancer Res* **63:** 7926–7935.

Kang Y. 2016. Dissecting tumor-stromal interactions in breast cancer bone metastasis. *Endocrinol Metab* **31:** 206–212.

Kang Y, Siegel PM, Shu W, Drobnjak M, Kakonen SM, Cordon-Cardo C, Guise TA, Massague J. 2003. A multigenic program mediating breast cancer metastasis to bone. *Cancer Cell* **3:** 537–549.

Korpal M, Yan J, Lu X, Xu S, Lerit DA, Kang Y. 2009. Imaging transforming growth factor-β signaling dynamics and therapeutic response in breast cancer bone metastasis. *Nat Med* **15:** 960–966.

Lu X, Wang Q, Hu G, Van Poznak C, Fleisher M, Reiss M, Massague J, Kang Y. 2009. ADAMTS1 and MMP1 proteolytically engage EGF-like ligands in an osteolytic signaling cascade for bone metastasis. *Genes Dev* **23:** 1882–1894.

Lu X, Mu E, Wei Y, Riethdorf S, Yang Q, Yuan M, Yan J, Hua Y, Tiede BJ, Lu X, et al. 2011. VCAM-1 promotes osteolytic expansion of indolent bone micrometastasis of breast cancer by engaging α4β1-positive osteoclast progenitors. *Cancer Cell* **20:** 701–714.

Melton LJ III, Kyle RA, Achenbach SJ, Oberg AL, Rajkumar SV. 2005. Fracture risk with multiple myeloma: A population-based study. *J Bone Miner Res* **20:** 487–493.

Morrison SJ, Scadden DT. 2014. The bone marrow niche for haematopoietic stem cells. *Nature* **505:** 327–334.

Muller A, Homey B, Soto H, Ge N, Catron D, Buchanan ME, McClanahan T, Murphy E, Yuan W, Wagner SN, et al. 2001. Involvement of chemokine receptors in breast cancer metastasis. *Nature* **410:** 50–56.

Mundy GR. 2002. Metastasis to bone: Causes, consequences and therapeutic opportunities. *Nat Rev Cancer* **2:** 584–593.

Park SI, Liao J, Berry JE, Li X, Koh AJ, Michalski ME, Eber MR, Soki FN, Sadler D, Sud S, et al. 2012. Cyclophosphamide creates a receptive microenvironment for prostate cancer skeletal metastasis. *Cancer Res* **72:** 2522–2532.

Ren G, Esposito M, Kang Y. 2015. Bone metastasis and the metastatic niche. *J Mol Med* **93:** 1203–1212.

Sethi N, Kang Y. 2011. Unravelling the complexity of metastasis-molecular understanding and targeted therapies. *Nat Rev Cancer* **11:** 735–748.

Sethi N, Dai X, Winter CG, Kang Y. 2011. Tumor-derived JAGGED1 promotes osteolytic bone metastasis of breast cancer by engaging notch signaling in bone cells. *Cancer Cell* **19:** 192–205.

Shiozawa Y, Pedersen EA, Havens AM, Jung Y, Mishra A, Joseph J, Kim JK, Patel LR, Ying C, Ziegler AM, et al. 2011. Human prostate cancer metastases target the hematopoietic stem cell niche to establish footholds in mouse bone marrow. *J Clin Invest* **121:** 1298–1312.

Wan L, Pantel K, Kang Y. 2013. Tumor metastasis: Moving new biological insights into the clinic. *Nat Med* **19:** 1450–1464.

Wang J, Loberg R, Taichman RS. 2006. The pivotal role of CXCL12 (SDF-1)/CXCR4 axis in bone metastasis. *Cancer Metastasis Rev* **25:** 573–587.

Wang H, Yu C, Gao X, Welte T, Muscarella AM, Tian L, Zhao H, Zhao Z, Du S, Tao J, et al. 2015. The osteogenic niche promotes early-stage bone colonization of disseminated breast cancer cells. *Cancer Cell* **27:** 193–210.

Weilbaecher KN, Guise TA, McCauley LK. 2011. Cancer to bone: A fatal attraction. *Nat Rev Cancer* **11:** 411–425.

Wu H, Bui MM, Leston DG, Shao H, Sokol L, Sotomayor EM, Zhang L. 2014. Clinical characteristics and prognostic factors of bone lymphomas: Focus on the clinical significance of multifocal bone involvement by primary bone large B-cell lymphomas. *BMC Cancer* **14:** 900.

Zhang J, Niu C, Ye L, Huang H, He X, Tong WG, Ross J, Haug J, Johnson T, Feng JQ, et al. 2003. Identification of the haematopoietic stem cell niche and control of the niche size. *Nature* **425:** 836–841.

Cancer, Oxidative Stress, and Metastasis

Jennifer G. Gill,[1,2] Elena Piskounova,[2] and Sean J. Morrison[2,3]
[1]*Department of Dermatology, University of Texas Southwestern Medical Center, Dallas, Texas 75390*
[2]*Department of Pediatrics, Children's Research Institute, University of Texas Southwestern Medical Center, Dallas, Texas 75390*
[3]*Howard Hughes Medical Institute, University of Texas Southwestern Medical Center, Dallas, Texas 75390*
Correspondence: Sean.Morrison@UTSouthwestern.edu

Reactive oxygen species (ROS) are highly reactive molecules that arise from a number of cellular sources, including oxidative metabolism in mitochondria. At low levels they can be advantageous to cells, activating signaling pathways that promote proliferation or survival. At higher levels, ROS can damage or kill cells by oxidizing proteins, lipids, and nucleic acids. It was hypothesized that antioxidants might benefit high-risk patients by reducing the rate of ROS-induced mutations and delaying cancer initiation. However, dietary supplementation with antioxidants has generally proven ineffective or detrimental in clinical trials. High ROS levels limit cancer cell survival during certain windows of cancer initiation and progression. During these periods, dietary supplementation with antioxidants may promote cancer cell survival and cancer progression. This raises the possibility that rather than treating cancer patients with antioxidants, they should be treated with pro-oxidants that exacerbate oxidative stress or block metabolic adaptations that confer oxidative stress resistance.

Reactive oxygen species (ROS), including superoxide radicals ($O_2\cdot^-$), hydroxyl radicals (OH•), and hydrogen peroxide (H_2O_2), are generated as by-products of aerobic metabolism as well as from a number of other sources. These ROS display varying reactivities toward different targets but share the ability to damage cells by oxidizing proteins, lipids, and DNA (Martinez-Cayuela 1995). The ability of ROS to mutate DNA and to damage cells raised the possibility that cellular aging and cancer initiation reflect accumulated ROS damage over time (Harman 1956). Although damage from ROS does contribute to aging and cancer initiation, ROS damage does not seem to provide a general explanation for the morbidities of aging, and antioxidant administration has so far failed to increase longevity outside of certain mutant genetic backgrounds (Finkel and Holbrook 2000). Nonetheless, decades of work in this area gave rise to the idea that antioxidants from the diet or from dietary supplements might slow aging and reduce cancer incidence by neutralizing cellular ROS (Greenwald et al. 1990). Results from clinical trials generally have not supported this idea.

SOURCES AND EFFECTS OF ROS

ROS are produced in various subcellular compartments by nonenzymatic and enzymatic reactions (Hernandez-Garcia et al. 2010). Nonenzymatic mechanisms include single-electron reduction of O_2 to produce superoxide in the mitochondria. Enzymatic mechanisms are numerous and include NADPH oxidases, nitric oxide synthases, xanthine oxidase, cytochrome P450 enzymes, cyclooxygenases, and lipoxygenases (Gorrini et al. 2013b). The endoplasmic reticulum also serves as a source of ROS during protein folding via protein disulfide isomerase, endoplasmic reticulum oxidoreductin, NADPH oxidase (especially NOX4), and other mechanisms (Malhotra et al. 2008; Bhandary et al. 2012; Higa and Chevet 2012). Peroxisomes produce ROS through β-oxidation of fatty acids and flavin oxidase activity (Schrader and Fahimi 2006). Hypoxia, sustained mitochondrial respiration, ER stress, and oncogenes all contribute to high ROS levels in some cancer cells (Szatrowski and Nathan 1991; Gorrini et al. 2013b). Because ROS are so highly reactive, ROS generally oxidize targets within, or adjacent to, the intracellular compartment in which they are generated.

Cellular ROS levels can also increase as a result of UV irradiation, ionizing radiation, toxins such as heavy metals, chemotherapy, and neighboring inflammatory cells, although the mechanisms vary widely (Federico et al. 2007; Azzam et al. 2012; Vera-Ramirez et al. 2012). For example, the chemotherapeutic doxorubicin forms a complex with topoisomerase and DNA that leads to double-strand breaks, increasing ROS levels and potentiating cellular damage (Lyu et al. 2007; Rowe et al. 2008; Zhang et al. 2012). Doxorubicin also promotes mitochondrial dysfunction through multiple mechanisms including changes in iron metabolism and releasing electrons from the electron transport chain (Granados-Principal et al. 2010).

Published by Cold Spring Harbor Laboratory Press; doi: 10.1101/sqb.2016.81.030791

At lower levels, ROS activate signaling that can be advantageous for cells, promoting proliferation, survival, or oxidative stress resistance (Ranjan et al. 2006; Rhee 2006; Schieber and Chandel 2014). Redox-sensitive signaling pathways include the PI3-kinase and MAP kinase signaling pathways in which ROS regulates signaling by epidermal growth factor (EGF) (Bae et al. 1997), Ras (Lander et al. 1997), platelet-derived growth factor (PDGF) (Sundaresan et al. 1995), and phosphatase and tensin homolog (PTEN) (Lee et al. 2002; Leslie et al. 2003). These pathways can be indirectly regulated by ROS or directly activated by "redox switches." For example, nitric oxide production results in S-nitrosylation of $p21^{Ras}$ at Cys118, increasing guanine nucleotide exchange and Ras signaling (Lander et al. 1997).

ANTIOXIDANT MECHANISMS FOR ROS DEFENSE

Cells must maintain homeostasis by limiting ROS production and having antioxidant mechanisms to neutralize ROS or mitigate oxidative stress. Antioxidant enzymes include superoxide dismutases (SODs), catalase, peroxiredoxins (PRDXs), thioredoxins, glutathione peroxidase, and heme oxygenase (Sabharwal and Schumacker 2014). Some of these enzymes, such as SOD, have different isoforms in mitochondria and in the cytoplasm. These enzymes work together to neutralize ROS (Fig. 1). For example, SOD can convert $O_2{\cdot}^-$ to O_2 or H_2O_2. Catalase and glutathione peroxidase subsequently convert the H_2O_2 to H_2O and O_2. Nonenzymatic antioxidants include proteins and metabolites produced by cells, such as thioredoxin, glutathione, and nicotinamide adenine dinucleotide phosphate (NADPH), as well as dietary components, such as vitamins A, C, and E, selenium, and β-carotene. Thioredoxin and glutathione are abundant endogenous redox buffers that serve as electron donors to peroxidases, which convert H_2O_2 to H_2O (Fig. 1). They can then be regenerated from their oxidized, disulfide forms, using NADPH as an electron donor.

Nuclear factor-erythroid 2-related factor 2 (NRF2) is a transcriptional master regulator of cellular redox status. It promotes the transcription of genes that encode antioxidant and detoxification enzymes in response to redox stress (Hayes and McMahon 2009). NRF2 is negatively regulated by Kelch-like ECH-associated protein 1 (KEAP1), which sequesters NRF2 in the cytoplasm (Hayes and Dinkova-Kostova 2014). However, ROS can oxidize and inactivate KEAP1, allowing NRF2 levels to increase in the nucleus and triggering the transcription of genes with antioxidant response elements in their promoters (Rushmore et al. 1991; Nioi et al. 2003). These genes include key components of the endogenous antioxidant response systems that import cystine into cells (Lewerenz et al. 2013) and promote the synthesis of glutathione (Higgins et al. 2009) and thioredoxin (Hayes and Dinkova-Kostova 2014). NRF2 also promotes the expression of glutathione peroxidase, glutathione reductase, and thioredoxin reductase (Fig. 1; Hayes and McMahon 2009; Abbas et al. 2011; Jeong et al. 2012; Hawkes et al. 2014; Lu and Holmgren 2014), as well as proteins such as ferritin, which blocks the formation of free radicals, and NADPH:quinone oxidoreductase 1, which inhibits the

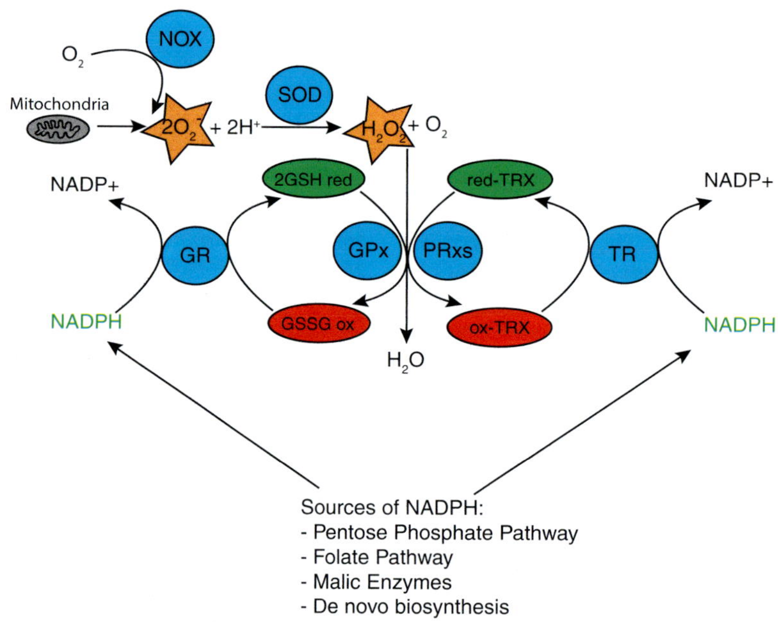

Figure 1. Endogenous antioxidant mechanisms. Reactive oxygen species (ROS) such as superoxide can be produced either in the cytoplasm or in mitochondria. Two main arms of the antioxidant response are shown: thioredoxin (TRX) and glutathione (GSH). Enzymes are highlighted in blue. GR, glutathione reductase; GSH red, reduced glutathione; GSSG ox, oxidized glutathione; GPx, glutathione peroxidase; SOD, superoxide dismutase; PRxs, peroxiredoxins; red-TRX, reduced thioredoxin; ox-TRX, oxidized thioredoxin; NOX, NADPH oxidase; NADPH, nicotinamide adenine dinucleotide phosphate.

formation of free radicals by quinones (Nioi and Hayes 2004). Finally, NRF2 promotes the expression of NADPH-generating enzymes to produce the NADPH required for glutathione and thioredoxin regeneration (Thimmulappa et al. 2002; Lee et al. 2003; Wu et al. 2011; Mitsuishi et al. 2012; Singh et al. 2013). This concerted upregulation of approximately 200 genes enables the adaptation of cells to increased ROS levels (Hayes and Dinkova-Kostova 2014).

Tumor suppressors also help to control ROS. For example, BRCA1 and fumarate hydratase both promote NRF2 function through different mechanisms (Gorrini et al. 2013b). p53 decreases ROS levels by inhibiting glycolysis and promoting the generation of NADPH via the pentose phosphate pathway (Green and Chipuk 2006). The increased DNA damage in the absence of p53 can be partially rescued by treatment with the antioxidant *N*-acetylcysteine (NAC), suggesting that the effects of p53 on redox homeostasis are a significant component of its functions that promote genomic stability (Liu et al. 2004; Sablina et al. 2005).

CLINICAL TRIALS OF ANTIOXIDANTS IN CANCER PREVENTION

Based on the hypothesis that cellular damage from ROS is a major driver of aging and oncogenic mutations, dietary supplementation with antioxidants was proposed to prevent and/or treat cancer (Greenwald et al. 1990). The Nutritional Prevention of Cancer Study Group tested whether dietary supplementation with selenium (a component of glutathione peroxidase and thioredoxin reductase) could reduce the incidence of skin cancers in patients with a history of basal and/or squamous cell carcinomas (Clark et al. 1996). Although selenium did not protect participants from the development of additional skin cancers, selenium supplementation was associated with a nearly 40% reduction in total cancer incidence, particularly in prostate cancer. The Linxian General Population Nutrition Intervention Trial found that a combination of selenium, vitamin E, and β-carotene reduced overall mortality as well as cancer rates (Blot et al. 1993; Qiao et al. 2009).

These promising findings spawned many additional trials to assess the efficacy of antioxidant use for cancer prevention (see Table 1 for a summary of randomized, placebo-controlled trials with more than 10,000 participants). Subsequent trials not only failed to reproduce these findings (Hennekens et al. 1996; Lee et al. 1999, 2005; Hercberg et al. 2004, 2007; Gaziano et al. 2009, 2012) but suggested that in some cases, antioxidants may actually promote cancer initiation and progression. The Alpha-Tocopherol Beta Carotene (ATBC) Cancer Prevention Study treated male smokers with vitamin E, β-carotene, both, or neither for 5–8 yr (ATBC 1994). Patients receiving β-carotene had an 18% increase in lung cancer incidence. These results were recapitulated in the CARET trial, which also found an increase in lung cancer incidence and mortality in participants taking β-carotene and vitamin A (Omenn et al. 1996a,b).

The SELECT trial was a randomized, double-blind, placebo-controlled trial of older males given vitamin E, selenium, both, or neither for 7–12 yr (Lippman et al. 2009; Klein et al. 2011). Men taking vitamin E alone were significantly more likely to develop prostate cancer, though overall cancer incidence did not significantly differ between groups. Among patients who received selenium, those with high baseline selenium levels had an increased risk of high-grade prostate carcinoma whereas those with low baseline selenium did not (Kristal et al. 2014).

Some trials also tested the ability of antioxidants to prevent the formation of new primary tumors in patients who had already been treated for cancer. The EUROSCAN trial treated patients with a history of head and neck cancer or lung cancer for 2 yr with vitamin A (retinyl palmitate), NAC, both, or neither (van Zandwijk et al. 2000). Patients receiving neither vitamin A nor NAC had the lowest incidence of new primary tumors, although this was not statistically significant. A Phase III randomized, double-blind, placebo-controlled trial tested the effect of selenium versus placebo for patients with resected non–small cell lung cancer (Karp et al. 2013). The trial was stopped early as selenium treatment was associated with a trend toward increased second primary tumors.

The U.S. Preventative Services Task Force (USPSTF) reviewed many of the above trials and concluded that there was insufficient data for or against the use of most nutrient supplements for cancer prevention (Moyer and U.S. Preventive Services Task Force 2014). Two exceptions included vitamin E and β-carotene. Vitamin E consistently showed no impact on cancer prevention. β-carotene showed an increased risk for lung cancer in smokers.

The other important question is whether dietary supplementation with antioxidants benefits healthy people. Many clinical trials have explored whether antioxidant use affects mortality due to heart disease, aging, or diseases of aging. A meta-analysis examining all-cause mortality in 68 randomized trials of antioxidants for many different indications included 232,606 participants and found a significant increase in mortality in patients taking β-carotene, vitamin A, and vitamin E (Bjelakovic et al. 2007). An additional meta-analysis showed that antioxidant supplements significantly increased the risk of bladder cancer (Myung et al. 2010).

CLINICAL TRIALS OF ANTIOXIDANTS IN CANCER TREATMENT

In addition to cancer prevention, some trials have included antioxidants as adjuvant therapy in the treatment of patients with cancer. A systematic review found no benefit of supplemental ascorbate (vitamin C) for overall or progression-free survival in cancer patients, most of whom had breast or colorectal cancer (Jacobs et al. 2015). Another systematic review examined treatment efficacy and patient survival with the use of any adjuvant antioxidants during chemotherapy or radiotherapy (Yasueda et al. 2016). The trials showed contradictory results, but

Table 1. Large, randomized, controlled trials designed to assess antioxidant impact on cancer incidence

Clinical trial	Reference(s)	Patient cohort	# enrollees	Treatment arms	Results
LINXIAN	Blot et al. 1993; Qiao et al. 2009	40–69-yr-old Linxian residents with no history of cancer or major disease	29,584	1. Retinol (200% RDI) + zinc (150% RDI) 2. Riboflavin (188% RDI) + niacin (200% RDI) 3. Vitamin C (200% RDI) + molybdenum (40% RDI) 4. Selenium (71% RDI) + vitamin E (111% RDI) + β-carotene (100% RDI)	Decrease in mortality and lower cancer incidences (especially stomach) in those taking selenium + vitamin E + β-carotene. This finding was durable at follow-up 10 yr after the study's end.
ATBC	ATBC Study Group 1994	Male smokers 50–69 yr of age in Finland	29,133	1. Vitamin E (185% RDI) 2. β-carotene (133% RDI) 3. Vitamin E + β-carotene (above doses) 4. Neither	Increased incidence of lung cancer with β-carotene. Slight increase in mortality in those who received vitamin E.
CARET	Omenn et al. 1996a,b	Male and female current or former smokers OR male workers exposed to asbestos	18,314	1. β-Carotene (100%–200% RDI) 2. Vitamin A (1000% RDI) 3. β-carotene + vitamin A (above doses) 4. Placebo	Increased incidence of lung cancer and mortality in β carotene + vitamin A group.
PHS I	Hennekens et al. 1996	U.S. male physicians, 40–84 yr old	22,071	1. β-carotene (167% RDI) 2. Placebo	No difference in cancer incidence or overall mortality.
WHS	Lee et al. 1999, 2005	Female health professionals >age 45	39,876	1. β-carotene (167% RDI) 2. Vitamin E (1000% RDI) 3. Placebo	No difference in cancer incidence or overall mortality.
SU.VI.MAX	Hercberg et al. 2004, 2007	French females age 35–60 and males age 45–60	13,017	1. Vitamin C (200% RDI) + vitamin E (111% RDI) + β-carotene (40% RDI) + selenium (142% RDI) + zinc (133% RDI) 2. Placebo	Lower cancer incidence and mortality seen in men, but not women, taking supplements. Increased incidence of skin cancer in women taking supplements.
PHS II	Gaziano et al. 2009, 2012	Male physicians ≥50 yr old	14,641	1. Vitamin E (667% RDI) + vitamin C (833% RDI) 2. Vitamin E (667% RDI) + placebo, 3. Placebo + vitamin C (833% RDI), 4. Multivitamin daily 5. Placebo	No effect on prostate and total cancer incidence (primary outcome measure). No effect on lung, colorectal, or other cancers. Modest decrease in incidence of total cancer with multivitamin.
SELECT	Lippman et al. 2009, Klein et al. 2011	Males >50–55 yr old w/PSA ≤4 ng/mL and normal DRE	35,533	1. Selenium (286% RDI) 2. Vitamin E (1333% RDI) 3. Selenium + vitamin E (above doses) 4. Placebo	No difference in prostate cancer incidence at 5.5 yr. At 7 yr (1.5 yr after trial ended), men taking vitamin E alone had a higher incidence of prostate cancer.
[META-ANALYSIS]	Bjelakovic et al. 2007	Varied; cohort consisted of 68 unique randomized trials	232,606	Varied; cohort consisted of 68 unique randomized trials	Statistically significant increase in mortality in patients taking β-carotene, vitamin A, and vitamin E.

β-Carotene and retinol do not have an official FDA recommended daily intake (RDI); for this table, RDI calculated is based on vitamin A RDI (5000 IU) with half of daily intake from β-carotene (2500 IU) and half from retinol (2500 IU).

Several of the above trials (e.g., LINXIAN and PHS II) had a factorial design, meaning that participants were assigned to more than one treatment arm.

PSA, prostate-specific antigen; DRE, digital rectal exam.

one trial showed that use of vitamin E and β-carotene in head and neck cancer patients receiving radiotherapy increased the risk of recurrence (RR = 2.41, 95% CI: 1.25–4.64) and cancer mortality (RR = 3.38, 95% CI: 1.11–10.34) specifically in smokers (Meyer et al. 2008).

Despite strong designs and very large numbers of participants, the above clinical trials (and other smaller trials) failed to yield clear evidence that antioxidants could reduce cancer development or progression. Contrary to what was expected, antioxidant use often appeared to increase cancer incidence, particularly in individuals at high risk.

ROS IN CANCER INITIATION AND PROGRESSION

ROS is a mutagen that promotes tumor initiation. ROS can oxidize guanine in DNA and RNA to form 8-hydroxyguanine (8-OHG) (Floyd 1990). 8-OHG can pair with adenine during DNA replication, resulting in G to T and C to A substitutions, potentially introducing missense mutations (Cheng et al. 1992). A large body of work has found a strong correlation between the formation of 8-OHG and carcinogenesis (Feig et al. 1994).

Consistent with the idea that ROS is carcinogenic, antioxidant enzymes are tumor suppressors. SOD is a family of three enzymes that are major scavengers of superoxide in the cytoplasm (SOD1), mitochondria (SOD2), and extracellularly (SOD3). *Sod1*-deficient mice develop liver cancer marked by extensive oxidative and DNA damage (Elchuri et al. 2005). Mice heterozygous for a null allele of *Sod2* also form tumors, particularly lymphoma and pituitary adenoma (Van Remmen et al. 2003). *Sod3*-deficient mice do not form tumors, but SOD3 overexpression reduces tumor formation by 50% in a skin carcinogenesis model (Kim et al. 2005). *Peroxiredoxin1* (*Prdx1*)-deficient mice develop lymphomas, sarcomas, and carcinomas (Neumann et al. 2003).

Some other tumor suppressors also act partly by suppressing ROS generation. Loss-of-function mutations in multiple tumor suppressors that promote genomic integrity, including ataxia telangiectasia mutated (ATM), P53, and BRCA1, lead to the generation of ROS (Bae et al. 2004; Reliene et al. 2004; Sablina et al. 2005; Reliene and Schiestl 2006; Esteve et al. 2010; Gorrini et al. 2013a). This may reflect the leakage of damaged DNA into the cytoplasm in these cells, inducing an interferon-mediated innate immune response (as would be stimulated by viral DNA) that promotes the generation of ROS (Santos et al. 2014; Tasdogan et al. 2016; A Tasdogan and H Fehling, pers. comm.). *Atm*-deficient cells show genomic instability, oxidative stress, hematopoietic stem cell depletion, and lymphoid neoplasia (Reliene et al. 2004; Reliene and Schiestl 2006). Treatment of *Atm*-deficient mice with NAC largely rescues these phenotypes, reducing ROS levels, DNA damage, and cancer incidence (Ito et al. 2004). NAC treatment of *p53*-deficient mice has similar effects (Sablina et al. 2005). Although clinical trials in patients at high risk of cancer tended to show that antioxidants were often ineffective or deleterious for cancer risk, these results in mice raise the possibility that antioxidants might reduce cancer risk in certain sensitized genetic backgrounds.

Numerous studies have also demonstrated a link between oncogene signaling and oxidative stress, though the mechanisms by which oncogenes increase ROS levels are murky. Ras activation increases the generation of superoxide (Irani et al. 1997). BCR-ABL-transformed cells show increased intracellular ROS, as well as oxidative DNA damage and chromosomal fragmentation (Sattler et al. 2000; Nowicki et al. 2004). c-Myc overexpression increases ROS levels, DNA damage, and genomic instability (Felsher and Bishop 1999; Vafa et al. 2002). The increase in ROS levels as a result of oncogene signaling may contribute to ongoing mutagenesis and genomic instability in cancer cells, promoting cancer progression. To balance these potentially toxic effects of ROS, several oncogenes also promote the expression of NRF2, which reduces ROS levels and promotes tumorigenesis (DeNicola et al. 2011).

Consistent with the idea that ROS can promote cancer initiation by promoting mutagenesis but impair cancer progression by causing oxidative damage, antioxidant enzymes have bimodal effects on cancer initiation and progression. *Prdx6* overexpression in keratinocytes can reduce the initiation of skin tumors, but once they arise, cancer progression is accelerated by *Prdx6* overexpression (Rolfs et al. 2013). Similarly, NRF2 confers resistance to chemical carcinogens but also promotes cancer progression by protecting cancer cells from oxidative stress and DNA damage (Ramos-Gomez et al. 2001; Iida et al. 2004; Hayes and McMahon 2006; Hu et al. 2006; Xu et al. 2006; Ma 2013; Satoh et al. 2013). Increased NRF2 expression in human cancers correlates with a poor prognosis (Moon and Giaccia 2015). Deletion of *NRF2* in pancreatic cancer cells increases DNA damage and decreases tumorigenesis (DeNicola et al. 2011). High levels of ROS are thus detrimental to cancer cells and cancer progression depends on endogenous antioxidants that attenuate oxidative stress.

ROS AND METASTASIS

Metastasis is a multistep process involving invasion, migration, intravasation into the blood, survival in circulation, extravasation into distant organs, and proliferation (Vanharanta and Massague 2013). Circulating cancer cells are commonly observed in the blood of patients and mice with various cancers (Nagrath et al. 2007; Stott et al. 2010; Yu et al. 2013, 2014; Sullivan et al. 2014). Nevertheless, metastasis is a very inefficient process (Vanharanta and Massague 2013) as very few metastasizing cancer cells survive and even fewer proliferate to form micrometastases (Luzzi et al. 1998; Cameron et al. 2000; Kienast et al. 2010). Accumulating evidence suggests that oxidative stress kills cancer cells at multiple stages of the metastatic process, contributing to the inefficiency of the process.

Metastasis begins with detachment from the local extracellular matrix. Epithelial cells undergo cell death when they detach from extracellular matrix in culture as a result of reduced glucose uptake, ATP depletion, and oxidative stress (Debnath et al. 2002; Debnath and Brugge 2005). Oncogenic signaling promotes the survival of detached breast epithelial cells by increasing glucose uptake and flux through the pentose phosphate pathway, which generates NADPH and regenerates glutathione (Schafer et al. 2009). Multiple transcription factors also cooperate to induce an antioxidant response that promotes survival, including NRF2 and ATF4. NRF2 and ATF4 promote the expression of serine/glycine biosynthesis enzymes to increase glutathione synthesis (DeNicola et al. 2015) as well as heme oxygenase 1 (Dey et al. 2015), each of which reduces oxidative stress, blocks anoikis, and promotes survival during metastasis.

Cancer cells are more sensitive than normal cells to elevated ROS levels (Raj et al. 2011). Cancer cells rely on glutathione and thioredoxin to protect them from ROS during cancer initiation and cancer progression (Harris et al. 2015). Combined inhibition of glutathione and thioredoxin synergistically induces the death of cancer cells (Harris et al. 2015). Antioxidant treatment of mouse models of lung cancer increases tumor progression and reduces mouse survival by reducing ROS levels, DNA damage, and p53 expression in the cancer cells (Sayin et al. 2014). Oxidative stress also impairs cancer progression by globally suppressing protein translation: *NRF2*-deficient cancer cells show an increase in oxidized cysteine residues in components of the translational initiation complex, globally reducing translation (Chio et al. 2016). This phenotype can be rescued by antioxidant treatment. Consistent with the critical role of NRF2 in redox regulation, some cancers suppress ROS by mutations in *KEAP1* or *NRF2* that prevent NRF2 from being sequestered in the cytoplasm, constitutively activating NRF2 (Singh et al. 2006; Ohta et al. 2008; Shibata et al. 2008a,b).

Circulating melanoma cells in the blood of xenografted mice as well as metastatic nodules have higher levels of ROS relative to primary subcutaneous tumors (Piskounova et al. 2015). Oxidative stress is a barrier to distant metastasis in these melanomas as treatment with the antioxidant NAC increases the frequency of circulating melanoma cells in the blood, as well as metastatic disease burden, without significantly affecting the growth of primary subcutaneous tumors (Piskounova et al. 2015). The finding that distant metastasis is limited by oxidative stress is not an artifact of xenotransplantation into immunocompromised mice as similar results were observed in immunocompetent mice with autochthanous melanomas: Treatment with NAC or vitamin E promoted distant metastasis without affecting the growth of subcutaneous tumors (Le Gal et al. 2015). Consistent with this, cancer cells depend on NRF2 (Wang et al. 2016), thioredoxin-like 2 (Qu et al. 2011), superoxide dismutase (Kamarajugadda et al. 2013; Glasauer et al. 2014), and glutamate cysteine ligase (the rate-limiting step of glutathione synthesis) (Nguyen et al. 2016) to survive during metastasis.

A number of studies have thus indicated that reducing oxidative stress is critical for metastasis. Nonetheless, other studies have reported that ROS can promote metastasis. Antioxidants inhibit the metastasis of some cell lines (Ferraro et al. 2006; Ishikawa et al. 2008; Porporato et al. 2014). Mouse melanoma cells in an aged microenvironment show decreased APE1 expression as a result of changes in Wnt signaling, which increases ROS, metastasis, and therapy resistance (Kaur et al. 2016). Mouse lung carcinoma cells containing mitochondrial DNA with mutations in NADH dehydrogenase subunit 6 (ND6) displayed higher ROS levels and increased metastasis when compared with wild-type mitochondrial DNA and the increase in metastasis could be inhibited by NAC treatment (Ishikawa et al. 2008). ROS can also act cell-extrinsically in the tumor microenvironment to promote cancer progression (Jezierska-Drutel et al. 2013), either by influencing the properties of tumor stromal cells (Cat et al. 2006; Toullec et al. 2010) or by attenuating the activity of inflammatory cells (Satoh et al. 2010). These studies are a reminder that ROS can induce signaling that provides a selective advantage to cells in certain circumstances and that the net effect of ROS on cancer reflects a complex combination of adaptive and maladaptive consequences within the cells and their environment.

CANCER CELLS UNDERGO METABOLIC CHANGES TO MANAGE ROS

Because oxidative stress limits cancer progression, the rare cancer cells that successfully metastasize may undergo metabolic changes that allow them to cope with oxidative stress. Consistent with this, melanoma cells undergo reversible metabolic changes during metastasis that increase their fitness to form tumors after metastasis (Piskounova et al. 2015).

NADPH is central to oxidative stress resistance. In cells in which glutathione and thioredoxin have been depleted by oxidative stress, NADPH must be diverted to regenerate the reduced forms of these redox buffers (Fig. 2). Cancer cells use multiple metabolic pathways to generate NADPH, including the pentose phosphate (Patra and Hay 2014), folate (Fan et al. 2014), and malic enzyme pathways (Fig. 2). The pentose phosphate pathway is the first line of defense against oxidative stress in many human cells and can promote the survival of cells during neoplastic transformation and during detachment from extracellular matrix (Boada et al. 2000; Debnath et al. 2002; Debnath and Brugge 2005; Bensaad et al. 2006; Sukhatme and Chan 2012; Hu et al. 2013; Kuehne et al. 2015).

De novo serine synthesis and the folate pathway are another major source of NADPH for cancer cells (Fan et al. 2014; Lewis et al. 2014; Ye et al. 2014). De novo serine synthesis can be limiting for flux through the folate pathway, and reduced serine hydroxymethyltransferase expression reduces the cellular $NADPH/NADP^+$ ratio while increasing ROS levels and cell death (Ye et al. 2014). Phosphoglycerate dehydrogenase (PHGDH) catalyzes the first step in serine biosynthesis. This enzyme is

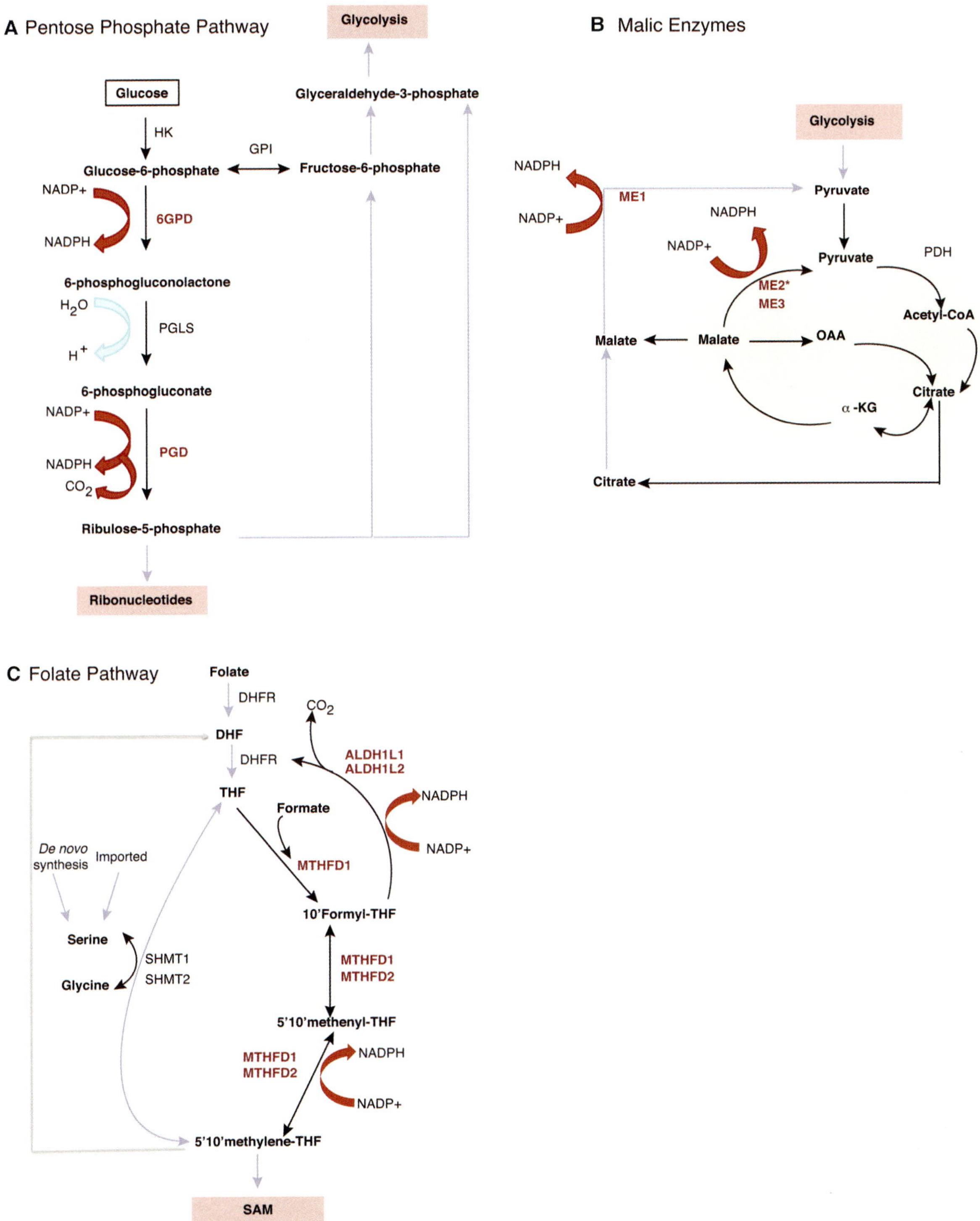

Figure 2. NADPH regenerating pathways. Cancer cells depend on different metabolic pathways to produce NADPH to regenerate endogenous antioxidants, such as GSH and TXN. (*A*) The pentose phosphate pathway. In the pentose phosphate pathway, NADPH is generated in two reactions catalyzed by glucose-6-phosphate dehydrogenase (G6PD) and 6-phosphogluconate dehydrogenase (PGD), shown in red. G6PD catalyzes the rate-limiting step in the oxidative branch of the pentose phosphate pathway. Its activity is regulated by the NADP+/NADPH ratio, where NADPH is the negative regulator and NADP+ is required for proper enzymatic function. (*B*) Malic enzymes. Three malic enzymes have been identified in mammalian cells. They differ with respect to their localization in the cytosol (ME1) versus mitochondria (ME2, ME3) and with respect to their use of NADP (ME1, ME3) versus NAD or NADP (ME2*) as electron acceptors. Malic enzymes mediate the conversion of malate to pyruvate, accompanied by the production of NADPH. (*C*) The folate pathway. The folate pathway uses one-carbon groups that come from serine to generate either nucleotides or NADPH. Serine can either be synthesized de novo by cells or imported. Several enzymes within the folate pathway, aldehyde dehydrogenase L1 (ALDH1L1) and 2 (ALDH1L2), and methylenetetrahydrofolate dehydrogenase 1 (MTHFD1) and 2 (MTHFD2) are able to regenerate NADPH.

increased in expression in many breast cancers and melanomas, and is necessary for the proliferation of those cells (Locasale et al. 2011; Mullarky et al. 2011; Possemato et al. 2011). The folate pathway uses one-carbon groups from serine to generate either nucleotides or NADPH. Several enzymes within the folate pathway generate NADPH including methyltetrahydrofolate dehydrogenases 1 and 2 (MTHFD1 and MTHFD2) and aldehyde dehydrogenase-like 1 and 2 (ALDH1L1 and ALDH1L2) (Fig. 2; Fan et al. 2014). ALDH1L2 expression reversibly increases in melanoma cells during metastasis and knockdown of ALDH1L2 or MTHFD1 reduces the metastasis of patient-derived melanoma xenografts in vivo without affecting the growth of primary tumors in the same mice (Piskounova et al. 2015).

Finally, malic enzymes mediate the conversion of malate to pyruvate, accompanied by NADPH production (Fig. 2). Malic enzymes promote the growth of several cancers (Son et al. 2013; Ren et al. 2014). ME2 expression is associated with reduced ROS levels in melanoma cells and promotes cutaneous melanoma proliferation and invasion in culture (Jiang et al. 2013).

Cancer cells sometimes benefit from metabolic changes that preserve NADPH for the regeneration of glutathione and thioredoxin. AMP-activated protein kinase (AMPK) is activated in response to ROS as well as during energy stress and can promote the survival of cells partly through redox regulation (Schafer et al. 2009). AMPK inhibits acetyl-CoA carboxylases, inhibiting the consumption of NADPH by fatty acid synthesis and promoting the generation of NADPH by fatty acid oxidation (Jeon et al. 2012). High ROS levels in cancer cells also decrease oxidative phosphorylation by promoting glycolysis by stabilizing hypoxia-inducible factor-1 (HIF-1) (Chandel et al. 2000; Semenza 2011) or by oxidizing and inhibiting pyruvate kinase M2, diverting glucose into the pentose phosphate pathway (Anastasiou et al. 2011). These changes may reflect a broader need to shut down anabolic pathways that generate ROS or consume NADPH in cancer cells experiencing oxidative stress.

Consistent with these observations, some cancer cells undergo metabolic changes during invasion in vitro and metastasis in vivo that would be expected to reduce the generation of ROS (Chen et al. 2007; Lu et al. 2010; Qu et al. 2011; Kamarajugadda et al. 2012, 2013; Dong et al. 2013; Shi et al. 2014). For example, HIF-1 activity is transiently increased during metastasis due to high ROS levels (Montagner et al. 2012; Vanharanta et al. 2013; Zhao et al. 2014). HIF-1 activation metabolically reprograms metastasizing cells away from oxidative phosphorylation to glycolysis and lactic acid production, through increased expression of lactate dehydrogenase and pyruvate dehydrogenase. These metabolic changes reduce ROS levels and promote survival during metastasis.

The metabolic plasticity of cancer cells allows them to undergo dynamic changes in mitochondrial mass and mitochondrial function that facilitate their ability to cope with energy stress and oxidative stress (Senft and Ronai 2016; Vyas et al. 2016). Both glycolysis and oxidative phosphorylation can be used in cancer cells in complementary strategies to enhance metabolic plasticity to overcome changes in the tumor environment or in energy demands (Jose et al. 2011). PGC-1α, which promotes mitochondrial biogenesis, is dynamically expressed by cancer cells. Although metastasizing cells can increase mitochondrial biogenesis and respiration by increasing PGC-1α expression (LeBleu et al. 2014), PGC-1α inhibits metastasis in other contexts (Luo et al. 2016). Because PGC-1α would be expected to increase mitochondrial mass and the generation of ROS, the observation that PGC-1α^{low} cells have more metastatic potential in some cancers (Luo et al. 2016) is consistent with the observation that oxidative stress limits distant metastasis (Piskounova et al. 2015). Cancer cells also dynamically control mitochondrial fusion and fission to regulate oxidative phosphorylation and ROS levels (Hagenbuchner et al. 2013) and to promote invasion (Zhao et al. 2013). The question of whether metastasizing cancer cells benefit from increased or decreased mitochondrial function may depend on the tissue of origin as mitochondrial DNA (mtDNA) copy numbers vary widely across tumor types (Reznik et al. 2016).

CONCLUSION

ROS promote cancer initiation by promoting mutagenesis and perhaps by activating signaling pathways that promote proliferation, survival, and stress resistance. However, ROS also limits cancer initiation and progression by causing oxidative stress that kills many cancer cells. For this reason, cancer cells depend on a variety of mechanisms to suppress ROS and to cope with oxidative stress. Antioxidants promote cancer initiation and progression in experimental mouse models as well as in clinical trials. Cancer may be more effectively treated with pro-oxidants that exacerbate the oxidative stress experienced by cancer cells or that prevent metabolic adaptations that confer oxidative stress resistance.

ACKNOWLEDGMENTS

S.J.M. is a Howard Hughes Medical Institute (HHMI) Investigator, the Mary McDermott Cook Chair in Pediatric Genetics, the Kathryn and Gene Bishop Distinguished Chair in Pediatric Research, the director of the Hamon Laboratory for Stem Cells and Cancer, and a Cancer Prevention and Research Institute of Texas Scholar. We thank Alpaslan Tasdogan and Kati Ahlqvist for discussion and critical comments. This work was supported by the Cancer Prevention and Research Institute of Texas and the National Institutes of Health (R37 AG024945 and R01 DK100848).

REFERENCES

Abbas K, Breton J, Planson AG, Bouton C, Bignon J, Seguin C, Riquier S, Toledano MB, Drapier JC. 2011. Nitric oxide activates an Nrf2/sulfiredoxin antioxidant pathway in macrophages. *Free Radic Biol Med* **51:** 107–114.

Anastasiou D, Poulogiannis G, Asara JM, Boxer MB, Jiang JK, Shen M, Bellinger G, Sasaki AT, Locasale JW, Auld DS, et al. 2011. Inhibition of pyruvate kinase M2 by reactive oxygen species contributes to cellular antioxidant responses. *Science* **334:** 1278–1283.

ATBC. 1994. The effect of vitamin E and β carotene on the incidence of lung cancer and other cancers in male smokers. The Alpha-Tocopherol Beta Carotene Cancer Prevention Study Group. *N Engl J Med* **330:** 1029–1035.

Azzam EI, Jay-Gerin JP, Pain D. 2012. Ionizing radiation-induced metabolic oxidative stress and prolonged cell injury. *Cancer Lett* **327:** 48–60.

Bae YS, Kang SW, Seo MS, Baines IC, Tekle E, Chock PB, Rhee SG. 1997. Epidermal growth factor (EGF)-induced generation of hydrogen peroxide. Role in EGF receptor-mediated tyrosine phosphorylation. *J Biol Chem* **272:** 217–221.

Bae I, Fan S, Meng Q, Rih JK, Kim HJ, Kang HJ, Xu J, Goldberg ID, Jaiswal AK, Rosen EM. 2004. BRCA1 induces antioxidant gene expression and resistance to oxidative stress. *Cancer Res* **64:** 7893–7909.

Bensaad K, Tsuruta A, Selak MA, Vidal MN, Nakano K, Bartrons R, Gottlieb E, Vousden KH. 2006. TIGAR, a p53-inducible regulator of glycolysis and apoptosis. *Cell* **126:** 107–120.

Bhandary B, Marahatta A, Kim HR, Chae HJ. 2012. An involvement of oxidative stress in endoplasmic reticulum stress and its associated diseases. *Int J Mol Sci* **14:** 434–456.

Bjelakovic G, Nikolova D, Gluud LL, Simonetti RG, Gluud C. 2007. Mortality in randomized trials of antioxidant supplements for primary and secondary prevention: Systematic review and meta-analysis. *JAMA* **297:** 842–857.

Blot WJ, Li JY, Taylor PR, Guo W, Dawsey S, Wang GQ, Yang CS, Zheng SF, Gail M, Li GY, et al. 1993. Nutrition intervention trials in Linxian, China: Supplementation with specific vitamin/mineral combinations, cancer incidence, and disease-specific mortality in the general population. *J Natl Cancer Inst* **85:** 1483–1492.

Boada J, Roig T, Perez X, Gamez A, Bartrons R, Cascante M, Bermudez J. 2000. Cells overexpressing fructose-2,6-bisphosphatase showed enhanced pentose phosphate pathway flux and resistance to oxidative stress. *FEBS Lett* **480:** 261–264.

Cameron MD, Schmidt EE, Kerkvliet N, Nadkarni KV, Morris VL, Groom AC, Chambers AF, MacDonald IC. 2000. Temporal progression of metastasis in lung: Cell survival, dormancy, and location dependence of metastatic inefficiency. *Cancer Res* **60:** 2541–2546.

Cat B, Stuhlmann D, Steinbrenner H, Alili L, Holtkotter O, Sies H, Brenneisen P. 2006. Enhancement of tumor invasion depends on transdifferentiation of skin fibroblasts mediated by reactive oxygen species. *J Cell Sci* **119:** 2727–2738.

Chandel NS, McClintock DS, Feliciano CE, Wood TM, Melendez JA, Rodriguez AM, Schumacker PT. 2000. Reactive oxygen species generated at mitochondrial complex III stabilize hypoxia-inducible factor-1α during hypoxia: A mechanism of O_2 sensing. *J Biol Chem* **275:** 25130–25138.

Chen EI, Hewel J, Krueger JS, Tiraby C, Weber MR, Kralli A, Becker K, Yates JR III, Felding-Habermann B. 2007. Adaptation of energy metabolism in breast cancer brain metastases. *Cancer Res* **67:** 1472–1486.

Cheng KC, Cahill DS, Kasai H, Nishimura S, Loeb LA. 1992. 8-Hydroxyguanine, an abundant form of oxidative DNA damage, causes G→T and A→C substitutions. *J Biol Chem* **267:** 166–172.

Chio II, Jafarnejad SM, Ponz-Sarvise M, Park Y, Rivera K, Palm W, Wilson J, Sangar V, Hao Y, Ohlund D, et al. 2016. NRF2 promotes tumor maintenance by modulating mRNA translation in pancreatic cancer. *Cell* **166:** 963–976.

Clark LC, Combs GF Jr, Turnbull BW, Slate EH, Chalker DK, Chow J, Davis LS, Glover RA, Graham GF, Gross EG, et al. 1996. Effects of selenium supplementation for cancer prevention in patients with carcinoma of the skin. A randomized controlled trial. Nutritional Prevention of Cancer Study Group. *JAMA* **276:** 1957–1963.

Debnath J, Brugge JS. 2005. Modelling glandular epithelial cancers in three-dimensional cultures. *Nat Rev Cancer* **5:** 675–688.

Debnath J, Mills KR, Collins NL, Reginato MJ, Muthuswamy SK, Brugge JS. 2002. The role of apoptosis in creating and maintaining luminal space within normal and oncogene-expressing mammary acini. *Cell* **111:** 29–40.

DeNicola GM, Karreth FA, Humpton TJ, Gopinathan A, Wei C, Frese K, Mangal D, Yu KH, Yeo CJ, Calhoun ES, et al. 2011. Oncogene-induced Nrf2 transcription promotes ROS detoxification and tumorigenesis. *Nature* **475:** 106–109.

DeNicola GM, Chen PH, Mullarky E, Sudderth JA, Hu Z, Wu D, Tang H, Xie Y, Asara JM, Huffman KE, et al. 2015. NRF2 regulates serine biosynthesis in non–small cell lung cancer. *Nat Genet* **47:** 1475–1481.

Dey S, Sayers CM, Verginadis II, Lehman SL, Cheng Y, Cerniglia GJ, Tuttle SW, Feldman MD, Zhang PJ, Fuchs SY, et al. 2015. ATF4-dependent induction of heme oxygenase 1 prevents anoikis and promotes metastasis. *J Clin Invest* **125:** 2592–2608.

Dong C, Yuan T, Wu Y, Wang Y, Fan TW, Miriyala S, Lin Y, Yao J, Shi J, Kang T, et al. 2013. Loss of FBP1 by Snail-mediated repression provides metabolic advantages in basal-like breast cancer. *Cancer Cell* **23:** 316–331.

Elchuri S, Oberley TD, Qi W, Eisenstein RS, Jackson Roberts L, Van Remmen H, Epstein CJ, Huang TT. 2005. CuZnSOD deficiency leads to persistent and widespread oxidative damage and hepatocarcinogenesis later in life. *Oncogene* **24:** 367–380.

Esteve JM, Armengod ME, Knecht E. 2010. BRCA1 negatively regulates formation of autophagic vacuoles in MCF-7 breast cancer cells. *Exp Cell Res* **316:** 2618–2629.

Fan J, Ye J, Kamphorst JJ, Shlomi T, Thompson CB, Rabinowitz JD. 2014. Quantitative flux analysis reveals folate-dependent NADPH production. *Nature* **510:** 298–302.

Federico A, Morgillo F, Tuccillo C, Ciardiello F, Loguercio C. 2007. Chronic inflammation and oxidative stress in human carcinogenesis. *Int J Cancer* **121:** 2381–2386.

Feig DI, Reid TM, Loeb LA. 1994. Reactive oxygen species in tumorigenesis. *Cancer Res* **54:** 1890s–1894s.

Felsher DW, Bishop JM. 1999. Transient excess of MYC activity can elicit genomic instability and tumorigenesis. *Proc Natl Acad Sci* **96:** 3940–3944.

Ferraro D, Corso S, Fasano E, Panieri E, Santangelo R, Borrello S, Giordano S, Pani G, Galeotti T. 2006. Pro-metastatic signaling by c-Met through RAC-1 and reactive oxygen species (ROS). *Oncogene* **25:** 3689–3698.

Finkel T, Holbrook NJ. 2000. Oxidants, oxidative stress and the biology of ageing. *Nature* **408:** 239–247.

Floyd RA. 1990. The role of 8-hydroxyguanine in carcinogenesis. *Carcinogenesis* **11:** 1447–1450.

Gaziano JM, Glynn RJ, Christen WG, Kurth T, Belanger C, MacFadyen J, Bubes V, Manson JE, Sesso HD, Buring JE. 2009. Vitamins E and C in the prevention of prostate and total cancer in men: The Physicians' Health Study II randomized controlled trial. *JAMA* **301:** 52–62.

Gaziano JM, Sesso HD, Christen WG, Bubes V, Smith JP, MacFadyen J, Schvartz M, Manson JE, Glynn RJ, Buring JE. 2012. Multivitamins in the prevention of cancer in men: The Physicians' Health Study II randomized controlled trial. *JAMA* **308:** 1871–1880.

Glasauer A, Sena LA, Diebold LP, Mazar AP, Chandel NS. 2014. Targeting SOD1 reduces experimental non-small-cell lung cancer. *J Clin Invest* **124:** 117–128.

Gorrini C, Baniasadi PS, Harris IS, Silvester J, Inoue S, Snow B, Joshi PA, Wakeham A, Molyneux SD, Martin B, et al. 2013a. BRCA1 interacts with Nrf2 to regulate antioxidant signaling and cell survival. *J Exp Med* **210:** 1529–1544.

Gorrini C, Harris IS, Mak TW. 2013b. Modulation of oxidative stress as an anticancer strategy. *Nat Rev Drug Discov* **12:** 931–947.

Granados-Principal S, Quiles JL, Ramirez-Tortosa CL, Sanchez-Rovira P, Ramirez-Tortosa MC. 2010. New advances in molecular mechanisms and the prevention of adriamycin toxicity by antioxidant nutrients. *Food Chem Toxicol* **48:** 1425–1438.

Green DR, Chipuk JE. 2006. p53 and metabolism: Inside the TIGAR. *Cell* **126:** 30–32.

Greenwald P, Nixon DW, Malone WF, Kelloff GJ, Stern HR, Witkin KM. 1990. Concepts in cancer chemoprevention research. *Cancer* **65:** 1483–1490.

Hagenbuchner J, Kuznetsov AV, Obexer P, Ausserlechner MJ. 2013. BIRC5/Survivin enhances aerobic glycolysis and drug resistance by altered regulation of the mitochondrial fusion/fission machinery. *Oncogene* **32:** 4748–4757.

Harman D. 1956. Aging: A theory based on free radical and radiation chemistry. *J Gerontol* **11:** 298–300.

Harris IS, Treloar AE, Inoue S, Sasaki M, Gorrini C, Lee KC, Yung KY, Brenner D, Knobbe-Thomsen CB, Cox MA, et al. 2015. Glutathione and thioredoxin antioxidant pathways synergize to drive cancer initiation and progression. *Cancer Cell* **27:** 211–222.

Hawkes HJ, Karlenius TC, Tonissen KF. 2014. Regulation of the human thioredoxin gene promoter and its key substrates: A study of functional and putative regulatory elements. *Biochim Biophys Acta* **1840:** 303–314.

Hayes JD, Dinkova-Kostova AT. 2014. The Nrf2 regulatory network provides an interface between redox and intermediary metabolism. *Trends Biochem Sci* **39:** 199–218.

Hayes JD, McMahon M. 2006. The double-edged sword of Nrf2: Subversion of redox homeostasis during the evolution of cancer. *Mol Cell* **21:** 732–734.

Hayes JD, McMahon M. 2009. *NRF2* and *KEAP1* mutations: Permanent activation of an adaptive response in cancer. *Trends Biochem Sci* **34:** 176–188.

Hennekens CH, Buring JE, Manson JE, Stampfer M, Rosner B, Cook NR, Belanger C, LaMotte F, Gaziano JM, Ridker PM, et al. 1996. Lack of effect of long-term supplementation with β carotene on the incidence of malignant neoplasms and cardiovascular disease. *N Engl J Med* **334:** 1145–1149.

Hercberg S, Galan P, Preziosi P, Bertrais S, Mennen L, Malvy D, Roussel AM, Favier A, Briancon S. 2004. The SU.VI.MAX Study: A randomized, placebo-controlled trial of the health effects of antioxidant vitamins and minerals. *Arch Intern Med* **164:** 2335–2342.

Hercberg S, Ezzedine K, Guinot C, Preziosi P, Galan P, Bertrais S, Estaquio C, Briancon S, Favier A, Latreille J, et al. 2007. Antioxidant supplementation increases the risk of skin cancers in women but not in men. *J Nutr* **137:** 2098–2105.

Hernandez-Garcia D, Wood CD, Castro-Obregon S, Covarrubias L. 2010. Reactive oxygen species: A radical role in development? *Free Radic Biol Med* **49:** 130–143.

Higa A, Chevet E. 2012. Redox signaling loops in the unfolded protein response. *Cell Signal* **24:** 1548–1555.

Higgins LG, Kelleher MO, Eggleston IM, Itoh K, Yamamoto M, Hayes JD. 2009. Transcription factor Nrf2 mediates an adaptive response to sulforaphane that protects fibroblasts in vitro against the cytotoxic effects of electrophiles, peroxides and redox-cycling agents. *Toxicol Appl Pharmacol* **237:** 267–280.

Hu R, Xu C, Shen G, Jain MR, Khor TO, Gopalkrishnan A, Lin W, Reddy B, Chan JY, Kong AN. 2006. Gene expression profiles induced by cancer chemopreventive isothiocyanate sulforaphane in the liver of C57BL/6J mice and C57BL/6J/Nrf2 (−/−) mice. *Cancer Lett* **243:** 170–192.

Hu T, Zhang C, Tang Q, Su Y, Li B, Chen L, Zhang Z, Cai T, Zhu Y. 2013. Variant G6PD levels promote tumor cell proliferation or apoptosis via the STAT3/5 pathway in the human melanoma xenograft mouse model. *BMC Cancer* **13:** 251.

Iida K, Itoh K, Kumagai Y, Oyasu R, Hattori K, Kawai K, Shimazui T, Akaza H, Yamamoto M. 2004. Nrf2 is essential for the chemopreventive efficacy of oltipraz against urinary bladder carcinogenesis. *Cancer Res* **64:** 6424–6431.

Irani K, Xia Y, Zweier JL, Sollott SJ, Der CJ, Fearon ER, Sundaresan M, Finkel T, Goldschmidt-Clermont PJ. 1997. Mitogenic signaling mediated by oxidants in Ras-transformed fibroblasts. *Science* **275:** 1649–1652.

Ishikawa K, Takenaga K, Akimoto M, Koshikawa N, Yamaguchi A, Imanishi H, Nakada K, Honma Y, Hayashi J. 2008. ROS-generating mitochondrial DNA mutations can regulate tumor cell metastasis. *Science* **320:** 661–664.

Ito K, Hirao A, Arai F, Matsuoka S, Takubo K, Hamaguchi I, Nomiyama K, Hosokawa K, Sakurada K, Nakagata N, et al. 2004. Regulation of oxidative stress by ATM is required for self-renewal of haematopoietic stem cells. *Nature* **431:** 997–1002.

Jacobs C, Hutton B, Ng T, Shorr R, Clemons M. 2015. Is there a role for oral or intravenous ascorbate (vitamin C) in treating patients with cancer? A systematic review. *Oncologist* **20:** 210–223.

Jeon SM, Chandel NS, Hay N. 2012. AMPK regulates NADPH homeostasis to promote tumour cell survival during energy stress. *Nature* **485:** 661–665.

Jeong W, Bae SH, Toledano MB, Rhee SG. 2012. Role of sulfiredoxin as a regulator of peroxiredoxin function and regulation of its expression. *Free Radic Biol Med* **53:** 447–456.

Jezierska-Drutel A, Rosenzweig SA, Neumann CA. 2013. Role of oxidative stress and the microenvironment in breast cancer development and progression. *Adv Cancer Res* **119:** 107–125.

Jiang P, Du W, Mancuso A, Wellen KE, Yang X. 2013. Reciprocal regulation of p53 and malic enzymes modulates metabolism and senescence. *Nature* **493:** 689–693.

Jose C, Bellance N, Rossignol R. 2011. Choosing between glycolysis and oxidative phosphorylation: A tumor's dilemma? *Biochim Biophys Acta* **1807:** 552–561.

Kamarajugadda S, Stemboroski L, Cai Q, Simpson NE, Nayak S, Tan M, Lu J. 2012. Glucose oxidation modulates anoikis and tumor metastasis. *Mol Cell Biol* **32:** 1893–1907.

Kamarajugadda S, Cai Q, Chen H, Nayak S, Zhu J, He M, Jin Y, Zhang Y, Ai L, Martin SS, et al. 2013. Manganese superoxide dismutase promotes anoikis resistance and tumor metastasis. *Cell Death Dis* **4:** e504.

Karp DD, Lee SJ, Keller SM, Wright GS, Aisner S, Belinsky SA, Johnson DH, Johnston MR, Goodman G, Clamon G, et al. 2013. Randomized, double-blind, placebo-controlled, phase III chemoprevention trial of selenium supplementation in patients with resected stage I non-small-cell lung cancer: ECOG 5597. *J Clin Oncol* **31:** 4179–4187.

Kaur A, Webster MR, Marchbank K, Behera R, Ndoye A, Kugel CH III, Dang VM, Appleton J, O'Connell MP, Cheng P, et al. 2016. sFRP2 in the aged microenvironment drives melanoma metastasis and therapy resistance. *Nature* **532:** 250–254.

Kienast Y, von Baumgarten L, Fuhrmann M, Klinkert WE, Goldbrunner R, Herms J, Winkler F. 2010. Real-time imaging reveals the single steps of brain metastasis formation. *Nat Med* **16:** 116–122.

Kim SH, Kim MO, Gao P, Youm CA, Park HR, Lee TS, Kim KS, Suh JG, Lee HT, Park BJ, et al. 2005. Overexpression of extracellular superoxide dismutase (EC-SOD) in mouse skin plays a protective role in DMBA/TPA-induced tumor formation. *Oncol Res* **15:** 333–341.

Klein EA, Thompson IM Jr, Tangen CM, Crowley JJ, Lucia MS, Goodman PJ, Minasian LM, Ford LG, Parnes HL, Gaziano JM, et al. 2011. Vitamin E and the risk of prostate cancer: The Selenium and Vitamin E Cancer Prevention Trial (SELECT). *JAMA* **306:** 1549–1556.

Kristal AR, Darke AK, Morris JS, Tangen CM, Goodman PJ, Thompson IM, Meyskens FL Jr, Goodman GE, Minasian LM, Parnes HL, et al. 2014. Baseline selenium status and effects of selenium and vitamin E supplementation on prostate cancer risk. *J Natl Cancer Inst* **106:** djt456.

Kuehne A, Emmert H, Soehle J, Winnefeld M, Fischer F, Wenck H, Gallinat S, Terstegen L, Lucius R, Hildebrand J, et al. 2015. Acute activation of oxidative pentose phosphate pathway as first-line response to oxidative stress in human skin cells. *Mol Cell* **59:** 359–371.

Lander HM, Hajjar DP, Hempstead BL, Mirza UA, Chait BT, Campbell S, Quilliam LA. 1997. A molecular redox switch on p21^{ras}. Structural basis for the nitric oxide-p21^{ras} interaction. *J Biol Chem* **272:** 4323–4326.

LeBleu VS, O'Connell JT, Gonzalez Herrera KN, Wikman H, Pantel K, Haigis MC, de Carvalho FM, Damascena A, Domingos Chinen LT, Rocha RM, et al. 2014. PGC-1α mediates mitochondrial biogenesis and oxidative phosphorylation in cancer cells to promote metastasis. *Nat Cell Biol* **16:** 992–1003, 1001–1015.

Lee IM, Cook NR, Manson JE, Buring JE, Hennekens CH. 1999. β-Carotene supplementation and incidence of cancer and cardiovascular disease: The Women's Health Study. *J Natl Cancer Inst* **91:** 2102–2106.

Lee SR, Yang KS, Kwon J, Lee C, Jeong W, Rhee SG. 2002. Reversible inactivation of the tumor suppressor PTEN by H_2O_2. *J Biol Chem* **277:** 20336–20342.

Lee JM, Calkins MJ, Chan K, Kan YW, Johnson JA. 2003. Identification of the NF-E2-related factor-2-dependent genes conferring protection against oxidative stress in primary cortical astrocytes using oligonucleotide microarray analysis. *J Biol Chem* **278:** 12029–12038.

Lee IM, Cook NR, Gaziano JM, Gordon D, Ridker PM, Manson JE, Hennekens CH, Buring JE. 2005. Vitamin E in the primary prevention of cardiovascular disease and cancer: The Women's Health Study: A randomized controlled trial. *JAMA* **294:** 56–65.

Le Gal K, Ibrahim MX, Wiel C, Sayin VI, Akula MK, Karlsson C, Dalin MG, Akyurek LM, Lindahl P, Nilsson J, et al. 2015. Antioxidants can increase melanoma metastasis in mice. *Sci Transl Med* **7:** 308re308.

Leslie NR, Bennett D, Lindsay YE, Stewart H, Gray A, Downes CP. 2003. Redox regulation of PI 3-kinase signalling via inactivation of PTEN. *EMBO J* **22:** 5501–5510.

Lewerenz J, Hewett SJ, Huang Y, Lambros M, Gout PW, Kalivas PW, Massie A, Smolders I, Methner A, Pergande M, et al. 2013. The cystine/glutamate antiporter system x_c^- in health and disease: From molecular mechanisms to novel therapeutic opportunities. *Antioxid Redox Signal* **18:** 522–555.

Lewis CA, Parker SJ, Fiske BP, McCloskey D, Gui DY, Green CR, Vokes NI, Feist AM, Vander Heiden MG, Metallo CM. 2014. Tracing compartmentalized NADPH metabolism in the cytosol and mitochondria of mammalian cells. *Mol Cell* **55:** 253–263.

Lippman SM, Klein EA, Goodman PJ, Lucia MS, Thompson IM, Ford LG, Parnes HL, Minasian LM, Gaziano JM, Hartline JA, et al. 2009. Effect of selenium and vitamin E on risk of prostate cancer and other cancers: The Selenium and Vitamin E Cancer Prevention Trial (SELECT). *JAMA* **301:** 39–51.

Liu G, Parant JM, Lang G, Chau P, Chavez-Reyes A, El-Naggar AK, Multani A, Chang S, Lozano G. 2004. Chromosome stability, in the absence of apoptosis, is critical for suppression of tumorigenesis in Trp53 mutant mice. *Nat Genet* **36:** 63–68.

Locasale JW, Grassian AR, Melman T, Lyssiotis CA, Mattaini KR, Bass AJ, Heffron G, Metallo CM, Muranen T, Sharfi H, et al. 2011. Phosphoglycerate dehydrogenase diverts glycolytic flux and contributes to oncogenesis. *Nat Genet* **43:** 869–874.

Lu J, Holmgren A. 2014. The thioredoxin antioxidant system. *Free Radic Biol Med* **66:** 75–87.

Lu X, Bennet B, Mu E, Rabinowitz J, Kang Y. 2010. Metabolomic changes accompanying transformation and acquisition of metastatic potential in a syngeneic mouse mammary tumor model. *J Biol Chem* **285:** 9317–9321.

Luo C, Lim JH, Lee Y, Granter SR, Thomas A, Vazquez F, Widlund HR, Puigserver P. 2016. A PGC1α-mediated transcriptional axis suppresses melanoma metastasis. *Nature* **537:** 422–426.

Luzzi KJ, MacDonald IC, Schmidt EE, Kerkvliet N, Morris VL, Chambers AF, Groom AC. 1998. Multistep nature of metastatic inefficiency: Dormancy of solitary cells after successful extravasation and limited survival of early micrometastases. *Am J Pathol* **153:** 865–873.

Lyu YL, Kerrigan JE, Lin CP, Azarova AM, Tsai YC, Ban Y, Liu LF. 2007. Topoisomerase IIβ mediated DNA double-strand breaks: Implications in doxorubicin cardiotoxicity and prevention by dexrazoxane. *Cancer Res* **67:** 8839–8846.

Ma Q. 2013. Role of nrf2 in oxidative stress and toxicity. *Annu Rev Pharmacol Toxicol* **53:** 401–426.

Malhotra JD, Miao H, Zhang K, Wolfson A, Pennathur S, Pipe SW, Kaufman RJ. 2008. Antioxidants reduce endoplasmic reticulum stress and improve protein secretion. *Proc Natl Acad Sci* **105:** 18525–18530.

Martinez-Cayuela M. 1995. Oxygen free radicals and human disease. *Biochimie* **77:** 147–161.

Meyer F, Bairati I, Fortin A, Gelinas M, Nabid A, Brochet F, Tetu B. 2008. Interaction between antioxidant vitamin supplementation and cigarette smoking during radiation therapy in relation to long-term effects on recurrence and mortality: A randomized trial among head and neck cancer patients. *Int J Cancer* **122:** 1679–1683.

Mitsuishi Y, Taguchi K, Kawatani Y, Shibata T, Nukiwa T, Aburatani H, Yamamoto M, Motohashi H. 2012. Nrf2 redirects glucose and glutamine into anabolic pathways in metabolic reprogramming. *Cancer Cell* **22:** 66–79.

Montagner M, Enzo E, Forcato M, Zanconato F, Parenti A, Rampazzo E, Basso G, Leo G, Rosato A, Bicciato S, et al. 2012. SHARP1 suppresses breast cancer metastasis by promoting degradation of hypoxia-inducible factors. *Nature* **487:** 380–384.

Moon EJ, Giaccia A. 2015. Dual roles of NRF2 in tumor prevention and progression: Possible implications in cancer treatment. *Free Radic Biol Med* **79:** 292–299.

Moyer VA; U.S. Preventive Services Task Force. 2014. Vitamin, mineral, and multivitamin supplements for the primary prevention of cardiovascular disease and cancer: U.S. Preventive Services Task Force recommendation statement. *Ann Intern Med* **160:** 558–564.

Mullarky E, Mattaini KR, Vander Heiden MG, Cantley LC, Locasale JW. 2011. PHGDH amplification and altered glucose metabolism in human melanoma. *Pigment Cell Melanoma Res* **24:** 1112–1115.

Myung SK, Kim Y, Ju W, Choi HJ, Bae WK. 2010. Effects of antioxidant supplements on cancer prevention: Meta-analysis of randomized controlled trials. *Ann Oncol* **21:** 166–179.

Nagrath S, Sequist LV, Maheswaran S, Bell DW, Irimia D, Ulkus L, Smith MR, Kwak EL, Digumarthy S, Muzikansky A, et al. 2007. Isolation of rare circulating tumour cells in cancer patients by microchip technology. *Nature* **450:** 1235–1239.

Neumann CA, Krause DS, Carman CV, Das S, Dubey DP, Abraham JL, Bronson RT, Fujiwara Y, Orkin SH, Van Etten RA. 2003. Essential role for the peroxiredoxin Prdx1 in erythrocyte antioxidant defence and tumour suppression. *Nature* **424:** 561–565.

Nguyen A, Loo JM, Mital R, Weinberg EM, Man FY, Zeng Z, Paty PB, Saltz L, Janjigian YY, de Stanchina E, et al. 2016. PKLR promotes colorectal cancer liver colonization through induction of glutathione synthesis. *J Clin Invest* **126:** 681–694.

Nioi P, Hayes JD. 2004. Contribution of NAD(P)H:quinone oxidoreductase 1 to protection against carcinogenesis, and regulation of its gene by the Nrf2 basic-region leucine zipper and the arylhydrocarbon receptor basic helix-loop-helix transcription factors. *Mutat Res* **555:** 149–171.

Nioi P, McMahon M, Itoh K, Yamamoto M, Hayes JD. 2003. Identification of a novel Nrf2-regulated antioxidant response element (ARE) in the mouse NAD(P)H:quinone oxidoreductase 1 gene: Reassessment of the ARE consensus sequence. *Biochem J* **374:** 337–348.

Nowicki MO, Falinski R, Koptyra M, Slupianek A, Stoklosa T, Gloc E, Nieborowska-Skorska M, Blasiak J, Skorski T. 2004. BCR/ABL oncogenic kinase promotes unfaithful repair of the reactive oxygen species–dependent DNA double-strand breaks. *Blood* **104:** 3746–3753.

Ohta T, Iijima K, Miyamoto M, Nakahara I, Tanaka H, Ohtsuji M, Suzuki T, Kobayashi A, Yokota J, Sakiyama T, et al. 2008. Loss of Keap1 function activates Nrf2 and provides advantages for lung cancer cell growth. *Cancer Res* **68:** 1303–1309.

Omenn GS, Goodman GE, Thornquist MD, Balmes J, Cullen MR, Glass A, Keogh JP, Meyskens FL Jr, Valanis B, Williams JH Jr, et al. 1996a. Risk factors for lung cancer and for intervention effects in CARET, the β-Carotene and Retinol Efficacy Trial. *J Natl Cancer Inst* **88:** 1550–1559.

Omenn GS, Goodman GE, Thornquist MD, Balmes J, Cullen MR, Glass A, Keogh JP, Meyskens FL, Valanis B, Williams JH, et al. 1996b. Effects of a combination of β carotene and vitamin A on lung cancer and cardiovascular disease. *N Engl J Med* **334:** 1150–1155.

Patra KC, Hay N. 2014. The pentose phosphate pathway and cancer. *Trends Biochem Sci* **39:** 347–354.

Piskounova E, Agathocleous M, Murphy MM, Hu Z, Huddlestun SE, Zhao Z, Leitch AM, Johnson TM, DeBerardinis RJ, Morrison SJ. 2015. Oxidative stress inhibits distant metastasis by human melanoma cells. *Nature* **527:** 186–191.

Porporato PE, Payen VL, Perez-Escuredo J, De Saedeleer CJ, Danhier P, Copetti T, Dhup S, Tardy M, Vazeille T, Bouzin C, et al. 2014. A mitochondrial switch promotes tumor metastasis. *Cell Rep* **8:** 754–766.

Possemato R, Marks KM, Shaul YD, Pacold ME, Kim D, Birsoy K, Sethumadhavan S, Woo HK, Jang HG, Jha AK, et al. 2011. Functional genomics reveal that the serine synthesis pathway is essential in breast cancer. *Nature* **476:** 346–350.

Qiao YL, Dawsey SM, Kamangar F, Fan JH, Abnet CC, Sun XD, Johnson LL, Gail MH, Dong ZW, Yu B, et al. 2009. Total and cancer mortality after supplementation with vitamins and minerals: Follow-up of the Linxian General Population Nutrition Intervention Trial. *J Natl Cancer Inst* **101:** 507–518.

Qu Y, Wang J, Ray PS, Guo H, Huang J, Shin-Sim M, Bukoye BA, Liu B, Lee AV, Lin X, et al. 2011. Thioredoxin-like 2 regulates human cancer cell growth and metastasis via redox homeostasis and NF-κB signaling. *J Clin Invest* **121:** 212–225.

Raj L, Ide T, Gurkar AU, Foley M, Schenone M, Li X, Tolliday NJ, Golub TR, Carr SA, Shamji AF, et al. 2011. Selective killing of cancer cells by a small molecule targeting the stress response to ROS. *Nature* **475:** 231–234.

Ramos-Gomez M, Kwak MK, Dolan PM, Itoh K, Yamamoto M, Talalay P, Kensler TW. 2001. Sensitivity to carcinogenesis is increased and chemoprotective efficacy of enzyme inducers is lost in *nrf2* transcription factor–deficient mice. *Proc Natl Acad Sci* **98:** 3410–3415.

Ranjan P, Anathy V, Burch PM, Weirather K, Lambeth JD, Heintz NH. 2006. Redox-dependent expression of cyclin D1 and cell proliferation by Nox1 in mouse lung epithelial cells. *Antioxid Redox Signal* **8:** 1447–1459.

Reliene R, Schiestl RH. 2006. Antioxidant *N*-acetyl cysteine reduces incidence and multiplicity of lymphoma in *Atm* deficient mice. *DNA Repair (Amst)* **5:** 852–859.

Reliene R, Fischer E, Schiestl RH. 2004. Effect of *N*-acetyl cysteine on oxidative DNA damage and the frequency of DNA deletions in *Atm*-deficient mice. *Cancer Res* **64:** 5148–5153.

Ren JG, Seth P, Clish CB, Lorkiewicz PK, Higashi RM, Lane AN, Fan TW, Sukhatme VP. 2014. Knockdown of malic enzyme 2 suppresses lung tumor growth, induces differentiation and impacts PI3K/AKT signaling. *Sci Rep* **4:** 5414.

Reznik E, Miller ML, Senbabaoglu Y, Riaz N, Sarungbam J, Tickoo SK, Al-Ahmadie HA, Lee W, Seshan VE, Hakimi AA, et al. 2016. Mitochondrial DNA copy number variation across human cancers. *Elife* **5:** e10769.

Rhee SG. 2006. Cell signaling. H_2O_2, a necessary evil for cell signaling. *Science* **312:** 1882–1883.

Rolfs F, Huber M, Gruber F, Bohm F, Pfister HJ, Bochkov VN, Tschachler E, Dummer R, Hohl D, Schafer M, et al. 2013. Dual role of the antioxidant enzyme peroxiredoxin 6 in skin carcinogenesis. *Cancer Res* **73:** 3460–3469.

Rowe LA, Degtyareva N, Doetsch PW. 2008. DNA damage-induced reactive oxygen species (ROS) stress response in *Saccharomyces cerevisiae*. *Free Radic Biol Med* **45:** 1167–1177.

Rushmore TH, Morton MR, Pickett CB. 1991. The antioxidant responsive element. Activation by oxidative stress and identification of the DNA consensus sequence required for functional activity. *J Biol Chem* **266:** 11632–11639.

Sabharwal SS, Schumacker PT. 2014. Mitochondrial ROS in cancer: Initiators, amplifiers or an Achilles' heel? *Nat Rev Cancer* **14:** 709–721.

Sablina AA, Budanov AV, Ilyinskaya GV, Agapova LS, Kravchenko JE, Chumakov PM. 2005. The antioxidant function of the p53 tumor suppressor. *Nat Med* **11:** 1306–1313.

Santos MA, Faryabi RB, Ergen AV, Day AM, Malhowski A, Canela A, Onozawa M, Lee JE, Callen E, Gutierrez-Martinez P, et al. 2014. DNA-damage-induced differentiation of leukaemic cells as an anti-cancer barrier. *Nature* **514:** 107–111.

Satoh H, Moriguchi T, Taguchi K, Takai J, Maher JM, Suzuki T, Winnard PT Jr, Raman V, Ebina M, Nukiwa T, et al. 2010. Nrf2-deficiency creates a responsive microenvironment for metastasis to the lung. *Carcinogenesis* **31:** 1833–1843.

Satoh H, Moriguchi T, Takai J, Ebina M, Yamamoto M. 2013. Nrf2 prevents initiation but accelerates progression through the Kras signaling pathway during lung carcinogenesis. *Cancer Res* **73:** 4158–4168.

Sattler M, Verma S, Shrikhande G, Byrne CH, Pride YB, Winkler T, Greenfield EA, Salgia R, Griffin JD. 2000. The BCR/ABL tyrosine kinase induces production of reactive oxygen species in hematopoietic cells. *J Biol Chem* **275:** 24273–24278.

Sayin VI, Ibrahim MX, Larsson E, Nilsson JA, Lindahl P, Bergo MO. 2014. Antioxidants accelerate lung cancer progression in mice. *Sci Transl Med* **6:** 221ra215.

Schafer ZT, Grassian AR, Song L, Jiang Z, Gerhart-Hines Z, Irie HY, Gao S, Puigserver P, Brugge JS. 2009. Antioxidant and oncogene rescue of metabolic defects caused by loss of matrix attachment. *Nature* **461:** 109–113.

Schieber M, Chandel NS. 2014. ROS function in redox signaling and oxidative stress. *Curr Biol* **24:** R453–R462.

Schrader M, Fahimi HD. 2006. Peroxisomes and oxidative stress. *Biochim Biophys Acta* **1763:** 1755–1766.

Semenza GL. 2011. Regulation of metabolism by hypoxia-inducible factor 1. *Cold Spring Harb Symp Quant Biol* **76:** 347–353.

Senft D, Ronai ZA. 2016. Regulators of mitochondrial dynamics in cancer. *Curr Opin Cell Biol* **39:** 43–52.

Shi Z, Wang B, Chihanga T, Kennedy MA, Weber GF. 2014. Energy metabolism during anchorage-independence. Induction by osteopontin-c. *PLoS ONE* **9:** e105675.

Shibata T, Kokubu A, Gotoh M, Ojima H, Ohta T, Yamamoto M, Hirohashi S. 2008a. Genetic alteration of Keap1 confers constitutive Nrf2 activation and resistance to chemotherapy in gallbladder cancer. *Gastroenterology* **135:** 1358–1368, 1368 e1351–1354.

Shibata T, Ohta T, Tong KI, Kokubu A, Odogawa R, Tsuta K, Asamura H, Yamamoto M, Hirohashi S. 2008b. Cancer related mutations in NRF2 impair its recognition by Keap1-Cul3 E3 ligase and promote malignancy. *Proc Natl Acad Sci* **105:** 13568–13573.

Singh A, Misra V, Thimmulappa RK, Lee H, Ames S, Hoque MO, Herman JG, Baylin SB, Sidransky D, Gabrielson E, et al. 2006. Dysfunctional KEAP1-NRF2 interaction in non-small-cell lung cancer. *PLoS Med* **3:** e420.

Singh A, Happel C, Manna SK, Acquaah-Mensah G, Carrerero J, Kumar S, Nasipuri P, Krausz KW, Wakabayashi N, Dewi R, et al. 2013. Transcription factor NRF2 regulates miR-1 and miR-206 to drive tumorigenesis. *J Clin Invest* **123:** 2921–2934.

Son J, Lyssiotis CA, Ying H, Wang X, Hua S, Ligorio M, Perera RM, Ferrone CR, Mullarky E, Shyh-Chang N, et al. 2013. Glutamine supports pancreatic cancer growth through a KRAS-regulated metabolic pathway. *Nature* **496:** 101–105.

Stott SL, Lee RJ, Nagrath S, Yu M, Miyamoto DT, Ulkus L, Inserra EJ, Ulman M, Springer S, Nakamura Z, et al. 2010. Isolation and characterization of circulating tumor cells from patients with localized and metastatic prostate cancer. *Sci Transl Med* **2:** 25ra23.

Sukhatme VP, Chan B. 2012. Glycolytic cancer cells lacking 6-phosphogluconate dehydrogenase metabolize glucose to induce senescence. *FEBS Lett* **586:** 2389–2395.

Sullivan JP, Nahed BV, Madden MW, Oliveira SM, Springer S, Bhere D, Chi AS, Wakimoto H, Rothenberg SM, Sequist LV, et al. 2014. Brain tumor cells in circulation are enriched for mesenchymal gene expression. *Cancer Discov* **4:** 1299–1309.

Sundaresan M, Yu ZX, Ferrans VJ, Irani K, Finkel T. 1995. Requirement for generation of H_2O_2 for platelet-derived growth factor signal transduction. *Science* **270:** 296–299.

Szatrowski TP, Nathan CF. 1991. Production of large amounts of hydrogen peroxide by human tumor cells. *Cancer Res* **51:** 794–798.

Tasdogan A, Kumar S, Allies G, Bausinger J, Beckel F, Hofemeister H, Mulaw M, Madan V, Scharfetter-Kochanek K, Feuring-Buske M, et al. 2016. DNA damage-induced HSPC malfunction depends on ROS accumulation downstream of IFN-1 signaling and bid mobilization. *Cell Stem Cell* **19:** 752–767.

Thimmulappa RK, Mai KH, Srisuma S, Kensler TW, Yamamoto M, Biswal S. 2002. Identification of Nrf2-regulated genes induced by the chemopreventive agent sulforaphane by oligonucleotide microarray. *Cancer Res* **62:** 5196–5203.

Toullec A, Gerald D, Despouy G, Bourachot B, Cardon M, Lefort S, Richardson M, Rigaill G, Parrini MC, Lucchesi C, et al. 2010. Oxidative stress promotes myofibroblast differentiation and tumour spreading. *EMBO Mol Med* **2:** 211–230.

Vafa O, Wade M, Kern S, Beeche M, Pandita TK, Hampton GM, Wahl GM. 2002. c-Myc can induce DNA damage, increase reactive oxygen species, and mitigate p53 function: A mechanism for oncogene-induced genetic instability. *Mol Cell* **9:** 1031–1044.

Vanharanta S, Massague J. 2013. Origins of metastatic traits. *Cancer Cell* **24:** 410–421.

Vanharanta S, Shu W, Brenet F, Hakimi AA, Heguy A, Viale A, Reuter VE, Hsieh JJ, Scandura JM, Massague J. 2013. Epigenetic expansion of VHL-HIF signal output drives multiorgan metastasis in renal cancer. *Nat Med* **19:** 50–56.

Van Remmen H, Ikeno Y, Hamilton M, Pahlavani M, Wolf N, Thorpe SR, Alderson NL, Baynes JW, Epstein CJ, Huang TT, et al. 2003. Life-long reduction in MnSOD activity results in increased DNA damage and higher incidence of cancer but does not accelerate aging. *Physiol Genomics* **16:** 29–37.

van Zandwijk N, Dalesio O, Pastorino U, de Vries N, van Tinteren H. 2000. EUROSCAN, a randomized trial of vitamin A and *N*-acetylcysteine in patients with head and neck cancer or lung cancer. For the European Organization for Research and Treatment of Cancer Head and Neck and Lung Cancer Cooperative Groups. *J Natl Cancer Inst* **92:** 977–986.

Vera-Ramirez L, Ramirez-Tortosa M, Perez-Lopez P, Granados-Principal S, Battino M, Quiles JL. 2012. Long-term effects of systemic cancer treatment on DNA oxidative damage: The potential for targeted therapies. *Cancer Lett* **327:** 134–141.

Vyas S, Zaganjor E, Haigis MC. 2016. Mitochondria and cancer. *Cell* **166:** 555–566.

Wang H, Liu X, Long M, Huang Y, Zhang L, Zhang R, Zheng Y, Liao X, Wang Y, Liao Q, et al. 2016. NRF2 activation by antioxidant antidiabetic agents accelerates tumor metastasis. *Sci Transl Med* **8:** 334ra351.

Wu KC, Cui JY, Klaassen CD. 2011. Beneficial role of Nrf2 in regulating NADPH generation and consumption. *Toxicol Sci* **123:** 590–600.

Xu C, Huang MT, Shen G, Yuan X, Lin W, Khor TO, Conney AH, Kong AN. 2006. Inhibition of 7,12-dimethylbenz(*a*)anthracene-induced skin tumorigenesis in C57BL/6 mice by sulforaphane is mediated by nuclear factor E2–related factor 2. *Cancer Res* **66:** 8293–8296.

Yasueda A, Urushima H, Ito T. 2016. Efficacy and interaction of antioxidant supplements as adjuvant therapy in cancer treatment: A systematic review. *Integr Cancer Ther* **15:** 17–39.

Ye J, Fan J, Venneti S, Wan YW, Pawel BR, Zhang J, Finley LW, Lu C, Lindsten T, Cross JR, et al. 2014. Serine catabolism regulates mitochondrial redox control during hypoxia. *Cancer Discov* **4:** 1406–1417.

Yu M, Bardia A, Wittner BS, Stott SL, Smas ME, Ting DT, Isakoff SJ, Ciciliano JC, Wells MN, Shah AM, et al. 2013. Circulating breast tumor cells exhibit dynamic changes in epithelial and mesenchymal composition. *Science* **339:** 580–584.

Yu M, Bardia A, Aceto N, Bersani F, Madden MW, Donaldson MC, Desai R, Zhu H, Comaills V, Zheng Z, et al. 2014. Cancer therapy. Ex vivo culture of circulating breast tumor cells for individualized testing of drug susceptibility. *Science* **345:** 216–220.

Zhang S, Liu X, Bawa-Khalfe T, Lu LS, Lyu YL, Liu LF, Yeh ET. 2012. Identification of the molecular basis of doxorubicin-induced cardiotoxicity. *Nat Med* **18:** 1639–1642.

Zhao J, Zhang J, Yu M, Xie Y, Huang Y, Wolff DW, Abel PW, Tu Y. 2013. Mitochondrial dynamics regulates migration and invasion of breast cancer cells. *Oncogene* **32:** 4814–4824.

Zhao T, Zhu Y, Morinibu A, Kobayashi M, Shinomiya K, Itasaka S, Yoshimura M, Guo G, Hiraoka M, Harada H. 2014. HIF-1-mediated metabolic reprogramming reduces ROS levels and facilitates the metastatic colonization of cancers in lungs. *Sci Rep* **4:** 3793.

RON Signaling Is a Key Mediator of Tumor Progression in Many Human Cancers

NAJME FAHAM AND ALANA L. WELM
Department of Oncological Sciences, Huntsman Cancer Institute, University of Utah, Salt Lake City, Utah 84112
Correspondence: alana.welm@hci.utah.edu

With an increasing body of literature covering RON receptor tyrosine kinase function in different types of human cancers, it is becoming clear that RON has prominent roles in both cancer cells and in the tumor-associated microenvironment. RON not only activates several oncogenic signaling pathways in cancer cells, leading to more aggressive behavior, but also promotes an immunosuppressive, alternatively activated phenotype in macrophages and limits the antitumor immune response. These two unique functions of this oncogene, the strong correlation between RON expression and poor outcomes in cancer, and the high tolerability of a new RON inhibitor make it an exciting therapeutic target, the blocking of which offers an advantage toward improving the survival of cancer patients. Here, we discuss recent findings on the role of RON signaling in cancer progression and its potential in cancer therapy.

Approximately 20 receptor tyrosine kinase (RTK) families have been identified so far in humans (Lemmon and Schlessinger 2010). In addition to their similar structure (sharing extracellular, transmembrane, and intracellular domains), tyrosine kinases conduct similar functions by regulating target protein posttranslational modification through transfer of phosphates from ATP to the hydroxyl group of a tyrosine (Manning et al. 2002). The result of this phosphorylation is activation of intrinsic tyrosine kinase activity, which eventually leads to signal transduction via multiple downstream signaling cascades (Schlessinger 2000).

One of these RTKs with various biological activities is receptor d'origine nantais (RON), which has gained a lot of attention since its discovery in 1993 (Ronsin et al. 1993). RON belongs to the MET proto-oncogene family that, together with the other prototype RTK, MET, constitute the only members of this family in humans (Wang et al. 2006). In animals, though, different orthologs have been identified that are highly homologous to human RON: STK in mice and Sea in chickens, pointing to the conservation of RON through evolution in different species (Hayman 1987; Iwama et al. 1994). Shortly after the discovery of RON, macrophage-stimulating protein (MSP) was identified as the only ligand for RON (Gaudino et al. 1994; Wang et al. 1994). Physiological roles of RON include regulation of the innate immune response during inflammation and promotion of wound healing (Gaudino et al. 1995; Quantin et al. 1995; Correll et al. 1997; Morrison and Correll 2002; Wang et al. 2002).

RON is expressed at low levels in healthy adult tissues of epithelial origin (such as skin, colon, breast, lung, and kidney) and at various levels in macrophages, hematopoietic cells, and osteoclasts (Wang et al. 2006; Meyer et al. 2009; Kretschmann et al. 2010; Fialin et al. 2013). Studies indicate that RON can also be expressed in fibroblasts during pathological conditions (Tong et al. 2011; Benight and Waltz 2012). Expression of RON, however, becomes high in several epithelial tumors such as colon, lung, breast, stomach, ovary, pancreas, and bladder cancers (Maggiora et al. 1998, 2003; Willett et al. 1998; Chen et al. 2000; Okino et al. 2001; Camp et al. 2005; Cheng et al. 2005; Lee et al. 2005). In tumors with mesenchymal origin, such as sarcoma, RON has been less studied; childhood Ewing sarcoma and rhabdomyosarcoma are the only two examples in which RON has been shown to be expressed and activated (Potratz et al. 2010). Within the last few years, data have accumulated to unveil more and more about the roles of RON both in cancer and noncancer settings. Here we discuss recent findings regarding the pathological relevance of RON in cancer progression.

STRUCTURAL AND BIOCHEMICAL PROPERTIES OF HUMAN RON

The *RON* gene, residing on chromosome 3p21, contains 20 exons and 19 introns that encode 1400 amino acids to generate the full-length protein (Ronsin et al. 1993; Angeloni et al. 2000). RON is first translated as a

Published by Cold Spring Harbor Laboratory Press; doi: 10.1101/sqb.2016.81.031377

single-chain cytoplasmic pro-protein. Proteolytic processing results in presentation of the receptor at the cell surface as a disulfide-linked heterodimeric receptor consisting of a 45-kDa extracellular α chain and a 150-kDa transmembrane-spanning β chain that includes a highly conserved intracellular catalytic domain (Gaudino et al. 1994; Wang et al. 1994; Waltz et al. 2001). The extracellular sequences of RON contain several unique structures including the semaphorin (sema) domain, followed by the plexin–semaphorin–integrin (PSI) domain and four immunoglobulin–plexin–transcription (IPT) domains (Chang et al. 2015). The sema domain is responsible for recognizing the receptor ligand, MSP (Wang et al. 2006). All other extracellular domains have specific roles and are often subjected to deletion or truncation under pathological conditions, which results in generation of multiple isoforms with different oncogenic activities. For example, the IPT units are required to maintain the integrity of RON—its activity, proper maturation, and cell surface localization (Collesi et al. 1996; Lu et al. 2007; Zhang et al. 2010). Accordingly, deletion of the first IPT domain causes conformational changes and spontaneous dimerization of RON. This dimerization leads to constitutive activation, resistance to proteolytic degradation, and increased oncogenic potential of RON, as indicated by promotion of epithelial-to-mesenchymal transition (EMT), cell migration, and colony formation (Ma et al. 2010a). Alterations in the IPT units of RON occur with high frequencies in several types of cancer such as colon, pancreas, glioma, and breast (Wang et al. 2000; Zhou et al. 2003; Xu et al. 2004, 2005; Eckerich et al. 2009; Yao et al. 2013b; Chakedis et al. 2016b). Specific oncogenic variants of RON are discussed in detail below.

The intracellular region of Ron consists of the juxtamembrane (JM) domain, the highly conserved kinase domain, and the noncatalytic carboxy-terminal tail (Danilkovitch et al. 1999). The JM domain is involved in regulation of RON stability through recruitment of ubiquitin–protein ligase c-Cbl to phosphorylated Y1017, which results in subsequent ubiquitination and degradation of RON (Penengo et al. 2003; Thien and Langdon 2005; Germano et al. 2006). Alteration of this region, either through mutation or partial deletion, makes RON resistant to c-Cbl-mediated degradation and results in increased stability, constitutive activation, and enhanced oncogenic potential of RON (Wei et al. 2005a). In addition to the regulatory role of Y1017 in the JM domain, an acidic JM-C region has been identified in this domain that is critical for RON autoinhibition. Experimental deletion of this region results in increased receptor autophosphorylation (Wang et al. 2014).

The kinase domain consists of two lobes with several subdomains. The N-lobe contains the αC helix and the P loop, which are essential for ATP recruitment. Y1238 and Y1239 are the two essential tyrosines that reside in the activation loop of the C-lobe and their phosphorylation is the hallmark of RON activity (Gaudino et al. 1994; Danilkovitch and Leonard 1999). These two tyrosines are persistently phosphorylated in oncogenic RON variants to sustain active downstream signal transduction. There are some other essential residues in the N-lobe, such as K1114 and E1130, which are required for ATP binding to RON; mutation of these residues converts RON to a kinase dead receptor (van den Akker et al. 2004; Wang et al. 2014). In addition to Y1238 and Y1239, there are other tyrosines that contribute to regulation of RON activity. For example, Y1198 is highly conserved among RTKs and its phosphorylation is directly associated with RON activity (Jeffers et al. 1997; Wei et al. 2005b). Mutation of this tyrosine to phenylalanine dramatically reduces RON autophosphorylation (Wang et al. 2014). Y1317 is another conserved tyrosine in the consensus sequence of the kinase domain that forms a putative docking site for SH2-containing molecules. Mutation of this tyrosine abolishes the level of receptor trans-autophosphorylation, pointing to its importance in receptor activity (Santoro et al. 2000).

The carboxy-terminal domain of RON contains two conserved tandem tyrosines, Y1353 and Y1360, known as docking sites (Ponzetto et al. 1994; Iwama et al. 1996; Wang et al. 2002). These sites serve as bridges to transduce signals from the kinase domain to downstream effectors of RON through binding to the SH2 domain of several adaptors, including Gab1 as the key adaptor (Ponzetto et al. 1994; Germano et al. 2006; Chaudhuri et al. 2011). The role of these tyrosines and their necessity for RON signal transduction is somewhat controversial, however. It has been shown that an experimental mutation in the kinase domain of RON (M1254 T) is able to overcome the requirement for Y1353 and Y1360 in promoting oncogenic activity (Santoro et al. 2000). The docking tyrosines in murine RON/Stk have also been shown to be dispensable for constitutive activation of mitogen-activated protein kinase (MAPK) (Wei et al. 2005b). However, in a separate study, these tyrosines were shown to have inhibitory function on the kinase activity of RON when expressed as a competing peptide, suggesting they are important for RON activity (Yokoyama et al. 2005). In our hands, mutation of these two tyrosines to phenylalanine makes the receptor even more competent for activating phosphoinositide 3-kinase (PI3K) and MAPK pathways (N Faham, unpubl.). The apparent discrepancy in the requirement for Y1353 and Y1360 is likely due to the presence of several critical tyrosines in the kinase domain that can compensate as docking sites for transducing the signal to effectors in the absence of the two carboxy-terminal tyrosines (Hanks and Quinn 1991; Songyang et al. 1995; Santoro et al. 2000; Wei et al. 2005a). These observations are in line with a study showing that the multisubstrate docking site of MET is not required for transmitting the signal to RAS and inducing cell scattering (Tulasne et al. 1999). Based on these investigations, it has been suggested that not all signaling and biological activities of MET are mediated through its carboxy-terminal tyrosine residues. In the case of RON, it is clear that deletion of the entire carboxy-terminal tail abrogates downstream signaling and eliminates RON-mediated tumorigenic activities (Lu et al. 2007).

LIGAND-DEPENDENT VERSUS LIGAND-INDEPENDENT RON SIGNALING

RON activation under physiological conditions, where there is basal level of RON, occurs through binding of its ligand, MSP (Danilkovitch et al. 1999; Wang et al. 2006, 2013a; Chaudhuri et al. 2013; Chaudhuri 2014). Hepatocytes are the major source of MSP, which activates RON in a paracrine manner (Bezerra et al. 1993; Yoshimura et al. 1993). Other organs like lungs, adrenal glands, placenta, and kidney can also express MSP, but to a lesser extent (Chang et al. 2015). Under pathological conditions, (e.g., in certain cancers), MSP is overexpressed in cancer cells along with RON and can lead to autocrine activation of RON (Riggins et al. 2006). MSP circulates in the blood in its biologically inactive form (Nanney et al. 1998; Rampino et al. 2002; Kawaguchi et al. 2009), and cleavage by matriptase or other serine-like proteases gives rise to conversion of pro-MSP to the mature dimeric active form (Bhatt et al. 2007; Kawaguchi et al. 2009). Binding of active MSP to the sema domains of RON induces receptor dimerization and subsequent conformational changes in the extracellular domain, which increases MSP binding affinity (Wang et al. 1997; Carafoli et al. 2005; Chao et al. 2012). These conformational changes are critical to trigger RON activation (Yokoyama et al. 2005) and are followed by sequential phosphorylation of regulatory tyrosines in the kinase domain and docking tyrosines in the carboxy-terminal tail (Ponzetto et al. 1994; Danilkovitch et al. 1999).

In the case of RON overexpression (e.g., in tumors), activation can occur independently of MSP by formation of RON homodimers via the sema domain (Chao et al. 2012). It seems that the close proximity of densely expressed RON molecules allows them to transphosphorylate each other without requiring ligand (Danilkovitch-Miagkova 2003). It is unclear, however, whether activation of RON via MSP versus ligand-independent activation due to RON overexpression leads to exposure of different tyrosines and, therefore, differential phosphorylation patterns on the RON receptor. In one study, MSP binding caused increased phosphorylation of Y1360 and not Y1353, compared with auto-activation of RON (Feres et al. 2009). This is in contrast with another study indicating Y1353 as the critical tyrosine for MSP-induced RON function (Chaudhuri et al. 2011). The reason for these discrepancies might be due to analysis of different cell types.

Even though both MSP-dependent and MSP-independent RON activation result in up-regulation of kinase activity and enhanced downstream signaling (Gaudino et al. 1994), there are some differences in these two types of activation with regard to biological outcomes. Based on the evidence so far and examples below, RON does not require MSP for all of its intrinsic functions in cancer cells, but there are certain functions that are MSP-dependent. On the other hand, to our knowledge thus far, MSP appears to be completely dependent on RON to mediate its function (Wang et al. 1994, 1995). For instance, with regard to the intrinsic effect of RON on cancer cells, it has been shown that MSP is necessary to prevent anoikis in RON-overexpressing epithelial cells under serum-starved conditions. Neither expression of RON alone nor presence of MSP without RON was enough to prevent anoikis in these cells, implying ligand-dependent function (Danilkovitch et al. 2000). We have found that RON overexpression in breast cancer cell lines that contain no detectable MSP is sufficient to promote cancer progression and metastasis when injected orthotopically in immunocompromised mice (N Faham, unpubl.). This is presumably a ligand-independent function because murine MSP is widely believed not to be able to bind and activate human RON. MSP expression in these cells results in a higher incidence of metastasis suggesting that the metastatic potential of RON is enhanced through ligand-dependent mechanisms (Zinser et al. 2006; Welm et al. 2007; Cunha et al. 2014). Similarly, Feres et al. (2009) reported that expression of RON in immortalized human breast MCF-10A cells resulted in protection from cell death and increased spreading and migratory potential, independent of MSP. However, MSP stimulation was required to boost RON-mediated cell migration and proliferation. Another example of MSP-independent function of RON was reported in bladder cancer cells under serum-starved conditions, where RON translocated to the nucleus in complex with EGFR and operated as a transcriptional activator of approximately 134 different target genes. Interestingly, neither MSP stimulation nor RON homodimerization or phosphorylation was required for nuclear translocation of RON (Liu et al. 2010). This group has suggested that cancer cells can bypass regular mechanisms of RON activation under stress conditions to confer a survival advantage. In yet separate study, phosphorylation of RON through interaction with integrins and as a consequence of adhesion to extracellular matrix is another example of MSP independent activation of RON, which was mediated by c-Src-FAK pathway (Danilkovitch-Miagkova et al. 2000).

MECHANISMS OF RON ACTIVATION IN CANCER

Activation of RON in tumors is most often due to receptor overexpression, rather than classical MSP binding (Camp et al. 2007; Wang et al. 2007; Yao et al. 2013a). Mutations in the *RON* gene, generation of splicing variants/truncated forms, and, very rarely, increased gene copy number are also documented mechanisms of RON activation in different cancers (Wang et al. 2013a).

Mutation of *RON* (resulting in R1018G, located in the JM domain) has been reported in 11% of gastroesophageal adenocarcinoma, but *RON* is rarely mutated in other cancer types (Catenacci et al. 2011; Yao et al. 2013a). As a result of this mutation, which disrupts the conserved c-Cbl binding motif, RON protein stays stable and active. In addition to mutation, increased copy number is another mechanism of RON activation in gastroesophageal cancer, which has been shown to be a prognostic factor for worse survival (Catenacci et al. 2011).

In contrast to rare gene mutation and amplification events as causes of RON activation, generation of oncogenic splicing variants and truncated forms are common in multiple epithelial cancers (Wang et al. 2006). Approximately nine different nonmutated RON isoforms have been described in various epithelial cancer types including colorectal cancer, glioma, and breast cancer (Gaudino et al. 1994; Zhou et al. 2003; Eckerich et al. 2009; Liu et al. 2011; Yao et al. 2013b). Exon skipping is the most common event leading to splice variants, termed RONΔ170, RONΔ165, RONΔ160, and RONΔ155 (Collesi et al. 1996; Zhou et al. 2003; Lu et al. 2007). Except for RONΔ170 which is kinase domain defective, and soluble RONΔ85, which acts as a dominant-negative isoform without any scattering activity, the rest of the isoforms are constitutively active in the absence of MSP and some have been shown to have cell-transforming potential (Angeloni et al. 2003, 2004; Wang et al. 2006; Ma et al. 2010a,b).

A unique example of a potent, constitutively active, variant of RON is RONΔ55 (referred to as short form, or sfRON). sfRON lacks most of the extracellular domains, but retains the transmembrane and intracellular domains. In contrast to most of the other RON variants that are generated by exon skipping, RONΔ55 is produced by an alternative transcription start site within exon 10 of *RON* (Bardella et al. 2004). We have shown that overexpression of this isoform converts slow-growing, nonmetastatic breast tumors derived from MCF7 cells into fast-growing tumors that spontaneously metastasizes from the mammary gland to liver and bones. Mechanistic studies revealed that this phenotype occurs through strong activation of PI3K pathway (Liu et al. 2011). Overexpression of sfRON in ovarian cancer cells also causes activation of PI3K/PDK1 pathway, resulting in EMT and aggressive tumor growth and metastasis (Moxley et al. 2016). Along with these findings, overexpression of sfRON in T47D cells increased their motility and growth rate which was accompanied by increase in EMT-related SLUG expression (Bardella et al. 2004). Oncogenic signaling of sfRON in acute myeloid leukemia, however, has been reported through activation of the Src family kinase Lyn as well as Bcl-2 (Fialin et al. 2013).

Recently, a novel splice variant of RON (termed P5P6) was found in the majority of pancreatic cancers as a result of partial skipping of exon 5 and 6. This isoform is constitutively phosphorylated (in the absence of MSP), activates AKT, and has transforming ability in immortalized pancreatic epithelial cells (Chakedis et al. 2016a). This isoform, together with sfRON, comprises most of the RON protein in pancreatic cancer specimens. Expression of these variants induces markedly different patterns of gene expression than wild-type RON (Chakedis et al. 2016b). It is interesting that even though RON P5P6 is similar to RONΔ160 (both resulting from complete skipping of exon 5 and 6), RON P5P6 signals through AKT, whereas RONΔ160 preferentially activates the β-catenin pathway (Xu et al. 2005). Likewise, most RON variants can activate multiple signaling cascades, but with different substrate specificity than that of wild-type RON. These distinct features arise from increased catalytic efficiency of the kinase domain, which changes the substrate profile (Santoro et al. 1998; Xu et al. 2005). It would be interesting to characterize signaling pathways downstream from different RON isoforms, either individually or as heterodimer with RON or other RTKs, because not all of these isoforms have been analyzed in detail. Lack of partial or complete domains in these variants leads to generation of different three-dimensional structures that either reside on the cell surface or remain intracellular because of their inability to convert to a two-chain form (Iwama et al. 1996). Structural characterization of these isoforms and their effect on downstream signaling cascades in different tumors may be critical for designing treatment strategies. As some of these RON variants are constitutively active because of either partial or complete deletion of the extracellular domain, they do not require MSP for activation. Therefore, monoclonal antibodies that prevent binding of MSP to RON may be ineffective in cancers harboring these isoforms. Using small molecule inhibitors or antibodies with a different mode of action might have a broader efficacy.

SIGNALING PATHWAYS DOWNSTREAM FROM RON IN CANCER PROGRESSION

It is well accepted that RON mediates several oncogenic functions in cancer cells, including cell proliferation and survival; cell adhesion and spreading; cell dissociation and migration; EMT and matrix invasion; and establishment of metastasis. These features point to the importance of RON in later stages of cancer and RON expression correlates strongly with invasion, tumor stage, and poor prognosis in most, but not all, cancers (Ronsin et al. 1993; Wang et al. 2003, 2006; Lee et al. 2005; O'Toole et al. 2006; Thomas et al. 2007; Welm et al. 2007; Catenacci et al. 2011; Benight and Waltz 2012; Song et al. 2012; Tactacan et al. 2012).

Aggressive cancer phenotypes caused by RON are the result of activation of complex downstream signaling networks including PI3K/AKT, MAPK, JNK, β-catenin, and STAT pathways (Danilkovitch-Miagkova 2003; Yao et al. 2013a). Even though numerous signaling pathways downstream from RON have been described, few studies have focused on the detailed regulation of these pathways in each phenotype.

Regarding tumorigenesis, overexpression of RON in mammary epithelium using a transgenic mouse model induced a tumorigenic phenotype associated with high metastatic burden in lung and liver. Expression analysis of molecules involved in cell cycle progression revealed that RON overexpression led to increased phosphorylation of MAPK and β-catenin, and up-regulation of β-catenin target genes such as cyclin D1 and c-Myc (Zinser et al. 2006). Conditional deletion of β-catenin in the context of RON overexpression resulted in delayed onset of mammary hyperplastic nodules and tumorigenesis, decreased tumor growth, and decreased liver metastasis at the studied time point, revealing a contributing, but not essential, role of β-catenin in mammary tumorigenesis downstream from RON (Wagh et al. 2012). Gene expression analysis of the A2780 ovarian cell line has shown up-regulation of several MAPK target genes in response to MSP (Chaudhuri et al. 2011). In leukemia and multiple myeloma, RON-induced IL-6 secretion seemed to underlie constitutive activation of Jak/Stat3 pathway and poor prognosis (Del Gatto et al. 1995; Danilkovitch-Miagkova 2003). Signaling in RON-overexpressing gastroesophageal adenocarcinoma cell lines and tissues has been reported to be through STAT3, the inhibition of which resulted in decreased viability of cell lines (Catenacci et al. 2011).

Regarding cell motility and EMT in vitro, activation of RON by MSP in noncancerous MDCK cells caused enhanced migration and cell motility through activation of MAPK pathway (Xiangming et al. 2011). In a further detailed in vitro investigation, RSK2, a principle effector of the Ras-ERK1/2 pathway, was reported as the main molecule bridging RON signaling to EMT-like biochemical changes such as migration and spindle-like morphology. In this investigation, MSP stimulation of RON in MDCK cells led to phosphorylation of ERK1/2 and RSK2, which resulted in enhanced migration along with diminished E-cadherin and increased vimentin expression—classic features of EMT. Inhibition of RSK2 in these settings reversed the spindle-like morphology and restored E-cadherin expression. Induction of EMT in these cells seems to be collaborative with TGF-β signaling (Wang et al. 2004; Ma et al. 2011). Feres et al. have shown that c-Src activity was essential for MSP-independent RON-mediated migration, cell spreading, and survival in MCF-10A cells; this might be attributed to the involvement of c-Src signaling in adhesion. MAPK and PI3K/AKT pathways, however, contributed to proliferation in the presence of MSP (Riggins et al. 2006; Feres et al. 2009). In a separate study, despite simultaneous activation of MAPK, FAK, and c-Src pathways, MSP exerted its anti-anoikis effect via PI3K pathway (Danilkovitch et al. 2000).

Besides signaling downstream from RON as an individual RTK, cross talk with other RTKs can further diversify the cellular pathways used or can enhance signaling intensity through particular pathways. This phenomenon can happen with or without the associated ligand and can be unidirectional or bidirectional. Interactions of RON with several RTKs including MET, EGFR, IGFR, and PDGFR have been reported (Follenzi et al. 2000; Danilkovitch-Miagkova and Leonard 2001; Peace et al. 2003; Thomas and Theodorescu 2006; Kobayashi et al. 2009; Liu et al. 2010; Benvenuti et al. 2011; Jaquish et al. 2011; Keller et al. 2013). Results of these interactions are sustained active signaling, enhanced cell survival, and increased invasiveness of tumors. These interactions are nicely reviewed elsewhere (Yao et al. 2013a) and are major causes of compensatory signaling and acquired resistance to single targeted therapies in cancer (Potratz et al. 2010; Wang et al. 2013b; Zhao et al. 2013). As mentioned earlier, different RON isoforms can signal distinctly from wild-type RON, and it is possible that individual or coexpression of these variants further affects other RTK interactions and downstream signaling during tumor progression. So far, it does not appear that a particular signaling pathway can be marked as the "Achilles heel" of RON signaling in all cancers, or even in a single cancer type. Activation of signaling downstream from RON differs greatly based on tissue availability of adaptor and signaling proteins in various cancers, the presence or absence of MSP, generation of RON isoforms, and presence of other RTKs.

CELL-AUTONOMOUS VERSUS NON-CELL-AUTONOMOUS FUNCTIONS OF RON IN CANCER PROGRESSION

Further complexity regarding RON function in cancer arises from its function in the tumor microenvironment. The contributing roles of tumor microenvironment in cancer progression and metastasis are increasingly realized, and the prominent effects of RON in resolving inflammation and promotion of wound healing are opening an exciting new opportunity to consider in cancer therapy. RON is expressed at low levels in certain types of terminally differentiated resident macrophages such as microglia, dermal macrophages, and alveolar macrophages and at very high levels in peritoneal macrophages (Brunelleschi et al. 2001; Suzuki et al. 2008; Okabe and Medzhitov 2014). RON activation promotes polarization of macrophages toward the immunosuppressive alternatively activated (also known as "M2") phenotype (Correll et al. 1997; Liu et al. 1999; Morrison and Correll 2002; Morrison et al. 2004; Kretschmann et al. 2010; Eyob et al. 2013b). RON does this in part by down-regulating IL-12, and therefore decreasing IFN-γ production by natural killer cells (Morrison and Correll 2002; Wilson et al. 2008). However, there are many other RON-mediated events that suppress inflammation, such as reduction of MHC class II surface expression, decreased STAT1 phosphorylation (in response to decrease in IFN-γ production), up-regulation of STAT3 phosphorylation, increased production of IL-10 and IL-6, and reduction of COX-2 expression through activation of NF-κB (Zhou et al. 2002; Gunella et al. 2006; Ni et al. 2007; Yu et al. 2009). Interestingly, RON was recently reported to play a protective role in obesity-induced chronic inflammation, through exerting a repair phenotype in a subpopulation of macrophages (Yu et al. 2016). A detailed review on the role of RON in inflammation can be found elsewhere (Wang and Hankey 2013).

Based on reports so far, the non-cell-autonomous, or extrinsic, functions of RON on the tumor microenvironment appear to be dependent on MSP (Morrison and Correll 2002; Morrison et al. 2004; Wilson et al. 2008; Sharda et al. 2011). As a result of RON-mediated immunosuppressive effects, "M2" macrophages fail to present antigen to T cells and instead produce cytokines that inhibit expansion of lymphocytes. They also secrete particular cytokines, angiogenic factors and growth factors that support proliferation and migration of epithelial cancer cells (Wyckoff et al. 2004; Mosser and Edwards 2008; Solinas et al. 2009). In some cases, tumors appear to hijack RON function by up-regulating MSP to influence the inflammatory process. For example, we have shown that loss of host RON activity in the presence of tumor specifically prevents suppression of the peripheral $CD8^+$ T-cell population. Increased $CD8^+$ T-cell activity in this model, followed by production of antitumor factors like TNF-α, leads to reduction in metastatic burden by inhibiting conversion of micrometastases to macrometastases (Eyob et al. 2013a). Inhibition of RON kinase activity by a selective small molecule inhibitor (BMS-777607) caused $CD8^+$ T-cell-mediated clearance of micrometastatic tumors and reduced outgrowth of established metastatic nodules in the mouse lungs (Eyob et al. 2013a). Similar results have been shown in a prostate cancer model (Gurusamy et al. 2013).

Another MSP-dependent, extrinsic effect of RON in the cancer setting is promotion of osteolysis through activation of osteoclasts, which are macrophage-derived cells in bones. This feature of RON might play an important role in several pathological conditions, such as osteoporosis and certain metastatic cancers like breast cancer, non–small cell lung cancer, prostate cancer, and multiple myeloma (Kozlow and Guise 2005; Roodman 2010; Sturge et al. 2011; Ibrahim et al. 2013). There are several tumor-derived factors that can mediate osteolytic lesions including TGF-β and PTHrP (Yin et al. 2005; Azim et al. 2012). RON is highly expressed on osteoclasts, but the role of MSP/RON pathway in osteolysis is just now being investigated (Kretschmann et al. 2010; Andrade et al. 2017). Our current work shows that MSP/RON signaling is responsible for bone resorption in breast cancer–mediated bone metastasis and also in osteoporosis. The mechanism of enhanced osteoclasts activity in these settings appears to be through activation of Src signaling and parallel to RANK pathway. This effect of RON strongly relies on the presence of MSP, as RON overexpression in mammary tumors by itself did not cause bone metastasis (AL Welm, unpubl.; Zinser et al. 2006), whereas MSP overexpression in these tumors led to spontaneous bone metastasis through host RON. Furthermore, activation of osteoclasts in vitro by MSP-expressing breast cancer cells versus non-MSP-expressing cells has been shown previously (Welm et al. 2007). We also confirmed this exciting novel role of RON/MSP using a RON inhibitor, BMS-777607, which reversed osteolytic lesions in breast cancer mouse models and improved bone turn over markers in human subjects (Andrade et al. 2017). A summary of the current understanding of extrinsic and intrinsic roles of RON, as well as its MSP-dependent and -independent functions in cancer, is presented in Figure 1.

STRATEGIES FOR TARGETING RON FOR CANCER THERAPY

Based on strong associations between RON activity and poor outcomes in cancer patients, the simultaneous intrinsic and extrinsic roles of RON in promoting cancer progression, and its relatively dispensable function in most normal adult tissues, targeting RON is a ripe opportunity for cancer therapy, particularly in patients whose cancers are strongly addicted to this oncogene. In addition, studies have shown that activation of RON is responsible for emergence of resistance in response to single-agent therapies targeting other RTKs. For example, resistance to lapatinib in HER2-positive breast cancer cells and resistance to IGF1R inhibitor in childhood sarcomas have both been attributed to RON activity (Potratz et al. 2010; Wang et al. 2013b). In pancreatic cancers, RON overexpression is also associated with resistance to gemcitabine (Logan-Collins et al. 2010). In all of these examples, inhibition or silencing of RON restored sensitivity to the original treatment. In another study, knockdown/inhibition of RON sensitized pancreatic cells to histone deacetylase (HDAC) inhibitors (Zou et al. 2013). These data provide further rationale to consider RON inhibition to maximize treatments efficacy in certain cancers.

Various strategies have been reported for targeted therapy based on RON, including monoclonal antibodies and small molecule inhibitors (SMIs), some of which are already in clinical trials (refer to Wang et al. 2010, Chang et al. 2015 and Yao et al. 2013a for detailed descriptions). Monoclonal antibodies have shown efficacy in some preclinical models (O'Toole et al. 2006; Li et al. 2010; Padhye et al. 2011; Yao et al. 2011). However, SMIs might have potential to be used in a more widespread way, given that RON activity does not always depend on MSP (Yao et al. 2013a; Bieniasz et al. 2015). Likewise, it is unclear how anti-RON antibodies might affect signaling when another RTK is involved. Due to the high degree of similarity in the kinase domains of MET and RON, available SMIs often show dual inhibitory effect with slightly different IC_{50} values (Yan et al. 2013; Yao et al. 2013a; Chang et al. 2015). To our knowledge, BMS-777607 and LCRF-0004 are the only SMIs reported to inhibit RON at lower IC_{50} than Met (Schroeder et al. 2009; Raeppel et al. 2010, 2015). BMS-777607/ASLAN002 has shown very promising results in preclinical models of breast, pancreatic, prostate, and colorectal cancer and has recently finished phase I clinical trials (Dai and Siemann 2010; Eyob et al. 2013a; Zeng et al. 2014; Bieniasz et al. 2015; Andrade et al. 2017). However, as experienced with other single targeted therapies (Alexander and Wang 2015), resistance has been reported with targeting RON alone, even in cancers that were highly addicted to RON signaling (Sharma et al. 2013; Zhao et al. 2013; Kang et al. 2014). The mechanism lies in compensatory signaling coming from other RTKs like MET, as well as multikinase inhibitory potential of this inhibitor which can lead to induction of polyploidy and resistance to chemotherapeutics (Sharma et al. 2013; Zhao et al. 2013). Mechanistic studies, however, have revealed that this phenotype was attributed to the usage of higher doses of BMS-777607 required for targeting RON, and were due to inhibition of aurora B kinase (Sharma et al. 2013; Zhao et al. 2013; Zeng et al. 2014). More work is needed to determine how and when RON inhibitors should be used and how to design combinatorial treat-

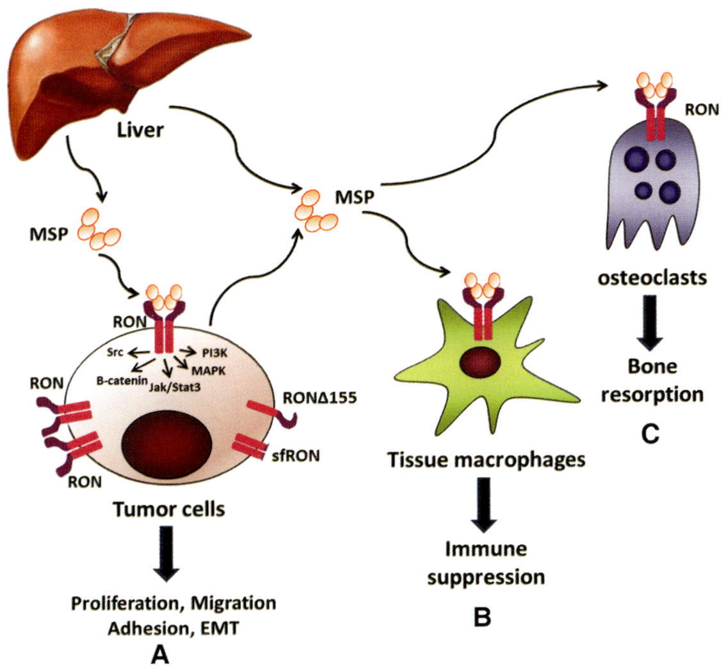

Figure 1. Schematic of intrinsic and extrinsic roles of RON in cancer progression. (*A*) Secreted macrophage-stimulating protein (MSP) from the liver or tumor cells results in activation of RON in cancer cells in a paracrine or autocrine manner, respectively. Cell-autonomous or intrinsic functions of RON in cancer cells leads to activation of several downstream signaling pathways including phosphoinositide 3-kinase (PI3K), mitogen-activated protein kinase (MAPK), β-catenin, Src, and Jak/Stat3 (Danilkovitch-Miagkova 2003; Yao et al. 2013a). Activation of RON in cancer cells can be both MSP-dependent or MSP-independent. However, two major types of RON activation in cancers occurs through protein overexpression and/or generation of splice variants, which are constitutively active and do not depend on MSP for activity (Wang et al. 2006, 2007; Camp et al. 2007; Yao et al. 2013a). Signaling downstream from each variant can be different from that of full-length wild-type RON (Xu et al. 2005; Chakedis et al. 2016a). sfRON and RONΔ155 are two examples of RON variants that are constitutively active in different cancers (Zhou et al. 2003; Liu et al. 2011; Fialin et al. 2013). Non-cell-autonomous, or extrinsic, functions of RON include (*B*) immune suppression in tumor microenvironment, through promotion of an alternatively activated phenotype in macrophages (Morrison and Correll 2002; Morrison et al. 2004; Eyob et al. 2013b) and (*C*) bone resorption through activation of osteoclasts (Kretschmann et al. 2010; Andrade et al. 2017). Based on the reports so far, these extrinsic effects of RON in macrophages and osteoclasts are MSP-dependent (Morrison et al. 2004; Wilson et al. 2008; Sharda et al. 2011; Andrade et al. 2017).

ment strategies in each cancer subtype to circumvent possible resistance.

CONCLUSION

RON is a multifunctional mediator of cancer progression and metastasis for many tumor types. Its dual role in the cancer cells and in the tumor microenvironment, while having no known critically important functions in normal adult tissues, makes it an ideal target for cancer therapy. Many years of basic science research have revealed the function of RON in cancer cells and in the immune system. Published data from multiple laboratories now indicate that inhibiting RON reduces tumor burden and metastatic growth but also boosts the immune response to the tumors. Several inhibitors are now in clinical development, and a continuing understanding of RON function in multiple cancer types should inform thoughtful design of clinical trials on the horizon.

ACKNOWLEDGMENTS

The unpublished work discussed in this review was supported by the U.S. Department of Defense (DOD) Breast Cancer Research Program (W81XWH0810109 to A.L.W.), the National Cancer Institute (R01CA166422 to A.L.W.), and the Susan G. Komen Foundation (PDF14301461 to N.F.).

REFERENCES

Alexander PB, Wang XF. 2015. Resistance to receptor tyrosine kinase inhibition in cancer: Molecular mechanisms and therapeutic strategies. *Front Med* **9:** 134–138.

Andrade K, Fornetti J, Zhao L, Miller SC, Randall RL, Anderson N, Waltz SE, McHale M, Welm AL. 2017. RON kinase: A target for treatment of cancer-induced bone destruction and osteoporosis. *Sci Transl Med* **9:** eaai9338.

Angeloni D, Danilkovitch-Miagkova A, Ivanov SV, Breathnach R, Johnson BE, Leonard EJ, Lerman MI. 2000. Gene structure of the human receptor tyrosine kinase RON and mutation analysis in lung cancer samples. *Genes Chromosomes Cancer* **29:** 147–156.

Angeloni D, Duh FM, Moody M, Dean M, Zabarovsky ER, Sentchenko V, Braga E, Lerman MI. 2003. C to A single nucleotide polymorphism in intron 18 of the human MST1R (RON) gene that maps at 3p21.3. *Mol Cell Probes* **17:** 55–57.

Angeloni D, Danilkovitch-Miagkova A, Miagkov A, Leonard EJ, Lerman MI. 2004. The soluble sema domain of the RON receptor inhibits macrophage-stimulating protein-induced receptor activation. *J Biol Chem* **279:** 3726–3732.

Azim HA, Kamal NS, Azim HA Jr. 2012. Bone metastasis in breast cancer: The story of RANK-ligand. *J Egypt Natl Canc Inst* **24:** 107–114.

Bardella C, Costa B, Maggiora P, Patane S, Olivero M, Ranzani GN, De Bortoli M, Comoglio PM, Di Renzo MF. 2004. Truncated RON tyrosine kinase drives tumor cell progression and abrogates cell-cell adhesion through E-cadherin transcriptional repression. *Cancer Res* **64:** 5154–5161.

Benight NM, Waltz SE. 2012. Ron receptor tyrosine kinase signaling as a therapeutic target. *Expert Opin Ther Targets* **16:** 921–931.

Benvenuti S, Lazzari L, Arnesano A, Li Chiavi G, Gentile A, Comoglio PM. 2011. Ron kinase transphosphorylation sustains MET oncogene addiction. *Cancer Res* **71:** 1945–1955.

Bezerra JA, Witte DP, Aronow BJ, Degen SJ. 1993. Hepatocyte-specific expression of the mouse hepatocyte growth factor-like protein. *Hepatology* **18:** 394–399.

Bhatt AS, Welm A, Farady CJ, Vasquez M, Wilson K, Craik CS. 2007. Coordinate expression and functional profiling identify an extracellular proteolytic signaling pathway. *Proc Natl Acad Sci* **104:** 5771–5776.

Bieniasz M, Radhakrishnan P, Faham N, De La OJ, Welm AL. 2015. Preclinical efficacy of Ron kinase inhibitors alone and in combination with PI3K inhibitors for treatment of sfRon-expressing breast cancer patient-derived xenografts. *Clin Cancer Res* **21:** 5588–5600.

Brunelleschi S, Penengo L, Lavagno L, Santoro C, Colangelo D, Viano I, Gaudino G. 2001. Macrophage stimulating protein (MSP) evokes superoxide anion production by human macrophages of different origin. *Br J Pharmacol* **134:** 1285–1295.

Camp ER, Liu W, Fan F, Yang A, Somcio R, Ellis LM. 2005. RON, a tyrosine kinase receptor involved in tumor progression and metastasis. *Ann Surg Oncol* **12:** 273–281.

Camp ER, Yang A, Gray MJ, Fan F, Hamilton SR, Evans DB, Hooper AT, Pereira DS, Hicklin DJ, Ellis LM. 2007. Tyrosine kinase receptor RON in human pancreatic cancer: Expression, function, and validation as a target. *Cancer* **109:** 1030–1039.

Carafoli F, Chirgadze DY, Blundell TL, Gherardi E. 2005. Crystal structure of the β-chain of human hepatocyte growth factor-like/macrophage stimulating protein. *FEBS J* **272:** 5799–5807.

Catenacci DV, Cervantes G, Yala S, Nelson EA, El-Hashani E, Kanteti R, El Dinali M, Hasina R, Bragelmann J, Seiwert T, et al. 2011. RON (MST1R) is a novel prognostic marker and therapeutic target for gastroesophageal adenocarcinoma. *Cancer Biol Ther* **12:** 9–46.

Chakedis J, French R, Babicky M, Jaquish D, Howard H, Mose E, Lam R, Holman P, Miyamoto J, Walterscheid Z, et al. 2016a. A novel protein isoform of the RON tyrosine kinase receptor transforms human pancreatic duct epithelial cells. *Oncogene* **35:** 3249–3259.

Chakedis J, French R, Babicky M, Jaquish D, Mose E, Cheng P, Holman P, Howard H, Miyamoto J, Porras P, et al. 2016b. Characterization of RON protein isoforms in pancreatic cancer: implications for biology and therapeutics. *Oncotarget* **7:** 45959–45975.

Chang K, Karnad A, Zhao S, Freeman JW. 2015. Roles of c-Met and RON kinases in tumor progression and their potential as therapeutic targets. *Oncotarget* **6:** 3507–3518.

Chao KL, Tsai IW, Chen C, Herzberg O. 2012. Crystal structure of the Sema-PSI extracellular domain of human RON receptor tyrosine kinase. *PLoS One* **7:** e41912.

Chaudhuri A. 2014. Regulation of macrophage polarization by RON receptor tyrosine kinase signaling. *Front Immunol* **5:** 546.

Chaudhuri A, Xie MH, Yang B, Mahapatra K, Liu J, Marsters S, Bodepudi S, Ashkenazi A. 2011. Distinct involvement of the Gab1 and Grb2 adaptor proteins in signal transduction by the related receptor tyrosine kinases RON and MET. *J Biol Chem* **286:** 32762–32774.

Chaudhuri A, Wilson NS, Yang B, Paler Martinez A, Liu J, Zhu C, Bricker N, Couto S, Modrusan Z, French D, et al. 2013. Host genetic background impacts modulation of the TLR4 pathway by RON in tissue-associated macrophages. *Immunol Cell Biol* **91:** 451–460.

Chen YQ, Zhou YQ, Angeloni D, Kurtz AL, Qiang XZ, Wang MH. 2000. Overexpression and activation of the RON receptor tyrosine kinase in a panel of human colorectal carcinoma cell lines. *Exp Cell Res* **261:** 229–238.

Cheng HL, Liu HS, Lin YJ, Chen HH, Hsu PY, Chang TY, Ho CL, Tzai TS, Chow NH. 2005. Co-expression of RON and MET is a prognostic indicator for patients with transitional-cell carcinoma of the bladder. *Br J Cancer* **92:** 1906–1914.

Collesi C, Santoro MM, Gaudino G, Comoglio PM. 1996. A splicing variant of the RON transcript induces constitutive tyrosine kinase activity and an invasive phenotype. *Mol Cell Biol* **16:** 5518–5526.

Correll PH, Iwama A, Tondat S, Mayrhofer G, Suda T, Bernstein A. 1997. Deregulated inflammatory response in mice lacking the STK/RON receptor tyrosine kinase. *Genes Funct* **1:** 69–83.

Cunha S, Lin YC, Goossen EA, DeVette CI, Albertella MR, Thomson S, Mulvihill MJ, Welm AL. 2014. The RON receptor tyrosine kinase promotes metastasis by triggering MBD4-dependent DNA methylation reprogramming. *Cell Rep* **6:** 141–154.

Dai Y, Siemann DW. 2010. BMS-777607, a small-molecule met kinase inhibitor, suppresses hepatocyte growth factor-stimulated prostate cancer metastatic phenotype in vitro. *Mol Cancer Ther* **9:** 1554–1561.

Danilkovitch A, Leonard EJ. 1999. Kinases involved in MSP/RON signaling. *J Leukoc Biol* **65:** 345–348.

Danilkovitch A, Miller M, Leonard EJ. 1999. Interaction of macrophage-stimulating protein with its receptor. Residues critical for β chain binding and evidence for independent α chain binding. *J Biol Chem* **274:** 29937–29943.

Danilkovitch A, Donley S, Skeel A, Leonard EJ. 2000. Two independent signaling pathways mediate the antiapoptotic action of macrophage-stimulating protein on epithelial cells. *Mol Cell Biol* **20:** 2218–2227.

Danilkovitch-Miagkova A. 2003. Oncogenic signaling pathways activated by RON receptor tyrosine kinase. *Curr Cancer Drug Targets* **3:** 31–40.

Danilkovitch-Miagkova A, Leonard EJ. 2001. Cross-talk between RON receptor tyrosine kinase and other transmembrane receptors. *Histol Histopathol* **16:** 623–631.

Danilkovitch-Miagkova A, Angeloni D, Skeel A, Donley S, Lerman M, Leonard EJ. 2000. Integrin-mediated RON growth factor receptor phosphorylation requires tyrosine kinase activity of both the receptor and c-Src. *J Biol Chem* **275:** 14783–14786.

Del Gatto F, Gilbert E, Ronsin C, Breathnach R. 1995. Structure of the promoter for the human macrophage stimulating protein receptor gene. *Biochim Biophys Acta* **1263:** 93–95.

Eckerich C, Schulte A, Martens T, Zapf S, Westphal M, Lamszus K. 2009. RON receptor tyrosine kinase in human gliomas: Expression, function, and identification of a novel soluble splice variant. *J Neurochem* **109:** 969–980.

Eyob H, Ekiz HA, Derose YS, Waltz SE, Williams MA, Welm AL. 2013a. Inhibition of Ron kinase blocks conversion of micrometastases to overt metastases by boosting antitumor immunity. *Cancer Discov* **3:** 751–760.

Eyob H, Ekiz HA, Welm AL. 2013b. RON promotes the metastatic spread of breast carcinomas by subverting antitumor immune responses. *Oncoimmunology* **2:** e25670.

Feres KJ, Ischenko I, Hayman MJ. 2009. The RON receptor tyrosine kinase promotes MSP-independent cell spreading and survival in breast epithelial cells. *Oncogene* **28:** 279–288.

Fialin C, Larrue C, Vergez F, Sarry JE, Bertoli S, Mansat-De Mas V, Demur C, Delabesse E, Payrastre B, Manenti S, et al. 2013. The short form of RON is expressed in acute myeloid leukemia and sensitizes leukemic cells to cMET inhibitors. *Leukemia* **27:** 325–335.

Follenzi A, Bakovic S, Gual P, Stella MC, Longati P, Comoglio PM. 2000. Cross-talk between the proto-oncogenes Met and Ron. *Oncogene* **19:** 3041–3049.

Gaudino G, Follenzi A, Naldini L, Collesi C, Santoro M, Gallo KA, Godowski PJ, Comoglio PM. 1994. RON is a heterodimeric tyrosine kinase receptor activated by the HGF homologue MSP. *EMBO J* **13:** 3524–3532.

Gaudino G, Avantaggiato V, Follenzi A, Acampora D, Simeone A, Comoglio PM. 1995. The proto-oncogene RON is involved in development of epithelial, bone and neuro-endocrine tissues. *Oncogene* **11:** 2627–2637.

Germano S, Barberis D, Santoro MM, Penengo L, Citri A, Yarden Y, Gaudino G. 2006. Geldanamycins trigger a novel Ron degradative pathway, hampering oncogenic signaling. *J Biol Chem* **281:** 21710–21719.

Gunella G, Bardelli C, Amoruso A, Viano I, Balbo P, Brunelleschi S. 2006. Macrophage-stimulating protein differently affects human alveolar macrophages from smoker and non-smoker patients: Evaluation of respiratory burst, cytokine release and NF-κB pathway. *Br J Pharmacol* **148:** 478–489.

Gurusamy D, Gray JK, Pathrose P, Kulkarni RM, Finkleman FD, Waltz SE. 2013. Myeloid-specific expression of Ron receptor kinase promotes prostate tumor growth. *Cancer Res* **73:** 1752–1763.

Hanks SK, Quinn AM. 1991. Protein kinase catalytic domain sequence database: Identification of conserved features of primary structure and classification of family members. *Methods Enzymol* **200:** 38–62.

Hayman MJ. 1987. The sea oncogene of the avian erythroblastosis virus S13. *Pathol Immunopathol Res* **6:** 390–399.

Ibrahim T, Mercatali L, Amadori D. 2013. A new emergency in oncology: Bone metastases in breast cancer patients (Review). *Oncol Lett* **6:** 306–310.

Iwama A, Okano K, Sudo T, Matsuda Y, Suda T. 1994. Molecular cloning of a novel receptor tyrosine kinase gene, STK, derived from enriched hematopoietic stem cells. *Blood* **83:** 3160–3169.

Iwama A, Yamaguchi N, Suda T. 1996. STK/RON receptor tyrosine kinase mediates both apoptotic and growth signals via the multifunctional docking site conserved among the HGF receptor family. *EMBO J* **15:** 5866–5875.

Jaquish DV, Yu PT, Shields DJ, French RP, Maruyama KP, Niessen S, Hoover H, Cheresh DA, Cravatt B, Lowy AM. 2011. IGF1-R signals through the RON receptor to mediate pancreatic cancer cell migration. *Carcinogenesis* **32:** 1151–1156.

Jeffers M, Schmidt L, Nakaigawa N, Webb CP, Weirich G, Kishida T, Zbar B, Vande Woude GF. 1997. Activating mutations for the met tyrosine kinase receptor in human cancer. *Proc Natl Acad Sci* **94:** 11445–11450.

Kang CM, Babicky ML, Lowy AM. 2014. The RON receptor tyrosine kinase in pancreatic cancer pathogenesis and its potential implications for future targeted therapies. *Pancreas* **43:** 183–189.

Kawaguchi M, Orikawa H, Baba T, Fukushima T, Kataoka H. 2009. Hepatocyte growth factor activator is a serum activator of single-chain precursor macrophage-stimulating protein. *FEBS J* **276:** 3481–3490.

Keller J, Nimnual AS, Shroyer KR, Joy C, Ischenko I, Chandler CS, Dong LM, Hayman MJ, Chan EL. 2013. Ron tyrosine kinase receptor synergises with EGFR to confer adverse features in head and neck squamous cell carcinoma. *Br J Cancer* **109:** 482–492.

Kobayashi T, Furukawa Y, Kikuchi J, Ito C, Miyata Y, Muto S, Tanaka A, Kusano E. 2009. Transactivation of RON receptor tyrosine kinase by interaction with PDGF receptor β during steady-state growth of human mesangial cells. *Kidney Int* **75:** 1173–1183.

Kozlow W, Guise TA. 2005. Breast cancer metastasis to bone: Mechanisms of osteolysis and implications for therapy. *J Mammary Gland Biol Neoplasia* **10:** 169–180.

Kretschmann KL, Eyob H, Buys SS, Welm AL. 2010. The macrophage stimulating protein/Ron pathway as a potential therapeutic target to impede multiple mechanisms involved in breast cancer progression. *Curr Drug Targets* **11:** 1157–1168.

Lee WY, Chen HH, Chow NH, Su WC, Lin PW, Guo HR. 2005. Prognostic significance of co-expression of RON and MET receptors in node-negative breast cancer patients. *Clin Cancer Res* **11:** 2222–2228.

Lemmon MA, Schlessinger J. 2010. Cell signaling by receptor tyrosine kinases. *Cell* **141:** 1117–1134.

Li Z, Yao H, Guin S, Padhye SS, Zhou YQ, Wang MH. 2010. Monoclonal antibody (mAb)-induced down-regulation of RON receptor tyrosine kinase diminishes tumorigenic activities of colon cancer cells. *Int J Oncol* **37:** 473–482.

Liu QP, Fruit K, Ward J, Correll PH. 1999. Negative regulation of macrophage activation in response to IFN-γ and lipopolysaccharide by the STK/RON receptor tyrosine kinase. *J Immunol* **163:** 6606–6613.

Liu HS, Hsu PY, Lai MD, Chang HY, Ho CL, Cheng HL, Chen HT, Lin YJ, Wu TJ, Tzai TS, et al. 2010. An unusual function of RON receptor tyrosine kinase as a transcriptional regulator in cooperation with EGFR in human cancer cells. *Carcinogenesis* **31:** 1456–1464.

Liu X, Zhao L, Derose YS, Lin YC, Bieniasz M, Eyob H, Buys SS, Neumayer L, Welm AL. 2011. Short-form Ron promotes spontaneous breast cancer metastasis through interaction with phosphoinositide 3-kinase. *Genes Cancer* **2:** 753–762.

Logan-Collins J, Thomas RM, Yu P, Jaquish D, Mose E, French R, Stuart W, McClaine R, Aronow B, Hoffman RM, et al. 2010. Silencing of RON receptor signaling promotes apoptosis and gemcitabine sensitivity in pancreatic cancers. *Cancer Res* **70:** 1130–1140.

Lu Y, Yao HP, Wang MH. 2007. Multiple variants of the RON receptor tyrosine kinase: Biochemical properties, tumorigenic activities, and potential drug targets. *Cancer Lett* **257:** 157–164.

Ma Q, Zhang K, Guin S, Zhou YQ, Wang MH. 2010a. Deletion or insertion in the first immunoglobulin-plexin-transcription (IPT) domain differentially regulates expression and tumorigenic activities of RON receptor tyrosine kinase. *Mol Cancer* **9:** 307.

Ma Q, Zhang K, Yao HP, Zhou YQ, Padhye S, Wang MH. 2010b. Inhibition of MSP-RON signaling pathway in cancer cells by a novel soluble form of RON comprising the entire sema sequence. *Int J Oncol* **36:** 1551–1561.

Ma Q, Guin S, Padhye SS, Zhou YQ, Zhang RW, Wang MH. 2011. Ribosomal protein S6 kinase (RSK)-2 as a central effector molecule in RON receptor tyrosine kinase mediated epithelial to mesenchymal transition induced by macrophage-stimulating protein. *Mol Cancer* **10:** 66.

Maggiora P, Marchio S, Stella MC, Giai M, Belfiore A, De Bortoli M, Di Renzo MF, Costantino A, Sismondi P, Comoglio PM. 1998. Overexpression of the RON gene in human breast carcinoma. *Oncogene* **16:** 2927–2933.

Maggiora P, Lorenzato A, Fracchioli S, Costa B, Castagnaro M, Arisio R, Katsaros D, Massobrio M, Comoglio PM, Flavia Di Renzo M. 2003. The RON and MET oncogenes are co-expressed in human ovarian carcinomas and cooperate in activating invasiveness. *Exp Cell Res* **288:** 382–389.

Manning G, Whyte DB, Martinez R, Hunter T, Sudarsanam S. 2002. The protein kinase complement of the human genome. *Science* **298:** 1912–1934.

Meyer SE, Waltz SE, Goss KH. 2009. The Ron receptor tyrosine kinase is not required for adenoma formation in $Apc^{Min/+}$ mice. *Mol Carcinog* **48:** 995–1004.

Morrison AC, Correll PH. 2002. Activation of the stem cell-derived tyrosine kinase/RON receptor tyrosine kinase by macrophage-stimulating protein results in the induction of arginase activity in murine peritoneal macrophages. *J Immunol* **168:** 853–860.

Morrison AC, Wilson CB, Ray M, Correll PH. 2004. Macrophage-stimulating protein, the ligand for the stem cell-derived tyrosine kinase/RON receptor tyrosine kinase, inhibits IL-12 production by primary peritoneal macrophages stimulated with IFN-γ and lipopolysaccharide. *J Immunol* **172:** 1825–1832.

Mosser DM, Edwards JP. 2008. Exploring the full spectrum of macrophage activation. *Nat Rev Immunol* **8:** 958–969.

Moxley KM, Wang L, Welm AL, Bieniasz M. 2016. Short-form Ron is a novel determinant of ovarian cancer initiation and progression. *Genes Cancer* **7:** 169–181.

Nanney LB, Skeel A, Luan J, Polis S, Richmond A, Wang MH, Leonard EJ. 1998. Proteolytic cleavage and activation of pro-macrophage-stimulating protein and upregulation of its receptor in tissue injury. *J Invest Dermatol* **111:** 573–581.

Ni S, Zhao C, Feng GS, Paulson RF, Correll PH. 2007. A novel Stat3 binding motif in Gab2 mediates transformation of primary hematopoietic cells by the Stk/Ron receptor tyrosine kinase in response to Friend virus infection. *Mol Cell Biol* **27:** 3708–3715.

Okabe Y, Medzhitov R. 2014. Tissue-specific signals control reversible program of localization and functional polarization of macrophages. *Cell* **157:** 832–844.

Okino T, Egami H, Ohmachi H, Takai E, Tamori Y, Nakagawa A, Nakano S, Sakamoto O, Suda T, Ogawa M. 2001. Immunohistochemical analysis of distribution of RON receptor tyrosine kinase in human digestive organs. *Dig Dis Sci* **46:** 424–429.

O'Toole JM, Rabenau KE, Burns K, Lu D, Mangalampalli V, Balderes P, Covino N, Bassi R, Prewett M, Gottfredsen KJ, et al. 2006. Therapeutic implications of a human neutralizing antibody to the macrophage-stimulating protein receptor tyrosine kinase (RON), a c-MET family member. *Cancer Res* **66:** 9162–9170.

Padhye SS, Guin S, Yao HP, Zhou YQ, Zhang R, Wang MH. 2011. Sustained expression of the RON receptor tyrosine kinase by pancreatic cancer stem cells as a potential targeting moiety for antibody-directed chemotherapeutics. *Mol Pharm* **8:** 2310–2319.

Peace BE, Hill KJ, Degen SJ, Waltz SE. 2003. Cross-talk between the receptor tyrosine kinases Ron and epidermal growth factor receptor. *Exp Cell Res* **289:** 317–325.

Penengo L, Rubin C, Yarden Y, Gaudino G. 2003. c-Cbl is a critical modulator of the Ron tyrosine kinase receptor. *Oncogene* **22:** 3669–3679.

Ponzetto C, Bardelli A, Zhen Z, Maina F, dalla Zonca P, Giordano S, Graziani A, Panayotou G, Comoglio PM. 1994. A multifunctional docking site mediates signaling and transformation by the hepatocyte growth factor/scatter factor receptor family. *Cell* **77:** 261–271.

Potratz JC, Saunders DN, Wai DH, Ng TL, McKinney SE, Carboni JM, Gottardis MM, Triche TJ, Jurgens H, Pollak MN, et al. 2010. Synthetic lethality screens reveal RPS6 and MST1R as modifiers of insulin-like growth factor-1 receptor inhibitor activity in childhood sarcomas. *Cancer Res* **70:** 8770–8781.

Quantin B, Schuhbaur B, Gesnel MC, Dollé P, Breathnach R. 1995. Restricted expression of the ron gene encoding the macrophage stimulating protein receptor during mouse development. *Dev Dyn* **204:** 383–390.

Raeppel S, Gaudette F, Mannion M, Claridge S, Saavedra O, Isakovic L, Deziel R, Beaulieu N, Beaulieu C, Dupont I, et al. 2010. Identification of a novel series of potent RON receptor tyrosine kinase inhibitors. *Bioorg Med Chem Lett* **20:** 2745–2749.

Raeppel SL, Raeppel F, Therrien E. 2015. Design and synthesis of close analogs of LCRF-0004, a potent and selective RON receptor tyrosine kinase inhibitor. *Bioorg Med Chem Lett* **25:** 2527–2531.

Rampino T, Collesi C, Gregorini M, Maggio M, Soccio G, Guallini P, Dal Canton A. 2002. Macrophage-stimulating protein is produced by tubular cells and activates mesangial cells. *J Am Soc Nephrol* **13:** 649–657.

Riggins RB, Thomas KS, Ta HQ, Wen J, Davis RJ, Schuh NR, Donelan SS, Owen KA, Gibson MA, Shupnik MA, et al. 2006. Physical and functional interactions between Cas and c-Src induce tamoxifen resistance of breast cancer cells through pathways involving epidermal growth factor receptor and signal transducer and activator of transcription 5b. *Cancer Res* **66:** 7007–7015.

Ronsin C, Muscatelli F, Mattei MG, Breathnach R. 1993. A novel putative receptor protein tyrosine kinase of the met family. *Oncogene* **8:** 1195–1202.

Roodman GD. 2010. Pathogenesis of myeloma bone disease. *J Cell Biochem* **109:** 283–291.

Santoro MM, Penengo L, Minetto M, Orecchia S, Cilli M, Gaudino G. 1998. Point mutations in the tyrosine kinase domain release the oncogenic and metastatic potential of the Ron receptor. *Oncogene* **17:** 741–749.

Santoro MM, Penengo L, Orecchia S, Cilli M, Gaudino G. 2000. The Ron oncogenic activity induced by the MEN2B-like substitution overcomes the requirement for the multifunctional docking site. *Oncogene* **19:** 5208–5211.

Schlessinger J. 2000. Cell signaling by receptor tyrosine kinases. *Cell* **103:** 211–225.

Schroeder GM, An Y, Cai ZW, Chen XT, Clark C, Cornelius LA, Dai J, Gullo-Brown J, Gupta A, Henley B, et al. 2009. Discovery of N-(4-(2-amino-3-chloropyridin-4-yloxy)-3-fluorophenyl)-4-ethoxy-1-(4-fluorophenyl)-2-oxo-1,2-dihydropyridine-3-carboxamide (BMS-777607), a selective and orally efficacious inhibitor of the Met kinase superfamily. *J Med Chem* **52:** 1251–1254.

Sharda DR, Yu S, Ray M, Squadrito ML, De Palma M, Wynn TA, Morris SM Jr, Hankey PA. 2011. Regulation of macrophage arginase expression and tumor growth by the Ron receptor tyrosine kinase. *J Immunol* **187:** 2181–2192.

Sharma S, Zeng JY, Zhuang CM, Zhou YQ, Yao HP, Hu X, Zhang R, Wang MH. 2013. Small-molecule inhibitor BMS-777607 induces breast cancer cell polyploidy with increased resistance to cytotoxic chemotherapy agents. *Mol Cancer Ther* **12:** 725–736.

Solinas G, Germano G, Mantovani A, Allavena P. 2009. Tumor-associated macrophages (TAM) as major players of the cancer-related inflammation. *J Leukoc Biol* **86:** 1065–1073.

Song YA, Park YL, Kim KY, Myung E, Chung CY, Cho SB, Lee WS, Jung YD, Kweon SS, Joo YE. 2012. RON is associated with tumor progression via the inhibition of apoptosis and cell cycle arrest in human gastric cancer. *Pathol Int* **62:** 127–136.

Songyang Z, Carraway KL III, Eck MJ, Harrison SC, Feldman RA, Mohammadi M, Schlessinger J, Hubbard SR, Smith DP, Eng C, et al. 1995. Catalytic specificity of protein-tyrosine kinases is critical for selective signalling. *Nature* **373:** 536–539.

Sturge J, Caley MP, Waxman J. 2011. Bone metastasis in prostate cancer: Emerging therapeutic strategies. *Nat Rev Clin Oncol* **8:** 357–368.

Suzuki Y, Funakoshi H, Machide M, Matsumoto K, Nakamura T. 2008. Regulation of cell migration and cytokine production by HGF-like protein (HLP)/macrophage stimulating protein (MSP) in primary microglia. *Biomed Res* **29:** 77–84.

Tactacan CM, Chang DK, Cowley MJ, Humphrey ES, Wu J, Gill AJ, Chou A, Nones K, Grimmond SM, Sutherland RL, et al. 2012. RON is not a prognostic marker for resectable pancreatic cancer. *BMC Cancer* **12:** 395.

Thien CB, Langdon WY. 2005. c-Cbl and Cbl-b ubiquitin ligases: Substrate diversity and the negative regulation of signalling responses. *Biochem J* **391:** 153–166.

Thomas CY, Theodorescu D. 2006. Collaboration of RON and epidermal growth factor receptor in human bladder carcinogenesis. *J Urol* **176:** 1909–1910.

Thomas RM, Toney K, Fenoglio-Preiser C, Revelo-Penafiel MP, Hingorani SR, Tuveson DA, Waltz SE, Lowy AM. 2007. The RON receptor tyrosine kinase mediates oncogenic phenotypes in pancreatic cancer cells and is increasingly expressed during pancreatic cancer progression. *Cancer Res* **67:** 6075–6082.

Tong XM, Wang JC, Shen Y, Xie JJ, Zhang JY, Jin J. 2011. Inhibition of inflammatory mediators and related signaling pathways by macrophage-stimulating protein in rheumatoid arthritis synovial fibroblasts. *Inflamm Res* **60:** 823–829.

Tulasne D, Paumelle R, Weidner KM, Vandenbunder B, Fafeur V. 1999. The multisubstrate docking site of the MET receptor is dispensable for MET-mediated RAS signaling and cell scattering. *Mol Biol Cell* **10:** 551–565.

van den Akker E, van Dijk T, Parren-van Amelsvoort M, Grossmann KS, Schaeper U, Toney-Earley K, Waltz SE, Lowenberg B, von Lindern M. 2004. Tyrosine kinase receptor RON functions downstream of the erythropoietin receptor to induce expansion of erythroid progenitors. *Blood* **103:** 4457–4465.

Wagh PK, Zinser GM, Gray JK, Shrestha A, Waltz SE. 2012. Conditional deletion of β-catenin in mammary epithelial cells of Ron receptor, Mst1r, overexpressing mice alters mammary tumorigenesis. *Endocrinology* **153:** 2735–2746.

Waltz SE, Eaton L, Toney-Earley K, Hess KA, Peace BE, Ihlendorf JR, Wang MH, Kaestner KH, Degen SJ. 2001. Ron-mediated cytoplasmic signaling is dispensable for viability but is required to limit inflammatory responses. *J Clin Invest* **108:** 567–576.

Wang X, Hankey PA. 2013. The ron receptor tyrosine kinase: A key regulator of inflammation and cancer progression. *Crit Rev Immunol* **33:** 549–574.

Wang MH, Ronsin C, Gesnel MC, Coupey L, Skeel A, Leonard EJ, Breathnach R. 1994. Identification of the ron gene product as the receptor for the human macrophage stimulating protein. *Science* **266:** 117–119.

Wang MH, Iwama A, Skeel A, Suda T, Leonard EJ. 1995. The murine stk gene product, a transmembrane protein tyrosine kinase, is a receptor for macrophage-stimulating protein. *Proc Natl Acad Sci* **92:** 3933–3937.

Wang MH, Julian FM, Breathnach R, Godowski PJ, Takehara T, Yoshikawa W, Hagiya M, Leonard EJ. 1997. Macrophage stimulating protein (MSP) binds to its receptor via the MSP β chain. *J Biol Chem* **272:** 16999–17004.

Wang MH, Kurtz AL, Chen Y. 2000. Identification of a novel splicing product of the RON receptor tyrosine kinase in human colorectal carcinoma cells. *Carcinogenesis* **21:** 1507–1512.

Wang MH, Zhou YQ, Chen YQ. 2002. Macrophage-stimulating protein and RON receptor tyrosine kinase: Potential regulators of macrophage inflammatory activities. *Scand J Immunol* **56:** 545–553.

Wang MH, Wang D, Chen YQ. 2003. Oncogenic and invasive potentials of human macrophage-stimulating protein receptor, the RON receptor tyrosine kinase. *Carcinogenesis* **24:** 1291–1300.

Wang D, Shen Q, Chen YQ, Wang MH. 2004. Collaborative activities of macrophage-stimulating protein and transforming growth factor-β1 in induction of epithelial to mesenchymal transition: Roles of the RON receptor tyrosine kinase. *Oncogene* **23:** 1668–1680.

Wang MH, Yao HP, Zhou YQ. 2006. Oncogenesis of RON receptor tyrosine kinase: A molecular target for malignant epithelial cancers. *Acta Pharmacol Sin* **27:** 641–650.

Wang MH, Lee W, Luo YL, Weis MT, Yao HP. 2007. Altered expression of the RON receptor tyrosine kinase in various epithelial cancers and its contribution to tumourigenic phenotypes in thyroid cancer cells. *J Pathol* **213:** 402–411.

Wang MH, Padhye SS, Guin S, Ma Q, Zhou YQ. 2010. Potential therapeutics specific to c-MET/RON receptor tyrosine kinases for molecular targeting in cancer therapy. *Acta Pharmacol Sin* **31:** 1181–1188.

Wang MH, Zhang R, Zhou YQ, Yao HP. 2013a. Pathogenesis of RON receptor tyrosine kinase in cancer cells: Activation mechanism, functional crosstalk, and signaling addiction. *J Biomed Res* **27:** 345–356.

Wang Q, Quan H, Zhao J, Xie C, Wang L, Lou L. 2013b. RON confers lapatinib resistance in HER2-positive breast cancer cells. *Cancer Lett* **340:** 43–50.

Wang X, Yennawar N, Hankey PA. 2014. Autoinhibition of the Ron receptor tyrosine kinase by the juxtamembrane domain. *Cell Commun Signal* **12:** 28.

Wei X, Hao L, Ni S, Liu Q, Xu J, Correll PH. 2005a. Altered exon usage in the juxtamembrane domain of mouse and human RON regulates receptor activity and signaling specificity. *J Biol Chem* **280:** 40241–40251.

Wei X, Ni S, Correll PH. 2005b. Uncoupling ligand-dependent and -independent mechanisms for mitogen-activated protein kinase activation by the murine Ron receptor tyrosine kinase. *J Biol Chem* **280:** 35098–35107.

Welm AL, Sneddon JB, Taylor C, Nuyten DS, van de Vijver MJ, Hasegawa BH, Bishop JM. 2007. The macrophage-stimulating protein pathway promotes metastasis in a mouse model for breast cancer and predicts poor prognosis in humans. *Proc Natl Acad Sci* **104:** 7570–7575.

Willett CG, Wang MH, Emanuel RL, Graham SA, Smith DI, Shridhar V, Sugarbaker DJ, Sunday ME. 1998. Macrophage-stimulating protein and its receptor in non-small-cell lung tumors: Induction of receptor tyrosine phosphorylation and cell migration. *Am J Respir Cell Mol Biol* **18:** 489–496.

Wilson CB, Ray M, Lutz M, Sharda D, Xu J, Hankey PA. 2008. The RON receptor tyrosine kinase regulates IFN-γ production and responses in innate immunity. *J Immunol* **181:** 2303–2310.

Wyckoff J, Wang W, Lin EY, Wang Y, Pixley F, Stanley ER, Graf T, Pollard JW, Segall J, Condeelis J. 2004. A paracrine loop between tumor cells and macrophages is required for tumor cell migration in mammary tumors. *Cancer Res* **64:** 7022–7029.

Xiangming X, Yun Q, Guoliang Z, Jianjiang L, Lisong T. 2011. Mechanisms of RON-mediated epithelial-mesenchymal transition in MDCK cells through the MAPK pathway. *Braz J Med Biol Res* **44:** 634–641.

Xu XM, Wang D, Shen Q, Chen YQ, Wang MH. 2004. RNA-mediated gene silencing of the RON receptor tyrosine kinase alters oncogenic phenotypes of human colorectal carcinoma cells. *Oncogene* **23:** 8464–8474.

Xu XM, Zhou YQ, Wang MH. 2005. Mechanisms of cytoplasmic {β}-catenin accumulation and its involvement in tumorigenic activities mediated by oncogenic splicing variant of the receptor originated from Nantes tyrosine kinase. *J Biol Chem* **280:** 25087–25094.

Yan SB, Peek VL, Ajamie R, Buchanan SG, Graff JR, Heidler SA, Hui YH, Huss KL, Konicek BW, Manro JR, et al. 2013. LY2801653 is an orally bioavailable multi-kinase inhibitor with potent activity against MET, MST1R, and other oncoproteins, and displays anti-tumor activities in mouse xenograft models. *Invest New Drugs* **31:** 833–844.

Yao HP, Zhou YQ, Ma Q, Guin S, Padhye SS, Zhang RW, Wang MH. 2011. The monoclonal antibody Zt/f2 targeting RON receptor tyrosine kinase as potential therapeutics against tumor growth-mediated by colon cancer cells. *Mol Cancer* **10:** 82.

Yao HP, Zhou YQ, Zhang R, Wang MH. 2013a. MSP-RON signalling in cancer: Pathogenesis and therapeutic potential. *Nat Rev Cancer* **13:** 466–481.

Yao HP, Zhuang CM, Zhou YQ, Zeng JY, Zhang RW, Wang MH. 2013b. Oncogenic variant RON160 expression in breast cancer and its potential as a therapeutic target by small molecule tyrosine kinase inhibitor. *Curr Cancer Drug Targets* **13:** 686–697.

Yin JJ, Pollock CB, Kelly K. 2005. Mechanisms of cancer metastasis to the bone. *Cell Res* **15:** 57–62.

Yokoyama N, Ischenko I, Hayman MJ, Miller WT. 2005. The C terminus of RON tyrosine kinase plays an autoinhibitory role. *J Biol Chem* **280:** 8893–8900.

Yoshimura T, Yuhki N, Wang MH, Skeel A, Leonard EJ. 1993. Cloning, sequencing, and expression of human macrophage stimulating protein (MSP, MST1) confirms MSP as a member of the family of kringle proteins and locates the MSP gene on chromosome 3. *J Biol Chem* **268:** 15461–15468.

Yu H, Pardoll D, Jove R. 2009. STATs in cancer inflammation and immunity: A leading role for STAT3. *Nat Rev Cancer* **9:** 798–809.

Yu S, Allen JN, Dey A, Zhang L, Balandaram G, Kennett MJ, Xia M, Xiong N, Peters JM, Patterson A, et al. 2016. The Ron receptor tyrosine kinase regulates macrophage heterogeneity and plays a protective role in diet-induced obesity, atherosclerosis, and hepatosteatosis. *J Immunol* **197:** 256–265.

Zeng JY, Sharma S, Zhou YQ, Yao HP, Hu X, Zhang R, Wang MH. 2014. Synergistic activities of MET/RON inhibitor BMS-777607 and mTOR inhibitor AZD8055 to polyploid cells derived from pancreatic cancer and cancer stem cells. *Mol Cancer Ther* **13:** 37–48.

Zhang K, Zhou YQ, Yao HP, Wang MH. 2010. Alterations in a defined extracellular region of the RON receptor tyrosine kinase promote RON-mediated motile and invasive phenotypes in epithelial cells. *Int J Oncol* **36:** 255–264.

Zhao S, Cao L, Freeman JW. 2013. Knockdown of RON receptor kinase delays but does not prevent tumor progression while enhancing HGF/MET signaling in pancreatic cancer cell lines. *Oncogenesis* **2:** e76.

Zhou YQ, Chen YQ, Fisher JH, Wang MH. 2002. Activation of the RON receptor tyrosine kinase by macrophage-stimulating protein inhibits inducible cyclooxygenase-2 expression in murine macrophages. *J Biol Chem* **277:** 38104–38110.

Zhou YQ, He C, Chen YQ, Wang D, Wang MH. 2003. Altered expression of the RON receptor tyrosine kinase in primary human colorectal adenocarcinomas: Generation of different splicing RON variants and their oncogenic potential. *Oncogene* **22:** 186–197.

Zinser GM, Leonis MA, Toney K, Pathrose P, Thobe M, Kader SA, Peace BE, Beauman SR, Collins MH, Waltz SE. 2006. Mammary-specific Ron receptor overexpression induces highly metastatic mammary tumors associated with β-catenin activation. *Cancer Res* **66:** 11967–11974.

Zou Y, Howell GM, Humphrey LE, Wang J, Brattain MG. 2013. Ron knockdown and Ron monoclonal antibody IMC-RON8 sensitize pancreatic cancer to histone deacetylase inhibitors (HDACi). *PLoS One* **8:** e69992.

Physical and Chemical Gradients in the Tumor Microenvironment Regulate Tumor Cell Invasion, Migration, and Metastasis

MADELEINE J. OUDIN[1] AND VALERIE M. WEAVER[2,3,4,5]

[1]*Koch Institute for Integrative Cancer Research, Massachusetts Institute of Technology, Cambridge, Massachusetts 02139*

[2]*Department of Surgery, Center for Bioengineering and Tissue Regeneration, University of California, San Francisco, San Francisco, California 94143*

[3]*UCSF Comprehensive Cancer Center, Helen Diller Family Cancer Research Center, University of California, San Francisco, San Francisco, California 94143*

[4]*Department of Anatomy, Department of Bioengineering and Therapeutic Sciences, and Department of Radiation Oncology, University of California San Francisco, San Francisco, California 94143*

[5]*Eli and Edythe Broad Center of Regeneration Medicine and Stem Cell Research and The Helen Diller Family Comprehensive Cancer Center, University of California, San Francisco, San Francisco, California 94143*

Correspondence: Valerie.Weaver@ucsf.edu

Cancer metastasis requires the invasion of tumor cells into the stroma and the directed migration of tumor cells through the stroma toward the vasculature and lymphatics where they can disseminate and colonize secondary organs. Physical and biochemical gradients that form within the primary tumor tissue promote tumor cell invasion and drive persistent migration toward blood vessels and the lymphatics to facilitate tumor cell dissemination. These microenvironment cues include hypoxia and pH gradients, gradients of soluble cues that induce chemotaxis, and ions that facilitate galvanotaxis, as well as modifications to the concentration, organization, and stiffness of the extracellular matrix that produce haptotactic, alignotactic, and durotactic gradients. These gradients form through dynamic interactions between the tumor cells and the resident fibroblasts, adipocytes, nerves, endothelial cells, infiltrating immune cells, and mesenchymal stem cells. Malignant progression results from the integrated response of the tumor to these extrinsic physical and chemical cues. Here, we first describe how these physical and chemical gradients develop, and we discuss their role in tumor progression. We then review assays to study these gradients. We conclude with a discussion of clinical strategies used to detect and inhibit these gradients in tumors and of new intervention opportunities. Clarifying the role of these gradients in tumor evolution offers a unique approach to target metastasis.

Tumors are highly heterogeneous and this feature compromises treatment efficacy and promotes tumor cell dissemination and metastasis to reduce patient survival. Tumor heterogeneity is driven in part by genetic alterations induced through genomic instability and maintained by the evolutionary selection of these genetically modified cells (Alizadeh et al. 2015). Epigenetic, transcriptional, and posttranslational modifications also contribute significantly to tumor cell diversity. Furthermore, cancer develops within a complex tissue microenvironment that itself fosters tumor heterogeneity through clonal selection as well as via epigenetic reprogramming and modifications of the tumor phenotype. The tumor microenvironment is composed of cellular constituents that include cells of the vasculature, infiltrating immune cells, and resident fibroblasts, adipocytes and nerves, and a noncellular component composed of diverse soluble factors and a highly variable and evolving insoluble extracellular matrix (ECM) (Joyce and Pollard 2009). The composition and organization of the tumor microenvironment vary significantly within and across tumors. One key feature of the noncellular microenvironment is the existence of local chemical and physical gradients that exert profound effects on the tumor phenotype, including its predilection to disseminate. In this regard, metastasis is facilitated by efficient local tumor cell invasion that is fostered by ECM and chemokine/cytokine/growth factor gradients which cooperate to promote the directional migration of the tumor cells toward the vasculature and lymphatics and facilitates their extravasation from the primary tumor site into the circulation and their intravasation into the distal tissue. In this review, we discuss the role of directional migration in cancer progression and metastasis in the context of local microenvironmental gradients that develop in the primary and metastatic tumor tissue. We begin by reviewing the generation of endogenous gradients within both the primary and metastatic tumor tissues. We then summarize some of the techniques currently used to study directional migration in cancer and discuss putative signaling mechanisms regulating directional migration in response to external cues and the intrinsic state of the tumor cell. We have

Published by Cold Spring Harbor Laboratory Press; doi: 10.1101/sqb.2016.81.030817

chosen to focus exclusively on methods applied to the study of cancer progression and mechanisms driving directional migration of tumor cells. Finally, we discuss existing approaches available to interfere with directional migration that could prevent tumor metastasis and outline potential future avenues for therapeutic development.

TUMOR HETEROGENEITY LEADS TO ENDOGENOUS GRADIENT GENERATION IN TUMORS

Local Gradients of Secreted Factors

Chemokines and growth factors (GFs) are key drivers of tumor cell invasion and facilitate tumor cell extravasation and intravasation to promote metastasis. Tumor cells secrete chemoattractants for macrophages, neutrophils, lymphocytes, fibroblasts, and mesenchymal stem cells that once recruited secrete chemokines and GFs that stimulate tumor cell migration (Roussos et al. 2011). The specific, directed migration of stromal cells into the tumor and the reciprocal persistent migration of the tumor cells into and through the stroma are reinforced by the restricted expression of specific receptors for each of the plethora of secreted factors. For example, tumor-associated macrophages (TAMs) release growth factors, including epidermal growth factor (EGF), that activate the EGFR receptor expressed on the breast tumor cells to stimulate and direct their migration (Wyckoff et al. 2004). The breast tumor cells in turn secrete colony stimulating factor 1 (CSF-1), which stimulates and recruits TAMs through TAM-specific expression of the CSF-1 receptor. This vicious feed-forward paracrine loop between the invading tumor cells and infiltrating macrophages ensures the persistent migration of the tumor cells toward the vasculature that is key for tumor cell dissemination and metastasis. The chemokine CXCL12, produced by pericytes and tumor cells, can also drive in breast tumor cell chemotaxis in vivo (Müller et al. 2001). Although the molecular details remain to be elaborated, several in vitro studies have attested to the existence of similar local reciprocal gradients of GFs such as EGF ligands, fibroblast growth factor (FGF), platelet-derived growth factor (PDGF), transforming growth factor β (TGF-β), vascular endothelial growth factor (VEGF), and CSF-1, in a variety of tumor types, investigating how this interplay stimulates stromal cell infiltration to potentiate tumor cell migration and dissemination (Roussos et al. 2011). Interestingly, tumor cells themselves can also create local gradients to facilitate their directed migration as has been shown through their matrix metalloproteinase-mediated degradation of the ECM or as was recently illustrated for melanoma tumor cells that create their own lipid chemoattractant lysophosphatidic acid gradient to foster their migration (Muinonen-Martin et al. 2014; Tweedy et al. 2016). Furthermore, soluble gradients have been implicated in tissue tropism within the context of metastasis and distal organ colonization. Indeed, chemokines secreted by cancer-associated fibroblasts (CAFs) have been associated with local stromal cells secreting a chemokine whose receptor is located on the tumor cell. For example, CAF-secreted CXCL12/SDF-1 drives metastasis of CXCR4-expressing breast cancer cells to the lymph nodes and lung (Müller et al. 2001), with CCL27 driving CCR10-dependent skin metastasis of melanoma and CCL21/CCL19 activating CCR7 to promote lymph node metastasis (Ben-Baruch 2008) (Figs. 1 and 2).

Changes in ECM Abundance and Organization

The ECM is a complex scaffold of proteins that provide structural support to the surrounding tissue. Increased deposition and accelerated post-translational modifications including cross-linking of the ECM are characteristic features of the tumor desmoplastic response that is associated with poor patient prognosis (Walker 2001). Recently, proteomics of the decellularized, soluble ECM fraction of the tumor tissue identified core matrisome proteins and found that this signature contains as many as 274 proteins composed of glycoproteins, collagens, and proteoglycans (Naba et al. 2012). Intriguingly, using this approach and relying on analysis that uses species-specific peptides, it was revealed that many of the tumor ECM proteins including a number of abundant laminins (LAMA3, B3, C2) and Snedl are secreted predominantly by the tumor cells, whereas other ECM components such as tenascin C (TNC) and vitronectin are secreted exclusively by resident fibroblasts and infiltrating stromal cells (Naba et al. 2014). Complementing these studies, advanced proteomic approaches have been able to profile the insoluble ECM proteins within the tumor, revealing distinct insoluble protein profiles within a tumor type that not only reflect the tumor stage but also the tumor genotype (Goddard et al. 2016; Laklai et al. 2016). Data revealed that the most aggressive pancreatic tumors that harbor SMAD4 mutations and that are associated with the shortest patient survival deposited quantities of collagen 12, TNC, and fibronectin (FN) nests surrounding the invading ductal lesions (Laklai et al. 2016). Overall, multiple proteomic studies have permitted researchers to conclude that the ECM is temporally modulated, spatially distinct, and influenced by the tumor genotype, and that its phenotype is dictated by both the tumor cells themselves as well as the resident and infiltrating stromal cells (Naba et al. 2014, 2016). For instance, FN, which is typically elevated in tumors, is highly heterogeneous with the highest levels found at the tissue periphery, near the vasculature and associated with thick collagen bundles (Astrof and Hynes 2009; Zhang et al. 2014; Miroshnikova et al. (in review). Peripheral solid tumors are typically fibrotic and characterized by elevated reorganized collagens that are often lysyl oxidase cross-linked into bundles upon which tumor cells have been seen migrating (Wang et al. 2004; Wyckoff et al. 2006; Erler and Weaver 2009). Using two-photon imaging has made it possible to characterize tumor collagen organization, and using structured illumination polarized imaging informs on the degree of alignment

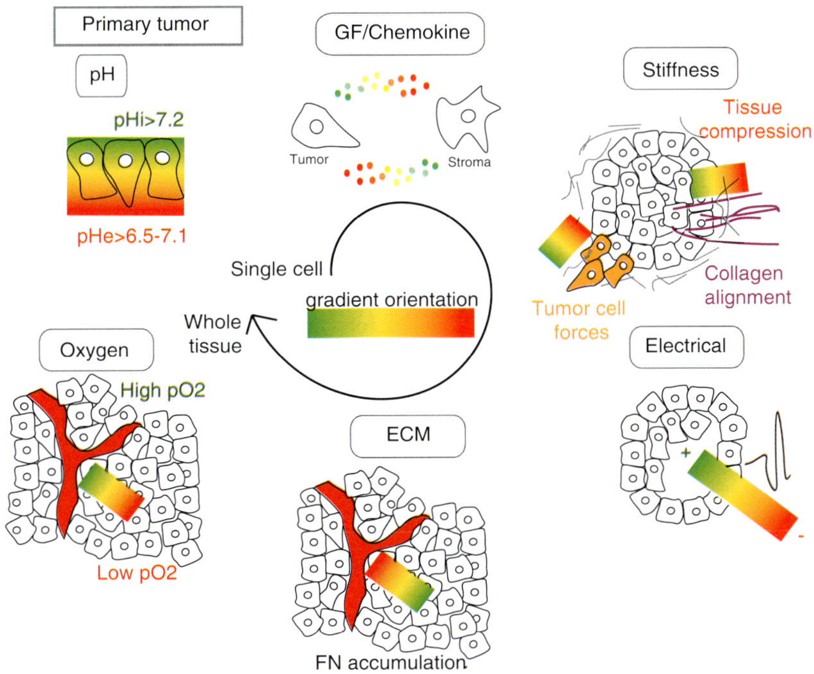

Figure 1. Endogenous gradients are present in primary breast tumors. Tumorigenesis is associated with changes in the tumor microenvironment that lead to the development of endogenous gradients at the single-cell level to whole tissue (summarized in part 1). pH: At the single-cell level, pH gradients develop between the intracellular and extracellular compartments. GF/Chemokine: Individual tumor cells can also respond to soluble gradients secreted by surrounding stromal cells. Stiffness: Tissue compression by the tumor and external forces generated by tumor cells or tumor cell–driven ECM remodeling of collagen can all lead to the development of endogenous stiffness gradients at the tumor–stroma interface. Electrical: Tumorigenesis is accompanied by changes in membrane potential of luminal structures. ECM: The ECM of tumor versus normal tissues changes and some ECM cues, in particular FN, are heterogeneously distributed throughout the tumor, leading to the development of endogenous gradients. Oxygen: It is well established that tumor tissues become hypoxic as they grow due to poor vascularization, leading to whole-tissue oxygen gradients. GF, growth factor; ECM, extracellular matrix; FN, fibronectin.

(Acerbi et al. 2015). These two approaches have revealed that the thick, highly aligned, fibrillar collagens typically observed in a primary tumor are the result of the dramatic reorganization of the relaxed, coiled short fibers found in a normal tissue (Levental et al. 2009). Provocatively, the organization of the tumor-associated fibrillar collagens is highly heterogeneous with prominent oriented fibers frequently localized to the invasive front, associated with invading blood vessels, and around the tumor periphery (Provenzano et al. 2006; Levental et al. 2009; Lopez et al. 2011; Acerbi et al. 2015). The clinical relevance of collagen reorganization at the primary tumor was emphasized by a retrospective study in which the presence of perpendicular collagen bundles in invasive human breast tumors was found to associate with poor patient outcome (Egeblad et al. 2010; Conklin and Keely 2012). Intriguingly, recent work revealed that several matricellular proteins, such as FN and TNC, can also accumulate in the premetastatic niche and may serve to attract and subsequently enhance tumor cell growth and survival, particularly because they are deposited as a highly endogenous gradient in the niche (Kaplan et al. 2005; Costa-Silva et al. 2015). Evidence to support this possibility was provided by experimental data showing that TNC is a major gene that is significantly up-regulated in breast tumor cells that preferentially colonize the lung (Minn et al. 2005).

Stiffness Gradients

Tumors are mechanically corrupted tissues whose physical attributes typically include high tumor solid stress that reflects the tumor genotype and is promoted by the expanding tumor cell mass, the elevated interstitial pressure, and the resistance exerted by the surrounding stiffened, fibrotic ECM (Padera et al. 2004; Paszek et al. 2005; Willipinski-Stapelfeldt et al. 2005). ECM stiffening reflects an increase in protein abundance; particularly of fibrillar proteins such as collagens I and III, as well as their reorganization and posttranslation modifications that include lysyl oxidase–mediated cross-linking (Erler and Weaver 2009; Levental et al. 2009; Egeblad et al. 2010; Cox et al. 2016). Although imaging elastography and unconfined compression analysis have consistently revealed an incremental increase in tissue stiffness as a function of tumor stage, grade atomic force microscopy (AFM) indentation analysis has also emphasized the heterogeneity of the ECM stiffness within a tumor (Evans et al. 2012; Plodinec et al. 2012; Yi et al. 2013; Fenner et al. 2014). For instance, AFM indentation revealed that the invasive front of human tumors is by far the stiffest (Acerbi et al. 2015) and determined that the vasculature within the core of the tumor is softer than the vasculature at the periphery (Lopez et al. 2011); possibly reflecting

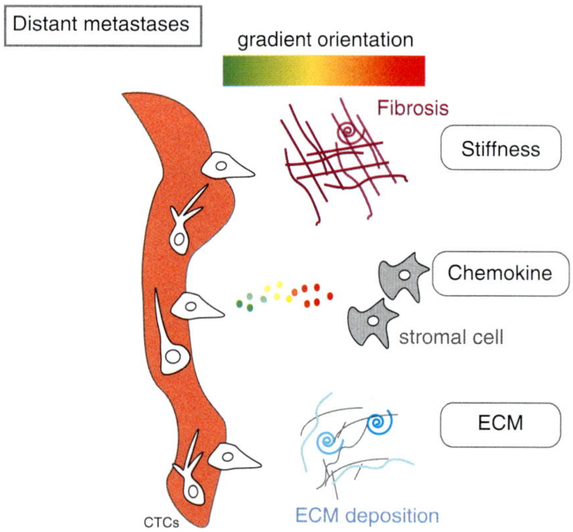

Figure 2. Endogenous gradients at sites of distant metastases. Local changes in the tumor microenvironment contribute to colonization and the formation of distant metastases. Fibrotic tissues, such as in the lung or liver, show increased stiffness and increased colonization. Secretion of chemokines from local stromal cells present in secondary sites can promote extravasation to tumor cells. Accumulation of ECM proteins in the premetastatic niche also leads to the development of endogenous gradients of haptotactic cues. ECM, extracellular matrix; CTCs, circulating tumor cells.

differences in pericyte coverage (Ribeiro and Okamoto 2015). These distinct regions of ECM stiffness result in endogenous stiffness gradients that can foster tumor cell migration through a process termed durotaxis (Lo et al. 2000; Isenberg et al. 2009; Plotnikov et al. 2012). Interestingly, two common organs of secondary metastasis are the liver and the lung (Wynn 2008; Steeg 2016), both of which suffer from age, disease, and obesity-mediated chronic fibrosis (Zeisberg and Kalluri 2013). Chronic liver and lung fibrosis are characterized by tissue stiffening through elevated levels of fibrillar-type collagens, TNC and FN, and more proteoglycan and lysyl oxidase-mediated cross-linking, which would comprise a strong stiffness differential gradient toward which disseminated tumor cells would be attracted (Levental et al. 2009; Chang et al. 2017).

Establishment of Hypoxia

Tumors are hypoxic (Wilson and Hay 2011) as a result of insufficient or dysfunctional vasculature that is often leaky, heterogeneous, and tortuous. This irregular branching and abnormal shunting lead to heterogeneous blood flow (Carmeliet and Jain 2011), which in turn creates endogenous hypoxia gradients within the tumor that can include extremely low pO_2 levels in some areas of the tumor (Li et al. 2007). The tumors adapt to the reduced oxygen by adjusting their intrinsic metabolism, which creates a layered metabolite gradient within the tissue. Thus, vascular heterogeneity creates gradients of extracellular oxygen levels or pO_2, which alter the intravascular pO_2 and the metabolic consumption rate of the tumor and stromal cells within the tissue. The combination of these events creates pO_2 and metabolite gradients within the tissue that can have a profound effect on tumor cell behavior.

Inversion of Normal pH Gradient

In differentiated epithelial cells, the intracellular pH (pHi, 7.2) is lower than the extracellular pH (pHe, 7.4). However, in a tumor tissue the pH can change dramatically depending on the pO_2 and metabolite availability, resulting in a reversal of the normal pH gradient (Webb et al. 2011), with cancer cells developing a higher pHi of 7.4 and much lower pHe of 6.5–7.1(Stüwe et al. 2007; Gallagher et al. 2008; Hashim et al. 2011). Changes in pH in tumor cells can occur as a consequence of up-regulation of expression and/or activity of ion pumps and transporters, such as H^+ ATPases (Martinez-Zaguilan et al. 1993), Na^+–H^+ exchanger NHE1 of the SLC9A family (McLean et al. 2000), and monocarboxylate–H^+ efflux cotransporters MCT1 and MCT4 or the SLC16A family (Pinheiro et al. 2008; Chiche et al. 2012) that drive H^+ efflux, leading to a high pHe. Furthermore, high activity of carbonic anhydrases CAIX and CAXII driven by hypoxia can accelerate the hydration of extracellular CO_2 to HCO_3^- and H^+ (Pastorekova et al. 2006; Chiche et al. 2009; Swietach et al. 2009). Finally, reduced blood flow and subsequent oxygen depletion at the early "carcinoma in situ" stage will drive tumor cells to undergo a metabolic switch toward a glycolytic phenotype (Gatenby and Gillies 2004). The increased flux of carbons resulting from glycolysis leads to an acidification of the extracellular pH of the tumors that is particularly pronounced in luminal structures (Barathova et al. 2008). Importantly, similar to pO_2 and metabolite gradients, tumors can also develop pH gradients that can influence their behavior and additionally contribute to tumor heterogeneity.

Alterations in Electrical Fields

Electrical signaling plays a fundamental role in cell and tissue homeostasis. The membrane potential of a cell is maintained by the unequal distribution of ions across the plasma membrane, which creates voltage differences between the cytoplasm and the extracellular environment (Yang and Brackenbury 2013). In epithelial tissues, naturally occurring electrical fields (EFs) are maintained by the spatial organization of ion pumps, with the apical domain of the cell being especially enriched in Na^+ channels and Cl^- transporters, and the basolateral domain containing Na^+–K^+ ATPases (McCaig et al. 2009). The generation of ionic gradients leads to extracellular ionic current flow and the establishment of endogenous voltage gradients. Normal epithelial cells are highly polarized; however, the membrane potential of tumor cells is often depolarized, and this difference results in a drop in transepithelial potential difference (James et al. 1956; Faupel et al. 1997). Depolarized tumor cells display increased intracellular Na^+ levels, stable K^+ levels, and Cl^- and Ca^{2+} influx (Yang and Brackenbury 2013). Tumor depolarization results in an extracellular voltage gradient between normal and tumor cells (Cuzick et al. 1998) that can vary widely within one tumor and differs between tumor types. For example, across the luminal wall of rat prostate glands, a steady direct current electrical field of 500 mV/mm exists, with a transepithelial electrical potential difference of −10 mV, the lumen being negatively charged (Szatkowski et al. 2000; Djamgoz et al. 2001). In contrast, in the mammary duct, the transepithelial electrical potential is +30 mV, with duct lumen being positively charged (Pu et al. 2007).

TECHNIQUES TO STUDY DIRECTIONAL MIGRATION

The transwell or Boyden chamber assay (Boyden 1962) is the original assay used to study directional migration, in which cells are plated on a porous filter membrane in between two cell culture chambers. By layering a soluble or ECM cue in only one of the two chambers, it is possible to study directional migration through the filter toward a cue of interest. Although this assay has unequivocally enhanced our understanding of directional migration, the approach is fraught with many limitations that include most importantly an inability to control the local environment and to generate gradient slopes. Furthermore, the technique does not easily lend itself to live imaging. In recent years, the convergence of bioengineering and cancer research has helped facilitate the development of more accurate and reliable methods to study directional migration, both in vitro and in vivo (see Fig. 3).

Micropipette/Micropatterning Assays for Local Two-Dimensional Gradients In Vitro

The first directional migration assay that was developed to generate local quantifiable gradients involved the use of a micromanipulator-controlled micropipette, loaded with a given concentration of a soluble cue that was slowly released in proximity to a two-dimensional (2D)-plated cell. In this assay, micropipettes are placed within 1–2 cell diameters of each cell, and a pump is used to control the flow of the chemoattractant out of the pipette and, in turn, the gradient (Mouneimne et al. 2004; Soon et al. 2005). This type of assay has been used successfully to examine carcinoma cell responses to soluble GF gradients (Carmona et al. 2016). However, this method is low-throughput, technically tedious, and requires expensive equipment. Moreover, the gradients generated by the micropipette approach are transient, making it difficult to quantify chronic cellular changes, and lend themselves only to the study of the effect of soluble cues on cell behavior. To address this issue, 2D micropatterning has been developed to study durotactic gradients. Although these 2D durotactic gradients have enjoyed varying degrees of success, they are notoriously technically challenging and thus not easily adapted to most cancer cell biology laboratories. To facilitate these studies, new durotactic gradient assays have been developed using mechanically tunable 2D polymer hydrogels conjugated with different ECM ligands (Roca-Cusachs et al. 2013). In these assays, stiffness gradients are generated by vary-

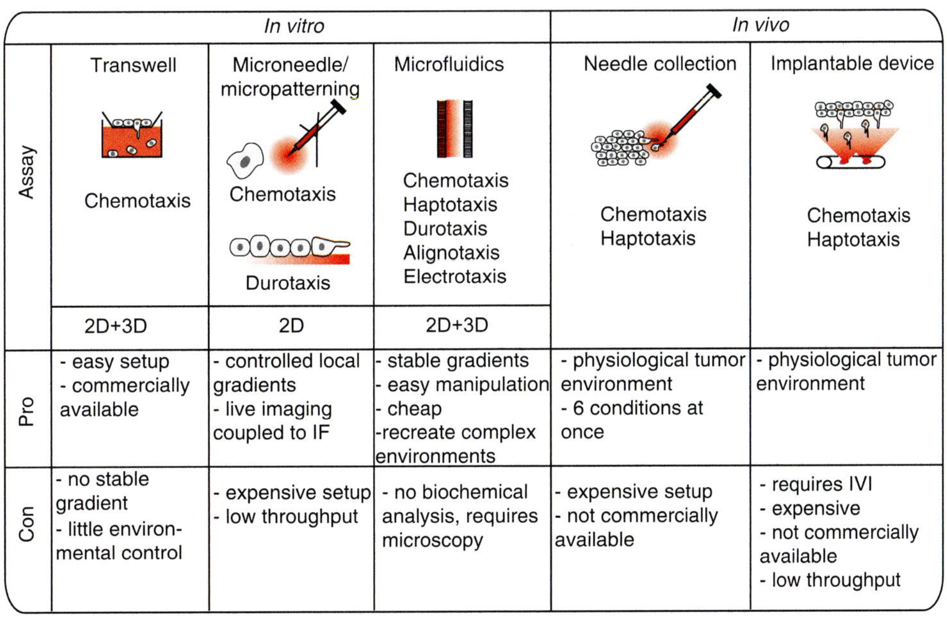

Figure 3. In vitro and in vivo methods to study directional migration: commonly used directional migration assays as well as pros and cons associated with each assay. In vitro study of directional migration relies on three main types of assays: the traditional transwell assay, microneedles, or micropatterning and microfluidics. To study in vivo directional migration, the main assays currently used are the needle collection assays and implantable devices. IF, immunofluorescence; IVI, intervertebral implant.

ing the cross-linking density of polyacrylamide hydrogels (Kuo et al. 2012), controlled by the amount of the cross-linker or light, for photoinitiated polymerization of gels (Sunyer et al. 2012). Alternately, a three-dimensional (3D) durotaxis gradient can be generated using a novel 3D collagen/FN/basement membrane hydrogel bioreactor that rapidly induces a 3D stiffness gradient by varying the angle of uniaxial stretch (Cassereau et al. 2015).

New Advances in Microfluidics

Advances in microfluidic technology have paved the way for its application to the study of directional migration. These tailor-made microfluidic devices permit precise spatiotemporal control over the microenvironment the cells are in and the gradients toward which cells will migrate. At present the vast majority of these microfluidic devices use polydimethyl-siloxane (PDMS), which facilitates the generation of microfabricated patterns and structures. Microfluidic devices are easily adopted in a cell biology platform and have the added benefit that they are reasonably cheap to generate. They also use small volumes of reagents, can be easily manipulated, and lend themselves to high-resolution time-lapse imaging (Bersini et al. 2014). To this end, microfluidic devices have been used to study the impact of soluble gradients of growth factors on cancer cell migration with good success (Li and Lin 2011). Some of the newer generation microfluidic devices incorporate pumps that provide a continuous flow of the soluble gradient and permit intermittent fluid analysis (Wu et al. 2012), whereas others generate long-term stable gradients and have revealed that subtle differences in gradients can exert significant biological effects (Suraneni et al. 2012). Stable ECM haptotactic gradients have also been developed using microfluidics and used to study 2D and 3D responses to multiple ECM cues such as FN, laminin, and vitronectin (Chan et al. 2014; Oudin et al. 2016a). Microfluidic channel width can also be used to directly modulate collagen alignment. In these iterations, a flow-based 3D microchannel assay that plated collagen in wide channels (3 mm) produced a random matrix, whereas when the collagen was added to a narrow channel (1 mm), it produced an aligned collagen matrix (Riching et al. 2014). Finally, microfluidics have significantly improved the study of galvanotaxis, which requires stable, integrated, tightly sealed cell culture chambers to maintain the provision of multiple stable electric fields (Huang et al. 2009; Li and Lin 2011).

Visualizing and Measuring Directional Migration In Vivo

Generating any sustained gradient—whether it is a soluble factor, a mechanical constraint, or a biochemical cue—within an endogenous tumor microenvironment in vivo is extremely challenging. One assay developed to study tumor cell migration responses to local gradients in vivo is the needle collection or in vivo invasion assay. In these assays, stainless steel needles containing an attractant are placed within an endogenous tumor into an anesthetized mouse that has developed a genetically engineered or a xenografted tumor (Hernandez et al. 2009). Intravital imaging studies in mice containing needles with EGF have shown that cells enter the needle by active migration only (Wyckoff et al. 2004). Since its inception this assay has been used to study both chemotactic and haptotactic responses (Oudin et al. 2016a). More recently, a new microscale device was developed which incorporates multiple reservoirs in the delivery system to permit the time-resolved delivery of several compounds. This new device permits the slow diffusion of drugs over time (Jonas et al. 2015). These devices can be placed on the edges of tumors and, coupled with intravital imaging, can permit real-time imaging of the tumor cells response to large sustained local gradients in vivo (Oudin et al. 2016a). Other implantable devices have been developed, such as the iNanivid (Williams et al. 2016b), to visualize cell responses in vivo directly, and these approaches are likely to become far more powerful in the near future. Overall, in vivo directional migration studies are technically difficult, require specific noncommercially available equipment, and are expensive. Nevertheless, they serve to validate the importance of microenvironmental gradients as a key regulator of the tumor phenotype.

MECHANISMS OF DIRECTIONAL MIGRATION

Directional migration favors the invasion of tumor cells into the parenchyma and their metastatic dissemination and colonization. The basic mechanisms driving cell migration, as well as the different modes a cell can use to migrate, have been described in detail elsewhere (Friedl and Wolf 2003; Petrie et al. 2009). Directional migration occurs through a series of defined steps that include cell polarization via asymmetric protrusion formation, focal adhesion formation and adhesion, and cell contraction leading to rear end detachment. Here, we focus specifically on mechanisms that have been shown to be involved in the directed migration of cancer cells and that have been implicated in cancer progression. The pathways described below have been summarized in Figure 4.

Chemotaxis

Numerous soluble cues drive chemotaxis by stimulating actin polymerization through RhoGTPases through the cells' cognate receptor tyrosine kinase (RTK) receptors or Gαi protein-coupled receptors (GPCRs) depending on the ligand (for review, see Roussos et al. 2011). In breast cancer cells, multiple GFs such as EGF, heparin-binding (HB)-EGF, TGF-α, insulin-like growth factor (IGF), and hepatocyte growth factor (HGF), but not PDGF and heregulin, induce strong protrusion responses that are predicted to drive 3D invasion in vitro (Meyer et al. 2012). Interestingly, the in vivo invasion assay revealed that EGF, TGF-α, and CSF-1, but not heregulin, VEGFα, fibroblast growth factor 1 (FGF-1), and PDGF, can induce the directed invasion of PyMT tumor cells in

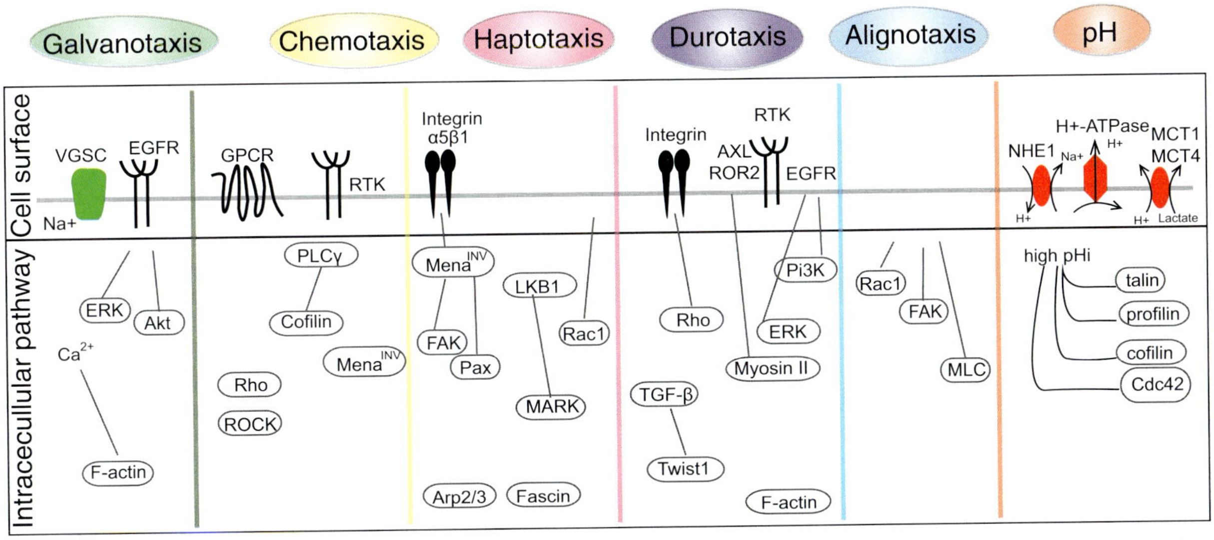

Figure 4. Cellular mechanisms known to drive directional migration: a summary of some of the main receptor families and signaling pathways that have been involved in the multiple modes of directional migration. These data are explained and cited in the text in more detail. It is important to note that some modes of directional migration, like galvanotaxis and pH-mediated migration, have their own individual cell surface receptors that facilitate responses to these cues. However, some of the intracellular signaling pathways mediating the responses overlap with other modes of migration. Chemotaxis, haptotaxis, and durotaxis have all been shown to require overlapping cell surface receptors. Similarly, many of the intracellular pathways, such as kinases and actin-myosin machinery, are required for response to a variety of different cues. VGSC, voltage-gated sodium channel; EGFR, epidermal growth factor receptor; GPCR, Gαi protein-coupled receptor; RTK, receptor tyrosine kinase; PLCγ, phospholipase C γ; TGF-β, transforming growth factor β.

mammary tumors formed in polyoma middle T (PyMT)-mouse mammary tumor virus (MMTV) mice and identified a set of core signaling pathways including those that influence Rho/ROCK activity, the capping protein cofilin, and the Arp2/3 pathway (Wang et al. 2004; Wyckoff et al. 2004; Patsialou et al. 2012). These studies also identified the actin regulatory protein Mena, a member of the Ena/VASP family whose alternately spliced isoform $Mena^{INV}$ has been strongly implicated in breast metastasis as well as poor patient prognosis (Goswami et al. 2009; Roussos et al. 2010; Gertler and Condeelis 2011; Oudin et al. 2016a). Importantly, $Mena^{INV}$ deregulates the phosphatase PTP1B to enhance RTK phosphorylation and potentiate actin polymerization via the phospholipase Cγ (PLCγ)–cofilin pathway greatly potentiating the tumor cells' response to chemotactic GFs such as EGF, IGF, and HGF (Philippar et al. 2008; Hughes et al. 2015). Chemotaxis downstream from GPCRs also contributes to tumor cell migration. For instance, GIV/Girdin, a non-receptor guanine nucleotide exchange factor (GEF) that triggers trimeric G-protein activation, binds to RTKs to promote metastasis (Garcia-Marcos et al. 2015), likely by fostering chemotaxis in response to insulin through enhanced Akt signaling. Similarly, chemokines such as CXCL12 (or SDF-1) can also drive chemotaxis via their cognate chemokine receptors.

Haptotaxis

The characteristic increase and heterogeneity in the distribution of the newly deposited and remodeled ECM protein in the tumor is consistent with the notion that haptotaxis contributes to the heterogeneity of the tumor phenotype. Haptotaxis is defined as the directed migration of cells on gradients of substrate-bound cues, in which cells are migrating on or within an ECM (Wu et al. 2012). In this respect, tumor haptotaxis toward gradients of the glycoprotein FN has been well studied. Studies using models of triple-negative breast cancer revealed that the actin binding protein Mena promotes FN haptotaxis through its ability to directly interact with the FN receptor integrin α5β1 (Oudin et al. 2016a). In this work, levels of the invasive isoform of Mena, $Mena^{INV}$, whose expression associates with a poor breast cancer patient outcome, drove tumor cell haptotaxis toward high concentrations of FN by increasing focal adhesion kinase (FAK)-dependent outside-in signaling by enhancing inside-out signaling through ECM remodeling. The concept was validated in vivo using an implantable device to create FN gradients and intravital imaging to monitor and quantify effects on tumor cell migration. Importantly, these studies showed not only that FN can function as a bona fide directional cue, but also that a haptotaxis phenotype fosters tumor metastasis. Interestingly, in metastatic melanoma, loss of the tumor suppressor LKB1 and its downstream target, the MARK family kinases, compromises FN haptotaxis (Chan et al. 2014). Moreover, inserting a P29S mutation into the proinvasion, prometastatic Rac1 molecule also abrogates fibroblast haptotaxis in vitro (King et al. 2016). Therefore, these data suggest context-specific effects of FN haptotaxis during tumor progression.

In addition to the haptotactic behavior of FN, Boyden chamber experiments have attested to the haptotactic potential of thrombospondin for human melanoma (Taraboletti et al. 1987). Given that thrombospondin is a glycoprotein localized at basement membrane and vessel walls, this raises the possibility that circulating melanoma cells could feasibly encounter gradients of thrombospondin that could facilitate their intravasation into secondary organs to form metastatic lesions. Similarly, vitronectin, an ECM component involved in melanoma metastasis and present in serum, can act as haptotactic cue for A2058 melanoma cells by ligating $\alpha v\beta 3$ integrin to promote the tyrosine phosphorylation of paxilin (Aznavoorian et al. 1996). Collagen, one of the most abundant ECM components in solid tumors, can also drive haptotaxis. Indeed, pancreatic tumor cells can drive haptotaxis in response to collagen by binding $\alpha 2\beta 1$ integrin to enhance FAK activity (Lu et al. 2014). GIV/Girdin, which promotes metastasis, is a nonreceptor GEF that triggers trimeric G-protein activation that binds to RTKs (Garcia-Marcos et al. 2015) and also activates integrins to promote haptotaxis toward collagen (Leyme et al. 2015).

Durotaxis

Cells preferentially migrate toward and remain associated with a stiffer ECM through a process termed durotaxis (Lo et al. 2000). Over the past several years, the original observations defining durotaxis have been elaborated upon using 2D and 3D natural and synthetic biomaterials with ECM stiffness gradients and have illustrated how cells persistently migrate up a stiffness gradient (Isenberg et al. 2009; Plotnikov et al. 2012; Cassereau et al. 2015). The importance of durotaxis to malignancy has also been shown by a series of in vitro studies using 3D hydrogels with calibrated stiffness and in vivo experiments in transgenic mouse models in which ECM rigidity was reduced using cross-linking inhibitors or enhanced through ECM cross-linking (for review, see Kai et al. 2016). These studies showed that a stiffened ECM promotes breast and squamous cell invasion by promoting integrin focal adhesions to enhance growth factor and GPCR signaling that activates RhoGTPases, Wnt, mitogen-activated protein (MAP), and phosphoinositide 3-kinases (PI3Ks) (Paszek et al. 2005; Levental et al. 2009). The relevance of ECM stiffness to breast tumor metastasis was further demonstrated by transgenic mouse studies in which collagen cross-linking and ECM stiffness were manipulated. These studies identified TGF-β and microRNAs that target tumor suppressors like the PI3K repressor phosphatase and tensin homolog (PTEN) as critical molecular modulators of the durotactic phenotype (Pickup et al. 2013; Mouw et al. 2014). Recently, a role for ECM rigidity in promoting pancreatic tumor progression and glioblastoma aggression that expand upon in vitro studies was demonstrated (Ulrich et al. 2009; Laklai et al. 2016; Miroshnikova et al. 2016). Building upon these observations data generated using MDA-MB-231 mesenchymal breast cancer cells defined an integrin-associated myosin-dependent molecular mechanism that regulates cellular rigidity sensing downstream from RTK signaling that is corrupted in tumors through AXL and ROR2 kinase perturbations (Lo et al. 2000; Hoffman et al. 2011; Plotnikov et al. 2012; Raab et al. 2012; Yang et al. 2016).

Alignotaxis

The reorganization of collagen from curly, coiled fibers to strait aligned fibers promotes the directed, rapid, and persistent migration of tumor cells, which we describe as alignotaxis. Alignotaxis was illustrated by studies conducted using the metastatic MDA-MB-231 breast cancer cells (Provenzano et al. 2008). Furthermore, aligned collagen fibers that are assembled through coordinated tumor colony interactions foster the rapid and directed migration of tumor cells (Shi et al. 2013). Consistently, intravital imaging studies in breast tumor models revealed that tumor cells located in close proximity to collagen fibers migrate faster than those located within the core of the tumor mass (Wang et al. 2002). Using fluorescently tagged myosin light chain (MLC), one study showed that MLC is recruited to the protrusions of migrating cells, with second-harmonic generation (SHG) imaging showing concomitant deformation in collagen fibers at the site of the protrusion (Wyckoff et al. 2006). Computational modeling followed by experimental validation showed that migratory persistence is enhanced by collagen fiber alignment because it limits the number of protrusions and restricts their formation to the direction of the alignment (Riching et al. 2014). Additional studies revealed that collagen fiber alignment can predict cell motility, relative to the pore size and density (Fraley et al. 2015), and that an increase in protrusion frequency, persistence, and lengthening along the direction of the fibers depends on FAK and Rac1 signaling (Carey et al. 2016). Indeed, preventing collagen alignment by inhibiting the activity of the collagen cross-linker LOXL2 (Grossman et al. 2016), or through knockdown of the proteoglycan syndecan1 (Yang et al. 2011), but not by inhibiting protease activity (matrix metalloproteinases [MMPs] and serine and cysteine proteases), reduced the directional migration of tumor cells in vitro and in vivo (Wyckoff et al. 2006; Levental et al. 2009; Rubashkin et al. 2014). Thus compelling in vitro and in vivo data attest to the importance of alignotaxis as a regulator of the local invasion of tumor cells.

Galvanotaxis

Galvanotaxis (or electrotaxis) describes the directional movement driven by different EFs. Multiple studies have illustrated a correlation between the metastatic potential of a tumor cell and its sensitivity to galvanotaxis (Djamgoz et al. 2001), including the galvanotaxis-mediated directional migration of prostate cancer cells (Djamgoz et al. 2001), breast cancer (Fraser et al. 2005), lung adenocarcinoma (Sun et al. 2012), and glioblastoma cells (Huang et al. 2016). Interestingly, although prostate, lung, and glioblastoma multiforme (GBM) tumor cells show cathodal galvanotaxis, the breast tumor cells migrate toward the anode, suggesting that different tumor microenvironments might contribute to EF-driven invasion. Although the primary stimulus differs, galvanotaxis likely exploits many of the same cytoskeletal pathways that mediate chemotaxis, haptotaxis, and durotaxis

(Mycielska and Djamgoz 2004). For instance, EFs increase intracellular Ca^{2+} to induce actin polymerization and actomyosin contractility, which are key for the formation of localized cellular protrusions that permit directional migration (Mycielska and Djamgoz 2004). More recently, voltage-gated Na^{+} channels (VGSCs) were shown to control the galvanotaxis behavior of tumor cells, and their levels correlated with the tumor cells metastatic potential (Djamgoz et al. 2001; Fraser et al. 2005). More recently, A549 lung carcinoma galvanotaxing toward a cathode polarized their EGFRs and assembled filamentous actin cables in the direction of the cathode. These galvanotaxing cells also showed oriented ERK and Akt signaling that lined up with the EF and apparently was critical for directional migration (Yan et al. 2009). Consistently, one study showed that the galvanotaxis of brain tumor initiating cells from three distinct GBM subtypes also depended upon the oriented activity of PI3K and ERK signaling (Huang et al. 2016). Although these findings suggest galvanotaxis can promote the directed migration of tumor cells, data clearly suggest that the nature of the behavior including the anodal versus the cathodal effect is context-dependent and likely cell type–dependent.

pH

The tumor-associated reversal of the normal pH gradient can profoundly influence intracellular signaling and modify the extracellular microenvironment to influence cell migration. Thus, the abnormally elevated intracellular pHi in the tumor cell can activate several intracellular signaling pathways linked to integrins, ion channel activity, and actin polymerization that will promote cell proliferation and migration. For instance, several guanine nucleotide exchange factors, such as DBS, that stimulate CDC42 become activated when the pH rises (Grillo-Hill et al. 2015). Other molecules that bind to actin and promote polymerization, such as cofilin (Frantz et al. 2008), profilin (McLachlan et al. 2007), and talin (Srivastava et al. 2008), also have increased activity at higher pHi. The extrusion of H^{+} from tumor cells is accompanied by an influx of water and subsequent osmotic swelling and opening of aquaporin channels (Saadoun et al. 2005), effects which contribute to driving cell migration. Consistently, the low pHe induced in tumor cells can significantly alter the ECM to create a prometastatic environment (Webb et al. 2011). Indeed, the lower pHe in the tumor microenvironment can activate multiple proteases such as MMP3 and MMP9 (Johnson et al. 2000), which result in ECM remodeling to foster the creation of migratory tracks and the release of ECM-bound soluble factors that stimulate cell motility.

STRATEGIES TO TARGET DIRECTIONAL MIGRATION AND PREVENT METASTASIS

Exploiting Directional Migration to Predict Metastasis

The ability to predict which patients' tumor has a high probability of metastasis would provide critical clinical insight to guide treatment and would likely influence the patient outcome. In this regard, a number of commercially available tests have recently been developed to estimate metastatic risk in cancer patients. One such test is the MammaPrint DX, which is a 70-gene signature that can identify breast cancer patients who are most likely to develop metastasis within 5 years of diagnosis with reasonable accuracy (van't Veer et al. 2002; van de Vijver et al. 2002). Similarly, the 15-gene signature DecisionDx-UM test is able to accurately predict which patients with uveal melanoma are most likely to metastasize and has been tested in large multicenter trials with excellent results (Harbour 2014). Exploiting the concept of tumor gradients as key regulators of tumor migration and metastasis is the derivation of a new metastasis biomarker that quantifies the number of tumor microenvironment of metastasis (TMEM) sites within a tumor. TMEMs are a tripartite structure composed of a macrophage, a Mena-expressing tumor cell, and an endothelial cell that result from and reflect chemotactic gradients within the tumor (Robinson et al. 2009). Recent clinical work suggests that as the number of TMEMs within an ER^{+}/PR^{+} breast tumor increases, metastasis increases; hence, patient prognosis decreases. The success of these new approaches to predict patient prognosis emphasizes the potential utility of developing directional migration biomarkers to predict metastatic disease (Rohan et al. 2014).

Blocking Cell Surface Receptors and Signaling to Impede Directional Migration

Therapies targeting the main chemotactic signaling pathways such as GFs/RTK and chemokine/GPCR have been approved for patient administration, with many more in clinical trials. Although most of these are aimed at inhibiting tumor growth and angiogenesis, they have the potential to impact metastasis-relevant processes. Yet, despite the fact that a role of EGF chemotaxis has been well established in breast cancer metastasis in mouse models, EGFR inhibitors have had mixed clinical efficacy (Crown et al. 2012). Small-molecule tyrosine kinase inhibitors (TKIs) such as gefitinib or erlotinib have failed to show any significant improvement in the triple-negative breast cancer (TNBC) patient outcome over standard-of-care chemotherapy (Baselga et al. 2005). The monoclonal EGFR antibody cetuximab showed only a modest increase in the overall response rate and progression-free survival relative to cisplatin (Baselga et al. 2013). Inhibitors targeting other surface RTKs that can modulate directional migration are currently in clinical trials for breast cancer, such as Met, IGF1R, FGFR, and ERBB3, and await verification of their utility (Jin and Mu 2015). Many molecular mechanisms are proposed to account for the failure of these promising RTK targets, including the expansion of genetically mutated tumor cells and reliance on parallel survival pathways.

Integrins are the main cell surface receptors that allow cells to sense, respond to, and reorganize their tumor microenvironment and cells use integrins for haptotaxis, chemotaxis, and durotaxis. However, although in vitro and preclinical studies have shown promising results (Schaffner et al. 2013), many of the integrin targeting clinical trials have ultimately failed (Desgrosellier and Cheresh 2010). The high-affinity anti-α5β1 antibody volociximab stabilized disease in metastatic renal cell carcinoma (Conti et al. 2013); however, another monoclonal α5β1 integrin antibody PF-04605412 had no effect on a variety of solid tumors and has consequently been discontinued (Mateo et al. 2014). A phase 3 trial in KRas wild-type metastatic colorectal cancer found no improvement with abituzumub, an anti-αv integrin antibody, relative to standard of care (Élez et al. 2015). Similarly, the integrin antagonist cilengitide did not improve the outcome in glioblastoma patients compared with radiotherapy, and the development of this drug was also discontinued (Stupp et al. 2014). Again the reason for these clinical failures remains unclear and are likely complex, including cross talk with diverse RTK signaling pathways.

An alternative to directly targeting cell surface receptors is identifying and targeting the signaling pathways that become activated downstream from cell surface receptors that mediate responses to multiple directional cues. Examples of such downstream targets toward which potent clinical inhibitors could be directed include FAK and Src kinase. Several clinical trials are ongoing that treat tumors by targeting FAK, although conclusive results are yet to be obtained. Interestingly, multiple preclinical studies have shown that inhibition of FAK may synergize well with therapies targeting other pathways or components of the microenvironment. Indeed, inhibition of FAK with VS-4718 delayed tumor progression in the p48-Cre;LSL-Kras^{G12D};Trp53$^{flox/+}$ KPC model for human pancreatic ductal adenocarcinoma (PDAC) and also rendered these tumors more sensitive to T-cell immunotherapy and PD-1 antagonists' PDAC (Jiang et al. 2016). In metastatic melanoma, treatment with a FAK inhibitor prevented resistance to Braf inhibitors caused by fibroblast-mediated ECM deposition, which activated integrin signaling in tumor cells (Hirata et al. 2015).

Galvanatoxis, regulated mainly by VGSCs, can drive directional migration that contributes to metastasis (Fraser et al. 2005), suggesting that blocking VGSC activity can be a strategy to block metastasis. Indeed, inhibition of VGSCs by tetrodotoxin has been used in a prostate cancer in vivo model to reduce metastatic burden (Yildirim et al. 2012). Targeting pathways that allow cells to respond to EFs may be useful in targeting directional migration. Elucidating the mechanisms driving directional migration in response to multiple cues has increased our understanding of the pathways to target to inhibit metastasis. However, it is important to note how challenging it is to develop antimetastatic therapies from a clinical perspective (for a review, see Steeg 2016).

Disrupting the Gradient

An alternate approach to target and prevent directional migration in tumors is by directly disrupting the endogenous gradients that drive tumor cell invasion (Fig. 5). Toward this objective, several stromal cells have been identified that can contribute to the generation of local growth factor or chemokine gradients that drive directional migration in tumors. The same stromal compart-

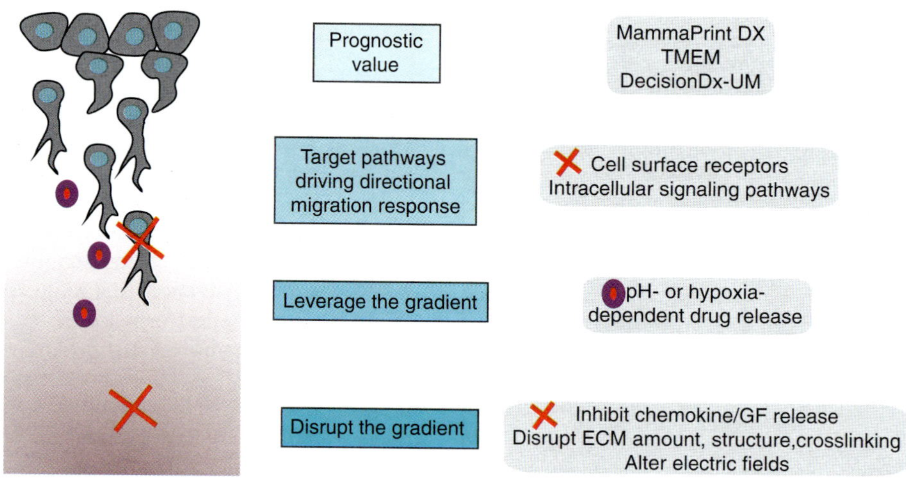

Figure 5. Strategies to target directional migration. We have identified four strategies to use directional migration to target tumor growth, local invasion, dissemination, and colonization (outlined in the four boxes in this figure). First, studying the mechanisms driving directional migration can lead to the development prognostic markers that can be used to predict likelihood of disease progression. Second, identifying the core signaling pathways that drive the cell responses to endogenous gradients will allow for use of already available drugs or the development of drugs against new targets. Third, the physical and biochemical properties of gradients can be leveraged to facilitate local drug delivery and drug release. Fourth, strategies to directly disrupt the gradient can be developed to remove the local cues that promote aggressive tumor phenotypes in the first place. TMEM, tumor microenvironment of metastasis; GF, growth factor; ECM, extracellular matrix.

ment also plays an important role in the deposition of ECM, as well as its organization, and thereby regulates haptotaxis, alignotaxis, and durotaxis in tumors (Conklin and Keely 2012; Naba et al. 2012). Macrophages in particular, which secrete promigratory growth factors, were identified as viable therapeutic targets in cancer. However, despite these provocative findings, depletion of macrophages, through genetic or pharmacological perturbations, does reduce tumor cell motility and intravasation in breast cancer tumors (Wyckoff et al. 2004; Harney et al. 2015). Nevertheless, several pharmacological agents are currently in clinical trials that should block macrophage precursor recruitment, deplete tumor-associated macrophages, or reprogram their function that might prove more efficacious (Williams et al. 2016a). Inhibition of the CSF-1 receptor on macrophages in particular offers one promising strategy to reduce directed cell migration and tumor dissemination and is currently in clinical trials in advanced refractory breast or prostate cancer.

Oriented fibrillary collagen can promote directional migration in tumors through alignotaxis and durotaxis, suggesting strategies to prevent collagen alignment and stiffening, and may provide a viable therapeutic strategy to prevent tumor progression and metastasis. Consistent with this concept, several preclinical studies have attested to the utility of targeting collagen-processing enzymes to inhibit malignancy in vivo. For instance, inhibition of LOX through pharmacological inhibition or injection with lysyl oxidase (LOX)-neutralizing antibodies reduced tumor metastasis from solid tumors (Levental et al. 2009; Cox et al. 2013; Pickup et al. 2013). Similarly, studies using function-blocking antibodies to inhibit lysyl oxidase-like 2 (LOXL2) reduced collagen alignment and decreased tumor burden in a breast tumor xenograft (Grossman et al. 2016); these antibodies have been associated with invasiveness and poor outcome in breast cancer (Ahn et al. 2013). Higher LOXL2 levels are also associated with tumor invasion and poor outcome in pancreatic cancer patients and showed some efficacy in mouse models of PDAC (Park et al. 2016). However, when a LOXL2 inhibitor simtuzamab (Gilead) was tested in combination with gemcitabine in a cohort of previously untreated advanced pancreatic cancer patients, there was no observed significant clinical benefit over placebo. Alternately, FN is required for LOX-dependent collagen cross-linking, and treatment of mice with liver fibrosis with a peptide that blocks FN fibril formation (pUR4) significantly decreased their fibrosis, suggesting this type of treatment may be useful to reduce ECM accumulation in solid tumors (Altrock et al. 2015). Similarly, mice treated with the antifibrotic agent pirfenidone showed a significant reduction in lung metastasis (Takai et al. 2016).

Intriguingly, galvanotaxis has been extensively characterized in the context of wound healing (Zhao et al. 2006), and studies have shown that the movement of epithelial sheets can be easily and tightly controlled through manipulation of electric currents (Cohen et al. 2014). Such findings suggest that application of low-magnitude EFs could locally manipulate cell migration within tissues, particularly in the context of nonhealing and chronic wounds, and could be developed as a noninvasive approach to control the local invasive behavior of primary tumors (Wu and Lin 2014).

Taking Advantage of the Gradient

Chemotherapy remains the standard of care for most cancers despite its nonspecificity and high toxicity. One approach is to concentrate the chemical and target its delivery through microfabricated nanoparticles generated with diverse polymeric carriers that can be functionalized to specifically target the tumor itself by exploiting unique properties of the tumor microenvironment (Kwon et al. 2015). In this case, the inverted pH gradient present in a variety of solid tumors that acidifies the extracellular pH has been exploited by developing pH-sensitive carrier systems that permit optimal cargo release only in the presence of the low pH within the tumor (Tian and Bae 2012). For example, a layer-by-layer pH-responsive sheddable nanoparticle shell exists in which the neuronal layers that encapsulate the drug are only shed in response to the low pH in the tumor, exposing the nanoparticle surface loaded with a chemotherapeutic agent that can be taken up by the cells (Poon et al. 2011). Similarly, the hypoxic regions of the tumor have been leveraged to optimize therapeutic delivery to tumor regions using multiple mechanisms (Patel and Sant 2016). One such example is provided by a bioactive prodrug, such as the N-oxide drug tirapazamine (TPZ), which induces DNA damage, works synergistically with chemotherapy and radiation therapy, and is only activated under hypoxia (von Pawel et al. 2000). These data suggest that the exploitation of endogenous gradients (hypoxia, pH) within a tumor could be used to enhance drug-targeting strategies that would potentiate tumor treatment and reduce many undesirable side effects.

CONCLUSION AND OUTSTANDING QUESTIONS

How Do Different Gradients Affect Each Other?

Out of necessity, most cancer studies focus on the tumor cells responses to a single cue. Yet, in vivo, tumor cells simultaneously face a variety of endogenous gradients, many of which interact positively and negatively with each other (Fig. 6). This raises the important point that to understand tumor evolution it will be critical to design studies that can interrogate these complex physical and chemical interactions. For example, changes in ECM deposition and organization simultaneously promote haptotaxis, alignotaxis, and durotaxis. Furthermore, a recent study showed that GBM galvanotaxis proceeds toward the anode on a poly-L-ornithine/laminin 2D surface but switches to a cathode bias when the tumor cells are assayed with a 3D hyaluronic acid and collagen gel (Huang et al. 2016). This underscores the importance of using in vitro assays that can accurately model multiple aspects of the tumor microenvironment to gain a clear and accurate

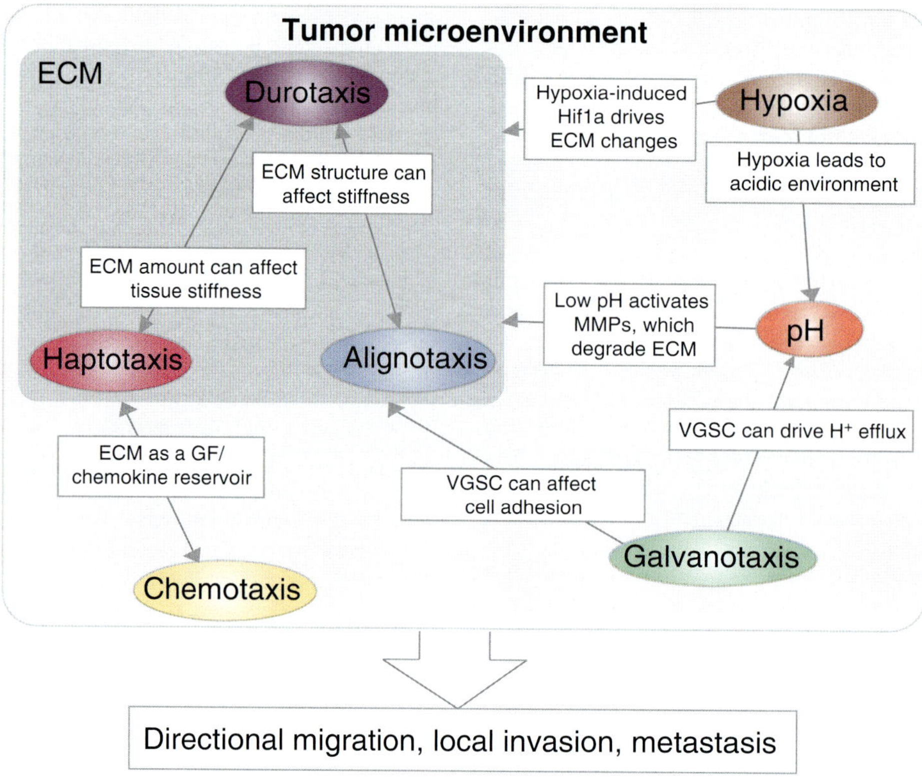

Figure 6. Potential for cross talk in response to the abundance of cues provided by the tumor microenvironment. The changes in local microenvironments that occur as a consequence of tumor formation lead to the generation of multiple endogenous gradients simultaneously, which each can promote their own mode of directional migration. However, is not clear that changes in one cue can affect another and that common signaling pathways might regulate responses to multiple cues. The *inset* highlights the potential for cross talk in the context of directional migration in response to extracellular matrix (ECM) cues: given that both ECM amount and structure can affect stiffness, haptotaxis, durotaxis, and alignotaxis and could each occur simultaneously, and it is likely that pathways driving one response may be involved in the other. The ECM is well known to act as a reservoir for growth factor (GF), and several signaling pathways have been shown to be involved in response to both chemotactic and haptotactic cues. Hypoxia has been shown to drive changes in ECM via up-regulation of Hif1α, implying that endogenous hypoxia gradients could lead to the generation of haptotactic gradients. Hypoxia leads to the development of an acidic environment, which in turn leads to matrix metalloproteinase (MMP) activation, which is known to degrade and reorganize ECM. Therefore, the establishment of pH gradients could affect response to ECM gradients. Finally, changes in cellular electrical potential, in part mediated by voltage-gated sodium channels (VGSCs), can affect cell adhesion, as well as drive H^+ efflux, suggesting that galvanotaxis may affect responses to ECM gradients and the maintenance of pH gradient.

understanding of the role of directional migration of tumors in vivo.

How Does One Cell Integrate Multiple Directional Cues?

The same cell surface receptors and signaling pathways have been shown to regulate directional responses to multiple cues, suggesting that there may be core signaling pathways that coordinate directional responses in several contexts. For example, the invasive isoform of the actin regulator $Mena^{INV}$ drives haptotaxis to high FN gradients (Oudin et al. 2016a) while also sensitizing cells to low EGF concentrations (Hughes et al. 2015). Interestingly, when breast tumor cells are incubated with both cues, expression of $Mena^{INV}$ renders them hyperinvasive, driving synergy between the signaling pathways downstream from both integrins and RTKs, leading to an increased migratory response that is stronger than each cue alone (Oudin et al. 2016b). Altogether, these studies highlight the importance of studying how different cues cooperate mechanistically to promote metastasis.

Attraction versus Repulsion?

In this review, we focused our discussion on how attractive cues promote local invasion. However, repulsive cues can also contribute to directional cell migration as has been consistently shown for neural guidance in the developing brain. For instance, the soluble ligand Slit and its transmembrane receptor Round-about (Robo) are well-established regulators of axon growth (Tessier-Lavigne and Goodman 1996); recent work suggests that this pathway may also play a role in cancer (Mehlen et al. 2011). This raises the intriguing possibility that repulsion may also play a role in modulating tumor cell guidance and metastasis.

How Do Current Therapies Affect Directional Migration?

Tumor cells that are particularly sensitive to physical and chemical directional cues are often the same cells that are intrinsically highly metastatic. This observation merits investigation because such heightened sensitivity could influence their treatment response and promote their dissemination. In this regard, Mena^{INV}, which drives chemotaxis and haptotaxis (Hughes et al. 2015; Oudin et al. 2016a), also promotes resistance to paclitaxel, a chemotherapeutic drug that is standard of care for the treatment of metastatic breast cancer (Oudin et al. 2017).

ACKNOWLEDGMENTS

This work was supported by K99-CA207866-01 to M.J.O. from the National Cancer Institute (NCI), the National Institutes of Health (NIH) NCI R01 grants CA138818-01A1, CA192914-01, CA174929-01, and CA085482 and 1U01CA202241-01 to V.M.W., and a Department of Defense (DOD) Breast Cancer Research Program (BCRP) grant BC122990 to V.M.W.

We apologize to the authors whose work we were unable to cite because of space limitations.

REFERENCES

Acerbi I, Cassereau L, Dean I, Shi Q, Au A, Park C, Chen YY, Liphardt J, Hwang ES, Weaver VM. 2015. Human breast cancer invasion and aggression correlates with ECM stiffening and immune cell infiltration. *Integr Biol (Camb)* **7:** 1120–1134.

Ahn SG, Dong SM, Oshima A, Kim WH, Lee HM, Lee SA, Kwon S-H, Lee J-H, Lee JM, Jeong J, et al. 2013. LOXL2 expression is associated with invasiveness and negatively influences survival in breast cancer patients. *Breast Cancer Res Treat* **141:** 89–99.

Alizadeh AA, Aranda V, Bardelli A, Blanpain C, Bock C, Borowski C, Caldas C, Califano A, Doherty M, Elsner M, et al. 2015. Toward understanding and exploiting tumor heterogeneity. *Nat Med* **21:** 846–853.

Altrock E, Sens C, Wuerfel C, Vasel M, Kawelke N, Dooley S, Sottile J, Nakchbandi IA. 2015. Inhibition of fibronectin deposition improves experimental liver fibrosis. *J Hepatol* **62:** 625–633.

Astrof S, Hynes RO. 2009. Fibronectins in vascular morphogenesis. *Angiogenesis* **12:** 165–175.

Aznavoorian S, Stracke ML, Parsons J, McClanahan J, Liotta LA. 1996. Integrin $\alpha v \beta 3$ mediates chemotactic and haptotactic motility in human melanoma cells through different signaling pathways. *J Biol Chem* **271:** 3247–3254.

Barathova M, Takacova M, Holotnakova T, Gibadulinova A, Ohradanova A, Zatovicova M, Hulikova A, Kopacek J, Parkkila S, Supuran CT, et al. 2008. Alternative splicing variant of the hypoxia marker carbonic anhydrase IX expressed independently of hypoxia and tumour phenotype. *Br J Cancer* **98:** 129–136.

Baselga J, Albanell J, Ruiz A, Lluch A, Gascón P, Guillém V, González S, Sauleda S, Marimón I, Tabernero JM, et al. 2005. Phase II and tumor pharmacodynamic study of gefitinib in patients with advanced breast cancer. *J Clin Oncol* **23:** 5323–5333.

Baselga J, Gómez P, Greil R, Braga S, Climent MA, Wardley AM, Kaufman B, Stemmer SM, Pêgo A, Chan A, et al. 2013. Randomized phase II study of the anti-epidermal growth factor receptor monoclonal antibody cetuximab with cisplatin versus cisplatin alone in patients with metastatic triple-negative breast cancer. *J Clin Oncol* **31:** 2586–2592.

Ben-Baruch A. 2008. Organ selectivity in metastasis: Regulation by chemokines and their receptors. *Clin Exp Metastasis* **25:** 345–356.

Bersini S, Jeon JS, Moretti M, Kamm RD. 2014. In vitro models of the metastatic cascade: From local invasion to extravasation. *Drug Discov Today* **19:** 735–742.

Boyden S. 1962. The chemotactic effect of mixtures of antibody and antigen on polymorphonuclear leucocytes. *J Exp Med* **115:** 453–466.

Carey SP, Goldblatt ZE, Martin KE, Romero B, Williams RM, Reinhart-King CA. 2016. Local extracellular matrix alignment directs cellular protrusion dynamics and migration through Rac1 and FAK. *Integr Biol (Camb)* **8:** 821–835.

Carmeliet P, Jain RK. 2011. Principles and mechanisms of vessel normalization for cancer and other angiogenic diseases. *Nat Rev Drug Discov* **10:** 417–427.

Carmona G, Perera U, Gillett C, Naba A, Law A-L, Sharma VP, Wang J, Wyckoff J, Balsamo M, Mosis F, et al. 2016. Lamellipodin promotes invasive 3D cancer cell migration via regulated interactions with Ena/VASP and SCAR/WAVE. *Oncogene* **35:** 5155–5169.

Cassereau L, Miroshnikova YA, Ou G, Lakins J, Weaver VM. 2015. A 3D tension bioreactor platform to study the interplay between ECM stiffness and tumor phenotype. *J Biotechnol* **193:** 66–69.

Chan KT, Asokan SB, King SJ, Bo T, Dubose ES, Liu W, Berginski ME, Simon JM, Davis IJ, Gomez SM, et al. 2014. LKB1 loss in melanoma disrupts directional migration toward extracellular matrix cues. *J Cell Biol* **207:** 299–315.

Chang TT, Thakar D, Weaver VM. 2017. Force-dependent breaching of the basement membrane. *Matrix Biol* **57-58:** 178–189.

Chiche J, Ilc K, Laferriere J, Trottier E, Dayan F, Mazure NM, Brahimi-Horn MC, Pouyssegur J. 2009. Hypoxia-inducible carbonic anhydrase IX and XII promote tumor cell growth by counteracting acidosis through the regulation of the intracellular pH. *Cancer Res* **69:** 358–368.

Chiche J, Le Fur Y, Vilmen C, Frassineti F, Daniel L, Halestrap AP, Cozzone PJ, Pouysségur J, Lutz NW. 2012. In vivo pH in metabolic-defective Ras-transformed fibroblast tumors: Key role of the monocarboxylate transporter, MCT4, for inducing an alkaline intracellular pH. *Int J Cancer* **130:** 1511–1520.

Cohen DJ, James Nelson W, Maharbiz MM. 2014. Galvanotactic control of collective cell migration in epithelial monolayers. *Nat Mater* **13:** 409–417.

Conklin MW, Keely PJ. 2012. Why the stroma matters in breast cancer: Insights into breast cancer patient outcomes through the examination of stromal biomarkers. *Cell Adh Migr* **6:** 249–260.

Conti A, Santoni M, Amantini C, Burattini L, Berardi R, Santoni G, Cascinu S, Muzzonigro G. 2013. Progress of molecular targeted therapies for advanced renal cell carcinoma. *Biomed Res Int* **2013:** 419176.

Costa-Silva B, Aiello NM, Ocean AJ, Singh S, Zhang H, Thakur BK, Becker A, Hoshino A, Mark MT, Molina H, et al. 2015. Pancreatic cancer exosomes initiate pre-metastatic niche formation in the liver. *Nat Cell Biol* **17:** 816–826.

Cox TR, Bird D, Baker AM, Barker HE, Ho MWY, Lang G, Erler JT. 2013. LOX-mediated collagen crosslinking is responsible for fibrosis-enhanced metastasis. *Cancer Res* **73:** 1721–1732.

Cox TR, Gartland A, Erler JT. 2016. Lysyl oxidase, a targetable secreted molecule involved in cancer metastasis. *Cancer Res* **76:** 188–192.

Crown J, O'Shaughnessy J, Gullo G. 2012. Emerging targeted therapies in triple-negative breast cancer. *Ann Oncol* **23**(Suppl 6): vi56–vi65.

Cuzick J, Holland R, Barth V, Davies R, Faupel M, Fentiman I, Frischbier HJ, LaMarque JL, Merson M, Sacchini V, et al. 1998. Electropotential measurements as a new diagnostic modality for breast cancer. *Lancet* **352:** 359–363.

Desgrosellier JS, Cheresh DA. 2010. Integrins in cancer: Biological implications and therapeutic opportunities. *Nat Rev Cancer* **10:** 9–22.

Djamgoz MBA, Mycielska M, Madeja Z, Fraser SP, Korohoda W. 2001. Directional movement of rat prostate cancer cells in direct-current electric field: Involvement of voltage gated Na+ channel activity. *J Cell Sci* **114:** 2697–2705.

Egeblad M, Rasch MG, Weaver VM. 2010. Dynamic interplay between the collagen scaffold and tumor evolution. *Curr Opin Cell Biol* **22:** 697–706.

Élez E, Kocáková I, Höhler T, Martens UM, Bokemeyer C, Van Cutsem E, Melichar B, Smakal M, Csőszi T, Topuzov E, et al. 2015. Abituzumab combined with cetuximab plus irinotecan versus cetuximab plus irinotecan alone for patients with KRAS wild-type metastatic colorectal cancer: The randomised phase I/II POSEIDON trial. *Ann Oncol* **26:** 132–140.

Erler JT, Weaver VM. 2009. Three-dimensional context regulation of metastasis. *Clin Exp Metastasis* **26:** 35–49.

Evans A, Whelehan P, Thomson K, McLean D, Brauer K, Purdie C, Baker L, Jordan L, Rauchhaus P, Thompson A. 2012. Invasive breast cancer: Relationship between shear-wave elastographic findings and histologic prognostic factors. *Radiology* **263:** 673–677.

Faupel M, Vanel D, Barth V, Davies R, Fentiman IS, Holland R, Lamarque JL, Sacchini V, Schreer I. 1997. Electropotential evaluation as a new technique for diagnosing breast lesions. *Eur J Radiol* **24:** 33–38.

Fenner J, Stacer AC, Winterroth F, Johnson TD, Luker KE, Luker GD. 2014. Macroscopic stiffness of breast tumors predicts metastasis. *Sci Rep* **4:** 5512.

Fraley SI, Wu P-H, He L, Feng Y, Krisnamurthy R, Longmore GD, Wirtz D. 2015. Three-dimensional matrix fiber alignment modulates cell migration and MT1-MMP utility by spatially and temporally directing protrusions. *Sci Rep* **5:** 14580.

Frantz C, Barreiro G, Dominguez L, Chen X, Eddy R, Condeelis J, Kelly MJS, Jacobson MP, Barber DL. 2008. Cofilin is a pH sensor for actin free barbed end formation: Role of phosphoinositide binding. *J Cell Biol* **183:** 865–879.

Fraser SP, Diss JKJ, Chioni A-M, Mycielska ME, Pan H, Yamaci RF, Pani F, Siwy Z, Krasowska M, Grzywna Z, et al. 2005. Voltage-gated sodium channel expression and potentiation of human breast cancer metastasis. *Clin Cancer Res* **11:** 5381–5389.

Friedl P, Wolf K. 2003. Tumour-cell invasion and migration: Diversity and escape mechanisms. *Nat Rev Cancer* **3:** 362–374.

Gallagher FA, Kettunen MI, Day SE, Hu D-E, Ardenkjaer-Larsen JH, Zandt R, Jensen PR, Karlsson M, Golman K, Lerche MH, et al. 2008. Magnetic resonance imaging of pH in vivo using hyperpolarized 13C-labelled bicarbonate. *Nature* **453:** 940–943.

Garcia-Marcos M, Ghosh P, Farquhar MG. 2015. GIV/Girdin transmits signals from multiple receptors by triggering trimeric G protein activation. *J Biol Chem* **290:** 6697–6704.

Gatenby RA, Gillies RJ. 2004. Why do cancers have high aerobic glycolysis? *Nat Rev Cancer* **4:** 891–899.

Gertler F, Condeelis J. 2011. Metastasis: Tumor cells becoming MENAcing. *Trends Cell Biol* **21:** 81–90.

Goddard ET, Hill RC, Barrett A, Betts C, Guo Q, Maller O, Borges VF, Hansen KC, Schedin P. 2016. Quantitative extracellular matrix proteomics to study mammary and liver tissue microenvironments. *Int J Biochem Cell Biol* **81:** 223–232.

Goswami S, Philippar U, Sun D, Patsialou A, Avraham J, Wang W, Di Modugno F, Nistico P, Gertler FB, Condeelis JS. 2009. Identification of invasion specific splice variants of the cytoskeletal protein Mena present in mammary tumor cells during invasion in vivo. *Clin Exp Metastasis* **26:** 153–159.

Grillo-Hill BK, Choi C, Jimenez-Vidal M, Barber DL. 2015. Increased H^+ efflux is sufficient to induce dysplasia and necessary for viability with oncogene expression. *Elife* **4:** e03270.

Grossman M, Ben-Chetrit N, Zhuravlev A, Afik R, Bassat E, Solomonov I, Yarden Y, Sagi I. 2016. Tumor cell invasion can be blocked by modulators of collagen fibril alignment that control assembly of the extracellular matrix. *Cancer Res* **76:** 4249–4258.

Harbour JW. 2014. A prognostic test to predict the risk of metastasis in uveal melanoma based on a 15-gene expression profile. *Methods Mol Biol* **1102:** 427–440.

Harney AS, Arwert EN, Entenberg D, Wang Y, Guo P, Qian B-Z, Oktay MH, Pollard JW, Jones JG, Condeelis JS. 2015. Real-time imaging reveals local, transient vascular permeability, and tumor cell intravasation stimulated by TIE2hi macrophage-derived VEGFA. *Cancer Discov* **5:** 932–943.

Hashim AI, Zhang X, Wojtkowiak JW, Martinez GV, Gillies RJ. 2011. Imaging pH and metastasis. *NMR Biomed* **24:** 582–591.

Hernandez L, Smirnova T, Wyckoff J, Condeelis J, Segall JE. 2009. In vivo assay for tumor cell invasion. *Methods Mol Biol* **571:** 227–238.

Hirata E, Girotti MR, Viros A, Hooper S, Spencer-Dene B, Matsuda M, Larkin J, Marais R, Sahai E. 2015. Intravital imaging reveals how BRAF inhibition generates drug-tolerant microenvironments with high integrin β1/FAK signaling. *Cancer Cell* **27:** 574–588.

Hoffman BD, Grashoff C, Schwartz MA. 2011. Dynamic molecular processes mediate cellular mechanotransduction. *Nature* **475:** 316–323.

Huang C-W, Cheng J-Y, Yen M-H, Young T-H. 2009. Electrotaxis of lung cancer cells in a multiple-electric-field chip. *Biosens Bioelectron* **24:** 3510–3516.

Huang Y-J, Hoffmann G, Wheeler B, Schiapparelli P, Quinones-Hinojosa A, Searson P, et al. 2016. Cellular microenvironment modulates the galvanotaxis of brain tumor initiating cells. *Sci Rep* **6:** 21583.

Hughes SK, Oudin MJ, Tadros J, Neil J, Del Rosario A, Joughin BA, Ritsma L, Wyckoff J, Vasile E, Eddy R, et al. 2015. PTP1B-dependent regulation of receptor tyrosine kinase signaling by the actin-binding protein Mena. *Mol Biol Cell* **26:** 3867–3878.

Isenberg BC, DiMilla PA, Walker M, Kim S, Wong JY. 2009. Vascular smooth muscle cell durotaxis depends on substrate stiffness gradient strength. *Biophys J* **97:** 1313–1322.

James AM, Ambrose EJ, Lowick JH. 1956. Differences between the electrical charge carried by normal and homologous tumour cells. *Nature* **177:** 576–577.

Jiang H, Hegde S, Knolhoff BL, Zhu Y, Herndon JM, Meyer MA, Nywening TM, Hawkins WG, Shapiro IM, Weaver DT, et al. 2016. Targeting focal adhesion kinase renders pancreatic cancers responsive to checkpoint immunotherapy. *Nat Med* **22:** 851–860.

Jin X, Mu P. 2015. Targeting breast cancer metastasis. *Breast Cancer (Auckl)* **9:** 23–34.

Johnson LL, Pavlovsky AG, Johnson AR, Janowicz JA, Man CF, Ortwine DF, Purchase CF, White AD, Hupe DJ. 2000. A rationalization of the acidic pH dependence for stromelysin-1 (Matrix metalloproteinase-3) catalysis and inhibition. *J Biol Chem* **275:** 11026–11033.

Jonas O, Landry HM, Fuller JE, Santini JT, Baselga J, Tepper RI, Cima MJ, Langer R. 2015. An implantable microdevice to perform high-throughput in vivo drug sensitivity testing in tumors. *Sci Transl Med* **7:** 284ra57.

Joyce JA, Pollard JW. 2009. Microenvironmental regulation of metastasis. *Nat Rev Cancer* **9:** 239–252.

Kai FB, Laklai H, Weaver VM. 2016. Force matters: Biomechanical regulation of cell invasion and migration in disease. *Trends Cell Biol* **26:** 486–497.

Kaplan RN, Riba RD, Zacharoulis S, Bramley AH, Vincent L, Costa C, MacDonald DD, Jin DK, Shido K, Kerns SA, et al. 2005. VEGFR1-positive haematopoietic bone marrow progenitors initiate the pre-metastatic niche. *Nature* **438:** 820–827.

King SJ, Asokan SB, Haynes EM, Zimmerman SP, Rotty JD, Alb JG, Tagliatela A, Blake DR, Lebedeva IP, Marston D, et al. 2016. Lamellipodia are crucial for haptotactic sensing and response. *J Cell Sci* **129:** 2329–2342.

Kuo C-HR, Xian J, Brenton JD, Franze K, Sivaniah E. 2012. Complex stiffness gradient substrates for studying mechanotactic cell migration. *Adv Mater* **24:** 6059–6064.

Kwon EJ, Lo JH, Bhatia SN. 2015. Smart nanosystems: Bio-inspired technologies that interact with the host environment. *Proc Natl Acad Sci* **112:** 14460–14466.

Laklai H, Miroshnikova YA, Pickup MW, Collisson EA, Kim GE, Barrett AS, Hill RC, Lakins JN, Schlaepfer DD, Mouw JK, et al. 2016. Genotype tunes pancreatic ductal adenocarcinoma tissue tension to induce matricellular fibrosis and tumor progression. *Nat Med* **22:** 497–505.

Levental KR, Yu H, Kass L, Lakins JN, Egeblad M, Erler JT, Fong SFT, Csiszar K, Giaccia A, Weninger W, et al. 2009. Matrix crosslinking forces tumor progression by enhancing integrin signaling. *Cell* **139:** 891–906.

Leyme A, Marivin A, Perez-Gutierrez L, Nguyen LT, Garcia-Marcos M. 2015. Integrins activate trimeric G proteins via the nonreceptor protein GIV/Girdin. *J Cell Biol* **210:** 1165–1184.

Li J, Lin F. 2011. Microfluidic devices for studying chemotaxis and electrotaxis. *Trends Cell Biol* **21:** 489–497.

Li X-F, Carlin S, Urano M, Russell J, Ling CC, O'Donoghue JA. 2007. Visualization of hypoxia in microscopic tumors by immunofluorescent microscopy. *Cancer Res* **67:** 7646–7653.

Lo CM, Wang HB, Dembo M, Wang YL. 2000. Cell movement is guided by the rigidity of the substrate. *Biophys J* **79:** 144–152.

Lopez JI, Kang I, You W-K, McDonald DM, Weaver VM. 2011. In situ force mapping of mammary gland transformation. *Integr Biol* **3:** 910–921.

Lu J, Zhou S, Siech M, Habisch H, Seufferlein T, Bachem MG. 2014. Pancreatic stellate cells promote hapto-migration of cancer cells through collagen I-mediated signalling pathway. *Br J Cancer* **110:** 409–420.

Martinez-Zaguilan R, Lynch RM, Martinez GM, Gillies RJ. 1993. Vacuolar-type H(+)-ATPases are functionally expressed in plasma membranes of human tumor cells. *Am J Physiol* **265:** C1015–C1029.

Mateo J, Berlin J, De Bono JS, Cohen RB, Keedy V, Mugundu G, Zhang L, Abbattista A, Davis C, Gallo Stampino C, et al. 2014. A first-in-human study of the anti-α5β1 integrin monoclonal antibody PF-04605412 administered intravenously to patients with advanced solid tumors. *Cancer Chemother Pharmacol* **74:** 1039–1046.

McCaig CD, Song B, Rajnicek AM. 2009. Electrical dimensions in cell science. *J Cell Sci* **122:** 4267–4276.

McLachlan GD, Cahill SM, Girvin ME, Almo SC. 2007. Acid-induced equilibrium folding intermediate of human platelet profilin. *Biochemistry* **46:** 6931–6943.

McLean LA, Roscoe J, Jorgensen NK, Gorin FA, Cala PM. 2000. Malignant gliomas display altered pH regulation by NHE1 compared with nontransformed astrocytes. *Am J Physiol Cell Physiol* **278:** C676–C688.

Mehlen P, Delloye-Bourgeois C, Chédotal A. 2011. Novel roles for Slits and netrins: Axon guidance cues as anticancer targets? *Nat Rev Cancer* **11:** 188–197.

Meyer AS, Hughes-Alford SK, Kay JE, Castillo A, Wells A, Gertler FB, Lauffenburger DA. 2012. 2D protrusion but not motility predicts growth factor-induced cancer cell migration in 3D collagen. *J Cell Biol* **197:** 721–729.

Minn AJ, Gupta GP, Siegel PM, Bos PD, Shu W, Giri DD, Viale A, Olshen AB, Gerald WL, Massague J. 2005. Genes that mediate breast cancer metastasis to lung. *Nature* **436:** 518–524.

Miroshnikova YA, Mouw JK, Barnes JM, Pickup MW, Lakins JN, Kim Y, Lobo K, Persson AI, Reis GF, McKnight TR, et al. 2016. Tissue mechanics promote IDH1-dependent HIF1α-tenascin C feedback to regulate glioblastoma aggression. *Nat Cell Biol* **18:** 1336–1345.

Mouneimne G, Soon L, DesMarais V, Sidani M, Song X, Yip S-C, Ghosh M, Eddy R, Backer JM, Condeelis J. 2004. Phospholipase C and cofilin are required for carcinoma cell directionality in response to EGF stimulation. *J Cell Biol* **166:** 697–708.

Mouw JK, Yui Y, Damiano L, Bainer RO, Lakins JN, Acerbi I, Ou G, Wijekoon AC, Levental KR, Gilbert PM, et al. 2014. Tissue mechanics modulate microRNA-dependent PTEN expression to regulate malignant progression. *Nat Med* **20:** 360–367.

Muinonen-Martin AJ, Susanto O, Zhang Q, Smethurst E, Faller WJ, Veltman DM, Kalna G, Lindsay C, Bennett DC, Sansom OJ, et al. 2014. Melanoma cells break down LPA to establish local gradients that drive chemotactic dispersal. *PLoS Biol* **12:** e1001966.

Müller A, Homey B, Soto H, Ge N, Catron D, Buchanan ME, McClanahan T, Murphy E, Yuan W, Wagner SN, et al. 2001. Involvement of chemokine receptors in breast cancer metastasis. *Nature* **410:** 50–56.

Mycielska ME, Djamgoz MBA. 2004. Cellular mechanisms of direct-current electric field effects: Galvanotaxis and metastatic disease. *J Cell Sci* **117:** 1631–1639.

Naba A, Clauser KR, Hoersch S, Liu H, Carr SA, Hynes RO. 2012. The matrisome: In silico definition and in vivo characterization by proteomics of normal and tumor extracellular matrices. *Mol Cell Proteomics* **11:** M111.014647.

Naba A, Clauser KR, Lamar JM, Carr SA, Hynes RO. 2014. Extracellular matrix signatures of human mammary carcinoma identify novel metastasis promoters. *Elife* **3:** e01308.

Naba A, Clauser KR, Ding H, Whittaker CA, Carr SA, Hynes RO. 2016. The extracellular matrix: Tools and insights for the "omics" era. *Matrix Biol* **49:** 10–24.

Oudin MJ, Jonas O, Kosciuk T, Broye LC, Guido BC, Wyckoff J, Riquelme D, Lamar JM, Asokan SB, Whittaker C, et al. 2016a. Tumor cell-driven extracellular matrix remodeling drives haptotaxis during metastatic progression. *Cancer Discov* **6:** 516–531.

Oudin MJ, Miller MA, Klazen JAZ, Kosciuk T, Lussiez A, Hughes SK, Tadros J, Bear JE, Lauffenburger DA, Gertler FB. 2016b. MenaINV mediates synergistic cross-talk between signaling pathways driving chemotaxis and haptotaxis. *Mol Biol Cell* **27:** 3085–3094.

Oudin MJ, Barbier L, Schäfer C, Kosciuk T, Miller MA, Han S, Jonas O, Lauffenburger DA, Gertler F. 2017. Mena confers resistance to Paclitaxel in triple-negative breast cancer. *Mol Cancer Ther* **16:** 143–155.

Padera TP, Stoll BR, Tooredman JB, Capen D, di Tomaso E, Jain RK. 2004. Pathology: Cancer cells compress intratumour vessels. *Nature* **427:** 695.

Park JS, Lee J-H, Lee YS, Kim JK, Dong SM, Yoon DS. 2016. Emerging role of LOXL2 in the promotion of pancreas cancer metastasis. *Oncotarget* **7:** 42539–42552.

Pastorekova S, Parkkila S, Zavada J. 2006. Tumor-associated carbonic anhydrases and their clinical significance. *Adv Clin Chem* **42:** 167–216.

Paszek MJ, Zahir N, Johnson KR, Lakins JN, Rozenberg GI, Gefen A, Reinhart-King CA, Margulies SS, Dembo M, Boettiger D, et al. 2005. Tensional homeostasis and the malignant phenotype. *Cancer Cell* **8:** 241–254.

Patel A, Sant S. 2016. Hypoxic tumor microenvironment: Opportunities to develop targeted therapies. *Biotechnol Adv* **34:** 803–812.

Patsialou A, Wang Y, Lin J, Whitney K, Goswami S, Kenny PA, Condeelis JS. 2012. Selective gene-expression profiling of migratory tumor cells in vivo predicts clinical outcome in breast cancer patients. *Breast Cancer Res* **14:** R139.

Petrie RJ, Doyle AD, Yamada KM. 2009. Random versus directionally persistent cell migration. *Nat Rev Mol Cell Biol* **10:** 538–549.

Philippar U, Roussos ET, Oser M, Yamaguchi H, Kim H-D, Giampieri S, Wang Y, Goswami S, Wyckoff JB, Lauffenburger DA, et al. 2008. A Mena invasion isoform potentiates EGF-induced carcinoma cell invasion and metastasis. *Dev Cell* **15:** 813–828.

Pickup MW, Laklai H, Acerbi I, Owens P, Gorska AE, Chytil A, Aakre M, Weaver VM, Moses HL. 2013. Stromally derived lysyl oxidase promotes metastasis of transforming growth factor-β-deficient mouse mammary carcinomas. *Cancer Res* **73:** 5336–5346.

Pinheiro C, Longatto-Filho A, Ferreira L, Pereira SMM, Etlinger D, Moreira MAR, Jubé LF, Queiroz GS, Schmitt F, Baltazar F. 2008. Increasing expression of monocarboxylate transporters 1 and 4 along progression to invasive cervical carcinoma. *Int J Gynecol Pathol* **27:** 568–574.

Plodinec M, Loparic M, Monnier CA, Obermann EC, Zanetti-Dallenbach R, Oertle P, Hyotyla JT, Aebi U, Bentires-Alj M, Lim RYH, et al. 2012. The nanomechanical signature of breast cancer. *Nat Nanotechnol* **7:** 757–765.

Plotnikov SV, Pasapera AM, Sabass B, Waterman CM. 2012. Force fluctuations within focal adhesions mediate ECM-rigidity sensing to guide directed cell migration. *Cell* **151:** 1513–1527.

Poon Z, Chang D, Zhao X, Hammond PT. 2011. Layer-by-layer nanoparticles with a pH-sheddable layer for in vivo targeting of tumor hypoxia. *ACS Nano* **5:** 4284–4292.

Provenzano PP, Eliceiri KW, Campbell JM, Inman DR, White JG, Keely PJ. 2006. Collagen reorganization at the tumor-stromal interface facilitates local invasion. *BMC Med* **4:** 38.

Provenzano PP, Inman DR, Eliceiri KW, Trier SM, Keely PJ. 2008. Contact guidance mediated three-dimensional cell migration is regulated by Rho/ROCK-dependent matrix reorganization. *Biophys J* **95:** 5374–5384.

Pu J, McCaig CD, Cao L, Zhao Z, Segall JE, Zhao M. 2007. EGF receptor signalling is essential for electric-field-directed migration of breast cancer cells. *J Cell Sci* **120:** 3395–3403.

Raab M, Swift J, Dingal PCDP, Shah P, Shin J-W, Discher DE. 2012. Crawling from soft to stiff matrix polarizes the cytoskeleton and phosphoregulates myosin-II heavy chain. *J Cell Biol* **199:** 669–683.

Ribeiro AL, Okamoto OK. 2015. Combined effects of pericytes in the tumor microenvironment. *Stem Cells Int* **2015:** 868475.

Riching KM, Cox BL, Salick MR, Pehlke C, Riching AS, Ponik SM, Bass BR, Crone WC, Jiang Y, Weaver AM, et al. 2014. 3D collagen alignment limits protrusions to enhance breast cancer cell persistence. *Biophys J* **107:** 2546–2558.

Robinson BD, Sica GL, Liu Y-F, Rohan TE, Gertler FB, Condeelis JS, Jones JG. 2009. Tumor microenvironment of metastasis in human breast carcinoma: A potential prognostic marker linked to hematogenous dissemination. *Clin Cancer Res* **15:** 2433–2441.

Roca-Cusachs P, Sunyer R, Trepat X. 2013. Mechanical guidance of cell migration: Lessons from chemotaxis. *Curr Opin Cell Biol* **25:** 543–549.

Rohan TE, Xue X, Lin H-M, D'Alfonso TM, Ginter PS, Oktay MH, Robinson BD, Ginsberg M, Gertler FB, Glass AG, et al. 2014. Tumor microenvironment of metastasis and risk of distant metastasis of breast cancer. *J Natl Cancer Inst* **106:** dju136.

Roussos ET, Wang Y, Wyckoff JB, Sellers RS, Wang W, Li J, Pollard JW, Gertler FB, Condeelis JS. 2010. Mena deficiency delays tumor progression and decreases metastasis in polyoma middle-T transgenic mouse mammary tumors. *Breast Cancer Res* **12:** R101.

Roussos ET, Condeelis JS, Patsialou A. 2011. Chemotaxis in cancer. *Nat Rev Cancer* **11:** 573–587.

Rubashkin MG, Cassereau L, Bainer R, DuFort CC, Yui Y, Ou G, Paszek MJ, Davidson MW, Chen YY, Weaver VM. 2014. Force engages vinculin and promotes tumor progression by enhancing PI3K activation of phosphatidylinositol (3,4,5)-triphosphate. *Cancer Res* **74:** 4597–4611.

Saadoun S, Papadopoulos MC, Hara-Chikuma M, Verkman AS. 2005. Impairment of angiogenesis and cell migration by targeted aquaporin-1 gene disruption. *Nature* **434:** 786–792.

Schaffner F, Ray AM, Dontenwill M. 2013. Integrin α5β1, the fibronectin receptor, as a pertinent therapeutic target in solid tumors. *Cancers (Basel)* **5:** 27–47.

Shi Q, Ghosh RP, Engelke H, Rycroft CH, Cassereau L, Sethian JA, Weaver VM, Liphardt JT. 2013. Rapid disorganization of mechanically interacting systems of mammary acini. *Proc Natl Acad Sci* **111:** 658–663.

Soon L, Mouneimne G, Segall J, Wyckoff J, Condeelis J. 2005. Description and characterization of a chamber for viewing and quantifying cancer cell chemotaxis. *Cell Motil Cytoskeleton* **62:** 27–34.

Srivastava J, Barreiro G, Groscurth S, Gingras AR, Goult BT, Critchley DR, Kelly MJS, Jacobson MP, Barber DL. 2008. Structural model and functional significance of pH-dependent talin-actin binding for focal adhesion remodeling. *Proc Natl Acad Sci* **105:** 14436–14441.

Steeg PS. 2016. Targeting metastasis. *Nat Rev Cancer* **16:** 201–218.

Stupp R, Hegi ME, Gorlia T, Erridge SC, Perry J, Hong YK, Aldape KD, Lhermitte B, Pietsch T, Grujicic D, et al. 2014. Cilengitide combined with standard treatment for patients with newly diagnosed glioblastoma with methylated MGMT promoter (CENTRIC EORTC 26071-22072 study): A multicentre, randomised, open-label, phase 3 trial. *Lancet Oncol* **15:** 1100–1108.

Stüwe L, Müller M, Fabian A, Waning J, Mally S, Noël J, Schwab A, Stock C. 2007. pH dependence of melanoma cell migration: Protons extruded by NHE1 dominate protons of the bulk solution. *J Physiol* **585:** 351–360.

Sun Y-S, Peng S-W, Lin K-H, Cheng J-Y. 2012. Electrotaxis of lung cancer cells in ordered three-dimensional scaffolds. *Biomicrofluidics* **6:** 14102–14102-14.

Sunyer R, Jin AJ, Nossal R, Sackett DL. 2012. Fabrication of hydrogels with steep stiffness gradients for studying cell mechanical response. *PLoS One* **7:** e46107.

Suraneni P, Rubinstein B, Unruh JR, Durnin M, Hanein D, Li R. 2012. The Arp2/3 complex is required for lamellipodia extension and directional fibroblast cell migration. *J Cell Biol* **197:** 239–251.

Swietach P, Patiar S, Supuran CT, Harris AL, Vaughan-Jones RD. 2009. The role of carbonic anhydrase 9 in regulating extracellular and intracellular pH in three-dimensional tumor cell growths. *J Biol Chem* **284:** 20299–20310.

Szatkowski M, Mycielska M, Knowles R, Kho AL, Djamgoz MBA. 2000. Electrophysiological recordings from the rat prostate gland in vitro: Identified single-cell and transepithelial (lumen) potentials. *BJU Int* **86:** 1068–1075.

Takai K, Le A, Weaver VM, Werb Z. 2016. Targeting the cancer-associated fibroblasts as a treatment in triple-negative breast cancer. *Oncotarget* **7:** 82889–82901.

Taraboletti G, Roberts DD, Liotta LA. 1987. Thrombospondin-induced tumor cell migration: Haptotaxis and chemotaxis are mediated by different molecular domains. *J Cell Biol* **105:** 2409–2415.

Tessier-Lavigne M, Goodman CS. 1996. The molecular biology of axon guidance. *Science (80-)* **274:** 1123–1133.

Tian L, Bae YH. 2012. Cancer nanomedicines targeting tumor extracellular pH. *Colloids Surf B Biointerfaces* **99:** 116–126.

Tweedy L, Susanto O, Insall RH. 2016. Self-generated chemotactic gradients—Cells steering themselves. *Curr Opin Cell Biol* **42:** 46–51.

Ulrich TA, de Juan Pardo EM, Kumar S. 2009. The mechanical rigidity of the extracellular matrix regulates the structure, motility, and proliferation of glioma cells. *Cancer Res* **69:** 4167–4174.

van de Vijver MJ, He YD, van't Veer LJ, Dai H, Hart AA, Voskuil DW, Schreiber GJ, Peterse JL, Roberts C, Marton MJ, et al. 2002. A gene-expression signature as a predictor of survival in breast cancer. *N Engl J Med* **347:** 1999–2009.

van't Veer L, Dai H, van de Vijver M, He YD, Hart AA, Mao M, Peterse HL, van der Kooy K, Marton MJ, Witteveen AT, et al 2002. Gene expression profiling predicts clinical outcome of breast cancer. *Nature* **415:** 530–536.

von Pawel J, von Roemeling R, Gatzemeier U, Boyer M, Elisson LO, Clark P, Talbot D, Rey A, Butler TW, Hirsh V, et al. 2000. Tirapazamine plus cisplatin versus cisplatin in advanced non-small-cell lung cancer: A report of the international CATAPULT I study group. Cisplatin and Tirapazamine in subjects with advanced previously untreated non-small-cell lung tumors. *J Clin Oncol* **18:** 1351–1359.

Walker RA. 2001. The complexities of breast cancer desmoplasia. *Breast Cancer Res* **3:** 143–145.

Wang W, Wyckoff JB, Frohlich VC, Oleynikov Y, Hüttelmaier S, Zavadil J, Cermak L, Bottinger EP, Singer RH, White JG, et al. 2002. Single cell behavior in metastatic primary mammary tumors correlated with gene expression patterns revealed by molecular profiling. *Cancer Res* **62:** 6278–6288.

Wang W, Goswami S, Lapidus K, Wells AL, Wyckoff JB, Sahai E, Singer RH, Segall JE, Condeelis JS. 2004. Identification and testing of a gene expression signature of invasive carcinoma cells within primary mammary tumors. *Cancer Res* **64:** 8585–8594.

Webb BA, Chimenti M, Jacobson MP, Barber DL. 2011. Dysregulated pH: A perfect storm for cancer progression. *Nat Rev Cancer* **11:** 671–677.

Williams CB, Yeh ES, Soloff AC. 2016a. Tumor-associated macrophages: Unwitting accomplices in breast cancer malignancy. *NPJ Breast Cancer* **2:** 15025.

Williams JK, Entenberg D, Wang Y, Avivar-Valderas A, Padgen M, Clark A, Aguirre-Ghiso JA, Castracane J, Condeelis JS. 2016b. Validation of a device for the active manipulation of the tumor microenvironment during intravital imaging. *IntraVital* **5:** e1182271.

Willipinski-Stapelfeldt B, Riethdorf S, Assmann V, Woelfle U, Rau T, Sauter G, Heukeshoven J, Pantel K. 2005. Changes in cytoskeletal protein composition indicative of an epithelial-mesenchymal transition in human micrometastatic and primary breast carcinoma cells. *Clin Cancer Res* **11:** 8006–8014.

Wilson WR, Hay MP. 2011. Targeting hypoxia in cancer therapy. *Nat Rev Cancer* **11:** 393–410.

Wu J, Lin F. 2014. Recent developments in electrotaxis assays. *Adv Wound Care* **3:** 149–155.

Wu C, Asokan SB, Berginski ME, Haynes EM, Sharpless NE, Griffith JD, Gomez SM, Bear JE. 2012. Arp2/3 is critical for lamellipodia and response to extracellular matrix cues but is dispensable for chemotaxis. *Cell* **148:** 973–987.

Wyckoff J, Wang W, Lin EY, Wang Y, Pixley F, Stanley ER, Graf T, Pollard JW, Segall J, Condeelis J. 2004. A paracrine loop between tumor cells and macrophages is required for tumor cell migration in mammary tumors. *Cancer Res* **64:** 7022–7029.

Wyckoff JB, Pinner SE, Gschmeissner S, Condeelis JS, Sahai E. 2006. ROCK- and myosin-dependent matrix deformation enables protease-independent tumor-cell invasion in vivo. *Curr Biol* **16:** 1515–1523.

Wynn TA. 2008. Cellular and molecular mechanisms of fibrosis. *J Pathol* **214:** 199–210.

Yan X, Han J, Zhang Z, Wang J, Cheng Q, Gao K, Ni Y, Wang Y. 2009. Lung cancer A549 cells migrate directionally in DC electric fields with polarized and activated EGFRs. *Bioelectromagnetics* **30:** 29–35.

Yang M, Brackenbury WJ. 2013. Membrane potential and cancer progression. *Front Physiol* **4:** 185.

Yang N, Mosher R, Seo S, Beebe D, Friedl A. 2011. Syndecan-1 in breast cancer stroma fibroblasts regulates extracellular matrix fiber organization and carcinoma cell motility. *Am J Pathol* **178:** 325–335.

Yang B, Lieu ZZ, Wolfenson H, Hameed FM, Bershadsky AD, Sheetz MP. 2016. Mechanosensing controlled directly by tyrosine kinases. *Nano Lett* **16:** 5951–5961.

Yi A, Moon WK, Cho N, Chang JM, Bae MS, Kim SJ, Han W, Park IA. 2013. Association of tumour stiffness on sonoelastography with axillary nodal status in T1 breast carcinoma patients. *Eur Radiol* **23:** 2979–2987.

Yildirim S, Altun S, Gumushan H, Patel A, Djamgoz MBA. 2012. Voltage-gated sodium channel activity promotes prostate cancer metastasis in vivo. *Cancer Lett* **323:** 58–61.

Zeisberg M, Kalluri R. 2013. Cellular mechanisms of tissue fibrosis. 1. Common and organ-specific mechanisms associated with tissue fibrosis. *Am J Physiol Cell Physiol* **304:** C216–C225.

Zhang Y, Lin Z, Foolen J, Schoen I, Santoro A, Zenobi-Wong M, Vogel V. 2014. Disentangling the multifactorial contributions of fibronectin, collagen and cyclic strain on MMP expression and extracellular matrix remodeling by fibroblasts. *Matrix Biol* **40:** 62–72.

Zhao M, Song B, Pu J, Wada T, Reid B, Tai G, Wang F, Guo A, Walczysko P, Gu Y, et al. 2006. Electrical signals control wound healing through phosphatidylinositol-3-OH kinase-γ and PTEN. *Nature* **442:** 457–460.

Pathways Involved in Formation of Mammary Organoid Architecture Have Keys to Understanding Drug Resistance and to Discovery of Druggable Targets

SAORI FURUTA[1,2] AND MINA J. BISSELL[1]
[1]*Life Sciences Division, Lawrence Berkeley National Laboratory, Berkeley, California 94720*
[2]*Department of Biochemistry and Cancer Biology, College of Medicine and Life Sciences, University of Toledo Health Science, Toledo, Ohio 43614*
Correspondence: saori.furuta@utoledo.edu; MJBissell@lbl.gov

Signals from the extracellular matrix (ECM) are received at the cell surface receptor, transmitted to the cytoskeletons, and transferred to the nucleus and chromatin for tissue- and context-specific gene expression. Cells, in return, modulate the cell shape and ECM, allowing for the maintenance of tissue homeostasis as well as for coevolution and adaptation to the environmental signals. We postulated the existence of dynamic and reciprocal interactions between the ECM and the nucleus more than three decades ago, but now these pathways have been proven experimentally thanks to the advances in imaging and cell/molecular biology techniques. In this review, we will introduce some of our recent work that has validated the critical roles of the three-dimensional (3D) tissue architecture in determining mammary biology, therapeutic response, and druggable targets. We describe a novel screen based on reversion of the malignant phenotype in 3D assays. We will also summarize our recent discoveries of the integration of feedback signaling for mammary acinar formation and phenotypic reversion of tumor cells in the LrECM. Lastly, we will introduce our exciting discovery of the physical linkages between the cell surface and cytofibers within a tunnel deep inside of the nucleus, enabling interaction with nuclear lamin and SUN proteins.

The microenvironment exerts control over the genome in both normal and malignant cells (Bissell and Labarge 2005; Bissell and Hines 2011). Our laboratory has spent more than three decades on understanding how the tissue microenvironment in the breast establishes and maintains tissue specificity and how this regulation is compromised in breast cancer. Such information is essential for therapeutic targeting of the tumor and the microenvironment for treatment and prevention of all forms of cancers, although the data described here deal exclusively with breast and breast cancer.

More than three decades ago, we speculated that the extracellular matrix (ECM) and nuclear machinery reciprocally and dynamically interact to regulate growth, gene expression, and phenotype of the mammary gland and other organs (Bissell et al. 1982). We postulated that the information emanating from the ECM is transferred to the cell surface receptors, is transmitted to the cytoskeleton, and is received by the nuclear matrix through the contiguous physical linkages and biochemical signaling to chromatin to allow tissue-specific gene expression. This dynamic and reciprocal signaling confers changes in cell shape, polarity, and functions that are tissue-specific and context-dependent and, ultimately, triggers the de novo synthesis of molecules. Cell and tissue polarity allow secretion of tissue-specific molecules including milk proteins and basement membrane (BM) molecules such as laminins, which, when incorporated into a structurally correct BM, will in turn give the signal for final architecture of the tissue. Almost 40 years later, the enormous advances of superresolution imaging and biomechanical/biochemical techniques have unveiled the bona fide existence of physical connections that we believe "hard-wire" the ECM to the nucleus (Wang et al. 2009; Jorgens et al. 2017). Such networks of cytoskeletal structures decisively control not only the architecture and mechanical environment but also all biological activities of the tissues. In particular, dysregulation of cytoskeletal dynamics largely contributes to malignant behavior of cancer cells (Mokady and Meiri 2015).

We summarized some of our work with human cells, three-dimensional (3D) models, reversion of the malignant phenotype, and integration of signaling, which occurs only in 3D, in a previous volume of *Cold Spring Harbor Symposia on Quantitative Biology* (Bissell et al. 2005). In this current brief review, we will not dwell on much of our earlier experimental work, which has been reviewed a number of times (Nelson and Bissell 2006; Radisky and Bissell 2006; Xu et al. 2009a; Spencer et al. 2010; Bissell and Hines 2011). Here we will introduce some of our recent work, in which we have taken advantage of old and new assays and a screen based on 3D tissue architecture to highlight the translational aspects of our studies. We will also introduce our recent studies that have shed light on how acinus is formed and how the signaling pathways integrate to close the laminin/nucleus loop in formation and maintenance of the mammary acini.

Published by Cold Spring Harbor Laboratory Press; doi: 10.1101/sqb.2016.81.030825

FORMATION OF MAMMARY ORGANOIDS IN 3D SUBSTRATA

A brief historical chronology of the relation between substrata, cell shape, and function was published very recently (Simian and Bissell 2017). But it has been clear from decades of literature that when cells were removed from the tissues and placed in culture, both form and function were lost (the much older but substantial literature was summarized in Bissell et al. 1982). A breakthrough occurred when Kleinman et al. (1987) analyzed the composition of what was then referred to as the EHS (Engelbreth–Holm–Swarm) matrix, a laminin-rich gel produced by a mouse tumor with striking similarity to the composition of the BM in glandular tissues. Later the gel was popularized and sold as Matrigel. The linkage between the ECM, cell shape, and gene expression was based on our finding that normal mammary epithelial cells (MECs), when isolated from the tissues and cultured in a laminin-rich gel (LrECM), form 3D alveoli-like structures. These structures are apicobasally polarized and capable of secreting milk into the lumen (Barcellos-Hoff et al. 1989). Furthermore, these structures were indeed the first reported "organoids," and our experiments and models laid the foundation for what now is generating much excitement (Simian and Bissell 2017). Of particular interest to the field are the findings that have much relevance to the fundamental properties of the tissues in vivo (i.e., apicobasal polarity of cell and tissue). This is best shown with the following work (Xu et al. 2009b): We wondered why mouse mammary cells in 2D substrata do not respond to prolactin (PR) by making milk proteins despite the fact that the cells were expressing the PR receptor. We showed that, whereas normal MECs on 2D monolayers maintain a degree of apicobasal polarity and express the PR receptors, they do not produce milk upon addition of prolactin. This is because, in these 2D-cultured cells, the PR receptors are on the basal surface that is attached to the rigid plastic substrata and not available to the medium added to the top of the monolayer (Fig. 1). Thus, in effect, the receptor and ligand are spatially segregated in 2D. On the other hand, in 3D LrECM cultures, the PR receptor is exposed to the ligand, because now the basal surface is bound to the softer ECM and becomes accessible. If the experiment is repeated with cells in PolyHEMA, a synthetic hydrogel, however, the basal surface is exposed, but the cells do not make milk when they receive PR. Such a defect indicates that, in addition to exposure of the basal receptor, we need the presence of biological molecules within the laminin-containing gel. We showed that laminin-induced polarity triggers activation of the transcription factor STAT5, which then transactivates the expression of the milk protein β-casein (Fig. 1; Xu et al. 2009b).

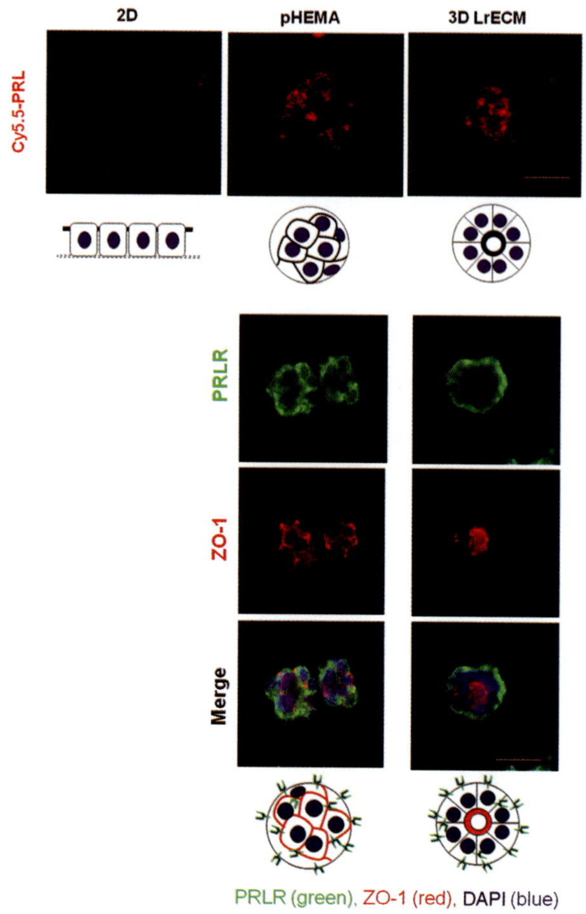

Figure 1. Prolactin receptor is exposed on the basolateral surface when mammary epithelial cells form apicobasally polarized colonies in three-dimensional (3D) laminin-rich extracellular matrix (LrECM). (*Top*) Representative micrograms of mouse mammary epithelial cells EpH4 incubated with Cy5.5-labeled prolactin (PRL). Cells were cultured in 2D plastic, on poly-HEMA (pHEMA), or in 3D LrECM. (*Bottom*) EpH4 cultures on pHEMA or LrECM were stained for prolactin receptor (PRLR), apical marker ZO-1, and nuclei (DAPI). Scale bar, 25 μm. (Reprinted, with permission, from Xu et al. 2009b, the Rockefeller University Press.)

Because the nonmalignant cells remember to be "normal" in 3D, we explored the possibility that the tumor cells may also remember to look like a tumor in 3D! In collaboration with Ole Petersen, we developed the first human cell assay, where the two populations could be distinguished easily and robustly in 3D within less than a week instead of months in mice (Petersen et al. 1992). Thus, 3D phenotype could be used to distinguish normal from malignant cells. We took advantage of the isogenic HMT3522 breast cell lines that were established from reduction mammoplasty and propagated from nonmalignant cells (S1), to premalignant cells (S2), and after many tries, to malignant cells (T4-2) (Briand et al. 1987). We showed that, whereas nonmalignant S1 cells form quiescent, apicobasally polarized spheroids in 3D LrECM cultures, their malignant counterparts, T4-2 cells, form proliferative, disorganized masses in the same culturing condition (Fig. 2; Petersen et al. 1992; Weaver et al. 1997; Lee et al. 2007). Such differential response to LrECM between normal and malignant cells to the same cues suggested that the pathways responding to laminin 111 (LN)-1 are altered in malignant cells (Petersen et al. 1992).

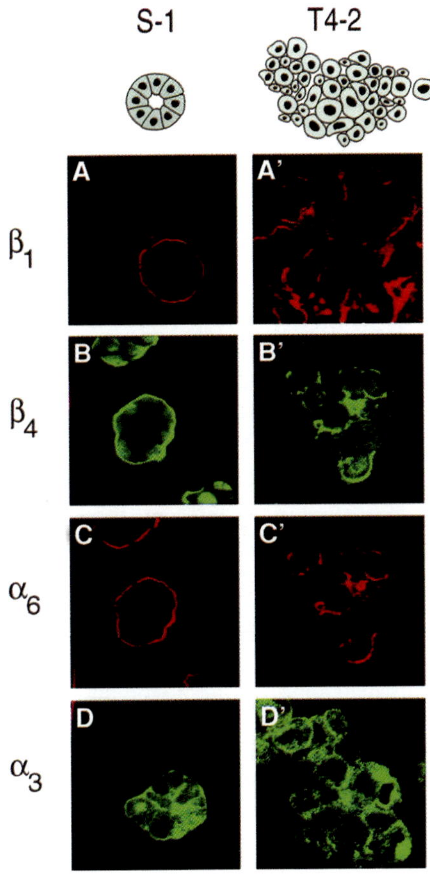

Figure 2. Normal and malignant mammary epithelial cells form distinct colonies in three-dimensional (3D) laminin-rich extracellular matrix (LrECM) cultures. Immunofluorescence characterization of the HMT-3522 cells in 3D cultures after 10–12 d inside the Engelbreth–Holm–Swarm (EHS) matrix. (*A–D*) S-1 acini and (*A′–D′*) T4-2 colonies immunostained β1- (*A* and *A′*), β4- (*B* and *B′*), α6- (*C* and *C′*), and α3-integrin (*D* and *D′*). Note that β1-, β4-, and α6-integrins were targeted to the cell–ECM junction in the S-1 acini (*A–C*), whereas in T4-2 colonies (*A′–C′*) this polarized-basal distribution was lost. S-1 acini showed basolateral α-3 integrins (*D*), whereas T4-2 colonies (*D′*) showed disorganized plasma membrane and cytosolic expression of this integrin. (Reprinted, with permission, from Weaver et al. 1997, the Rockefeller University Press.)

We explored the potential mechanisms by which nonmalignant MECs and breast cancer cells (both from the same individual) show such profound phenotypic differences in LrECM. To address the mechanism of growth factor–independent, autonomous growth of cancer cells, we measured the levels of receptors expressed on the surface of normal versus malignant cells and found that, although these cells appeared to have similar receptors, the level of integrins, including β1 and α/β6, as well as epidermal growth factor receptor (EGFR) were elevated between four- and sixfold in malignant cells in 3D LrECM (Fig. 2; Weaver et al. 1997). Remarkably, reduction of a number of these pathways to the level of nonmalignant S1 led to a phenotypic reversion of morphology and growth, allowing T4-2 cells to resemble the S1 cells but only in 3D (summarized in Bissell et al. 2005; see below for more details). To get insight into events that led to conversion of S1 cells to premalignant S2 and to malignant T4-2, we found that once EGF, the only growth factor in the medium, was removed, cells became adapted to the absence of EGF by up-regulating the expression of a matrix metalloproteinase (MMP), TACE. TACE promotes shedding of an autocrine factor, amphiregulin (ARG), from the cell surface. ARG then binds the EGFR on cancer cells to induce oncogenic signaling, without requirement of exogenous EGF (Kenny and Bissell 2007).

We asked: How is it that the nonmalignant cells, with many mutations, deletions, and amplifications, know to still form an acinus in 3D, but once malignant, they no longer do? In a data-rich and ambitious paper (Rizki et al. 2008), we summarized years of work and showed that the structure of the acini is compromised even before malignancy sets in (please see the upper panel in Fig. 5). To understand how each step is executed to induce formation of a complete acinus in the LrECM (pure laminin 111, as a 20% rat tale collagen gel can substitute for Matrigel and, of course, it is much better defined) (Alcaraz et al. 2008), we started with putting single primary human breast cells from reduction mammoplasty, or even a breast cell line inside the gel, and imaged live cells continuously for 10 days when all cells formed acini. Remarkably, the cells began to rotate continuously through a process we termed "coherent angular motion" (CAMo) (Tanner et al. 2012). Comparing cellular motions between normal MECs and breast cancer cells, we found that nonmalignant cells undergo CAMo to form a polarized spherical cluster, whereas they are tightly tethered to each other and remain polar during cell division. The coherent rotating movement of acinus forming MECs resembles that of *Drosophila* follicles during embryogenesis (Haigo and Bilder 2011). In contrast, malignant cells undergo a random movement and form a disorganized mass as they divide and their interactions are quite loose (Fig. 3; Tanner et al. 2012). The ability of nonmalignant MECs to undergo CAMo was dependent on the presence of intact cell–cell junctions, because inhibition of PAR3 (tight junction and polarity) or E-cadherin (adherin junctions) completely abrogated CAMo in nonmalignant MECs. This result again highlights the fact that the same mechanical and biochemical cues from the ECM are transduced differently in normal versus malignant cells, leading to different cellular motions and, thus, formation of different tissue structures. When tumor cells are treated to revert to a normal phenotype, the movement becomes coherent again (Tanner et al. 2012).

By using the 3D LrECM cultures, we explored whether altering the extracellular growth signaling pathways could convert the 3D phenotype of malignant cells to that of normal cells. We discovered that suppression of extracellular growth signals, including ECM or growth factor receptors (e.g., β1 integrin or EGFR), induces tumor reversion and restores a normal-like, polarized, and quiescent phenotype in malignant cells (Figs. 2, 4, and 5; Weaver et al. 1997, 2002; Bissell et al. 2005; Lee et al. 2007, 2012; Beliveau et al. 2010). We have found

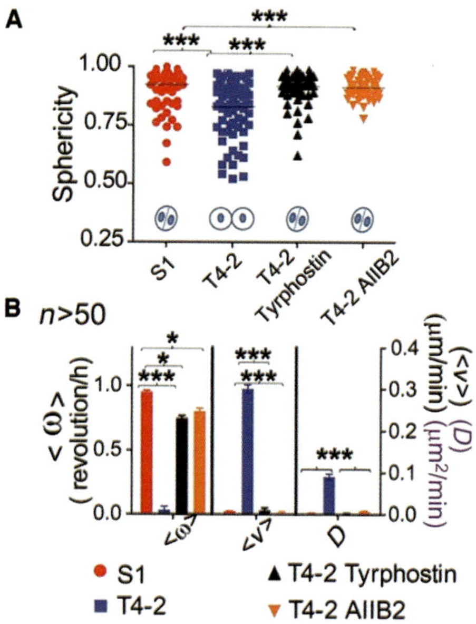

Figure 3. Different movements of normal versus malignant mammary epithelial cells during morphogenesis in three-dimensional (3D) laminin-rich extracellular matrix (LrECM). (*A*) Comparison of cell adhesion measured by sphericity of the cell aggregates. (*B*) Graph depicting different types of motility: ω, angular motion; *v*, linear displacement; *D*, diffusion; $^{*}P < 0.05$; $^{**}P < 0.01$; $^{***}P < 0.0001$. (Reprinted, with permission, from Tanner et al. 2012, the National Academy of Sciences of the United States of America.)

more than a dozen different pathways involved in tumor reversion (Weaver et al. 1997, 2002; Bissell et al. 2005; Lee et al. 2007, 2012), suggesting the presence of common regulators that integrate these different signals to produce essentially the same architecture—namely, a reverted structure that resembles a nonmalignant acinus. For example, up-regulation of integrins, EGFR, and other growth pathways converge on activation of MMP-9 (Beliveau et al. 2010). Inhibition of any of these pathways to the level of S1 (nonmalignant) inactivates MMP-9, leading to phenotypic reversion (Beliveau et al. 2010).

The phenomenon of tumor reversion provides us with a robust tool to determine critical signals that are involved in formation of normal tissue architecture and are dysregulated in cancer. Such an approach allows us to identify novel tumor suppressors or oncogenes, based on cells' ability or inability, respectively, to undergo phenotypic reversion (Chen et al. 2009; Lee et al. 2012). Utilization of tumor reversion also contributes to development of a new therapeutic regimen that corrects microenvironmental cues (Weaver et al. 1997; Wang et al. 1998; Kenny and Bissell 2003; Beliveau et al. 2010; Lee et al. 2012; Onodera et al. 2013).

USE OF 3D MORPHOGENESIS ASSAY TO IDENTIFY NOVEL TUMOR SUPPRESSORS AND "ONCOGENES" AND TO DISCOVER POTENTIAL TREATMENTS FOR BREAST CANCER

It has long been known that early pregnancy may have protective effects for breast cancer (Britt et al. 2007). The underlying mechanism, however, is still speculative. We postulated that such protection is mediated by formation of mammary alveoli during pregnancy that may produce certain antitumor factors to suppress the growth of premalignant and malignant cells in the mammary gland. Whereas this may appear to be a wild and crazy guess, evolution has had to protect the continuation of the species by developing lush and intricate mechanisms that we already know are built-in to assure that babies are conceived and thrive once born. So a healthy and robust acinar structure is necessary for producing milk and secreting it into the lumen so the baby can be properly fed! We had already developed a robust, physiologically relevant model to examine how acini are formed and how they produce milk vectorially (Aggeler et al. 1991). To test this possibility, we used 3D LrECM cultures that allow for formation of polarized acini recapitulating mammary alveologenesis in vivo during pregnancy. We collected the conditioned media (CM) of nonmalignant human MECs during acinar formation and used the media to culture malignant MECs in 3D LrECM. We found that tumor cells underwent phenotypic reversion when cultured in the presence of the CM of acinar-forming nonmalignant MECs. We then analyzed the components of the CM by mass spectrometry and identified IL25 as the major factor that showed antitumor effects (Furuta et al.

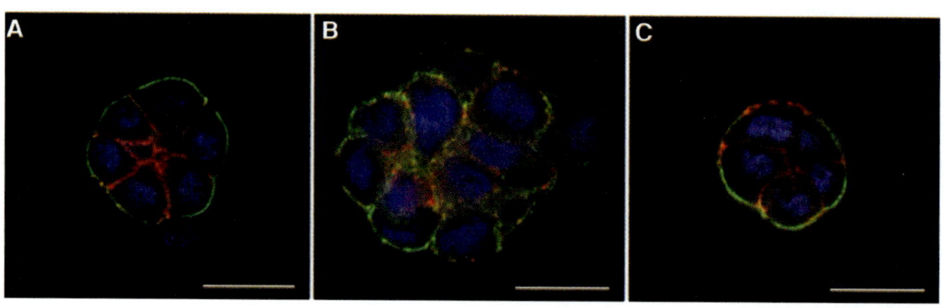

Figure 4. Epidermal growth factor receptor (EGFR) inhibition restores S1-like quiescent, polarized three-dimensional (3D) structures of malignant T4-2 cells. (*A–C*) HMT-3522 S1 (*A*), HMT-3522 T4-2 (*B*), and HMT-3522 T4-2 treated with an EGFR inhibitor, AG1478 (*C*) were cultured in the 3D on-top assay for 4 d. Colonies were then extracted and immunostained against α_6 integrin (green) and β-catenin (red). Nuclei were counterstained with DAPI (blue). Confocal sections through the centers of the colonies are shown. Scale bar, 20 μm. (Reprinted, with permission, from Lee et al. 2007, Nature Publishing Group.)

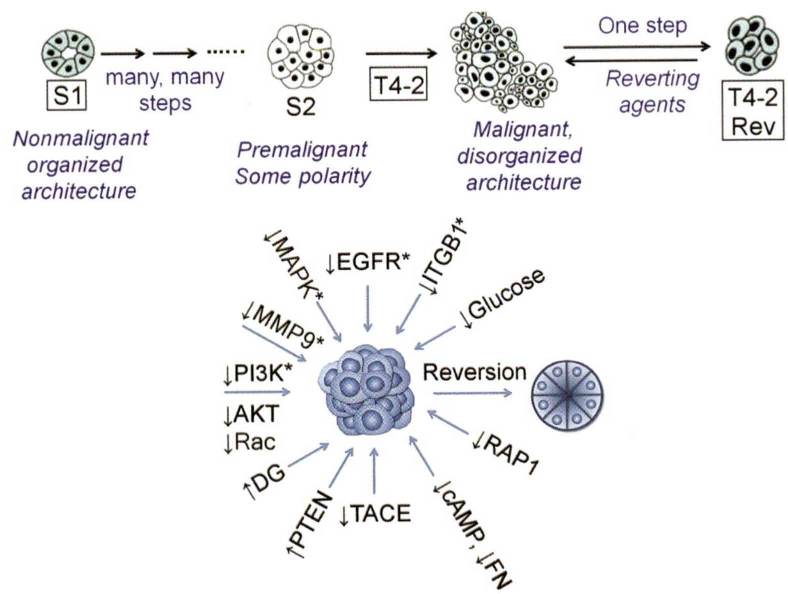

Figure 5. Inhibition of diverse signaling pathways leads to phenotypic reversion of tumor cells. (*Top*) Scheme of tumor progression and reversion. (*Bottom*) Different means to induce tumor reversion. The methods used in gene and microRNA expression profiling are indicated by an asterisk (*). EGFR, epidermal growth factor receptor; ITGB1, integrin beta 1; cAMP, cyclic adenosine monophosphate; FN, fibronectin; PTEN, phosphatase and tensin homolog; DG, dystroglycan; PI3K, phosphoinositide 3-kinase; MMP9, matrix metalloproteinase 9; MAPK, mitogen-activated protein kinase.

2011). We further explored the mechanism of antitumor effect of IL-25 and found that the majority of cancer cells analyzed up-regulated the receptor IL25R, which could bind both IL25 and IL17B. IL-25-IL25R interactions activated death receptor signaling leading to caspase activation. In contrast, IL17B–IL25R interaction led to activation of NF-κB and proliferative pathway. Interestingly, nonmalignant MECs produced IL25 only during acinar formation, whereas cancer cells produced only IL17B. Furthermore, IL25R is present only in cancer cells. These results uncover a novel mechanism of normal breast tissue exerting natural defense against cancer cells. This unexpected discovery was possible because of normal cells' ability to undergo in vivo–like tissue morphogenesis in 3D LrECM cultures (Fig. 6; Furuta et al. 2011).

We have often wondered why so many cancer patients, despite high levels of EGFR in their primary tumors, do not respond to a number of current tyrosine kinase inhibitors (TKIs) such as lapatinib, and why so often they respond and within a short period they become resistant? We hypothesized that there may be additional intermediates in the EGFR and related pathways that still have not been discovered because many of the screens rely only on monolayer cultures and normal cells are hardly ever included in toxicity studies. The 3D LrECM cultures can be "used" to test these ideas. We often had noticed that a small fraction of T4-2 cancer cells in our 3D assays displayed resistance to signaling inhibitors that induce phenotypic reversion in the bulk of tumor cells. We suspected that the existence of these "resistant colonies" to our EGFR inhibitors, "reverting agents," was due to up-regulation of some novel oncogenic intermediate in the resistant cells. We generated the cDNA library from malignant T4-2 cells, overexpressed the library in T4-2 cells, and determined which cDNA made T4-2 cells resistant to the EGFR inhibitor in 3D cultures. We identified FAM83A as the suspected oncogene that confers resistance to EGFR inhibitor (Fig. 7; Lee et al. 2012). During a chance encounter, we discovered that another laboratory, using an entirely different screen by using breast cell lines from Martha Stampher's 184 breast cancer progression series, had identified FAM83B in their screen (Cipriano et al. 2012). In the ensued collaboration, together we have uncovered that FAM83A and FAM83B are among the eight members of a new class of onco-

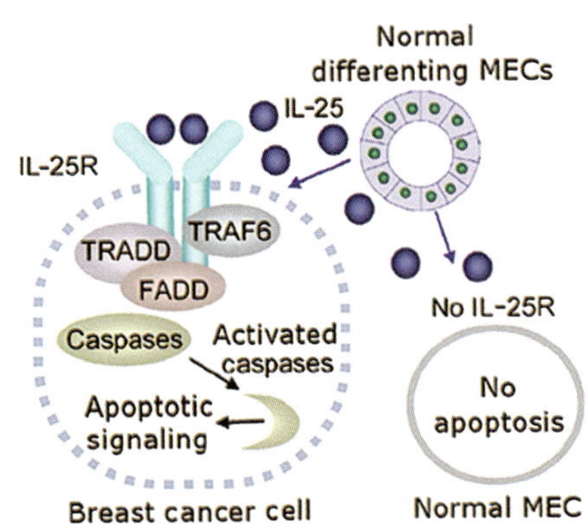

Figure 6. Schematic for the cytotoxic activity of IL-25 specific to breast cancer cells that express IL25R. Nonmalignant mammary epithelial cells (MECs) do not express IL-25R and are resistant to apoptosis induced by IL-25. Breast cancer cells that express IL-25R are susceptible to IL-25-induced apoptosis. (Reprinted, with permission, from Furuta et al. 2011, AAAS.)

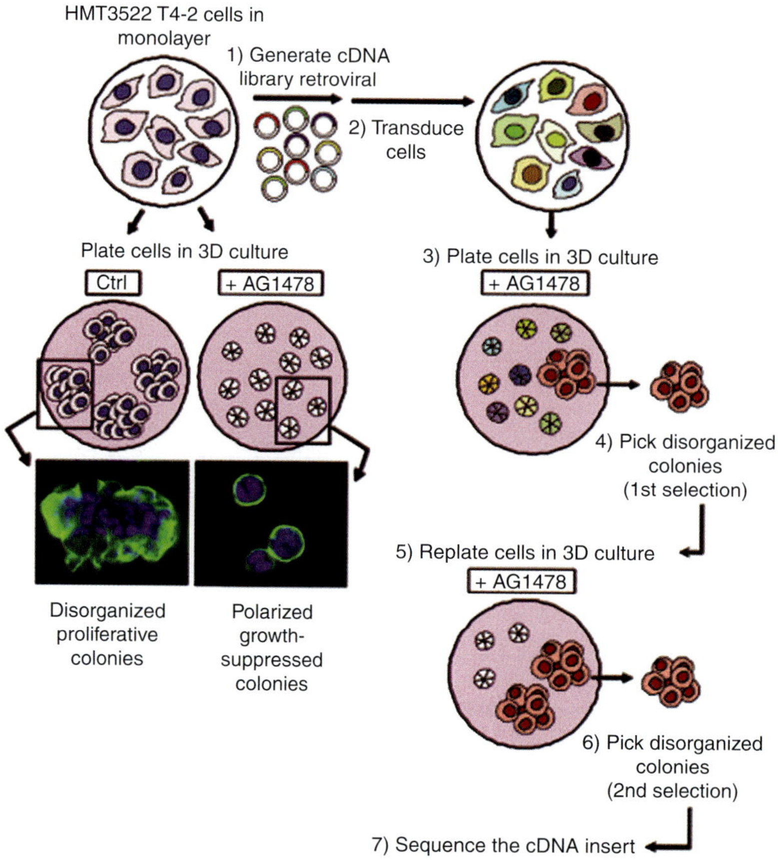

Figure 7. Process for the screen to identify FAM83A. Treatment of T4-2 cells with a small-molecule epidermal growth factor receptor (EGFR) inhibitor, AG1478, induced reversion to a polarized acinar phenotype in three-dimensional (3D) laminin-rich extracellular matrix (LrECM) culture, as detected by α_6 integrin staining (green) to delineate basal polarity (*left*); after transduction with the cDNA library, a subpopulation of cells disorganized in 3D culture after AG1478 treatment was isolated, expanded, and searched for the cDNA insert (*right*). (Reprinted, with permission, from Lee et al. 2012, American Society for Clinical Investigation.)

genes, FAM83A through FAM83H. We found that FAM83A interacts with both c-RAF and the p85 subunit of phosphoinositide 3-kinase (PI3K), facilitating their activation. We further showed that treatment of mammary tumors with lapatinib, a clinically used EGFR and EGFR2 inhibitor, while killing the majority of cells, up-regulated FAM83A expression in a small fraction of surviving cells in vivo leading to resistance to TKIs. This result strongly supports that FAM83A up-regulation is a cause of lapitinib resistance of breast tumors (Lee et al. 2012). These FAMs indeed are highly expressed in many tumor types including lung, stomach, and breast and are druggable targets in the pathways of EGFR and PI3K. They are induced to high levels in vivo after mice with human tumors are given lapatinib (Snijders et al. 2017). Such findings also suggest that the degree of phenotypic reversion of tumor cells in 3D LrECM cultures could be used as the index of the efficacy of cancer treatment in vivo.

We have examined also the contribution of 3D LrECM environment to chemo-resistance of breast cancer cells. We treated different human epidermal growth factor receptor 2 (HER2)-amplified breast cancer cell lines (AU565, SKBR3, HCC1569, and BT549) cultured in either 2D plastic or 3D LrECM with a clinically used HER2-targeting drug (trastuzumab, pertuzumab, and lapatinib). We found that the drug responses of these cell lines differ significantly between 2D and 3D conditions because of differential activation of PI3K and mitogen-activated protein kinase (MAPK) pathways in response to the drug (Weigelt et al. 2010). The cotreatment of β1 integrin inhibitory antibody (AIIBII), however, augmented drug effect of all the conditions, suggesting that integrin-mediated cell–ECM anchorage had reduced drug efficacy for these cells tested (Weigelt et al. 2010).

To apply this finding to therapy, we xenografted different breast cancer cell lines subcutaneously into immunocompromised mice and applied β1 integrin inhibitory antibody (AIIBII). We observed significant suppression of tumor take and growth by AIIBII. In contrast, normal breast tissue, as well as nonmalignant cells undergoing acinar morphogenesis in 3D LrECM cultures, were not affected by AIIBII (Park et al. 2006). This result suggests that polarized nonmalignant cells were resistant to the growth inhibitory effects of the β1 integrin blocking antibody. Further, the study strongly indicates that the β1 integrin inhibitory antibody could be used as a cancer-specific therapy with minimal side effects on both normal cells in 3D cultures and in vivo (Park et al. 2008). We tested the efficacy of adjuvant use of AIIBII for radiation treatment of xenografted breast cancer cells. We found that AIIBII, which drastically inhibited AKT activation, significantly enhanced radiation-induced apoptosis of breast cancer cells. This allowed the radiation dosage to be reduced by fourfold (8 Gy to 2 Gy) to obtain the same apoptotic effect (Park et al. 2008). These results suggested strongly that targeting the β1 integrin-mediated cell–ECM linkage could be a promising method for breast cancer treatment. The results of using the humanized inhibitory antibodies against β1 integrin are encouraging.

LINKAGE BETWEEN THE ECM AND NUCLEUS

Three-dimensional environments that enable tissues to express tissue-specific genes for functional differentiation have allowed us also to unravel important signaling pathways that are integrated to define the tissue architecture.

One of the first evidences of the linkage between the ECM and the nucleus from our laboratory was our finding of the ECM-response element. The promoter region of bovine β-casein gene was found to contain 160-bp bovine casein response element (BCE)-1 transcriptional enhancer at ~1.5 kb upstream of the transcription start site. BCE-1 controls transcriptional activation of the milk protein, β-casein gene (Schmidhauser et al. 1992). BCE-1 is activated not by binding of transcription factors that have consensus sequences within that region but by removing chromatin inactivation exerted by histone deacetylase in response to the ECM protein laminin (Streuli et al. 1995; Myers et al. 1998).

Another example is Seid et al.'s finding of the promoter region of LpS1 gene (called proximal G-string) in sea urchin that is activated in response to collagen. A collagen cross-linking inhibitor (β-aminopropionitrile) impairs activation of proximal G-string, whereas a mutation in that region causes constitutive activation, suggesting that G-string activation is mediated by removal of repression in response to collagen (Seid et al. 1997).

Lastly, we recently have discovered that the promoter region of the laminin α3 chain (LAMA3) of laminin 332 (also known as laminin 5) is activated by p53 in response to the ECM laminins in the normal breast. The LAMA3 gene promoter contains numerous (more than 20) consensus sequences of p53 within 1 kb upstream of the transcription start site, whereas inactivation of p53 abrogates LAMA3 transcription. LAMA3 expression is essential for formation of normal mammary acini (milk secretory unit) in 3D ECM cultures (S Furuta and MJ Bissell, in prep.).

Another evidence for the ECM–nucleus linkage is the shuttling of nuclear actin between the nucleus and cytosol in response to ECM cues. Although polymeric actin is an essential cytoskeletal component in the cytosol, monomeric actin is shuttled between the cytosol and nucleus (Sen et al. 2015). In the nucleus, actin plays critical roles in nuclear matrix organization, chromatin remodeling, mRNA processing, and transcriptional regulation (Visa and Percipalle 2010). Nuclear actin is part of the nuclear matrix (Kiseleva et al. 2004) and interacts with the nuclear lamina, pre-mRNA splicing proteins, and chromatin remodeling complexes for transcriptional activation of tissue-specific genes (Clubb and Locke 1998; Lamond and Spector 2003; Gruenbaum et al. 2005; Farrants 2008; Visa and Percipalle 2010; Sen et al. 2015). Conversely, polymeric actin in the nucleus is thought to play roles in regulating nuclear shape and positioning in response to mechanical stress (Plessner et al. 2015). Actin translocation into the nucleus is facilitated by importin 9 in response to stress (e.g., heat shock) (Nishida et al. 1987; Pendleton et al. 2003; Dopie et al. 2012; Belin et al. 2015). Conversely, we found that, when cells are exposed to laminin 111 or LrECM, nuclear actin is transported from the nucleus to the cytosol (Fig. 8; Spencer et al. 2011). This destabilizes RNA polymerase II/III complex at transcription site and inhibits transcription and DNA synthesis, allowing for cells to enter the quiescent state for functional differentiation (Spencer et al. 2011). In contrast, overexpression of actin harboring nuclear localization signal inhibits export of nuclear actin and growth arrest (Spencer et al. 2011). Translocation of nuclear actin to the cytosol is mediated by exportin 6 (Stüven et al. 2003). Exportin 6 levels determine cells' abilities to translocate nuclear actin. Exportin 6 expression is elevated in normal breast epithelial cells in response to laminin or LrECM, triggering translocation of nuclear actin. In contrast, laminin-responsive exportin 6 up-regulation is abrogated in breast cancer cells, impairing actin translocation (A Bruni-Cardoso and MJ Bissell, in prep.).

Most recently, we have elucidated the detailed mechanism that integrates different signaling cascades for mammary morphogenesis using the 3D LrECM system. These signals emanate from reciprocal interactions between different cellular and acellular components, forming integrated feedback loops (Xu et al. 2009a; Mammoto and Ingber 2010; Patwari and Lee 2010; Basson 2012). We compared gene expression profiles of nonmalignant S1, malignant T4-2 cells, and T4-2 cells reverted with six different reverting agents, including inhibitors against EGFR, β1 integrin, and MMP9, after culturing in LrECM for 10 days. We found that there is a cluster of 70 genes that were all up-regulated in T4-2 cells, but down-modulated in S1 and "reverted" T4-2 cells (Bissell et al. 2005). To test whether these genes are under the control of common transcriptional regulators, we analyzed the consensus sequences in the promoter regions of these genes. We found that all of them possess multiple binding sites of the RelA subunit of NF-κB. NF-κB was highly up-regulated in malignant T4-2 cells and activated the expression of these 70 genes. We then inhibited NF-κB in T4-2 cells and found that this led to phenotypic reversion of malignant cells in LrECM culture (Becker-Weimann et al. 2013). This result suggests that NF-κB plays a critical role in disintegration of diverse signaling involved in mammary morphogenesis and must be silenced to allow phenotypic reversion in response to LrECM. To further

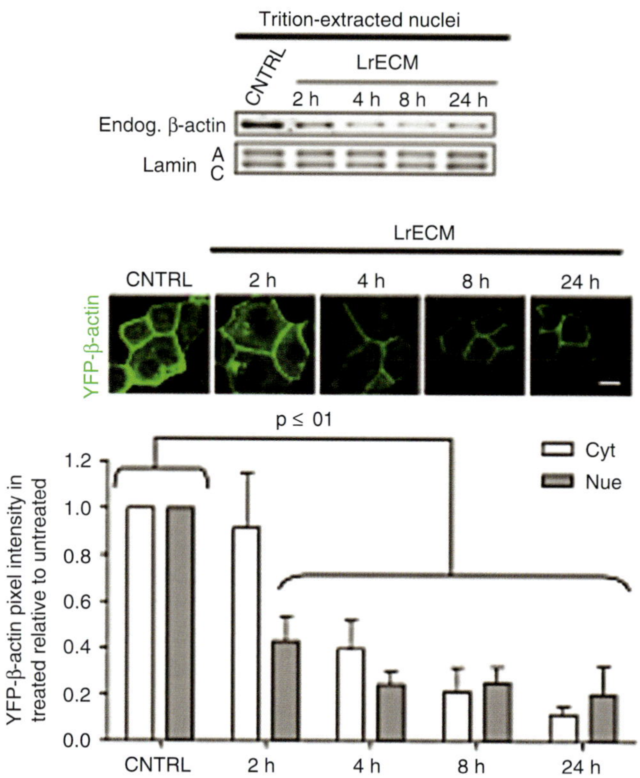

Figure 8. Laminin-rich extracellular matrix (LrECM) triggers the export of actin from the nucleus in normal mammary epithelial cells. (*Top*) Western blot analysis of actin level in the nucleus after addition of LrECM. (*Middle*) Micrograms of cells expressing yellow fluorescent protein (YFP)-β-actin. (*Bottom*) Quantification of the YFP signal in the cytosol versus nucleus after addition of LrECM. Cyt, cytosol; Nuc, nucleus. (Reprinted, with permission, from Spencer et al. 2011, The Company of Biologists Limited.)

explore the detailed molecular mechanism of the signal integration, we performed microRNA expression profiling of S1, T4-2 cells, and T4-2 cells reverted with the same conditions as those used for gene expression profiling (Fig. 5). By combining the results of gene expression profiling and microRNA expression profiling, we identified a group of microRNAs that could regulate the cluster of genes identified in gene expression profiling. While studying the mechanism of regulation of these microRNAs, we unraveled the signaling circuit that is activated in response to the basement membrane proteins laminins and responsible for the morphogenetic program of MECs (Fig. 9) (S Furuta and MJ Bissell, in prep.). We found that LrECM treatment in nonmalignant MECs activates p53 as the earliest responses (within 30 min), which then binds and transactivates the expression of LAMA3 (α3 chain of laminin-332) for exerting breast tissue specificity. Concomitantly, activated p53 up-regulates expression of a group of microRNAs that inactivate the expression of MMP9, the enzyme that degrades laminins, for protection of de novo synthesized laminins. Up-regulated laminin-332 then repeats the whole morphogenetic signaling cascade (S Furuta and MJ Bissell, in prep.). During proper morphogenesis, cytoskeletons are reorganized, and cell junctions and polarity are reinforced, leading to formation of apicobasally polarized structures. We showed that de novo synthesis of laminin-332 by MECs is the ultimate determinant of these processes. We also showed that this mechanism, conversely, is defective in cancer cells, which accounts for their inability to form quiescent, polarized structures in LrECM. Nevertheless, modulation of signals "downstream" of p53 (e.g., NF-κB, EGFR, β1 integrin, or MMP9) in cancer could reciprocally restore the upstream p53 activation, allowing for resumption of a normal morphogenetic program (Fig. 9). The overall findings of this study not only attest to the fact that morphogenetic signaling cascades establish integrated feedback loops but also show the rapid and intimate communications between the ECM and nucleus (S Furuta and MJ Bissell, in prep.).

Lastly, we visualized cytoskeletal structures that serve as the physical and mechanical connections between the ECM and nucleus using superresolution imaging techniques. Recent studies by other groups revealed the detailed mechanism of mechanotransduction that transmits the ECM stimuli to the nucleus. It is found that force received at the cell–ECM interphase is transmitted, through cytoskeletons and then to the linker of nucleoskeleton and cytoskeleton (LINC) complex located between the outer and inner nuclear membranes (Chang et al. 2015). LINC mechanically links cytoskeleton to the nucleus (Osmanagic-Myers et al. 2015) and plays roles in force transduction, as well as DNA damage repair and chromosome anchoring (Chang et al. 2015). Defects of LINC lead to nuclear deformation and impaired polarization of cells (Osmanagic-Myers et al. 2015). To further analyze these cytoskeletal structures deep inside of the nucleus, we collaborated with Manfred Auer and visualized the ultrastructures of mammary acini formed by nonmalignant MECs in LrECM cultures. Our work showed that, within the nucleus, cytoskeletal elements comprised of actin and keratin invaginate and

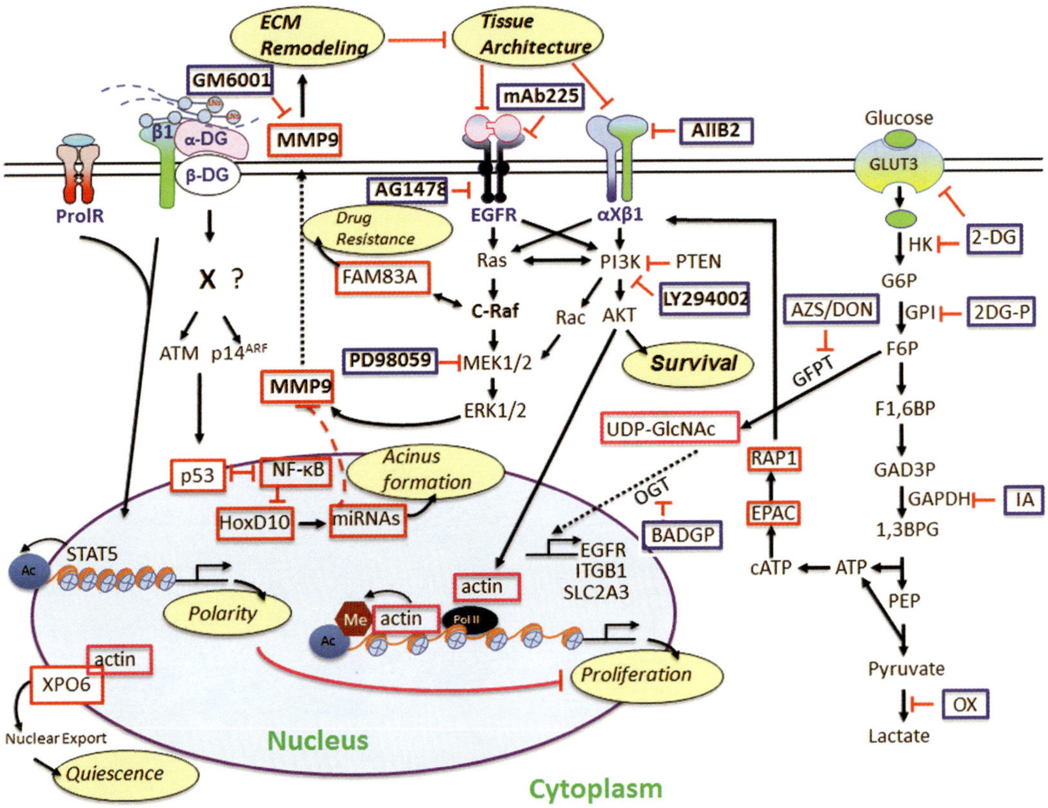

Figure 9. Scheme of signal integration and feedback loops during mammary acinar formation and tumor cell reversion. The different signaling cascades that we have reported are involved in acinar formation and tumor reversion are shown. ECM, extracellular matrix; MMP9, matrix metalloproteinase 9; DG, dystroglycan; EGFR, epidermal growth factor receptor; ITGB1, integrin beta 1; PI3K, phoshoinositide 3-kinase; PTEN, phosphatase and tensin homolog; cATP, cyclic adenosine triphosphate; PEP, phosphoenolpyruvate; GAPDH, glyceraldehyde 3-phosphatedehydrogenase; GPI, glucose 6-phosphate isomerase. (Courtesy of Sun-Young Lee.)

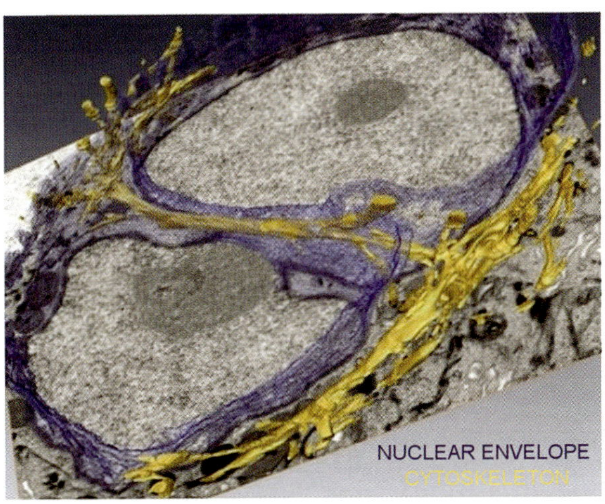

Figure 10. Cytoskeletal structures (yellow) in the nucleus are completely enclosed by the nuclear envelope (purple) and transverse the nuclear tunnel. The image was captured by serial block face scanning electron microscopy. Note that this image shows the ultrastructure of a single nucleus. (Reprinted, with permission, from Jorgens et al. 2017, The Company of Biologists Limited.)

transverse the nuclear tunnel. This structure ultimately forms the bridge between the adhesion complexes on the cell surface (desmosomes and hemidesmosomes) and nucleus (Fig. 10; Jorgens et al. 2017). These findings provide the visual support for the physical and mechanical links from the cell surface to the deep inside of the nucleus. The finding serves as the evidence for the physical path of dynamic reciprocity between the ECM and the nucleus by which ECM stimuli are transmitted into the nucleus for tissue-specific and context-dependent gene expression. Future investigation of whether this system is impaired in cancer cells is of great importance and is under investigation in our laboratory.

ACKNOWLEDGMENTS

The work from the Furuta and Bissell laboratories was supported by the breast cancer program of the U.S. Department of Defense (DOD), Breast Cancer Research Foundation, and the National Cancer Institute (NCI) to M.J.B. S.F. was supported by the Startup fund from the

Department of Biochemistry & Cancer Biology at the University of Toledo.

REFERENCES

Aggeler J, Ward J, Blackie LM, Barcellos-Hoff MH, Streuli CH, Bissell MJ. 1991. Cytodifferentiation of mouse mammary epithelial cells cultured on a reconstituted basement membrane reveals striking similarities to development in vivo. *J Cell Sci* **99:** 407–417.

Alcaraz J, Xu R, Mori H, Nelson CM, Mroue R, Spencer VA, Brownfield D, Radisky DC, Bustamante C, Bissell MJ. 2008. Laminin and biomimetic extracellular elasticity enhance functional differentiation in mammary epithelia. *EMBO J* **27:** 2829–2838.

Barcellos-Hoff MH, Aggeler J, Ram TG, Bissell MJ. 1989. Functional differentiation and alveolar morphogenesis of primary mammary cultures on reconstituted basement membrane. *Development* **105:** 223–235.

Basson MA. 2012. Signaling in cell differentiation and morphogenesis. *Cold Spring Harb Perspect Biol* **4:** a008151.

Becker-Weimann S, Xiong G, Furuta S, Han J, Kuhn I, Akavia UD, Pe'er D, Bissell MJ, Xu R. 2013. NF-κB disrupts tissue polarity in 3D by preventing integration of microenvironmental signals. *Oncotarget* **4:** 2010–2020.

Belin BJ, Lee T, Mullins RD. 2015. DNA damage induces nuclear actin filament assembly by Formin-2 and Spire-1/2 that promotes efficient DNA repair. [corrected]. *Elife* **4:** e07735.

Beliveau A, Mott JD, Lo A, Chen EI, Koller AA, Yaswen P, Muschler J, Bissell MJ. 2010. Raf-induced MMP9 disrupts tissue architecture of human breast cells in three-dimensional culture and is necessary for tumor growth in vivo. *Genes Dev* **24:** 2800–2811.

Bissell MJ, Hines WC. 2011. Why don't we get more cancer? A proposed role of the microenvironment in restraining cancer progression. *Nat Med* **17:** 320–329.

Bissell MJ, Labarge MA. 2005. Context, tissue plasticity, and cancer: Are tumor stem cells also regulated by the microenvironment? *Cancer Cell* **7:** 17–23.

Bissell MJ, Hall HG, Parry G. 1982. How does the extracellular matrix direct gene expression? *J Theor Biol* **99:** 31–68.

Bissell MJ, Kenny PA, Radisky DC. 2005. Microenvironmental regulators of tissue structure and function also regulate tumor induction and progression: The role of extracellular matrix and its degrading enzymes. *Cold Spring Harb Symp Quant Biol* **70:** 343–356.

Briand P, Petersen OW, Van Deurs B. 1987. A new diploid nontumorigenic human breast epithelial cell line isolated and propagated in chemically defined medium. *In Vitro Cell Dev Biol* **23:** 181–188.

Britt K, Ashworth A, Smalley M. 2007. Pregnancy and the risk of breast cancer. *Endocr Relat Cancer* **14:** 907–933.

Chang W, Worman HJ, Gundersen GG. 2015. Accessorizing and anchoring the LINC complex for multifunctionality. *J Cell Biol* **208:** 11–22.

Chen A, Cuevas I, Kenny PA, Miyake H, Mace K, Ghajar C, Boudreau A, Bissell MJ, Boudreau N. 2009. Endothelial cell migration and vascular endothelial growth factor expression are the result of loss of breast tissue polarity. *Cancer Res* **69:** 6721–6729.

Cipriano R, Graham J, Miskimen KL, Bryson BL, Bruntz RC, Scott SA, Brown HA, Stark GR, Jackson MW. 2012. FAM83B mediates EGFR- and RAS-driven oncogenic transformation. *J Clin Invest* **122:** 3197–3210.

Clubb BH, Locke M. 1998. Peripheral nuclear matrix actin forms perinuclear shells. *J Cell Biochem* **70:** 240–251.

Dopie J, Skarp KP, Rajakylä EK, Tanhuanpää K, Vartiainen MK. 2012. Active maintenance of nuclear actin by importin 9 supports transcription. *Proc Natl Acad Sci* **109:** E544–E552.

Farrants AK. 2008. Chromatin remodelling and actin organization. *FEBS Lett* **582:** 2041–2050.

Furuta S, Jeng YM, Zhou L, Huang L, Kuhn I, Bissell MJ, Lee WH. 2011. IL-25 causes apoptosis of IL-25R-expressing breast cancer cells without toxicity to nonmalignant cells. *Sci Transl Med* **3:** 78ra31.

Gruenbaum Y, Margalit A, Goldman RD, Shumaker DK, Wilson KL. 2005. The nuclear lamina comes of age. *Nat Rev Mol Cell Biol* **6:** 21–31.

Haigo S, Bilder D. 2011. Global tissue revolutions in a morphogenetic movement controlling elongation. *Science* **331:** 1071–1074.

Jorgens DM, Inman JL, Wojcik M, Robertson C, Palsdottir H, Tsai WT, Huang H, Bruni-Cardoso A, López CS, Bissell MJ, et al. 2017. Deep nuclear invaginations are linked to cytoskeletal filaments - integrated bioimaging of epithelial cells in 3D culture. *J Cell Sci* **130:** 177–189.

Kenny PA, Bissell MJ. 2003. Tumor reversion: Correction of malignant behavior by microenvironmental cues. *Int J Cancer* **107:** 688–695.

Kenny PA, Bissell DM. 2007. Targeting TACE-dependent EGFR ligand shedding in breast cancer. *J Clin Invest* **117:** 337–345.

Kiseleva E, Drummond SP, Goldberg MW, Rutherford SA, Allen TD, Wilson KL. 2004. Actin- and protein-4.1-containing filaments link nuclear pore complexes to subnuclear organelles in *Xenopus* oocyte nuclei. *J Cell Sci* **117:** 2481–2490.

Kleinman HK, McGarvey ML, Hassell JR, Star VL, Cannon FB, Laurie GW, Martin GR. 1987. Support of cultured hepatocytes by a laminin-rich gel. Evidence for a functionally significant subendothelial matrix in normal rat liver. *Biochemistry* **25:** 312–318.

Lamond AI, Spector DL. 2003. Nuclear speckles: A model for nuclear organelles. *Nat Rev Mol Cell Biol* **4:** 605–612.

Lee GY, Kenny PA, Lee EH, Bissell MJ. 2007. Three-dimensional culture models of normal and malignant breast epithelial cells. *Nat Methods* **4:** 359–365.

Lee S-Y, Meier R, Furuta S, Lenburg ME, Kenny PA, Xu R, Bissell MJ. 2012. FAM83A confers EGFR-TKI resistance in breast cancer cells and in mice. *J Clin Invest* **122:** 3211–3220.

Mammoto T, Ingber DE. 2010. Mechanical control of tissue and organ development. *Development* **137:** 1407–1420.

Mokady D, Meiri D. 2015. RhoGTPases—A novel link between cytoskeleton organization and cisplatin resistance. *Drug Resist Updat* **19:** 22–32.

Myers CA, Schmidhauser C, Mellentin-Michelotti J, Fragoso G, Roskelley CD, Casperson G, Mossi R, Pujuguet P, Hager G, Bissell MJ. 1998. Characterization of BCE-1, a transcriptional enhancer regulated by prolactin and extracellular matrix and modulated by the state of histone acetylation. *Mol Cell Biol* **18:** 2184–2195.

Nelson CM, Bissell MJ. 2006. Of extracellular matrix, scaffolds, and signaling: Tissue architecture regulates development, homeostasis, and cancer. *Annu Rev Cell Dev Biol* **22:** 287–309.

Nishida E, Iida K, Yonezawa N, Koyasu S, Yahara I, Sakai H. 1987. Cofilin is a component of intranuclear and cytoplasmic actin rods induced in cultured cells. *Proc Natl Acad Sci* **84:** 5262–5266.

Onodera Y, Nam JM, Bissell MJ. 2013. Increased sugar uptake activates oncogenic signaling via EPAC-Rap1 and O-GlcNAc pathways. *J Clin Invest* **124:** 367–384.

Osmanagic-Myers S, Dechat T, Foisner R. 2015. Lamins at the crossroads of mechanosignaling. *Genes Dev* **29:** 225–237.

Park CC, Zhang H, Pallavicini M, Gray JW, Baehner F, Park CJ, Bissell MJ. 2006. β1 Integrin inhibitory antibody induces apoptosis of breast cancer cells, inhibits growth, and distinguishes malignant from normal phenotype in three dimensional cultures and in vivo. *Cancer Res* **66:** 1526–1535.

Park CC, Zhang HJ, Yao ES, Park CJ, Bissell MJ. 2008. β1 integrin inhibition dramatically enhances radiotherapy efficacy in human breast cancer xenografts. *Cancer Res* **658:** 4398–4405.

Patwari P, Lee RT. 2010. Mechanical control of tissue morphogenesis. *Circ Res* **103:** 234–243.

Pendleton A, Pope B, Weeds A, Koffer A. 2003. Latrunculin B or ATP depletion induces cofilin-dependent translocation of actin into nuclei of mast cells. *J Biol Chem* **278:** 14393–14400.

Petersen OW, Ronnov-Jessen L, Howlett AR, Bissell MJ. 1992. Interaction with basement membrane serves to rapidly distinguish growth and differentiation pattern of normal and malignant human breast epithelial cells. *Proc Natl Acad Sci* **89:** 9064–9068.

Plessner M, Melak M, Chinchilla P, Baarlink C, Grosse R. 2015. Nuclear F-actin formation and reorganization upon cell spreading. *J Biol Chem* **290:** 11209–11216.

Radisky DC, Bissell MJ. 2006. Matrix metalloproteinase-induced genomic instability. *Curr Opin Genet Dev* **16:** 45–50.

Rizki A, Weaver VM, Lee S-Y, Rozenberg G, Chin K, Myers CA, Bascom JL, Mott JD, Semeiks JR, Grate LR, et al. 2008. A human breast cell model of preinvasive to invasive transition. *Cancer Res* **68:** 1378–1387.

Schmidhauser C, Casperson GF, Myers CA, Sanzo KT, Bolten S, Bissell MJ. 1992. A novel transcriptional enhancer is involved in the prolactin- and extracellular matrix–dependent regulation of β-casein gene expression. *Mol Biol Cell* **3:** 699–709.

Seid CA, Ramachandran RK, George JM, Govindarajan V, González-Rimbau MF, Flytzanis CN, Tomlinson CR. 1997. An extracellular matrix response element in the promoter of the *LpS1* genes of the sea urchin *Lytechinus pictus*. *Nucleic Acids Res* **25:** 3175–3182.

Sen B, Xie Z, Uzer G, Thompson WR, Styner M, Wu X, Rubin J. 2015. Intranuclear actin regulates osteogenesis. *Stem Cell* **33:** 3065–3076.

Simian M, Bissell MJ. 2017. Organoids: A historical perspective of thinking in three dimensions. *J Cell Biol* **216:** 31.

Snijders AM, Lee SY, Hang B, Hao W, Bissell MJ, Mao JH. 2017. FAM83 family oncogenes are broadly involved in human cancers: An integrative multi-omics approach. *Mol Oncol* **11:** 167–179.

Spencer VA, Xu R, Bissell MJ. 2010. Gene expression in the third dimension: The ECM-nucleus connection. *J Mammary Gland Biol Neoplasia* **15:** 65–71.

Spencer VA, Costes S, Inman JL, Xu R, Chen J, Hendzel MJ, Bissell MJ. 2011. Depletion of nuclear actin is a key mediator of quiescence in epithelial cells. *J Cell Sci* **124:** 123–132.

Streuli CH, Schmidhauser C, Bailey N, Yurchenco P, Skubitz AP, Roskelley C, Bissell MJ. 1995. Laminin mediates tissue-specific gene expression in mammary epithelia. *J Cell Biol* **129:** 591–603.

Stüven T, Hartmann E, Görlich D. 2003. Exportin 6: A novel nuclear export receptor that is specific for profilin.actin complexes. *EMBO J* **22:** 5928–5940.

Tanner K, Mori H, Mroue R, Bruni-Cardoso A, Bissell MJ. 2012. Coherent angular motion in the establishment of multicellular architecture of glandular tissues. *Proc Natl Acad Sci* **109:** 1973–1978.

Visa N, Percipalle P. 2010. Nuclear functions of actin. *Cold Spring Harb Perspect Biol* **2:** a000620.

Wang F, Weaver VM, Petersen OW, Larabell CA, Dedhar S, Briand P, Lupu R, Bissell MJ. 1998. Reciprocal interactions between β1-integrin and epidermal growth factor receptor in three-dimensional basement membrane breast cultures: A different perspective in epithelial biology. *Proc Natl Acad Sci* **95:** 14821–14826.

Wang N, Tytell JD, Ingber DE. 2009. Mechanotransduction at a distance: Mechanically coupling the extracellular matrix with the nucleus. *Nat Rev Mol Cell Biol* **10:** 75–82.

Weaver VM, Petersen OW, Wang F, Larabell CA, Briand P, Damsky C, Bissell MJ. 1997. Reversion of the malignant phenotype of human breast cells in three-dimensional culture and in vivo by integrin blocking antibodies. *J Cell Biol* **137:** 231–245.

Weaver VM, Lelièvre S, Lakins JN, Chrenek MA, Jones JC, Giancotti F, Werb Z, Bissell MJ. 2002. β4 integrin-dependent formation of polarized three-dimensional architecture confers resistance to apoptosis in normal and malignant mammary epithelium. *Cancer Cell* **2:** 205–216.

Weigelt B, Lo AT, Park CC, Gray JW, Bissell MJ. 2010. HER2 signaling pathway activation and response of breast cancer cells to HER2-targeting agents is dependent strongly on the 3D microenvironment. *Breast Cancer Res Treat* **122:** 35–43.

Xu R, Boudreau A, Bissell MJ. 2009a. Tissue architecture and function: Dynamic reciprocity via extra- and intra-cellular matrices. *Cancer Metastasis Rev* **28:** 167–176.

Xu R, Nelson CM, Muschler JL, Veiseh M, Vonderhaar BK, Bissell MJ. 2009b. Sustained activation of STAT5 is essential for chromatin remodeling and maintenance of mammary-specific function. *J Cell Biol* **184:** 57–66.

Explaining the Paucity of Intratumoral T Cells: A Construction Out of Known Entities

DOUGLAS T. FEARON[1,2,3]
[1]Cold Spring Harbor Laboratory, Cold Spring Harbor, New York 11724
[2]Weill Cornell Medicine, New York City, New York 10065
[3]University of Cambridge, Cambridge CB2 1TN, United Kingdom
Correspondence: dfearon@cshl.edu

This essay addresses the question of how tumors escape control by the immune system. The literature strongly points to inadequate accumulation of T cells among cancer cells as being the proximate cause, but this observation has no acceptable explanation as yet. An approach to this problem is adopted wherein the chemokines and chemokine receptors that normally mediate the trafficking of T cells to inflamed tissues are reviewed and considered in the context of their relative levels of expression in a transplanted colorectal tumor model. This method of reasoning—consistent with Bertrand Russell's (1985) advice, "Whenever possible, substitute constructions out of known entities for inferences to unknown entities"—leads to the proposal that signaling via the chemokine receptor, CXCR4, impairs the function of CXCR3 on the immune cells that are responsible for suppressing the growth of cancers.

The immune system has the capacity to control the growth of cancer. Clinical trials in which patients with certain types of solid tumors were administered neutralizing antibodies to membrane proteins that are expressed on activated effector and regulatory T cells, such as CTLA-4 and PD-1, or to PD-L1, the ligand for PD-1 that is expressed by some myelomonocytic cells and cancer cells, have shown that some patients show striking and often durable remissions (Hodi et al. 2010; Brahmer et al. 2012; Topalian et al. 2012; Hamid et al. 2013; Wolchok et al. 2013; Herbst et al. 2014; Powles et al. 2014; Tumeh et al. 2014; Motzer et al. 2015; Ferris et al. 2016). Other studies in which in vitro expanded tumor-infiltrating T cells have been adoptively transferred to cancer patients also have documented regressions of tumors (Tran et al. 2016). Finally, adoptive transfer of T cells expressing chimeric antigen receptors, so-called CAR T cells, has been an effective therapy for CD19-expressing B-cell malignancies (Porter et al. 2011). Thus, as predicted by earlier studies, such as that showing better long-term outcomes in colorectal cancer patients whose tumors were characterized by "$CD8^+$ T cells infiltrated within cancer cell nests" (Naito et al. 1998), the immune system, through the cytotoxic activities of the $CD8^+$ T cell, has the capacity to kill immunogenic cancer cells. These successes present tumor immunologists with an urgent responsibility now to understand why only subsets of patients with only some types of cancers respond to these forms of immunotherapy. That is to say, "Why is immunotherapy ineffective in most patients?"

Potential contemporary answers to this question range from being self-evident, although nonetheless important, such as whether cancer cells have induced an adaptive immune response, to others for which mechanistic explanations are still evolving, such as T-cell "exhaustion" and immune inhibition by "myeloid-derived suppressor cells". Although the former can be definitively resolved by experimental approaches like the development of assays of anticancer T-cell immunity, the latter explanations will be more difficult to affirm or disprove. The complexity of the vertebrate immune system presents the tumor immunologist with almost too many experimental opportunities to "validate" hypotheses that may have conceptual rather than molecular definitions. More relevant to this question of tumoral immune escape is our knowledge that although the immune system evolved to defend against an extraordinary array of pathogens, it also developed "brakes" that prevent attack of semiallogenic fetuses and autoimmune diseases. Namely, it is "normal" that the adaptive immune system may not always damage immunogenic tissue. For this reason, we should not assume that the capacity of an immunogenic tumor to escape immune control is a consequence of an abnormal immunological process, like a hostile tumor microenvironment (TME) causing "T-cell exhaustion." Instead, this unfortunate clinical circumstance may be explained by an immune pathway that benefits the host in biological circumstances other than cancer.

T-CELL EXCLUSION—THE SHARED CHARACTERISTIC OF TUMORS ESCAPING IMMUNE CONTROL

The most striking difference between a tissue immune reaction eliminates epithelial cells expressing immuno-

Published by Cold Spring Harbor Laboratory Press; doi: 10.1101/sqb.2016.81.030783

genic antigens, as in influenza virus–infected pulmonary epithelium, and an ineffective tumor immune reaction is the relative numbers of infiltrating T cells. Although informative and discerning metrics have been devised, such as "immunoscore" (Galon et al. 2016), and scientifically vague, but descriptive terms, such as "hot" and "cold" tumors, have been promoted to exemplify this difference, the most unambiguous characteristic of an immunologically uncontrolled tumor is the relative paucity of T cells that are juxtaposed to cancer cells (Joyce and Fearon 2015). In contrast, tumors that are being effectively eliminated by an immune response have dense infiltrates of $CD8^+$ T cells, as do the islets of Langerhans of the pancreas in autoimmune type I diabetes mellitus, organs undergoing cell-mediated immune rejection, and, indeed, virally infected tissues, the last being a biological circumstance that selected this $CD8^+$ T-cell response. Therefore, if the relative intensity of the intratumoral accumulation of $CD8^+$ T cells is the critical variable that distinguishes between effective and ineffective antitumor immune reactions, one should determine whether the principles that have been discovered to regulate these other examples of $CD8^+$ T-cell accumulation also pertain to cancer. This approach would be consistent with the admonition of Bertrand Russell: "Whenever possible, substitute constructions out of known entities for inferences to unknown entities" (Russell 1985).

THE NETWORK OF CHEMOKINE AND CHEMOKINE RECEPTORS: CXCR3 AS THE MEDIATOR OF INTRATUMORAL T-CELL ACCUMULATION

We have known for many years that an extensive network of chemokines and chemokine receptors regulates the trafficking of immune cells (Griffith et al. 2014). The first chemokine receptor, which is thought to be CXCR4, appears in the jawless fish, lamprey, an ancient example of the vertebrate lineage in which the adaptive immune system that is typified by the *RAG* genes is absent (Kuroda et al. 2003). With the further evolution of vertebrates, there is an explosion of chemokines and chemokine receptors, such that the zebrafish has 63 genes encoding chemokines and 24 genes encoding chemokine receptors (DeVries et al. 2006). The relevance of this system for directing the cellular migration is not limited to the immune system and is essential also for the development of many organs. However, in the postnatal period, the primary role of this system may well be immunological. The cells of the immune system most often mediate their effector functions in tissues that are remote from their sites of development and residence, the bone marrow, thymus, and secondary lymphoid organs. With respect to the effector $CD8^+$ T cell, a remarkably consistent model has been developed. After their initial clonal expansion in secondary lymphoid organs, effector $CD8^+$ T cells are released into the peripheral blood. Although these effector cells up-regulate their expression of a number of chemokine receptors during differentiation from naïve to effector $CD8^+$ T cell, a single receptor, CXCR3, appears to be most often responsible for mediating homing to sites of inflammation. These CXCR3-mediated $CD8^+$ T-cell responses depend on the production in the inflamed tissues of the chemokines, CXCL9, CXCL10, and CXCL11, the levels of which are regulated by type I and type II interferon (IFN) (Griffith et al. 2014). A dominant role for CXCR3 in T-cell homing has been found for a range of biological circumstances in which target cells express major histocompatibility complex (MHC) class I/peptide complexes bearing neoepitopes, including allograft rejection (Seung et al. 2011), viral infections (Klein et al. 2005; Hsieh et al. 2006), autoimmunity (Frigerio et al. 2002), and cancer (Mikucki et al. 2015), the last example being especially relevant to the subject of this discourse. In the B16 mouse model of melanoma, CXCR3 was shown to have a nonredundant role in mediating the intratumoral accumulation of adoptively transferred, cancer-specific, effector $CD8^+$ T cells, which resulted in control of tumor growth by two complementary approaches—use of $CD8^+$ T cells in which the *CXCR3* gene had been disabled and administration of neutralizing anti-CXCR3 antibody. Two other receptors, CCR2 and CCR5, did not support the intratumoral trafficking of $CD8^+$ T cells in this experimental model, serving, in effect, as negative controls. Therefore, despite the multiplicity of chemokine receptors that an effector $CD8^+$ T cell expresses, and regardless of a diversity of immunological tissue reactions, CXCR3 has a dominant, positive role in the trafficking of these cells. In this regard, it is interesting to note that other immune cell types that should be considered important for orchestrating an effective antitumor immune reaction—TH1 cells, the subset of dendritic cells (DCs) with antigen cross-presenting capability, and natural killer (NK) cells—also express CXCR3, whereas other cell types that may even have tissue protective functions, such as monocytes and macrophages, do not.

Because the response of the preclinical B16 melanoma mouse model to anti-CTLA-4 immunotherapy (Leach et al. 1996) has been predictive of clinical immunotherapeutic responses of human melanoma (Hodi et al. 2010), the absence of a clinical response and the paucity of intratumoral T cells in those cancers that do not respond to immunotherapy may be considered to be caused by an absence of signaling by CXCR3 on $CD8^+$ T cells within the TME. This deficiency may have one obvious explanation, which would be insufficient level of intratumoral CXCL9, CXCL10, or CXCL11, the ligands for CXCR3, but a second explanation is also possible, which is that the TME impairs signaling by CXCR3. There is evidence in support of the first explanation with the finding of epigenetic silencing of CXCL9 and CXCL10 as a means by which a preclinical tumor model circumvents immune control (Peng et al. 2015). Epigenetic silencing of the expression of these chemokines has been proposed also as the mechanism by which the semiallogeneic fetus escapes rejection by the maternal adaptive immune response (Nancy et al. 2012). Moreover, the finding that the type I IFN response is necessary for immune control of immunogenic mouse tumors (Diamond et al. 2011; Fuertes et al. 2011) may be related to the capacity of this cytokine to induce the transcription of CXCL10. Thus, it is reasonable to propose that insufficient levels of the CXCR3-related chemokines account for the impaired intratumoral accumulation of T cells. However, there may also be evidence in support of impaired CXCR3 function on immune cells in the TME.

IMPAIRED FUNCTION OF CXCR3 IN THE C26 COLORECTAL TUMOR MODEL

The C26 tumor is a chemically induced, transplantable mouse colon carcinoma that has been used in studies of cancer-induced cachexia. Its origin as a mutagen-induced cancer may indicate that it expresses neoantigens for presentation via the MHC class I pathway when transplanted and grown in mice. Therefore, it may serve as an appropriate model to explore the mechanism of impaired intratumoral accumulation of cancer-specific $CD8^+$ T cells. In our studies of cancer-induced cachexia, we found relatively high mRNA levels for CXCL9 and CXCL10 in analyses of transplanted C26 tumors (Flint et al. 2016), which indicated that these tumors had two elements of an antitumor immune response, immunogenicity and CXCR3-related chemokines, that might be anticipated to be sufficient for immune control tumor growth. However, control obviously does not happen and ectopic C26 tumors continue to grow in immunocompetent, syngeneic mice. To gain insight into the reasons for this escape from immune control, we performed RNA-seq on whole C26 tumors to assess the expression of genes that mark specific immune cell types (Table 1). The C26 tumor has abundant monocytes and/or macrophages, as exemplified by the high mRNA levels of CSF1R, ITGAM, and F4/80. In contrast, and assuming that the expression of each gene per cell is relatively equivalent, the frequency of T cells is only 1%–5% that of monocytes/macrophages, based on the low mRNA levels of CD3e, CD4, and CD8a. Also, the mRNA levels of *BATF3*, *CD103*, *XCR1*, *CLEC9a*, and *FLT3*, five genes expressed by the antigen crossing-presenting subset of DCs, are similarly low, as is the mRNA level for the gene, *NCR1* (NKp46), that encodes a receptor that is uniquely expressed by NK cells. Therefore, in the C26 tumor, there is a relative paucity of precisely those immune cells that would be capable of promoting the killing of cancer cells: CTLs, NK cells, and the subset of DCs that direct clonal expansion of cancer-specific $CD8^+$ T cells within the draining lymph node and within the tumor itself.

Table 1. Expression of genes denoting immune cell types in the C26 tumor

Gene	Cell type	RPKM
CSF1R	Mono, Mac	2652
ITGAM	Mono, Mac	2901
F4/80	Mac	1427
CD3e	T cell	33
CD4	T cell	69
FOXP3	Treg cell	14
CD8a	T cell, $BATF3^+$ DC	117
BATF3	$BATF3^+$ DC	29
CD103	$BATF3^+$ DC	22
XCR1	$BATF3^+$ DC	31
CLEC9a	$BATF3^+$ DC	22
FLT3	$BATF3^+$ DC	15
NCR1	NK cell	35

RPKM, reads per kilobase of transcript per million mapped reads; Mono, monocyte; Mac, macrophage; Treg, T regulatory; DC, dendritic cell; NK, natural killer.

An examination of the expression of chemokines and chemokine receptors in the C26 tumor reveals a pattern that could account for the marked difference in immune cell types that accumulate in the C26 tumor. The mRNA for CXCR4, which is expressed by all immune cell types, is relatively abundant, presumably reflecting the presence of monocytes and macrophages (Table 2). In accord with this surmise, the mRNA levels for CCR2 and CCR5, two chemokine receptors that are expressed by monocytes and macrophages, are high. In contrast, mRNA levels for CXCR3 are drastically lower. This finding is consistent with the presumed low frequency of T cells, $BATF3^+$ DCs, and NK cells in the C26 tumor because these three cell types, in contrast to monocytes and macrophages, express CXCR3. The paucity of CXCR3-expressing immune cells cannot be explained by low tumoral expression of the ligands for this receptor because the combined mRNA level of CXCL9, CXCL10, and CXCL11 is even higher than that of CCL2 mRNA, the ligand for CCR2. This circumstance, assuming that these transcripts are translated, excludes the possibility that the relative deficiency of CXCR3-expressing cells is secondary to low levels if the receptor's cognate chemokines. Indeed, if

Table 2. Expression of chemokine and chemokine receptor genes in the C26 tumor

Receptor	Cell types	RPKM	Chemokine	RPKM	Chemokine-R/chemokine ratio
CCR2	Teff, Mono, Mac	825	CCL2	1125	0.7
CCR5	Teff, iNKT, NK, Mac, pDC	1775	CCL3	42	7.7
			CCL4	55	
			CCL5	133	
CXCR3	TH1, CTL, pDC, $Batf3^+$DC, iNKT, NK, ILC1	10	CXCL9	1384	0.005
			CXCL10	473	
			CXCL11	68	
CXCR4	All	308	CXCL12	187	1.7

RPKM, reads per kilobase of transcript per million mapped reads; Teff, effector T-cell; Mono, monocyte; Mac, macrophage; iNKT, invariant natural killer T; pDC, plasmacytoid dendritic cell; CTL, cytotoxic T lymphocyte.

one calculates the ratio of the mRNA levels of each chemokine receptor to its associated chemokine(s) as a measure of the activity of the receptor, one finds that the function of CXCR3 may be <1% that of CCR2 in the C26 tumor. This finding is compelling evidence in support of the possibility that the CXCR3-expressing cells may not accumulate in the C26 tumor even when CXCL9-11 are expressed and strongly suggests that some aspect of the TME impairs CXCR3 function and causes an intratumoral immune phenotype that is equivalent to that of B16 melanoma-bearing mice treated in which cancer-specific $CD8^+$ T cells are lacking CXCR3.

CXCR4 AS THE CHEMOKINE RECEPTOR THAT INHIBITS CXCR3

By conforming to the philosophical axiom of Bertrand Russell and focusing on the "known entity" of the chemokine/chemokine receptor system, we have identified a disparity between the intratumoral mRNA levels of CXCR3 and its ligands that strongly suggests that the TME suppresses the function of this chemokine receptor. Again, we turn to the literature in seeking a potential explanation of how the TME might alter the function of CXCR3. A spontaneously immunogenic, autochthonous mouse model of pancreatic ductal adenocarcinoma (PDAC) (Hingorani et al. 2005) escapes immune control by excluding T cells from the vicinity of cancer cells (Feig et al. 2013). Several findings suggested the possibility that CXCR4, a chemokine receptor found on all immune cells, had a role in this T-cell exclusion phenomenon: Conditional depletion of the cancer-associated fibroblasts enabled T-cell-dependent suppression of tumor growth; the cancer-associated fibroblasts were the major, intratumoral cellular source of CXCL12, the ligand for CXCR4; and the PDAC cancer cells appeared to be "coated" with the chemokine, despite the absence of CXCL12 mRNA in these cells (Kraman et al. 2010; Feig et al. 2013). A potential immunosuppressive role for CXCR4 was assessed by treating PDAC-bearing mice with AMD3100, a small-molecule inhibitor of CXCR4. Within 24 h of treatment initiation, T cells had accumulated among cancer cells, and within 48 h, in combination with a T-cell checkpoint antagonist, anti-PD-L1 antibody, to which mouse and human PDACs are consistently resistant, the mouse tumor volume had decreased by 15%. If the many publications attesting to the inhibitory specificity of AMD3100 for CXCR4 are correct, and if this PDAC tumor model obeys the same, nonredundant requirement for CXCR3 expression and function by $CD8^+$ T cells in the B16 mouse melanoma (Mikucki et al. 2015), one is led to the conclusion that CXCR4 signaling suppresses the intratumoral function of CXCR3 in mouse PDAC to mediate the exclusion of T cells. The finding that AMD3100 also led to immune control of an ectopic Lewis lung tumor expressing the antigen, ovalbumin (Feig et al. 2013), indicates that this conclusion is not restricted to mouse PDAC. (See Fig. 1.)

CXCR4 has several characteristics that distinguish it from the other leukocyte, Gαi protein-coupled receptors (GPCRs) that mediate chemotaxis. An ortholog of CXCR4 exists in the lamprey (Kuroda et al. 2003), a member of the Agnatha class. Therefore, this chemokine receptor, uniquely among all chemokine receptors, evolved before the vertebrate adaptive immune system of immunoglobulin and T-cell receptors. CXCR4 and its CXCL12 ligand demonstrate the highest degree of evolutionary conservation when chemokines and chemokine receptor sequences of human, mouse, chicken, frog, zebrafish, and pufferfish are compared (DeVries et al. 2006). Furthermore, unlike other chemokine receptors, CXCR4 has important nonimmunological roles, such as in the development of neural (Zou et al. 1996; Chalasani et al. 2003; Haas and Gilmour 2006) and cardiovascular systems (Tachibana et al. 1998), germ cell migration (Knaut et al. 2003; Molyneaux et al. 2003), and hematopoiesis (Zou et al. 1996). These features have led some investigators to classify CXCR4 and its ligand, CXCL12, as having "homeostatic" functions, in contrast to the immune and inflammatory functions of the other leukocyte chemokine receptors. Seen from this perspective, it is reasonable to consider the role of CXCR4 in other organ systems to obtain hints concerning whether and how it may alter the function of CXCR3 on immune cells.

An early demonstration of CXCR4 signaling altering the function of another GPCR was the finding that the chemotactic activities of both the μ- and δ-opioid GPCRs in several cell types were desensitized following activation of CXCR4 in the responding cells (Szabo et al. 2002). Interestingly, this study also found that pretreatment with CXCL12 followed by opioid administration into the periaqueductal gray matter of the brain increased a behavioral response to a painful stimulus, consistent with CXCR4 signaling suppressing opioid receptor signaling of a cellular response distinct from motility. This conclusion was confirmed and extended by the demonstration that CXCR4, CXCL12, and the μ-opioid receptor were coexpressed by individual neurons in several regions of the brain areas (Heinisch et al. 2011). Moreover,

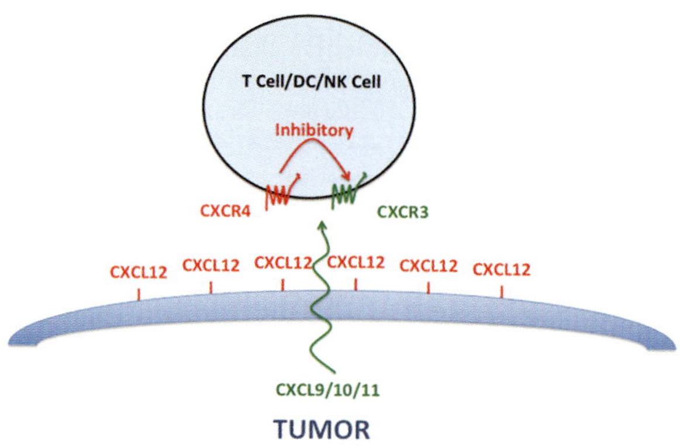

Figure 1. Proposed basis for the exclusion of T cells from tumors. CXCL12 "coats" tumor cells and stimulates CXCR4 on T cells, dendritic cells (DCs), and natural killer (NK) cells. CXCR4 then inhibits the capacity of CXCR3 to respond to tumor-derived CXCL9, CXCL10, and CXCL11.

whole patch-clamp recordings of the periaqueductal grey neurons showed that CXCL12 blocked morphine-induced electrophysiological effects. This effect of CXCL12 was suppressed by AMD3100, demonstrating that it was mediated by CXCR4.

CXCR4 may mediate these inhibitory effects on opioid receptor function as a consequence of these two receptors forming heteromeric complexes (Stephens and Handel 2013). Therefore, in the context of this discourse on tumor immunity, it is of some interest that CXCR4 forms hetero-oligomeric complexes with CCR2 and CCR5 (Sohy et al. 2007, 2009). In the heterodimer formed between CXCR4 and CCR2, there is negative binding cooperativity between the subunits such that ligands that are specific for one receptor suppress the binding of the ligand to the other receptor. A similar negative binding cooperativity has been demonstrated for the chemokines interacting with heterodimers formed between CXCR4 and CCR5. The basis for such negative binding cooperativity was not determined but was suggested to be a consequence of allosteric modulation across the receptor heteromer interface. Although there is no evidence that CCR2 or CCR5 mediate the intratumoral accumulation of T cells, and, in fact, there is a report that CCR5 does not mediate this process (Mikucki et al. 2015), these findings do show the capacity of CXCL12 to cause CXCR4 to suppress the signaling by another chemokine receptor. In this respect, then, it may be relevant that CXCR4 and CXCR3 have been shown to form heteromeric complexes when ectopically expressed in HEK293T cells (Watts et al. 2013). Negative binding cooperativity between CXCR4 and CXCR3, however, was not observed in intact HEK293T cells, although the analysis was not performed with cells that normally express these two receptors.

These studies of the interaction of ligands with CXCR4–CCR2–CCR5 chemokine receptor heteromers also revealed that the effects of inhibitors of CXCR4 are more complex than has been concluded based on standard experiments using cells expressing only single types of chemokine receptors. For example, treatment of cells expressing only CCR2 with the CXCR4 antagonist, AMD3100, does not affect the binding of CCL2, and treating cells expressing only CXCR4 inhibits binding of CXCL12. With cells expressing CXCR4 in addition to CCR2, however, AMD3100 treatment suppresses not only binding of CXCL12 to CXCR4, but also that of CCL2 to CCR2 (Sohy et al. 2009). Similarly, AMD3100 does not alter the binding of CCL4 to cells expressing only CCR5, but the CXCR4 antagonist does alter the kinetics of the interaction between CCR5 and CCL4 with cells that also express CXCR4. These effects, which may be secondary to the partial agonistic activity of AMD3100 on CXCR4 (Zhang et al. 2002), emphasize the necessity to conduct in vitro experiments that reflect at least some aspects of the realities of in vivo biology—namely, that cells rarely express only a single chemokine receptor. For example, populations of activated effector $CD8^+$ T cells may express CCR2, CCR5, CXCR3, CXCR4, and CXCR6. Studies of chemokine receptor antagonists must determine which inhibitors are truly "specific" for their target receptor and do not alter the function of potential heteromeric partners.

Finally, CXCL12, the ligand of CXCR4, also shows a complexity that distinguishes it from other chemokines, especially with respect to the form in which it is present in carcinomas. In the autochthonous mouse model of PDAC, cancer cells stain with anti-CXCL12 (Feig et al. 2013). This staining with anti-CXCL12 of cancer cells has been reported also for human ovarian cancer (Scotton et al. 2002; Jiang et al. 2006), colorectal cancer (Akishima-Fukasawa et al. 2009), and PDAC (Liang et al. 2010), and we have observed it in mouse and human non–small cell lung cancer (NSCLC). These observations may be relevant to the means by which these cancers escape immune control because they are all characterized by exclusion of T cells from the vicinity of cancer cells (Naito et al. 1998; Zhang et al. 2003; Salmon et al. 2012; Feig et al. 2013) except for relatively infrequent examples of NSCLC and colorectal cancer that are associated with a high somatic mutational burden (Le et al. 2015; Rizvi et al. 2015). The means by which CXCL12 associates with cancer cells is not known, but it is reasonable to consider the possibility that this cell surface CXCL12 will be presented to CXCR4-expressing cells in a multimeric form rather than the monomeric form that is typically used in in vitro studies. Recent studies suggest that monomeric CXCL12 may not be the only or even most prevalent form of the chemokine. For example, dimeric CXCL12 is induced by the interaction of monomeric CXCL12 with negatively charged glycosaminoglycans (Veldkamp et al. 2005), which are a ubiquitous constituent of the TME. Dimeric CXCL12, in contrast to monomeric CXCL12, has been found to be resistant to proteolytic inactivation (Takekoshi et al. 2012), suggesting a means by which the half-life of the chemokine may be extended in a tissue-specific manner. Furthermore, dimeric CXCL12 elicits CXCR4-dependent Ca^{2+} flux, adenylyl cyclase inhibition, and activation of ERK1/2, but not chemotaxis (Drury et al. 2011). The interaction of dimeric CXCL12 with CXCR4 is also influenced by posttranslational sulfation of extracellular tyrosines (Veldkamp et al. 2006, 2008; Seibert et al. 2008). Thus, the physical state of CXCL12 may be dictated by the TME, and this may be an additional way that a tumor determines the type of intracellular signals that are elicited by CXCR4, adding an additional element of biological flexibility to this unusual chemokine/chemokine receptor system.

CONCLUSION

Immunogenic carcinomas escape immune control by creating a microenvironment that prevents the accumulation of T cells, DCs, and NK cells. Signaling by CXCR3 may usually mediate the accumulation of these immune cell types in tissue responses associated with the killing of cells expressing antigens, suggesting that the relative absence of these immune cell types in tumors is secondary either to a deficiency of the CXCR3-associated chemokines or to the malfunction of the chemokine receptor. An example is presented of a TME in which there is abundant expression of the CXCR3-associated chemokines, CXCL9, CXCL10, and CXCL11, but a relative paucity of T cells, DCs, and NK cells. The absence of CXCR3-expresssing immune cells in this tumor, then, must be secondary to inhibition of CXCR3 function. Because suppressing signaling by CXCR4 leads to intratumoral accumulation of T cells in this and other tumor models, because CXCL12 is present in tumors, and because the depletion of the intratumoral stromal cells that produce CXCL12 leads to control of tumor growth, one may reason that CXCR4 is responsible for the inhibition of CXCR3 function. This inhibition may be mediated through a physical association of CXCR4 with CXCR3 because ligation of CXCR4 in heteromeric complexes containing another GPCR may inhibit the function of the partner chemokine receptor. This type of cross-desensitization may occur also with CXCR4 “antagonists” that have partial agonistic activity, which may compromise the therapeutic efficacy of these antagonists in overcoming CXCR4-mediated restriction on the intratumoral accumulation of T cells.

This essay opened with a quote from Bertrand Russell and, in closing, one is reminded of another comment from this philosopher/mathematician: “The method of ‘postulating’ what we want has many advantages; they are the same as the advantages of theft over honest toil. Let us leave them to others and proceed with our honest toil” (Russell 1919). “Honest toil” in this essay refers to the experiments that have been performed by many thoughtful scientists whose experiments have provided the careful measurements of biological reactions that have enabled “constructions of known entities.” Whether the author has deviated too far in the direction of “postulating” what he “wants” is a question that, thankfully, can be answered by experimentation.

ACKNOWLEDGMENTS

The author and his laboratory are supported by the Lustgarten Foundation. The author is grateful to his postdoctoral fellows and PhD students who, over the years, have enthusiastically taken on the difficult and rewarding challenges that immunology provides.

REFERENCES

Akishima-Fukasawa Y, Nakanishi Y, Ino Y, Moriya Y, Kanai Y, Hirohashi S. 2009. Prognostic significance of CXCL12 expression in patients with colorectal carcinoma. *Am J Clin Pathol* **132:** 202–210.

Brahmer JR, Tykodi SS, Chow LQ, Hwu WJ, Topalian SL, Hwu P, Drake CG, Camacho LH, Kauh J, Odunsi K, et al. 2012. Safety and activity of anti-PD-L1 antibody in patients with advanced cancer. *N Engl J Med* **366:** 2455–2465.

Chalasani SH, Sabelko KA, Sunshine MJ, Littman DR, Raper JA. 2003. A chemokine, SDF-1, reduces the effectiveness of multiple axonal repellents and is required for normal axon pathfinding. *J Neurosci* **23:** 1360–1371.

DeVries ME, Kelvin AA, Xu L, Ran L, Robinson J, Kelvin DJ. 2006. Defining the origins and evolution of the chemokine/chemokine receptor system. *J Immunol* **176:** 401–415.

Diamond MS, Kinder M, Matsushita H, Mashayekhi M, Dunn GP, Archambault JM, Lee H, Arthur CD, White JM, Kalinke U, et al. 2011. Type I interferon is selectively required by dendritic cells for immune rejection of tumors. *J Exp Med* **208:** 1989–2003.

Drury LJ, Ziarek JJ, Gravel S, Veldkamp CT, Takekoshi T, Hwang ST, Heveker N, Volkman BF, Dwinell MB. 2011. Monomeric and dimeric CXCL12 inhibit metastasis through distinct CXCR4 interactions and signaling pathways. *Proc Natl Acad Sci* **108:** 17655–17660.

Feig C, Jones JO, Kraman M, Wells RJ, Deonarine A, Chan DS, Connell CM, Roberts EW, Zhao Q, Caballero OL, et al. 2013. Targeting CXCL12 from FAP-expressing carcinoma-associated fibroblasts synergizes with anti-PD-L1 immunotherapy in pancreatic cancer. *Proc Natl Acad Sci* **110:** 20212–20217.

Ferris RL, Blumenschein G Jr, Fayette J, Guigay J, Colevas AD, Licitra L, Harrington K, Kasper S, Vokes EE, Even C, et al. 2016. Nivolumab for recurrent squamous-cell carcinoma of the head and neck. *N Engl J Med* **375:** 1856–1867.

Flint TR, Janowitz T, Connell CM, Roberts EW, Denton AE, Coll AP, Jodrell DI, Fearon DT. 2016. Tumor-induced IL-6 reprograms host metabolism to suppress anti-tumor immunity. *Cell Metab* **24:** 672–684.

Frigerio S, Junt T, Lu B, Gerard C, Zumsteg U, Holländer GA, Piali L. 2002. β cells are responsible for CXCR3-mediated T-cell infiltration in insulitis. *Nat Med* **8:** 1414–1420.

Fuertes MB, Kacha AK, Kline J, Woo SR, Kranz DM, Murphy KM, Gajewski TF. 2011. Host type I IFN signals are required for antitumor $CD8^+$ T cell responses through $CD8\alpha^+$ dendritic cells. *J Exp Med* **208:** 2005–2016.

Galon J, Fox BA, Bifulco CB, Masucci G, Rau T, Botti G, Marincola FM, Ciliberto G, Pages F, Ascierto PA, et al. 2016. Immunoscore and Immunoprofiling in cancer: An update from the melanoma and immunotherapy bridge 2015. *J Transl Med* **14:** 273–279.

Griffith JW, Sokol CL, Luster AD. 2014. Chemokines and chemokine receptors: Positioning cells for host defense and immunity. *Annu Rev Immunol* **32:** 659–702.

Haas P, Gilmour D. 2006. Chemokine signaling mediates self-organizing tissue migration in the zebrafish lateral line. *Dev Cell* **10:** 673–680.

Hamid O, Robert C, Daud A, Hodi FS, Hwu WJ, Kefford R, Wolchok JD, Hersey P, Joseph RW, Weber JS, et al. 2013. Safety and tumor responses with lambrolizumab (anti-PD-1) in melanoma. *N Engl J Med* **369:** 134–144.

Heinisch S, Palma J, Kirby LG. 2011. Interactions between chemokine and μ-opioid receptors: Anatomical findings and electrophysiological studies in the rat periaqueductal grey. *Brain Behav Immun* **25:** 360–372.

Herbst RS, Soria JC, Kowanetz M, Fine GD, Hamid O, Gordon MS, Sosman JA, McDermott DF, Powderly JD, Gettinger SN, et al. 2014. Predictive correlates of response to the anti-PD-L1 antibody MPDL3280A in cancer patients. *Nature* **515:** 563–567.

Hingorani SR, Wang L, Multani AS, Combs C, Deramaudt TB, Hruban RH, Rustgi AK, Chang S, Tuveson DA. 2005. Trp53R172H and KrasG12D cooperate to promote chromosomal instability and widely metastatic pancreatic ductal adenocarcinoma in mice. *Cancer Cell* **7:** 469–483.

Hodi FS, O'Day SJ, McDermott DF, Weber RW, Sosman JA, Haanen JB, Gonzalez R, Robert C, Schadendorf D, Hassel JC, et al. 2010. Improved survival with ipilimumab in patients with metastatic melanoma. *N Engl J Med* **363:** 711–723.

Hsieh MF, Lai SL, Chen JP, Sung JM, Lin YL, Wu-Hsieh BA, Gerard C, Luster A, Liao F. 2006. Both CXCR3 and CXCL10/IFN-inducible protein 10 are required for resistance to primary infection by dengue virus. *J Immunol* **177:** 1855–1863.

Jiang YP, Wu XH, Shi B, Wu WX, Yin GR. 2006. Expression of chemokine CXCL12 and its receptor CXCR4 in human epithelial ovarian cancer: An independent prognostic factor for tumor progression. *Gynecol Oncol* **103:** 226–233.
Joyce JA, Fearon DT. 2015. T cell exclusion, immune privilege, and the tumor microenvironment. *Science* **348:** 74–80.
Klein RS, Lin E, Zhang B, Luster AD, Tollett J, Samuel MA, Engle M, Diamond MS. 2005. Neuronal CXCL10 directs $CD8^+$ T-cell recruitment and control of West Nile virus encephalitis. *J Virol* **79:** 11457–11466.
Knaut H, Werz C, Geisler R, Nüsslein-Volhard C; Tübingen 2000 Screen Consortium. 2003. A zebrafish homologue of the chemokine receptor Cxcr4 is a germ-cell guidance receptor. *Nature* **421:** 279–282.
Kraman M, Bambrough PJ, Arnold JN, Roberts EW, Magiera L, Jones JO, Gopinathan A, Tuveson DA, Fearon DT. 2010. Suppression of antitumor immunity by stromal cells expressing fibroblast activation protein-α. *Science* **330:** 827–830.
Kuroda N, Uinuk-ool TS, Sato A, Samonte IE, Figueroa F, Mayer WE, Klein J. 2003. Identification of chemokines and a chemokine receptor in cichlid fish, shark, and lamprey. *Immunogenetics* **54:** 884–895.
Le DT, Uram JN, Wang H, Bartlett BR, Kemberling H, Eyring AD, Skora AD, Luber BS, Azad NS, Laheru D, et al. 2015. PD-1 blockade in tumors with mismatch-repair deficiency. *N Engl J Med* **372:** 2509–2520.
Leach DR, Krummel MF, Allison JP. 1996. Enhancement of antitumor immunity by CTLA-4 blockade. *Science* **271:** 1734–1736.
Liang JJ, Zhu S, Bruggeman R, Zaino RJ, Evans DB, Fleming JB, Gomez HF, Zander DS, Wang H. 2010. High levels of expression of human stromal cell-derived factor-1 are associated with worse prognosis in patients with stage II pancreatic ductal adenocarcinoma. *Cancer Epidemiol Biomarkers Prev* **19:** 2598–2604.
Mikucki ME, Fisher DT, Matsuzaki J, Skitzki JJ, Gaulin NB, Muhitch JB, Ku AW, Frelinger JG, Odunsi K, Gajewski TF, et al. 2015. Non-redundant requirement for CXCR3 signalling during tumoricidal T-cell trafficking across tumour vascular checkpoints. *Nat Commun* **6:** 7458–7472.
Molyneaux KA, Zinszner H, Kunwar PS, Schaible K, Stebler J, Sunshine MJ, O'Brien W, Raz E, Littman D, Wylie C, et al. 2003. The chemokine SDF1/CXCL12 and its receptor CXCR4 regulate mouse germ cell migration and survival. *Development* **130:** 4279–4286.
Motzer RJ, Escudier B, McDermott DF, George S, Hammers HJ, Srinivas S, Tykodi SS, Sosman JA, Procopio G, Plimack ER, et al. 2015. Nivolumab versus everolimus in advanced renal-cell carcinoma. *N Engl J Med* **373:** 1803–1813.
Naito Y, Saito K, Shiiba K, Ohuchi A, Saigenji K, Nagura H, Ohtani H. 1998. $CD8^+$ T cells infiltrated within cancer cell nests as a prognostic factor in human colorectal cancer. *Cancer Res* **58:** 3491–3494.
Nancy P, Tagliani E, Tay CS, Asp P, Levy DE, Erlebacher A. 2012. Chemokine gene silencing in decidual stromal cells limits T cell access to the maternal–fetal interface. *Science* **336:** 1317–1321.
Peng D, Kryczek I, Nagarsheth N, Zhao L, Wei S, Wang W, Sun Y, Zhao E, Vatan L, Szeliga W, et al. 2015. Epigenetic silencing of TH1-type chemokines shapes tumour immunity and immunotherapy. *Nature* **527:** 249–253.
Porter DL, Levine BL, Kalos M, Bagg A, June CH. 2011. Chimeric antigen receptor-modified T cells in chronic lymphoid leukemia. *N Engl J Med* **365:** 725–733.
Powles T, Eder JP, Fine GD, Braiteh FS, Loriot Y, Cruz C, Bellmunt J, Burris HA, Petrylak DP, Teng SL, et al. 2014. MPDL3280A (anti-PD-L1) treatment leads to clinical activity in metastatic bladder cancer. *Nature* **515:** 558–562.
Rizvi NA, Hellmann MD, Snyder A, Kvistborg P, Makarov V, Havel JJ, Lee W, Yuan J, Wong P, Ho TS, et al. 2015. Cancer immunology. Mutational landscape determines sensitivity to PD-1 blockade in non-small cell lung cancer. *Science* **348:** 124–128.
Russell B. 1985. Logical atomism. In *The philosophy of logical atomism* (ed. Pears DF), pp. 157–234. Open Court, La Salle.
Salmon H, Franciszkiewicz K, Damotte D, Dieu-Nosjean MC, Validire P, Trautmann A, Mami-Chouaib F, Donnadieu E. 2012. Matrix architecture defines the preferential localization and migration of T cells into the stroma of human lung tumors. *J Clin Invest* **122:** 899–910.
Scotton CJ, Wilson JL, Scott K, Stamp G, Wilbanks GD, Fricker S, Bridger G, Balkwill FR. 2002. Multiple actions of the chemokine CXCL12 on epithelial tumor cells in human ovarian cancer. *Cancer Res* **62:** 5930–5938.
Seibert C, Veldkamp CT, Peterson FC, Chait BT, Volkman BF, Sakmar TP. 2008. Sequential tyrosine sulfation of CXCR4 by tyrosylprotein sulfotransferases. *Biochemistry* **47:** 11251–11262.
Seung E, Cho JL, Sparwasser T, Medoff BD, Luster AD. 2011. Inhibiting CXCR3-dependent $CD8^+$ T cell trafficking enhances tolerance induction in a mouse model of lung rejection. *J Immunol* **186:** 6830–6838.
Sohy D, Parmentier M, Springael JY. 2007. Allosteric transinhibition by specific antagonists in CCR2/CXCR4 heterodimers. *J Biol Chem* **282:** 30062–30069.
Sohy D, Yano H, de Nadai P, Urizar E, Guillabert A, Javitch JA, Parmentier M, Springael JY. 2009. Hetero-oligomerization of CCR2, CCR5, and CXCR4 and the protean effects of "selective" antagonists. *J Biol Chem* **284:** 31270–31279.
Stephens B, Handel TM. 2013. Chemokine receptor oligomerization and allostery. *Prog Mol Biol Transl Sci* **115:** 375–420.
Szabo I, Chen XH, Xin L, Adler MW, Howard OM, Oppenheim JJ, Rogers TJ. 2002. Heterologous desensitization of opioid receptors by chemokines inhibits chemotaxis and enhances the perception of pain. *Proc Natl Acad Sci* **99:** 10276–10281.
Tachibana K, Hirota S, Iizasa H, Yoshida H, Kawabata K, Kataoka Y, Kitamura Y, Matsushima K, Yoshida N, Nishikawa S, et al. 1998. The chemokine receptor CXCR4 is essential for vascularization of the gastrointestinal tract. *Nature* **393:** 591–594.
Takekoshi T, Ziarek JJ, Volkman BF, Hwang ST. 2012. A locked, dimeric CXCL12 variant effectively inhibits pulmonary metastasis of CXCR4-expressing melanoma cells due to enhanced serum stability. *Mol Cancer Ther* **11:** 2516–2525.
Topalian SL, Hodi FS, Brahmer JR, Gettinger SN, Smith DC, McDermott DF, Powderly JD, Carvajal RD, Sosman JA, Atkins MB, et al. 2012. Safety, activity, and immune correlates of anti-PD-1 antibody in cancer. *N Engl J Med* **366:** 2443–2454.
Tran E, Robbins PF, Lu YC, Prickett TD, Gartner JJ, Jia L, Pasetto A, Zheng Z, Ray S, Groh EM, et al. 2016. T-cell transfer therapy targeting mutant KRAS in cancer. *N Engl J Med* **375:** 2255–2262.
Tumeh PC, Harview CL, Yearley JH, Shintaku IP, Taylor EJ, Robert L, Chmielowski B, Spasic M, Henry G, Ciobanu V, et al. 2014. PD-1 blockade induces responses by inhibiting adaptive immune resistance. *Nature* **515:** 568–571.
Veldkamp CT, Peterson FC, Pelzek AJ, Volkman BF. 2005. The monomer-dimer equilibrium of stromal cell-derived factor-1 (CXCL 12) is altered by pH, phosphate, sulfate, and heparin. *Protein Sci* **14:** 1071–1081.
Veldkamp CT, Seibert C, Peterson FC, Sakmar TP, Volkman BF. 2006. Recognition of a CXCR4 sulfotyrosine by the chemokine stromal cell-derived factor-1α (SDF-1α/CXCL12). *J Mol Biol* **359:** 1400–1409.
Veldkamp CT, Seibert C, Peterson FC, De la Cruz NB, Haugner JCIII, Basnet H, Sakmar TP, Volkman BF. 2008. Structural basis of CXCR4 sulfotyrosine recognition by the chemokine SDF-1/CXCL12. *Sci Signal* **1:** ra4.
Watts AO, van Lipzig MM, Jaeger WC, Seeber RM, van Zwam M, Vinet J, van der Lee MM, Siderius M, Zaman GJ, Boddeke HW, et al. 2013. Identification and profiling of CXCR3-CXCR4 chemokine receptor heteromer complexes. *Br J Pharmacol* **168:** 1662–1674.
Wolchok JD, Kluger H, Callahan MK, Postow MA, Rizvi NA, Lesokhin AM, Segal NH, Ariyan CE, Gordon RA, Reed K,

et al. 2013. Nivolumab plus ipilimumab in advanced melanoma. *N Engl J Med* **369:** 122–133.

Zhang WB, Navenot JM, Haribabu B, Tamamura H, Hiramatu K, Omagari A, Pei G, Manfredi JP, Fujii N, Broach JR, et al. 2002. A point mutation that confers constitutive activity to CXCR4 reveals that T140 is an inverse agonist and that AMD3100 and ALX40-4C are weak partial agonists. *J Biol Chem* **277:** 24515–24521.

Zhang L, Conejo-Garcia JR, Katsaros D, Gimotty PA, Massobrio M, Regnani G, Makrigiannakis A, Gray H, Schlienger K, Liebman MN, et al. 2003. Intratumoral T cells, recurrence, and survival in epithelial ovarian cancer. *N Engl J Med* **348:** 203–213.

Zou YR, Kottmann AH, Kuroda M, Taniuchi I, Littman DR. 1996. Function of the chemokine receptor CXCR4 in haematopoiesis and in cerebellar development. *Nature* **393:** 595–599.

Modeling Breast Cancer Intertumor and Intratumor Heterogeneity Using Xenografts

ALEJANDRA BRUNA,[1] OSCAR M. RUEDA,[1] AND CARLOS CALDAS[1,2]
[1]*Department of Oncology and Cancer Research UK Cambridge Institute, Li Ka Shing Centre, University of Cambridge, Cambridge CB2 0RE, United Kingdom*
[2]*Breast Cancer Programme, Cambridge Cancer Centre, Cambridge CB2 2QQ, United Kingdom*
Correspondence: carlos.caldas@cruk.cam.ac.uk

Breast cancer is a heterogeneous disease that can be stratified in at least 10 different subtypes. We present here a platform for derivation of preclinical models based on patient-derived tumor xenografts (PDTXs) that represent these subgroups. These models preserve the transcriptome, methylome, copy-number, and mutational landscape features of the tumor of origin through different passaging. Furthermore, the intratumoral composition of these models is formed by communities of clones very similar to the ones present in the originating tumor. Finally, we show that short-term cultures of cells from these models (PDTX-derived tumor cells, PDTCs) also preserve the molecular features of the tumor and can be used for high-throughput drug testing of single compounds or combinations, with high reproducibility and clinical predictive power.

Breast cancer is a heterogeneous disease. Consequently, precise breast cancer stratification is a crucial step toward more efficacious treatment. For many years patients have been stratified based on clinical–pathological features including staging, tumor grade, and the estrogen receptor (ER), progesterone receptor (PR), and Her2 status. Molecular stratification based on gene expression profiling defined the five intrinsic subtypes of breast cancer: basal-like, Her2, luminal A, luminal B, and normal-like (Perou et al. 2000).

However, these classifications are not optimal. Clinical and histological stratification still fails to assign patients to the right treatment; in particular within the largest subset of patients, which are $ER^+/Her2^-$, where hormone therapy, with Tamoxifen or aromatase inhibitors, frequently fails, and where use of chemotherapy is fraught with poor prediction of benefit. This problem persists even using the gene expression–based intrinsic subtype classification, where luminal A and luminal B are very broad groups that only differ in proliferation status. We therefore hypothesized that a taxonomy of breast cancer based on genomic driver events would be an improvement, which would ultimately lead to better treatment decisions. The METABRIC (Molecular Taxonomy of Breast Cancer International Consortium) project (Curtis et al. 2012) used copy-number and expression profiling of 997 primary breast cancer samples to identify the top 1000 candidate driver genes based on their expression being driven in *cis* by copy-number aberrations. Using these 1000 *cis*-driven genes, enriching for oncogenes and tumor-suppressor genes, we showed that breast cancers could be divided into 10 integrative clusters (IntClusts). This molecular stratification was validated in a further 983 samples from the same cohort (Curtis et al. 2012) and subsequently in several external data sets (Ali et al. 2014). Some of the IntClusts were characterized by a very clear copy-number driver event, like the *CCND1/PAK1* amplification in IntClust 2, the *HER2* amplification in IntClust 5, or the *MYC* amplicon in IntClust9, but in some other groups drivers were more elusive. To gain insight into different molecular events, we also performed targeted sequencing of 173 genes (Ali et al. 2014) that are relevant in breast cancer, finding that 40 of them are drivers (mut-driver genes), with different frequencies in each IntClust. This new layer of molecular data contributed to refining the molecular landscape of breast cancer. Significantly, some of these mutation drivers seem to have IntClust specific prognostic value. For example, *PIK3CA* is associated with worse survival in only three $ER^+/Her2^-$ subgroups. We also quantified the level of intratumoral heterogeneity in individual tumors, revealing its distribution is distinctly different between IntClusts. Importantly we also showed that intratumoral heterogeneity is a biomarker of disease-specific survival.

However, all of these important biological associations that we can infer with a well-annotated and well-characterized cohort like METABRIC need a set of representative preclinical models in order to be validated. With this aim in mind, we developed a classifier to assign existing breast cancer cell lines into one of the IntClusts (Curtis et al. 2012). We observed the fit was poor because the genomic profiles of the cell lines did not reflect well those found in breast cancer samples in the clinic. This is a consequence of the nature of cell lines, which not only fail to capture the intratumor and intertumor heterogeneity seen in the clinic but also have other limitations probably reflecting adaptation to long-term culture in plastic. These drawbacks are thought to have strongly impacted

Published by Cold Spring Harbor Laboratory Press; doi: 10.1101/sqb.2016.81.031112

cancer drug development, which has suffered from high attrition rates, mainly because of lack of efficacy in the clinic (Harrison 2016).

DEVELOPING A COLLECTION OF PATIENT-DERIVED TUMOR XENOGRAFTS

Patient-derived tumor xenografts (PDTXs) have emerged as tools with the potential to improve cancer drug development (Tentler et al. 2012; Hidalgo et al. 2014; Aparicio et al. 2015; Gao et al. 2015). We therefore set out to develop PDTXs as a better model to study drug sensitivity but also to reflect tumor evolution under the selective pressure produced by treatment, enabling the study of mechanisms of sensitivity and resistance (Fig. 1). We have now optimized a framework to efficiently engraft breast cancer samples from the hospital into immunocompromised mice (NSG). With time we have substantially improved this platform to (1) ensure rapid and efficient implantation of new breast cancer samples, (2) expand the successful engrafted samples through serial transplants for biobanking purposes, and (3) provide these models as a research tool. The breast cancer PDTX biobank is now one of the largest in the academic setting and is composed by almost equal number of ER^+ and ER^- samples, most of which are from breast tumor primaries.

We have now robustly shown that the PDTXs represent the breast cancer patient population and preserve most of the features of the originating tumor sample. This has included full characterization at the transcriptome, methylome, copy-number, and mutational landscape levels, demonstrating similarity with the originating human tumor material and its preservation through several passages in the mouse.

We soon realized the downstream analysis of sequencing data was complicated by the nature of PDTX samples, which are unknown mixtures of mouse stromal and human epithelial malignant cells. This prompted the development of a bioinformatics pipeline that allowed for the identification with 99.9% accuracy of mouse and human DNA sequences. Significantly, we have observed the proportion of mouse stromal cells in the PDTXs remains stable through serial passages. These data were confirmed by orthogonal methods such as fluorescence in situ hybridization (FISH) and immunohistochemistry (Bruna et al. 2016).

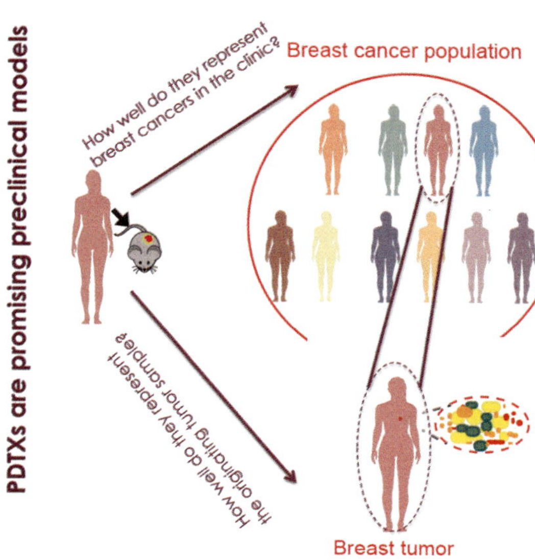

Figure 1. Modeling breast cancer intertumor and intratumor heterogeneity. PDTXs, patient-derived tumor xenografts.

We used exome sequencing to define single-nucleotide variations (SNVs) and small insertions and deletions (indels), shallow whole-genome sequencing to obtain copy-number profiles, Illumina expression microarrays (and RNA-seq) to measure gene expression and infer pathway activation, and reduced representation bisulfite sequencing (RRBS) to characterize DNA methylation.

We classified PDTXs into the IntClusts using the method we previously described (Ali et al. 2014). The copy-number profiles of PDTXs classified into each IntClust are similar to those reported in primary tumors, contrasting to what we observed in cell lines (Curtis et al. 2012; Ali et al. 2014). The subtype distribution of engrafted PDTX samples in the biobank is biased toward ER^- and ER^+ tumors with worse prognosis (Curtis et al. 2012; Bruna et al. 2016), reflecting the favored engraftment of more aggressive breast cancer subtypes. Hence, the ER^- PDTX samples in the biobank have a distribution of most frequently mutated genes similar to that found in ER^- breast cancers from the The Cancer Genome Atlas (TCGA) cohort (Cancer Genome Atlas 2012). In contrast, frequencies of mutated genes in ER^+ PDTX samples mirrored those found in more aggressive ER^+ subtypes (Eirew et al. 2015).

When comparing PDTX to their matched originating sample and to later passages, we observed that overall the molecular profiles are remarkably similar. The biggest drift occurs with engrafting, although it is model-specific, suggesting there is not a common mechanism underlying changes observed when implanting. It is hence more likely that patient- and tumor-specific determinants, in combination with growing in a new environment in the immunodeficient mouse host, affects the observed drifts. Significantly, all the molecular traits examined remained remarkably stable through serial passages in the mouse, even after more than 10 passages (Bruna et al. 2016).

We further investigated the intratumoral composition and architecture of PDTX samples and observed xenografts are communities of clones with varying degrees of complexity similar to those observed in the breast cancer clinical population, and these communities share most of the clones found in the matched originating cancer sample (Bruna et al. 2016; Pereira et al. 2016). Because cancer genomes constantly evolve under selective pressure, we investigated the extent of the human-to-mouse and mouse-to-mouse clonal dynamics using PyClone (Roth et al. 2014; Eirew et al. 2015), a method for estimating the proportion of cells in a tumor carrying each mutation and clustering them into subclones (groups of cells with the same genotype). We observed that most of the clones remained stable after engraftment and throughout passag-

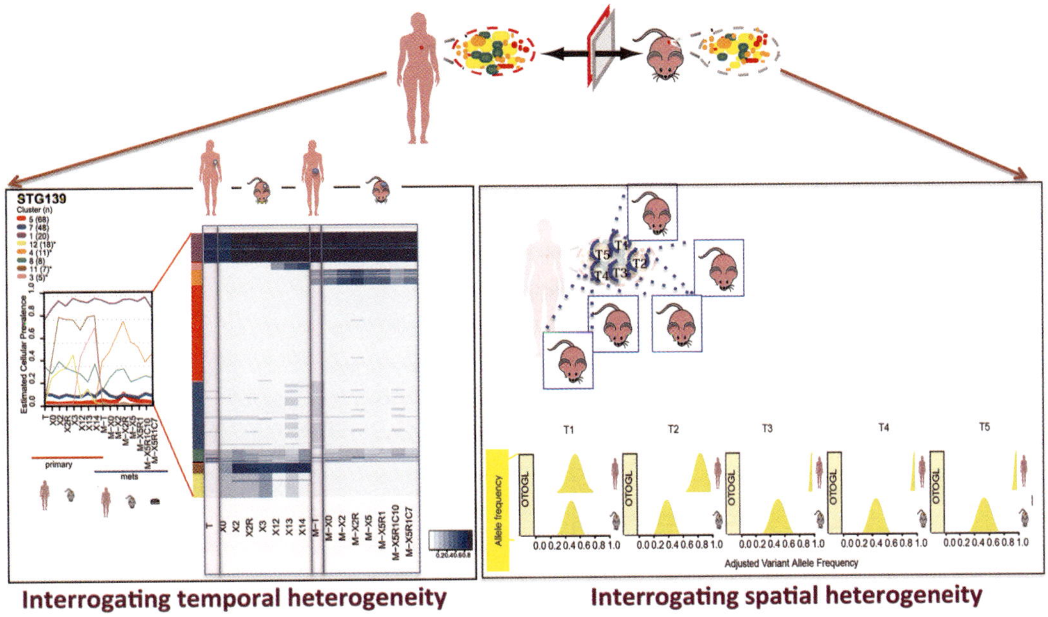

Figure 2. Patient-derived tumor xenografts (PDTXs) retain most of the molecular features of their originating tumor samples, including the intratumor heterogeneity, and remain very stable in the mouse. (Portions adapted from Bruna et al. 2016.)

ing, although in certain instances some subclones increased or decreased their prevalence or even appeared or disappeared. In fact, only 20% of the almost 200 clonal clusters detected changed significantly. Importantly, only four of these contained known breast cancer drivers. We further investigated the temporal and spatial intratumoral heterogeneity in a few examples. By comparing the clonal cluster composition and allele frequency distributions in matched primary, metastatic, and PDTX samples we observed both common and private clones. We further engrafted five spatially separated biopsies from a patient tumor sample and observed minor spatial intratumor heterogeneity. However, clonal dynamics were remarkably similar in independently engrafted mice regardless of their prevalence in the originating human tumor sample. Altogether, these results significantly added to previous data showing clones follow deterministic evolutionary trajectories (Fig. 2; Eirew et al. 2015).

The potential value of PDTXs gained further momentum with a study by Novartis showing xenografts captured the interpatient drug response variability (Gao et al. 2015). This suggested PDTXs would reach their full potential if their use would be possible for high-content studies. However, it is not feasible to do this in mice for ethical and economic reasons. This led us to develop a pipeline that would maintain and expand cancer samples in vivo in mice and then use the xenografts to generate single cells for short-term culture and high-throughput drug screening. These short-term cultures from PDTXs (PDTX-derived tumor cells or PDTCs) retain similar molecular features, including the intratumor heterogeneity, as in the originating PDTX sample (Fig. 3; Bruna et al. 2016). We then did a proof of concept screen, using the Welcome Trust Sanger Institute setup for high-throughput drug testing in cell lines (Garnett et al. 2012), in 22 PDTX models at different passages. We tested 102 compounds relevant to cancer, either as single agents or as drug:drug combinations in a 5 × 5 matrix of standard of care agents cisplatin and paclitaxel with six clinically relevant targeted compounds. We observed the technical and biological reproducibility was similar to that reported in established cell lines and organoids (Garnett et al. 2012; van de Wetering et al. 2015). To further verify the robustness of the data, we tested eight PDTC models using in parallel three end point DNA content- or ATP-dependent assays and observed highly correlated drug responses independently of the assay used. We also observed a very high correlation of responses in all models tested to pairs of compounds targeting the same pathway or sharing a similar mechanism of action. Furthermore, as an example of successful drug:drug combination approaches, we validated synergy between an HSP90 inhibitor and paclitaxel in Her2^{+} models (Bruna et al. 2016). These results support the idea that the PDTX/PDTC platform generates biologically robust data and shows its potential of becoming a resourceful intermediate in drug discovery. To further support the value of the PDTX/PDTC platform as an improved preclinical tool with clinical predictive power, we validated a set of ex vivo drug responses in vivo. Out of the 40 selected ex vivo drug responses, 33 were validated in vivo, even though different compounds with similar specificity had to be used in some cases.

In summary, we have redefined the molecular taxonomy of breast cancer using a genomic-driver-based stratification, and we have now generated a set of xenografts that model the intertumor and intratumor heterogeneity seen in the clinic. We have demonstrated these models can be used as a robust preclinical drug testing platform, and we will use them also for other perturbations to test functional heterogeneity.

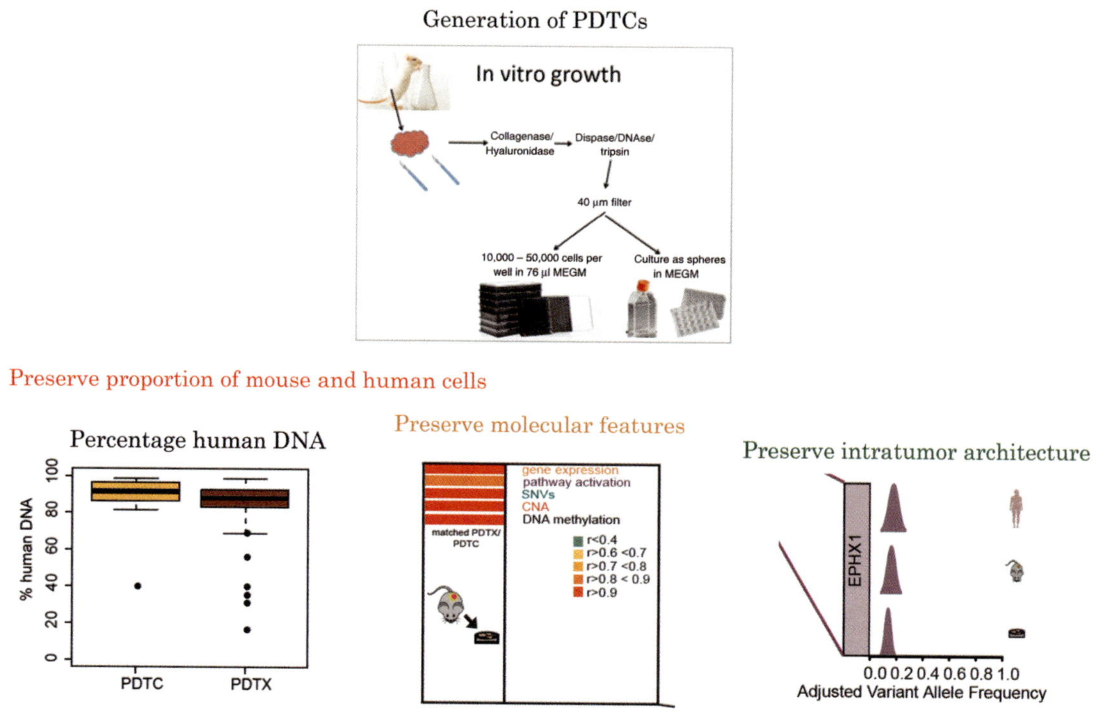

Figure 3. Generation of short-term cultures of PDTX-derived tumor cells (PDTCs) that preserve the proportion of mouse and human cells, the molecular features, and the intratumor heterogeneity of its originating sample. (Portions adapted from Bruna et al. 2016.)

REFERENCES

Ali HR, Rueda OM, Chin SF, Curtis C, Dunning MJ, Aparicio SA, Caldas C. 2014. Genome-driven integrated classification of breast cancer validated in over 7,500 samples. *Genome Biol* **15:** 431.

Aparicio S, Hidalgo M, Kung AL. 2015. Examining the utility of patient-derived xenograft mouse models. *Nat Rev Cancer* **15:** 311–316.

Bruna A, Rueda OM, Greenwood W, Batra AS, Callari M, Batra RN, Pogrebniak K, Sandoval J, Cassidy JW, Tufegdzic-Vidakovic A, Sammut SJ, et al. 2016. A biobank of breast cancer explants with preserved intra-tumor heterogeneity to screen anticancer compounds. *Cell* **167:** 260–274 e222.

Cancer Genome Atlas Network. 2012. Comprehensive molecular portraits of human breast tumours. *Nature* **490:** 61–70.

Curtis C, Shah SP, Chin SF, Turashvili G, Rueda OM, Dunning MJ, Speed D, Lynch AG, Samarajiwa S, Yuan Y, et al. 2012. The genomic and transcriptomic architecture of 2,000 breast tumours reveals novel subgroups. *Nature* **486:** 346–352.

Eirew P, Steif A, Khattra J, Ha G, Yap D, Farahani H, Gelmon K, Chia S, Mar C, Wan A, et al. 2015. Dynamics of genomic clones in breast cancer patient xenografts at single-cell resolution. *Nature* **518:** 422–426.

Gao H, Korn JM, Ferretti S, Monahan JE, Wang Y, Singh M, Zhang C, Schnell C, Yang G, Zhang Y, et al. 2015. High-throughput screening using patient-derived tumor xenografts to predict clinical trial drug response. *Nat Med* **21:** 1318–1325.

Garnett MJ, Edelman EJ, Heidorn SJ, Greenman CD, Dastur A, Lau KW, Greninger P, Thompson IR, Luo X, Soares J, et al. 2012. Systematic identification of genomic markers of drug sensitivity in cancer cells. *Nature* **483:** 570–575.

Harrison RK. 2016. Phase II and phase III failures: 2013–2015. *Nat Rev Drug Discov* **15:** 817–818.

Hidalgo M, Amant F, Biankin AV, Budinská E, Byrne AT, Caldas C, Clarke RB, de Jong S, Jonkers J, Mælandsmo GM, et al. 2014. Patient-derived xenograft models: An emerging platform for translational cancer research. *Cancer Discov* **4:** 998–1013.

Pereira B, Chin SF, Rueda OM, Vollan HK, Provenzano E, Bardwell HA, Pugh M, Jones L, Russell R, Sammut SJ, et al. 2016. The somatic mutation profiles of 2,433 breast cancers refines their genomic and transcriptomic landscapes. *Nat Commun* **7:** 11479.

Perou CM, Sørlie T, Eisen MB, van de Rijn M, Jeffrey SS, Rees CA, Pollack JR, Ross DT, Johnsen H, Akslen LA, et al. 2000. Molecular portraits of human breast tumors. *Nature* **406:** 747–752.

Roth A, Khattra J, Yap D, Wan A, Laks E, Biele J, Ha G, Aparicio S, Bouchard-Côté A, Shah SP. 2014. PyClone: Statistical inference of clonal population structure in cancer. *Nat Methods* **11:** 396–398.

Tentler JJ, Tan AC, Weekes CD, Jimeno A, Leong S, Pitts TM, Arcaroli JJ, Messersmith WA, Eckhardt SG. 2012. Patient-derived tumour xenografts as models for oncology drug development. *Nat Rev Clin Oncol* **9:** 338–350.

van de Wetering M, Francies HE, Francis JM, Bounova G, Iorio F, Pronk A, van Houdt W, van Gorp J, Taylor-Weiner A, Kester L, et al. 2015. Prospective derivation of a living organoid biobank of colorectal cancer patients. *Cell* **161:** 933–945.

Challenges and Opportunities in Modeling Pancreatic Cancer

Michael E. Feigin[1,2] and David A. Tuveson[1,2,3]
[1]*Cold Spring Harbor Laboratory, Cold Spring Harbor, New York 11724*
[2]*Lustgarten Foundation Pancreatic Cancer Research Laboratory, Cold Spring Harbor, New York 11724*
[3]*Rubenstein Center for Pancreatic Cancer Research, Memorial Sloan Kettering Cancer Center, New York, New York 10065*
Correspondence: feiginm@cshl.edu; dtuveson@cshl.edu

The ability to faithfully model complex processes lies at the heart of experimental biology. Although a reductionist approach necessarily reduces this complexity, it is nevertheless required for untangling the contributions and interactions of the various system components. It has long been appreciated that cancer is a complex process that involves positive and negative interactions between tumor cells, normal host tissue, and the associated cells of the tumor microenvironment. However, accurate models for studying these complex interactions in vitro have remained elusive. We seek to generate models of mouse and human pancreatic cancer that are relevant to disease biology and useful for elucidating poorly understood facets of this deadly disease. The ability to model, manipulate, and predict the therapeutic response of an individual's disease outside their body represents the promise of precision medicine. Therefore, these models are patient-specific and allow the identification of new biomarkers and novel treatment modalities for rapid translation to the clinic. In this perspective we will discuss recent advances in modeling pancreatic cancer in vitro, the discoveries these models have enabled, and future challenges and opportunities awaiting further investigation.

Pancreatic ductal adenocarcinoma (PDA) is among the most highly lethal malignancies, with a median survival rate of <8% (Siegel et al. 2016). Several factors account for this poor outcome, including the advanced stage at which most patients are diagnosed and the minimal effectiveness of currently utilized therapeutic options (Ying et al. 2016). Although surgical intervention can be curative, the vast majority of patients (>80%) present at diagnosis with metastases and are considered ineligible for surgery. In these patients, standard of care generally consists of chemotherapy, with minimal long-term benefit. The modest activity of systemic chemotherapy is thought to be a consequence of the highly desmoplastic, poorly vascularized nature of PDA, where up to 90% of the total tumor volume can consist of a dense extracellular matrix, fibroblasts, and cells of the immune system (Moir et al. 2015). Understanding the contribution of the individual components of the tumor microenvironment to tumor progression or control is imperative to developing more effective treatment options. Therefore, in vivo and in vitro models that accurately reflect this complexity are required for defining and manipulating the signals that mediate tumor–stromal interactions.

Human and mouse tumor-derived cell lines grown on tissue culture plastic in two dimensions (2D) have been a mainstay for understanding the genes and pathways critical for cancer cell growth and oncogenic processes such as cell migration. They are easily cultured and can be genetically manipulated, transplanted into mice for tumor growth and metastasis assays, and grown in large quantities for high-throughput chemical and genetic screens. However, several important drawbacks exist with this culture method. Importantly, normal, nontransformed human or mouse cells rarely proliferate under these conditions, hindering the ability to directly compare gene expression or the results of experimental manipulations between normal and tumor. When nontransformed cells can be cultured in a monolayer, this oftentimes results in the loss of cell polarity, a critical property required for epithelial cell function. It is also challenging to coculture distinct cell types in monolayers, hindering the ability to study tumor–stromal cell interactions. Finally, transplantation of 2D-cultured cancer cells into the mouse generally results in tumors that do not recapitulate the complex pathophysiology, microenvironment and therapeutic response of the primary tumor. Therefore, models that overcome these limitations are required.

Genetically engineered mouse models (GEMMs) of PDA have demonstrated the ability to accurately model disease progression, from the early stage pancreatic intraepithelial neoplasm (PanIN) stages through desmoplastic tumors with stroma characteristic of the human disease (Perez-Mancera et al. 2012). In particular, the "KC" mouse, harboring a mutant KRas allele (*KRas*^{G12D}), and the "KPC" mouse, harboring both mutant KRas and mutant p53 (*Trp53*^{R172H}), driven by *Pdx1-Cre* in the pancreas, have provided opportunities for preclinical studies with novel therapeutics, generating a rationale for

Published by Cold Spring Harbor Laboratory Press; doi: 10.1101/sqb.2016.81.031104

clinical trials and changes in PDA clinical intervention (Hingorani et al. 2003, 2005). Although these mouse models are useful for understanding the genetic drivers of PDA and performing therapeutic trials, they are expensive and time-consuming to maintain. Furthermore, the inherent tumor microenvironmental complexity makes it difficult to untangle interactions between the epithelial-derived tumor cells, stromal fibroblasts, and immune components.

To address the deficiencies of both standard 2D cell culture and GEMMs, we developed an in vitro three-dimensional (3D) organoid model, allowing the growth and manipulation of normal, premalignant, and malignant pancreatic tissue from mouse and human (Boj et al. 2015). The organoids can be passaged indefinitely in semisolid matrices, stored as a frozen pellet, and resuscitated in a manner similar to traditional cell culture methods, genetically manipulated, grown in large quantities for therapeutic or genetic screens, and orthotopically transplanted into mice where they recapitulate the phenotype of the tissue from which they were derived. Importantly, the tumors derived from organoid transplants recapitulate many properties of human PDA, including desmoplasia and poor vascularization. This system provides several advantages over standard cell culture, including the ability to grow normal and malignant pancreatic ductal cells under identical conditions, allowing direct comparisons of differential gene expression, cell signaling networks, metabolism, and therapeutic response. We have also developed a coculture model whereby organoids can be grown alongside pancreatic stellate cells (PSCs, the resident fibroblast of the pancreas), allowing detailed investigations into the positive and negative interactions that occur between these cell types during tumorigenesis.

Understanding the complex genetic underpinnings of PDA has been hampered by the low tumor cellularity and difficulty in obtaining samples from patients with advanced disease (Bailey et al. 2016). The organoid model overcomes these obstacles by allowing the selective culture and expansion of the epithelial-derived tumor cells, allowing deeper sequencing without stromal contamination. Furthermore, organoids can be generated from endoscopic ultrasound-guided fine-needle biopsies, including from those patients with inoperable, metastatic disease, allowing for the first time a large-scale analysis of advanced PDA. In this perspective, we will highlight recent discoveries in PDA biology made possible by the organoid model and propose new challenges and opportunities for studying this deadly disease.

ELUCIDATING THE ROLE OF Nrf2 IN PDA PROGRESSION

Nearly all patients with PDA display an activating mutation in the *KRAS* oncogene, a small GTPase with a wide range of downstream effectors (Javle et al. 2016). Genetic ablation of oncogenic *KRas* in mouse models leads to tumor regression, demonstrating the importance of therapeutically targeting this protein in patients with PDA (Ying et al. 2012). However, effective drugs targeting KRas directly have yet to be developed. Therefore, much research has focused on understanding the downstream mechanisms by which oncogenic KRas promotes tumorigenesis. We previously showed that oncogenic KRas increased the expression of the redox regulator transcription factor Nrf2 (nuclear factor erythroid-derived 2-like 2, *Nfe2l2/Nrf2*), lowering intracellular reactive oxygen species (ROS) (DeNicola et al. 2011). Disruption of this pathway inhibited KRas-driven tumorigenesis, demonstrating that Nrf2-mediated ROS detoxification is a critical mediator of oncogenesis. However, as directly targeting transcription factors therapeutically has proved difficult, we endeavored to understand the mechanisms by which Nrf2 promoted tumorigenesis.

Accordingly, we utilized the organoid model system to determine the mechanism by which Nrf2 maintains PDA proliferation (Chio et al. 2016). Knockdown of Nrf2 in human tumor organoids reduced proliferation, whereas knockdown had little effect on normal human organoids. Mouse organoids harboring oncogenic KRas were able to form tumors upon orthotopic transplantation in athymic nude mice; in contrast, engraftment was impaired in mouse organoids lacking Nrf2. In concordance with previous results, activation of oncogenic KRas decreased levels of ROS, which was reversed upon knockdown of Nrf2. We hypothesized that Nrf2-regulated changes in ROS levels may alter patterns of protein cysteine oxidation, thereby modifying protein function and driving cancer cell proliferation. Therefore, we developed a proteomic platform to specifically detect reduced cysteine-containing peptides, independent of protein abundance differences. This method revealed that loss of Nrf2, in the context of oncogenic KRas and mutant p53 ("KP"), modified the oxidized cysteine content of proteins regulating mRNA translation. Organoids derived from KP mice displayed up-regulation of translation when compared with those derived from normal pancreata, and protein translation was attenuated in these organoids following Nrf2 loss. To determine if reversible oxidation of the translation machinery controlled translation activity, we expressed cysteine to aspartic acid oxidation mimics of several translational regulatory proteins and measured protein synthesis. Expression of one such mutant, of the elongation factor eEF2, decreased the rate of nascent protein synthesis as compared with wild-type eEF2, showing that eEF2 is a bona fide redox sensor and supporting the overall premise that Nrf2 protects PDA cells from oxidative stress to promote tumorigenesis.

In addition to the effects of Nrf2 on mediating translation through redox regulation, we observed that Nrf2-deficient cells displayed defects in cap-dependent translation, driven by changes in PI3K (phosphoinositide 3-kinase)/Akt/mTOR (mammalian target of rapamycin) signaling. Cells lacking Nrf2 had decreased EGFR (epidermal growth factor receptor) activation, leading to decreased Akt and Erk phosphorylation, known regulators of cap-dependent translation. The lack of EGFR activity resulted from decreased shedding of EGF by the metalloprotease Adam10. Adam10 was shown to contain a reduced cysteine residue sensitive to changes in Nrf2 levels, and an Adam10 cysteine oxidation mimic reduced nascent protein synthesis in KP organoids. Therefore, we hypothesize that Nrf2 regulates Adam10 activity in a redox-dependent manner, mediating EGF autocrine signaling and cap-dependent translation in PDA.

As Akt is a critical regulator of KRas signaling and tumor proliferation, Akt inhibitors are under clinical investigation in pancreatic cancer (Akinleye et al. 2015) (Clinical Trials Identifier: NCT01783171). However, KRas mutant cells overcome Akt inhibition through reactivation of EGFR, in a process that requires autocrine EGF signaling (Mendoza et al. 2011). Because this autocrine EGF signaling is disrupted in tumor cells lacking Nrf2, we hypothesized that Akt inhibition would synergize with the pro-exidant buthionine sulfoximine (BSO), an inhibitor of intracellular glutathione (GSH) synthesis. KRas mutant organoids treated with both inhibitors failed to reactivate EGFR, and the combination was synergistic in reducing proliferation in organoids and human PDA cells. Finally, the combination of Akt inhibiton and BSO treatment was more effective than either drug alone in reducing tumor growth in KPC mice. Therefore, we propose that modulating redox regulation in combination with Akt inhibition should be considered for clinical translation (Chio et al. 2016).

UNTANGLING THE COMPLEXITY OF THE TUMOR MICROENVIRONMENT

The complex role of the tumor microenvironment as a regulator of PDA pathogenesis is an area of intense investigation. Cancer-associated fibroblasts (CAFs) are responsible for secretion of extracellular matrix (ECM) and production of tumor-promoting growth factors, resulting in tumor cell growth, decreased drug delivery, and drug resistance (Bachem et al. 2005; Hwang et al. 2008; Straussman et al. 2012; Jacobetz et al. 2013). Therefore, disruption of the ECM, either through enzymatic digestion or targeting CAFs directly, has been carried out in both mouse models and human clinical trials (Froeling et al. 2011; Olive et al. 2009; Hingorani et al. 2016). However, depletion of smooth muscle actin (αSMA)-expressing CAFs (Ozdemir et al. 2014) and long-term pharmacological inhibition of the CAFs that respond to sonic hedgehog ligand (Lee et al. 2014; Rhim et al. 2014) reduced survival in mice and produced highly undifferentiated PDA tumors. These divergent results may reflect the presence of CAF heterogeneity, as evidenced by the expression of differing patterns of fibroblast markers in vivo (Öhlund et al. 2014). Therefore, identifying these different CAF subpopulations, and probing their effects on tumorigenesis, is required for developing more specific CAF-targeted therapeutics.

To address the role of CAFs in PDA, we developed a coculture model allowing direct interactions between organoids and pancreatic stellate cells (PSCs), the resident fibroblast of the pancreas, which can be activated to form CAFs. Upon coculture in Matrigel with tumor-derived organoids, the PSCs acquired an activated phenotype, including cellular extensions in close contact with organoids, and deposited a collagen I–containing ECM organized into collagen fibrils. Coculture enhanced the proliferative rate of both PSCs and organoids, as well as allowed indefinite culture of organoids in growth factor–reduced media. Therefore, this model accurately reflects the desmoplastic stroma of PDA and provides opportunities to study the reciprocal interactions driving cell proliferation of each cell type. Furthermore, CAF heterogeneity was observed in the coculture model, as PSCs expressing high levels of αSMA were found exclusively in contact with organoids (termed myCAFs for myofibroblastic CAFs), whereas PSCs not in direct contact with organoids were αSMA^{low} and secreted high levels of the inflammatory cytokine interleukin 6 (IL6) (termed iCAFs for inflammatory CAFs). These observations were extended to mouse and human PDA, where SMA expression was restricted to fibroblasts in direct proximity to neoplastic cells, whereas fibroblasts expressing IL6 were found far from the neoplastic cells and surrounded by stroma. Although contact between organoids and PSCs was required for myCAF formation, spatial separation of organoids and PCSs through transwell culture increased iCAF formation exclusively. Although these two PSC populations are mutually exclusive in terms of gene expression profiles and secretory activity, they can switch phenotype in response to organoid contact or secreted factors (Öhlund et al. 2017). Therefore, the PDA stroma is more complex than previously appreciated and may inform our understanding of previous experiments disrupting CAF activity and the resultant effects on tumor progression. Importantly, this organoid-PSC coculture model should continue to provide fundamental insights into the molecular and cellular details of the PDA tumor microenvironment and guide us toward new and more fruitful therapeutic avenues for stromal targeting.

CHALLENGES AND OPPORTUNITIES IN PDA RESEARCH

The development of accurate models presents exciting opportunities for making impactful discoveries. The pancreatic organoids have provided insight into the role of Nrf2 in PDA progression and the complexity of the tumor microenvironment, generating hypotheses for new therapeutic interventions. We have adapted the human organoid model as a screening platform for therapeutic discovery, allowing the unbiased identification of drugs or drug combinations that are efficacious in PDA. We have also utilized both mouse and human organoids to identify and validate new tumor-specific biomarkers. Finally, we envision using organoids to study poorly understood facets of cancer cell biology, including the role of cell polarity, noncoding mutations, and perineural invasion.

Many patients diagnosed with PDA are ineligible for surgery because of the presence of advanced metastatic disease. As surgical samples provide the raw material for large-scale sequencing efforts, interrogating the genetics of late-stage, metastatic PDA has been technically challenging. Furthermore, the low cellularity of PDA requires deep sequencing to ensure detection of even the most common mutations. The organoid model overcomes these challenges in several ways. First, we are able to successfully culture organoids with high reliability (>80%) from endoscopic ultrasound-guided fine-needle biopsies, a diagnostic procedure performed on patients with advanced disease. By growing these samples as organoids, we are able to selectively culture the epithelial component, allowing high-resolution DNA and RNA sequencing at lower coverage. In parallel, we have developed a high-throughput, 384-well-based therapeutics platform to determine the sensitivity of an individual patient's organoids to standard of care chemotherapy, FDA-approved investigational agents, or drugs targeting proteins of interest. The primary goal of such an approach is to identify the most active drug or combination of drugs that can be applied directly to the patient. Combining the genotype, gene expression, and therapeutic sensitivity information, we hope to develop algorithms for rapidly choosing the optimal course of action for future patients. This approach to precision medicine necessitates the generation and characterization of a large-scale, living biobank of PDA samples, an undertaking currently in progress.

Among the reasons that PDA is often diagnosed at a late stage is the absence of specific biomarkers of disease. The existing PDA biomarker, the carbohydrate antigen CA-19-9, cannot be used as a screening tool because of high false-positive and false-negative rates (Swords et al. 2016). Furthermore, CA-19-9 can be elevated in the absence of cancer, including in patients with pancreatitis. Current approaches to identifying novel biomarkers are hampered by their inability to compare cell surface markers or secreted factors from neoplastic cells versus those from the normal ductal epithelium. The organoid system overcomes this "lack of the best control" complication by allowing direct comparison between panels of normal or tumor organoids from individual patients. Accordingly, we have used mass spectrometric approaches to identify differentially modified protein carriers of CA-19-9 that may improve the specificity of this biomarker, and our preliminary results are encouraging. The ability to culture and manipulate normal human pancreatic ductal cells has been critical to these endeavors.

Finally, another hallmark of PDA is perineural invasion (PNI), the process by which tumor cells migrate along the surface of nerves and invade the neuronal sheath (Amit et al. 2016). The presence of PNI correlates with poor prognosis and is associated with extreme pain. Intriguingly, genes regulating axon guidance are frequently mutated in PDA, providing a possible cell-intrinsic mechanism for this process (Biankin et al. 2012). It has been well-established that tumor cells can promote neural sprouting and Schwann cell recruitment (Demir et al. 2016), and that neuron and Schwann cell–derived signals can regulate tumor cell invasion (Liang et al. 2016). However, models faithfully recapitulating neuronal–epithelial interactions are required for understanding the mechanisms driving PNI and for screening compounds that can inhibit this process. Existing models for neural invasion rely on cancer cells grown in 2D or cell suspension or utilize nonphysiological neural tissue. Therefore, we propose to develop a coculture model containing tumor organoids and nerves from the celiac ganglia that innervate the pancreas. Such a model would allow interrogation of the signals regulating PNI in PDA and directly test the hypothesis that the axon guidance pathway mediates this process.

CONCLUSION

PDA is a highly lethal malignancy, with numerous outstanding challenges standing in the way of improved outcomes for patients. First, more specific biomarkers are needed for detecting disease at an early stage where surgical intervention is possible. Second, a deeper understanding of the complexities in the PDA tumor microenvironment is required if we hope to target only those interactions driving tumor progression. Third, we must overcome the difficulties in drug delivery, stemming from an abnormal vasculature, as well as determine optimal drug combinations and dosing strategies. Fourth, we need to develop a means to effectively target the oncogene KRas, either directly, or indirectly, or through disruption of downstream signaling pathways, or synthetic lethal interactions. Finally, it is imperative to identify additional molecular events that promote PDA progression and metastasis, as these logically serve as ideal therapeutic targets. We have sought to address these challenges through the development and utilization of in vivo and in vitro models that reflect the complex genetic and pathophysiological characteristics of human PDA. These models present new opportunities for a deep molecular understanding of PDA and rapid translation of findings into the clinic for patient benefit.

ACKNOWLEDGMENTS

D.A.T. is a distinguished scholar of the Lustgarten Foundation and Director of the Lustgarten Foundation–designated laboratory of Pancreatic Cancer Research. D.A.T. is also supported by the Cold Spring Harbor Laboratory Association, the Carcinoid Foundation, and the David Rubinstein Center for Pancreatic Cancer Research at Memorial Sloan Kettering Cancer Center (MSKCC). In addition, we are grateful for support from the Starr Foundation (I7-A718 for D.A.T.), the Department of Defense (DOD) (W81XWH-13-PRCRP-IA for D.A.T.), the Louis Morin Charitable Trust (M.E.F.), and the National Institutes of Health (NIH) (5P30CA45508-26, 5P50CA 101955-07, 1U10CA180944-01, 5U01CA168409-3, 1R01CA190092-03, and 1R01CA188134 for D.A.T.).

REFERENCES

Akinleye A, Iragavarapu C, Furqan M, Cang S, Liu D. 2015. Novel agents for advanced pancreatic cancer. *Oncotarget* **6:** 39521–39537.

Amit M, Na'ara S, Gil Z. 2016. Mechanisms of cancer dissemination along nerves. *Nat Rev Cancer* **16:** 399–408.

Bachem MG, Schunemann M, Ramadani M, Siech M, Beger H, Buck A, Zhou S, Schmid-Kotsas A, Adler G. 2005. Pancreatic carcinoma cells induce fibrosis by stimulating proliferation and matrix synthesis of stellate cells. *Gastroenterology* **128:** 907–921.

Bailey P, Chang DK, Nones K, Johns AL, Patch AM, Gingras MC, Miller DK, Christ AN, Bruxner TJ, Quinn MC, et al. 2016. Genomic analyses identify molecular subtypes of pancreatic cancer. *Nature* **531:** 47–52.

Biankin AV, Waddell N, Kassahn KS, Gingras MC, Muthuswamy LB, Johns AL, Miller DK, Wilson PJ, Patch AM, Wu J, et al. 2012. Pancreatic cancer genomes reveal aberrations in axon guidance pathway genes. *Nature* **491:** 399–405.

Boj SF, Hwang CI, Baker LA, Chio II, Engle DD, Corbo V, Jager M, Ponz-Sarvise M, Tiriac H, Spector MS, et al. 2015. Organoid models of human and mouse ductal pancreatic cancer. *Cell* **160:** 324–338.

Chio II, Jafarnejad SM, Ponz-Sarvise M, Park Y, Rivera K, Palm W, Wilson J, Sangar V, Hao Y, Ohlund D, et al. 2016. NRF2 promotes tumor maintenance by modulating mRNA translation in pancreatic cancer. *Cell* **166:** 963–976.

Demir IE, Kujundzic K, Pfitzinger PL, Saricaoglu ÖC, Teller S, Kehl T, Reyes CM, Ertl LS, Miao Z, Schall TJ, et al. 2016. Early pancreatic cancer lesions suppress pain through CXCL12-mediated chemoattraction of Schwann cells. *Proc Natl Acad Sci* **114:** E85–E94.

DeNicola GM, Karreth FA, Humpton TJ, Gopinathan A, Wei C, Frese K, Mangal D, Yu KH, Yeo CJ, Calhoun ES, et al. 2011. Oncogene-induced Nrf2 transcription promotes ROS detoxification and tumorigenesis. *Nature* **475:** 106–109.

Froeling FE, Feig C, Chelala C, Dobson R, Mein CE, Tuveson DA, Clevers H, Hart IR, Kocher HM. 2011. Retinoic acid-induced pancreatic stellate cell quiescence reduces paracrine Wnt-β-catenin signaling to slow tumor progression. *Gastroenterology* **141:** 1486–1497, 1497 e1481–1414.

Hingorani SR, Petricoin EF, Maitra A, Rajapakse V, King C, Jacobetz MA, Ross S, Conrads TP, Veenstra TD, Hitt BA, et al. 2003. Preinvasive and invasive ductal pancreatic cancer and its early detection in the mouse. *Cancer Cell* **4:** 437–450.

Hingorani SR, Wang L, Multani AS, Combs C, Deramaudt TB, Hruban RH, Rustgi AK, Chang S, Tuveson DA. 2005. Trp53R172H and KrasG12D cooperate to promote chromosomal instability and widely metastatic pancreatic ductal adenocarcinoma in mice. *Cancer Cell* **7:** 469–483.

Hingorani SR, Harris WP, Beck JT, Berdov BA, Wagner SA, Pshevlotsky EM, Tjulandin SA, Gladkov OA, Holcombe RF, Korn R, et al. 2016. Phase Ib study of PEGylated recombinant human hyaluronidase and gemcitabine in patients with advanced pancreatic cancer. *Clin Cancer Res* **22:** 2848–2854.

Hwang RF, Moore T, Arumugam T, Ramachandran V, Amos KD, Rivera A, Ji B, Evans DB, Logsdon CD. 2008. Cancer-associated stromal fibroblasts promote pancreatic tumor progression. *Cancer Res* **68:** 918–926.

Jacobetz MA, Chan DS, Neesse A, Bapiro TE, Cook N, Frese KK, Feig C, Nakagawa T, Caldwell ME, Zecchini HI, et al. 2013. Hyaluronan impairs vascular function and drug delivery in a mouse model of pancreatic cancer. *Gut* **62:** 112–120.

Javle M, Golan T, Maitra A. 2016. Changing the course of pancreatic cancer—Focus on recent translational advances. *Cancer Treat Rev* **44:** 17–25.

Lee JJ, Perera RM, Wang H, Wu DC, Liu XS, Han S, Fitamant J, Jones PD, Ghanta KS, Kawano S, et al. 2014. Stromal response to Hedgehog signaling restrains pancreatic cancer progression. *Proc Natl Acad Sci* **111:** E3091–E3100.

Liang D, Shi S, Xu J, Zhang B, Qin Y, Ji S, Xu W, Liu J, Liu L, Liu C, et al. 2016. New insights into perineural invasion of pancreatic cancer: More than pain. *Biochim Biophys Acta* **1865:** 111–122.

Mendoza MC, Er EE, Blenis J. 2011. The Ras-ERK and PI3K-mTOR pathways: Cross-talk and compensation. *Trends Biochem Sci* **36:** 320–328.

Moir JA, Mann J, White SA. 2015. The role of pancreatic stellate cells in pancreatic cancer. *Surg Oncol* **24:** 232–238.

Öhlund D, Elyada E, Tuveson D. 2014. Fibroblast heterogeneity in the cancer wound. *J Exp Med* **211:** 1503–1523.

Öhlund D, Handly-Santana A, Biffi G, Elyada E, Almeida AS, Ponz-Sarvise M, Corbo V, Oni TE, Hearn SA, Lee EJ, et al. 2017. Distinct populations of inflammatory fibroblasts and myofibroblasts in pancreatic cancer. *J Exp Med* doi: 10.1084/jem.20162024.

Olive KP, Jacobetz MA, Davidson CJ, Gopinathan A, McIntyre D, Honess D, Madhu B, Goldgraben MA, Caldwell ME, Allard D, et al. 2009. Inhibition of Hedgehog signaling enhances delivery of chemotherapy in a mouse model of pancreatic cancer. *Science* **324:** 1457–1461.

Ozdemir BC, Pentcheva-Hoang T, Carstens JL, Zheng X, Wu CC, Simpson TR, Laklai H, Sugimoto H, Kahlert C, Novitskiy SV, et al. 2014. Depletion of carcinoma-associated fibroblasts and fibrosis induces immunosuppression and accelerates pancreas cancer with reduced survival. *Cancer Cell* **25:** 719–734.

Perez-Mancera PA, Guerra C, Barbacid M, Tuveson DA. 2012. What we have learned about pancreatic cancer from mouse models. *Gastroenterology* **142:** 1079–1092.

Rhim AD, Oberstein PE, Thomas DH, Mirek ET, Palermo CF, Sastra SA, Dekleva EN, Saunders T, Becerra CP, Tattersall IW, et al. 2014. Stromal elements act to restrain, rather than support, pancreatic ductal adenocarcinoma. *Cancer Cell* **25:** 735–747.

Siegel RL, Miller KD, Jemal A. 2016. Cancer statistics, 2016. *CA Cancer J Clin* **66:** 7–30.

Straussman R, Morikawa T, Shee K, Barzily-Rokni M, Qian ZR, Du J, Davis A, Mongare MM, Gould J, Frederick DT, et al. 2012. Tumour micro-environment elicits innate resistance to RAF inhibitors through HGF secretion. *Nature* **487:** 500–504.

Swords DS, Firpo MA, Scaife CL, Mulvihill SJ. 2016. Biomarkers in pancreatic adenocarcinoma: Current perspectives. *Onco Targets Ther* **9:** 7459–7467.

Ying H, Kimmelman AC, Lyssiotis CA, Hua S, Chu GC, Fletcher-Sananikone E, Locasale JW, Son J, Zhang H, Coloff JL, et al. 2012. Oncogenic Kras maintains pancreatic tumors through regulation of anabolic glucose metabolism. *Cell* **149:** 656–670.

Ying H, Dey P, Yao W, Kimmelman AC, Draetta GF, Maitra A, DePinho RA. 2016. Genetics and biology of pancreatic ductal adenocarcinoma. *Genes Dev* **30:** 355–385.

Functional Genomic Characterization of Cancer Genomes

Thomas P. Howard,[1,2] Francisca Vazquez,[1,2] Aviad Tsherniak,[2] Andrew L. Hong,[1,2,3] Mik Rinne,[1,2] Andrew J. Aguirre,[1,2] Jesse S. Boehm,[2] and William C. Hahn[1,2]

[1]*Department of Medical Oncology, Dana-Farber Cancer Institute, Harvard Medical School, Boston, Massachusetts 02215*

[2]*Broad Institute of Harvard and Massachusetts Institute of Technology, Cambridge, Massachusetts 02142*

[3]*Boston Children's Hospital, Boston, Massachusetts 02115*

Correspondence: William_Hahn@dfci.harvard.edu

International efforts to sequence cancer genomes now provide an overview of the major genetic alterations that occur in most human cancers. These studies have identified many highly recurrent alterations in specific cancer subtypes but have also identified mutations that occur at lower frequency and unstudied variants of known cancer-associated genes. To elucidate the function of such cancer alleles, we have developed several approaches to systematically interrogate genomic changes found in human tumors. In general, we have taken two complementary approaches. In the first approach, we focus on perturbing genes identified as mutated, amplified, or deleted by cancer genome annotation efforts, whereas in the second, we have taken an unbiased approach to identify genes that are essential for cancer cell proliferation or survival in cell lines that are extensively annotated to identify context-specific essential genes. These studies begin to allow us to define a cancer dependencies map.

Major advances in sequencing technologies over the last decade have led to an exponential increase in our ability to analyze and describe cancer genomes (Stratton et al. 2009; Garraway and Lander 2013; Vogelstein et al. 2013). The massive throughput of these technologies has facilitated international studies to identify genetic alterations and define gene programs in specific types of human cancers. Indeed, we now have a draft view of the major alterations that occur in most human cancers.

However, these studies have also made it clear that most adult human cancers harbor hundreds if not thousands of genetic alterations, probably as a consequence of past or ongoing genomic instability (Hanahan and Weinberg 2011; Vogelstein et al. 2013). In addition, efforts to profile human cancers as part of the clinical evaluation of tumors have led to the identification of a large number of cancer-associated alleles that have never been studied. Two major questions need to be answered for every genomic alteration found in cancer: (i) Does the gene initiate cancer or is it required for tumor maintenance? (ii) Does the cancer-specific variation yield a therapeutic opportunity? Yet, validating each and every observed allele, expression change, and copy number alteration is a dauntingly time-consuming task if studied one event at a time.

To complement the knowledge gained in these genome characterization studies, we and others have developed systematic methods to manipulate gene expression in human cells with the goal of elucidating the function of these cancer alleles (Fig. 1). Gain-of-function screens for increases in cell transformation, proliferation, and viability take advantage of expression libraries composed of cDNA or open reading frames (ORFs) (Gerhard et al. 2004; Rual et al. 2004; Yang et al. 2011; ORFeome Collaboration 2016). Identifying genes that, when overexpressed or mutated, drive an oncogenic phenotype allows for the prioritization of functional candidates from the many genes identified through genomic technologies.

Conversely, loss-of-function screens allow for the interrogation of candidate tumor suppressor genes. Depletion by RNA interference (RNAi) or inactivation with the clustered regularly interspaced short palindromic repeats (CRISPR)–Cas9 system mimics loss-of-function mutations observed in cancer. Unbiased screens using these technologies have discovered novel tumor suppressors both in vitro and in vivo (Berns et al. 2004; Westbrook et al. 2005; Zender et al. 2008; Bric et al. 2009; Cheung et al. 2011a; Hsu et al. 2014; Chen et al. 2015; Wang et al. 2015). In light of the high rates of passenger mutations, these loss-of-function approaches provide additional information that permits the identification of bona fide tumor suppressor genes.

Loss-of-function technologies also allow for the assessment of essential genes on which cancer cells harboring particular genetic alterations are dependent. An understanding of cancer dependencies, defined here as genes that when depleted by RNAi, inactivated by CRISPR–Cas9, or inhibited by a small molecule lead to decreased proliferation or viability of cancer cells, will help inform therapeutic decisions and future drug development.

Published by Cold Spring Harbor Laboratory Press; doi: 10.1101/sqb.2016.81.031070

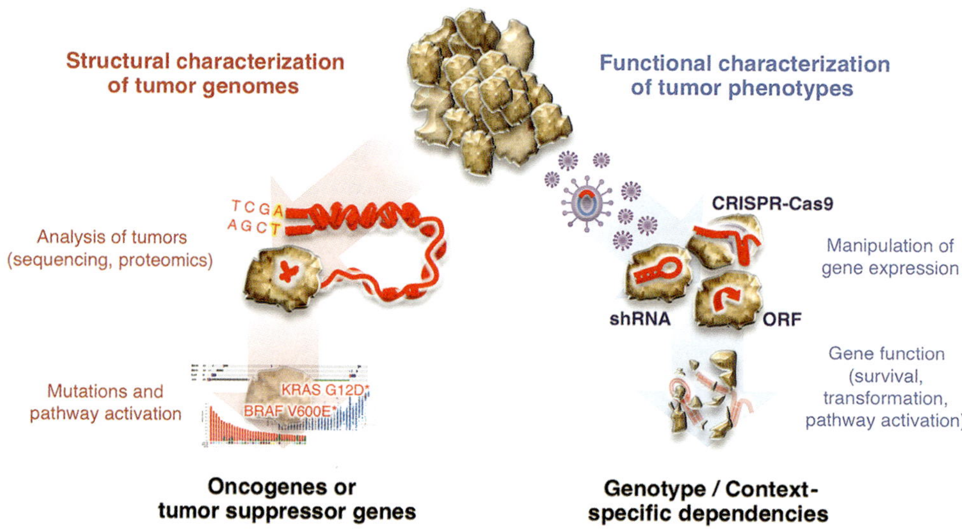

Figure 1. Convergence of structural and functional characterizations provides actionable information on tumor genomes. Genomic feature plots were generated from cBioPortal (Cerami et al. 2012; Gao et al. 2013). shRNA, short-hairpin RNA; ORF, open reading frame.

Here, we review recent advances in the use of these technologies to decipher cancer gene function. We discuss genomic tools to identify functional alterations as well as technologies used to define cancer dependencies.

SCREENS FOR FUNCTIONAL CANCER ALTERATIONS

Interrogation of Copy Number Alterations

The two major types of expression libraries are composed of cDNA or ORF constructs (Boehm and Hahn 2011). cDNA libraries are generated from the reverse transcription of mRNA from a particular cell type or tissue. Thus, cDNA libraries are biased toward genes that are highly expressed in the template tissue and remain incomplete because of unexpressed genes and incomplete reverse transcription of large transcripts. In contrast, ORF expression constructs are produced by individually cloning polymerase chain reaction (PCR) products of cDNA to remove untranslated region (UTR) sequences. Thousands of clones are then pooled to obtain a library with equal representation across the genome. Even when ORF libraries are fully sequenced, these collections contain alleles that vary from database sequences, and the largest libraries, exceeding 17,000 genes, currently do not include all transcript variants and isoforms for many genes (Yang et al. 2011; ORFeome Collaboration 2016).

In the context of cancer, ORF overexpression can mimic copy number amplifications that increase the prevalence of driving oncogenes. Screens for genes that, when overexpressed, drive an increase in proliferation or transformation potential permits the identification of functional copy number changes from the large number of passenger alterations caused by genomic instability (Table 1). For example, Dunn et al. introduced a selected library of ORFs corresponding to genes frequently amplified in ovarian cancer, identifying *GAB2* as a novel ovarian cancer oncogene (Dunn et al. 2014). Injection of liver progenitor cells overexpressing a number of candidates into mice identified two coamplified genes, *BIRC2 (cIAP1)* and *YAP1*, as capable of transforming cells (Zender et al. 2006). Given that large amplifications often include multiple genes, these studies can hone in on the specific genes responsible for oncogenesis or tumor maintenance, such as *TLOC1* and *SKIL* in 3q26 (Hagerstrand et al. 2013), *KCNK9* in 8q24.3 (Mu et al. 2003), and *PPM1D* in 17q23 amplified cancers (Li et al. 2002).

To complement ORF expression, one can also perform gene suppression studies using RNAi (discussed in detail below) to identify amplified genes that are essential for the survival of cells that harbor such amplifications (Table 1). Amplified genes required for survival are more likely to be drivers in those cells than nonessential amplified genes. RNAi screens in large numbers of cell lines identified *PAX8* as oncogenic in many ovarian cancers (Cheung et al. 2011a), whereas in vivo RNAi delivery confirmed *ID4* as a novel ovarian oncogene (Ren et al. 2012).

Independent validation by ORF expression and RNAi methods establish oncogene candidates as capable of both transforming normal cells and maintaining cancer cell survival. Targets confirmed by both experiments are high-confidence oncogenes and include *CRKL* in non–small cell lung cancer (Cheung et al. 2011b) and *FRS2* in ovarian cancer (Luo et al. 2015).

Characterizing Variant Alleles

cDNA and ORF constructs can also express variant alleles of genes to assess their oncogenic potential (Table 1). For example, introduction of oncogenic *PIK3CA* into noncancerous cells drives an increase in proliferation and cancer-related phenotypes (Isakoff et al. 2005).

Table 1. Functional assessment of cancer genomic alterations

Genomic observation	Functional genomics tool(s)	Model	Phenotype(s)	Conclusion
Copy number amplification; mRNA overexpression	ORF library of wild-type alleles	Noncancerous, immortalized cells	↑ proliferation; ↑ transformation potential	Increased gene function drives cancer
	RNAi, CRISPR–Cas9 targeting amplified genes	Cancer cell lines	↓ proliferation; ↓ viability	Increased gene function required for survival
Hotspot point mutations; variants of unknown significance in validated oncogenes	ORF library of variant alleles	Noncancerous, immortalized cells	↑ proliferation; ↑ transformation potential	Oncogenic driver mutations
Inactivating mutations; copy number loss	RNAi, CRISPR–Cas9 targeting wild-type alleles of mutated genes	Noncancerous, immortalized cells	↑ proliferation; ↑ transformation potential	Loss of function drives cancer
Mutation, gene expression, and copy number profiles	RNAi, CRISPR–Cas9 targeting entire genome	Cancer cell lines	↓ proliferation; ↓ viability	Genomic alteration leads to dependencies on specific genes

mRNA, microRNA; ORF, open reading frame; RNAi, RNA interference.

Although experiments with these tools have validated many of the most common oncogenic mutations, genome sequencing has revealed that a majority of variants found in cancer are present at low frequencies in the population. This finding makes it difficult to gauge functionality of variants even in known oncogenes and tumor suppressors (Garraway and Lander 2013; Vogelstein et al. 2013).

Recently, Kim et al. (2016) reported a method to assess variants of unknown significance (VUSs) through introduction of a pooled variant allele ORF library into immortalized cells and testing for transformation potential in mice (Fig. 2). This proof-of-concept experiment reported a dozen previously unknown transforming alleles, including some that were present at very low frequency, demonstrating that even rare alleles may contribute to cell transformation (Kim et al. 2016). Complementing the results of transformation potential, gene expression signatures were derived for each variant allele to firmly identify VUSs that resemble profiles of known oncogenic alleles of the same genes (Fig. 2; Kim et al. 2016). Expansion of this technology would inform both basic research and clinical decision-making as genomic characterization of patients becomes more prevalent.

LARGE-SCALE CANCER DEPENDENCY SCREENS

RNAi

RNAi can be mediated by two types of reagents: small interfering RNAs (siRNAs), which are transiently introduced into cells, and short-hairpin RNAs (shRNAs), which are stably expressed following genomic integration via viral vectors (Rao et al. 2009). In both cases, binding of si/shRNA to complementary messenger RNA (mRNA) transcripts of interest leads to a decrease in mRNA levels and their protein products. This depletion, suppression, or knockdown of the target gene then allows for functional assessment of a phenotype.

Genome-scale RNAi experiments can be performed in an arrayed or pooled format (Echeverri and Perrimon 2006). In an arrayed experiment, a small number of cells are individually exposed to each reagent, usually in a 96- or 384-well plate format. siRNA work best in arrays, as they induce potent depletion over 2–4 d, allowing for the assessment of short-term phenotypes (Rao et al. 2009). In a pooled experiment, a population of cells is exposed to a library of reagents targeting many genes in a massively parallel manner. shRNA are preferable for pooled experiments, as the DNA encoding each shRNA is incorporated into the infected cells' genomes, serving as a barcode for later detection of relative changes in representation (Echeverri and Perrimon 2006). If an shRNA decreases proliferation of cells, it will be relatively depleted in the final pool after a growth period, whereas an shRNA that increases proliferation will be relatively enriched in the final pool (Cowley et al. 2014). Additionally, shRNA expression is permanent, allowing for observation of long-term phenotypes (Rao et al. 2009).

RNAi has been used to determine whether loss of a particular gene can drive proliferation and transformation, allowing for the assessment of the function of inactivating mutations observed in patients (Table 1; Berns et al. 2004; Westbrook et al. 2005; Bric et al. 2009). In addition, RNAi can be used to assess dependencies on a genome scale (Luo et al. 2009). Project Achilles is an effort to screen hundreds of cancer cell lines with pooled, genome-wide shRNA libraries to determine which genes are required for the growth and viability of many cancer cell lines (Cheung et al. 2011a; Cowley et al. 2014). The large number of cell lines is designed to cover a wide spectrum of cell lineages and genomic alterations. All cell lines screened through Project Achilles are also profiled as part of the Cancer Cell Line Encyclopedia (CCLE) (Barretina et al. 2012), which facilitates the integration of genomic and transcriptomic information with dependency findings.

Although genome-wide RNAi experiments have identified several gene dependencies (Barbie et al. 2009; Luo

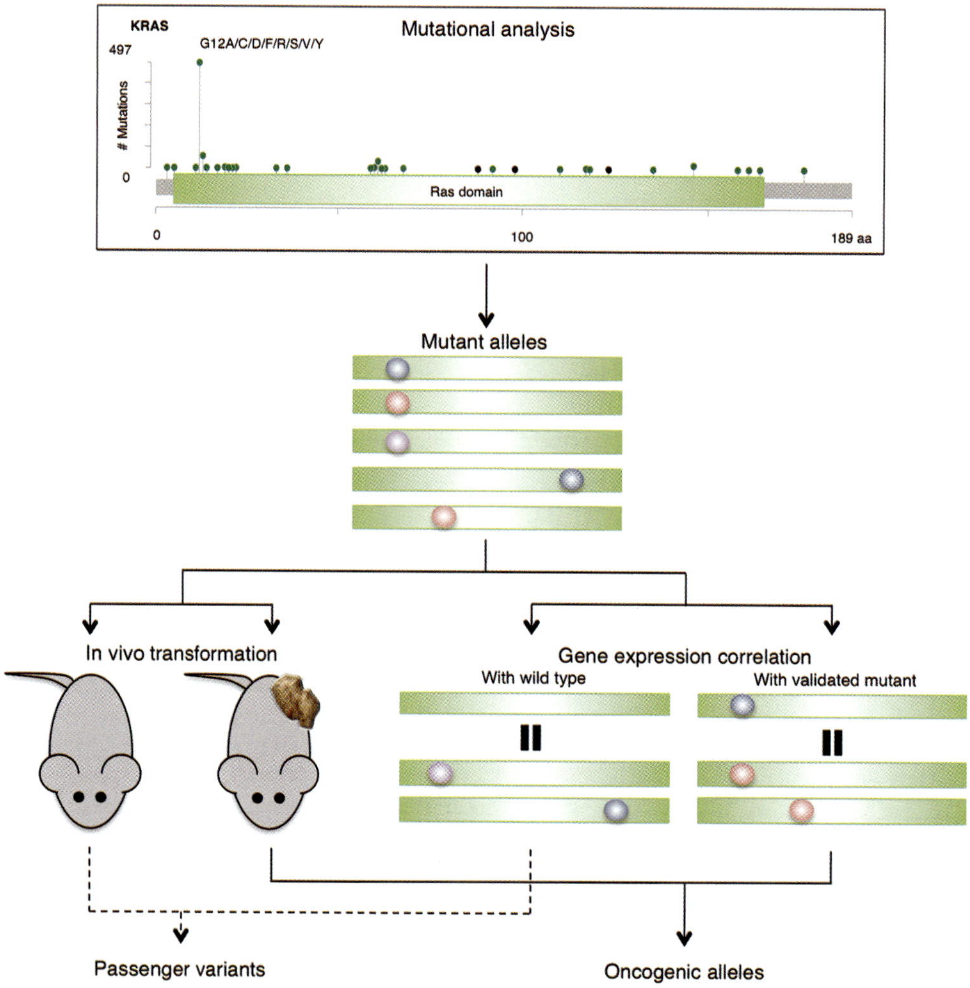

Figure 2. Functional interrogation of cancer variants. The KRAS mutation plot was adapted from cBioPortal (Cerami et al. 2012; Gao et al. 2013).

et al. 2009; Scholl et al. 2009), this technique has some important limitations. First, the vast majority of RNAi reagents induce only partial loss of gene expression, preventing the identification of dependent genes that require full inactivation. Furthermore, there is a significant variation in the quality of reagents even with advanced design tools, creating many false-negative results due to ineffective suppression (Booker et al. 2011).

More importantly, off-target effects of si/shRNAs, defined as phenotypic consequences of interactions with mRNAs other than the intended complementary sequence, can lead to frequent false-positive results (Echeverri et al. 2006). Screens for cancer dependencies, where the output phenotype is simply a loss of cellular viability and proliferation, are particularly susceptible to false positives given the many housekeeping genes that, when inadvertently depleted, could copy this phenotype (Kaelin 2012). Off-target effects are mediated by seed sequence complementarity, defined as bases 2–8 of an si/shRNA that can mimic the native miRNA pathways (Jackson et al. 2003, 2006). Paired control reagents with changes of bases 9–11 (C911) abrogate on-target effects, which require perfect complementarity but retain most off-target effects through the seed sequence (Buehler et al. 2012). Although these controls are very helpful in targeted screens and validation experiments, paired controls are not feasible at the genome scale necessary for cancer dependency screens.

For large-scale experiments, the most practical method to correct for off-target effects is to employ redundant reagents designed to target the same mRNA at independent locations and then computationally correct for outliers. Multiple si/shRNAs that target the same gene and produce the same phenotype are more likely producing on-target effects than a single reagent or those that produce disparate phenotypes (Echeverri et al. 2006). One algorithm, ATARiS, considers shRNA dependency data (approximately five shRNAs per gene) across many cell lines, identifying the reagents that behave similarly to each other in most contexts as on-target and eliminating shRNAs that do not cause the same phenotypes (Shao et al. 2013). Dependency data of approved shRNAs are then collapsed into a gene solution that reports the relative dependency of a particular cell line on that gene.

CRISPR–Cas9

The CRISPR–Cas9 bacterial immune system has been adapted for genomic editing in mammalian cells (Jinek et al. 2012, 2013; Cho et al. 2013; Cong et al. 2013; Mali et al. 2013; Hsu et al. 2014). Small-guide RNAs (sgRNAs) direct the Cas9 endonuclease in a sequence-specific manner to genomic sites, allowing Cas9 to induce a double-strand DNA break (Cong et al. 2013; Hsu et al. 2014). These breaks are generally repaired in an error-prone fashion, creating inactivating mutations with high frequencies at target sites (Shalem et al. 2014; Wang et al. 2014). Pooled sgRNA screens are performed similarly to pooled shRNA screens, with a program such as ATARiS or MAGeCK collapsing multiple sgRNAs targeting the same gene into a single score (Shao et al. 2013; Li et al. 2014).

Gene editing with CRISPR–Cas9 has several advantages over RNAi (Shalem et al. 2014; Wang et al. 2014, 2015). First, biallelic gene inactivation at the DNA level bypasses issues in which partial depletion of mRNA is insufficient to cause a phenotype (Shalem et al. 2014; Wang et al. 2015). Second, genome editing with the CRISPR–Cas9 system leads to substantially fewer off-target effects when compared to RNAi (Shalem et al. 2014; Wang et al. 2014). Even when using first-generation libraries, editing was observed at likely alternative sites at very low frequencies (<2.5%) (Wang et al. 2014). Overall, the increased depletion strength and specificity by CRISPR–Cas9 gene editing should reduce both false-negative and false-positive results from genome-wide loss-of-function screens.

CRISPR–Cas9 screens are not limited to the total inactivation of protein coding genes (Fig. 3). Shi et al. used sgRNAs targeting individual protein domains to identify preferentially important protein domains and potential drug targets (Shi et al. 2015). At a genome-wide level, CRISPR–Cas9 has been used to disrupt miRNAs in myeloid leukemia, identifying specific dependencies in these cells that would be invisible to RNAi (Wallace et al. 2016). As a proof of concept, CRISPR–Cas9 was used to identify functional p53 and estrogen receptor enhancer elements from a large set of expected binding sites determined by chromatin immunoprecipitation with sequenc-

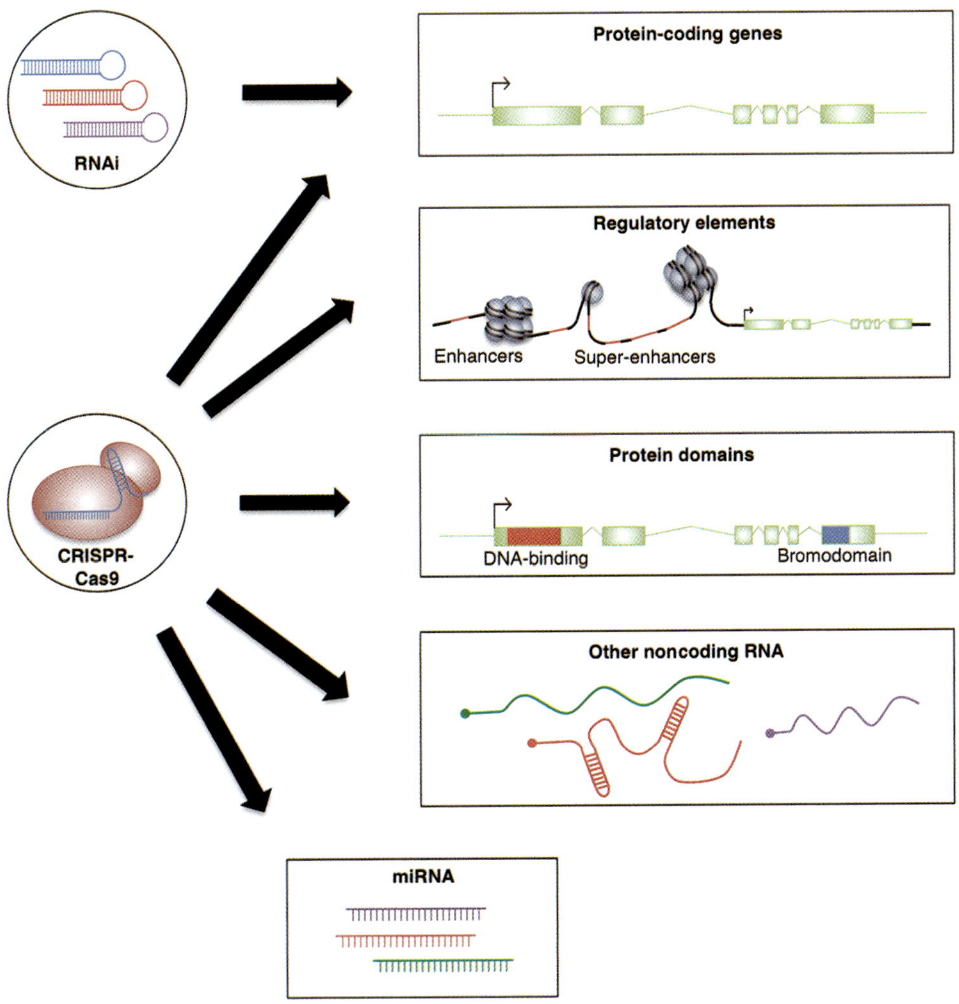

Figure 3. Comparison of genetic elements targeted by RNA interference (RNAi) and CRISPR–Cas9, miRNA, microRNA.

ing (ChIP-seq) and consensus motif analysis (Korkmaz et al. 2016). Using all possible sgRNAs to achieve saturated mutagenesis, Canver et al. determined key functional bases in the BCL11A enhancer (Canver et al. 2015). The creation of activation (Konermann et al. 2015) or catalytically inactive (Braun et al. 2016) Cas9 complexes allows for gene expression perturbations in both positive and negative directions on a genome scale. All of these advances extend our ability to functionally interrogate cancer genomes and expand the cancer dependency map.

Although CRISPR–Cas9-mediated gene editing is quite specific, we and others have identified a gene-independent effect induced by CRISPR–Cas9. In multiple cell lines, sgRNAs targeted to amplified regions tend to score as dependencies, even if the sgRNAs match intergenic sequences or nonexpressed genes (Wang et al. 2015; Aguirre et al. 2016). Additionally, sgRNAs from first-generation, imperfect genome-wide libraries predicted to target multiple locations throughout the genome also scored as false positives (Aguirre et al. 2016). Further investigation discovered an additive loss of cellular viability with every Cas9-induced cut, independent of the inactivated target (Aguirre et al. 2016). The decrease in viability appears to be independent of the type of structural rearrangement, although multiple cuts made on independent chromosomes have a greater impact than the same number of cuts on a single chromosome (Aguirre et al. 2016). Importantly, this dependency on generally amplified regions is not seen in RNAi screens of the very same cell lines (Aguirre et al. 2016). Thus, alterations in copy number—the most frequent structural abnormalities in human cancer—have a significant impact on results from genome-wide CRISPR–Cas9 screens where the output phenotype is a loss of viability.

Overall, these initial findings highlight the importance of considering copy number alterations when analyzing CRISPR–Cas9 screening results. In fact, genomic copy number variation data are a requirement for any CRISPR–Cas9-based screen in cancer cell lines. For low number alterations, a computational correction should suffice to filter out false positives (manuscript in preparation). For high number alterations (>10–20), any signal due to gene impacts is likely obscured by the cutting effects, necessitating that these genes be excluded. Well-designed negative-control sgRNAs are important to set a baseline for filtering. Finally, even though CRISPR–Cas9 technologies have many advantages, it is important to recognize that additional weaknesses will likely be discovered as the technology continues to be implemented more broadly.

Small-Molecule Screens

Although not a true functional genomics tool, small-molecule screens are useful in complementing the technologies introduced above. The major advantage of compound-based screens is that drugs already approved by the U.S. Federal Drug Administration (FDA) or in clinical trials can more easily be repurposed for a newly investigated cancer type or mutational profile, shortening the time from discovery to treatment. The major disadvantage is that most small molecules have off-target effects and many inhibitors target a number of related proteins, making it more difficult to assess mechanism. A number of large small-molecule screening projects exist (Seashore-Ludlow et al. 2015; Rees et al. 2016; Yu et al. 2016), and comparisons with genome-wide RNAi and CRISPR–Cas9 screens can reveal important similarities, with hits present in all three screening modalities being the most promising (Hong et al. 2016).

Types of Cancer Dependencies

Findings from Project Achilles and other genome-wide screens have identified a number of cancer dependencies. Of particular interest are genes that are preferentially dependent in limited cancer contexts, such as cancers driven by a particular mutation or from a single-cell lineage as opposed to dependencies ubiquitous to all cells, such as on housekeeping genes, as there is more likely to be a therapeutic window that could be exploited in patients.

The simplest dependencies are upon the driving oncogenic mutations underlying that particular cancer. Unsurprisingly, a number of well-characterized oncogenes have been found as preferential dependencies, including *KRAS* in pancreatic (Cheung et al. 2011a) and lung cancer (Sunaga et al. 2011) and *BCR* and *ABL1* in chronic myelogenous leukemia (Luo et al. 2008; Wang et al. 2015).

These genetic findings mimic clinical observations in patients, where therapies targeting oncogenes have been effective treatments for cancers that depend on these driving mutations. However, many of these mutations are currently thought to be "undruggable," such as oncogenic *KRAS* and amplified *MYC* (Verdine and Walensky 2007; Cox et al. 2014). Furthermore, resistance to oncogene-targeted inhibitors, as has been observed in both *BCR-ABL* (Branford et al. 2003) and BRAF V600E (Shi et al. 2014) driven cancers, highlights the importance of finding additional dependencies even in cancers that harbor "targetable" mutations.

In these cases, there is a need to discover synthetic lethal dependencies, in which inactivation of one gene in a cancer context can sensitize to loss of a second gene (Kaelin 2005). Whereas mutation or loss of expression of a single gene has no impact, loss of both genes is lethal. From a therapeutics perspective, a patient's noncancerous tissues should retain function of the gene mutated in the tumor, allowing for a preferential dependency in the cancer on the second gene.

Screens have identified a number of such dependencies. For example, the mammalian SWI/SNF (BAF) chromatin-remodeling complex is made up of more than a dozen subunits, a number of which are incorporated into the complex in a mutually exclusive fashion (Wilson and Roberts 2011). Two sets of these pairs have been identified as synthetic lethal partners: Mutations in *ARID1A*, the most frequently mutated subunit of the complex, sensitize to loss of *ARID1B* (Helming et al. 2014b); similarly, mutation or loss of expression of *SMARCA4* sensitizes to loss of *SMARCA2* (Oike et al. 2013; Hoffman et al. 2014;

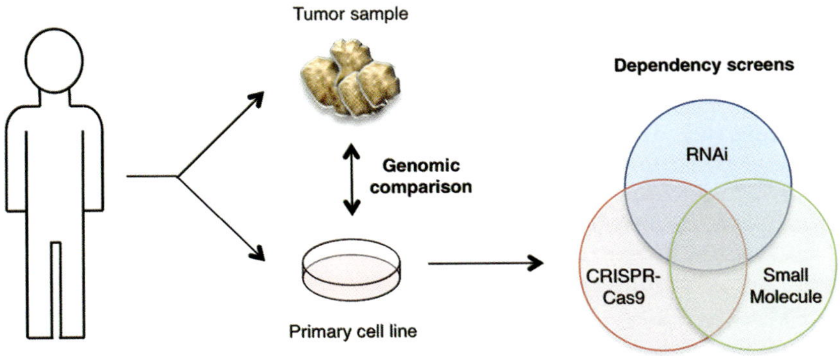

Figure 4. Primary cell lines resemble patient tumors and are amenable to dependency screens. RNAi, RNA interference.

Wilson et al. 2014). Cells require either ARID1A or ARID1B and either SMARCA2 or SMARCA4 to build functional complexes, but cannot survive loss of both paired subunits (Helming et al. 2014a).

Finally, deletion and loss of heterozygosity of tumor suppressor genes can delete other proximal genes. In some cases, the resultant hemizygosity of essential genes creates a preferential dependency for further depletion of those same genes. Termed CYCLOPS (copy number alterations yielding cancer liabilities owing to partial loss), these dependencies tend toward components of essential complexes, such as the ribosome, proteasome, and spliceosome (Nijhawan et al. 2012). In a related case, hemizygous deletion of *MTAP*, frequent given its genomic proximity to the tumor suppressor *CDKN2A*, confers dependency on an independent gene, *PRMT5*, because of accumulation of an inhibiting metabolite (Kryukov et al. 2016).

REMAINING CHALLENGES

All of the screening efforts described above are dependent on cell lines as models for cancer. Despite decades of significant advancements driven by research in cell lines, two major questions regarding established cell line use remain. First, how well do established cell lines reflect cancer biology seen in patients? Comparisons of cell lines characterized as part of the CCLE and patient tumors from The Cancer Genome Atlas (TCGA) have shown a significant variation in genomic markers between some cell line models and primary tumors of the same lineage (Domcke et al. 2013). Cells grown in culture for extended periods of time tend to genomically resemble each other more than primary cancers from the same tissue, suggesting both clonal selection for growth on plastic and evolution due to a variety of pressures not present in vivo (Gillet et al. 2011). Second, do we have enough cell lines available to fully encompass the extraordinary diversity in cancer lineages, mutations, and changes to cellular homeostasis observed in patients? Rare cancers (incidence of <6/100,000/year) account for almost 25% of all cancers, highlighting the need to develop as many models as possible to study cancer broadly (Gatta et al. 2011).

The establishment of new, recently derived cell lines aims to overcome some of the disadvantages of established lines while building on model diversity (Fig. 4; Hong et al. 2016). A pediatric undifferentiated sarcoma cell line recapitulated the histology, copy number profile, and genomic fusion events detected in a relapsed tumor from the same patient and was amenable to high-throughput RNAi, CRISPR–Cas9, and small-molecule screens (Hong et al. 2016). Although it is impossible to apply genomic technologies to compare older cell lines to their original tumors, freshly established cell lines can be validated as accurate models through genomic sequencing. Increasing both the number and quality of models will greatly improve our ability to establish a comprehensive cancer dependency map.

CONCLUSION

By intersecting genomic characterization with functional screens, we are approaching an era in which sequencing information from patients' tumors can be used to accurately predict dependencies and inform treatment for that specific cancer. The continued use of CRISPR–Cas9 screens to supplement the dependency data already produced by RNAi will discover many valuable dependencies. It is crucial that more cancer types, with a wide range of mutations and genomic features, continue to be screened using functional genomics technologies in order to capture all of the variation that we observe in patients' tumors.

Although the last decade has been defined by massive advancements in cancer genome characterization, we envision that the next decade will build on this knowledge with high-throughput functional studies to understand the clinically actionable information held in the cancer genome. With continued resources and effort, a cancer dependency map could usher in an unprecedented era of directed medicine for cancer treatment.

ACKNOWLEDGMENTS

We thank all members of the Hahn Laboratory for insightful discussions. This work was supported in part by grants from the U.S. National Cancer Institute (NCI) (U01

CA176058, U01 CA199253). T.P.H. was supported by award numbers T32GM007753 and T32GM007226 from the National Institute of General Medical Sciences. The content is solely the responsibility of the authors and does not necessarily represent the official views of the National Institute of General Medical Sciences or the National Institutes of Health. We apologize to authors whose work we did not cite because of length requirements.

REFERENCES

Aguirre AJ, Meyers RM, Weir BA, Vazquez F, Zhang CZ, Ben-David U, Cook A, Ha G, Harrington WF, Doshi MB, et al. 2016. Genomic copy number dictates a gene-independent cell response to CRISPR–Cas9 targeting. *Cancer Discov* **6:** 914–929.

Barbie DA, Tamayo P, Boehm JS, Kim SY, Moody SE, Dunn IF, Schinzel AC, Sandy P, Meylan E, Scholl C, et al. 2009. Systematic RNA interference reveals that oncogenic KRAS-driven cancers require TBK1. *Nature* **462:** 108–112.

Barretina J, Caponigro G, Stransky N, Venkatesan K, Margolin AA, Kim S, Wilson CJ, Lehar J, Kryukov GV, Sonkin D, et al. 2012. The Cancer Cell Line Encyclopedia enables predictive modelling of anticancer drug sensitivity. *Nature* **483:** 603–607.

Berns K, Hijmans EM, Mullenders J, Brummelkamp TR, Velds A, Heimerikx M, Kerkhoven RM, Madiredjo M, Nijkamp W, Weigelt B, et al. 2004. A large-scale RNAi screen in human cells identifies new components of the p53 pathway. *Nature* **428:** 431–437.

Boehm JS, Hahn WC. 2011. Towards systematic functional characterization of cancer genomes. *Nat Rev Genet* **12:** 487–498.

Booker M, Samsonova AA, Kwon Y, Flockhart I, Mohr SE, Perrimon N. 2011. False negative rates in *Drosophila* cell-based RNAi screens: A case study. *BMC Genomics* **12:** 50.

Branford S, Rudzki Z, Walsh S, Parkinson I, Grigg A, Szer J, Taylor K, Herrmann R, Seymour JF, Arthur C, et al. 2003. Detection of *BCR-ABL* mutations in patients with CML treated with imatinib is virtually always accompanied by clinical resistance, and mutations in the ATP phosphate-binding loop (P-loop) are associated with a poor prognosis. *Blood* **102:** 276–283.

Braun CJ, Bruno PM, Horlbeck MA, Gilbert LA, Weissman JS, Hemann MT. 2016. Versatile in vivo regulation of tumor phenotypes by dCas9-mediated transcriptional perturbation. *Proc Natl Acad Sci* **113:** E3892–E3900.

Bric A, Miething C, Bialucha CU, Scuoppo C, Zender L, Krasnitz A, Xuan Z, Zuber J, Wigler M, Hicks J, et al. 2009. Functional identification of tumor-suppressor genes through an in vivo RNA interference screen in a mouse lymphoma model. *Cancer Cell* **16:** 324–335.

Buehler E, Chen YC, Martin S. 2012. C911: A bench-level control for sequence specific siRNA off-target effects. *PLoS One* **7:** e51942.

Canver MC, Smith EC, Sher F, Pinello L, Sanjana NE, Shalem O, Chen DD, Schupp PG, Vinjamur DS, Garcia SP, et al. 2015. BCL11A enhancer dissection by Cas9-mediated in situ saturating mutagenesis. *Nature* **527:** 192–197.

Cerami E, Gao J, Dogrusoz U, Gross BE, Sumer SO, Aksoy BA, Jacobsen A, Byrne CJ, Heuer ML, Larsson E, et al. 2012. The cBio cancer genomics portal: An open platform for exploring multidimensional cancer genomics data. *Cancer Discov* **2:** 401–404.

Chen S, Sanjana NE, Zheng K, Shalem O, Lee K, Shi X, Scott DA, Song J, Pan JQ, Weissleder R, et al. 2015. Genome-wide CRISPR screen in a mouse model of tumor growth and metastasis. *Cell* **160:** 1246–1260.

Cheung HW, Cowley GS, Weir BA, Boehm JS, Rusin S, Scott JA, East A, Ali LD, Lizotte PH, Wong TC, et al. 2011a. Systematic investigation of genetic vulnerabilities across cancer cell lines reveals lineage-specific dependencies in ovarian cancer. *Proc Natl Acad Sci* **108:** 12372–12377.

Cheung HW, Du J, Boehm JS, He F, Weir BA, Wang X, Butaney M, Sequist LV, Luo B, Engelman JA, et al. 2011b. Amplification of CRKL induces transformation and epidermal growth factor receptor inhibitor resistance in human non–small cell lung cancers. *Cancer Discov* **1:** 608–625.

Cho SW, Kim S, Kim JM, Kim JS. 2013. Targeted genome engineering in human cells with the Cas9 RNA-guided endonuclease. *Nat Biotechnol* **31:** 230–232.

Cong L, Ran FA, Cox D, Lin S, Barretto R, Habib N, Hsu PD, Wu X, Jiang W, Marraffini LA, et al. 2013. Multiplex genome engineering using CRISPR/Cas systems. *Science* **339:** 819–823.

Cowley GS, Weir BA, Vazquez F, Tamayo P, Scott JA, Rusin S, East-Seletsky A, Ali LD, Gerath WF, Pantel SE, et al. 2014. Parallel genome-scale loss of function screens in 216 cancer cell lines for the identification of context-specific genetic dependencies. *Sci Data* **1:** 140035.

Cox AD, Fesik SW, Kimmelman AC, Luo J, Der CJ. 2014. Drugging the undruggable RAS: Mission possible? *Nat Rev Drug Discov* **13:** 828–851.

Domcke S, Sinha R, Levine DA, Sander C, Schultz N. 2013. Evaluating cell lines as tumour models by comparison of genomic profiles. *Nat Commun* **4:** 2126.

Dunn GP, Cheung HW, Agarwalla PK, Thomas S, Zektser Y, Karst AM, Boehm JS, Weir BA, Berlin AM, Zou L, et al. 2014. In vivo multiplexed interrogation of amplified genes identifies GAB2 as an ovarian cancer oncogene. *Proc Natl Acad Sci* **111:** 1102–1107.

Echeverri CJ, Perrimon N. 2006. High-throughput RNAi screening in cultured cells: A user's guide. *Nat Rev Genet* **7:** 373–384.

Echeverri CJ, Beachy PA, Baum B, Boutros M, Buchholz F, Chanda SK, Downward J, Ellenberg J, Fraser AG, Hacohen N, et al. 2006. Minimizing the risk of reporting false positives in large-scale RNAi screens. *Nat Methods* **3:** 777–779.

Gao J, Aksoy BA, Dogrusoz U, Dresdner G, Gross B, Sumer SO, Sun Y, Jacobsen A, Sinha R, Larsson E, et al. 2013. Integrative analysis of complex cancer genomics and clinical profiles using the cBioPortal. *Sci Signal* **6:** 11.

Garraway LA, Lander ES. 2013. Lessons from the cancer genome. *Cell* **153:** 17–37.

Gatta G, van der Zwan JM, Casali PG, Siesling S, Dei Tos AP, Kunkler I, Otter R, Licitra L, Mallone S, Tavilla A, et al. 2011. Rare cancers are not so rare: The rare cancer burden in Europe. *Eur J Cancer* **47:** 2493–2511.

Gerhard DS, Wagner L, Feingold EA, Shenmen CM, Grouse LH, Schuler G, Klein SL, Old S, Rasooly R, Good P, et al. 2004. The status, quality, and expansion of the NIH full-length cDNA project: The Mammalian Gene Collection (MGC). *Genome Res* **14:** 2121–2127.

Gillet JP, Calcagno AM, Varma S, Marino M, Green LJ, Vora MI, Patel C, Orina JN, Eliseeva TA, Singal V, et al. 2011. Redefining the relevance of established cancer cell lines to the study of mechanisms of clinical anti-cancer drug resistance. *Proc Natl Acad Sci* **108:** 18708–18713.

Hagerstrand D, Tong A, Schumacher SE, Ilic N, Shen RR, Cheung HW, Vazquez F, Shrestha Y, Kim SY, Giacomelli AO, et al. 2013. Systematic interrogation of 3q26 identifies *TLOC1* and *SKIL* as cancer drivers. *Cancer Discov* **3:** 1044–1057.

Hanahan D, Weinberg RA. 2011. Hallmarks of cancer: The next generation. *Cell* **144:** 646–674.

Helming KC, Wang X, Roberts CW. 2014a. Vulnerabilities of mutant SWI/SNF complexes in cancer. *Cancer Cell* **26:** 309–317.

Helming KC, Wang X, Wilson BG, Vazquez F, Haswell JR, Manchester HE, Kim Y, Kryukov GV, Ghandi M, Aguirre AJ, et al. 2014b. ARID1B is a specific vulnerability in *ARID1A*-mutant cancers. *Nat Med* **20:** 251–254.

Hoffman GR, Rahal R, Buxton F, Xiang K, McAllister G, Frias E, Bagdasarian L, Huber J, Lindeman A, Chen D, et al. 2014.

Functional epigenetics approach identifies BRM/SMARCA2 as a critical synthetic lethal target in BRG1-deficient cancers. *Proc Natl Acad Sci* **111:** 3128–3133.
Hong AL, Tseng YY, Cowley GS, Jonas O, Cheah JH, Kynnap BD, Doshi MB, Oh C, Meyer SC, Church AJ, et al. 2016. Integrated genetic and pharmacologic interrogation of rare cancers. *Nat Commun* **7:** 11987.
Hsu PD, Lander ES, Zhang F. 2014. Development and applications of CRISPR–Cas9 for genome engineering. *Cell* **157:** 1262–1278.
Isakoff SJ, Engelman JA, Irie HY, Luo J, Brachmann SM, Pearline RV, Cantley LC, Brugge JS. 2005. Breast cancer–associated *PIK3CA* mutations are oncogenic in mammary epithelial cells. *Cancer Res* **65:** 10992–11000.
Jackson AL, Bartz SR, Schelter J, Kobayashi SV, Burchard J, Mao M, Li B, Cavet G, Linsley PS. 2003. Expression profiling reveals off-target gene regulation by RNAi. *Nat Biotechnol* **21:** 635–637.
Jackson AL, Burchard J, Schelter J, Chau BN, Cleary M, Lim L, Linsley PS. 2006. Widespread siRNA "off-target" transcript silencing mediated by seed region sequence complementarity. *RNA* **12:** 1179–1187.
Jinek M, Chylinski K, Fonfara I, Hauer M, Doudna JA, Charpentier E. 2012. A programmable dual-RNA-guided DNA endonuclease in adaptive bacterial immunity. *Science* **337:** 816–821.
Jinek M, East A, Cheng A, Lin S, Ma E, Doudna J. 2013. RNA-programmed genome editing in human cells. *Elife* **2:** e00471.
Kaelin WG Jr. 2005. The concept of synthetic lethality in the context of anticancer therapy. *Nat Rev Cancer* **5:** 689–698.
Kaelin WG Jr. 2012. Molecular biology. Use and abuse of RNAi to study mammalian gene function. *Science* **337:** 421–422.
Kim E, Ilic N, Shrestha Y, Zou L, Kamburov A, Zhu C, Yang X, Lubonja R, Tran N, Nguyen C, et al. 2016. Systematic functional interrogation of rare cancer variants identifies oncogenic alleles. *Cancer Discov* **6:** 714–726.
Konermann S, Brigham MD, Trevino AE, Joung J, Abudayyeh OO, Barcena C, Hsu PD, Habib N, Gootenberg JS, Nishimasu H, et al. 2015. Genome-scale transcriptional activation by an engineered CRISPR–Cas9 complex. *Nature* **517:** 583–588.
Korkmaz G, Lopes R, Ugalde AP, Nevedomskaya E, Han R, Myacheva K, Zwart W, Elkon R, Agami R. 2016. Functional genetic screens for enhancer elements in the human genome using CRISPR–Cas9. *Nat Biotechnol* **34:** 192–198.
Kryukov GV, Wilson FH, Ruth JR, Paulk J, Tsherniak A, Marlow SE, Vazquez F, Weir BA, Fitzgerald ME, Tanaka M, et al. 2016. MTAP deletion confers enhanced dependency on the PRMT5 arginine methyltransferase in cancer cells. *Science* **351:** 1214–1218.
Li J, Yang Y, Peng Y, Austin RJ, van Eyndhoven WG, Nguyen KC, Gabriele T, McCurrach ME, Marks JR, Hoey T, et al. 2002. Oncogenic properties of PPM1D located within a breast cancer amplification epicenter at 17q23. *Nat Genet* **31:** 133–134.
Li W, Xu H, Xiao T, Cong L, Love MI, Zhang F, Irizarry RA, Liu JS, Brown M, Liu XS. 2014. MAGeCK enables robust identification of essential genes from genome-scale CRISPR/Cas9 knockout screens. *Genome Biol* **15:** 554.
Luo B, Cheung HW, Subramanian A, Sharifnia T, Okamoto M, Yang X, Hinkle G, Boehm JS, Beroukhim R, Weir BA, et al. 2008. Highly parallel identification of essential genes in cancer cells. *Proc Natl Acad Sci* **105:** 20380–20385.
Luo J, Emanuele MJ, Li D, Creighton CJ, Schlabach MR, Westbrook TF, Wong KK, Elledge SJ. 2009. A genome-wide RNAi screen identifies multiple synthetic lethal interactions with the Ras oncogene. *Cell* **137:** 835–848.
Luo LY, Kim E, Cheung HW, Weir BA, Dunn GP, Shen RR, Hahn WC. 2015. The tyrosine kinase adaptor protein FRS2 is oncogenic and amplified in high-grade serous ovarian cancer. *Mol Cancer Res* **13:** 502–509.
Mali P, Yang L, Esvelt KM, Aach J, Guell M, DiCarlo JE, Norville JE, Church GM. 2013. RNA-guided human genome engineering via Cas9. *Science* **339:** 823–826.
Mu D, Chen L, Zhang X, See LH, Koch CM, Yen C, Tong JJ, Spiegel L, Nguyen KC, Servoss A, et al. 2003. Genomic amplification and oncogenic properties of the KCNK9 potassium channel gene. *Cancer Cell* **3:** 297–302.
Nijhawan D, Zack TI, Ren Y, Strickland MR, Lamothe R, Schumacher SE, Tsherniak A, Besche HC, Rosenbluh J, Shehata S, et al. 2012. Cancer vulnerabilities unveiled by genomic loss. *Cell* **150:** 842–854.
Oike T, Ogiwara H, Tominaga Y, Ito K, Ando O, Tsuta K, Mizukami T, Shimada Y, Isomura H, Komachi M, et al. 2013. A synthetic lethality-based strategy to treat cancers harboring a genetic deficiency in the chromatin remodeling factor BRG1. *Cancer Res* **73:** 5508–5518.
ORFeome Collaboration. 2016. The ORFeome Collaboration: A genome-scale human ORF-clone resource. *Nat Methods* **13:** 191–192.
Rao DD, Vorhies JS, Senzer N, Nemunaitis J. 2009. siRNA vs. shRNA: Similarities and differences. *Adv Drug Deliv Rev* **61:** 746–759.
Rees MG, Seashore-Ludlow B, Cheah JH, Adams DJ, Price EV, Gill S, Javaid S, Coletti ME, Jones VL, Bodycombe NE, et al. 2016. Correlating chemical sensitivity and basal gene expression reveals mechanism of action. *Nat Chem Biol* **12:** 109–116.
Ren Y, Cheung HW, von Maltzhan G, Agrawal A, Cowley GS, Weir BA, Boehm JS, Tamayo P, Karst AM, Liu JF, et al. 2012. Targeted tumor-penetrating siRNA nanocomplexes for credentialing the ovarian cancer oncogene ID4. *Sci Transl Med* **4:** 147ra112.
Rual JF, Hirozane-Kishikawa T, Hao T, Bertin N, Li S, Dricot A, Li N, Rosenberg J, Lamesch P, Vidalain PO, et al. 2004. Human ORFeome version 1.1: A platform for reverse proteomics. *Genome Res* **14:** 2128–2135.
Scholl C, Frohling S, Dunn IF, Schinzel AC, Barbie DA, Kim SY, Silver SJ, Tamayo P, Wadlow RC, Ramaswamy S, et al. 2009. Synthetic lethal interaction between oncogenic KRAS dependency and STK33 suppression in human cancer cells. *Cell* **137:** 821–834.
Seashore-Ludlow B, Rees MG, Cheah JH, Cokol M, Price EV, Coletti ME, Jones V, Bodycombe NE, Soule CK, Gould J, et al. 2015. Harnessing connectivity in a large-scale small-molecule sensitivity dataset. *Cancer Discov* **5:** 1210–1223.
Shalem O, Sanjana NE, Hartenian E, Shi X, Scott DA, Mikkelsen TS, Heckl D, Ebert BL, Root DE, Doench JG, et al. 2014. Genome-scale CRISPR–Cas9 knockout screening in human cells. *Science* **343:** 84–87.
Shao DD, Tsherniak A, Gopal S, Weir BA, Tamayo P, Stransky N, Schumacher SE, Zack TI, Beroukhim R, Garraway LA, et al. 2013. ATARiS: Computational quantification of gene suppression phenotypes from multisample RNAi screens. *Genome Res* **23:** 665–678.
Shi H, Hugo W, Kong X, Hong A, Koya RC, Moriceau G, Chodon T, Guo R, Johnson DB, Dahlman KB, et al. 2014. Acquired resistance and clonal evolution in melanoma during BRAF inhibitor therapy. *Cancer Discov* **4:** 80–93.
Shi J, Wang E, Milazzo JP, Wang Z, Kinney JB, Vakoc CR. 2015. Discovery of cancer drug targets by CRISPR–Cas9 screening of protein domains. *Nat Biotechnol* **33:** 661–667.
Stratton MR, Campbell PJ, Futreal PA. 2009. The cancer genome. *Nature* **458:** 719–724.
Sunaga N, Shames DS, Girard L, Peyton M, Larsen JE, Imai H, Soh J, Sato M, Yanagitani N, Kaira K, et al. 2011. Knockdown of oncogenic KRAS in non–small cell lung cancers suppresses tumor growth and sensitizes tumor cells to targeted therapy. *Mol Cancer Ther* **10:** 336–346.
Verdine GL, Walensky LD. 2007. The challenge of drugging undruggable targets in cancer: Lessons learned from targeting BCL-2 family members. *Clin Cancer Res* **13:** 7264–7270.
Vogelstein B, Papadopoulos N, Velculescu VE, Zhou S, Diaz LA Jr, Kinzler KW. 2013. Cancer genome landscapes. *Science* **339:** 1546–1558.
Wallace J, Hu R, Mosbruger TL, Dahlem TJ, Stephens WZ, Rao DS, Round JL, O'Connell RM. 2016. Genome-wide

CRISPR–Cas9 screen identifies microRNAs that regulate myeloid leukemia cell growth. *PLoS One* **11:** e0153689.

Wang T, Wei JJ, Sabatini DM, Lander ES. 2014. Genetic screens in human cells using the CRISPR–Cas9 system. *Science* **343:** 80–84.

Wang T, Birsoy K, Hughes NW, Krupczak KM, Post Y, Wei JJ, Lander ES, Sabatini DM. 2015. Identification and characterization of essential genes in the human genome. *Science* **350:** 1096–1101.

Westbrook TF, Martin ES, Schlabach MR, Leng Y, Liang AC, Feng B, Zhao JJ, Roberts TM, Mandel G, Hannon GJ, et al. 2005. A genetic screen for candidate tumor suppressors identifies REST. *Cell* **121:** 837–848.

Wilson BG, Roberts CW. 2011. SWI/SNF nucleosome remodellers and cancer. *Nat Rev Cancer* **11:** 481–492.

Wilson BG, Helming KC, Wang X, Kim Y, Vazquez F, Jagani Z, Hahn WC, Roberts CW. 2014. Residual complexes containing SMARCA2 (BRM) underlie the oncogenic drive of *SMARCA4* (*BRG1*) mutation. *Mol Cell Biol* **34:** 1136–1144.

Yang X, Boehm JS, Yang X, Salehi-Ashtiani K, Hao T, Shen Y, Lubonja R, Thomas SR, Alkan O, Bhimdi T, et al. 2011. A public genome-scale lentiviral expression library of human ORFs. *Nat Methods* **8:** 659–661.

Yu C, Mannan AM, Yvone GM, Ross KN, Zhang YL, Marton MA, Taylor BR, Crenshaw A, Gould JZ, Tamayo P, et al. 2016. High-throughput identification of genotype-specific cancer vulnerabilities in mixtures of barcoded tumor cell lines. *Nat Biotechnol* **34:** 419–423.

Zender L, Spector MS, Xue W, Flemming P, Cordon-Cardo C, Silke J, Fan ST, Luk JM, Wigler M, Hannon GJ, et al. 2006. Identification and validation of oncogenes in liver cancer using an integrative oncogenomic approach. *Cell* **125:** 1253–1267.

Zender L, Xue W, Zuber J, Semighini CP, Krasnitz A, Ma B, Zender P, Kubicka S, Luk JM, Schirmacher P, et al. 2008. An oncogenomics-based in vivo RNAi screen identifies tumor suppressors in liver cancer. *Cell* **135:** 852–864.

How Cancer Genomics Drives Cancer Biology: Does Synthetic Lethality Explain Mutually Exclusive Oncogenic Mutations?

HAROLD VARMUS,[1] ARUN M. UNNI,[1] AND WILLIAM W. LOCKWOOD[2]
[1]*Meyer Cancer Center, Weill Cornell Medicine, New York, New York 10065*
[2]*BC Cancer Agency and University of British Columbia, Vancouver, British Columbia V5Z 1L3, Canada*
Correspondence: varmus@med.cornell.edu

Large-scale analyses of cancer genomes are revealing patterns of mutations that suggest biologically significant ideas about many aspects of cancer, including carcinogenesis, classification, and preventive and therapeutic strategies. Among those patterns is "mutual exclusivity," a phenomenon observed when two or more mutations that are commonly observed in samples of a type of cancer are not found combined in individual tumors. We have been studying a striking example of mutual exclusivity: the absence of coexisting mutations in the *KRAS* and *EGFR* proto-oncogenes in human lung adenocarcinomas, despite the high individual frequencies of such mutations in this common type of cancer. Multiple lines of evidence suggest that toxic effects of the joint expression of *KRAS* and *EGFR* mutant oncogenes, rather than loss of any selective advantages conferred by a second oncogene that operates through the same signaling pathway, are responsible for the observed mutational pattern. We discuss the potential for understanding the physiological basis of such toxicity, for exploiting it therapeutically, and for extending the studies to other examples of mutual exclusivity.

For more than three decades, the study of carcinogenesis has focused increasingly on the mutations that drive the conversion of cells that populate normal tissues into cells that display the hallmarks of cancer: excessive growth, an ability to invade, a failure to die or differentiate, and many other features of malignancy (Hanahan and Weinberg 2000, 2011). Such mutations may arise stochastically as a result of errors in DNA replication or cell division, or they may be attributable to environmental, genetic, or behavior factors that promote mutations by damaging DNA directly or by impeding DNA repair (Stratton et al. 2009). Databases cataloging these mutations have been expanding exponentially under the influence of much more rapid methods for DNA sequencing and other modes of characterization and with investments in large-scale cancer genome projects, such as such as The Cancer Genome Atlas (TCGA; https://cancergenome.nih.gov/) or the International Consortium for the Genomics of Cancer (ICGC; http://icgc.org/).

GENERAL FEATURES OF ONCOGENIC MUTATIONS

Individual somatic mutations identified in human cancers may be classified based on the chemical and physical changes in genetic material (base substitutions; small or large deletions or insertions; rearrangements such as DNA amplifications or chromosomal translocations) or on the functional consequences of the mutations (gain or loss of function) (Futreal et al. 2004; Watson et al. 2013). Gain-of-function (GOF) mutations that convert proto-oncogenes into active oncogenes often result from specific base substitutions, chromosomal translocations, or gene amplifications that strengthen gene function by increasing the abundance, changing the catalytic properties, or eliminating regulatory control of the gene product. Loss-of-function (LOF) mutations, such as inactivating missense or nonsense mutations or deletions, often combined with loss of heterozygosity, usually by intrachromosomal deletion or chromosomal loss, affect tumor-suppressor genes (TSGs); similar functional effects can be achieved by selected nucleotide substitutions that confer a "dominant-negative" effect. It is generally assumed that such LOF and GOF mutations offer a selective advantage, such as faster growth or less frequent cell death. When they occur in a cancer's founding cell, they will be present in every cell in the cancerous clone; when they occur subsequently, they will be found in all cells in an evolutionary "branch" or subclone of the tumor.

Many oncogenic mutations are associated with certain types of cancers at only relatively low frequencies, but some mutations—such as GOF alleles of mutant *KRAS* in pancreatic adenocarcinomas, LOF mutations in the *RB* gene in small cell lung cancers, and the Philadelphia chromosome, harboring the BCR-ABL oncogene in chronic myeloid leukemia—are highly associated with a pathological diagnosis. Such findings suggest that cells in

Published by Cold Spring Harbor Laboratory Press; doi: 10.1101/sqb.2016.81.030866

different tissues are susceptible to the cancer-inducing effects of specific mutations, a proposal that needs validation by experiments that expose cells at different stages of development in different organs to various mutations associated with established cancers.

Other mutations may occur coincidentally in the founding clone, without making a functional contribution to the cancer phenotype; those are considered "passenger" mutations and are rarely found recurrently in multiple tumors. Other biochemical or "epigenetic" changes in chromosomes (e.g., altered patterns of DNA methylation or modification of chromatin proteins) are also common in cancers (Meyerson et al. 2010), as are alterations in gene expression in adjacent, noncancerous cells that constitute a tumor's microenvironment; but the potentially important roles of the epigenome (Baylin and Jones 2016) and the microenvironment (Hanahan and Coussens 2012) in the initiation and maintenance of the oncogenic state have not been clearly defined.

GENOMIC COMPLEXITIES

Even when an analysis is confined to changes in the cancer cell's DNA sequence (genotype), without an effort to discern functional consequences, the situation is complicated by two fundamental features: Tumors are usually evident only after multiple mutations have occurred, and many cells in a tumor undergo subsequent mutations that contribute to the evolution of the cancer. These features mean that cancer genomes are complex, unique, internally heterogeneous, and changeable over time.

Such complexities are daunting. Still, recognizable genotypic patterns have emerged as new methods in genomics, such as "next-generation sequencing" (NGS), have been applied to many tumor samples from a wide variety of histopathological types (Garraway and Lander 2013). As a result, both the individual mutations and combinations of lesions found at significant frequency in various histological types of cancers are increasingly used as criteria for making accurate diagnoses, which inform therapeutic decisions and prognostic advice (Garraway 2013). Further, analysis of the genetic alterations in a tumor (the "mutational signature") is a useful guide to exogenous and endogenous factors that contribute to tumorigenesis by chemically damaging or misreplicating DNA (Lawrence et al. 2013). Even mutant alleles present at low frequencies in a tumor or mutations that appear during tumor growth may be useful clinically as indicators of subclones resistant to "targeted" drugs or endowed with augmented pathogenic potential.

MUTATIONAL PATTERNS AND MUTUAL EXCLUSIVITY

In addition to associating individual mutations with certain types of cancer and cancer phenotypes, the genotyping of many cancers has identified patterns of apparently cooperating mutations that are provocative but poorly understand. One such pattern is the co-appearance of certain mutations in certain tumor types—such as the dual LOF mutations in *TP53* and *RB* genes in small cell lung cancers or the constellation of mutations that promote Wnt signaling (LOF mutations of *APC* or GOF mutations in β-catenin, R-spondin, or Axin), inactivate *TP53*, and activate *KRAS* in colorectal cancers (Cancer Genome Atlas Network 2012; George et al. 2015).

Another striking pattern—to be discussed at greater length here—is the absence of certain pairs of mutations that are individually frequent in certain tumor types. This pattern of "mutual exclusivity" is most easily recognized by displaying the well-recognized, recurrent mutations in a set of one type of tumors, using software such as that used at the cBioPortal website (http://www.cbioportal.org) (Cerami et al. 2012). We have used statistical methods to determine the likelihood of such mutual exclusivities and judge their functional significance (Fig. 1).

Several interpretations of such exclusivities have been proposed. When the mutated genes are thought likely to have similar consequences—for example, by encoding proteins that act in the same signaling pathway—a mutation in one could, in principle, obviate any selective advantage that might normally be conferred by the other. But there are more interesting possible explanations for exclusivity—a combination of mutations might result in synthetically lethal or at least growth-impairing phenotypes, by providing a toxic signal or impairing an essential cellular function.

Consider one example of a striking but still unexplained mutual exclusivity. Mutations affecting one of four different RNA splicing factors are commonly found in myelodysplasias and, less frequently, in certain solid tumors (Yoshida et al. 2011; Ogawa 2012; Lee et al. 2016). These mutations change the transcriptome by favoring certain alternative choices during RNA processing, so they are fundamentally GOF mutations. Because the second (normal) copy of each gene is retained—and indeed cannot be eliminated without loss of viability (Fei et al. 2016)—the mutant version also appears to be deficient in some essential functions of the normal gene. But heterozygous mutations affecting two splicing factors have not been observed in a single tumor. This could be accounted for in either of two ways. Mutations affecting two factors might prevent a sufficient level of normal splicing. Alternatively, two mutant splicing factors might produce more aberrantly processed RNAs than growing cells could tolerate. In either view—the loss of normal function or a gain of aberrant function—effects on cell growth or viability could disadvantage a cell with two splicing factor mutations in comparison with other cells in the same cancer.

EXPLAINING A FREQUENTLY ENCOUNTERED EXAMPLE OF MUTUAL EXCLUSIVITY

We have been studying a common (and commonly noted) example of mutual exclusivity: the rarity of en-

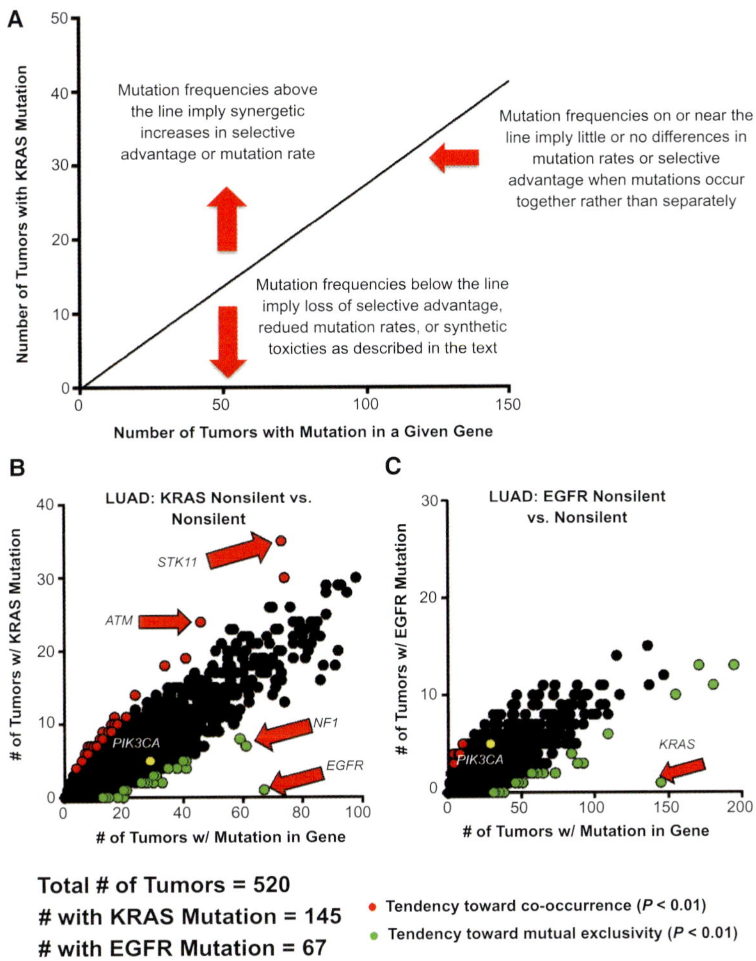

Figure 1. Assessing mutational relationships in lung adenocarcinoma. (*A*) The diagram demonstrates the method used to identify gene pairs in which mutations are negatively associated (mutually exclusive) in cancer types, as well as those that are positively associated and those for which the frequency of detection is not influenced by coexistence in the tumor. With the assumptions regarding mutational frequencies and selective advantages or disadvantages, as described in the text, the method is based on the prediction that the more frequently any gene is mutated in a population of lung adenocarcinomas (x-axis), the greater the probability that it will occur in a tumor with a mutation in *KRAS* (y-axis). 2×2 Fisher's Exact Tests are then used to determine nonrandom associations as previously described (Unni et al. 2015). (*B*,*C*) In these panels, each point represents a unique gene, but data points overlap when multiple genes are mutated in the same number of tumors. Red dots represent genes for which nonsilent mutations are positively associated with proto-oncogene mutations in the tumors (*Right side*) Fisher's Exact Test $P \leq 0.01$; green dots represent genes for which mutations are negatively associated with mutant proto-oncogenes (*Left side*) Fisher's Exact Test $P \leq 0.01$; yellow dots represent some important genes that are comutated at the "expected" rate. Genes of special interest are indicated by the arrows. (*B*) Co-occurrence of nonsilent mutations in other genes with missense *KRAS* mutations in lung adenocarcinoma (LUAD). The frequency may be changed from the expected number by differences in mutations rates or by changes in the selective advantages or disadvantages of the second mutation occurring in the presence of the first. Those changes may include toxic effects on the cell in which the second mutation occurs, as discussed extensively in the text. These factors will account for genes that are detected in mutant forms more or less frequently than expected with mutant *KRAS*. These include *STK11* and *ATM* (experimentally validated oncogenes that cooperate with mutant *KRAS* in tumorigenesis) and *EGFR* and *NF1* (both of which act in the RTK–RAS–RAF–ERK pathway). (*C*) A similar analysis for nonsilent mutations co-occurring with missense *EGFR* mutations, highlighting the negative association with *KRAS* mutations.

countering coexisting *EGFR* and *KRAS* GOF mutations in human lung adenocarcinomas, despite the fact that each kind of mutation is found frequently (in 10%–30% of tumors) without the other (Fig. 1; Cancer Genome Atlas Research Network 2014).

Interpretation of this observation has to be considered in the light of at least three important factors: the mutation rates in the relevant genes in a single cell (before, during, or after carcinogenesis)[1]; the effect of any mutations in that gene on the relative growth and survival rates of that cell; and the consequences of other coexisting mutation(s) that might confer a selective advantage or disad-

[1]Mutation rates vary among genes, based on the size of the gene, its base composition, and its transcriptional activity; further, the average mutation rate in any cell will be affected by its exposure to DNA damaging agents, the fidelity of its replication machinery, and the integrity of its DNA repair mechanisms.

vantage on that cell. These features obviously complicate any analysis of tumor genotypes, because they can all affect the likelihood of detecting coincident mutations when DNA from a tumor sample is studied. It is equally important to consider the number of cells that exist in a tumor or in a tumorigenic state—from one to billions—when a mutation occurs in a single cell. This will dramatically influence the frequency with which mutations are detected in tumor tissue.

One way to approach the problem is to remove the bias of selective pressure from the analysis, examining tumors for the coexistence of a mutation known to be oncogenic and another mutation that does not alter coding potential. Assuming that the oncogenic mutation has little or no effect on the somatic mutation rates and that the noncoding mutation can provide no selective advantage or disadvantage whether it occurs with or without the oncogenic mutation, the frequency at which coincident mutations are detected should be predictable from the frequencies at which either is detected alone: calculated simply as the product of the two mutation frequencies. Consistent with this idea, we have previously shown that such silent mutations are detected along with codon 12 KRAS mutations as the products of their individual frequencies (see Fig. 1B in Unni et al. 2015).

But the situation is more complicated when we ask whether two mutations that change the coding sequence—and particularly two mutations that affect a cell's selective advantage—are present in the same cell. If the mutation rate, starting cell number, and the selective advantages of either mutant gene are not affected by the presence of the other, the prediction is a simple one: coexistence of the two mutations should be detected at a frequency determined by multiplying the frequencies at which each mutation is detected singly. Thus if a mutation in gene A is found in 30% of tumors and a mutation in gene B is found in 10%, coincident detection of the two should then occur in ~3% of tumors. However, if the presence of the mutation in gene A eliminates any advantage conferred by the mutation in gene B or vice versa, as in the situation often referred to as "pathway redundancy," then the likelihood of finding both mutations in the tumor will be markedly reduced. Conversely, if the coexistence the two mutations has a synergistic effect that exceeds the sum of the selective advantages conferred by each, then tumors with both mutations will be more common that predicted from the product of individual frequencies. Finally, and relevant to the problem we address here, if coexistence of mutations confers toxicity—a selective disadvantage—on cells, then (as in the loss of a selective advantage) tumors will less commonly—and perhaps never—be found to harbor both mutations.

This kind of analysis is germane to an observation made by several laboratories and studied mechanistically here: Despite the relatively high frequencies of detection in human lung adenocarcinomas of codon 12 mutations in KRAS (~30%) and a few idiosyncratic mutations in EGFR (~10%–15%), coexistence of the two is rarely, if ever, found (Ding et al. 2008; Cancer Genome Atlas Research Network 2014; Unni et al. 2015). We have assessed exon sequence data for more than 500 lung adenocarcinomas, taking into consideration specific demographic factors associated with the incidence of these mutations such as smoking status, and we have found a significant negative association between the two mutant oncogenes.

An exome-wide survey for the co-occurrence of mutated genes with mutant KRAS in 520 lung adenocarcinomas revealed that mutations in EGFR deviated most dramatically from the frequency expected for situations in which mutation rates and selective advantages were unaltered by coexistence of the mutations (Fig. 1). Importantly, the analysis also revealed mutated genes, such as *STK11*, that occur together with mutant KRAS more commonly than anticipated by chance. This combination has been posited to cooperate during tumor progression (Ji et al. 2007), suggesting that the method can reveal mutated gene pairs that occur more and less frequently than would have been predicted if co-occurrence did not influence selective advantages or mutation rates. For simplicity, the analysis did not take into account gene copy numbers or methylation changes; in principle, either could create GOF or LOF. Further, we have not considered the variables of time and allele frequency; the sequenced DNAs were obtained from tumors that had probably evolved for different numbers of cell divisions, and not all of the recorded mutations were necessarily in all the cells of the tumor.

In an initial effort to understand why mutations in *EGFR* and *KRAS* were not found in the same tumor at detectable frequencies, we took advantage of transgenic mice that we had engineered to express mutant alleles of human *KRAS* (G12D) and human *EGFR* (a small deletion in exon 19) in lung epithelium under the control of a doxycycline-responsive transcriptional regulator, rtTA (Fisher et al. 2001; Politi et al. 2006). We made the simple prediction that in tritransgenic mice—carrying both oncogenic transgenes and a third encoding the regulator—tumors would develop more rapidly (and mice would be killed earlier) if the oncoproteins had additive activities; tumors would appear at the same rate as in the mice with only the more potent transgene if there were no additive effects; and tumors would occur more slowly or not at all if the combination produced detrimental consequences.

What we observed, however, was more complicated and surprising. All mice developed multiple tumors throughout the lungs. As indicated by the Kaplan–Meier plots shown in Figure 2 in Unni et al (2015), the tumors affected the survival of the tritransgenic animals at approximately the same rate as in the line of bitransgenic animals having only the more potent of the two oncogenic transgenes (*KRAS*-G12D). This seemed to imply that addition of the mutant *EGFR* transgene had no further effect, neither providing a selective advantage nor impairing cell growth or viability in the presence of the *KRAS* transgene. However, when we tested individual microdissected tumors for expression of the two genes, RNA specific for only one of the two transgenes was found in the great majority of the tumors. The most likely

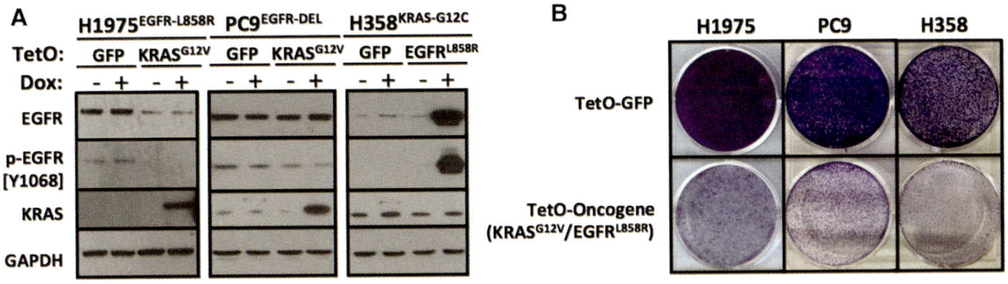

Figure 2. Coexpression of mutant KRAS and mutant EGFR in human lung adenocarcinoma cell lines decreases viability. (*A*) Induced expression of transduced genes in human lung cancer cell lines. H1975, PC9, and H358 cell lines (with endogenous mutations in either *EGFR* or *KRAS* as indicated) were transduced with the indicated tetracycline (TetO) responsive plasmids. Lysates of cells cultured in the presence or absence of doxycycline (Dox) for 24 h were prepared and assayed for the indicated protein expression by western blotting. (*B*) Coexpression of mutant *KRAS* and *EGFR* reduces cell viability. Cells with either a TetO-Oncogene ($KRAS^{G12V}$ for PC9 and H1975; $EGFR^{L858R}$ for H358) or TetO-GFP were grown in the presence of Dox for 7 d and cell density was assessed by crystal violet staining. Only the cells in which Dox induces expression of the second oncogene have reduced cell densities, consistent with the hypothesis that the coexpression of *KRAS* and *EGFR* leads to a less fit cellular state.

explanation of this observation is that detectable tumors arose only when cells expressed a single oncogenic transgene, not both. The tumors we harvested and characterized presumably developed from cells that never expressed both transgenes or from cells that may have expressed both transgenes initially but then extinguished the expression of one of them. Because we do not know the proportion of lung epithelial cells that turn on each transgene immediately or soon after administration of doxycycline, it is difficult to know which scenario is more likely. Similar findings with other pairs of *EGFR* and *KRAS* transgenic mice have recently been published by Ambrogio et al. (2016).

One of the shortcomings of the experiments with transgenic mice is that our claims for synthetic toxicity of the two expressed oncogenes are conjectural and not based on direct observation of noxious effects. Furthermore, the experiments were, of necessity, performed in mouse rather than human cells. To overcome these limitations, we placed inducible DNA vectors carrying a mutant *EGFR* or a mutant *KRAS* gene into human lung adenocarcinoma cell lines that express either of the two oncogenic drivers of the cancers from which the cell lines were derived. Thus PC9 cells, with an *EGFR*-exon 19 deletion mutation, were equipped with a doxycycline-inducible *KRAS* gene (G12D), and H358 cells, with a KRAS-G12C mutation, were modified by addition of an inducible *EGFR*-exon 19 mutation. PC9 and H358 lines carrying an inducible green fluorescent protein (GFP) cDNA served as controls. When the exogenous coding domains were induced in cultured cells, coexpression of a second oncogene (mutant *KRAS* in PC9 cells or mutant *EGFR* in H358 cells) caused a gradual decline in cell numbers, whereas expression of GFP produced no obvious changes. By 7–10 d after induction, only 10%–20% as many lung cancer cells remained in the cultures in which a second oncogene was induced as in the control cultures, suggesting that cell death mechanisms were activated, with or without restriction of cell growth (Fig. 2)

Several morphological, biochemical, and physiological properties were associated with the diminished numbers of cells in these cultures. Most obviously, many of the cells showed a marked change in shape and cytoplasmic contents. Electron micrographs revealed the kinds of vesicular structures associated with macropinocytosis, and the predicted augmented uptake of proteins was documented with fluorescent dextran (Unni et al. 2015). Because similar findings have been reported in cells with high levels of mutant KRAS proteins (Commisso et al. 2013), we also examined the abundance and phosphorylation status of several proteins implicated in transmission of signals downstream from KRAS and EGFR (Fig. 3A). Of special note are the increased levels of phosphorylation of four serine/threonine kinases—AKT, ERK, p38, and JNK—suggesting that enzymatic activity in these branches of the RAS signaling network was augmented by coexpression of mutant EGFR and mutant KRAS (Fig. 3B).

When the tumor cell lines expressing both oncogenes were examined for markers of cell cycle progression, apoptosis, and autophagy, we could document perturbations of all these processes, and each could help to explain the loss of cell viability. Combined with the morphological changes attributed to increased cell vacuolization, we suggest that a unique form of cell death, termed "methuosis," can be induced by coexpression of the mutant oncogenes (Maltese and Overmeyer 2014). However, whether this is true only for these experimental systems or is indeed a feature specifically associated with the synthetic lethality associated with mutant KRAS and EGFR remains to be explored.

One major factor that could influence the phenotype observed is the level at which the mutant oncogenes are expressed in the experimental systems. For example, our work used doxycycline inducible systems in mice and cell culture, and these can produce proteins at high levels (Unni et al. 2015). A recent study exploring the relationship of mutant *BRAF* and *KRAS* in lung adenocarcinoma, however, expressed the mutant genes from their endogenous promoters, and senescence rather than methuosis was observed upon coexpression (Cisowski et al. 2016). Although the levels of the oncoproteins may influence the

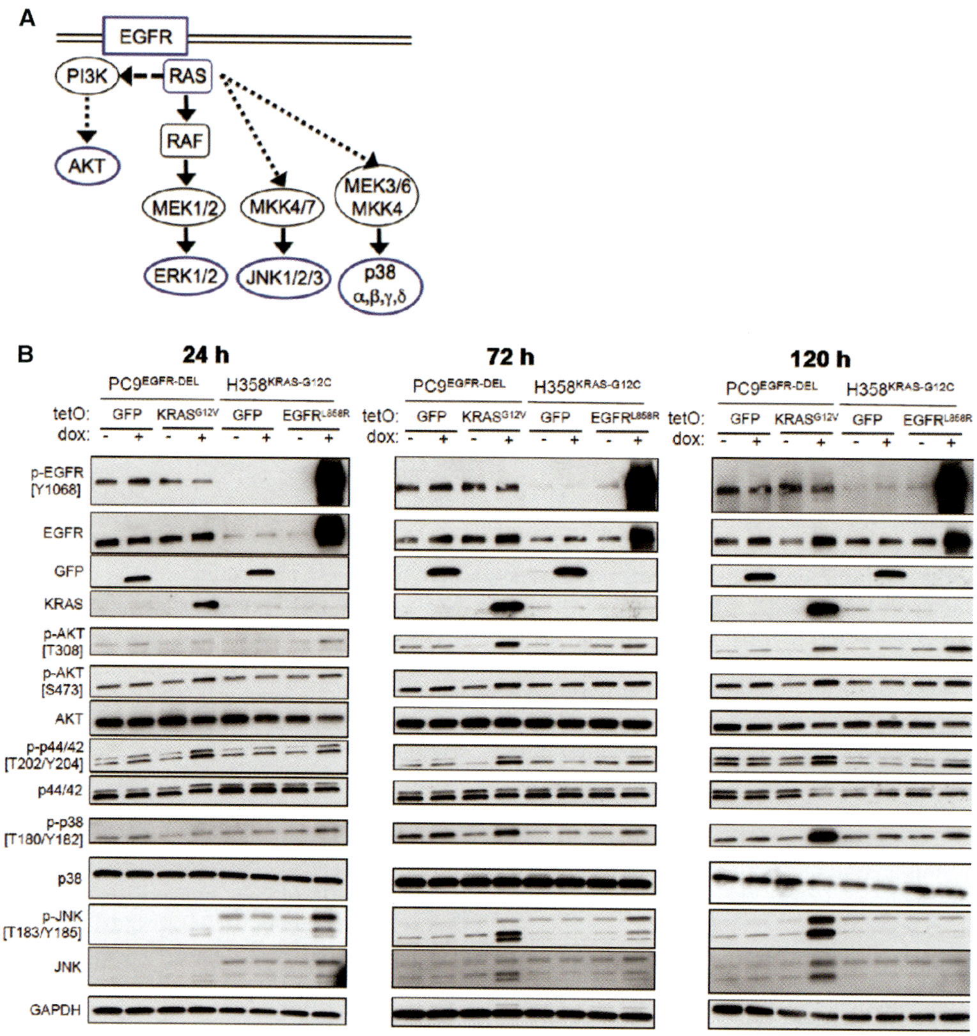

Figure 3. Downstream components of the EGFR-KRAS signaling axis are activated at different time points after induction of a second mutant oncogene. (*A*) EGFR and KRAS mainly mediate their oncogenic effects through induction of the PI3K-AKT and RAF-ERK pathways. However, KRAS has also been shown to activate additional MAPK signaling pathways including JNK and p38 under specific conditions—namely, through association and activation of RAC (Coso et al. 1995; Minden et al. 1995; Bhanot et al. 2010). (*B*) Temporal effects of mutant *KRAS* and *EGFR* coexpression on downstream signaling pathways over time. Increased MAPK signaling is observed in cells coexpressing mutant oncogenes; however, the times at which they are activated vary depending on the line and the specific MAPK pathway components (ERK, JNK, or p38). The indicated cells were cultured for 24, 72, and 120 h with or without Dox; lysates were assayed by western blotting for the indicated proteins and phospho-proteins. Where relevant, the phosphorylated tyrosine (Y) or threonine (T) residue that determines reactivity with phospho-specific antibodies is shown.

specific physiological effects, the premise that mutually exclusive mutant oncogenes lead to a less fit cellular state upon coexpression remains consistent.

SEARCHING FOR THE MECHANISM AND UTILITY OF AGONISTIC SYNTHETIC LETHALITY

To understand and exploit the apparent synthetic lethality resulting from coexpression of two oncogenes, it is necessary to identify the factors that mediate the toxic effects and to learn how to recapitulate them. We have considered and, in some instances, initiated attempts to do this in several ways.

1. An unbiased screen with reagents that mutate or inhibit the function of thousands of individual genes—through gene editing, inhibitory RNAs, or chemical compounds—might identify genes required for synthetic lethality. To that end, we have infected cultures of H358 cells carrying an uninduced *EGFR* (L858R) transgene, in addition to an endogenous somatic *KRAS* mutation, with a lentivirus library carrying CRISPR–Cas9 and guide RNAs targeting about 20,000 genes (Shalem et al. 2014). The infected cells were then treated with doxycycline to induce mutant *KRAS*, expecting that after a few passages the cultures would be enriched for cells in which gene editing had inactivated any gene required to mediate the toxic effects of oncogene coexpression. NGS sequencing of DNA

from surviving cells revealed enrichment for cells in which CRISPR had inactivated *EGFR* (presumably the mutant transgene) or a variety of endogenous genes, including two genes encoding Mediator proteins.

2. Enough is now known about the proteins involved in EGFR and KRAS signaling to make educated guesses about the components that might be critical for synthetic toxicity. A set of inhibitory RNAs has been developed to "knock down" the relevant gene products individually to test whether they are required to impair the function of cells expressing both oncogenes. Furthermore, we have recently observed a dynamic relationship in the activation of MAPK pathway components upon coinduction of mutant *EGFR* and *KRAS* in different cell lines (Fig. 3). This has led us to explore the kinetics of signaling upon coinduction of the mutant oncogenes using quantitative phospho-proteomics to identify the key components that mediate synthetic lethality
3. Compounds that inhibit enzymes, such as protein kinases or phosphatases, that regulate the activity of components of EGFR and KRAS signaling networks could provide information about elements that contribute to—or protect cells from—the detrimental effects of EGFR plus KRAS on cell growth and survival. In principle, some compounds might be instructive about ways to use drug treatment of tumors to mimic the synthetic lethality produced by combinations of oncogenic mutations. Others might help to identify proteins that mediate the toxicity of the genetic combinations.

To attempt to understand synthetic agonistic lethality more generally, we encourage studies of recognized examples of mutually exclusive oncogenes and efforts to find oncogenic combinations that are detrimental to cell growth or survival. Among the several known oncogenic mutations that appear to be mutually exclusive, perhaps the best known, other than the *EGFR*/*KRAS* combination, is the rarity of an *NRAS* mutation coincident with a *BRAF* mutation in melanoma, despite the high frequency of each mutation in this tumor type. Further, Petti et al. (2006) have reported that forced expression of mutant *NRAS* in a melanoma clone with mutant *BRAF* causes the cells to become senescent. We have surveyed TCGA databases for other missing combinations of mutant genes in many types of cancer, and some of those are listed in Table 1; direct tests of the effects of such combinations in appropriate cell types are likely to be informative. Another approach to identifying toxic combinations relies on the use of a library of genetic vectors carrying a large selection of activated oncogenes, preferably linked to barcodes. Propagation of cells into which such libraries have been introduced, combined with DNA sequencing to look for clones that become underrepresented during growth, should produce candidate combinations for synthetic toxicity.

A review of mutually exclusive oncogenic mutations shows that such pairs are not confined to GOF mutations in proto-oncogenes (Table 1). Several, such as mutations

Table 1. Mutual exclusivities in selected cancer types

Cancer type	Reported mutual exclusivity[a]	Reference
Myelodysplasia syndromes	U2AF35, ZRSR2, SRSF2, SF3B1	Yoshida et al. 2011
Lung adenocarcinoma	KRAS, EGFR, BRAF, ROS1, ALK, RET, MAP2K1, HRAS, NRAS, MET, ERBB2, RIT1, NF1	Cancer Genome Atlas Research Network 2014
Melanoma	HRAS, KRAS, NRAS, BRAF, NF1	Cancer Genome Atlas Network 2015
Colorectal adenocarcinoma	IGF2, IRS2, PIK3CA, PIK3R1, PTEN; KRAS, NRAS, BRAF	Cancer Genome Atlas Network 2012

[a]Genes belonging to pairs or sets of mutated alleles that are claimed by the authors of the cited reports to have a "mutually exclusive" relationship are listed as groups, separated by a semicolon. An allele may be "mutated" as a result of single base changes, short insertions or deletions, amplification, large deletions, or gene fusions. The criteria used to invoke "mutual exclusivity" vary and are described in the cited literature.

of *NF1* and *EGFR* in glioblastomas, involve a LOF mutation in a tumor-suppressor gene plus an activated proto-oncogene. Of course, mutational loss of a tumor-suppressor function may lead to a functional gain in a signaling protein that drives growth; examples might include the augmented signaling by a RAS protein after loss of a GAP (GTPase activator protein) or the increased HIF (hypoxia-inducible factor) activity after loss of the von Hippel–Lindau (VHL) protein. Further experiments to investigate why such combinations occur and to seek other detrimental pairings might enrich our understanding of the exclusivities and suggest additional strategies for therapies based on them.

CONCLUSION: INTERPRETATIONS AND FURTHER PROSPECTS

Mutations in *EGFR* and *KRAS*, each found at high frequency in human lung adenocarcinomas, rarely if ever co-occur in a single tumor. Our efforts thus far to understand this phenomenon are most consistent with the notion that coexpression of the two mutations is synthetically lethal or at least toxic. It is, however, difficult to study the mechanism that explains the mutual exclusivity in an intact organism, even in mouse models, let alone human patients.

The most productive approach thus far has been to induce expression of the second oncogene in human adenocarcinoma cell lines that carry mutant *EGFR* or mutant *KRAS* as a driver oncogene after introduction of the second oncogene regulated by a doxycycline-responsive transcriptional activator. This provides a setting in which the effects of adding the second oncogene can be observed acutely, but it may not accurately recapitulate events in vivo, because the production of the two oncoproteins, especially the induced protein, may exceed levels encountered in the intact organism. Moreover, the growth conditions in tissue culture and the absence of the natural microenvironment may further alter the outcome. Thus it remains conceivable that the explanation for mutual exclusivity of certain oncogenic mutations in human cancers is a manifestation of other features. For example, tumor cells expressing both oncogenes might experience a growth disadvantage (e.g., senescence) without undergoing a lethal process. There might also be situations in which mutual exclusivity is, in fact, explained solely by functional redundancy of the relevant cancer genes, without a selective disadvantage conferred by the second gene contributing to the exclusivity.

Establishing alternative explanations for mutually exclusive pairs of oncogenic mutations will be difficult, and we have chosen to focus on efforts to understand the mechanisms by which cancer cells die in culture when the second oncogene is expressed. We have done this because the synthetic toxicity of *EGFR* and *KRAS* mutations is unexpected and its mechanism, even in cell culture, might be instructive about novel approaches to therapy. Many efforts have been made to screen for lethal outcomes in cancer cells (e.g., those in which common mutant *KRAS* alleles are oncogenic drivers) by using libraries of inhibitory RNAs, drugs that inhibit gene products, or CRISPR–Cas9-based libraries that inactivate genes (Downward 2015). In such screens, the intent is to combine a LOF event with a GOF mutation (e.g., in *KRAS*) and to cause cell toxicity by eliminating or severely reducing the functionality of a second gene that is specifically required for the viability of a cancer cell driven by a certain oncogene. Our findings imply that it might also be important to seek GOF changes—using agonists, not just antagonists, of a second gene's function—to kill or disable cancer cells.

We refer here to GOF and LOF (or agonistic or antagonistic) approaches as a way of characterizing the action taken experimentally to perturb cancer cell behavior or viability. But the outcome may involve biochemical steps that are complex and difficult to classify as losses or gains of function. Cell signaling is still poorly understood, involving branching pathways, feedback events, and dosage. In this light, inhibition of a negative regulator (an apparent LOF) would have a positive effect on the regulated protein (a GOF). Further, by viewing events in cell signaling in a binary language—largely "on or off" or "activated or inactivated"—we may be ignoring important effects of variations in signal intensity (e.g., imperceptible when the signal is weak, oncogenic when moderate, and toxic when intense).

In addition to exploring the basis of the common mutual exclusivities, such as mutant *EGFR* and mutant *KRAS*, we encourage the further study of other combinations of oncogenic mutations that rarely or never co-occur. Those examples that have attracted attention include mutant *BRAF* and mutant *NRAS* (shown to cause cell senescence when forced to be expressed concurrently in melanoma cells (Petti et al. 2006)); the mutual exclusivity of splicing factor mutations; mutant *KRAS* and *BRAF* in lung adenocarcinoma (Cisowski et al. 2016); inactivation of *NF1* and activating *EGFR* insertions in glioblastoma; and mutant *KRAS* and mutant *BRAF* in colorectal adenocarcinoma (Table 1; Yoshida et al. 2011; Cancer Genome Atlas Network 2012, 2015; Cancer Genome Atlas Research Network 2014). The list of mutually exclusive mutations, however, may be considerably longer, based on analysis of large data sets, such as those produced by TCGA. Because some genes in these analyses will have been infrequently mutated, it may be difficult to provide convincing evidence that sets of mutations are truly mutually exclusive, unless subjected to the kinds of experimental tests described here. Clues to the possible significance of exclusivity may be obtained by grouping genes by functional classes. Such approaches have been used, for example, to predict the mutual exclusivity of mutations affecting genes that modify chromatin in glioblastomas (e.g., *IDH1, MLL3, SETD1A;* (Brennan et al. 2013)) or affecting genes that encode protein kinases in acute myeloid leukemias (e.g., *FLT3, KIT, ABL1*; Cancer Genome Atlas Research Network 2013). Elucidating the mechanisms by which exclusivity is enforced in a variety of cancer types may suggest therapeutic strategies and enable chemical or genetic screens to pursue those strategies.

One other inadequately studied aspect of mutual exclusivity is the possible contribution of germline variation to a cell's ability to tolerate certain mutational combinations. Most of the evidence for inherited factors in carcinogenesis has come from the identification of germline mutations in tumor-suppressor genes, in genes that participate in DNA repair, and (least frequently) in proto-oncogenes. However, it seems likely that other kinds of genetic variation, responsible for less obvious changes in cancer risk, may occur at loci that determine the penetrance of oncogenic mutations in various cell lineages. Similarly, such variation may determine the probability of synthetic lethality due to combinations of GOF and LOF changes.

ACKNOWLEDGMENTS

The authors are grateful for support from the National Institutes of Health (NIH) Intramural Research Program, the Meyer Cancer Center at Weill Cornell Medicine, the British Columbia Cancer Foundation, and the Canadian Institutes of Health Research (CIHR). W.W.L. also receives salary support from the CIHR New Investigator and Michael Smith Foundation for Health Research Scholar programs. We also thank our laboratory colleagues at the NIH, Weill Cornell Medicine (WCM), and the University of British Columbia (UBC) for helpful conversations.

REFERENCES

Ambrogio C, Barbacid M, Santamaria D. 2016. In vivo oncogenic conflict triggered by co-existing KRAS and EGFR activating mutations in lung adenocarcinoma. *Oncogene* doi: 10.1038/onc.2016.385.

Baylin SB, Jones PA. 2016. Epigenetic determinants of cancer. *Cold Spring Harb Perspect Biol* **8:** a019505.

Bhanot H, Young AM, Overmeyer JH, Maltese WA. 2010. Induction of nonapoptotic cell death by activated Ras requires inverse regulation of Rac1 and Arf6. *Mol Cancer Res* **8:** 1358–1374.

Brennan CW, Verhaak RG, McKenna A, Campos B, Noushmehr H, Salama SR, Zheng S, Chakravarty D, Sanborn JZ, Berman SH, et al. 2013. The somatic genomic landscape of glioblastoma. *Cell* **155:** 462–477.

Cancer Genome Atlas Network. 2012. Comprehensive molecular characterization of human colon and rectal cancer. *Nature* **487:** 330–337.

Cancer Genome Atlas Network. 2015. Genomic classification of cutaneous melanoma. *Cell* **161:** 1681–1696.

Cancer Genome Atlas Research Network. 2013. Genomic and epigenomic landscapes of adult de novo acute myeloid leukemia. *N Engl J Med* **368:** 2059–2074.

Cancer Genome Atlas Research Network. 2014. Comprehensive molecular profiling of lung adenocarcinoma. *Nature* **511:** 543–550.

Cerami E, Gao J, Dogrusoz U, Gross BE, Sumer SO, Aksoy BA, Jacobsen A, Byrne CJ, Heuer ML, Larsson E, et al. 2012. The cBio cancer genomics portal: An open platform for exploring multidimensional cancer genomics data. *Cancer Discov* **2:** 401–404.

Cisowski J, Sayin VI, Liu M, Karlsson C, Bergo MO. 2016. Oncogene-induced senescence underlies the mutual exclusive nature of oncogenic KRAS and BRAF. *Oncogene* **35:** 1328–1333.

Commisso C, Davidson SM, Soydaner-Azeloglu RG, Parker SJ, Kamphorst JJ, Hackett S, Grabocka E, Nofal M, Drebin JA, Thompson CB, et al. 2013. Macropinocytosis of protein is an amino acid supply route in Ras-transformed cells. *Nature* **497:** 633–637.

Coso OA, Chiariello M, Yu JC, Teramoto H, Crespo P, Xu N, Miki T, Gutkind JS. 1995. The small GTP-binding proteins Rac1 and Cdc42 regulate the activity of the JNK/SAPK signaling pathway. *Cell* **81:** 1137–1146.

Ding L, Getz G, Wheeler DA, Mardis ER, McLellan MD, Cibulskis K, Sougnez C, Greulich H, Muzny DM, Morgan MB, et al. 2008. Somatic mutations affect key pathways in lung adenocarcinoma. *Nature* **455:** 1069–1075.

Downward J. 2015. RAS synthetic lethal screens revisited: Still seeking the elusive prize? *Clin Cancer Res* **21:** 1802–1809.

Fei DL, Motowski H, Chatrikhi R, Prasad S, Yu J, Gao S, Kielkopf CL, Bradley RK, Varmus H. 2016. Wild-type U2AF1 antagonizes the splicing program characteristic of U2AF1-mutant tumors and is required for cell survival. *PLoS Genet* **12:** e1006384.

Fisher GH, Wellen SL, Klimstra D, Lenczowski JM, Tichelaar JW, Lizak MJ, Whitsett JA, Koretsky A, Varmus HE. 2001. Induction and apoptotic regression of lung adenocarcinomas by regulation of a K-Ras transgene in the presence and absence of tumor suppressor genes. *Genes Dev* **15:** 3249–3262.

Futreal PA, Coin L, Marshall M, Down T, Hubbard T, Wooster R, Rahman N, Stratton MR. 2004. A census of human cancer genes. *Nat Rev Cancer* **4:** 177–183.

Garraway LA. 2013. Genomics-driven oncology: Framework for an emerging paradigm. *J Clin Oncol* **31:** 1806–1814.

Garraway LA, Lander ES. 2013. Lessons from the cancer genome. *Cell* **153:** 17–37.

George J, Lim JS, Jang SJ, Cun Y, Ozretic L, Kong G, Leenders F, Lu X, Fernandez-Cuesta L, Bosco G, et al. 2015. Comprehensive genomic profiles of small cell lung cancer. *Nature* **524:** 47–53.

Hanahan D, Coussens LM. 2012. Accessories to the crime: Functions of cells recruited to the tumor microenvironment. *Cancer Cell* **21:** 309–322.

Hanahan D, Weinberg RA. 2000. The hallmarks of cancer. *Cell* **100:** 57–70.

Hanahan D, Weinberg RA. 2011. Hallmarks of cancer: The next generation. *Cell* **144:** 646–674.

Ji H, Ramsey MR, Hayes DN, Fan C, McNamara K, Kozlowski P, Torrice C, Wu MC, Shimamura T, Perera SA, et al. 2007. LKB1 modulates lung cancer differentiation and metastasis. *Nature* **448:** 807–810.

Lawrence MS, Stojanov P, Polak P, Kryukov GV, Cibulskis K, Sivachenko A, Carter SL, Stewart C, Mermel CH, Roberts SA, et al. 2013. Mutational heterogeneity in cancer and the search for new cancer-associated genes. *Nature* **499:** 214–218.

Lee SD K, Obeng EA, Kim E, Micol JB, Yoshimi A, Willekens C, Inoue D, Saada V, Cho H, Chung YR, et al. 2016. Synthetic lethal interactions of MDS-associated spliceosomal gene mutations identifies the basis for their mutual exclusivity. In *American Society for Hematology 58th Annual Meeting & Exposition, December 3–6*, p. 961. American Society for Hematology, Washington, DC.

Maltese WA, Overmeyer JH. 2014. Methuosis: Nonapoptotic cell death associated with vacuolization of macropinosome and endosome compartments. *Am J Pathol* **184:** 1630–1642.

Meyerson M, Gabriel S, Getz G. 2010. Advances in understanding cancer genomes through second-generation sequencing. *Nat Rev Genet* **11:** 685–696.

Minden A, Lin A, Claret FX, Abo A, Karin M. 1995. Selective activation of the JNK signaling cascade and c-Jun transcriptional activity by the small GTPases Rac and Cdc42Hs. *Cell* **81:** 1147–1157.

Ogawa S. 2012. Splicing factor mutations in myelodysplasia. *Int J Hematol* **96:** 438–442.

Petti C, Molla A, Vegetti C, Ferrone S, Anichini A, Sensi M. 2006. Coexpression of NRASQ61R and BRAFV600E in human melanoma cells activates senescence and increases susceptibility to cell-mediated cytotoxicity. *Cancer Res* **66:** 6503–6511.

Politi K, Zakowski MF, Fan PD, Schonfeld EA, Pao W, Varmus HE. 2006. Lung adenocarcinomas induced in mice by mutant EGF receptors found in human lung cancers respond to a tyrosine kinase inhibitor or to down-regulation of the receptors. *Genes Dev* **20:** 1496–1510.

Shalem O, Sanjana NE, Hartenian E, Shi X, Scott DA, Mikkelsen TS, Heckl D, Ebert BL, Root DE, Doench JG, et al. 2014. Genome-scale CRISPR–Cas9 knockout screening in human cells. *Science* **343:** 84–87.

Stratton MR, Campbell PJ, Futreal PA. 2009. The cancer genome. *Nature* **458:** 719–724.

Unni AM, Lockwood WW, Zejnullahu K, Lee-Lin SQ, Varmus H. 2015. Evidence that synthetic lethality underlies the mutual exclusivity of oncogenic KRAS and EGFR mutations in lung adenocarcinoma. *Elife* **4:** e06907.

Watson IR, Takahashi K, Futreal PA, Chin L. 2013. Emerging patterns of somatic mutations in cancer. *Nat Rev Genet* **14:** 703–718.

Yoshida K, Sanada M, Shiraishi Y, Nowak D, Nagata Y, Yamamoto R, Sato Y, Sato-Otsubo A, Kon A, Nagasaki M, et al. 2011. Frequent pathway mutations of splicing machinery in myelodysplasia. *Nature* **478:** 64–69.

A Pipeline for Drug Target Identification and Validation

EUSEBIO MANCHADO,[1,6] CHUN-HAO HUANG,[1,2,6] NILGUN TASDEMIR,[1,3]
DARJUS F. TSCHAHARGANEH,[1] JOHN E. WILKINSON,[4] AND SCOTT W. LOWE[1,2,5]

[1]*Department of Cancer Biology and Genetics, Memorial Sloan Kettering Cancer Center, New York, New York 10065*

[2]*Weill Cornell Graduate School of Medical Sciences, Cornell University, New York, New York 10065*

[3]*Watson School of Biological Sciences, Cold Spring Harbor Laboratory, Cold Spring Harbor, New York 11724*

[4]*ULAM/Department of Pathology, University of Michigan School of Medicine, Ann Arbor, Michigan 48109*

[5]*Howard Hughes Medical Institute, Memorial Sloan Kettering Cancer Center, New York, New York 10065*

Correspondence: lowes@mskcc.org

Rapid and affordable tumor profiling has led to an explosion of genomic data that is facilitating the development of new cancer therapies. The potential of therapeutic strategies aimed at inactivating the oncogenic lesions that contribute to the aberrant survival and proliferation of tumor cells has yielded remarkable success in some malignancies such as BRAF-mutant melanoma and BCR-ABL expressing chronic myeloid leukemia. However, the direct inhibition of several well-established oncoproteins in some of these cancers is not possible or produces only transient benefits. Functional genomics represents a powerful approach for the identification of vulnerabilities linked to specific genetic alterations and has provided substantial insights into cancer signaling networks. Still, as inhibition of gene function can have diverse effects on both tumor and normal tissues, information on the potency of target inhibition on tumor growth as well as the toxic side effects of target inhibition are also needed. Here, we discuss our RNA interference (RNAi) pipeline for cancer target discovery based on our optimized short-hairpin RNA (shRNA) tools for negative selection screens and inducible RNAi platform that, in combination with embryonic stem cell (ESC)-based genetically engineered mouse models (GEMMs), enable deep in vivo target validation.

TARGET IDENTIFICATION AND VALIDATION IN ONCOLOGY

Cancer genomes have complex landscapes of mutations and diverse types of genetic aberrations (Lawrence et al. 2013). Advances in sequencing technologies have greatly facilitated genome characterization, making it possible to compile a nearly complete catalog of genes altered in human tumors and the development of new therapies that rationally exploit these alterations. In a number of cases where a cancer driver has been functionally validated, particularly in the case of certain activated oncogenes, therapeutic strategies to inhibit the oncogenic event have proven remarkably effective in treating patients with these lesions. However, not all oncogenic events are druggable by traditional pharmacologic approaches, and this general approach cannot exploit loss of tumor suppressors. As a consequence, genome characterization alone has not identified obvious targets for a large fraction of human cancers. Therefore, it seems likely that the combination of genomic and functional approaches will be needed to identify new cancer vulnerabilities that lead to the rational development of effective therapies.

In principle, an ideal cancer target would be required for the ongoing proliferation and/or survival of cancer cells but not normal tissues. Such targets can be classified into three general categories: (1) the cancer "driver" itself, whose direct inhibition can produce an anticancer effect; (2) a "client" of a cancer driver that mediates its tumor-promoting effect and whose inhibition achieves a similar outcome as would targeting the driver itself; and (3) a "synthetic lethal" target that reflects an induced dependency of a cancer cell harboring a particular driving genetic event. Functional genomic approaches are ideal for identifying targets in classes 2 and 3. Still, even when a genotype-specific vulnerability is identified, it remains to be determined how robust target inhibition will be in an in vivo context and whether there may be deleterious effects to normal tissues. As discussed below, functional genomic techniques and new animal modeling approaches provide one path forward.

---

[6]Co-first authors.

Published by Cold Spring Harbor Laboratory Press; doi: 10.1101/sqb.2016.81.031096

RNA interference (RNAi) technology is a powerful tool for identifying genes whose suppression affects cell proliferation and viability. Although genome-wide RNAi screens have been widely used for the identification of new therapeutic targets in cancer, these screens are technically challenging, tedious, and error-prone because of the high complexity of the short-hairpin RNA (shRNA) libraries. During the past few years, we have incorporated an RNAi platform optimized for negative selection screens that consists of three modular components. First, we use miR30-based shRNA systems that enable robust gene knockdown from constitutive and inducible promoters and that link the shRNAs to fluorescent reporters, allowing the facile tracking and/or sorting of shRNA-expressing cells (Hemann et al. 2003; Dickins et al. 2005, 2007; Zuber et al. 2011a). This tracking system significantly increases the sensitivity of our screens by reducing the noise derived from cells that do not efficiently induce the shRNA. Second, we take a systematic approach to identify shRNAs optimized for potency and specificity (Fellmann et al. 2011). Third, we use focused shRNA libraries targeting genes hypothesized to be important for the desired phenotype, which reduces the library complexity and increases the likelihood of obtaining robust and reproducible results.

Using this conceptual approach, we performed several in vivo RNAi screens and identified a series of relevant tumor suppressors (Zender et al. 2008; Bric et al. 2009; Scuoppo et al. 2012). We also identified Brd4 as crucial for the maintenance of acute myeloid leukemia (AML) and platelet-derived growth factor receptor beta (PDGFRβ) as an antimetastasis target for tumors harboring mutant p53 (Zuber et al. 2011b; Weissmueller et al. 2014). More recently, we uncovered a synthetic lethal interaction between MYC-overexpressing hepatocellular carcinomas (HCCs) and CDK9 inhibition, and identified that fibroblast growth factor receptor 1 (FGFR1) inhibition confers hypersensitivity to MEK inhibition in KRAS-mutant lung and pancreatic cancer cells (Huang et al. 2014; Manchado et al. 2016). Importantly, all of the principles used in these studies can be directly applied to functional genomic approaches using variations on clustered regularly interspaced short palindromic repeat (CRISPR)–Cas9 genome editing technology.

A key aspect in the process of identifying a new cancer target is validating whether the target is required for the maintenance of the cancer phenotype in the complex physiological setting of in vivo tumor development. Mouse models represent one "gold standard" system for validating candidate targets. Tumor cell line xenograft models have been widely used for in vivo validation of candidate targets, because of their tractability, flexibility, and potential to reflect the genetic diversity of human tumors. However, the recipient mice in these subcutaneous graft models are immune-deficient and the tumor grows in an atypical site, thus not capturing the complex interaction between tumor cells, stroma, and immune system. To partially address these issues, our pipeline incorporates an approach to rapidly transit hits from a genetic screen to in depth in vivo target validation using genetically and pathologically accurate cancer mouse models.

Our validation approach is based on our inducible shRNA platform that enables suppressing gene function in established tumors instead of at the time of cell engraftment or, in genetically engineered models, at tumor initiation. The most elegant iteration of the method is based on the derivation of ESC lines that harbor multiple conditional cancer-predisposing alleles, tissue-specific Cre recombinases, and reverse tet-transactivators that, in combination, are used directly to generate a functional cancer mouse model without the extensive strain intercrossing required with conventional methods. These models allow us to not only genetically validate candidate targets in a physiologic context and with an unprecedented depth but also to anticipate potential toxicities associated with the therapeutic strategy, because the target is inhibited in the majority of tissue types and not just in the tumor cells.

Here, we summarize our RNAi pipeline that uses, in many cases, the same reagents to seamlessly go from genetic screens for target identification to deep in vivo validation approaches that stand in contrast to previous efforts that have simply looked at genetic dependencies in cell lines or xenograft models. We first discuss and illustrate our RNAi screening approach for target discovery by describing two shRNA screens aimed at identifying vulnerabilities in Myc-driven HCC and mechanisms of resistance to MEK inhibition in KRAS-mutant lung cancer. We then summarize our GEMM-ESC platform for in depth in vivo target validation and for anticipating toxicities associated to target inhibition.

RNAi SCREENS FOR TARGET IDENTIFICATION

Much of the current thinking in the identification and development of novel cancer targets involves the concept of *precision medicine*—specifically the concept of matching patients' tumors with specific genetic configurations and drug treatments predicted to target those lesions. The potential of such strategies has been revealed by the success of imatinib for BCR-ABL-expressing chronic myeloid leukemia and gefitinib or erlotinib for epidermal growth factor receptor (EGFR)-mutant lung adenocarcinoma. However, for some genetic lesions (e.g., amplified *MYC* or mutant *KRAS*), this has not been possible.

Functional genomics provides a nonbiased approach to uncover vulnerabilities linked to specific genetic alterations and has been widely applied toward this goal over the last decade. In its most basic iteration, RNAi or CRISPR libraries are introduced into cells and the representation of individual shRNAs or single-guide RNAs (sgRNAs) is assessed by next-generation sequencing, with the expectation that cells expressing shRNAs or sgRNAs targeting genes required for proliferation or survival will be depleted with time or under specific conditions (e.g., drug treatments).

Like many laboratories, we use functional genomics to identify cancer drug targets, focusing on exploiting vulnerabilities produced by important but undruggable cancer drivers. Below, we summarize our efforts to optimize shRNA screens for the identification of cancer therapeutic targets and illustrate this approach with two recent screens performed in our laboratory, which identified (i) CDK9—a key component of the positive transcription elongation factor b (P-TEFb) complex—as required for the proliferation of MYC-overexpressing HCC (Huang et al. 2014) and (ii) FGFR1 as a key mediator of a feedback signaling loop that induces adaptive resistance to MEK inhibition in KRAS-mutant lung cancer cells (Manchado et al. 2016).

Our screening approach strives to reduce library complexity and efficiently produces customized shRNA libraries in vectors appropriate for the screen. Using low complexity libraries makes screenings technically easier and more cost effective by requiring less cell culture, and increases the likelihood of identifying significant targets by allowing more replicates and enabling more accurate shRNA quantitation than when performing error-prone genome-wide screens. To achieve this, we generate focused lists of gene candidates based on the biology of the system that we would like to interrogate. For example, to identify new therapeutic strategies in MYC-overexpressing HCCs, we generated a customized shRNA library capable of suppressing proteins for which small molecule inhibitors are available (Huang et al. 2014). For identifying sensitizers of KRAS-mutant lung cancer cells to MEK inhibition, we screened a kinase library by hypothesizing that the activation of kinase signaling cascades mediates signaling reactivation and adaptive resistance to MEK inhibitors (Manchado et al. 2016). By taking this focused approach, we were able to follow up positive hits with relevant functional experiments, such as using small molecule inhibitors predicted to mimic genetic perturbations.

In addition to developing a biological hypothesis to focus our screen, we have invested heavily in optimizing tools and methods to produce robust results. Indeed, a multiplexed RNAi screen requires shRNAs that efficiently suppress their target and a system that allows the efficient delivery and expression of the shRNAs in the target cells. To meet this requirement, we developed an algorithm derived from analyzing data from a biological reporter assay that functionally evaluates shRNA potency on a large scale (Fellmann et al. 2011). To minimize false negatives, enhance the ability to study the loss of function of proliferation and survival genes, and provide a feasible system to efficiently deliver the shRNAs, we developed optimized systems such as the TRMPV tet-responsive miR30-based retroviral vector (Zuber et al. 2011a). This vector enables inducible shRNA expression and, by coupling the shRNA to a fluorescent reporter, allows only those cells expressing the shRNA to be analyzed, increasing the sensitivity of our screens. By using this approach, we produced kinase and drug target gene shRNA libraries that harbor remarkably accurate shRNA sequences of relatively equal distribution, which increases the likelihood of obtaining robust results (Huang et al. 2014; Manchado et al. 2016).

To identify targets specifically required for the proliferation/survival of Myc-overexpressing murine HCC cells or synthetic lethal interactions with the MEK inhibitor trametinib in human KRAS-mutant lung cancer cells, the vector libraries described above (human kinome and murine "drug target" genes) containing inducible shRNAs and a neomycin resistance gene were transduced in triplicate into either Myc-overexpressing or KRAS-mutant cancer cells under conditions that predominantly result in a single retroviral integration per cell. Our shRNA libraries included six informatically predicted shRNAs per gene and shRNAs targeting "neutral" and "essential" genes (Huang et al. 2014; Manchado et al. 2016), which provide negative and positive controls, respectively, help to calibrate sequencing results and normalize between independent experiments. Upon shRNA transduction and selection with neomycin, cells were treated with doxycycline (to allow shRNA induction) or with doxycycline together with trametinib to identify synthetic lethal interactions with MEK inhibition in KRAS-mutant cells, and were cultured for 10 population doublings under conditions such that each shRNA was represented in at least 1000 cells through the course of the experiment.

To determine shRNA representation, shRNA cassettes were amplified from genomic DNA and quantified by deep sequencing. Importantly, the use of a vector that links the shRNA to a fluorescent reporter allowed us to evaluate only those cells in which the shRNA was expressed (by fluorescence-activated cell sorting) and eliminated sequencing noise from silenced elements. Upon sequencing, data were analyzed using a custom script and the fold change of shRNA representation was calculated by comparing normalized shRNA reads at t_f to normalized reads at t_0, after eliminating from the analysis those shRNAs underrepresented in the initial screen samples. As a general rule, we require that candidate genes for further validation have at least two shRNAs that show more than fourfold average depletion and FDR < 0.05. Following this precise method we identified Cdk9, Psmb2, and Mcm6 as potential targets in Myc-overexpressing HCC and FGFR1 and other mitogen-activated protein kinase (MAPK) signaling effectors as sensitizers of KRAS-mutant lung cancer cells to MEK inhibition (Fig. 1; Huang et al. 2014; Manchado et al. 2016).

On the basis of the above experiences, we believe that our focused shRNA libraries and screening pipeline are capable of generating robust and reproducible results in both murine and human systems. By screening shRNA libraries targeting known drug target genes and human kinases in a murine HCC model driven by Myc overexpression and p53 loss and in a KRAS-mutant lung cancer cell line, respectively, we described a critical role for CDK9 in MYC-dependent transcription elongation and HCC maintenance and showed that the MEK inhibitor trametinib provokes a compensatory response involving FGFR1 that leads to adaptive drug resistance (Huang et al. 2014; Manchado et al. 2016).

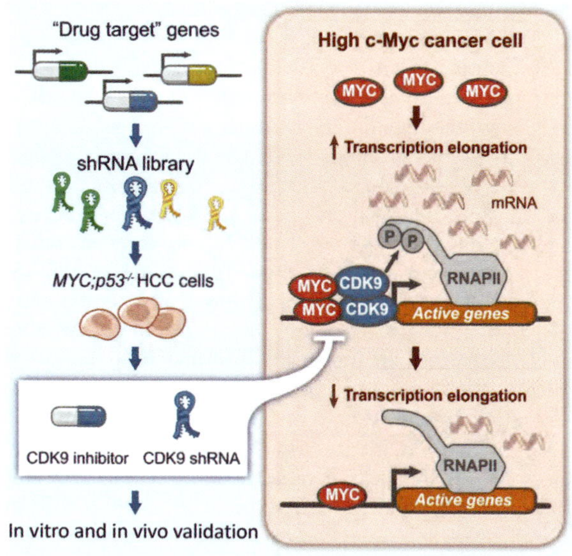

Figure 1. Schematic of a focused RNAi screen for genes encoding known drug targets. By screening a custom library of short-hairpin RNAs (shRNAs) directed toward known drug targets in a genetically defined Myc-driven hepatocellular carcinoma (HCC) model, we identified CDK9—a key component of the positive transcription elongation factor b (P-TEFb) complex—as required for MYC-overexpressing HCC. Results from shRNA-mediated and pharmacological CDK9 inhibition together highlight the relevance of transcription elongation in the addiction of cancer cells to MYC and suggest that targeting MYC-mediated transcription elongation can be an effective strategy for the treatment of MYC-overexpressing tumors.

It is noteworthy that the emergence of CRISPR–Cas9 technologies (Doudna and Charpentier 2014) has greatly facilitated functional genomics efforts (Shalem et al. 2015). Importantly, it is becoming increasingly clear that CRISPR–Cas9 technologies represent a powerful orthogonal approach to the RNAi methods described above (Deans et al. 2016). This method, in its simplest form, involves the CRISPR-associated Cas9 endonuclease and a sgRNA, can be experimentally harnessed to target and cleave DNA in an exquisite sequence-specific manner (Doudna and Charpentier 2014). One major advantage of CRISPR is that a much larger fraction of sgRNAs is effective when compared with shRNAs, making it less likely to have false-negative results (Morgens et al. 2016). Also, unlike RNAi, CRISPR–Cas can produce true null alleles. Additional iterations of the CRISPR–Cas methodology using nuclease-deficient Cas9 can enable inducible and reversible gene activation and repression (Shalem et al. 2015). Although these methods are still in their infancy, they will undoubtedly further enhance the capabilities of functional genomics screens.

Still, for many therapeutic targets, hypomorphic effects from RNAi are desirable and can help identify a realistic therapeutic window for inhibition of essential gene products. Recent efforts to increase the knockdown efficiency of miR-30-based shRNAs led to an enhanced microRNA-based hairpin, termed miR-E (Fellmann et al. 2013). The utility of single-copy miR-E shRNAs to facilitate the functional annotation of the genome has also been evaluated and validated in multiple independent studies (Huang et al. 2014; Schwickert et al. 2014; Watanabe et al. 2016). Moreover, new algorithms for microRNA-based shRNA technologies are also increasing the fraction of functional shRNAs in libraries (Pelossof et al., unpubl.). Consequently, our and other laboratories (Deans et al. 2016) are currently using RNAi and CRISPR as two orthogonal approaches that reinforce each other and believe that such an integrative approach will maximize the information obtainable from functional genomics screens.

RNAi FOR TARGET VALIDATION: IN VITRO AND IN VIVO APPROACHES

In Vitro Target Validation

A critical step in the process of target discovery involves the validation of the potential cancer targets. In our screening pipeline, target candidates identified following our scoring criteria are validated through several methods. First of all, candidate shRNAs are individually tested in competitive in vitro proliferation assays together with positive control shRNAs targeting genes such as *PLK1*, *RPA1*, and the oncogenic drivers (i.e., *Myc* and *KRAS*), and negative control shRNAs targeting Renilla luciferase. To partially recapitulate the settings of the screen, shRNA-transduced cells expressing a fluorescent reporter are mixed with nontransduced cells in an equal proportion and the percentage of each population is quantified over time by flow cytometry.

Next, to assess the absolute effect of target inhibition on cell proliferation and survival, we test the effect of candidate shRNAs in clonogenic and apoptosis assays in multiple cell lines. For shRNAs identified as synthetic lethal partners with a drug, we also examine whether the effects are additive or synergistic using different approaches such as the Chou–Talalay or Loewe methods (Loewe 1953; Chou 2010). For situations where small molecule inhibitors targeting the candidate gene exist, we test whether the drug recapitulates the effect of the shRNAs. We examine how the drug affects the proliferation and survival of cancer cells by using in vitro clonogenic assays and by measuring apoptosis (annexin V/propidium iodide), and we examine the synergistic activity of the drug combination using the methods mentioned above. Highlighting the robustness of our screening approach, we usually validate ~80% of the candidate genes identified following the scoring criteria described above.

As we aim to identify targets whose inhibition show selective antiproliferative effects in cancer cells or in cells from a certain tumor type, we examine the effect of suppressing the positive hits in nontransformed cells and in cancer cells from different lineages or with different genetic drivers. For example, we validated the antiproliferative effects of CDK9 inhibition in Myc-overexpressing HCC cells, but observed a limited effect in nontransformed liver cells (Huang et al. 2014). Similarly, FGFR1 inhibition showed combined activity with trametinib in KRAS-mutant pancreatic and lung cancer cells, but not in KRAS wild-type lung cancer cells or KRAS-mutant colorectal lines (Manchado et al. 2016).

An important aspect of target validation is confirming that the phenotypes induced by shRNAs are due to on-target effects. To do this, we use several orthogonal methods: (i) We validate our results in multiple cancer cell lines harboring different genetic drivers and derived from different tumor tissue types; (ii) we ensure that multiple shRNAs capable of knocking down the target protein produce similar results; (iii) we confirm that CRISPR–Cas9 genome editing targeting of the candidates recapitulates the shRNA effects; (iv) if available, we test and compare the effect of small molecule inhibitors targeting the hit; and (v) we perform rescue experiments by expressing the cDNA of the candidate gene containing a silent mutation in the shRNA target region. In the case where cDNA rescue experiments work, we also test whether introduction of a mutant cDNA defective in the druggable activity (e.g., kinase dead mutant) are unable to rescue the phenotype, thereby providing evidence that small molecule inhibitors might produce similar phenotypes.

In cases where multiple targets are identified and validated, we prioritize, for further in vivo validation, those candidate targets that are (1) validated across multiple human and murine systems, (2) targetable with selective and potent small molecule inhibitors, and/or (3) abnormally expressed (up-regulated/mutated) in tumor samples corresponding to the tumor type under study. Of course, all of the screening results are considered in the context of current knowledge, and specific hits can be pursued based on a potential understanding of the underlying mechanism.

In Vivo Target Validation

Assessing the relevance of a candidate target on tumor maintenance in vivo is a critical aspect of our target validation process. A powerful feature of our pipeline for identifying and validating therapeutic targets is the ability to seamlessly take candidate results identified through in vitro screens to rapid genetic validation in genetically and pathologically accurate mouse models of human cancer. Our in vivo validation method consists of two approaches that take advantage of the inducible shRNA platform described above. First, we test whether target knockdown has a substantial impact on tumor development in tumor cell line xenograft models. Second, we incorporate ESC GEMMs that harbor multiple conditional disease-associated alleles (Premsrirut et al. 2011; Dow et al. 2012). The prototypical ESC GEMM is produced from ESCs carrying at least four alleles: (1) one or more cancer-predisposing conditional alleles that can be activated by Cre recombinase (e.g., "lox-stop-lox" *Kras*^{G12D}; *Tp53*$^{flox/flox}$); (2) a tissue-specific Cre recombinase; (3) a Cre-activatable lox-stop-lox *rtTA3* allele to enable tissue-specific induction of a tet-responsive shRNA; and (4) a "homing cassette" that enables site-specific integration of the tet-responsive shRNA downstream from the *Col1a1* locus using recombination-mediated cassette exchange. Other iterations omit the tissue-specific Cre (enabling somatic activation using Cre-expressing adenoviruses or lentiviruses) or link the *rtTA3* allele to a fluorescent reporter (providing lineage tracing of Cre-recombined cells). For example, the *CAGs-LSL-rtTA3-IRES-KATE* (CAGs-RIK) transgene requires Cre activity to express the rtTA3, which is linked to the fluorescent reporter Kate and enables tissue-specific, Dox-dependent control of shRNA expression (Dow et al. 2014). These ES cells are then used to produce mice via tetraploid embryo complementation or blastocyst injection to generate multiallelic mouse models without extensive strain intercrossing.

Once established, multiallelic ES cell lines can be further manipulated in vitro (to introduce additional genetic complexity), for instance, by genome editing of other disease modifying genes or through recombination of a genetic element into the homing cassette. As additional genetic alterations are introduced in culture, multiple genes can be assessed in parallel without the need for large, expensive breeding colonies. Moreover, virtually all of the derived mice have a relevant genotype and can be used in experimental studies (Premsrirut et al. 2011). In one validation of the system, we generated an ESC GEMM lung adenocarcinoma model harboring LSL-*Kras*^{G12D}; CAGs-RIK; and tetracycline response element (TRE)-shKras transgenes and showed that induction of the Kras shRNA in established adenomas by doxycycline treatment (12 wk upon mutant Kras activation by intratracheal delivery of Adenovirus Cre) led to complete tumor regression compared with mice expressing a control shRNA (sh.REN) (Fig. 2). Similar results are observed in an ESC GEMM of Myc-overexpressing HCC, where induction of Cdk9 shRNAs blocked tumor progression when expressed in established liver tumors (data not shown).

One advantage of our ESC GEMM approach is that it produces genetically and physiologically accurate preclinical models for studying the impact of small molecule regimens in established tumors. In situations where a drug against the target exists, we use the in vivo models described above to compare and contrast the impact of genetic versus small molecule action, and begin to gain insights into factors that influence de novo or acquired resistance to such pharmacological treatments. Illustrating this point, we recently validated the combined inhibition of MEK and FGFR1 as a therapeutic strategy using a GEMM of *Kras*^{G12D} lung adenocarcinoma. Although trametinib and ponatinib, an FDA-approved multikinase inhibitors that inhibit FGFR1, had minor antitumor effects as single agents, the drug combination caused significant tumor regression characterized by massive necrosis (Fig. 3; Manchado et al. 2016).

The above-described ESC GEMMs enable deep and robust validation of potential cancer targets identified through shRNA screens. Of course, human-based models such as patient-derived xenograft systems are also required to establish human relevance (Manchado et al. 2016). By combining these and other orthogonal meth-

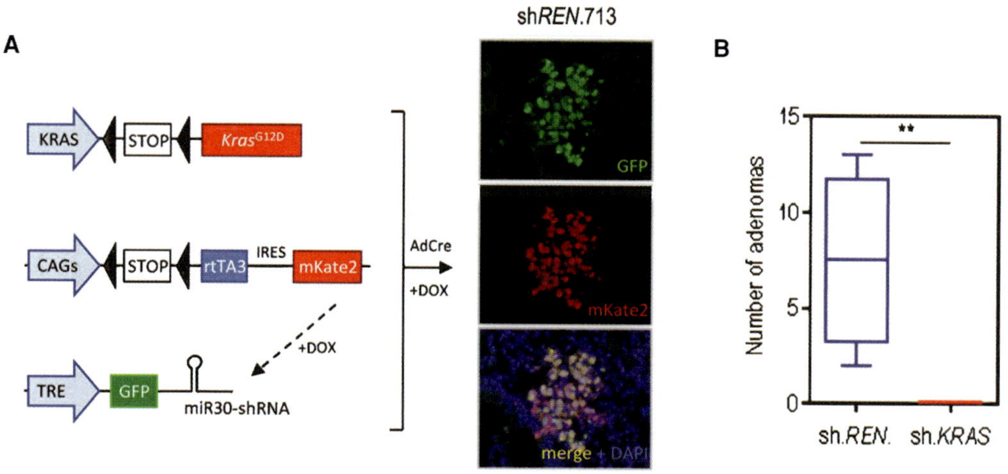

Figure 2. Embryonic stem cell (ESC)-based genetically engineered mouse models (GEMMs) for in vivo target validation. (*A*) Schematic representation of LSL-*Kras*^{G12D}; *CAGs-LSL-rtTA3-IRES-mKate* (RIK); and *TRE-GFP-miR30* alleles, before Cre-mediated recombination. Immunofluorescent stains for green fluorescent protein (GFP) and mKate2 in lung sections of triple transgenic mice (*CAGs-LSL-rtTA3-IRES-Kate*/sh-Ren.713/LSL-*Kras*^{G12D}). (*B*) Quantification of lung adenomas in LSL-*Kras*^{G12D} carrying either a control (Ren.713) or Kras-targeted short-hairpin RNA (shRNA) following 2 wk of treatment with doxycycline (**, $P < 0.01$). All mouse experiments were approved by the Memorial Sloan Kettering Cancer Center (MSKCC) Animal Care and Use Committee (protocol no. 12-04-006). Statistical significance was calculated by two-tailed Student's *t*-test (Prism 6 software). Significance values are *, $P < 0.05$; **, $P < 0.01$; and ***, $P < 0.001$.

ods, we strive to capture the genetic complexity and heterogeneity of the human disease, as well as the in vivo physiological tumor environment of genetically and pathologically accurate cancer mouse models.

ESTABLISHING A THERAPEUTIC INDEX USING IN VIVO RNAi

A crucial but often overlooked step of target validation is identifying the potential toxicities that can arise from target inhibition in normal tissues. Although many of the studies that strive to identify targets through functional genomics confirm a differential sensitivity between tumor and normal cells of the same tissue, they rarely address the critical issues that may produce dose-limiting toxicities in vivo (e.g., toxicity to the hematopoietic system or the gut). All too often, the liabilities of certain targets are not identified until toxicity studies are performed on clinical candidate compounds. In some instances, these toxicities can be "off-target," arising from a lack of drug specificity and the deleterious consequences of inhibiting an unintended target. Although such toxicities can be overcome by producing more selective agents, other toxicities are "on-target" and reflect the essential function of the target protein in certain normal tissues. Knowledge of potential "on-target" toxicities would be extremely valuable before initiating expensive drug development efforts.

Genetic studies in mouse models provide one way to reveal the function of a gene product in specific tissues or throughout the animal and, in principle, can identify pathologies associated with inhibiting a candidate drug target. In practice, these models are rarely used for this purpose. This underutilization is partially due to concerns that the consequences of gene deletion during development may not reflect the impact of acute protein inhibition in the adult. In addition, it has historically been extraordinarily time consuming and expensive to produce the appropriate models, and the relevant approaches are typically best suited for the academic setting where questions related to drug toxicity are not a priority. To address some of these limitations, we combine the optimized shRNA technology above with modeling approaches that allow acute and reversible gene suppression in normal tissues and used these to study the requirement for candidate drug targets across a wide range of adult tissues. Although recognizing that genetic studies in mice will never fully reflect the impact of drugs on their targets in normal human tissues, we believe that this approach will facilitate the assessment of on-target toxicities associated with target inhibition and modify the course of drug development. To illustrate our approach, we describe efforts using RNAi to explore the consequences of liver-specific Cdk9 suppression and systemic Brd4 suppression in vivo.

Liver Regeneration for Assessing Toxicity of Target Inhibition

As described above, the identification of CDK9 as an important maintenance gene in MYC-overexpressing cancers fueled interest in targeting transcription elongation as an anti-MYC strategy (Huang et al. 2014). CDK9, a key transcription elongation factor, plays an important role in transcriptional amplification and is involved in many essential processes. Still, there is evidence that some cells depend more on CDK9 than others, and as such it is considered not only as a promising drug target for cancer but also for diseases such as cardiac hypertro-

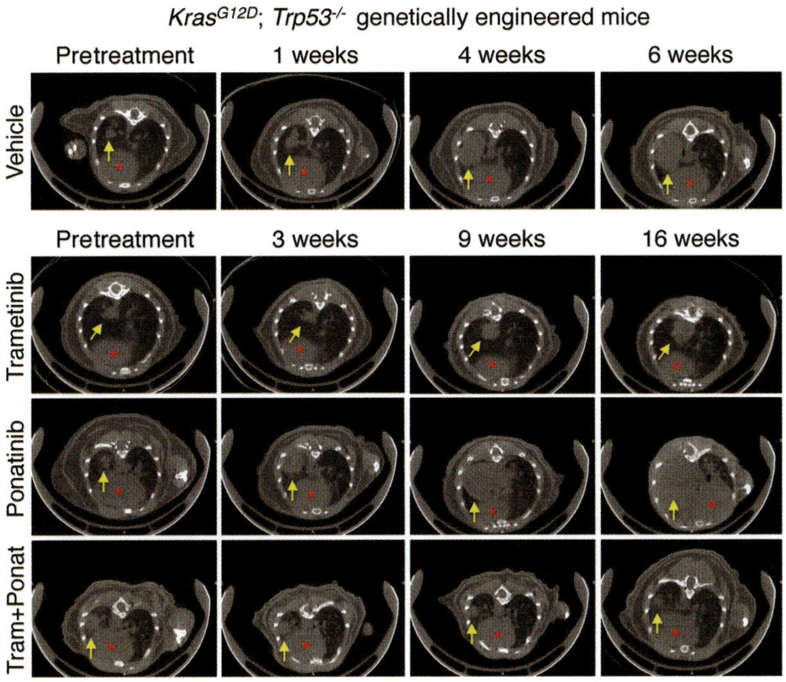

Figure 3. Ponatinib in combination with trametinib leads to tumor regression of KRAS-mutant lung tumors. Representative μCT images of the lungs of $Kras^{G12D}$; $Trp53^{-/-}$ genetically engineered mice treated with vehicle, trametinib (3 mg/kg), ponatinib (30 mg/kg), or both drugs in combination for the indicated times. Lung tumors are indicated by yellow arrows and red asterisks mark the hearts. All mouse experiments were approved by the Memorial Sloan Kettering Cancer Center (MSKCC) Animal Care and Use Committee (protocol no. 12-04-006).

phy (Krystof et al. 2010) and HIV/AIDS (Wei et al. 1998), among others (Wang and Fischer 2008). Although several CDK9 inhibitors are available, these agents are invariably toxic (Wang and Fischer 2008), most likely because of their lack of exquisite on-target selectivity to CDK9 and frequent off-target inhibition of other members of the CDK family. As a consequence, it is not clear whether these toxicities are on- or off-target, and the requirement for CDK9 function in normal adult tissues remains unknown.

Our studies suggested that CDK9 inhibitors would be particularly effective in MYC-overexpressing HCC (Huang et al. 2014). Therefore, as a first step to assess whether CDK9 inhibition in normal hepatocytes has deleterious side effects, we performed hydrodynamic tail vein injections of transposons expressing Cdk9 or Myc shRNAs followed by partial hepatectomy (Fig. 4A). This approach leads to suppression of target genes in ~20% of hepatocytes, as revealed by GFP immunofluorescence staining (Fig. 4B,C). Suppression of replication protein A, subunit 3 (Rpa3), an essential gene in adult mice (McJunkin et al. 2011), caused about 10% reduction of green fluorescent protein (GFP)-positive hepatocytes after partial hepatectomy, which led to lower liver to body weight ratios (Fig. 4B–E). In contrast, we found that suppression of Cdk9 or Myc did not show significant effects on the proliferation of hepatocytes and liver regeneration (Fig. 4B–E) but completely abrogated Myc-driven tumor formation (Huang et al. 2014). Interestingly, a previous study also reported that conditional deletion of Myc does not impair liver regeneration (Li et al. 2006), further supporting our results using RNAi. Together, these relatively straightforward genetic experiments enabled by tracking a reporter protein in vivo provide strong evidence for a therapeutic window by directly targeting CDK9 in livers harboring MYC-overexpressing HCC.

Studying the phenotypes associated with silencing CDK9 or its Cyclin partners (T1, T2, or K) systemically in vivo using genetic approaches has been challenging owing to embryonic lethality (Kohoutek 2009). Thus, further investigating the consequence of systemic Cdk9 suppression in adult mice using RNAi will be important to identify potential toxicities associated with CDK9 inhibition and understand its functions in vivo. Indeed, studies are underway to use ESC GEMMs to inducibly and systemically inhibit Cdk9 as discussed for Brd4 below.

Inducible and Reversible Transgenic RNAi Mice for Assessing Target Inhibition-Associated Toxicity

As described above, our laboratory has established an ESC GEMM platform for the production of transgenic mice harboring tetracycline/doxycycline (dox) responsive shRNAs that allows temporal and spatial control of endogenous gene expression (Premsrirut et al. 2011). Because shRNAs do not modify the target gene, this system enables reversible gene silencing (Premsrirut et al. 2011; Dow et al. 2012), thereby facilitating assessment of reversible and irreversible loss of function phe-

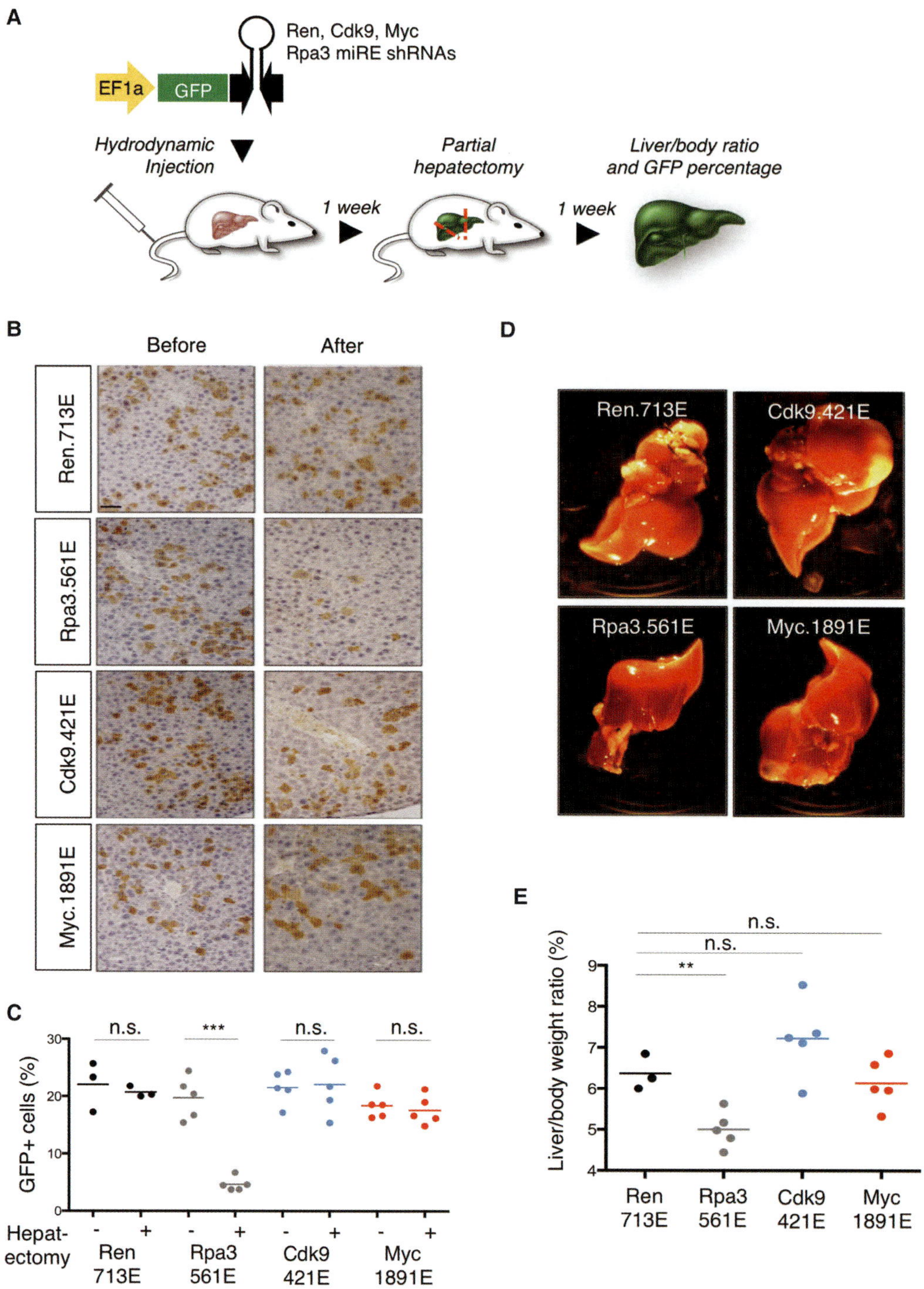

Figure 4. Liver regeneration for assessing toxicity of Cdk9 suppression in vivo. (*A*) Schematic representation of the liver regeneration. miR-E short-hairpin RNA (shRNA) transposon vectors targeting control Renilla luciferase, Cdk9, Myc, or Rpa3 were injected together with CMV-SB13 transposase by hydrodynamic tail vein injection into female C57BL/6 mice (6–8 wk of age) from Harlan Laboratories as previously described (Huang et al. 2014). Partial (2/3) hepatectomy (Taub 2004) was performed after 1 wk. Before being anesthetized with isoflurane, the animals received one subcutaneous dose of buprenorphine (0.5 mg/kg) and meloxicam (2 mg/kg). The median lobe, with its two central portions together with the left lateral lobe were mobilized, looped with a 4-0 vicryl ligation (Ethicon), and two-thirds of the liver volume were removed. Liver/body ratio and green fluorescent protein (GFP) percentage were examined after 2 wk. (*B*) Representative histological analysis of livers stained for GFP. Scale bar, 50 µm. (*C*) Cdk9 or Myc suppression does not show significant impact on the percentage of GFP^+ cells before and after partial hepatectomy. (*D*) Representative images of livers injected with the indicated shRNAs after partial hepatectomy. (*E*) Cdk9 or Myc suppression does not show significant impact on liver/body ratio compared with Ren.713E (neutral control shRNA). Rpa3.561E is used as a positive control shRNA. All mouse experiments were approved by the Memorial Sloan Kettering Cancer Center (MSKCC) Animal Care and Use Committee (protocol no. 11-06-011). Statistical significance was calculated by two-tailed Student's *t*-test (Prism 6 software). Significance values are *, $P < 0.05$; **, $P < 0.01$; and ***, $P < 0.001$.

notypes. For example, we have studied the functions of Apc (Premsrirut et al. 2011; Dow et al. 2015), Rpa3 (McJunkin et al. 2011), and Pten (Miething et al. 2014) under different conditions in vivo using inducible and reversible RNAi.

More recently, we began harnessing this platform to study drug target genes identified from the RNAi screening pipeline described earlier. When applied for target validation, such system can discover potential toxicities associated with on-target gene suppression, determine the kinetics of the toxicities upon target inhibition, and examine whether any or all are reversible. To illustrate the feasibility of this approach, we generated inducible Brd4 shRNA transgenic mice to identify the consequences of Brd4 inhibition in adult tissues (Bolden et al. 2014). Suppression of Brd4 in the hematopoietic system produced defects in hematopoiesis, with an impact on the proliferation of several lineages, and a general depletion of stem cells (Bolden et al. 2014). In addition, acute Brd4 inhibition disrupted the epidermis plus hair follicles and triggered the depletion of multiple cell types in the intestine (Bolden et al. 2014). Remarkably, these phenotypes were rapidly reversible, and mice completely recovered after restoration of normal Brd4 levels. Collectively, these data suggest that "on-target" Brd4 inhibition can produce deleterious pathologies but raise hope that these might be managed by dose and schedule adjustment.

The studies described above use mice in which the GFP/Brd4 shRNA cassette was expressed using the TRE-tight promoter, which we subsequently realized limited expression of the shRNAs to only a subset of tissues (Bolden et al. 2014). In more recent efforts, we have largely eliminated these tissue restrictions by expressing the shRNA from the conventional TRE promoter. To see whether this system would identify additional pathologies for Brd4 inhibition, ESC clones were targeted with two independent GFP-linked Brd4 miR-30 shRNAs expressed downstream from the TRE promoter (hereafter referred to as TG-Brd4.552 and TG-Brd4.1448) (Fig. 5A). Next, we examined the transgenic mice (CAGs-rtTA3; TG-shBrd4 mice) on a dox diet for 2 wk. Both dox-fed Brd4 shRNA mice showed significant weight loss over 2 wk (TG-Brd4.552: ~20%; TG-Brd4.1448: ~6%) compared to control TG-Ren.713 mice (Fig. 5B). Of note, transgenic mice harboring Brd4 shRNAs driven by TRE-tight promoter only showed reduced weight gain (Bolden et al. 2014), suggesting additional toxicities not observed with the TRE-tight system (data not shown). By evaluating the GFP-expressing organs from TG-Brd4.552 and TG-Brd4.1448 mice and comparing to control TG-Ren.713 mice after 2 wk on dox, we identified significant lesions in the pancreas (Fig. 5C). The lesions resulted from extensive scattered acinar cells undergoing necrosis and apoptosis (Fig. 5D), suggesting that Brd4 may play an important role in pancreas homeostasis and function. Together, this systemic RNAi model identified toxicity in the pancreas, in addition to epidermal hyperplasia, alopecia, and stem cell depletion in the small intestine, as described previously (Bolden et al. 2014).

The data from these Cdk9 and Brd4 studies highlight how the RNAi technology can be used to explore in vivo gene-specific toxicities, in principle, even in advance of therapeutic development, and to explore the tolerability of certain therapeutic strategies. Although RNAi-mediated protein knockdown may not always produce the same effects as small molecule protein inhibition, the systems above can examine the effects of acute, partial, and reversible consequences of target inhibition, providing a "first pass" assessment of therapeutic index to mimic phase I studies in clinical trials. Moreover, RNAi can be further used to precisely reveal the toxicities associated with recently developed phthalimide conjugation inhibitors for in vivo target protein degradation such as dBET, a bifunctional thalidomide-conjugated version of JQ1 (Winter et al. 2015). Indeed, clinical trials on various BET inhibitors are well underway, and early data suggest that hematologic and gut toxicities are dose limiting—toxicities that were predicted using our thorough RNAi target validation pipeline (Stathis et al. 2016). Still, to date, skin pathologies have not been reported. Although the basis for this discrepancy remains unknown, it is worth reiterating that the shRNA models show potent and sustained target inhibition 24/7, which may not be achieved with the clinical candidate. Regardless, these models can be effectively used to help anticipate any potential on-target toxicities that might be produced by a new agent.

CONCLUSION

In the past few years, we have made significant advances in understanding the molecular lesions that contribute to tumor formation. Although these efforts have led to the development of a number of successful molecularly targeted therapies, an increase in our molecular understanding of aberrant cancer cell signaling can lead to the functional identification of new vulnerabilities and, consequently, to more efficient therapies. RNAi, as well as emerging genome-editing technologies, represent a powerful way of revealing such cancer dependencies. The approaches described here uniquely exploit optimized RNAi screening systems for target identification as well as in vivo approaches for deep target validation. Together, these optimized RNAi approaches and additional orthogonal approaches, including new genome editing technologies, will facilitate the identification and validation of cancer targets whose further development is likely to produce real clinical benefit.

ACKNOWLEDGMENTS

We thank S. Houlihan, M. Taylor, D. Grace, and J. Simon for excellent technical assistance and F.J. Sánchez-Rivera for carefully editing the manuscript. We also thank members of the Lowe laboratory for stimulating discussions and, in particular, Dr. Charles J. Sherr for important suggestions. We thank Memorial Sloan Ketter-

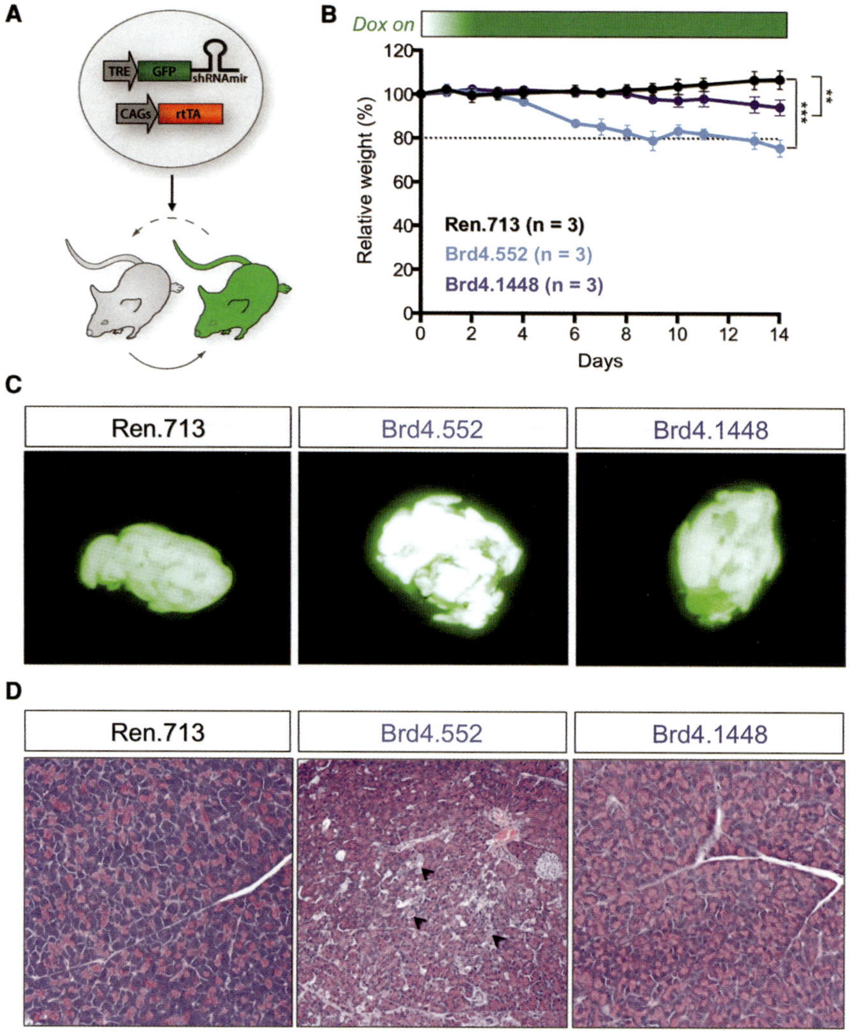

Figure 5. Inducible and reversible transgenic RNA interference (RNAi) mice for assessing toxicities associated with systemic Brd4 suppression. (*A*) Schematic representation of the generation and application of short-hairpin RNA (shRNA) transgenic mice. Tetracycline response element (TRE)-driven miR-30 shRNAs are targeted to the ColA1 locus to drive doxycycline (dox)-dependent genes knockdown in embryonic stem cells and embryonic and adult tissues of the mouse as previously described (Bolden et al. 2014). By further using a sensitive and widely expressed rtTA mice strain, CAGs-rtTA3, GFP-miR-30 shRNAs can be efficiently expressed in most tissues to study on-target toxicities. (*B*) Mean weight changes of male and female (combined) CAGs-rtTA3/+; TG-Ren.713, TG-Brd4.552, and TG-Brd4.1448 mice on dox diet, relative to day 0 of dox treatment. Error bars represent ± SD. (*C*) Green fluorescent protein (GFP) expression in pancreas from CAGs-rtTA3/+; TG-Ren.713, TG-Brd4.552, and TG-Brd4.1448 mice, maintained on dox diet for 2 wk. (*D*) Representative hemotoxylin and eosin (H&E) stains of pancreas from CAGs-rtTA3/+; TG-Ren.713, TG-Brd4.552, and TG-Brd4.1448 mice maintained on dox diet for 2 wk. Arrowheads indicate the extensive scattered acinar cells undergoing necrosis. All mouse experiments were approved by the Memorial Sloan Kettering Cancer Center (MSKCC) Animal Care and Use Committee (protocol no. 11-06-011). Statistical significance was calculated by two-tailed Student's *t*-test (Prism 6 software). Significance values are *, $P < 0.05$; **, $P < 0.01$; and ***, $P < 0.001$.

ing Cancer Center (MSKCC) Animal Facilities and MSKCC Genomics Core. E.M. was supported by The Jane Coffin Childs Memorial Fund for Medical Research and a K99/R00 grant from the National Institutes of Health (NIH)/National Cancer Institute (NCI). C.-H.H. was funded by the Ministry of Education of Taiwan. This work was supported by program project grants from the NCI (S.W.L.) and Cancer Target Discovery and Development Center (S.W.L.). S.W.L. is the Geoffrey Beene Chair for Cancer Biology and a Howard Hughes Medical Institute investigator.

REFERENCES

Bolden JE, Tasdemir N, Dow LE, van Es JH, Wilkinson JE, Zhao Z, Clevers H, Lowe SW. 2014. Inducible in vivo silencing of Brd4 identifies potential toxicities of sustained BET protein inhibition. *Cell Rep* **8:** 1919–1929.

Bric A, Miething C, Bialucha CU, Scuoppo C, Zender L, Krasnitz A, Xuan Z, Zuber J, Wigler M, Hicks J, et al. 2009. Functional identification of tumor-suppressor genes through an in vivo RNA interference screen in a mouse lymphoma model. *Cancer Cell* **16:** 324–335.

Chou TC. 2010. Drug combination studies and their synergy quantification using the Chou–Talalay method. *Cancer Res* **70:** 440–446.

Deans RM, Morgens DW, Okesli A, Pillay S, Horlbeck MA, Kampmann M, Gilbert LA, Li A, Mateo R, Smith M, et al. 2016. Parallel shRNA and CRISPR–Cas9 screens enable antiviral drug target identification. *Nat Chem Biol* **12:** 361–366.

Dickins RA, Hemann MT, Zilfou JT, Simpson DR, Ibarra I, Hannon GJ, Lowe SW. 2005. Probing tumor phenotypes using stable and regulated synthetic microRNA precursors. *Nat Genet* **37:** 1289–1295.

Dickins RA, McJunkin K, Hernando E, Premsrirut PK, Krizhanovsky V, Burgess DJ, Kim SY, Cordon-Cardo C, Zender L, Hannon GJ, et al. 2007. Tissue-specific and reversible RNA interference in transgenic mice. *Nat Genet* **39:** 914–921.

Doudna JA, Charpentier E. 2014. Genome editing. The new frontier of genome engineering with CRISPR–Cas9. *Science* **346:** 1258096.

Dow LE, Premsrirut PK, Zuber J, Fellmann C, McJunkin K, Miething C, Park Y, Dickins RA, Hannon GJ, Lowe SW. 2012. A pipeline for the generation of shRNA transgenic mice. *Nat Protoc* **7:** 374–393.

Dow LE, Nasr Z, Saborowski M, Ebbesen SH, Manchado E, Tasdemir N, Lee T, Pelletier J, Lowe SW. 2014. Conditional reverse tet-transactivator mouse strains for the efficient induction of TRE-regulated transgenes in mice. *PLoS One* **9:** e95236.

Dow LE, O'Rourke KP, Simon J, Tschaharganeh DF, van Es JH, Clevers H, Lowe SW. 2015. Apc restoration promotes cellular differentiation and reestablishes crypt homeostasis in colorectal cancer. *Cell* **161:** 1539–1552.

Fellmann C, Zuber J, McJunkin K, Chang K, Malone CD, Dickins RA, Xu Q, Hengartner MO, Elledge SJ, Hannon GJ, et al. 2011. Functional identification of optimized RNAi triggers using a massively parallel sensor assay. *Mol Cell* **41:** 733–746.

Fellmann C, Hoffmann T, Sridhar V, Hopfgartner B, Muhar M, Roth M, Lai DY, Barbosa IA, Kwon JS, Guan Y, et al. 2013. An optimized microRNA backbone for effective single-copy RNAi. *Cell Rep* **5:** 1704–1713.

Hemann MT, Fridman JS, Zilfou JT, Hernando E, Paddison PJ, Cordon-Cardo C, Hannon GJ, Lowe SW. 2003. An epi-allelic series of p53 hypomorphs created by stable RNAi produces distinct tumor phenotypes in vivo. *Nat Genet* **33:** 396–400.

Huang CH, Lujambio A, Zuber J, Tschaharganeh DF, Doran MG, Evans MJ, Kitzing T, Zhu N, de Stanchina E, Sawyers CL, et al. 2014. CDK9-mediated transcription elongation is required for MYC addiction in hepatocellular carcinoma. *Genes Dev* **28:** 1800–1814.

Kohoutek J. 2009. P-TEFb—The final frontier. *Cell Div* **4:** 19.

Krystof V, Chamrad I, Jorda R, Kohoutek J. 2010. Pharmacological targeting of CDK9 in cardiac hypertrophy. *Med Res Rev* **30:** 646–666.

Lawrence MS, Stojanov P, Polak P, Kryukov GV, Cibulskis K, Sivachenko A, Carter SL, Stewart C, Mermel CH, Roberts SA, et al. 2013. Mutational heterogeneity in cancer and the search for new cancer-associated genes. *Nature* **499:** 214–218.

Li F, Xiang Y, Potter J, Dinavahi R, Dang CV, Lee LA. 2006. Conditional deletion of c-myc does not impair liver regeneration. *Cancer Res* **66:** 5608–5612.

Loewe S. 1953. The problem of synergism and antagonism of combined drugs. *Arzneimittelforschung* **3:** 285–290.

Manchado E, Weissmueller S, Morris JP 4th, Chen CC, Wullenkord R, Lujambio A, de Stanchina E, Poirier JT, Gainor JF, Corcoran RB, et al. 2016. A combinatorial strategy for treating KRAS-mutant lung cancer. *Nature* **534:** 647–651.

McJunkin K, Mazurek A, Premsrirut PK, Zuber J, Dow LE, Simon J, Stillman B, Lowe SW. 2011. Reversible suppression of an essential gene in adult mice using transgenic RNA interference. *Proc Natl Acad Sci* **108:** 7113–7118.

Miething C, Scuoppo C, Bosbach B, Appelmann I, Nakitandwe J, Ma J, Wu G, Lintault L, Auer M, Premsrirut PK, et al. 2014. PTEN action in leukaemia dictated by the tissue microenvironment. *Nature* **510:** 402–406.

Morgens DW, Deans RM, Li A, Bassik MC. 2016. Systematic comparison of CRISPR/Cas9 and RNAi screens for essential genes. *Nat Biotechnol* **34:** 634–636.

Premsrirut PK, Dow LE, Kim SY, Camiolo M, Malone CD, Miething C, Scuoppo C, Zuber J, Dickins RA, Kogan SC, et al. 2011. A rapid and scalable system for studying gene function in mice using conditional RNA interference. *Cell* **145:** 145–158.

Schwickert TA, Tagoh H, Gultekin S, Dakic A, Axelsson E, Minnich M, Ebert A, Werner B, Roth M, Cimmino L, et al. 2014. Stage-specific control of early B cell development by the transcription factor Ikaros. *Nat Immunol* **15:** 283–293.

Scuoppo C, Miething C, Lindqvist L, Reyes J, Ruse C, Appelmann I, Yoon S, Krasnitz A, Teruya-Feldstein J, Pappin D, et al. 2012. A tumour suppressor network relying on the polyamine-hypusine axis. *Nature* **487:** 244–248.

Shalem O, Sanjana NE, Zhang F. 2015. High-throughput functional genomics using CRISPR–Cas9. *Nat Rev Genet* **16:** 299–311.

Stathis A, Zucca E, Bekradda M, Gomez-Roca C, Delord JP, de La Motte Rouge T, Uro-Coste E, de Braud F, Pelosi G, French CA. 2016. Clinical response of carcinomas harboring the BRD4-NUT oncoprotein to the targeted bromodomain inhibitor OTX015/MK-8628. *Cancer Discov* **6:** 492–500.

Taub R. 2004. Liver regeneration: From myth to mechanism. *Nat Rev Mol Cell Biol* **5:** 836–847.

Wang S, Fischer PM. 2008. Cyclin-dependent kinase 9: A key transcriptional regulator and potential drug target in oncology, virology and cardiology. *Trends Pharmacol Sci* **29:** 302–313.

Watanabe C, Cuellar TL, Haley B. 2016. Quantitative evaluation of first, second, and third generation hairpin systems reveals the limit of mammalian vector-based RNAi. *RNA Biol* **13:** 25–33.

Wei P, Garber ME, Fang SM, Fischer WH, Jones KA. 1998. A novel CDK9-associated C-type cyclin interacts directly with HIV-1 Tat and mediates its high-affinity, loop-specific binding to TAR RNA. *Cell* **92:** 451–462.

Weissmueller S, Manchado E, Saborowski M, Morris JP 4th, Wagenblast E, Davis CA, Moon SH, Pfister NT, Tschaharganeh DF, Kitzing T, et al. 2014. Mutant p53 drives pancreatic cancer metastasis through cell-autonomous PDGF receptor β signaling. *Cell* **157:** 382–394.

Winter GE, Buckley DL, Paulk J, Roberts JM, Souza A, Dhe-Paganon S, Bradner JE. 2015. Phthalimide conjugation as a strategy for in vivo target protein degradation. *Science* **348:** 1376–1381.

Zender L, Xue W, Zuber J, Semighini CP, Krasnitz A, Ma B, Zender P, Kubicka S, Luk JM, Schirmacher P, et al. 2008. An oncogenomics-based in vivo RNAi screen identifies tumor suppressors in liver cancer. *Cell* **135:** 852–864.

Zuber J, McJunkin K, Fellmann C, Dow LE, Taylor MJ, Hannon GJ, Lowe SW. 2011a. Toolkit for evaluating genes required for proliferation and survival using tetracycline-regulated RNAi. *Nat Biotechnol* **29:** 79–83.

Zuber J, Shi J, Wang E, Rappaport AR, Herrmann H, Sison EA, Magoon D, Qi J, Blatt K, Wunderlich M, et al. 2011b. RNAi screen identifies Brd4 as a therapeutic target in acute myeloid leukaemia. *Nature* **478:** 524–528.

Single-Cell Analysis of Circulating Tumor Cells as a Window into Tumor Heterogeneity

David T. Miyamoto,[1,2] David T. Ting,[1,3] Mehmet Toner,[4,5] Shyamala Maheswaran,[1,4] and Daniel A. Haber[1,3,6]

[1]*Massachusetts General Hospital Cancer Center, Boston, Massachusetts 02114*

[2]*Department of Radiation Oncology, Massachusetts General Hospital and Harvard Medical School, Boston, Massachusetts 02114*

[3]*Department of Medicine, Massachusetts General Hospital and Harvard Medical School, Boston, Massachusetts 02114*

[4]*Department of Surgery, Massachusetts General Hospital and Harvard Medical School, Boston, Massachusetts 02114*

[5]*Center for Engineering in Medicine, Massachusetts General Hospital and Shriners Hospitals for Children, Charlestown, Massachusetts 02129*

[6]*Howard Hughes Medical Institute, Chevy Chase, Maryland 20815*

Correspondence: dhaber@mgh.harvard.edu

Recent advances in microfluidic approaches have enabled the efficient isolation and detailed molecular characterization of circulating tumor cells (CTCs) in the peripheral blood of patients with cancer. Single-cell molecular analyses of CTCs reveal a tremendous degree of intracellular heterogeneity in CTC populations, reflective of heterogeneity across different patients as well as the underlying heterogeneity of tumors within each individual patient. These studies have enabled the identification of heterogeneous drug resistance mechanisms that can coexist in treatment refractory tumors. CTC analyses also enable serial noninvasive monitoring in patients and can capture the emergence of tumor heterogeneity over time, whether due to tumor evolution through genetic instability or through cellular plasticity. The presence and extent of intratumoral heterogeneity as revealed through the study of CTCs have important clinical implications for understanding and predicting the development of treatment resistance in a variety of solid tumors and for formulating appropriate therapeutic strategies in the effective treatment of cancer.

Major advances in our molecular understanding of cancer have led to the development of effective new therapies that target tumor-specific genetic alterations, often resulting in profound responses in subsets of patients with specific oncogenic drivers in their tumors (Haber et al. 2011; Hyman et al. 2017). However, acquired resistance to therapy remains a pressing problem, and the emergence of resistant subclones within heterogeneous tumors represents a significant barrier to the effective treatment of cancer. Detailed next-generation sequencing studies have showed dramatic spatial and temporal heterogeneity in metastatic and primary tumors from individual patients (Gerlinger et al. 2012; Wu et al. 2012), but such analyses present technical and logistical challenges to routine clinical implementation, particularly for cancers that frequently metastasize to the bone, brain, or lungs. Tumor heterogeneity is especially difficult to study in patients with multiple metastases, because capturing the full range of relevant tumor clones would require invasive biopsies of multiple different individual sites of metastatic disease. Furthermore, traditional tissue biopsies sample only a small portion of a tumor, and thus the range of clones within a heterogeneous tumor may not be fully represented.

The sampling of tumor cells that circulate in the peripheral blood, or "circulating tumor cells" (CTCs), provides an elegant solution to the study of tumor heterogeneity, because CTCs may be a more representative sample of invasive tumor cell populations derived from heterogeneous tumors in multiple sites within an individual patient. In addition, the ability to sample cells in the blood noninvasively at multiple time points allows for the longitudinal study of tumor evolution over time. Recent advances in microfluidic engineering have made possible the development of novel platform technologies that efficiently isolate rare CTCs at the single-cell level, thus enabling the study of single CTCs to gain insights into tumor heterogeneity and treatment resistance in a variety of cancers.

MICROFLUIDIC ISOLATION OF CIRCULATING TUMOR CELLS

Although the existence of circulating populations of tumor cells in the peripheral blood was first posited in 1869 (Ashworth 1869), the isolation and study of CTCs has been a challenging endeavor because of their rarity

Published by Cold Spring Harbor Laboratory Press; doi: 10.1101/sqb.2016.81.031120

and fragility. In the past several years, a variety of methods and technologies have been developed to isolate and enumerate CTCs. In this concise review, we focus on microfluidic CTC isolation technologies that were developed in our laboratories; other CTC isolation technologies have been reviewed in detail elsewhere (Alix-Panabieres and Pantel 2014, 2016; Haber and Velculescu 2014; Miyamoto et al. 2014; Li et al. 2015). In collaboration with our bioengineering colleagues at the Massachusetts General Hospital, we developed a series of microfluidic devices aimed at the gentle and efficient isolation of CTCs from blood. The first- and second-generation microfluidic devices, the $^{\mu p}$CTC-Chip and the ^{HB}CTC-Chip, process blood through microfluidic channels and rely on the physical capture and immobilization of CTCs onto microfluidic surfaces coated with antibodies directed against tumor-specific cell surface antigens (Nagrath et al. 2007; Stott et al. 2010). The third-generation device, the CTC-iChip, is a tumor antigen-independent CTC isolation technology based on the concept of negative depletion, in which leukocytes, erythrocytes, platelets, and noncellular objects are removed from the blood, resulting in the enrichment of an untagged population of CTCs (Ozkumur et al. 2013; Karabacak et al. 2014). Because the CTC-iChip enables the efficient, gentle isolation of intact, untagged CTCs in suspension, it has allowed for detailed single-cell molecular analyses of CTCs from cancer patients and mouse models, thus revealing new insights into tumor heterogeneity at the single-cell level.

The microfluidic CTC-iChip uses three integrated sequential steps, consisting of continuous deterministic lateral displacement for size-based separation of CTCs and leukocytes from whole blood, inertial focusing for precise alignment of cells in a microchannel, and microfluidic magnetophoresis for removal of leukocytes prebound to magnetic beads labeled with anti-CD45, anti-CD16, and anti-CD66b antibodies (Karabacak et al. 2014). This tumor antigen-independent purification method enables the isolation of CTCs without assumptions regarding the nature of cell surface epitopes present on the tumor cells. Using the integrated microfluidic CTC-iChip, up to 10^7 cells per second can be sorted, enabling the sensitive, efficient, and high-throughput isolation of CTCs. The purity of CTCs within the enriched product ranges between 0.01% and 10%, depending on the burden of CTCs present in the patient's circulation. Importantly, CTCs isolated using this method are intact and untagged and amenable to a variety of downstream biological and molecular analyses including immunofluorescence, RNA-in situ hybridization, single-cell RNA-seq, and in vitro culture (Ozkumur et al. 2013; Aceto et al. 2014; Ting et al. 2014; Yu et al. 2014; Miyamoto et al. 2015). These sensitive analyses are possible because the gentle microfluidic enrichment processes of the CTC-iChip preserves the integrity and viability of isolated cells, in contrast to other methods that require cellular fixation and antibody-mediated binding of magnetic beads to cell membrane epitopes. Of note, our single CTC RNA-seq studies suggest that approximately one-half of CTCs have RNA at various stages of degradation even when isolated using this gentle microfluidic method, indicating that many cells in the circulation are nonviable (Ting et al. 2014; Miyamoto et al. 2015). Nevertheless, the quality and integrity of RNA in CTCs isolated using the CTC-iChip are well-preserved compared with prior methods, thus enabling single-cell RNA expression analyses, including the study of intracellular heterogeneity.

INSIGHTS INTO INTRATUMOURAL HETEROGENEITY THROUGH CTCs

Prostate Cancer

Prostate cancer is a heterogeneous entity, with primary tumors that are frequently multifocal and arise from divergent cancer clones (Andreoiu and Cheng 2010; Cooper et al. 2015). Metastases in prostate cancer are likely established through polyclonal seeding of divergent clones (Gundem et al. 2015), although some studies point to the conservation of driver lesions in metastatic lesions within individuals (Kumar et al. 2016). Consistent with substantial intratumoral heterogeneity, we observed through single-cell immunofluorescence analysis heterogeneous and varying degrees of androgen receptor (AR) signaling in CTCs from patients with castration-resistant prostate cancer (CRPC), in striking contrast to relatively homogeneous CTC populations with activated AR signaling in patients with untreated prostate cancer (Miyamoto et al. 2012). Similar levels of heterogeneity have been observed in prostate CTCs with respect to cellular morphologic criteria, related to disease status and cell differentiation state (Chen et al. 2015; Beltran et al. 2016). Single-cell RNA-seq of CTCs isolated from CRPC patients revealed a tremendous degree of heterogeneity in expression profiles, both in CTC populations from individual patients and across different patients (Miyamoto et al. 2015). Indeed, upon unsupervised hierarchical clustering of transcriptional profiles, the mean correlation coefficient for single CTCs from individual patients was significantly lower than the mean correlation coefficient of single cells from prostate cancer cell lines, and it was similar to that of single cells across multiple different cell lines, suggesting a much higher level of heterogeneity in CTCs. Nevertheless, these single CTCs clustered based on their patient of origin, indicative of shared transcriptional programs among CTCs derived from any given individual patient.

Heterogeneity at the single CTC level was also observed in the acquisition of varied mechanisms of resistance to AR-targeted therapies (Miyamoto et al. 2015). In an analysis of specific molecular aberrations in CTCs, including previously reported AR splice variants and AR mutations (Antonarakis et al. 2014; Robinson et al. 2015), we noted that single CTCs from individual patients expressed remarkable heterogeneity in their expression of different AR splice variants. More than half of patients had multiple CTCs expressing different AR splice variants, and about one of six single CTCs had expression of multiple different AR splice variants simultaneously.

Through the analysis of transcriptional differences between CTCs from patients who were resistant to the antiandrogen therapy enzalutamide compared with patients who were enzalutamide naïve, we identified two AR-independent pathways associated with resistance to enzalutamide: we confirmed the previously reported activation of glucocorticoid receptor (GR) (Arora et al. 2013) and discovered elevated noncanonical Wnt (ncWnt) signaling, finding that both of these resistance pathways were activated in different subsets of cells. Thus, we found that complex and heterogeneous drug resistance mechanisms exist in advanced prostate cancer, as revealed by the study of heterogeneous populations of CTCs. The degree of heterogeneity observed in CTCs in CRPC patients is consistent with intratumoral heterogeneity observed in patients with advanced metastatic prostate cancer, suggestive of the polyclonal seeding of divergent clones (Lohr et al. 2014; Gundem et al. 2015; Jiang et al. 2015), and is reflective of the variety of molecular alterations that may occur in parallel in tumor cells during disease progression and the development of resistance to therapy.

Pancreatic Cancer

We observed similar intracellular heterogeneity in single CTCs isolated from the genetically engineered *LSL-Kras*^{G12D}, *Trp53*$^{flox/flox\ or\ +}$, *Pdx1-Cre* (KPC) mouse model of pancreatic cancer (Bardeesy et al. 2006; Ting et al. 2014). Although the genetic driver mutations were identical in CTCs from different mice because of their shared genetic background, there was nevertheless a significant level of heterogeneity observed within CTCs, with specific clustering of CTCs derived from different animals (Ting et al. 2014). These findings suggest the potential importance of somatically acquired genetic and epigenetic changes in defining the heterogeneity of CTC populations, despite a shared genetic background.

In the KPC pancreatic cancer mouse model, we noted three subsets of CTCs: a major "classical CTC" group with strong expression of epithelial markers, a group with enrichment of platelet markers, and a group associated with proliferation signatures (Ting et al. 2014). Notably, although the classical CTCs showed clear loss of the epithelial marker E-cadherin (*Cdh1*) consistent with epithelial-to-mesenchymal transition (EMT), they did not lose expression of other epithelial markers such as cytokeratins, and they showed heterogeneous expression of mesenchymal genes across single cells. Thus, many classical CTCs appear to be arrested in a biphenotypic EMT state. Nevertheless, they also showed great diversity in their transcriptional programs, including expression of cancer stem cell markers *Aldh1a1* and *Aldh1a2* in some cells, and surprising enrichment of extracellular matrix (ECM) transcripts in many CTCs. Indeed, this finding of ECM gene expression in CTCs was recapitulated in human pancreatic, breast, and prostate cancer patients, with enrichment of the core matrisome protein *SPARC* in 100% of pancreatic CTCs. Together, these studies of CTCs in the KPC genetic mouse model showed substantial intracellular heterogeneity in CTCs from mice despite a conserved genetic background and strong expression of ECM transcripts in a majority of pancreatic CTCs, suggestive of a remarkable ability of CTCs to make their own contributions to tumor stromal remodeling and establishment of a hospitable microenvironment at metastatic sites.

Breast Cancer

Considerable heterogeneity has also observed in single CTCs from breast cancer patients using a variety of methods, including microfluidic transcriptional profiling, targeted mutation detection, and next-generation sequencing (Powell et al. 2012; Deng et al. 2014; De Luca et al. 2016). We used an RNA-in situ hybridization (ISH) assay to observe a spectrum of expression of epithelial and mesenchymal markers in single breast CTCs (Yu et al. 2013), ranging from exclusively epithelial to exclusively mesenchymal CTCs, as well as CTCs in an intermediate state with dual expression of epithelial and mesenchymal markers, similar to the biphenotypic EMT state identified in the pancreatic cancer mouse model (Ting et al. 2014). Interestingly, the EMT features of CTCs varied according to histological subtype of breast cancer, where CTCs from ER^+/PR^+ and $HER2^+$ cancers were predominantly epithelial, and CTCs from triple-negative ($ER^-/PR^-/HER2^-$) breast cancer were predominantly mesenchymal. Examination of changes in EMT states before and after systemic therapy suggested that patients who responded to therapy showed a decrease in CTC numbers and/or a proportional decrease in mesenchymal compared with epithelial markers, and that those with progressive disease despite therapy showed an increase in mesenchymal markers in CTCs post-treatment. Thus, the EMT status of CTCs may serve as a potential biomarker of therapeutic response, with the degree of heterogeneity of cells reflective of their susceptibility to treatment.

Progressive disease in cancer patients is often accompanied by the appearance of multicellular clusters of CTCs (Molnar et al. 2001; Cho et al. 2012), and the presence of these CTC clusters in breast cancer patients is associated with a worse overall survival (Aceto et al. 2014). A detailed examination of CTC clusters in mouse models revealed that they have a critical role as mediators of cancer metastasis, because they have a higher propensity for seeding metastatic disease compared with single CTCs in circulation (Fidler 1973; Liotta et al. 1976; Aceto et al. 2014). Of note, cellular tagging and mixing studies in mice showed that nearly all CTC clusters were derived from oligoclonal precursor cells, indicative of their heterogeneous composition and origin (Aceto et al. 2014). Nevertheless, RNA-seq expression studies of breast cancer patient-derived single CTCs and CTC clusters showed a strong clustering pattern by patient of origin, and a high level of concordance in expression patterns between matched CTC clusters and single CTCs from individual breast cancer patients, with the exception of a few candidate cluster-related genes including XBP1, AGR2, HER3, and plakoglobin (Aceto et al. 2014). Indeed, knockdown of plakoglobin in mouse models suppressed CTC cluster formation and development of metastases, pointing to an important role of this protein in tumor dissemination. Thus, two distinct populations of CTCs appear to coexist in the circulation, including single CTCs and CTC clusters with much higher metastatic potential, derived from oligoclonal precursor cells representative of the inherent heterogeneity of the primary tumor.

MONITORING TUMOR EVOLUTION AND HETEROGENEITY WITH CTCs

Although the evolution of tumors in response to initially effective targeted therapies has become increasingly appreciated, either through the selection of rare preexisting subclonal populations or through molecular tumor evolution (Hata et al. 2016), the routine sampling of tumors at multiple time points remains a major technical challenge. CTCs provide a uniquely accessible mechanism to perform noninvasive serial sampling of tumors over time, and hence they could prove revolutionary in monitoring tumor evolution in patients undergoing treatments. Although circulating tumor DNA (ctDNA) is another type of "liquid biopsy" that may be used to assess for the emergence of specific tumor mutations (Bardelli and Pantel 2017), tumor evolution occurs not just through genetic changes, but also from phenotypic changes through epigenetic cellular plasticity (Fig. 1). Such functional changes cannot be easily ascertained using ctDNA but rather require the use of functional assays in live tumor cells. Thus, the analysis of CTCs, CTC-derived cell lines, and patient-derived xenograft models are invaluable tools in the study of tumor evolution.

In breast cancer, the emergence of HER2 expressing subpopulations has been observed in CTCs from patients with initially HER2$^-$ tumors after exposure to multiple courses of therapy (Fehm et al. 2010; Lindstrom et al. 2012). Analysis of CTCs in such patients revealed discrete HER2$^+$ and HER2$^-$ subpopulations, which had the ability to interconvert spontaneously when maintained in culture (Jordan et al. 2016). Thus, a dynamic equilibrium of HER2$^+$ and HER2$^-$ cell populations exist within a heterogeneous tumor cell population, driven by spontaneous interconversion between these phenotypes. HER2$^+$ CTCs were more rapidly proliferative with activation of multiple receptor tyrosine kinase pathways, whereas HER2$^-$ CTCs showed resistance to cytotoxic chemotherapy and activation of Notch and DNA damage pathways. These two cell populations had comparable tumor initiating potential and similar expression of the stem cell marker *ALDH1*, suggesting that an underlying cellular

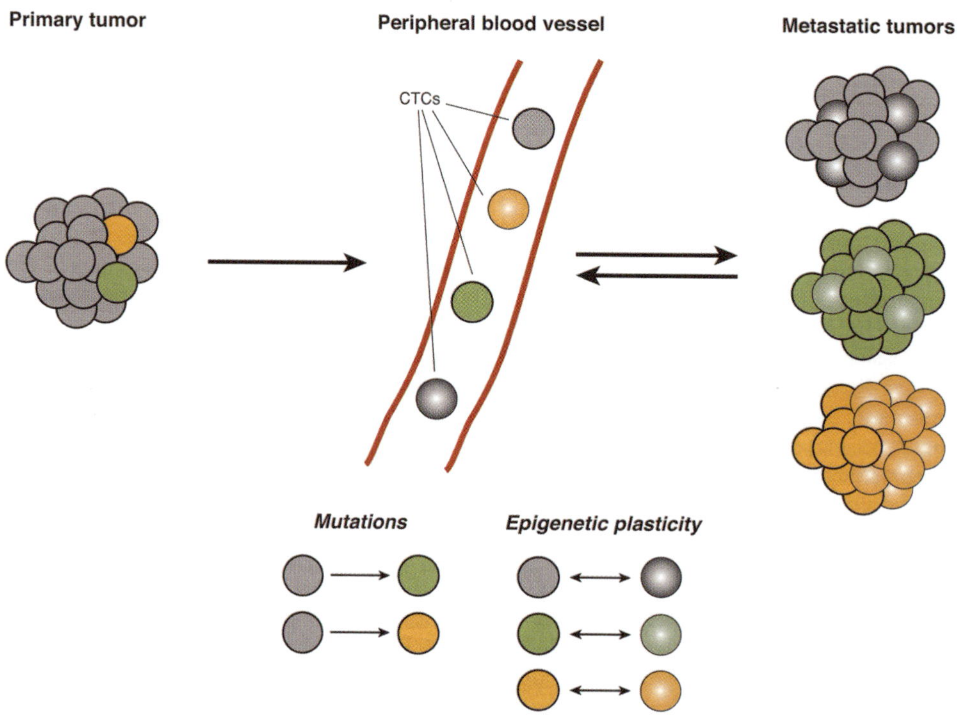

Figure 1. Schematic of heterogeneous tumors giving rise to circulating tumor cells (CTCs). CTCs arise from intravasation of cancer cells into peripheral blood vessels from primary or metastatic tumors. Single-cell heterogeneity can arise from genetic mutations (represented by gray circles becoming green or orange circles) or from epigenetic plasticity (represented by circles converting into spheres, and the reverse conversion of spheres to circles). Mutations and epigenetic changes can also occur simultaneously in the same cancer cell (represented by green and orange circles converting into green and orange spheres). Metastatic tumors may consist of heterogeneous groups of cancer cells that have undergone changes because of any combination of mutations, epigenetic plasticity, or both.

plasticity leads to disease progression and drug resistance in breast cancer, rather than preexisting drug-resistant subclones in a hierarchical cancer stem-cell model.

Whether resistant subclones arise within heterogeneous tumors through genetic instability or through cellular plasticity, their emergence portends the development of therapeutic resistance. The early detection of resistant subclonal populations may allow the early implementation of combination therapies that simultaneously target multiple different oncogenic pathways before a dominant resistant clone emerges. This strategy requires careful repeated monitoring of patients while on treatment, which may be accomplished using serial CTC analyses. In addition, the deployment of aggressive treatments earlier in the disease course before tumor evolution, such as adjuvant therapy following local therapy, may become an increasingly important approach guided by the monitoring of patients through CTC analyses.

CONCLUSION

In the new world of precision oncology, the development of novel therapeutics and diagnostics often merge with one another and are codependent. The emergence of heterogeneity in evolving tumors suggests the inadequacy of a "one-size-fits-all" approach to cancer therapy and necessitates the development of sophisticated molecular tests to guide the precision selection of appropriate therapeutic strategies. The driving force for the effective use of new targeted therapies is novel diagnostic strategies to identify the presence of specific molecular lesions in tumors, but at the same time there is no impetus to develop such diagnostic tests until clear therapeutic options have been developed that make these molecular lesions worth uncovering. This coevolution model in oncology is unprecedented and requires the integrated development of therapeutics and diagnostics at the levels of discovery, validation, and practical implementation in the clinic. Ultimately, the careful application of CTC-based diagnostic tools may help overcome treatment resistance mediated by tumor heterogeneity by guiding the rational selection and application of effective new targeted therapies.

ACKNOWLEDGMENTS

This work was supported by grants from the Prostate Cancer Foundation (D.T.M., D.A.H., S.M., M.T.), Charles Evans Foundation (D.A.H.), Department of Defense (D.T.M., D.T.T.), Howard Hughes Medical Institute (D.A.H.), Burroughs Wellcome Fund (D.T.T.), National Institute of Biomedical Imaging and Bioengineering (NIBIB) EB008047 (M.T.), and the MGH-Johnson & Johnson Center for Excellence in CTC Technologies (M.T., S.M.).

REFERENCES

Aceto N, Bardia A, Miyamoto DT, Donaldson MC, Wittner BS, Spencer JA, Yu M, Pely A, Engstrom A, Zhu H, et al. 2014. Circulating tumor cell clusters are oligoclonal precursors of breast cancer metastasis. *Cell* **158:** 1110–1122.

Alix-Panabieres C, Pantel K. 2014. Challenges in circulating tumour cell research. *Nat Rev Cancer* **14:** 623–631.

Alix-Panabieres C, Pantel K. 2016. Clinical applications of circulating tumor cells and circulating tumor DNA as liquid biopsy. *Cancer Discov* **6:** 479–491.

Andreoiu M, Cheng L. 2010. Multifocal prostate cancer: Biologic, prognostic, and therapeutic implications. *Hum Pathol* **41:** 781–793.

Antonarakis ES, Lu C, Wang H, Luber B, Nakazawa M, Roeser JC, Chen Y, Mohammad TA, Chen Y, Fedor HL, et al. 2014. AR-V7 and resistance to enzalutamide and abiraterone in prostate cancer. *N Engl J Med* **371:** 1028–1038.

Arora VK, Schenkein E, Murali R, Subudhi SK, Wongvipat J, Balbas MD, Shah N, Cai L, Efstathiou E, Logothetis C, et al. 2013. Glucocorticoid receptor confers resistance to antiandrogens by bypassing androgen receptor blockade. *Cell* **155:** 1309–1322.

Ashworth TR. 1869. A case of cancer in which cells similar to those in the tumors were seen in the blood after death. *Aust Med J* **14:** 146–149.

Bardeesy N, Aguirre AJ, Chu GC, Cheng KH, Lopez LV, Hezel AF, Feng B, Brennan C, Weissleder R, Mahmood U, et al. 2006. Both p16(Ink4a) and the p19(Arf)-p53 pathway constrain progression of pancreatic adenocarcinoma in the mouse. *Proc Natl Acad Sci* **103:** 5947–5952.

Bardelli A, Pantel K. 2017. Liquid biopsies, what we do not know (yet). *Cancer Cell* **31:** 172–179.

Beltran H, Jendrisak A, Landers M, Mosquera JM, Kossai M, Louw J, Krupa R, Graf RP, Schreiber NA, Nanus DM, et al. 2016. The initial detection and partial characterization of circulating tumor cells in neuroendocrine prostate cancer. *Clin Cancer Res* **22:** 1510–1519.

Chen JF, Ho H, Lichterman J, Lu YT, Zhang Y, Garcia MA, Chen SF, Liang AJ, Hodara E, Zhau HE, et al. 2015. Subclassification of prostate cancer circulating tumor cells by nuclear size reveals very small nuclear circulating tumor cells in patients with visceral metastases. *Cancer* **121:** 3240–3251.

Cho EH, Wendel M, Luttgen M, Yoshioka C, Marrinucci D, Lazar D, Schram E, Nieva J, Bazhenova L, Morgan A, et al. 2012. Characterization of circulating tumor cell aggregates identified in patients with epithelial tumors. *Phys Biol* **9:** 016001.

Cooper CS, Eeles R, Wedge DC, Van Loo P, Gundem G, Alexandrov LB, Kremeyer B, Butler A, Lynch AG, Camacho N, et al. 2015. Analysis of the genetic phylogeny of multifocal prostate cancer identifies multiple independent clonal expansions in neoplastic and morphologically normal prostate tissue. *Nat Genet* **47:** 367–372.

De Luca F, Rotunno G, Salvianti F, Galardi F, Pestrin M, Gabellini S, Simi L, Mancini I, Vannucchi AM, Pazzagli M, et al. 2016. Mutational analysis of single circulating tumor cells by next generation sequencing in metastatic breast cancer. *Oncotarget* **7:** 26107–26119.

Deng G, Krishnakumar S, Powell AA, Zhang H, Mindrinos MN, Telli ML, Davis RW, Jeffrey SS. 2014. Single cell mutational analysis of PIK3CA in circulating tumor cells and metastases in breast cancer reveals heterogeneity, discordance, and mutation persistence in cultured disseminated tumor cells from bone marrow. *BMC Cancer* **14:** 456.

Fehm T, Muller V, Aktas B, Janni W, Schneeweiss A, Stickeler E, Lattrich C, Lohberg CR, Solomayer E, Rack B, et al. 2010. HER2 status of circulating tumor cells in patients with metastatic breast cancer: A prospective, multicenter trial. *Breast Cancer Res Treat* **124:** 403–412.

Fidler IJ. 1973. The relationship of embolic homogeneity, number, size and viability to the incidence of experimental metastasis. *Eur J Cancer* **9:** 223–227.

Gerlinger M, Rowan AJ, Horswell S, Larkin J, Endesfelder D, Gronroos E, Martinez P, Matthews N, Stewart A, Tarpey P, et al. 2012. Intratumor heterogeneity and branched evolution revealed by multiregion sequencing. *N Engl J Med* **366:** 883–892.

Gundem G, Van Loo P, Kremeyer B, Alexandrov LB, Tubio JM, Papaemmanuil E, Brewer DS, Kallio HM, Hognas G, Annala M, et al. 2015. The evolutionary history of lethal metastatic prostate cancer. *Nature* **520:** 353–357.

Haber DA, Velculescu VE. 2014. Blood-based analyses of cancer: Circulating tumor cells and circulating tumor DNA. *Cancer Discov* **4:** 650–661.

Haber DA, Gray NS, Baselga J. 2011. The evolving war on cancer. *Cell* **145:** 19–24.

Hata AN, Niederst MJ, Archibald HL, Gomez-Caraballo M, Siddiqui FM, Mulvey HE, Maruvka YE, Ji F, Bhang HE, Krishnamurthy Radhakrishna V, et al. 2016. Tumor cells can follow distinct evolutionary paths to become resistant to epidermal growth factor receptor inhibition. *Nat Med* **22:** 262–269.

Hyman DM, Taylor BS, Baselga J. 2017. Implementing genome-driven oncology. *Cell* **168:** 584–599.

Jiang R, Lu YT, Ho H, Li B, Chen JF, Lin M, Li F, Wu K, Wu H, Lichterman J, et al. 2015. A comparison of isolated circulating tumor cells and tissue biopsies using whole-genome sequencing in prostate cancer. *Oncotarget* **6:** 44781–44793.

Jordan NV, Bardia A, Wittner BS, Benes C, Ligorio M, Zheng Y, Yu M, Sundaresan TK, Licausi JA, Desai R, et al. 2016. HER2 expression identifies dynamic functional states within circulating breast cancer cells. *Nature* **537:** 102–106.

Karabacak NM, Spuhler PS, Fachin F, Lim EJ, Pai V, Ozkumur E, Martel JM, Kojic N, Smith K, Chen PI, et al. 2014. Microfluidic, marker-free isolation of circulating tumor cells from blood samples. *Nat Protoc* **9:** 694–710.

Kumar A, Coleman I, Morrissey C, Zhang X, True LD, Gulati R, Etzioni R, Bolouri H, Montgomery B, White T, et al. 2016. Substantial interindividual and limited intraindividual genomic diversity among tumors from men with metastatic prostate cancer. *Nat Med* **22:** 369–378.

Li J, Gregory SG, Garcia-Blanco MA, Armstrong AJ. 2015. Using circulating tumor cells to inform on prostate cancer biology and clinical utility. *Crit Rev Clin Lab Sci* **52:** 191–210.

Lindstrom LS, Karlsson E, Wilking UM, Johansson U, Hartman J, Lidbrink EK, Hatschek T, Skoog L, Bergh J. 2012. Clinically used breast cancer markers such as estrogen receptor, progesterone receptor, and human epidermal growth factor receptor 2 are unstable throughout tumor progression. *J Clin Oncol* **30:** 2601–2608.

Liotta LA, Saidel MG, Kleinerman J. 1976. The significance of hematogenous tumor cell clumps in the metastatic process. *Cancer Res* **36:** 889–894.

Lohr JG, Adalsteinsson VA, Cibulskis K, Choudhury AD, Rosenberg M, Cruz-Gordillo P, Francis JM, Zhang CZ, Shalek AK, Satija R, et al. 2014. Whole-exome sequencing of circulating tumor cells provides a window into metastatic prostate cancer. *Nat Biotechnol* **32:** 479–484.

Miyamoto DT, Lee RJ, Stott SL, Ting DT, Wittner BS, Ulman M, Smas ME, Lord JB, Brannigan BW, Trautwein J, et al. 2012. Androgen receptor signaling in circulating tumor cells as a marker of hormonally responsive prostate cancer. *Cancer Discov* **2:** 995–1003.

Miyamoto DT, Sequist LV, Lee RJ. 2014. Circulating tumour cells-monitoring treatment response in prostate cancer. *Nat Rev Clin Oncol* **11:** 401–412.

Miyamoto DT, Zheng Y, Wittner BS, Lee RJ, Zhu H, Broderick KT, Desai R, Fox DB, Brannigan BW, Trautwein J, et al. 2015. RNA-Seq of single prostate CTCs implicates noncanonical Wnt signaling in antiandrogen resistance. *Science* **349:** 1351–1356.

Molnar B, Ladanyi A, Tanko L, Sreter L, Tulassay Z. 2001. Circulating tumor cell clusters in the peripheral blood of colorectal cancer patients. *Clin Cancer Res* **7:** 4080–4085.

Nagrath S, Sequist LV, Maheswaran S, Bell DW, Irimia D, Ulkus L, Smith MR, Kwak EL, Digumarthy S, Muzikansky A, et al. 2007. Isolation of rare circulating tumour cells in cancer patients by microchip technology. *Nature* **450:** 1235–1239.

Ozkumur E, Shah AM, Ciciliano JC, Emmink BL, Miyamoto DT, Brachtel E, Yu M, Chen PI, Morgan B, Trautwein J, et al. 2013. Inertial focusing for tumor antigen-dependent and -independent sorting of rare circulating tumor cells. *Sci Transl Med* **5:** 179ra147.

Powell AA, Talasaz AH, Zhang H, Coram MA, Reddy A, Deng G, Telli ML, Advani RH, Carlson RW, Mollick JA, et al. 2012. Single cell profiling of circulating tumor cells: Transcriptional heterogeneity and diversity from breast cancer cell lines. *PLoS One* **7:** e33788.

Robinson D, Van Allen EM, Wu YM, Schultz N, Lonigro RJ, Mosquera JM, Montgomery B, Taplin ME, Pritchard CC, Attard G, et al. 2015. Integrative clinical genomics of advanced prostate cancer. *Cell* **161:** 1215–1228.

Stott SL, Hsu CH, Tsukrov DI, Yu M, Miyamoto DT, Waltman BA, Rothenberg SM, Shah AM, Smas ME, Korir GK, et al. 2010. Isolation of circulating tumor cells using a microvortex-generating herringbone-chip. *Proc Natl Acad Sci* **107:** 18392–18397.

Ting DT, Wittner BS, Ligorio M, Vincent Jordan N, Shah AM, Miyamoto DT, Aceto N, Bersani F, Brannigan BW, Xega K, et al. 2014. Single-cell RNA sequencing identifies extracellular matrix gene expression by pancreatic circulating tumor cells. *Cell Rep* **8:** 1905–1918.

Wu X, Northcott PA, Dubuc A, Dupuy AJ, Shih DJ, Witt H, Croul S, Bouffet E, Fults DW, Eberhart CG, et al. 2012. Clonal selection drives genetic divergence of metastatic medulloblastoma. *Nature* **482:** 529–533.

Yu M, Bardia A, Wittner BS, Stott SL, Smas ME, Ting DT, Isakoff SJ, Ciciliano JC, Wells MN, Shah AM, et al. 2013. Circulating breast tumor cells exhibit dynamic changes in epithelial and mesenchymal composition. *Science* **339:** 580–584.

Yu M, Bardia A, Aceto N, Bersani F, Madden MW, Donaldson MC, Desai R, Zhu H, Comaills V, Zheng Z, et al. 2014. Cancer therapy. Ex vivo culture of circulating breast tumor cells for individualized testing of drug susceptibility. *Science* **345:** 216–220.

Discovery of Double-Stranded Genomic DNA in Circulating Exosomes

RAGHU KALLURI AND VALERIE S. LEBLEU
Department of Cancer Biology, Metastasis Research Center, University of Texas MD Anderson Cancer Center, Houston, Texas 77005
Correspondence: rkalluri@mdanderson.org; vlebleu@mdanderson.org

It is becoming increasingly clear that small vesicles released from cells (extracellular vesicles [EVs]) represent a heterogeneous population implicated in cell-to-cell communication. The classifications and nomenclature of EVs are evolving as enrichment strategies and specific characteristics are being unraveled. At present, physical properties of EVs—namely, size, shape, and density—are often used to identify subpopulations of EVs. A distinct group of EVs, termed exosomes, largely defined by their small size (~40–150 nm) and proposed subcellular origin, has been extensively studied in several aspects of cancer biology. Exosomes are implicated in modulating behavior of cancer cells as well as the immune and angiogenic responses in tumors, possibly contributing to cancer progression locally and systemically. Most intriguingly, the nucleic acid content of exosomes has been proposed to play a role in oncogenic transformation and transfer of cancer-specific genome to promote cancer pathogenesis. Here, we specifically focus on the discovery of exosomal DNA, studies related to the origin of genomic DNA in exosomes, and its utility in cancer diagnosis and disease monitoring.

Exosomes are extracellular vesicles (EVs) released via fusion of multivesicular bodies with the plasma membrane of cells. The content of the multivesicular bodies (intraluminal vesicles [ILVs]) is released in the form of exosomes in the extracellular space (Colombo et al. 2014; Kalluri 2016). This propositioned biogenesis is in contrast with that of "ectosomes," which emerge from the direct budding of the plasma membrane, giving rise to EVs of similar or greater size range when compared with exosomes (Fig. 1; Antonyak and Cerione 2015; Cocucci and Meldolesi 2015). Despite novel biology unraveling these processes, the exact molecular mechanism of exosome and ectosome formation is unknown. It remains challenging to definitively distinguish between exosomes and ectosomes, in particular when they share similar physical characteristics (size, shape, and density) and present with markers that can overlap between the two subpopulations. There are to date no definitive markers or processes to strictly enrich exosomes versus ectosomes, although proteins involved in their biogenesis are being rapidly unraveled (Colombo et al. 2014; Edgar 2016; Tkach and Thery 2016). The biogenesis of exosomes and ectosomes may play a role in defining their respective contents. Exosomes are implicated in intercellular exchange of macromolecules involving proteins, metabolites, lipids, mRNA, miRNA, lncRNA, viral RNA, mitochondrial, and chromosomal DNA (Kalluri 2016). Collectively, exosomes may present with a heterogeneous mixture of all cellular componenets (Fig. 1), raising an intriguing notion that exosomes may resemble "primordial"-like vesicles (Sinkovics 2015; Kalluri 2016). The evidence for the functional impact of cellular exchange of exosomal content, such as signaling molecules, messenger RNA (mRNA), and microRNA (miRNA), implicates exosomes in widespread multiorgan communication. The functional outcome of such exchange may not be limited to neoplastic events and tumor growth but also involve embryonic/organ development and non-cancer-associated pathologies (Colombo et al. 2014; Edgar 2016; McGough and Vincent 2016; Tkach and Thery 2016).

Most results related to cell–cell communication mediated by exosomes, although intriguing, have largely been collected from experiments performed employing cell culture system or employing experimental designs wherein animals are given bolus administration of large quantities of exogenously prepared exosomes from cultured cells. It is critical to note that the physiological function of exosomes remains largely unknown. Several factors have been implicated in the release of exosomes, and the loss of the *tumor suppressor-activated pathway 6* (*TSAP6*) gene in mice results in abnormal transferrin receptor down-regulation and reticulocyte maturation (Lespagnol et al. 2008)—processes that were first described in exosome biogenesis (Colombo et al. 2014; Edgar 2016). Interestingly, this process was linked to the DNA damage–induced TP53 pathway associated with exosome secretion (Lespagnol et al. 2008). It remains unknown whether such exosome secretion pathways are similar in all cell types. Emerging studies suggest distinct molecular mechanisms for exosome biogenesis, independent of the traditionally discussed endosomal sorting complexes required for transport (ESCRT)-dependent process (Stuffers et al. 2009).

Published by Cold Spring Harbor Laboratory Press; doi: 10.1101/sqb.2016.81.030932

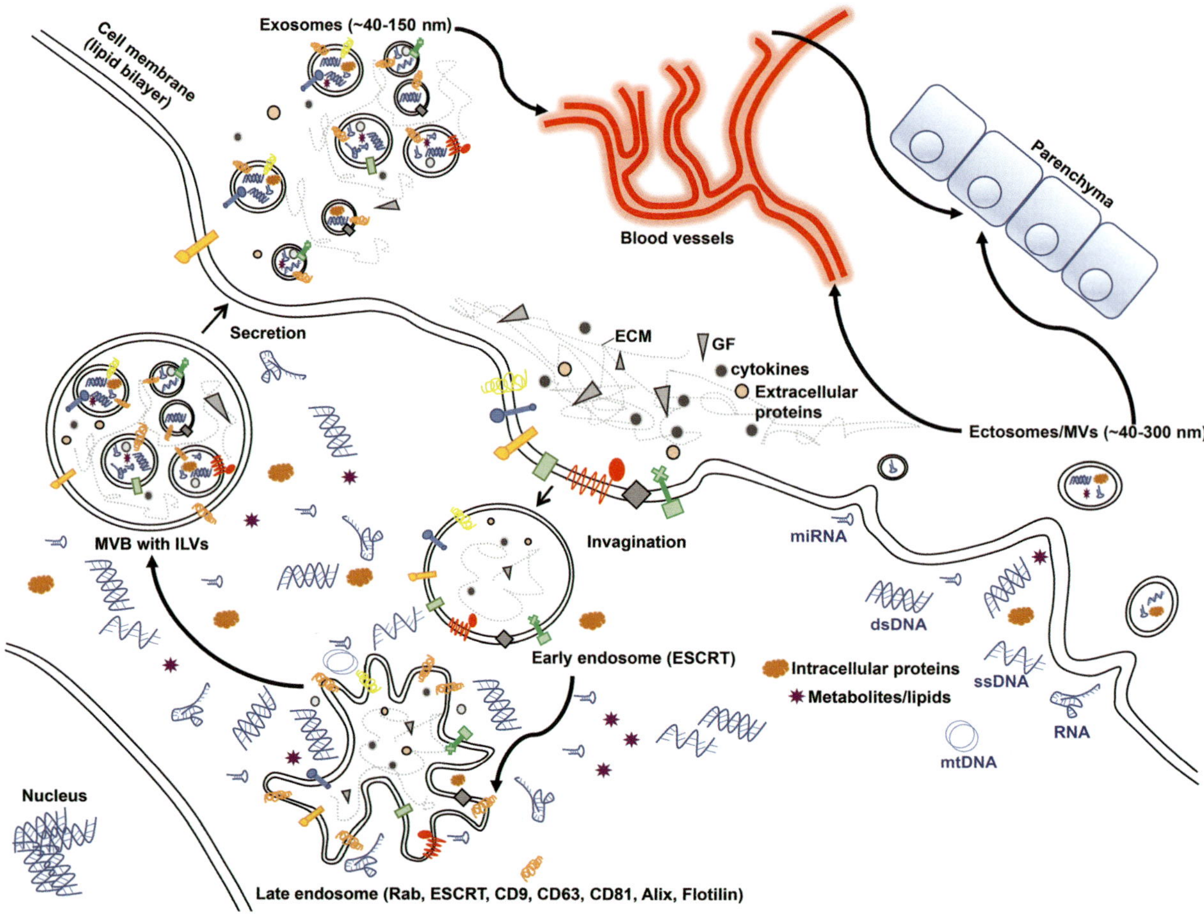

Figure 1. Invagination of the cell plasma membrane leads to the formation of intercellular vesicles encapsulating extracellular components (extracellular matrix [ECM], extracellular proteins, cytokines, and growth factors [GFs]). The intracellular vesicles mature into the early and late endosomes, implicating the endosomal sorting complexes required for transport (ESCRT) machinery and Rab proteins. Proteins (CD9, CD63, CD81, Alix, Flotilin) that decorate the late endosomes are often used as exosome markers following their secretion. The formation of the multivesicular body (MVB) containing the intraluminal vesicles (ILVs) follows a secondary invagination of the late endosome membranes, trapping within newly forming ILVs components of the cytosol (metabolites, intracellular proteins, nucleic acids). Following fusion of the MVs with the plasma membrane, the ILVs are released and referred to as exosomes. The surface proteins on the released exosomes are conformed similarly to their presentation on the plasma membrane. The exosomes enter the circulation or surrounding parenchyma. Ectosomes or microvesicles (MVs), in contrast, emerge from the budding of the plasma membrane and are often larger in size when compared to exosomes. miRNA, microRNA; dsDNA, double-stranded DNA; ssDNA, single-stranded DNA; mtDNA, mitochondrial DNA.

In cancer, the biology and function of exosomes is still largely unknown, and some suggestive evidence today points to their role in promoting cancer (Kahlert and Kalluri 2013; Yu et al. 2015; Kalluri 2016). Cancer cell–derived exosomes may elicit phenotypic and molecular responses in recipient cells that result in the activation of signaling pathways that may promote cancer cell proliferation, survival, invasion, and resistance to therapy (Yu et al. 2015). Tumor-derived exosomes may also remodel the tumor microenvironment to promote tumor angiogenesis and impair the immune control of tumor growth (Kahlert and Kalluri 2013). The content of cancer cell–derived exosomes may reflect the content of their cell of origin, perpetuating potential transfer of specific oncogenic components to recipient cells (Fujita et al. 2016; Choi et al. 2017). Interestingly, exosomes may also serve as a modulator of cancer cell heterogeneity, possibly impacting the populational equilibrium of diffuse B-cell lymphoma (Koch et al. 2014). Although these effects are largely attributed to signaling molecules and epigenetic regulators delivered into the recipient cells by exosomes, another intriguing mode of communication by exosomes entails the DNA present in them. Exosomes were reported to contain mitochondrial DNA (from culture supernatant of myoblasts [Guescini et al. 2010]) and chromosomal DNA (vide infra). Chromosomal DNA in exosomes was noted in cell culture supernatant as well as in human and mouse biological fluids, including blood, seminal fluid, and urine. Interestingly, the encapsulation of DNA within exosomes may confer enhanced stability to DNA when outside the cell (Jin et al. 2016). These discoveries have spurred a specific interest in exosomes DNA for use as liquid biopsies to facilitate the diagnosis and monitoring of cancer patients (Speicher and Pantel

2014; Wan et al. 2017; Yi et al. 2017). The exosomal DNA in the blood offers yet another level of complexity in the study of circulating DNA in cancer (Thierry et al. 2016) but also perhaps novel means to isolate cancer specific DNA, by possibly enriching for circulating exosomes derived from cancer cells (Kalluri 2016).

IDENTIFICATION OF DNA IN THE EXOSOMES

Using whole-genome sequencing, Kahlert et al. (2014) provided definitive identification of large fragments of double-stranded genomic DNA (cumulatively encompassing all chromosomes) in the circulating exosomes from patients with pancreatic cancer and also in exosomes derived from cancer cells. Several groups have also identified double-stranded DNA fragments in the exosomes ranging from 100 b to 17 kb in size (Cai et al. 2013; Kahlert et al. 2014; Lee et al. 2014; Thakur et al. 2014). Cai et al. (2013) reported the detection of chromosomal and mitochondrial DNA in exosomes derived from healthy donor human plasma and culture supernatant of human primary vascular smooth muscle cells, HEK293 cells (human embryonic kidney cells) and K562 cells (human immortalized myelogenous leukemia cells) (Cai et al. 2013). In their studies, DNase treatment supported that the long fragments of double-stranded DNA were predominantly found within the exosomes rather than outside. They further showed that the exosome's DNA, upon transfer to the recipient cells, localizes to the nucleus, recruits nuclear factor κB (NF-κB), and is transcribed. Further, exosome-derived DNA containing the BCR/ABL hybrid gene (from K562 cell supernatant) could be detected in normal recipient human neutrophils (Cai et al. 2013). Their results support the notion of functional horizontal transfer of DNA via exosomes (Cai et al. 2013, 2016). The detection of gene fusion in exosome's DNA sets the stage for further analyses of driver mutations in several cancer types.

DETECTION OF DNA MUTATIONS IN EXOSOMES

In the supernatants of pancreatic cancer cells and serum from pancreatic cancer patients, large fragments of double-stranded DNA were noted in exosomes, and Sanger sequencing analyses of polymerase chain reaction (PCR) amplicons revealed the presence of *KRAS* and *TP53* mutations (Kahlert et al. 2014). The double-stranded DNA in exosomes collected from prostate cancer cell supernatant also revealed mutations characteristic of the cell of origin (Lazaro-Ibanez et al. 2014), and double-stranded DNA was also identified in the plasma of prostate cancer patients (Lazaro-Ibanez et al. 2014) as well as in the seminal fluid of donors (Ronquist et al. 2012). The diagnostic value of exosome's DNA in liquid biopsies, including but not limited to blood, urine, and seminal fluid, remains to be determined because of the challenges associated with detection sensitivity and signal-to-noise ratio owing to fragments of DNA from cancer cells (mutation) mixed with DNA from normal cells (no mutation). Nonetheless, detection of actionable mutations (in *NOTCH1* and *BRCA2*) in exosome's DNA from pancreatic cancer patients has been reported (San Lucas et al. 2016). The most common mutations in pancreatic cancer patients (*KRAS* and *TP53*) are readily detected in circulating exosomes DNA (Allenson et al. 2017; Yang et al. 2017). These studies, however, provide caution because detection of *KRAS* mutation from an exosome's DNA in healthy individuals confounds the diagnostic utility. The exosome's DNA in cancer may indeed represent a new treasure trove for monitoring changes in the mutational landscape (e.g., in response to therapy); however, this may only be feasible upon enrichment of cancer cell–derived exosomes from the blood (Kalluri 2016).

BIOLOGICAL FUNCTION OF DNA IN THE EXOSOMES

The exact nature and relative abundance of the different species of DNA inside and outside of exosomes from cell culture supernatants and biological fluids is being actively investigated. It is likely that distinct types (single-stranded, double-stranded, mitochondrial; Fig. 2) and integrity of the exosome's DNA (fragment length, chromatin bound) are likely possible in the distinct type of exosomes based on their cell of origin. It is also plausible that the mechanisms enabling packaging of DNA within exosomes are cell type–specific and dynamically regulated. Analyses of exosomes from the supernatant of various cancer cell lines and non-cancer-associated fibroblast lines revealed exosomes DNA was more abundant in cancer cell–derived exosomes, and that double-stranded DNA was predominantly found within the exosomes (Kahlert et al. 2014; Thakur et al. 2014). DNase treatment of exosomes from cultured cell supernatants revealed that double-stranded DNA, rather than single-stranded DNA or RNA, was predominantly found inside exosomes, albeit DNA is also found outside exosomes (Kahlert et al. 2014; Thakur et al. 2014). This is in contrast with the reported single-stranded DNA in exosomes from the supernatant of medulloblastoma tumor cells (Balaj et al. 2011). Whole-genome sequence analyses revealed the entire coverage of the genome in exosomes and that there did not seem to be a notable bias for specific regions (Kahlert et al. 2014; Lee et al. 2014; Thakur et al. 2014). Using exosomes containing H-ras DNA sequence, derived from the culture supernatant of RAS-3 cells (tumorigenic c-H-ras transformed rat intestinal epithelial cells), the recipient RAT-1 cells (immortalized rat fibroblasts) showed transient H-ras DNA for up to 30 d following exposure to exosomes (Lee et al. 2014). Although this uptake did not transform RAT-1 cells, RAS-3-derived exosomes stimulated their proliferation (Lee et al. 2014).

BRCA1-KO human fibroblasts could be transformed into malignant cells when exposed to circulating exosomes from patients with breast, colorectal, pancreatic, and other

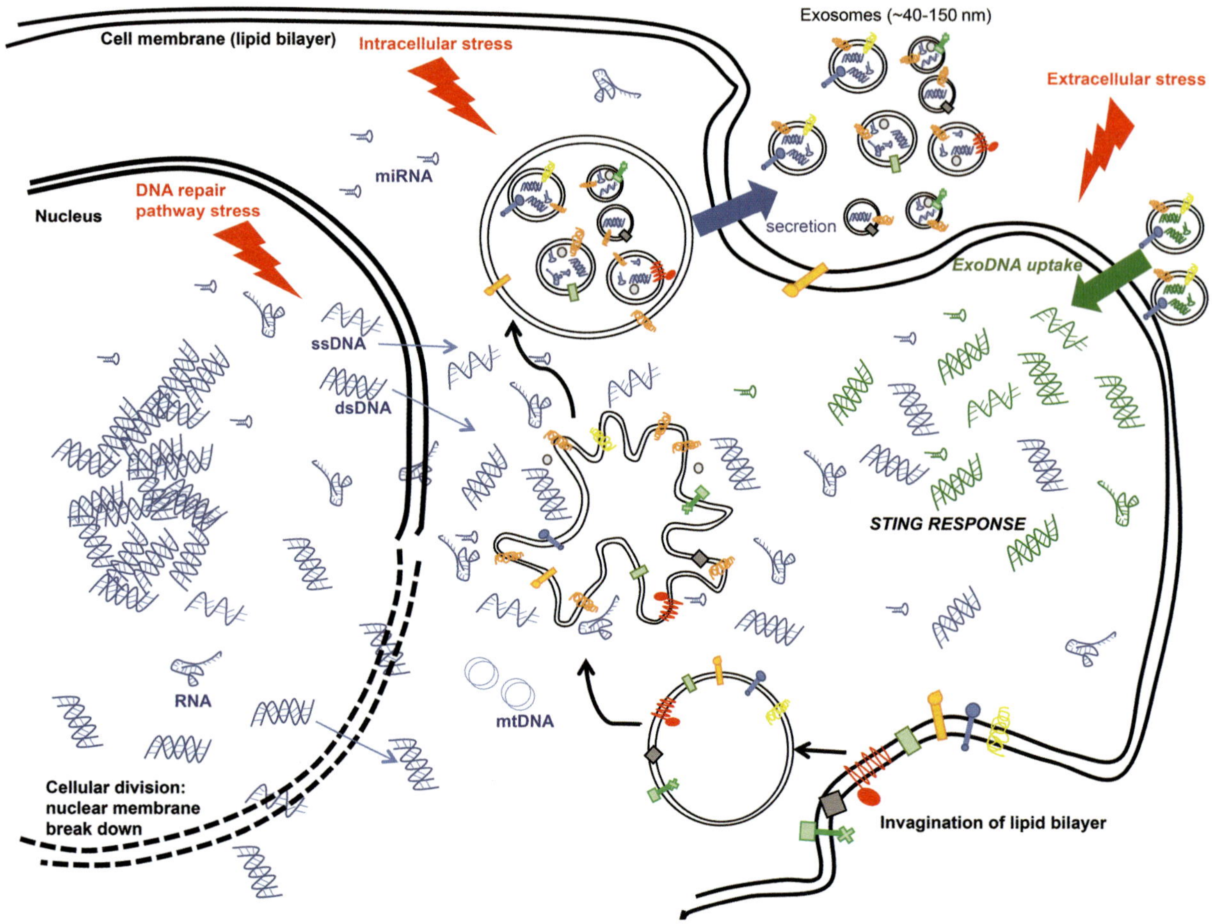

Figure 2. During the formation of intraluminal vesicles that will subsequently be released as exosomes, nucleic acids in the cytosol are encapsulated. These include messenger RNA (mRNA), microRNA (miRNA), mitochondrial DNA (mtDNA), and single-stranded and double-stranded chromosomal DNA (ssDNA and dsDNA, respectively). Increased cellular stress (extracellular stress, intracellular stress, and DNA repair pathway stress) may contribute to the increase in cytosolic nucleic acids that are shuttled into intraluminal vesicles (ILVs). Breakdown of the nuclear membrane during cellular division may also expose extraneous DNA to cytosolic compartments that would facilitate DNA uptake into ILVs. The increased content of cytosolic DNA, possibly via uptake of exosomal DNA, may elicit the stimulator of interferon genes (STING) pathway and stimulate an immune response.

cancer types (Hamam et al. 2016). Interestingly, studies using epidermal growth factor receptor (EGFR)-driven cancer cell lines subjected to EGFR kinase inhibitors yield an increase in the release of exosomes containing DNA (Montermini et al. 2015). Collectively these studies indicate that exosomal DNA cargo can be different but predominantly present as large fragments of double-stranded DNA. Moreover, DNA packaging may be influenced by exogenous agents acting on the cell of origin, yet with all regions of the genome appearing to be contained within a given pool of exosomes (billions). Further, the influence of exosomal DNA on the recipient cells may differ based on pre-existing genomic and microenvironmental stressors. The impact of an exosome's DNA may extend beyond cell–cell communications. Recent studies revealed that breast cancer cell–derived exosomal DNA, following treatment of the cancer cells with topoisomerase I inhibitor, may activate the stimulator of interferon genes (STING) pathway in the recipient dendritic cells (Fig. 2; Paludan and Bowie 2013; Kitai et al. 2017). Whether these events are rate-limiting for antitumor immunity remains to be determined, but it certainly opens a new means to modulate tumor immunity via exosomal DNA.

FUTURE DIRECTIONS

Among the fascinating, yet largely unanswered questions surrounding exosomal DNA are the mechanisms associated with chromosomal DNA packaging within ILVs, which eventually give rise to exosomes when released into the extracellular space. Because the biogenesis of ILVs could include the sampling of the cytosol, cytosolic chromosomal DNA likely can get encapsulated within the exosomes (Fig. 2). Cytosolic DNA accumulation in cancer cells may reflect their deregulated DNA repair pathways (Shen et al. 2015; Ho et al. 2016), and this could explain the higher abundance of exosomal DNA in cancer cells when compared with normal cells.

Alternatively, the different expression levels of cytosolic nucleases in distinct cell types, leading to enhanced cytosolic DNA content, could also account for different amounts of DNA within ILVs. Whether a correlation exists between cytosolic DNA content and abundance of exosomal DNA remains to be elucidated. Intriguingly, there may be a link between nuclear envelope breakdown in proliferating cells and vesicular packaging of aborted DNA replication forks (Fig. 2), although this hypothesis remains to be tested. The mechanism underlying the transfer of an exosome's DNA to its recipient cells is also largely unknown, although exosomal Connexin 43 (*GJA1* or Cx43) may enable docking and fusion of exosomes with the plasma membrane of Cx43 expressing recipient cells (Soares et al. 2015). The green fluorescent protein (GFP)-expressing plasmid incorporated into the $Cx43^+$ exosomes was more efficiently transcribed and translated when Cx43 was present on both exosomes and recipient cell surfaces (Soares et al. 2015). Interestingly, predicted DNA and RNA-binding motifs on Cx43 could suggest a role for Cx43 in exosome-mediated transfer of nucleic acids (Varela-Eirin et al. 2017).

In conclusion, many research groups validated the fascinating discovery that exosomes contain genomic double-stranded DNA. In the next decade, we will gain more knowledge with respect to the mechanism of how DNA enters the exosomes and how exosomal DNA can aid in diagnosis and therapy of cancer.

ACKNOWLEDGMENTS

The work in the laboratory of R.K. is supported by the Cancer Prevention and Research Institute of Texas. The V.S.L. laboratory is supported by the University of Texas MD Anderson Cancer Center (UT MDACC) Khalifa Bin Zayed Al Nahyan Foundation.

REFERENCES

Allenson K, Castillo J, San Lucas FA, Scelo G, Kim DU, Bernard V, Davis G, Kumar T, Katz M, Overman MJ, et al. 2017. High prevalence of mutant KRAS in circulating exosome-derived DNA from early stage pancreatic cancer patients. *Ann Oncol*. doi: 10.1093/annonc/mdx004.

Antonyak MA, Cerione RA. 2015. Emerging picture of the distinct traits and functions of microvesicles and exosomes. *Proc Natl Acad Sci* **112:** 3589–3590.

Balaj L, Lessard R, Dai L, Cho YJ, Pomeroy SL, Breakefield XO, Skog J. 2011. Tumour microvesicles contain retrotransposon elements and amplified oncogene sequences. *Nat Commun* **2:** 180.

Cai J, Han Y, Ren H, Chen C, He D, Zhou L, Eisner GM, Asico LD, Jose PA, Zeng C. 2013. Extracellular vesicle-mediated transfer of donor genomic DNA to recipient cells is a novel mechanism for genetic influence between cells. *J Mol Cell Biol* **5:** 227–238.

Cai J, Wu G, Jose PA, Zeng C. 2016. Functional transferred DNA within extracellular vesicles. *Exp Cell Res* **349:** 179–183.

Choi D, Lee TH, Spinelli C, Chennakrishnaiah S, D'Asti E, Rak J. 2017. Extracellular vesicle communication pathways as regulatory targets of oncogenic transformation. *Semin Cell Dev Biol*. doi: 10.1016/j.semcdb.2017.01.003.

Cocucci E, Meldolesi J. 2015. Ectosomes and exosomes: Shedding the confusion between extracellular vesicles. *Trends Cell Biol* **25:** 364–372.

Colombo M, Raposo G, Thery C. 2014. Biogenesis, secretion, and intercellular interactions of exosomes and other extracellular vesicles. *Annu Rev Cell Dev Biol* **30:** 255–289.

Edgar JR. 2016. Q&A: What are exosomes, exactly? *BMC Biol* **14:** 46.

Fujita Y, Yoshioka Y, Ochiya T. 2016. Extracellular vesicle transfer of cancer pathogenic components. *Cancer Sci* **107:** 385–390.

Guescini M, Guidolin D, Vallorani L, Casadei L, Gioacchini AM, Tibollo P, Battistelli M, Falcieri E, Battistin L, Agnati LF, et al. 2010. C2C12 myoblasts release micro-vesicles containing mtDNA and proteins involved in signal transduction. *Exp Cell Res* **316:** 1977–1984.

Hamam D, Abdouh M, Gao ZH, Arena V, Arena M, Arena GO. 2016. Transfer of malignant trait to BRCA1 deficient human fibroblasts following exposure to serum of cancer patients. *J Exp Clin Cancer Res* **35:** 80.

Ho SS, Zhang WY, Tan NY, Khatoo M, Suter MA, Tripathi S, Cheung FS, Lim WK, Tan PH, Ngeow J, et al. 2016. The DNA structure-specific endonuclease MUS81 mediates DNA sensor STING-dependent host rejection of prostate cancer cells. *Immunity* **44:** 1177–1189.

Jin Y, Chen K, Wang Z, Wang Y, Liu J, Lin L, Shao Y, Gao L, Yin H, Cui C, et al. 2016. DNA in serum extracellular vesicles is stable under different storage conditions. *BMC Cancer* **16:** 753.

Kahlert C, Kalluri R. 2013. Exosomes in tumor microenvironment influence cancer progression and metastasis. *J Mol Med (Berl)* **91:** 431–437.

Kahlert C, Melo SA, Protopopov A, Tang J, Seth S, Koch M, Zhang J, Weitz J, Chin L, Futreal A, et al. 2014. Identification of double-stranded genomic DNA spanning all chromosomes with mutated KRAS and p53 DNA in the serum exosomes of patients with pancreatic cancer. *J Biol Chem* **289:** 3869–3875.

Kalluri R. 2016. The biology and function of exosomes in cancer. *J Clin Invest* **126:** 1208–1215.

Kitai Y, Kawasaki T, Sueyoshi T, Kobiyama K, Ishii KJ, Zou J, Akira S, Matsuda T, Kawai T. 2017. DNA-containing exosomes derived from cancer cells treated with topotecan activate a STING-dependent pathway and reinforce antitumor immunity. *J Immunol* **198:** 1649–1659.

Koch R, Demant M, Aung T, Diering N, Cicholas A, Chapuy B, Wenzel D, Lahmann M, Guntsch A, Kiecke C, et al. 2014. Populational equilibrium through exosome-mediated Wnt signaling in tumor progression of diffuse large B-cell lymphoma. *Blood* **123:** 2189–2198.

Lazaro-Ibanez E, Sanz-Garcia A, Visakorpi T, Escobedo-Lucea C, Siljander P, Ayuso-Sacido A, Yliperttula M. 2014. Different gDNA content in the subpopulations of prostate cancer extracellular vesicles: Apoptotic bodies, microvesicles, and exosomes. *Prostate* **74:** 1379–1390.

Lee TH, Chennakrishnaiah S, Audemard E, Montermini L, Meehan B, Rak J. 2014. Oncogenic ras-driven cancer cell vesiculation leads to emission of double-stranded DNA capable of interacting with target cells. *Biochem Biophys Res Commun* **451:** 295–301.

Lespagnol A, Duflaut D, Beekman C, Blanc L, Fiucci G, Marine JC, Vidal M, Amson R, Telerman A. 2008. Exosome secretion, including the DNA damage-induced p53-dependent secretory pathway, is severely compromised in TSAP6/Steap3-null mice. *Cell Death Differ* **15:** 1723–1733.

McGough IJ, Vincent JP. 2016. Exosomes in developmental signalling. *Development* **143:** 2482–2493.

Montermini L, Meehan B, Garnier D, Lee WJ, Lee TH, Guha A, Al-Nedawi K, Rak J. 2015. Inhibition of oncogenic epidermal growth factor receptor kinase triggers release of exosome-like extracellular vesicles and impacts their phosphoprotein and DNA content. *J Biol Chem* **290:** 24534–24546.

Paludan SR, Bowie AG. 2013. Immune sensing of DNA. *Immunity* **38:** 870–880.

Ronquist GK, Larsson A, Stavreus-Evers A, Ronquist G. 2012. Prostasomes are heterogeneous regarding size and appearance but affiliated to one DNA-containing exosome family. *Prostate* **72:** 1736–1745.

San Lucas FA, Allenson K, Bernard V, Castillo J, Kim DU, Ellis K, Ehli EA, Davies GE, Petersen JL, Li D, et al. 2016. Minimally invasive genomic and transcriptomic profiling of visceral cancers by next-generation sequencing of circulating exosomes. *Ann Oncol* **27:** 635–641.

Shen YJ, Le Bert N, Chitre AA, Koo CX, Nga XH, Ho SS, Khatoo M, Tan NY, Ishii KJ, Gasser S. 2015. Genome-derived cytosolic DNA mediates type I interferon-dependent rejection of B cell lymphoma cells. *Cell Rep* **11:** 460–473.

Sinkovics JG. 2015. The cell survival pathways of the primordial RNA–DNA complex remain conserved in the extant genomes and may function as proto-oncogenes. *Eur J Microbiol Immunol (Bp)* **5:** 25–43.

Soares AR, Martins-Marques T, Ribeiro-Rodrigues T, Ferreira JV, Catarino S, Pinho MJ, Zuzarte M, Anjo SI, Manadas B, Sluijter JP, et al. 2015. Corrigendum: Gap junctional protein Cx43 is involved in the communication between extracellular vesicles and mammalian cells. *Sci Rep* **5:** 14888.

Speicher MR, Pantel K. 2014. Tumor signatures in the blood. *Nat Biotechnol* **32:** 441–443.

Stuffers S, Sem Wegner C, Stenmark H, Brech A. 2009. Multivesicular endosome biogenesis in the absence of ESCRTs. *Traffic* **10:** 925–937.

Thakur BK, Zhang H, Becker A, Matei I, Huang Y, Costa-Silva B, Zheng Y, Hoshino A, Brazier H, Xiang J, et al. 2014. Double-stranded DNA in exosomes: A novel biomarker in cancer detection. *Cell Res* **24:** 766–769.

Thierry AR, El Messaoudi S, Gahan PB, Anker P, Stroun M. 2016. Origins, structures, and functions of circulating DNA in oncology. *Cancer Metastasis Rev* **35:** 347–376.

Tkach M, Thery C. 2016. Communication by extracellular vesicles: Where we are and where we need to go. *Cell* **164:** 1226–1232.

Varela-Eirin M, Varela-Vazquez A, Rodriguez-Candela Mateos M, Vila-Sanjurjo A, Fonseca E, Mascarenas JL, Eugenio Vazquez M, Mayan MD. 2017. Recruitment of RNA molecules by connexin RNA-binding motifs: Implication in RNA and DNA transport through microvesicles and exosomes. *Biochim Biophys Acta* **1864:** 728–736.

Wan JC, Massie C, Garcia-Corbacho J, Mouliere F, Brenton JD, Caldas C, Pacey S, Baird R, Rosenfeld N. 2017. Liquid biopsies come of age: Towards implementation of circulating tumour DNA. *Nat Rev Cancer* **17:** 223–238.

Yang S, Che SP, Kurywchak P, Tavormina JL, Gansmo LB, Correa de Sampaio P, Tachezy M, Bockhorn M, Gebauer F, Haltom AR, et al. 2017. Detection of mutant KRAS and TP53 DNA in circulating exosomes from healthy individuals and patients with pancreatic cancer. *Cancer Biol Ther* **18:** 158–165.

Yi X, Ma J, Guan Y, Chen R, Yang L, Xia X. 2017. The feasibility of using mutation detection in ctDNA to assess tumor dynamics. *Int J Cancer*. doi: 10.1002/ijc.30620.

Yu S, Cao H, Shen B, Feng J. 2015. Tumor-derived exosomes in cancer progression and treatment failure. *Oncotarget* **6:** 37151–37168.

Moving Closer To Victory

Taru Muranen[1] and Joan S. Brugge[2]
[1]Cancer Center at Beth Israel Deaconess Medical Center, Department of Medicine, Division of Genetics, Boston, Massachusetts 02115
[2]Department of Cell Biology, Ludwig Center at Harvard, Harvard Medical School, Boston, Massachusetts 02115
Correspondence: Joan_brugge@hms.harvard.edu

The 81st Cold Spring Harbor Symposium on Quantitative Biology was the fifth symposium focused on cancer. Each of the five, separated by approximately a decade, has been marked by enormous leaps in our understanding of cancer and progress toward more effective cancer treatments. This year's symposium was no exception and was distinguished by extraordinary advances in the translation of this understanding to the development of new treatment strategies and a much deeper elaboration of the complex alterations that underlie all aspects of tumor behavior.

The first two cancer-focused symposia (1974 and 1979) were dominated by research on tumor viruses. During this time period, there were no technologies available to investigate the underpinnings of human cancer, and tumor viruses were the most accessible models to probe cancer mechanisms. With the development of recombinant DNA and sequencing technologies in the decade that followed, there was an explosion in the identification of bone fide human tumor oncogenes and tumor suppressors, many of which were homologs of genes captured from mammalian or avian genomes by tumor viruses. Indeed, the 1994 symposium was replete with presentations about the identification and initial functional characterization of these genes. The summary of this meeting, titled "Introduction to the Puzzle," reflected the opinion of the author Ed Harlow that although these findings provided clues to cancer mechanisms, they posed more questions than answers (Harlow 1994). He also noted the huge gap between the world of basic cancer research and clinical cancer research and treatments and referred to it as a "frustrating time."

Remarkably, in the next decade, many of the questions raised in the 1994 symposium summary were answered and the gap between basic and clinical research narrowed significantly. Tyler Jacks thus titled his summary of the 2005 symposium "The Beginning of the End of Frustration" (Jacks 2005). He concluded that this was a "turning point" in the field of cancer research, with the development of powerful genomic methods that facilitated the identification of an expanded array of cancer genes in human clinical specimens and the feasibility of guiding individual patient treatments based on their specific set of alterations (now known as "precision" medicine) and a continuum from basic research to clinical investigation. He also pointed out the promise of targeted therapies for cancer, highlighted by the development of multiple targeted therapies (e.g., gleevec, herceptin, Iressa, Tarceva, and Avastin). Although he referred to that time as a "new era" for targeted therapies, Dr. Jacks noted, quite presciently, that it was critical to understand the molecular context that confers drug sensitivity and resistance.

Questions also emerged from the new insights provided at the 2005 meeting. Central among these questions were ones related to the "context" or "microenvironment" of tumor cells—how tumor cells overcome their contextual suppressive signals in tumor initiation, the involvement of immune cells, vasculature, and other stromal cells in the biology of cancer, the role of stem/progenitor cells as "cells of origin" of different tumor types, and the contribution of epithelial-to-mesenchymal transition (EMT) in metastatic spread. In addition, exciting new avenues of cancer investigation emerged: novel approaches to cancer immunotherapy, cancer cell metabolism, cancer epigenetics, and noncoding microRNA regulation of gene/protein expression.

Since 2005, the field has made huge leaps once again. As we reviewed the 2016 meeting highlights, it was evident that many of the most exciting discoveries of the past 11 years came from the "emerging" fields that were highlighted in the 2005 review. Our understanding of the tumor microenvironment has vastly expanded. Most significantly, immunologists have developed effective strategies to prevent virus-induced cancers and awaken patient immune systems to induce long-term tumor regression. In addition, the deciphering of the complex mechanisms of gene expression and protein activity regulation—chromatin modifications, microRNAs, long noncoding RNAs, RNA editing, translation, and protein degradation regulation—has greatly enhanced our understanding of how these processes go awry in cancer. This decade also witnessed an enormous revitalization of investigations of metabolic pathways, leading to the recognition of the vast extent to which signaling pathways altered in cancer impact cancer metabolism and how metabolic pathways regulate epigenetic control of gene

Published by Cold Spring Harbor Laboratory Press; doi: 10.1101/sqb.2016.81.031179

expression and protein activity. We also learned how genetic and epigenetic alterations control stem and progenitor cell differentiation, thus contributing to the maintenance of a pool of undifferentiated stem-like cells within tumors. Furthermore, a more detailed understanding of alterations that contribute to metastatic spread of cancer cells emerged during this period. Lastly, we have witnessed a huge expansion of approved cancer treatments, most of which were developed based on the in-depth understanding of cancer that emerged over the last 40 years. This knowledge, together with technology improvements that allow cost-effective cancer gene detection, has refined our ability to guide patient treatment through "precision" medicine, matching the alterations in each patient's tumor to specific therapies. Although these advances are highly encouraging for cancer patients, this decade also revealed the enormous challenges in maintaining long-lasting cancer remission as a result of the ability of tumors to develop therapy resistance. The 2016 meeting provided some clues as to mechanisms of resistance but also raised critical questions that need to be addressed to prevent cancer recurrence and also to treat the most recalcitrant tumors that remain almost entirely untreatable.

It was not feasible to provide a comprehensive summary of the broad range of topics and presentations from the 2016 symposium; thus we have tried to capture the major new concepts and emerging questions in three overarching areas: (1) mechanisms of initiation, progression, and metastasis, (2) cancer therapies, and (3) therapy resistance.

MECHANISMS OF INITIATION, PROGRESSION, AND METASTASIS

Genetic and Epigenetic Alterations That Contribute to Tumorigenesis

Advances in sequencing and computational analysis tools have now provided unprecedented detail regarding tumor cell genetic alterations. Wigler's group has optimized DNA sequencing technology to allow analysis of single tumor cells, thus providing information on the extent of heterogeneity of genetic alterations in tumors. Novel mechanisms of epigenetic regulation of oncogene expression were described by Clark and coworkers using a three-dimensional mapping methodology (HiC) to define alterations in the topology of chromatin domains in tumors. They found large-scale alterations that cause long-range changes in chromatin silencing or activation of cancer genes. Stratton and colleagues developed algorithms to survey the types of genetic alterations in different types of tumors to predict the mutational processes involved in the development of mutations (e.g., smoking is associated with C>A transitions). They defined five different mutation signatures and classified tumors based on these mutations, thus providing clues to the nature of mutational drivers in different cancers. Amon described the effects of chromosome mis-segregation leading to tumor aneuploidy and the "aneuploidy paradox"—that is, how aneuploidy can both suppress and promote tumorigenesis. Aneuploidy leads to G_1 arrest in normal cells and to senescence in tumor cells that harbor severe mitotic aberrations. However, rare variants can escape this arrest because of favorable genetic alterations; in these cells, aneuploidy accelerates tumor evolution by increasing genomic instability.

This scale of genetic alterations in human cancer represents a significant challenge in modeling cancer in model organisms, such as mice, in which genetic modifications in specific oncogenes or tumor suppressors drive tumors so aggressively that there is no selection for genetic alterations that promote genomic instability. Tyler Jacks described new CRISPR technologies as "game changers" to address this problem by increasing the scale in which genetic alterations can be induced in mice using genetically engineered models with inducible or noninducible Cas9.

Mechanisms of Tumorigenesis Involving Differentiation Disruption

Tissue development involves hierarchical, branched differentiation of cells, with successive chromatin modifications that restrict the lineage potential as differentiation proceeds. Multiple presentations at the symposium described cancer-associated mechanisms that prevent the natural differentiation of cells and thus maintain cells in stem/progenitor states. These studies have provided a greater mechanistic understanding of the concept of "maturation arrest" of cancer cells that was originally proposed by Pierce and Dixon (1959). In some tumors, disruption of differentiation was associated with genetic alterations in chromatin modulators themselves. Armstrong and Melnick described leukemias in which there are two different mutated chromatin-modifying proteins (e.g., two histone lysine methyltransferases in AML: MLL1/KMT2A and DOT1L) or cooperating chromatin-modifying proteins and transcription factors (e.g., EZH2, a histone lysine methyltransferase, and BCL6, a zinc finger transcription factor corepressor in diffuse large B-cell lymphoma) that work together to suppress differentiation. Kadoch described the interesting paradox that mutations in the BAF chromatin-modifying complex are associated with either tumor-suppressor or oncogene activities, depending on which BAF complex genes are affected, thus highlighting the intricate regulatory aspects of these proteins. Feinberg brought up the important notions that stochastic epigenetic variation is a driving force in development and evolutionary adaptation and that cancer is an environmentally driven disease where repeated changes in the microenvironment may underlie this plasticity.

Another fascinating mechanism of epigenetic regulation was described by Thompson, who summarized evidence that mutations in certain metabolic enzymes—IDH1/2, succinate dehydrogenase (SDH), and fumarate hydrolase (FH)—lead to the generation of metabolites that inhibit multiple α-ketoglutarate (αKG)-dependent oxygenases that modify chromatin (histone demethylases, prolyl hydroxylases), collagen (prolyl-4-hydroxylases), and the TET (10–11 translocation) family of 5-methlycytosine (5mC) hydroxylases. These chromatin-modifying enzymes control differentiation, and most of the cancer-associated mutations in these enzymes maintain cells in stem/progenitor states. Interestingly, Morrison reported that ascorbate (vitamin C) negatively regulates hematopoietic stem cell renewal by promoting the activity of α-KG-dependent deoxygenases like TET, thus serving as a dietary tumor suppressant. Lastly, Neel described a novel association between the phosphatase PTP1B and this class of α-KG-dependent oxygenases, showing that loss of PTP1B delays the onset of HER2/Neu-driven breast tumors through a hypoxia-associated mechanism involving the Moyamoya disease gene RNF214 control of α-KG-dependent deoxygenases. These studies thus provided strong evidence for metabolic control of differentiation through epigenetic regulation of gene expression.

Three other presentations addressed the lineage plasticity of cancer cells. Weissman and Parada argued that in both glioblastomas and leukemias, tumor cells follow a unidirectional "natural" lineage differentiation, whereas Weinberg provided evidence that breast tumor cells could undergo "reverse" differentiation toward a more mesenchymal, stem-like phenotype with greater metastatic and tumor-initiating activity. Moreover, Weinberg showed that there are specific EMT programs that regulate normal versus cancer stem cells, and that it is feasible to target the cancer stem cells without affecting the normal stem cells. Parada provided evidence that the cells of origin in glioblastomas are $GFAP^{+}$ stem cells; however, additional genetic perturbations are required in a stem cell to develop malignant tumors. Weissman discussed the need to identify epigenetic states in single hematopoietic stem cells, arguing that in studies of pooled cells, potentially important epigenetic changes that lead to onset of cancer are masked. This highlights how technological advances are providing insights into stem cell biology that would not have been possible a decade ago.

Jacks described an interesting finding from their studies in a mouse model of lung adenocarcinoma where cancer stem-like cells as well as niche-like cells are coexistent. The niche-like cells are derived from cancer cells rather than stromal cells and were marked by the expression of porcupine, an enzyme required for post-translational modification and secretion of WNT ligands. Production of Wnt in the niche cells promoted tumor stem cell proliferation. Serrano provided a new twist on another cellular program that can contribute to cancer—senescence. His results support previous findings indicating that senescence leads to a strong inflammatory and remodeling response, which can promote cancer, but also revealed that in vivo expression of the cellular reprogramming factors OCT4, SOX2, KLF4, and c-MYC in mice leads to senescence and reprogramming, and that both processes coexist in close proximity.

Cancer Metabolism

Since the last cancer-focused CSHSQB, the field of cancer metabolism has undergone a dramatic revival. Cancer cells are highly proliferative and need to adapt metabolic pathways to fit these anabolic needs as well as adapt to alterations in oxygen, nutrients, and oxidative stress. In Thompson's keynote presentation, he pointed out that 10 years ago there was no overlap between metabolic pathways and cancer signaling pathways, yet we now understand that signaling and metabolic pathways are closely intertwined. Proteins such as PI3K, AKT, mTOR, MYC, JAK, STAT, and PKM2 (among others) are all important regulators of multiple metabolic programs through either direct modification of metabolic proteins or the induction of transcriptional programs. Shaw identified another link between a metabolic pathway and a tumor suppressor, showing that tumors with mutations in LKB are dependent on acetyl CoA carboxylase (ACC), and that LKB-mutant tumors are highly sensitive to a drug combination consisting of an ACC inhibitor and carboplatin. As described in the above section, cancer-associated mutations in metabolic genes such as IDH1/2, SDH, and FH have direct effects on chromatin modifications that affect stem/progenitor cell differentiation. Interestingly, Mak described another consequence of IDH mutation, independent of TET, on transcription of DNA repair genes. This leads to DNA damage and sensitivity to radiation and daunorubicin, thus revealing additional vulnerabilities of tumors with mutations in this metabolic pathway gene.

Thompson also discussed various ways that tumor cells can adapt to starvation. Under nutrient limiting conditions, particularly in RAS mutant tumors, tumor cells can uptake extracellular protein through macropinocytosis to fulfill their metabolic needs. He also showed that mTOR suppression in nutrient-limited tumor regions is required for this process since mTOR prevents macropinocytosis; thus, depending on nutrient availability, mTOR acts either as an oncogene (in nutrient-rich conditions) or as a tumor suppressor (in nutrient poor conditions). Another cellular process that allows adaptation to nutrient deficiencies during tumorigenesis is autophagy, which in normal cells maintains homoeostasis under starvation conditions by degrading proteins, lipids, carbohydrates, nucleic acids, and organelles to provide nutrients. White and coworkers reported that in mutant KRAS-driven tumors, autophagy is required for tumor growth, especially in hypoxic tumor regions where nutrient supply is low; inhibition of autophagy in these RAS-driven tumors leads to tumor regression. However, even in the autophagy-impaired tumors, autophagy in stromal cells can also aid tumor growth, supporting the concept that the microenvironment contributes to the metabolic needs of tumor cells. Interestingly, mTOR suppression promotes autophagy, so in nutrient poor KRAS tumors, the cancer cells can utilize both macropinocytosis and autophagy to overcome the limited nutrient availability and promote tumor growth.

Another metabolism-focused presentation revealed an interesting unrecognized function of the hypoxia response factor HIF2α. Simon showed that HIF2α modulates lipid storage in clear cell renal cell carcinomas (ccRCCs), where this protein is stabilized because of loss-of-function mutations in von Hippel–Lindau tumor-suppressor protein (pVHL), which is required for oxygen (O_2)-dependent suppression of HIF signaling. The lipid storage droplets sustain ER homeostasis, in particular under conditions of reduced nutrients and oxygen, thereby promoting tumor cell survival. Simon also described other alterations in ccRCC metabolism, including a consistent decrease in urea cycle enzymes, like ARG2/ASS1, involved in this pathway. Loss of ARG2/ASS1 was found to promote tumor cell growth by decreasing ornithine decarboxylase and ornithine aminotransferase metabolic activities, protecting the abundance of pyridoxal phosphate (a vitamin B_6 complex biosynthetic cofactor required for many other reactions involved in amino acid, lipid, and glucose metabolism). This novel mechanism of regulation of pyridoxal phosphate not only identified pyridoxal kinase as a potential therapeutic target in clear cell carcinoma but also provided a clear example of how understanding aberrations in tumor metabolism can reveal new insights into normal metabolism.

Several presentations highlighted the importance of adaptation to oxidative stress during tumorigenesis, adding to the mounting evidence that antioxidants can promote tumorigenesis. This refutes the prevailing dogma that antioxidants protect cells from tumorigenesis. Vousden provided a specific example, involving p53, the most commonly mutated gene in cancer (42%). She found that when serine is withheld from tumor cells, there is a strong induction of reactive oxygen species (ROS) before they shift into making their own serine. Mutant p53 induces an antioxidant response during the transition, allowing adaptation to this stress, and thus providing at least one explanation for why mutant p53 is favored over loss of p53 in some tumors. Another presentation on the importance of ROS adaptation in tumorigenesis by Tak Mak showed that estrogen induction of the antioxidant transcription factor NRF2 following homozygous loss of BRCA1 function rescues these cells from death, suggesting that estrogen promotes survival specifically in tissues responsive to estrogen. Morrison presented evidence that oxidative stress inhibits metastasis and that melanoma cells that are able to induce folate pathway NADPH production following dissemination from the primary tumor are more competent to survive in the circulation and metastasize. Lastly, Tuveson shared the unexpected findings that NRF2 antioxidant effects in pancreatic tumor cells are at least in part mediated through redox regulation of protein translation machinery leading to autocrine stimulation of growth factor pathways that promote tumorigenesis. Together, these new insights significantly reinforce the notion that although antioxidants may protect cells from ROS-induced alterations that promote tumor initiation, induction of antioxidant programs following tumor initiation can promote tumor progression through neutralization of the tumor-suppressive effects of ROS.

Tumor Microenvironment

There has been an explosive expansion of our knowledge about the broad and diverse effects of nonneoplastic cells that comprise the tumor microenvironment. We now understand that tumors are complex ecosystems composed of diverse populations of stromal and immune cells that support tumor growth and survival. Cancer cells hijack normal stromal cells and recruit immune cells to rescue tumor cell deficiencies associated with the intrinsic "costs" of mutations that allow cells to escape homeostatic controls that normally function to eliminate aberrantly proliferating cells outside their normal niches. Several of the presentations during the symposium described "vicious cycles" of reciprocal cross talk involving paracrine signaling between tumor cells and their microenvironment that promote cancer progression. Gerard Evans described how MYC expression in pancreatic tumors triggers rapid stromal expansion and immune cell infiltration through paracrine signaling from the epithelial cells to the stroma, thus hijacking the regenerative program of the stroma, creating the tumor microenvironment commonly associated with pancreatic tumors. Ron Evans added additional insights showing the critical importance of the vitamin D receptor (VDR) in regulating the multiple tumor promoting factors (the "secretome") produced by fibrotic cells in the microenvironment of pancreatic tumors. He showed that VDR knockout leads to fibrotic cell expansion in the pancreas and liver, thus raising the question whether vitamin D may regulate fibrosis that is commonly associated with pancreatic cancer. Indeed, treatment with a synthetic analog of vitamin D reduced the fibrotic response and secretome production in a pancreatic cancer model. Kang provided an example of tumor–stroma interactions associated with metastasis, showing how breast cancer bone metastasis involves cancer cell communication with multiple cells in the bone microenvironment. This involved osteoclast-mediated bone degradation leading to release of TGF-β, which induced tumor cell jagged1 expression that then stimulated Notch signaling in multiple bone stromal cells, thus promoting tumor growth as well as chemoresistance, and identifying jagged1 as a target in metastatic disease.

Weaver discussed another critical aspect of the tumor microenvironment—its physical state or "stiffness" caused by remodeling of extracellular matrix (ECM). Tumor cells "tune" their actomyosin cytoskeletal tension to the stiffness of the microenvironment. Stiff ECM promotes formation of focal adhesions and activation of signal transduction pathways involving RhoGTPases, ERK, PI3K, c-Myc, and WNT to reciprocally enhance matrix cross-linking and stiffness, as well as to enhance proliferation and survival, invasion, and angiogenesis. In addition, she presented evidence that in SMAD4-mutated pancreatic tumors (which have impaired TGF-β signaling), JAK-activated STAT3 is critically involved in tumor progression though a feedback loop whereby STAT3 signaling increases matricellular fibrosis and tissue tension, and STAT3 knockout attenuated tumor progression by reducing the stromal stiffening and epithelial contractility. These results implicate JAK kinases as targets in pancreatic cancer, and add to the growing body of evidence indicating that changes in the compliance of the tumor microenvironment affect tumor progression.

In another presentation that addressed the role of the microenvironment in tumor formation, Lozano compared the effects of expression of mutant p53 in the germline (where it is present in tumor and stroma) versus somatic (where it is only expressed in the tumor). She showed that mice with somatic tamoxifen-induction of the R172H p53 mutant allele had a shorter life span than mice carrying a germline R172H p53 mutation, suggesting that the tumor microenvironment's p53 status has an impact on the tumor evolution.

A microenvironment topic that was not discussed during the 2005 symposium is the role of exosomes in tumor progression. Exosomes are 40–150 nM vesicles secreted from normal and tumor cells that have been implicated in the regulation of angiogenesis, immunity, and metastasis through a previously unrecognized mechanism of cell–cell communication. At the 2016 symposium, Kalluri and Lyden shared insights on the diversity of exosomes and their cargos, tissue tropism of exosomes that influence metastatic honing of tumor cells to specific sites, and the possibility of exploiting this knowledge for diagnostic and therapeutic purposes.

CANCER THERAPIES

The main topics presented in this area dealt with the identification of new targets, development of new classes of inhibitors, immunotherapies, mechanisms of therapy resistance, and strategies to overcome resistance. In addition, patient-derived xenograft (PDX) models have been added to the arsenal of cancer models. These have been shown to recapitulate the heterogeneity of human tumors more meaningfully than in vitro models. During the symposium, Caldas described analysis of more than 80 breast cancer PDX models and the use of explants of these tumors to probe their drug sensitivities and to draw correlations between drug sensitivity and genetic alterations in the tumors.

Oncogenes with Multiple Distinct Mutations in Different Tumors

Rosen's presentation highlighted the complexities of targeting a single oncoprotein that is mutated at distinct residues in different tumors. Certain *BRAF* missense mutations activate BRAF monomers and are sensitive to vemurafenib, whereas others activate braf dimers and are insensitive to vemurafenib. Yet another class of mutations reduces or inactivates BRAF; these mutants effectively dimerize with and activate CRAF. Tumors with these mutations are resistant to the current BRAF inhibitors. Thus, not all BRAF-mutant melanomas have similar effects on BRAF activity, and cannot be treated identically.

New Targets

As new driver oncogenes have been identified, scientists both in academia and industry have made significant efforts to develop inhibitors that target these oncoproteins. One group of targets under intense drug development over the last decade is epigenetic regulators. During the 2016 symposium, several presentations described drugs developed against chromatin-modifying proteins for treatment of leukemias (EZH2, DOT1L, MEN1 [menin], and BRD4). Polyak also described impressive results from treatment of triple-negative breast cancer xenografts with JQ1, a BRD4 inhibitor. These new classes of drugs show significant promise for improving current cancer therapies. Cichowski also described new therapeutic approaches for two currently untreatable cancers by cotargeting both oncogenic and epigenetic vulnerabilities. In the first instance she found that BRD4 and MEK inhibitors synergize in Ras-driven nervous system tumors by co-operatively suppressing Ras transcriptional output. However, she also described a promising combination for castration-resistant prostate cancer, which frequently shows dramatic overexpression of the chromatin-modifying enzyme EZH2. Specifically, she identified agents that, together with EZH2 inhibitors, kill tumors by upregulating an apoptotic transcriptional program. In both tumor types, the therapeutic effects of these distinct combinations were due to the convergence of these epigenetic and oncogenic regulators on key nodal pathways.

Drugs targeting anti-apoptotic proteins have also emerged during the last decade. These drugs take advantage of tumor cell alterations or therapy-induced tumor states that lead to "apoptotic priming," a term coined by one of the speakers, Letai, to describe states in which pro-apoptotic proteins are balanced by anti-apoptotic proteins, such that perturbations that alter this balance determine the threshold in which cells will undergo apoptosis. Drugs that target anti-apoptotic proteins push cells toward a pro-apoptotic fate. Letai pointed out there has been great success with Bcl-2 inhibitors like ABT-199 in the clinic, with an 80% response rate in CLL, and in chemo-refractory AML, the combination of ABT-199 with hypomethylating agents increased the response rate to 93%.

In the last decade, much attention has been focused on targeting stroma proteins that support growth and survival of tumors. As mentioned above, Ron Evans described studies showing that the synthetic vitamin D analog, Paracalcitol, suppresses the aberrant stromal cell production of a panoply of secreted factors that enhance tumor cell growth and survival and, in combination with chemotherapy, significantly repress tumor growth.

Several speakers described a variety of genetic screens to identify the landscape of tumor cell vulnerabilities. Bill Hahn discussed large-scale shRNA and CRISPR screens of hundreds of tumor cell lines, enabling the definition of a distinct class of cancer-specific liabilities resulting from genome instability. Genes with single-copy loss in cancer, termed CYCLOPS genes for "copy-number alterations yielding cancer liabilities owing to partial loss," were particularly required for optimal cell growth. CYCLOPS were enriched in genes associated with large protein complexes (e.g., splicing complexes, proteasome, and ribosome). In addition, several transcription factors that are predicted to mark the cell-of-origin of the tumors cells (e.g., SOX10 in lung and PDEF in ovarian tumor cells) were found to be required for efficient growth in culture and considered to be "master regulators" of these tumor types. Lowe's laboratory examined specific kinase dependencies of liver tumors induced by MYC overexpression. They identified CDK9 as a MYC-specific dependency and inducible shRNAs targeting CDK9 reduced the tumor burden significantly; however, CDK9 inhibitors induced significant systemic toxicities, limiting translation of these findings to the clinic. Lastly, the studies described above by the Shaw group on the dependency of LKB1-deficient tumors on ACC were motivated by their pharmacologic screens to identify compounds that show specific "synthetic lethality" in LKB1-deficient cells.

Solit's presentation showed how tracking extraordinary responses to therapies can lead to the identification of prognostic markers of targeted therapies, using an example of a patient with bladder cancer who responded extraordinarily well to treatment with an mTOR inhibitor. Sequencing of the tumor showed that it carried mutations in two tumor suppressors—TSC1 and NF2—both of which are tumor suppressors that inhibit mTOR activation. Additional studies showed that bladder cancer patients with TSC mutations are highly sensitive to mTOR inhibitors and that down-regulation of NF2 in tumor cell lines with TSC1 mutations enhances the efficacy of mTOR inhibition. This highlighted the potential for studies of extraordinary "responders" to identify effective therapies. Solit also discussed "basket trials" that are target-directed rather than cancer-type directed—these include patients with diverse tumor types that share mutations in a common target gene. These trials are facilitated by programs in many major cancer centers that sequence oncogenes and tumor suppressors in most patient tumors; these data are now being aggregated and harmonized in Project Genie, which will be a valuable research for all cancer investigators and clinicians (accessible through http://www.cbioportal.org/genie).

Several other presentations provided interesting new findings relating to the context in which therapies are effective. For example, Dang showed how the circadian rhythm is important in determining the therapeutic window. He examined the circadian "metabolome" in the liver, where large fluctuations are observed throughout the day, in part because of fluctuations in MYC activation, which affects glucose/glutamine and NAD metabolism. Overexpression of MYC disrupts circadian fluctuations by directly inducing the transcription factor REV-ERBα, which dampens expression and oscillation of the clock protein BMAL1. Dang speculated that the differences between normal cellular circadian gene expression and the arrhythmic expression of circadian genes in cancer cells could potentially be exploited to optimize cancer therapy; for example, he raised the possibility that the induction of thrombocytopenia by the NAMPT inhibitor, which suppresses NAD production, could be reduced by dosing the inhibitor at a time NAMPT levels are low in the liver. Neel shared data showing how the effects of PTP1B inhibition are dependent on hypoxia since this state affects αKG-dependent deoxygenases (see Metabolism).

Immunotherapy

The first keynote presentation was delivered by Elaine Mardis, who covered the history of the immunotherapy field from the discovery of the first tumor antigens by Boon, Wolfel, and Schneider in the late '80s and early '90s, to the development of the highly effective cancer vaccines, immune checkpoint, and engineered chimeric antigen receptor T-cell (CAR-T) therapies. The most effective immunotherapies developed during the last decade are based on fundamental findings in immunology by Allison's laboratory, which indicated that there are homeostatic checkpoints that suppress immune cell activation to limit destruction of normal tissues. Multiple inhibitory immune checkpoints have been identified and therapeutic antibodies that block two of these checkpoints—CTLA4 and PDL/PDL1—have shown impressive efficacy in subsets of human cancers. A major focus of Mardis' presentation was on the development of personalized cancer vaccines targeting tumor "neoantigens"—that is, peptide sequences from mutated proteins that are presented on the major histocompatibility complex (MHC). She emphasized the importance of evaluating tumor mRNA expression to ensure that the mutated gene sequences are expressed in proteins. She also noted the difficulty in predicting neoantigens across the spectrum of different genetic variants, such as indels and fusions, and the ability to predict which of these neoantigens are actually processed and presented to MHC. Mardis suggested that the next steps would be to combine checkpoint therapy with neoantigen vaccines.

Carl June, whose laboratory first developed CAR-T therapies, discussed improved variants of CAR-T receptors as well as how to tackle the challenges currently facing CAR-T cell therapies, which involve treatment with genetically engineered T cells that produce chimeric receptors on their surface that allow them to kill tumor cells that express a specific protein (antigen). One of the difficulties with delivery of these therapies is the logistics associated with the need for personalization for each patient—the cells are harvested from a patient, shipped to a facility to be engineered, and then shipped back to the hospital where they will be administered. June is developing more generic CAR-T cells that will be compatible with different classes of histocompatibility antigens, thus making the therapy much simpler to deliver. Michel Sadelain also discussed how to engineer second-generation CAR-T cells, and how to augment the effectiveness of the existing CAR-T cell-based therapies by combining them with other small molecule inhibitors or checkpoint blockades. Furthermore, he described the development of CAR-T cells that are effective for solid tumors. One technological improvement he described was the use of CRISPR/Cas9 to engineer T cells with the chimeric receptor genes.

Fearon described two different mechanisms that limit effective immune clearance of tumor cells. One involves secretion of the chemokine CXCL12 by dense clusters of fibroblasts surrounding pancreatic tumors. This chemokine repulses effector T cells preventing them from entering the tumor. Elimination of fibroblasts or blocking CXCL12 receptor caused rapid T-cell accumulation in tumors and acted synergistically with α-PDL1 to induce regression of cancer cells, indicating that CXCL12's immune suppressive activity accounts for the failure of these T-cell checkpoint antagonists in this context. He also showed that tumor-associated cachexia can suppress immune cell function through impairment of hepatic PPARα-dependent mitochondrial β-oxidation of free fatty acids and ketogenesis, leading to induction of immunosuppressive glucocorticoid production, and suppression of the effects of immunotherapy.

Weissman described a novel therapy that targets CD47 on tumor cells. This protein prevents phagocytic destruction of tumor cells by innate immune cells and is referred to as a "don't eat me" signal. Inhibition of the CD47 interaction with its receptor SIRP-α on monocytes, macrophages, and dendritic cells increases tumor cell phagocytosis in both solid tumors and hematologic malignancies and is currently in clinical trials.

Although progress in immunotherapies is truly impressive, many challenges exist—among them are reducing the autoimmunity associated with checkpoint blockade immunotherapies, taming the toxicity of CAR-T cells, finding more generic therapies to reduce the complexity of patient-specific therapies, identifying immune-suppressive mechanisms in "immune-desert" tumors, and identifying mechanisms of resistance following effective initial immunotherapy treatments.

CANCER THERAPY RESISTANCE

Research over the past decade has revealed an enormous degree of genotypic and phenotypic heterogeneity associated with neoplastic cells within individual tumors. This heterogeneity plays a major role in the development of therapy resistance because distinct subpopulations of cells display different drug sensitivities. There are many factors that contribute to heterogeneity associated with therapy resistance; some of these are intrinsic to the neoplastic cells (e.g., genetic heterogeneity and adaptive responses to therapies), and some are caused by extrinsic factors from stromal and immune cells in the tumor microenvironment (ECM, paracrine factors). Presentations from several speakers provided insights into the mechanisms responsible for resistance, and therapy combinations to overcome or prevent resistance.

Some of the first FDA-approved drugs that targeted specific mutant oncoproteins, such as BCR-ABL, EGFR, and BRAF, have been in the clinic long enough that a great deal of information has accumulated on the mechanisms involved in drug resistance. As detailed in the introduction to Engelman's presentation, the most common mechanisms of resistance to this class of drugs are either mutations in the specific drug target (e.g., BCR-ABL, EGFR, BRAF) or "bypass" mutations that cause activation of critical downstream components from the pathway regulated by the targeted protein, e.g., the ERK pathway in the case of BRAF mutations. However, other mechanisms of resistance have also been elucidated and highlighted in some of the presentations.

Adaptive Resistance

Two talks addressed strategies to overcome resistance associated with adaptive responses to single-agent therapies. In general, these adaptive responses are due to loss of feedback inhibition mechanisms that are mediated by the target protein pathway or stress responses to the insult of the therapy. Lowe describes how MEK inhibition in KRAS mutant lung cancers leads to activation of FGFR1 and that an FGFR1 inhibitor increases the effectiveness of MEK inhibition. Brugge described adaptive responses to PI3K-mTOR inhibitors in PDX models of ovarian cancer, showing a significant treatment-induced up-regulation of both pro- and anti-apoptotic proteins. This new balance of pro- and anti-apoptotic proteins suppressed cell death, but also primed the cells to be sensitive to inhibitors of anti-apoptotic proteins. Most of the ovarian PDX models were sensitive to combined inhibition of the anti-apoptotic protein BCL-XL together with either PI3K-mTOR or MEK inhibitors; however, several of the ovarian PDX tumors required inhibition of other anti-apoptotic proteins, MCL1 and/or XIAP, for effective killing.

Lineage Switching

Several presentations described switches in the differentiation state of cancer cells associated with resistance. Engelman described how EGFR mutant lung tumor cells can become resistant to EGFR inhibitors through conversion to either an EMT- or small cell lung cancer/neuroendocrine-like state that is no longer dependent on mutant EGFR. The conversion to small cell lung cancer subtype was associated with mutations in p53 and loss or Rb. Interestingly, Sawyer described a set of androgen therapy-resistant prostate tumors in which mutations in p53 and Rb cause a lineage switch associated with loss of luminal identity and an increase in the stem cell transcription factor SOX2, while retaining inhibition of androgen-regulated genes. This resistance state could be recapitulated by genetic down-regulation of p53 and Rb and the switch can be toggled by modulation of p53 and Rb expression.

Haber described heterogeneity and lineage plasticity in estrogen receptor positive, $HER2^-$ (ER^+HER2^-) breast tumors that acquire a $HER2^+$ subpopulation after multiple courses of therapy. He showed that $HER2^+$ and $HER2^-$ circulating tumor cells interconvert spontaneously, with cells of either phenotype producing daughters of the opposite after several cell doublings. Both $HER2^+$ and $HER2^-$ circulating tumor cells have comparable tumor initiating potential; however HER2$^+$ cells are favored because of their higher proliferation rate. Oxidative stress or cytotoxic chemotherapy enhances transition to the HER2$^-$ phenotype because of the lower sensitivity of HER2$^-$ cells to these stresses, whereas Notch inhibition enriches for HER2$^+$ cells. This presentation highlighted how interconverting phenotypes within patient-derived circulating tumor cells can affect treatment outcome.

Tumor Microenvironment Factors

Several presentations described distinct factors in the microenvironment that can influence drug sensitivity. For example, Engelman showed that HGF produced in the tumor stroma can significantly reduce sensitivity to EGFR inhibition in EGFR mutant tumors through activation of its receptor MET, which can then bypass the mutant EGFR. Kang showed a distinct example of microenvironmental regulation of drug sensitivity, demonstrating that chemotherapy-induced Jagged2 expression in osteoblasts within the bone microenvironment of patients with disseminated breast cancer promotes resistance of tumor cells to chemotherapy. Lastly, Weaver showed how a reciprocal cross talk between pancreatic tumors and fibroblasts involving STAT3 leads to the development of fibrosis, which can both restrict drug access and activate resistance pathways in tumors cells.

Immunotherapy Resistance

Two questions posed during the immunotherapy presentations were "what are the mechanisms of resistance to immunotherapy, and what are the biomarkers of responsiveness?". At the time of the symposium, few answers to these questions were available. In acute lymphoblastic leukemia, escape variants were found to lack the epitope that is recognized by the CAR-T cells. As mentioned above, Fearon showed the cachexia is associated with suppression of the immune response through glucocorticoid production, suggesting this as one mechanism of resistance to immunotherapies. Lastly, Bergers discussed how tumor-associated VEGF can both promote angiogenesis as well as suppress immune responses, and how VEGF inhibition can enhance immune cell activation. These studies suggest that combining immunotherapies with VEGF-targeted therapies could both enhance immune cell activities as well as induce normalization of aberrant blood vessels, leading to more effective drug delivery.

CHALLENGES FOR THE FUTURE AND LOOKING AHEAD

The major challenges facing cancer research in the next decade are the development of (1) more effective prevention strategies, (2) early detection methodologies that reveal cancer subtype identities, (3) more effective therapies for recalcitrant cancers like pancreatic, lung, liver, and ovarian carcinomas, (4) effective therapies for the most common oncogenes (e.g., mutant p53, RAS, MYC), (5) a better understanding of immune escape of tumors, and (6) therapies with improved tolerabilities. These are complex and difficult challenges that require input from investigators from many areas of basic science (biology, chemistry, physics, mathematics, and engineering) and medicine.

Although the cancer therapy field is moving toward personalized cancer therapies to target specific alterations associated with each individual cancer, the complexity of this approach in terms of therapy delivery outside of major medical centers is enormous and not feasible in many contexts. Thus, there should be a significant effort devoted toward finding the "commonalities" of subsets of tumors—beyond those associated with common genetic alterations. Varmus suggested that we need to seek simpler solutions in the midst of all the "omics" data, to focus on pathways, patterns, and physiological changes instead of just individual genes, and to attack multiple targets by interpreting effects of mutations on pathways. Furthermore, he posed several questions, which when answered could be exploited therapeutically, such as what is the basis of oncogene addiction, what are the effects of mutations on cancer hallmarks, what is the significance of mutual exclusivity of mutations, and how can we take advantage of synthetic lethalities. We share this opinion and add that the answers to these questions could have a significant impact on overcoming therapy resistance since many different resistance mechanisms operative in any one tumor are likely to impinge on common downstream programs. By understanding the common programs that drive cancer and that mediate resistance, we will be able to rationally design more effective combination therapies and identify biomarkers for patient selection.

In his summary of the 2005 symposium, Tyler Jacks predicted that "before the next Cold Spring Harbor Symposium on this topic, many battles will be won, and the end of the war will be well within sight." The breadth of significant new insights in cancer mechanisms and effective therapies presented at the 81st CSHSQB testify to the accuracy of his prediction. Although there remain significant hurdles to overcome, we share Jacks' optimism today, and believe that, by the next cancer-focused CSHSQB, we will be that much closer to victory in the war on cancer.

ACKNOWLEDGMENTS

We thank Angie Martinez Gakidis for scientific editing.

REFERENCES

Harlow E. 1994. An introduction to the puzzle. *Cold Spring Harbor Symp Quant Biol* **59:** 709.

Jacks T. 2005. The beginning of the end of frustration. *Cold Spring Harbor Symp Quant Biol* **70:** 525.

Pierce GB, Dixon FJ. 1959. Testicular teratomas. I. Demonstration of teratogenesis by metamorphosis of multipotential cells. *Cancer* **12:** 573–583.

DORCAS CUMMINGS LECTURE

Video recording of the Dorcas Cummings Lecture is available on Cold Spring Harbor Laboratory's Leading Strand channel on YouTube (playlist 2016 Symposium Interview Series Targeting Cancer).

Dorcas Cummings Lecture

CHARLES SAWYERS

Dr. Charles Sawyers presented the Dorcas Cummings Lecture entitled "Reflections on Precision Medicine and Cancer Moonshots" to friends and neighbors of Cold Spring Harbor Laboratory and Symposium participants on Saturday, June 4, 2016. Dr. Sawyers is the Marie-Josée and Henry R. Kravis Chair of the Human Oncology and Pathogenesis Program at Memorial Sloan Kettering Cancer Center and an investigator at the Howard Hughes Medical Institute.

Thank you for that kind introduction, and thanks to all of you for giving up your late Saturday afternoon. I'm going to present part of a talk that the Howard Hughes Medical Institute asked me to do. It's called "The Holiday Lectures on Science" (http://www.hhmi.org/biointeractive/medicine-genomic-era). HHMI spends a lot of effort trying to disseminate science into the high schools in the Washington, D.C. area. The headquarters of the Howard Hughes Medical Institute is in Chevy Chase, so I want to give them a shout-out for this.

The term "precision medicine" was first coined in a report that we published in 2011. A lot of these reports sit on the shelf and gather dust and this one was doing that until January of 2015, when President Obama declared the Precision Medicine Initiative at the White House. It took a little time to get the wheels of government to buy into this. Now, the "Cancer Moonshot" thing is a more recent development. Many of you, I assume, watched at least part of Obama's State of the Union when, in the middle of it, with no prior announcement, he turned around and looked back at Vice President Biden and said, "It's time to cure cancer, and I'm putting you in charge of Mission Control." So, for the last 4 or 5 months, the vice president has been busy living up to that job, trying to get advice on what to do. I'm one of 36 people on a panel where we are on conference calls frequently.

I want to leave you with three main take-home messages today. The first one is that this is an incredibly exciting time for cancer research. The second is that this has been a long time in coming. It required years of commitment to understanding basic cell biology and cancer biology, largely due to work at places like this. The third thing is that the "moonshot" is *not* one project. There is not a *single* moonshot. This is a problem we're going to have to live with for a long time coming, and we have to avoid the hype of most media articles on a new discovery saying, "Oh, aren't we almost there?" I don't want to take the wind out of the sails—it's an incredibly exciting time—but we have to manage expectations.

Let's start with first principles. You all know the terms "benign" and "malignant." "Benign" means a tumor that starts in one cell and proliferates inappropriately, but it just sits there; it's noninvasive. It's not causing a problem unless it gets really big, and then you just ask the surgeon take it out. It starts to become a problem when the initial tumor cells acquire the ability to escape the "mother ship" and start invading into other tissues. The real problem is if it gets into the blood, then it can metastasize to other sites. Benign tumors are not cancer. The locally invasive and metastatic tumors are malignant. These are cancer.

How do we treat these? We rely on surgeons to cure cancer by physically removing the tumor cells. Obviously, that is the best way. We have to detect tumors early enough to have that work. The two other main therapies you've known about—and you've had friends who've had these therapies—are radiation and chemotherapy. These kill cancer cells by damaging the DNA in the cells to the point where the cells can't survive. But because normal cells and tumor cells both require DNA to replicate, these two therapies kill normal cells, too. We need more therapies because a surgeon can't see cells that have escaped the tumor, and neither can a radiologist. Therefore, surgeons can leave cells behind. Radiation and chemotherapies don't discriminate well enough between normal and tumor cells; therefore, there are a lot of side effects.

The excitement over the last 15-plus years is from new treatments that have emerged from our knowledge of the DNA sequence of tumor cells. We now know that cancers are caused by mutations in our normal genes. We call these cancer-causing genes, and we've now adopted the terminology of "drivers." These are mutations in a gene that now become the driving force of the cancer. There are two flavors. There are those that we call "oncogenes," which are "dominant," in the sense of "Mendel-and-the-peas" dominant versus recessive traits. The other category is composed of recessive "tumor-suppressor genes."

Cells, when they divide, go through something called the "cell cycle." An oncogene is a "Go" signal. A tumor-

Published by Cold Spring Harbor Laboratory Press; doi: 10.1101/sqb.2016.81.031971

suppressor gene is a "Stop" signal. If you get a mutation in one copy of a gene that's an oncogene, it sends a "Go" signal and the cell divides, and you form a tumor. For a tumor-suppressor gene, one mutation doesn't do it. You have to mutate both copies of the gene. Some of you might know that tumor-suppressor genes sometimes are inherited. The famous examples are the breast cancer genes, *BRCA1* and *BRCA2*. Families who have that mutation in their germline (meaning it's in every cell in their body and can pass on to their children) have one mutated gene already. Those people are at higher risk to get tumors because they only have to develop one additional mutation; the other one is already gone.

A very simple concept is, if you have a mutant oncogene and you know it's driving the cancer, why not make a drug that inhibits the mutant protein? That is what's called "targeted therapy." Perhaps the best example of this is a drug called Gleevec.

Leukemia is a cancer of the blood that results in the overproduction of white blood cells. The earliest symptoms are anemia, but mostly it's diagnosed by examining blood and bone marrow smears under a microscope. There are too many white blood cells in patients with chronic myeloid leukemia. You can grow these cells in the laboratory and look at the chromosomes, and if you use a technology called cytogenetics you can line up the pairs of somatic chromosomes—1 to 22—and then identify the sex chromosomes, XX or XY. If you're really good at looking at these chromosome portraits—these karyotypes—all the time, you can see that one of the pairs of chromosome 22 is too short. This discovery was made in 1960 in Philadelphia, and this is called "the Philadelphia chromosome." It's a pretty subtle change, but such is the power of observation.

It took another 10 to 12 years of developing technology to recognize that what was actually happening is that the missing little piece from chromosome 22 is actually now over on chromosome 9. What basically happened in one cell in the blood is that there was a break at the bottom of chromosome 9, there was a break at the bottom of chromosome 22, and these two pieces switched places. This is called "reciprocal translocation." It turns out that break is in the middle of a gene on chromosome 9 as well as one on chromosome 22. As a result, the gene on chromosome 22, which goes by the name *BCR* (breakpoint cluster region), and the one on 9, which goes by the name *ABL* (Abelson murine leukemia viral oncogene homolog), forms this hybrid gene called *BCR–ABL*, which creates a cancer-causing protein that is half-BCR and half-ABL. This hybrid gene does not exist in normal cells.

We have solved the crystal structure of this protein, so we actually know what the protein looks like at atomic scale. This protein is a molecular machine called an enzyme, which executes a function. If you look at the "skeleton" of the protein, there's this loop that can move back and forth between what's known as an active and inactive conformation, essentially a "flap" that opens and closes a pocket or recess in the protein structure where another molecule might bind. In this case, it binds a molecule called ATP (adenosine triphosphate). It then transfers one of ATP's phosphates—it has three—to another protein called a "substrate." That's a molecular signal. Think of this as a chain reaction, with a bucket brigade passing a signal from one protein to another, with the ultimate result of this chain of signal propagation being cell proliferation. That's what *BCR–ABL* does.

Now that we know exactly how this works, can we make a drug that would interfere with its activity? And can we do it very precisely so that it does not interfere with every protein that uses ATP? It turns out that using ATP is a process commonly used by many different proteins—over 500. In Basel, Switzerland, in the early 1990s, some chemists decided to look for such an inhibitor. When they started this project, we didn't know much of what I've just told you. That knowledge came later. You could imagine though, if you knew what I just told you, that maybe you could just draw on the board exactly what chemical would fit in that pocket where the ATP was binding. It turns out that even with what we know now, that's still incredibly challenging. The approach that was used more than 20 years ago is still the most efficient way to at least get a first sense of what the chemical will look like.

The successful approach came through a brute force strategy that, at the time, was only accessible to the pharmaceutical industry. Imagine searching through a whole library of chemical structures that chemists have made over the years, millions of them, to see if one might block *BCR–ABL*. That's exactly what happened, and that's how Gleevec popped out. Gleevec slips right into this position where the ATP would have been. What's really clever about it is that it slips in and locks this activation loop in what's known as the "closed" conformation. Even if the loop tries to move, it can't; Gleevec completely disables the signaling.

What happened when this drug went into the patients? I think you know a little bit of that story. Six patients who got the drug had white blood cell counts in the $50–100 \times 10^3$ cells/μL range before they started taking the drug (normally, it's in the $5–10 \times 10^3$ cells/μL range). All the blood counts came down to normal-or-below ranges within 30 days of treatment with Gleevec. Some patients with this disease who were first treated on clinical trials in the 1999–2000 timeframe are still alive and well, taking this pill once a day and living a normal life. That is what "targeted therapy" is. That is what "precision medicine" is: This pill works in patients who have the *BCR–ABL* mutation, but it doesn't work in patients who don't have the *BCR–ABL* mutation.

This one experiment led to a series of conversations among the whole field. In fact, I remember being in this auditorium about 15 years ago saying, "Is it time to do something really big?" Is there sort of an engineering approach to push this? How many more examples of this are there? It started in 1960 when Peter Nowell was looking through a microscope and saw the Philadelphia chromosome pattern. It was 40 years later when Gleevec was approved. We can go faster. The thing that enabled us to think about going at large scale was the ability to sequence DNA at great speed and low cost, as well as the computing power to maintain and process the massive amounts of data related to DNA and protein sequencing. Basically, the idea was to expand the example of Gleevec to other targeted therapies by defining *all* the cancer driver genes and their mutations.

New technology called "next-generation sequencing" has drastically increased the pace of cancer research. Using the old technology, it cost one billion dollars to sequence the first human genome. Using "NextGen" technology, that cost is down to 1000 dollars. The big idea is that, instead of relying on cytogeneticists to look in a crude way at chromosomes, let's sequence the DNA base by base—all three billion base pairs in the tumor—and compare it with the patient's normal DNA to see if we can find mutations specific to tumor cells. Let's do this for thousands of tumors and see what we learn.

I'll also remind you that the reason these mutations create a problem is because the DNA sequence, the so-called triplet codon, instructs the cell's machinery to insert a specific amino acid (for example, the triplet TAC encodes tyrosine) into a protein as it's being made. A mutation is a misspelling: The "spell-check" function didn't work. If the third base changes to a T, that's called a silent mutation because that TAT triplet sequence still codes for a tyrosine. Those silent mutations are screened out from this giant sequencing project to avoid confusion; those mutations are less likely to matter. We're more interested in the two other types of mutations. In missense mutations, changing the A to a G (TGC), for example, changes the tyrosine to a different amino acid, cysteine. That can have a very profound effect on a protein. The other type of mutation is if there's a codon change where the C turns to a G (i.e., TAG), it becomes a stop codon; the translation of protein from the messenger RNA that this derives from just completely stops. The protein doesn't get made properly. That's a very profound disruptive change.

Through a national effort run by the National Cancer Institute, as well as an international effort, the DNA-sequencing community and cancer biologists got together and sequenced well over 10,000 tumors from 23 different types of cancer. What did we learn? We can generate graphs where we can count the number of mutations we found in a tumor and we can now start to look at some patterns. There's something really interesting that just grabs you right off the bat, and that is that children with cancer have tumors with very few mutations: only four or five per tumor. Adults have more and there's a wide range, from 25 to hundreds to thousands, although most tumors have 50 to 100 mutations.

What's underlying this pattern? We think of lung cancer as lung cancer, and we know that smoking is associated with lung cancer. If you get lung cancer and you weren't a smoker, you have very few mutations. If you have lung cancer and you are a smoker, you have 150 to 200 mutations. This makes perfect sense. We know that carcinogens in tobacco smoke cause DNA mutations. We're not learning anything that we didn't suspect, but now we have real quantification of this. Patients with melanoma also have tumors with 100 to 200 mutations. Again, once we saw this, it made perfect sense. Skin cancer is caused by exposure to sunlight. Ultraviolet light causes DNA mutations. That's why we wear sunscreen. The tumors with the mutations in the thousands, it took a little while to sort this out, but this is primarily a subset of patients with colon cancer who have an underlying mutation in genes that are part of the "spell-check" (i.e., the cellular machinery responsible for maintaining the integrity and fidelity of the gene sequence). If you disrupt the "spell-check," the error rate is going to go way up, and that's exactly what happens here, and it turns out this makes a big difference for therapy as well.

What I've just described to you is the number of mutations, but how do we know if those mutations are in genes that are important in causing cancer? Are they drivers? How do we sort that out? We still don't know the best way to do this, but a very good way is to count them, determine where on the chromosome they are, and in what genes they are. The mutations can be laid out on a three-dimensional plot that displays each chromosome from 1 to 22 and then the sex chromosome from left to right (i.e., the x-axis). The length of each chromosome is displayed along the y-axis, with the start of each chromosome at the origin; the prevalence of mutations in a given gene on that chromosome is displayed vertically on the z-axis. You see that there are these skyscrapers of mutations, these towers. These are the genes that we know—and in many cases, already knew—are cancer genes. The stacks that are shorter in size are also clearly cancer genes because we keep seeing them again and again, but they are less common. In this way, as a statistical strategy, we can determine which genes are most likely to be the true drivers.

Now that we've analyzed so many thousands of tumors, where do we stand? It turns out there's quite a few driver genes. The most commonly mutated gene is called *TP53* (tumor protein 53). Some of you might have heard of other commonly occurring mutants: *BRAF*, *KRAS*, *PTEN*, *PIK3CA*. We can't really know ever if we have identified them all, but we certainly know we have the common ones, where "common" is defined statistically as present in at least 5% of the tumors of that type. If we sequenced another hundred thousand tumors, there's no question that we would find more, but these new ones will be less common across the population of cancer patients. The good news about this is that we know our enemy in way better detail than we did 10 years ago. It's challenging, but it's only several hundred genes, not thousands. As you'll see, we can bucket these hundreds into smaller groups to which, hopefully, common therapies can be applied. There's no doubt that cancer diagnosis will include—and is already including—sequencing to find out what's the status of these genes in a tumor, because it really matters.

One way of bucketing those hundreds of genes into just a few categories is to group them based on their function in normal cells. The smallest group consists of the genes involved in genome maintenance, which is the "spell-check" function in the cell. Of the other two major groups, one is simple to understand: It's the "Go" signal that I mentioned before for cell growth and survival. EGFR (epidermal growth factor receptor) is an example of a growth and survival gene. It's mutated in several different cancers, primarily lung cancer, and there are drugs that now have changed the treatment of lung cancer based on that. EGFR starts that ATP signaling process: When its ligand, epidermal growth factor, binds to the receptor (which is embedded in the cell membrane), it initiates a chain reaction of ATP phosphorylation of all these downstream proteins (RAS, BRAF, MEK), causing cells to grow. Other growth-related cancer genes are drivers on the oncogene side like cyclin D1 and *CDK4* that stimulate the cell cycle, and on the tumor-suppressor side, *P53* and *RB*, which inhibit the cell cycle. We understand the function of these in exquisite detail.

The other major category is genes that are normally involved in what's called "cell fate." All tissues—blood, pancreas, liver, prostate—begin with a single cell called a stem cell. These produce progenitor cells, which then differentiate. Blocking any of these steps can cause cells at a particular stage of development to overproliferate and form a tumor, possibly a malignant one. The idea is that there's a set of genes that control this process in a very regulated way. What we've learned from these sequencing projects is that, if one of the genes that control this goes awry and the cell can't differentiate, you've got a problem.

One of the best examples is colon cancer. The intestine is lined with a series of finger-like projections called villi, and around the base of each villus are "crypts," deep recesses in the lining of the intestine. The stem cells at the base of each crypt give rise to their "children"—the progenitor cells—that move upward out of the crypt, where they undergo their final differentiation to the cells covering the villi. These cells are critical; their function is primarily to absorb food nutrients. These cells actually only live for a couple of days; there's massive turnover. When they get to the top of the villus, they've hit the end of their useful life and they die in a process called programmed cell death. What happens if you have a mutation in one of these cell fate genes that's important in this process? There's a famous gene called *APC* (adenomatous polyposis coli) that does just that. If it's mutated, cells that are supposed to keep progressing out of the crypt never do. Because the APC gene is lost, they can't complete the migration and differentiation process and they pile up in the crypts, and you get colon cancer.

How many of these hundreds of mutations can we deal with? As an example, lung cancer has been a remarkable success story. For decades, lung cancer has been diagnosed through the microscope. We still do that. A surgeon takes a biopsy; we then look under the microscope and we see these different patterns, and those have been the basis for diagnosis and classification. Nearly half are defined as adenocarcinomas, followed by squamous and small-cell cancers, and finally, large-cell cancer. With DNA sequencing, we now know that just one histological pattern, adenocarcinoma, has numerous different types of mutant genes, so lung cancer is actually several diseases with different cancer-causing genes. There's still a portion we haven't yet figured out, but we know more than half of them. *KRAS* and *EGFR* each are mutated in about a third of adenocarcinoma tumors, with another third composed of a host of less common mutations (e.g., *ALK* fusions, *HER2*, *BRAF*, *PIK3CA*, *AKT1*, etc.). Clearly, "adenocarcinoma" is way too vague and broad a category; we need to subdivide these into the drivers responsible for them. The good news is that there are drugs either already available or that have been discovered over the last several years that inhibit many of these because they share that property of *BCR–ABL* of being an enzyme that requires ATP. In the same way that Gleevec was discovered, the inhibitors of many of these mutant genes have now also been discovered.

Does it really work? Here's an example of one of the first patients who received one of these pills that inhibits EGFR. This is a chest X-ray. All of this fluffy material is tumor and, within just a matter of a week, this patient's tumor was clearing away. This is a remarkable story. Now when we do clinical trials, we look at the results in graphs called waterfall plots. Each bar along the x-axis is a patient, and what's being measured on the y-axis is how big was the tumor before they started the therapy and how big was it a month or two later (essentially, the increase or decrease in tumor size); the patients are arranged from left to right along the x-axis from worst to best responders. With the EGFR inhibitor drug, erlotinib, there are patients who've now had almost 100% shrinkage of their tumor (and a lot of them), but there are others that haven't. A big part of what we're doing in the field is trying to understand, if all of these patients have the mutation, how come everyone didn't benefit? We're getting lots of new insights leading to the combinations of drugs that move this waterfall plot in a favorable direction (i.e., toward the right).

What I want to end with is this new form of therapy—I say "new," even though it was first tried a hundred years ago—called immunotherapy. To introduce this, I want to remind you how the immune system works. Imagine that you get a viral infection. It gets into the cell. It starts producing viral proteins in the cell. When a protein is made, it has a certain length of time that it's useful and then it degrades. When the cell degrades it, it turns out that these little peptide fragments of the viral protein then get presented by a molecule on the surface of the cell called a major histocompatibility complex (MHC) protein. There's a whole code of the shape of the peptide and the shape of the MHC that's very important in controlling the precision of the immune response.

In a similar way, you could imagine that a cancer cell that has a mutant driver, it's not normal. Could the immune system recognize that protein? Does it also get chewed up? Does it get presented on the outside of the cell? We now know that all of these are true. A T cell, which is part of the immune system, has a receptor that very specifically recognizes and binds to the viral peptide in the MHC on the surface of an infected cell. When that happens, that binding triggers the proliferation of the T cell so that it will kill the infected cell. That's how we get rid of an infection. If cancer were an infection, the same thing should happen: The immune cells should proliferate and kill the cancer cell before it spreads. We know that's not the case, so what happened? It turns out that, even in a normal situation, there's a built-in sort of fail-safe or backup system to control the immune response. If we didn't have a way to turn it off, the immune system, when overactive, can be lethal. There are a series of molecules on the surface of the cells called "immune checkpoint molecules" that turn the immune system off once it's been activated. There is exquisite detail in how this interaction works. We now know that the reason these cancers aren't eliminated quickly is that cancer cells have figured out a way to borrow that strategy and turn off the immune system before it even gets activated.

What can we do about that? Can we figure out a way to block this interaction and outsmart the tumor cell? There are a number of drugs that have recently been approved that go by the name "checkpoint inhibitor." One is an antibody against a checkpoint molecule called CTLA4, and another is against another checkpoint molecule called PD1. These two drugs—and there are many versions; different companies have developed them—are the most widely studied cancer drugs today. These drugs are being studied in every tumor type, and they are now approved as treatment for at least three or four different types. When you give the inhibitor, you disable the checkpoint molecule, and the T cells proliferate and the tumor cell is killed. The first example of that working was in melanoma. Remember, melanoma is caused by sun exposure. Those are tumors that happen to have a lot of mutations. This melanoma patient with metastases to the liver received CTLA4, and 36 weeks after the start of treatment, the melanoma disappeared, even in the liver. One of the most famous examples of this treatment being successful is former President Jimmy Carter. He had melanoma with metastasis to the brain. He got one of these drugs, and his disease completely cleared. What's really amazing about this is that some of these patients are out more than 10 years without a relapse. The immune system in this case has been "taught" to recognize cancer cells, and that "knowledge" is still there. It's like a drug that's continually active to pick up any further tumor cells.

Now, remember, the patients with lung cancers that didn't smoke have a low number of mutations, but the ones that did smoke have a high number of mutations. It turns out that these immune drugs work best in melanoma and in lung cancer, but only in the patients who have a smoking history. The other cancer types with very high numbers of mutations—the "spell-check" tumors, the colon cancers—also respond to this category of drugs. This is late-breaking in terms of why this is happening, but the leading hypothesis—and I think it's a very compelling one—is that, if you've got a lot of mutations, the probability that you've created a mutation in a peptide that's going to stimulate the immune system is higher. Again, having a lot of mutations is clearly bad, but your probability of responding to one of these drugs is increased if you have a lot of mutations.

So, now we've got surgery, radiation, chemotherapy, targeted therapy, and immunotherapy. We have an incredibly exciting arsenal. This is why all of us spent 5 days here talking about the nitty–gritty of how these drugs are working. The pace at which these drugs have been developed is startling. Gleevec was approved in 2001. It wasn't the first targeted therapy, but at the time, you could count them on one hand. It was pretty impressive just 2 years ago, when we were up to 35, but I just looked online yesterday, and there are 71 approved targeted therapies. It's a testament to this massive investment in biology, sequencing, etc. It's also a testament to the power of the biotechnology and pharmaceutical industry to react to this investment in public science and take it to the next level.

What's left to do? First, I think I've made it abundantly clear that this has been possible only because we understand the problem in the detail that we didn't have back when the "War on Cancer" was declared by President Nixon. We understand the taxonomy of how to classify tumors. We're using mutations to determine what cancer the patients have. We understand a lot more about how resistance is developed, etc. All this has led to an increasing number of drug and other treatment options, but we still have a lot of work ahead, and I want to finish with just a few thoughts on what that work looks like.

Together with my colleagues at Sloan Kettering, led by David Solit and Marc Ladanyi, about 2 years ago, we started sequencing the genomes of our patients who have metastatic cancer. We've done a little over 10,000. In two-thirds of those patients, we know what caused their cancer, but in a third, we still don't. We see a few mutations, but they're not on the driver list. We call that "cancer of unknown driver," and we need to figure that out. Otherwise, we're not going to have a good strategy to treat those tumors. We can talk about how we think we can do it, but it's not trivial. It's a challenge. The other problem is that, even of the two-thirds where we know exactly what's going on, we only have drugs for about half of them. The other half, which we call "undruggable," has not been neglected because these are rare cancers that do not attract the interest of the pharmaceutical community. These are super-challenging. These drivers are not these ATP-dependent enzymes. We actually don't know how to go after them chemically. Of course, we didn't know how to go after these ATP-dependent enzymes chemically until Gleevec. I'm optimistic. We just have to keep investing.

I'll leave you with the idea that there's not a single moonshot project that finishes the job. It's still complicated. I'm very grateful to the vice president for bringing so much attention to the problem. He has galvanized conversations among many branches of government that weren't happening before and among the pharmaceutical sector and among scientists. I'm an eternal optimist. We are going to continue to make progress in this fight.

Thank you very much.

CONVERSATIONS AT THE SYMPOSIUM

Video recording of the Symposium Conversations is available on Cold Spring Harbor Laboratory's Leading Strand channel on YouTube (playlist 2016 Symposium Interview Series Targeting Cancer).

A Conversation with Gabriele Bergers

INTERVIEWER: GEMMA ALDERTON

Senior Editor, Nature Reviews Cancer

Gabriele Bergers is a Group Leader at the Flanders Institute for Biotechnology (VIB), a Professor in the Department of Onocology at KU Leuven, Belgium, and a Professor in the Department of Neurological Surgery at the University of California, San Francisco.

Gemma Alderton: You work on the tumor microenvironment and in particular on angiogenesis. Could you give us some background as to where we're at with that?

Dr. Bergers: We have realized that the old idea of depleting tumor vessels to starve a tumor to death, as Judah Folkman had proposed, is not working. It actually can make it worse sometimes, because hypoxic conditions can lead to recurrence and even a more aggressive growth. Rakesh Jain flipped the whole idea and came up with the concept of vessel normalization to make blood vessels more functional; this enables better drug delivery and therefore better efficacy. The problem is that it's a transient process. You have good efficacy for a while, but then further vessel pruning will lead again to hypoxia and tumor relapse. This is a big hurdle. Indeed, antiangiogenic therapies, specifically those targeting vascular endothelial growth factor (VEGF) receptor signaling pathways, show transient efficacies and only work in a subset of tumors. In the beginning, nobody really understood why, if you just block blood vessels, that it works in some tumors but it doesn't work in others. Several laboratories, including ours, started to identify some of the resistance mechanisms. These are very distinct and different because it's not just classic resistance where the cells don't respond to the inhibitor. Here, they find evasive programs and still grow in the presence of the inhibitor. With these issues came a new concept: Can we do combination therapy that targets these evasive pathways and thereby sustain antiangiogenic efficacy?

We approached two different resistance mechanisms. One entailed the implication of myeloid cells in resistance causing a proangiogenic relapse. The other mechanism was independent of the reactivation of the angiogenic program; rather, it was an alternative way to grow. In this scenario, tumors became more invasive—simply expanding by co-opting and crawling along existing blood vessels. This appears to happen in a subset of gliomas. It's not that glioblastomas are not invasive, but they invade by different means and can enhance such angiotropic migration if necessary. Again, however, it only happens in a subset of tumors, so why does GBM A do it and GBM B doesn't do it?

That's when we started investigating the underpinning mechanisms in more detail. We realized that there is not just one prototypic vascular niche in tumors; it's really a heterogeneous group of distinct niches. You have the typical angiogenic niches with sprouting, abnormal tumor vessels, and you can have hypoxic niches in which tumor vessels are nonfunctional or regressed, as well as invasive, co-opted tumor niches in which tumor cells move along normal blood vessels. Similarly, you can have metastatic niches where tumor cells extravasate and co-opt blood vessels in distant organs. We then asked, how does antiangiogenic therapy affect these distinct vascular niches?

We realized that the tumor niches responded very differently. It depended not only on the composition of the different tumor cells, but also on the composition of the host cells. There's a substantial diversity: The heterogeneous tumor cells and heterogeneous host cells interacting dynamically.

With regard to targeting the vascular niche, we are aware that anti-VEGF therapy is not efficient enough. The question is, is it still worthwhile to target the tumor vasculature? By studying combination of immune modulatory and antiangiogenic therapies we realized that not only can we make blood vessels more functional and create an immunestimulatory environment, but you can actually turn some of those into vascular structures reminiscent of high endothelial venules (HEVs): These are specialized blood vessels that are normally found in lymphoid organs to enable lymphocyte trafficking and seem to help here to get T cells into the tumor. We found for the first time that we can therapeutically induce HEV formation in some of these tumors, which correlated with a better tumor response.

Gemma Alderton: Is that because of clearance from the T cells?

Published by Cold Spring Harbor Laboratory Press; doi: 10.1101/sqb.2016.81.031914

Dr. Bergers: Yes, we think so. More T cells infiltrate and are activated. For many years the immunology field and the angiogenesis field have taken mostly separate paths. For example, we have been thinking about VEGF as a proangiogenic factor, but we were not thinking about it as an immune-suppressive factor. By blocking VEGF we can induce angiostatic as well as immune-stimulatory effects. This comes back to the whole point that the immune system and the vasculature are functionally interregulated. During wounding, myeloid cells can become activated to attack and kill potential pathogens, but then they need to switch in the resolution phase to become immune-suppressive and angiogenic. It's a normal homeostasis program and the tumor takes advantage of this and can flip the cells to their own advantage. The good part is that you can flip the cells too, making them again angiostatic and immune stimulatory. This is an evolving theme in many laboratories. It will be important to see if one can induce these effects not only in primary tumors but also in metastatic lesions.

Tumor vessels themselves are highly immune-suppressive. They up-regulate PD-L1 (Programmed death ligand 1), for example, which serves as a negative checkpoint regulator on endothelial cells. If you have high PD-L1 levels on tumor vessels, and lower numbers of adhesion molecules, T cells do not sufficiently enter the tumor. That's the first barrier. These blood vessels become a gate. Instead of getting rid of tumor vessels, we need to turn them into vessels that can get lymphocytes into the tumor. Importantly though is to get the right cells in there and have them activated. This is where the reprogramming comes in. One needs to create an immunestimulatory environment with innate and adaptive immune cells.

There are a variety of clinical studies showing that spontaneous HEV formation in many human tumor types—breast cancer, melanomas, gastric cancer, colon cancer—mostly correlates with good prognosis and prolonged survival. The idea is, how can we use this therapeutically and target these blood vessels in a way that we can alter them and turn them into HEVs to enable more efficient T-cell infiltration in order to more efficiently attack the tumor.

Gemma Alderton: What are the real challenges in terms of your research? Is it modeling? What are the hurdles you need to overcome to move forward?

Dr. Bergers: It's always a problem when you start with mouse models and you need to show that it's really happening in patients. We would love to have tissue samples from patients who have been treated with these inhibitors to see if the hypothesis that we are trying to prove in mouse models really works in patients. This is often a major challenge because of the difficulties to obtain pre- and posttreated tumor samples from the same patient, or even just to obtain posttreated tumors because many patients will not undergo surgery anymore.

The other problem with antiangiogenic and immune modulatory therapy has been the lack of biomarkers. How do we know when response and relapse occurs? Many laboratories attempt to obtain signatures from peripheral blood of patients. We have looked into signatures of innate immune cells to test whether their changes within the tumor are also seen in those from peripheral blood. If one finds a signature, one could actually take blood from patients and oversee if the therapy is working and if, when it will stop working.

Gemma Alderton: That's challenging though, given the massively dynamic nature of those tumors.

Dr. Bergers: They're extremely dynamic, but that's what makes it interesting.

A Conversation with Joan Brugge

INTERVIEWER: RICHARD SEVER

Assistant Director, Cold Spring Harbor Laboratory Press

Joan Brugge is Director of the Ludwig Center and Chair of the Department of Cell Biology at Harvard Medical School.

Richard Sever: The first time you came to Cold Spring Harbor was 1974.

Dr. Brugge: That actually wasn't for a meeting. It was to learn how to do a procedure that was novel at the time, and Cold Spring Harbor was one of the only places that did it. It was a hybridization technique called C_0t analysis. I spent a week at the laboratory of Joe Sambrook and learned how to do it. My first meeting was probably the RNA tumor virus meeting when I moved here in '79.

Richard Sever: You've worked on so many things: the original identification of Src, EGFR (epidermal growth factor receptor) signaling, integrins, metabolism, the microenvironment of apoptosis. An old friend told me that this broad background meant that at one point you were appointed as the commissioner of the Oncogene Olympics at a meeting. Is that true? In refereeing, was it? Apparently, there was "Team Cytoplasm" and "Team Nucleus"?

Dr. Brugge: Yes, exactly. The meeting was so intense that I thought it would be great to have a break; a short break was scheduled in the afternoons, but everybody roamed around. They didn't know what to do, so I thought of an idea for something we could do that would allow us to "bond" and at the same time have a lot of fun—this became the Oncogene Olympics. We had many different events—soccer, softball, tennis—and we broke up into the teams for those people working on either nuclear or cytoplasmic oncogenes. Everyone enjoyed it so much that it became a tradition. It was lots of fun.

Richard Sever: Presumably, you were in "Team Cytoplasm."

Dr. Brugge: Definitely. I stayed away from the nucleus for many years.

Richard Sever: Your recent work remains very much in the cytoplasm, particularly with respect to high-grade ovarian cancer. Can you say why that's such an important cancer to look at, and what the challenges are?

Dr. Brugge: High-grade serous ovarian cancer is one of the most recalcitrant types of cancer. One problem is that the disease isn't detected until generally it's at stage 3 and 4, and the prognosis for patients at those stages is poor. Their 5-year survival is somewhere around 10% or 20%, and basically the treatments haven't changed for the last 20 years. Treatment's generally chemotherapy; the first-line treatment is a combination of a platinum- and a taxane-based drug. Eventually, all the tumor cells become resistant, so they basically throw every type of chemotherapy they can at the tumors. Patients go through a cycle of remission, relapse, remission, relapse, over and over, and eventually, the patient succumbs.

Richard Sever: It seemed like there was a lot of information that you learned from the DNA sequencing of tumors from TCGA (The Cancer Genome Atlas). What were the things that you picked up?

Dr. Brugge: The sequence didn't reveal any obvious "smoking gun" that we could go after right away. There weren't mutations in EGFR or ALK (anaplastic lymphoma kinase) or BRAF, or any of the other oncogenes that have drugs that are directed against them, but it did reveal some interesting things. One is that almost every single tumor had a mutation in the tumor suppressor p53, and it looks like that's the first event. That seems to trigger the initiation of the cancer. Additionally, there weren't any other mutations that were found at high frequency, but there were an enormous number of copy-number alterations.

The investigators that worked on the sequencing information were able to discern a few pathways that looked interesting for therapeutic intervention. One is that ~50% of the tumors had defects in some component of homologous recombination–mediated repair of DNA damage. Those tumors are predicted to be more sensitive to PARP (poly-ADP ribose polymerase) inhibitors, and PARP inhibitors are actually used in the clinic for patients that have tumors that are defective in DNA repair. Unfortunately, it's the same story as with chemo: There's initial

Published by Cold Spring Harbor Laboratory Press; doi: 10.1101/sqb.2016.81.033035

success, but then they undergo remission and relapse. They're trying to figure out what to do about that.

They also noticed that there were common alterations in multiple components of two other pathways: the phosphoinositide 3-kinase (PI3K)/AKT/mTOR pathway, and the RAS/ERK (extracellular-signal-regulated kinase) pathway. About 50% of the tumors had an amplification or deletion in regulators of those pathways, but then if you looked at single-copy deletion or amplification, they were in a huge percentage of tumors. It raised the question whether slight amplification of an oncogene coupled with slight or one-copy loss of a negative regulator of that pathway would be sufficient to activate the pathway.

Richard Sever: You were essentially seeing PI3K/AKT or ERK pathway signatures in this tumor?

Dr. Brugge: Yes, in terms of the alterations. Along with the genome information from the TCGA study, they also published proteomics data using the RPPA (reverse-phase protein array) platform that Gordon Mills developed at MD Anderson. That provided strong support that both of these pathways are activated. The phosphoprotein markers that are indicative of pathway activation were enriched in ~70%–75% of the tumors. That's what made us feel that it would be worth exploring these two pathways to see if there would be any therapeutic potential.

Richard Sever: What was the next stage in looking at the consequences of activation of the PI3K pathway?

Dr. Brugge: We decided to use tumor cell samples from patient-derived xenografts (PDXs) that had been developed at Dana-Farber. These match the representation of genomic alterations and pathway activation very nicely, relative to the human tumor data. They grow in mice, we harvest them, and then we have a pipeline for analysis so that we can get as reproducible data as possible. They're cultured exactly the same way for the exact number of days. We found a 10- to 15-fold variation in their sensitivity to PI3K/mTOR inhibitors. However, what we really found was that while inhibition of these pathways caused a cessation of proliferation, it didn't induce cell death.

We wanted to understand that. We had previous data that suggested there was an adaptive response, like a rewiring of pathways, for the cells to adapt to inhibition of this central pathway that regulates self-survival. Originally, we would have expected there to be an induction of death, but we found there's also this massive up-regulation of signaling pathways that provide survival protection. So anti-apoptotic pathways were activated significantly, like Bcl-2 (B-cell lymphoma 2) protein.

Richard Sever: Both pro- and anti-apoptotic pathways went up?

Dr. Brugge: Right. A new analysis by Gordon Mills includes twelve different pro- and anti-apoptotic proteins. They were almost all up-regulated. It wasn't just a survival response; there was also a response to inhibition of this survival pathway, which basically leads to a new balance. Pro-apoptotic mechanisms are up-regulated, but at the same time, there's an adaptive response to counter that, so the cells have to reach a new equilibrium with higher levels of both pro- and anti-apoptotic proteins. That suggested that if we could inhibit the anti-apoptotic proteins, it would unleash the elevated activity of the pro-apoptotic proteins to induce cell death.

Richard Sever: It's like the cells have one foot on the gas and one on the brake, and if you could kick the foot off the brake then you'd be okay.

Dr. Brugge: That's what we found. When we combined the Bcl-2/Bcl-xL (B-cell lymphoma, extra-large) combined inhibitor from AbbVie together with PI3K/mTOR inhibitor from Genentech, we got really nice synergistic death on all but a few of the models. There were three models that were much less sensitive, where we didn't see synergy with the Bcl-2/xL inhibitor. A bio-engineer in our lab, Ioannis Zervantonakis, did a systems-level regression analysis. He was able to predict which components of these pro- and anti-apoptotic proteins would be the best predictors for the response to the PI3K/combined inhibition. It worked out beautifully. The predictions made from his analysis were validated completely with the inhibition studies and with proteomic studies. For example, in those three PDX models that were the least sensitive, it was predicted that two other anti-apoptotic proteins, Mcl-1 (induced myeloid leukemia cell differentiation protein) and XIAP, were elevated and were furthermore predicted to correlate with resistance. Indeed, if we inhibited either one of those we got a very dramatic inhibition of survival in those models.

Richard Sever: So it really was a targeted therapy.

Dr. Brugge: Yes, but it goes a step further. The original inhibitor we used was a Bcl-2/xL combined inhibitor, but the regression analysis indicated that xL was much more predictive. We got separate inhibitors from AbbVie that were specific for Bcl-xL or Bcl-2, and found exactly that. The Bcl-xL inhibitor had great efficacy combining with the PI3K inhibitors, but Bcl-2 inhibitors basically didn't have any activity. We were able to parse out which of the components of the anti-apoptotic program are critical in different models. It looks like they will be good biomarkers if the therapy goes to the clinic.

Richard Sever: That was on the AKT side. You talked about this other pathway, the ERK/MAP-kinase pathway. Were you getting a similar regulation of these apoptosis regulators?

Dr. Brugge: Definitely. The ERK pathway had previously been shown to suppress expression of a very dominant, potent pro-apoptotic called BIM (Bcl-2-like protein 11). When we looked at the RPPA data, there were three different groups of the PDX models: one where ERK was high and BIM was low (which you'd predict), one where ERK was low and BIM was high, and two that were low for both. We suspected that by inhibiting ERK in the models that weren't already high for BIM we could up-regulate BIM and sensitize the cells, and that's exactly what happened. Those models in which ERK inhibition led to an increase in BIM were very sensitive to the combination of Bcl-2/xL and ERK or MEK (mitogen-activated protein kinase kinase) inhibition, but the others were not.

Richard Sever: MEK being the protein upstream of ERK.

Dr. Brugge: Right, MEK inhibitor will block the ERK pathway. Again, those models where we didn't see any synergy were the ones that had high Mcl-1, and where we used the Mcl-1 inhibitor combined with the MEK inhibitor we got beautiful synergy. Basically, the same PDX models that were sensitive to PI3K were sensitive to ERK. We found that PI3K inhibitor was inhibiting ERK, so essentially we saw the same activity because the PI3K was upstream of the ERK. When we inhibited PI3K, we inhibited ERK as well.

We've been talking to AbbVie and Genentech about clinical translation of this. They're most interested in a MEK inhibitor because they have less overall toxicity, and they think it would be easier to combine with the Bcl-2/xL inhibitor. We're setting up to do preclinical studies now to focus on that combination.

Richard Sever: Do you think that we're potentially at an inflection point that's equivalent to the triple cocktail moment with HIV?

Dr. Brugge: I think that's what we're likely going to have to work toward. It's not straightforward because of the complications in clinical trials. Generally, you can just add one at a time, and so you have to do layered trials to look at the addition of a second drug to the first one, and then yet another one. It's not going to be straightforward, but unless immunotherapy can clean up all of those resistant tumor cells from the initial therapies, then we're going to have to get better at killing, which is going to involve more inhibitors. We're hopeful that the immune system will do that. Those are orthogonal therapies so it might be more straightforward than to hit yet another signaling pathway or something.

A Conversation with Karen Cichowski

INTERVIEWER: PAULA KIBERSTIS

Senior Editor, Science

Karen Cichowski is a Professor of Medicine at Harvard Medical School.

Paula Kiberstis: *Ras* is particularly important in cancer because it's highly mutated in many different cancers, particularly in the most lethal cancers. How did you become interested in *ras*?

Dr. Cichowski: I became interested in *ras* because it is one of the most commonly deregulated genes and pathways in human cancer. Not only are there *ras* mutations in a broad spectrum of human tumors, but you often find genetic alterations that function upstream of or downstream from *ras*, highlighting the importance of this pathway in the majority of cancers.

Paula Kiberstis: When mutations in the gene were discovered 35 years ago, people thought it would be relatively easy to target because it was an activated protein. That has turned out not to be the case. Can you explain why?

Dr. Cichowski: It has to do with the chemistry, although people are making strides in targeting the Ras protein itself. However, there are many other approaches that one can take to develop therapies for *ras*-driven cancers. That's what we've been working on in the meantime.

Paula Kiberstis: You're targeting other proteins in the pathway. What sorts of things are you doing?

Dr. Cichowski: I firmly believe that to develop a therapy for any *ras*-driven tumor, you have to hit part of the *ras* pathway. You have to either hit the RAF/MEK/ERK pathway or the PI3 K/AKT/mTOR pathway. The key is trying to identify the other drugs that you can add onto these agents to cause durable tumor regression in humans. There are several approaches one can take to achieve this.

One approach that's been very successful is to think about other types of vulnerabilities—the nononcogenic vulnerabilities—that are in a cancer cell, and try to cotarget those. For example, perhaps you could develop a therapy by targeting part of the *ras* pathway and then capitalizing on responses to DNA damage or oxidative stress or proteotoxic stress, for example.

Paula Kiberstis: You've tested such therapies in mouse models of cancer?

Dr. Cichowski: Yes. We always start with human cells, and we develop a hypothesis and test as many human cell lines as possible, and sometimes human xenografts. Then we test our agents in genetically engineered mouse models. The models we use have been very powerful because every animal that develops a tumor is genetically distinct; just like tumors that develop in humans, the genetics of each tumor are different. Therefore, we are able to test specific drug combinations in a large number of mice to see which of these combinations work the best and most frequently.

The most important consideration is that we're not just looking to slow tumors down. We are looking for frank tumor regression. We want the tumors to melt away. I started off developing genetically engineered mouse models to try to create animals that got cancer, and that was very rewarding, but it is much more rewarding to see a tumor shrink.

Paula Kiberstis: Mouse models sometimes get a bad rap, especially from pharmaceutical companies who take on faith that drugs that have shown efficacy in mice are going to work out in the clinic. I know you're constantly improving the models, but what's your general reaction to the frustration that you hear sometimes from the industry?

Dr. Cichowski: To develop effective therapies, we need to take a variety of approaches. People perform cellular screens in a two-dimensional culture dish. That doesn't accurately mimic a tumor in a patient, but they are still important for target identification. Xenografts are valuable because they're human cells, they are easy to use, and we can manipulate them genetically to complement drug studies. However, because cell lines are cultured indefinitely, they can genetically and epigenetically drift from the original tumor. Patient-derived xenografts are as close to the source as we can get, but they are often difficult to obtain or grow and they lack an intact immune

Published by Cold Spring Harbor Laboratory Press; doi: 10.1101/sqb.2016.81.031955

system and microenvironment. Genetically engineered mouse models are valuable because they do have an intact immune system and tumors develop within the biologically relevant site of origin, such as the brain or the lung. The most powerful approach is to use multiple approaches.

I think that one of the reasons why drugs or specific therapies don't work in humans versus mouse models is because sometimes we're not able to dose humans in the same as we can in mouse models, often because of toxicity. Moreover, in many instances, we're not measuring pharmacodynamics in clinical trials so we don't even know if we have effectively hit the target in humans. To minimize false positives, when possible, we try to recapitulate the dosing that's used in humans. We perform extensive pharmacokinetic and pharmacodynamic studies, and then when we develop clinical trials with our clinical partners, we try to incorporate pharmacokinetics and pharmacodynamics in those human studies as well. My feeling is that, when things don't work, it's usually because the same experiment wasn't performed.

Paula Kiberstis: To get back to your combination therapies, that work has taken you into epigenetics. Tell us about that.

Dr. Cichowski: In many instances, oncogenes drive cancers; however, the ultimate transcriptional output of any oncogenic signal is likely to be determined and dictated by the epigenetic state of a cell. We're becoming more and more aware of the mutations in epigenetic regulatory enzymes. And we now know that there are mutations in both classical oncogenes, as well as in epigenetic enzymes. The question is do these defective proteins or pathways interact? And if so, how?

We noticed in several different tumor types that we would see a specific, interesting oncogenic mutation and a specific, interesting epigenetic mutation. We originally tried to cotarget these using what we thought was an "orthogonal" approach. However, in every instance where a specific drug combination works, there appears to be some nodal point of convergence. That, I think, is what's causing the vulnerability. That's what we're able to capitalize on in these combination therapies.

Paula Kiberstis: One of the drugs you're working with is a bromodomain inhibitor. How far along is that work?

Dr. Cichowski: There are bromodomain inhibitors that are in clinical trials for other indications. The hope is that they will have some clinical activity in individuals that have a very specific alteration that these inhibitors target. We're trying to develop a combination therapy now with the bromodomain inhibitors and MEK inhibitors, which works very well in our animal models. I'm very optimistic. There are many companies that have bromodomain inhibitors; now we have to see which one will be the most effective with the least toxicity. Clinical trials will determine that.

Paula Kiberstis: At Harvard Medical School you're surrounded by clinicians and near Kendall Square with all the drug companies.

Dr. Cichowski: Yes, that's quite valuable. The cancer center really makes an effort to bring in pharmaceutical companies on a regular basis so that we get that interface and that interaction, and we hear about new drugs as they're being developed. And our clinical partners are invaluable. It's that back and forth, bench-to-bedside, back-to-bench approach that has made the greatest impact.

A Conversation with Susan Clark

INTERVIEWER: MIRNA KVAJO

Scientific Editor, Cell

Susan Clark is a Senior Principal Research Fellow and Head of the Division of Genomics and Epigenetics at the Garvan Institute of Medical Research.

Mirna Kvajo: You've studied the role of epigenetics in normal cells' behavior and function and also in disease; it's fair to say that you are one of the pioneers in epigenetics. You're one of the first people who studied changes such as DNA methylation, and you also were the first to develop a sequencing method that allows one to look at methylation patterns. Can you tell me how you came into this field just as it was developing?

Dr. Clark: My research science journey began in the mid 1970s, when I started my undergraduate degree in Molecular Biology at the Australian National University, with the lofty goal to understand how genes are regulated. Restriction enzymes, ligation, and cloning had just been discovered. My Ph.D. in the late '70s at the University of Adelaide was to clone human histone genes for the first time and sequence them. We were using Maxam–Gilbert sequencing in those days, and you could read the bases on acrylamide gels. I noticed that the migration of the bands was different at CCG sites, and we wondered whether that difference in migration meant that the cytosines in this context were methylated.

In the 1980s restriction enzymes were found to be useful to detect DNA methylation as well, but clearly could only be used to detect methylation at a limited number of CpG sites. I joined the laboratory of Marianne Frommer at CSIRO Division of Molecular Biology, and she was also interested in DNA methylation. It took until the early '90s to solve the problem of how to assess methylation of individual CpG sites in a different way. The eureka moment for Marianne came from reading two new papers in the literature, one describing the chemistry of sodium bisulfite conversion of cytosine residues to uracil residues and the second the development of PCR (polymerase chain reaction). By combining these two technologies with sequencing, we showed that we could detect DNA methylation at single-molecule resolution. Simply sequencing amplified DNA after bisulfite conversion allowed you to read whether DNA was methylated or not and most importantly which CpG sites were modified in any different loci and cell type.

Bisulfite PCR sequencing totally changed what we thought about the role of DNA methylation. In 1975, Art Riggs and Robert Holliday proposed methylation might have a role in gene regulation. That hypothesis is what really got me into the field in the first place, but because the technology was limited to restriction enzymes, it was very difficult to address. Also, the field really didn't get much traction because in the '80s it was shown that the main models used for molecular biology research, *Drosophila* and worms, don't have DNA methylation, therefore how could it be important in the control of gene regulation? Suddenly, the hypothesis went out of favor.

It was therefore really fantastic to develop a new method that could be used to look at the methylation pattern of every cytosine residue in any sequence. That was the first time we and others could map methylation patterns in different cell types and demonstrate their importance in gene expression. Critically at the same time, the DNA methyltransferase enzymes were knocked out in mouse and these knockouts were shown to be lethal. The combination of showing methylation was essential for development, and a new method to detect individual cytosine sites really opened up interest in the field again.

Mirna Kvajo: You started by looking at this in a very fundamental way to understand this process of modifications through DNA, but at the same time there was the awareness that this could be important for pathological states. You're working on cancer now.

Dr. Clark: I am now, but that took a long time. Initially, we started looking at methylation profiles at CpG islands and showed that they are typically unmethylated regardless of the cell types, whereas tissue-specific promoters are differentially methylated in different cell types. Then we looked during embryo development to see how the methylation of candidate genes changes from fertilization to implantation. We did that work in the late 1990s. It was very difficult then to get funding to do basic research, as it

Published by Cold Spring Harbor Laboratory Press; doi: 10.1101/sqb.2016.81.031948

still is now. That's when we decided to see if methylation changes during the course of a disease.

That was a very good transition to make. I was inspired with the first report that the retinoblastoma protein is mutated in retinoblastoma tumors. I asked the question: Could the promoter CpG island also be methylated? One of the key investigators in the field sent us DNA from retinoblastoma tumors. We did bisulfite sequencing on the retinoblastoma promoter CpG island and showed that many of the retinoblastoma tumors had one allele methylated and one that was mutated. It demonstrated for the first time that it wasn't just genetic mistakes in these key driver tumor-suppressor genes; methylation was also a bona fide way of switching off a tumor-suppressor gene.

Mirna Kvajo: That complicated things for people who were hoping to look at cancer as a purely genetic disease.

Dr. Clark: There's still an active debate between genetic and epigenetic researchers about this! However from that first example, the field opened up. We and others showed that it's not just single genes that become methylated; many hundreds, even thousands, of genes get differentially methylated and repressed. The next question we asked, "Could there be regions of contiguous gene inactivation in cancer as a result of methylation?" It was very surprising at the time to show that not just single CpG island promoters were methylated, but megabase regions harboring neigboring CpG islands were hypermethylated in cancer. This result showed that there must be a program eliciting regional changes in epigenetic reprogramming across the cancer genome.

Mirna Kvajo: Did you look at other cancer types? Is there a general idea about how much methylation changes in cancer or in different cancer types?

Dr. Clark: Yes. Initally we were fairly opportunistic in selecting cancer types where we could collaborate with enthusiastic surgeons and oncologists. We moved from retinoblastoma tumors to leukemia and then to prostate and breast cancer. Now with the success of the ICGC (International Cancer Genome Consortium) and TCGA (The Cancer Genome Atlas) programs, it is well established that DNA methylation changes are widespread in cancer, even if each clinical sample and subtype of cancer might have a different level and profile of DNA methylation aberrations.

Mirna Kvajo: Do you work on other DNA modifications that have been implicated in the context of cancer?

Dr. Clark: One of the key questions that is still not solved in the cancer field is what elicits the opposing changes in DNA methylation? In cancer, you have global hypomethylation—so, loss of methylation—in addition to the gain: hypermethylation in CpG islands. We're interested in understanding the cause of that and the interrelation with histone marks and replication timing. It is now possible to be able to overlay the DNA methylation landscape with overarching histone modifications to see what is the correlation between methylation and chromatin remodeling, both in normal cells and in the cancerous state.

Mirna Kvajo: The field of epigenetics in general has a big influx of new techniques (and new adjustments and improvements of techniques) that allows this global analysis all the time. We could start a project at one point and, in the meantime, there are several iterations of how you could do it differently.

Dr. Clark: Well, recombinant DNA technology was driven by restriction enzymes and cloning. I think all fields are driven by innovations in technology, but this field of epigenomics in particular is really moving at such a fast pace. You think you've got a great result with a new technique, and suddenly it can be surpassed and the reviewers or referees will say "Why don't you try this as well?"

I think a game changer is a technology that allows you to manipulate or perturb the cell to induce malignancy in vivo in a controlled setting. The problem with looking at cancer is that by the time we can identify the cancer—by the time it's large enough for the surgeon to take it out—all of these epigenetic changes have already happened. It's very difficult. We really need a way to either follow the cell's journey from the premalignant state to the cancer state or a way of manipulating the normal cells with what we think are drivers of epigenetic change in cancer so we can catalog the temporal progression and understand what comes first—genetic or epigenetic lesions.

Personally, it would be preferable to avoid mouse models. Human cancers are really quite unique because it takes a long time for a cancer to grow, especially prostate cancer. Ideally, we need model systems in a human that we can follow, and that's where single-cell technology is a game changer. Methylation is heterogeneous in disease, and to be able to do single-cell technology and follow those cells as they evolve and migrate is really fantastic. CRISPR (clustered regularly interspaced short palindromic repeats) is another game-changing technique that allows you to manipulate epigenetic marks one at a time. Rather than just profiling and looking at patterns in our normal and cancer cell sequencing data, we now have the opportunity to be able to manipulate and hopefully to see what elicits these epigenetic patterns in the first place.

Mirna Kvajo: You've used another technique that moves a step beyond looking at DNA to the whole genome. Changes in chromatin conformation in cancer confer some properties that are different from the normal properties of the cell. Why should they care how the chromatin in the genome is folded, and why are cancer cells different?

Dr. Clark: We commonly just think of the genome as linear, because that's how the sequence is when we visualize it on a page. We have to remember that DNA isn't linear. It is folded in this incredibly complex and exquisite way inside the nucleus, and one of the challenges has been to determine what that structure is inside the nucleus. The development of Hi-C chromatin conformation sequencing technology has allowed us for the first time to get a picture, a snapshot, of how different regions of the chromatin interact with each other inside the nucleus. We were very interested to see if this conformation changes in cancer. We used Hi-C technology and compared normal cells and cancerous prostate cells to look at the topological domains and where the changes might be.

Mirna Kvajo: When it comes to their genome, cancer cells change on so many levels, and the way some transcriptional programs are going to happen or not really seems to depend on all of these factors.

Dr. Clark: It's all interdependent. Simply, the genome is quite de-arranged in cancer, so it's not surprising that has an impact on the conformation. One of the biggest surprises was that the basic structure of the core topological domains is intact. I would have predicted we would see a lot more disruption, but maintaining this core structure appears essential for cell viability. Even though the cancer has gone awry in many ways, I don't think the 3D genome can afford to totally disrupt its fundamental "origami" shape that defines a cell.

The other part of our finding is that where we do identify structural changes in cancer, these involve promoters and enhancers forming new contacts not found in normal cells. The new contacts facilitate ectopic expression of cancer genes. These new insights into alterations of the three-dimensional architecture in cancer have only been possible since the development of new ways of visualizing chromatin interactions.

A Conversation with Gerard Evan

INTERVIEWER: RICHARD SEVER

Assistant Director, Cold Spring Harbor Laboratory Press

Gerard Evan is the Sir William Dunn Professor of Biochemistry at the University of Cambridge.

Richard Sever: You've likened the evolution of a cancer to the Apollo 13 mission. Can you explain what you mean by that?

Dr. Evan: The language we use about cancer is loaded with terms that imply that it's got a plan or a purpose, and it's heading in a direction, which is basically to kill the patient. We use the word "progression" as the disease gets worse. Really though, cancers arise by random mutations, followed by natural selection for the fastest growing or most surviving variant. It's a very random process. There's this wonderful moment in the film *Apollo 13* when the astronauts have got too much carbon dioxide because they don't have scrubbers that can remove it and they don't know what to do. They tell the people at NASA that they've got to sort this out and find a solution. Then a guy comes in with a big garbage bag and empties it on the floor; you see that it's just this pile of debris. He says, "This is what they've got, and we've got to make something that does that out of this." And it's that process of coupling together bits and pieces that might not work particularly well, but that work well enough, that is the hallmark of evolution.

When we look at cancers, we see this chaotic process, yet remarkably, cancers of particular tissues tend to look the same as each other. That means that although they arise by a random process of mutation and selection, something constrains and channels the way that they form. For example, the diagnosis of pancreatic cancer is that it looks like other pancreatic cancers with a lot of connective tissue and stroma. Lung cancers look like other lung cancers with their abundant blood vessels and inflammation. Pancreas tumors look like each other but look different from lung tumors, even when they seem to be caused by the same general types of mutations in the same general types of genes.

Richard Sever: It's not a superficial similarity with different mutations?

Dr. Evan: That's what we don't know. Why is it that through random mutations pancreatic cancers all end up looking like each other, and lung cancers all end up looking like they do, and liver cancers end up looking like they do? It can't just be random. There has to be something that's constraining them.

There are two possible explanations, both of which have some merit. One is that the way cancers can evolve in particular tissues is constrained by the way in which that tissue works. That is, cancers look like they're hijacked, hacked versions of processes that occur normally in tissues when that tissue is damaged. When a tissue is damaged, the program that repairs it needs to be fairly specific for that tissue—this just reflects the fact that different tissues are built in different ways.

Richard Sever: To your "Apollo 13" analogy, this is because the parts available in that rubbish bag are different in different organs.

Dr. Evan: That's correct. The other idea, which I think is also probably true to some extent, is that there's some tissue-specific constraint as to which mutations and which processes can get mutated, and in which ways, to drive a cancer in each tissue type. There's probably some sort of top-down selection, i.e., certain types of mutation that give rise to faster growth in some tissues don't do so in another tissue, perhaps because they just don't have, say, the right blood supply in that tissue. They simply have no way of making it in that tissue. So my guess is that there are constraints from the bottom up, the canalization I referred to first, colliding with some constraints from the top that confer a selective advantage to specific mutations in one tissue versus another.

Published by Cold Spring Harbor Laboratory Press; doi: 10.1101/sqb.2016.81.031278

Richard Sever: Almost like internal as opposed to external constraints?

Dr. Evan: Correct. And we're left trying to sort this out. Of course, what we all want to know is if there are commonalities across cancers, or certain types of cancer. Are there general cancer principles? Yes, everybody's cancer is different from everybody else's cancer because they arise by these spontaneous, random mutations. And even within cancer, there's a huge amount of heterogeneity within the cells. But if cancers are really irreducible, complicated messes, then there's no way of generally describing the process or of identifying general therapies. On the other hand, if—as it seems to be—cancers are quite highly constrained, then we can guess that there must be general principles that would apply to, say, all cancers that arise in a particular tissue type. There could even be general principles that apply to all cancers in all tissues.

Richard Sever: That scenario is almost the opposite of personalized medicine.

Dr. Evan: Yes, I like the idea of impersonalized medicine. It's very easy to get bogged down in the differences between cancers, but just because there are differences doesn't mean to say they are different. All humans have huge differences one from the other, but they share a huge amount of commonality.

Richard Sever: Right. Antibiotics will work if you've got plague.

Dr. Evan: That's a perfect analogy. It doesn't matter who you are—unless you're allergic to the antibiotics! If we could identify these nodes, the shared common elements that go wrong in cancers, then there's no reason why we wouldn't be able to develop therapies that work against many cancers, perhaps all. It sounds like science fiction but to me it's a very real possibility.

Richard Sever: You've recently looked at Myc and Ras in pancreas and lung. What have you been doing there to look at this issue of constraints?

Dr. Evan: We know that these two archetypical oncogenes cooperate. Ras does something that Myc doesn't do, and Myc does something that Ras doesn't do, and you need to put them together to get the full process of tumorigenesis, or at least to provide the platform on which further evolution of tumors can arise. They are two cooperating cancer nodes whose deregulation is implicated in most, perhaps all, cancers. This Ras plus Myc cooperation is a platform on which all tumors depend, and upon which are erected minor secondary embellishments—giving this clone a bit more of an advantage to this, and perhaps a bit less of an advantage to that. The point is that targeting these secondary attributes is not going to knock out the platform and make the tumor go away. You're just going after the hydra head by head, when actually what you want to do is pour bleach on the thing and wipe it out completely. You want to hit the nodes.

We've built in vivo mouse models where we flick on Myc and Ras in different adult tissues and ask what happens. In all cases, the Myc–Ras combination drives tumorigenesis very effectively, almost instantaneously, but you have to have them both. Ras without Myc arrests cells permanently. Myc without Ras, the cells die. This is the prototypical way that biology solves the problem of how to make something incredibly easy—cell proliferation—when it needs to happen but incredibly difficult when it shouldn't happen. As an analogy, it's the same way that we make it easy to get into our house for us, but difficult for someone else. There's a key that turns a combination. You have to have the combination of Myc and Ras to unlock the proliferative machine and cells "cul-de-sac" if they are each activated separately by mutation. But when mutations activate the two together, then you get the full-blown cancer platform.

Remarkably, these Ras–Myc-codriven cancers looks like they are a hacked regenerative program from each tissue. If you activate Myc and Ras in pancreas, you immediately get a pancreatic adenocarcinoma: collapse of blood vessels, lots and lots of connective tissue, all the inflammatory and immune attributes of pancreatic cancer. You switch on the same Myc–Ras pair in lung, and you get a full-blown non–small cell lung cancer: huge numbers of blood vessels, highly inflammatory and devoid of immune cells. The tumors that Ras and Myc together induce in any one tissue look like each other in terms of behavior, dynamics, and appearance and like spontaneous tumors that emerge in that tissue. But they look very different from tumors driven by Ras and Myc in another tissue. So while cancers in differing tissues look very different they still can, and likely are, driven by the same mutated nodes.

It's rather like the way that the same cell cycle gives rise to all the distinct tissue types. If you look at the bewildering variety of tissue shapes and structures you might conclude that they all have different cell cycle machineries. But that's actually not the case. One size cell cycle fits all tissues. Another analogy that I find very useful is that long before people got old enough to ever get cancer, we were being invaded by highly genetically diverse, rapidly proliferating cells called bacteria. If we'd discovered infectious disease in the era of "-omics", I'm sure many would be saying, "You have to have a different way of targeting each different bacterial infection." It's not true. If you hit them where they share common dependencies, you can wipe them out. That's why we have antibiotics—we discovered antibiotics long before the "-omics" revolution.

A Conversation with Daniel Haber

INTERVIEWER: ELIZABETH MCKENNA

Associate Editor, Cancer Discovery

Daniel Haber is the Kurt J. Isselbacher Professor of Oncology at Harvard Medical School and Director of the Massachusetts General Hospital Cancer Center.

Elizabeth McKenna: Can you briefly introduce the topic of circulating tumor cells (CTCs)?

Dr. Haber: Circulating tumor cells are cancer cells that are found in the blood. This is how cancer spreads from a primary tumor in an organ to give rise to metastases. These cells are rare, and they're surrounded by many blood cells. We've spent a lot of time developing technologies to pull these cells out of the blood so we can study both metastasis, as well as cancer progression.

Some techniques use the fact that cancer cells are somewhat larger than blood cells so they try to filter them, but some cancer cells are not that large so you can miss a lot that way. Other technologies involve using antibodies against epitopes on tumor cells or epithelial cells to pull them out; those have become standard in the field. The challenge is that some epithelial cells become more mesenchymal and some cancer cells have different epitopes, so it's hard to get a reliable antibody to always pull out tumor cells.

The technologies that we've used in collaboration with Mehmet Toner, a bioengineer at Massachusetts General Hospital, purify out the normal cells. If you use an antibody against white blood cells to deplete them from a blood specimen and you remove the red blood cells and platelets, then you're left with the cells that don't belong in the blood: the tumor cells. That's hard to do; you need a technology that is really efficient at purifying away a large excess of normal blood cells. That's a big advantage in the field of microfluidics, where you can line up the cells in such a way that they pass single file through a channel. You can attach a magnetic particle with an antibody to the cells of interest and then pass them through a magnet so you can separate them. The technology has gotten more complex but also more powerful, to the point where you can reliably pull out these cells and ask the questions that you've always wanted to ask.

Elizabeth McKenna: Are CTCs similar to the parental tumor or have they diverged in any way? How are they generated and which ones are selected for metastasizing?

Dr. Haber: There's a lot of heterogeneity. Early cancers can shed cells into the blood way before any metastases occur. When you capture cells from a blood sample, maybe 90% or so are cells that are destined to die, but some of them could be alive and some of them could be endowed with the potential to give rise to metastasis. Recently, we've been able to culture some of them and the ones that grow in the laboratory look different. They grow in clusters, they grow under hypoxic conditions, and they're highly tumorigenic when you inject them back into mice. There's clearly a subpopulation in the blood that are bad actors.

Elizabeth McKenna: Can you learn anything about the development of the original tumor by collecting or culturing CTCs?

Dr. Haber: Technically, you can apply multiple assays to analyze CTCs at the single cell level. You can stain in situ for all kind of markers at the protein or RNA levels. You can pick individual cells and do RNA sequencing or do whole-genome amplification and look for mutations. The technologies have to be adapted for minute amounts of material, but those are available now and you can access the heterogeneity of CTCs. In a recent study looking at advanced prostate cancer, we could see how—even within an individual patient—different cells had different markers and different genetic abnormalities. It's a helpful way of looking at the complexity of cancer.

The most important thing from a clinical standpoint is that you can do this noninvasively with a blood test, and you can do this over and over again. Increasingly, we have these effective treatments for cancer and the more effective your treatment is the more the cancer shrinks, but when it grows back, it has different properties. You're

Published by Cold Spring Harbor Laboratory Press; doi: 10.1101/sqb.2016.81.031419

not going to be successful at treating it a second time with the drug that you used the first time. Cancers evolve and the great potential of liquid biopsies is that you can sample tumors repeatedly to see how the cancers change in response to therapies.

Elizabeth McKenna: What needs to be addressed to demonstrate that liquid biopsying is going to have clinical utility?

Dr. Haber: There are three different technologies, all at different stages. You can look at whole CTCs, where you can do anything from culture them to look at protein, RNA, or mutational changes. There's free plasma DNA, which you can analyze primarily looking at mutations at the DNA level. There are also technologies to try to isolate exosomes: little particles of cellular material, including mostly RNA, but other things as well. Ultimately, there will be different applications for each one in the clinic. There are some things that you'll do with one and some with another, depending on whether a particular cancer has recurrent mutations or RNA changes or other markers that are clinically relevant.

The most obvious application is trying to figure out the mechanism of resistance when you give a treatment and the tumor shrinks but then comes back. There are now second-, third-, and fourth-generation treatments and you want to guide those based on the mechanism of resistance. If blood-based biopsies give you a sense of that mechanism, you'll know how to treat the patient. As those methods become more applied, I think they'll also be applied earlier in diagnosis. There are many times when the primary biopsy isn't so good or when it takes time to get a primary biopsy. You could see liquid biopsies moving earlier in the course of treatment.

Also, some of these cells in the blood have different properties, and that raises the question of different targets. There's a lot of interest in why tumor cells in the blood are more sensitive to different stressors such as oxygen stressors or loss of anchorage dependence. We might have new treatments for cancer based on suppressing their survival in the blood. That would be really exciting.

Elizabeth McKenna: Are tests for CTCs or cell-free DNA being incorporated into clinical trial design? What bar needs to be met for this to be introduced into clinical practice?

Dr. Haber: Both circulating tumor DNA as well as CTCs are involved in a large number of clinical trials. For them to gain acceptance and really become part of clinical care you have to have a demonstrated clinical benefit, but also there is a need for better reimbursement structures that will make them clinically viable. A challenge for the field is that as treatments have gotten better, the diagnostics become more important to guide these treatments. We need to put more effort in supporting diagnostics so that we can direct therapeutics more effectively.

Elizabeth McKenna: Is there a particular disease where these liquid biopsies might change practices in the near future?

Dr. Haber: One is where you have targeted therapies and they give rise to a couple defined mechanisms of resistance. If you have mutation A, you need this treatment; if you have amplification B, you need that treatment. It really makes a difference. I could see that happening relatively quickly in cancers such as lung and breast cancer, potentially melanoma. Also, there are cancers like prostate and breast cancer, where tumors spread to bone, which is challenging to biopsy. For these cancers, especially where some can be fairly indolent and others can be aggressive and where there are different medical therapies available, that's where blood sampling would be really helpful.

Elizabeth McKenna: Moving back to the biology of CTCs, what are the next big unanswered questions in the field?

Dr. Haber: We're seeing new pathways that are relevant to CTC biology (i.e., tumor cells that are transiently present within the bloodstream, without their usual surroundings, survival signals, and basement membrane attachment) that we hadn't appreciated before based on tumor analyses. The idea that some therapeutics could target CTCs is really intriguing. Personally, I also want to see CTC studies enabling us to diagnose cancer at an earlier stage. If you can diagnose cancer when, instead of a trillion cells, there are only a million cells, then many treatment options become so much more effective, since there are so many fewer cancer cells to harbor drug resistance. The challenge is that most early diagnosis of cancer is fraught with risk: You pick up noninvasive cancers; you have false positives. Blood-based diagnostics will have a major contribution here, because if a cancer cell invades the blood it's probably unlikely to be an indolent "well behaved" cancer cell. If we can combine blood-based diagnostics with other tools that we now have for early detection of cancer, we could have a really big impact.

Elizabeth McKenna: You think that the presence of CTCs will occur before any change that would be detectable by imaging per se, so you could use it to track the disease before it might be detectable otherwise.

Dr. Haber: I think we will need a combination of assays. For example, in 20 patients with localized prostate cancer we could find CTCs in about a quarter to a fifth of them before surgery, and after surgery the CTCs were gone, and none of these patients have relapsed with 4 years follow-up. It's an example of how a subset of patients will have tumors that shed cells into the blood before metastases are established, and that offers a diagnostic opportunity. It could be the same in lung, in breast, or other types of cancers. Of course, it won't be true in every patient and we will have to combine blood-based assays with mammogram or a prostate-specific antigen (PSA) tests or lung CT scan screening. There's a lot to do to test and validate, but it's a whole new dimension of diagnostics.

Elizabeth McKenna: Are there ongoing clinical trials to see if the number of CTCs correlates with the response?

Do you need large number of patients to make it a clinical benchmark?

Dr. Haber: That's of course the challenge. In a patient with cancer who's being treated, that patient is his or her own control in testing whether CTCs follow or predict clinical course. On the other hand, if you're looking at early detection, even in high-risk individuals, the number of patients or healthy individuals at-risk you need to study is considerably higher. You have to design those studies very carefully. For early detection applications, the test has to be even more reliable and robust. You can't make mistakes, including false positives. It's a whole new dimension to the CTC field, but ultimately it could have a tremendous impact on the way we monitor for cancer.

A Conversation with William Kaelin

INTERVIEWER: JAN WITKOWSKI

Executive Director, Banbury Center at Cold Spring Harbor Laboratory

William Kaelin is a Professor in the Department of Medicine at the Dana–Farber Cancer Institute at Harvard Medical School and Associate Director for Basic Science at the Dana–Farber Harvard Cancer Center.

Jan Witkowski: How does your work relate to the topic of targeting cancer?

Dr. Kaelin: As a postdoctoral fellow, I worked on the retinoblastoma tumor-suppressor gene. Through that experience I learned the power of letting the cancer's genetics guide you in terms of which pathways and molecules they hold near and dear. When I started my own laboratory in the early 1990s, I came across a paper on the identification of the von Hippel–Lindau (VHL) tumor suppressor gene. As a clinician, I knew that the cancers that VHL gene mutation carriers develop, such as kidney cancers and hemangioblastomas, are rich in blood vessels and they also sometimes cause the patient to produce an excess number of red blood cells. What those two events have in common is that they would be normal responses if a tissue were starved of oxygen. You'd try to bring in more blood vessels and more red blood cells. We thought that by studying the VHL gene we'd learn something about oxygen sensing, but also learn something about the pathogenesis of kidney cancer and other cancers caused by VHL mutations.

In the course of that work we learned how the body senses changes in oxygen availability through a transcription factor called HIF (hypoxia-inducible factor), and the modifications of HIF that take place in an oxygen-dependent manner, and how that governs whether the VHL protein does or does not destroy the HIF protein. That led to some interesting new ways to approach diseases characterized by inadequate oxygen delivery. It also taught us that kidney cancers care very deeply about one particular HIF, namely HIF-2. HIF-2, in turn, turns on various genes, including the vascular endothelial growth factor (VEGF). That knowledge helped to accelerate the development of VEGF inhibitors for kidney cancer that have now been FDA-approved. We're also working with a biotech company called Peloton Therapeutics that's developing the first generation of small molecules that target HIF-2 directly. These have just entered clinical trials. It's very early days, but it looks like there could be some activity there, as well.

I'm a big believer in genetically validated pathways as fertile areas for looking for drug targets. Our work on VHL and oxygen sensing and, more broadly, metabolism led us to cancers that are caused by mutations in isocitrate dehydrogenase (IDH) 1 or 2. We learned from work done at Agios Pharmaceuticals that these mutant proteins actually overproduce an abnormal metabolite, a so-called oncometabolite, called 2-hydroxyglutarate (2HG). We, and many other laboratories, have begun to understand how 2HG causes cancer. This has helped provide a framework for the first generation of drugs that now prevent the production of 2HG. At least in leukemias caused by IDH mutations, this has led to some dramatic clinical responses.

Jan Witkowski: Your mention of blood vessels makes me think of Judah Folkman and his work on angiogenesis. How do his ideas fit with yours? You're saying that the increase in blood vessels is a consequence of lowered oxygen, whereas Folkman thought that the proliferation of blood vessels promoted tumor growth.

Dr. Kaelin: Some of the great German pathologists in the 19th century appreciated that all solid tumors have regions that are hypoxic and hypothesized, as did Folkman, that it would be a useful way to control tumor growth if we could block new blood vessels from supplying blood to them. I think the general idea that blocking angiogenesis would have anti-cancer activity is correct. What we have going for us in kidney cancer is a genetic "smoking gun": Here's a tumor that you can almost be sure is causing its angiogenesis by up-regulating HIF and, in particular, VEGF. That's important, because there are actually a host of factors that can induce angiogenesis. In many tumors, it's probably some "chicken soup" combination, whereas in kidney cancer you can be quite sure that

Published by Cold Spring Harbor Laboratory Press; doi: 10.1101/sqb.2016.81.031385

VEGF is doing most of the work. I think that's why there are other solid tumors that will respond to VEGF inhibitors, but the responses are much more significant in kidney cancer.

Although Folkman was correct in thinking we need to develop angiogenesis inhibitors, the inhibitors he developed were never really grounded in a firm molecular and genetic understanding of which growth factors are most essential for inducing blood vessels. That came later, when the work of Napoleone Ferrara and others taught us through mouse engineering and other studies that VEGF is clearly one of the most significant factors used by tumors to make blood vessels.

Jan Witkowski: If you block VEGF in kidney tumors, are there other pathways that it could use to induce angiogenesis?

Dr. Kaelin: Although VEGF inhibitors are very active in kidney cancer and some patients will go into durable remissions that can be measured in years, to my knowledge no one has been cured with a VEGF inhibitor. They inevitably relapse. As you suggest, it's quite likely that this results from workarounds or collateral pathways being activated that provide other means of inducing angiogenesis. If we understood more about those factors, we could perhaps build better combinations.

Jan Witkowski: You're very concerned that one of the factors involved in the length of time it takes from some fundamental discovery of mechanisms to having some molecule that the FDA approves is that the academic research is not—and we can discuss precisely what word to use—reproducible, at least not to a level that the pharmaceutical industry thinks that they can just take at face value.

Dr. Kaelin: There are several layers to that question. One thing that has impressed me watching the drug discovery and development process from the academic sidelines is that there's a big difference between the level of proof demanded for targets and strategies in academia versus pharma, with the latter often setting a higher bar. If you have a genetic smoking gun (preferably in people, but at least in a relevant animal model such as the mouse), this should be a target, as opposed to things that are based on circumstantial evidence. In terms of developing angiogenesis inhibitors, VEGF was the first strategy that was based on a molecular understanding of angiogenesis and where you had that genetic smoking gun. The process of going from target to an actual drug is accelerating, particularly when you have a high-value target where you have genetic evidence as well as corroborating lines of evidence. The basic findings also have to prove to be robust, meaning that other investigators in the field, either in academia or pharma or both, have been able to replicate the essence of the published work.

What you're hinting at is that our colleagues in the pharmaceutical and biotech industries are telling us (more frequently than we would like) that they find that either the published work related to targets emerging from academic laboratories is either not reproducible or, more commonly, reproducible but not robust, meaning it's true in a set of defined circumstances but those circumstances are sufficiently narrow that it's unlikely to translate into a meaningful new drug. We need a greater appreciation that for something to lead to a drug discovery effort we need to emphasize robustness. The key is whether there are corroborating lines of evidence to support the conclusion and if the conclusion can be used to make predictions that you otherwise would not have made that then turn out to be true. That's the heart of the scientific method and at the heart of scientific progress.

Unfortunately, the publication event has become the golden ring in this process. We have these very talented and ambitious trainees, and they see getting that publication as the end to the means. We have to step back and remember that it's really about whether or not the work will stand the test of time. Will it enable other people to build on it? That's what we have to do a better job emphasizing to our young people. Ideally, we would form a relationship with pharma and biotech that would be symbiotic and complementary and would build on each other's strengths. There are things that academia does better than they do, and there are things that they do better than we do. We have to understand that sort of hand-off.

Over the past 10 or 20 years, that appreciation is, to me, the secret sauce of American biomedicine: Understanding that we could rely on academic investigators to do classical, curiosity-driven research without putting a gun to their heads saying, "We want to see the application. We want to see the deliverables. You do your curiosity-driven research, learn the fundamental principles of the system, and if on occasion that leads to a translatable moment where you have something you could build on, do the hand-off to our colleagues in biotech and pharma." Unfortunately, implicitly or explicitly, we've coerced academic scientists to become more translational, to become more applied. One of the problems we've got is that we now have people in some cases doing science they probably shouldn't be doing. It's not what they would naturally be more inclined to do.

Jan Witkowski: Given that institutions, universities, and funding bodies require or expect it, perhaps people feel pressure to make claims about the applicability of their research?

Dr. Kaelin: Now we're getting close to the heart of the problem. When I read papers, maybe there's some interesting phenomenon reported in Figures 1 or 2 that in years past would have been developed more fully from a mechanistic and fundamental point of view. Now, though, you can almost see the race to get to that last figure, which is going to be some gratuitous attempt at clinical relevance. It's going to be a contrived therapeutic, preclinical experiment, or maybe some correlative data on patient outcomes, or some attempt to justify the work. It sometimes makes me want to cry, because it's people trying to justify what they're doing. I think it's actually having some unintended negative consequences.

Jan Witkowski: Do you see a way to deal with this issue?

Dr. Kaelin: We, as a scientific community, have to have a dialogue about this relationship: What we, in academia, should or should not be attempting to do, and what standards we should hold one another to, particularly when it comes to what constitutes a publishable set of observations. I've said only half jokingly that a paper 20 years ago was one thing proven six ways, and I think slowly we've moved toward papers being six things being proved one way. It becomes a house of cards and it leads to this robustness problem. We need to have a serious conversation among ourselves, because at the end of the day we're reviewing each other's papers. We have a lot of influence in terms of what gets published and what constitutes a publishable event. It's always easy to reflexively say, "Oh, I'd like to see one more experiment," but it does become a bridge too far. I think the NIH director and the NCI director and other people in Bethesda and Washington have to use their bully pulpits to say, "Enough is enough. This is what we now expect from the scientific community given these resources that we've been entrusted with."

A Conversation with Raghu Kalluri

INTERVIEWER: GEMMA ALDERTON

Senior Editor, Nature Reviews Cancer

Raghu Kalluri is Chair of the Department of Cancer Biology and Director of the Metastasis Research Center at the University of Texas MD Anderson Cancer Center.

Gemma Alderton: Previously, you've worked on fibrosis and fibroblasts, but you've moved recently into this field of exosomes.

Dr. Kalluri: I've worked in the area of tumor microenvironment for many years now. Work on exosomes is a logical extension because if you look at the tumor microenvironment, in addition to cells and extracellular proteins, there are also these little vesicles that are part of the extracellular milieu. Therefore we asked what these vesicles do and whether there is anything valuable we can learn from them to diagnose or treat cancer.

Gemma Alderton: Where are we at with regard to the diagnostic work in the field?

Dr. Kalluri: Obviously, the whole area of liquid biopsy is very big now. The idea that we can harness information about cancer from the bloodstream or any body fluid is very attractive because you don't have to keep probing tumor tissue or metastatic tissue. There's a lot of information that is carried in the blood via exosomes and free-circulating material. We are excited about exosomes because they come from live cells. That means that you get a reflection of what the cell is undergoing: whether it is stressed or not stressed, whether it has mutations in it causing it to have pathways that are altered. All of that will reflect in the exosomes coming from that cell, and that information can be useful to understand the behavior of the cell, diagnose cancer, and understand how that information could be used for therapy. In the next 5 years you will see serum genomics, body fluid genomics, and liquid-biopsy-associated biomarkers to be of great value to not only diagnose disease but also to be able to rapidly track if a therapy has worked or not.

Gemma Alderton: Could you detect cancers earlier than with imaging?

Dr. Kalluri: Potentially. Let's say you were to take blood from somebody who is 60 years old. If you have done enough studies, you could know, if you isolate the exosomes from this person's blood, which exosomes could predict if there is a certain type of cancer somewhere in the body. Even if you're not absolutely sure, it will give you some risk factor assessment. Then you can follow that up with imaging and other things. I think that's going to be the wave of the future. Of course, I think that early detection and prevention is probably the only way that we will beat cancer, not just treating an advanced disease.

Gemma Alderton: Cancer cells are not the only cells that produce exosomes, so a challenge in the field is determining which exosomes from a blood sample come from a cancer cell.

Dr. Kalluri: There is, of course, a need to distinguish anything coming from cancer cells from anything coming from other cells in the blood. Once we've done that though, when you identify those distinctions in a collection of everything, you know it's coming from cancer cells. That's one way we will overcome that. For example, you can take free-circulating DNA or even exosomes in the blood and do deep sequencing and you can find information. To make it really broadly available to everybody, you have to bring the cost down, and the technology should be very minimal to all centers that do this diagnosis. You're correct though, in that you have to find a way to enrich the sample for something coming from cancer cells, and we've done that. We found certain proteins that are present on exosomes coming from cancer cells but not from normal cells. That allows for an enrichment process that lets us look at that material inside the exosome and be more certain that it's coming from cancer cells. That allowed us to predict things earlier, at least in pancreatic cancer.

Gemma Alderton: Were these markers on the exosomes specific to pancreatic cancer? Do other types of cancer have different markers that would have to be identified separately?

Dr. Kalluri: My gut feeling, and not having any data to back it up right now, is that there will be some markers

Published by Cold Spring Harbor Laboratory Press; doi: 10.1101/sqb.2016.81.033043

coming from all types of cancer cells—just because there are certain common pathways that trigger in most cancer cells that will create proteins in those pathways—that will probably be on or in the exosomes. There are going to be some specific ones, but I'm more interested in the common ones. The protein we found seems to be a common one, but it tends to recognize exosomes coming from pancreatic cancer cells much better. That doesn't mean it doesn't do well with other cancers. It's just that it's not as robust and perfect as with pancreatic cancer samples. We have to do these studies systematically with other cancers to see if there's anything unique we can find. Overall, I'm optimistic that something more generic will be found.

The interest in exosomes is not just to have it on the surface so you can use it as a biomarker but also to enrich it so you can appreciate the microRNA in it, the mRNA in it, the DNA in it, the metabolites that are reflective of cancer cells. You're increasing the signal compared with background.

Gemma Alderton: A lot of questions today are about exosome biogenesis and how the huge plethora of proteins and nucleic acids found in exosomes arises.

Dr. Kalluri: The biogenesis of exosomes is pretty much an open field. There have been some studies done, some proteins identified. There is some good science there, but the bulk of it is unknown. The reason for that is that there was some simplification of the nomenclature, with exosomes being a particular type of entity, and extracellular vesicles being another entity and microvesicles yet another, all because of their size differences. We have a standard way of making exosomes. Somebody wants to call them "extracellular vesicles"? That's fine. One day, we will have to understand if there are in fact different populations doing different things within this collection of exosomes. For now, not having access to such tools, we collect all of them and we study them together. How all of these come about in a cell to be secreted, nobody knows. There could be five different ways of doing it, or ten, or maybe only one way.

Gemma Alderton: Might that be an interesting avenue to explore in terms of preventing biogenesis, particularly from cancer cells?

Dr. Kalluri: I would think so. If you prevent cancer cells from making exosomes, what would be the impact on tumor progression and in metastasis? That requires us to do more experiments, to understand what specific proteins or collection of proteins are important to suppress exosome production without altering other pathways, then asking that question. There are some proteins that tend to have some effect on biogenesis, but again, if it's a heterogeneous population, then you're only going to inhibit one way of producing exosomes but not the other way.

Gemma Alderton: There's been lots of discussion about using exosomes to target genes or proteins in cells.

Dr. Kalluri: That's an exciting area, because there's been enough research done to suggest that if we can enclose drugs or particles and get them to deliver that drug to a particular place in the body, it's a great way to have specific therapies for specific tissues and not have side effects. We have trillions, maybe even a quadrillion, of exosomes floating in our blood all through our body, and they tend to travel. They go into tissues. They enter cells. Why not use this natural property to send drugs? That's what we're pursuing. We're trying to understand how we can put drugs in exosomes and use their natural ability to float around the blood and enter the cells that they would enter naturally. If they enter a cell where there's some specific inhibition that needs to happen through a drug, then they'll have an impact.

Right now, the goal is to see if we can deliver drugs using exosomes. Once that happens, then we can ask "How efficient it is? Is it more efficient than liposomes or nanoparticles carrying these drugs?" It turns out that they are, because exosomes tend to enter cells more efficiently. They have so many other surface proteins on them that enable them to fuse with the plasma membrane, enter the cell, and dump whatever drug cargo they have. They are selected for cancer cells because cancer cells tend to take things in more readily than normal cells because of certain oncogenic events that are amplified. That particular property of cancer cells allows exosomes to enter them more readily.

Gemma Alderton: How do you make the exosomes to package them? Do you take them from the patients to individualize them?

Dr. Kalluri: You could take them from patients, but you could even take them from cells like fibroblasts, which, if you inject them in mice, don't do much. There doesn't seem to be any toxicity associated with them. You can make them from any cell type that you think is benign enough to not cause any problems. You can personalize it for the patient, and then send it back. For cells, it's easy. You could make billions of exosomes in a day, so you could create the drug payload very easily.

One of the most exciting things is that we already have ways like siRNAs or shRNAs to suppress bad transcripts that we don't want. The problem has been how to get them into the cancer cells. I think exosomes can solve that problem. Potentially, you could even deliver things that cells don't normally take in; exosomes will enter the cell and dump it in there. Certain drugs don't enter the cells, but exosomes could disguise them, take them inside, and deliver them to have an impact on something that is inside the cell. You could target the intracellular milieu with exosomes. You don't have to only target the surface of the cancer cell, like most monoclonal antibodies do.

Gemma Alderton: What sort of intracellular processes might you target?

Dr. Kalluri: Transcription factors. Many of them are critical for cancer and they're up-regulated. They're small molecules and notoriously difficult to target. You could use exosome technology to target them.

Gemma Alderton: What are some of the future challenges confronting the field?

Dr. Kalluri: With any young field like this you really need to do some fundamental biology. You need to understand why exosomes have what they have. Is it consistent? Does a cell in the morning make exosomes with a payload that is similar to a cell in the evening that is making exosomes? What is the variation? We need to understand some of the basic biology so that when we do experiments, we know that there's internal consistency. Moreover, it looks like pretty much everything that's found in the cell is found in the exosome, but we need to catalog that. For the next couple of years we're probably auditing and cataloging everything so that we know what these vesicles have and then what we can add to them to make them do what we need to do to treat cancer patients.

On the other hand, maybe exosomes coming from cancer cells can be inhibited. Understanding the biology and biogenesis of exosomes is probably going to be very important. That's a challenge because these are very small vesicles. They are not cells; it's hard to manipulate them. We have to come up with new imaging techniques, new way to manipulate them so that they can be generated in a way that we can understand how the composition of exosomes reflects specific manipulations of their cell of origin. These things are technology-dependent, but I think exosomes are going to be an exciting area of biology. In the next 5 to ten years, we'll be amazed by the things that exosomes are doing in our body and how they can be a reflection of some type of primordial particles that have evolved to provide complex functions in human beings.

A Conversation with Scott Lowe

INTERVIEWER: MIRNA KVAJO

Scientific Editor, Cell

Scott Lowe is Chair of the Cancer Biology & Genetics Program and Chair of the Geoffrey Beene Cancer Research Center at Memorial Sloan Kettering.

Mirna Kvajo: You've been studying oncogenes for a long time, and more broadly, the way genetic mutations affect the various processes of cells and how that results in oncogenesis and tumor progression. How did you start with this topic? Was there something specific that intrigued you?

Dr. Lowe: Like many people that end up in interesting fields, they come into it not expecting to study what they end up studying. When I was a graduate student, I started to work in a laboratory for an esoteric project that related to cholesterol metabolism: a hot topic today, but not so hot back then. The main focus of the laboratory, however, was on oncogenes and particularly how different oncogenes cooperate with each other to lead to transformation. My advisor, Earl Ruley, proposed that certain oncogenes could provoke antiproliferative responses that would suppress the ability of the cell to become cancerous and that a second event was needed to overcome this cellular response.

In the course of my graduate studies we began to understand that this response could involve the p53 tumor suppressor, and that this interplay between deregulated proliferation triggered by an oncogene and then a cellular response—first we studied apoptosis, then senescence—could be a very potent antitumor mechanism. That got me excited about what we often call "innate tumor suppressive mechanisms": the natural processes that normal cells possess to help combat tumor development. If we understand those processes, we might be able to target them.

Mirna Kvajo: You mentioned p53, but you've looked at many tumor suppressors. Are there some common denominators about these genes with respect to how they act or the kinds of processes they affect?

Dr. Lowe: There are several. All tumor suppressors in one way or another have an ability normally to limit the expansion of cell populations. p53 does this primarily by inhibiting proliferation or by promoting cell death, but other tumor suppressors have different mechanisms. The APC (adenomatous polyposis coli) tumor suppressor in colon cancer does this normally by affecting decisions of stem cells to proliferate or differentiate. When it's present, cells are much more likely to undergo normal processes of differentiation than when it is suppressed. Right now, we are quite interested in a class of tumor suppressors that affect chromatin biology, but at the end of the day, p53 is the big one. I think after all the studies and all the sequencing and all the things that have been done, it does all point to p53 really being a central tumor suppressor. That's very interesting, because you can make mice without p53 function even though this molecule plays such an important role in limiting tumor development.

Mirna Kvajo: When you look at a particular gene, you also look at the whole network of oncogenes and the molecules and processes they control as they change in the cells. When you started your research, what could you accomplish and how did that affect the way in which you were thinking and analyzing the function of oncogenes, and how has the field progressed?

Dr. Lowe: There has been a huge change over the last two decades. When I started my laboratory we used cells in culture, and we still use those models extensively today. My laboratory focused on normal cells, often fibroblasts. They aren't particularly a great model of many human cancers, but they can be cultured easily. We were trying to create combinations of genetic changes by activating different oncogenes and inactivating different tumor suppressor genes to create different genotypes to study the different behavior: Why a certain genotype might proliferate better, or another one might die more. We learned a lot about the mechanisms by which molecules such as p53 worked and oncogenes such as ras transformed cells. The challenge in the late '90s and early 2000s was that a lot of the phenotypes we study in cancer can't be studied in a Petri dish.

Published by Cold Spring Harbor Laboratory Press; doi: 10.1101/sqb.2016.81.031351

That led us to follow in the footsteps of others to explore gene function in animal models. We immediately faced a challenge because the standard ways in which one can manipulate genes in mammals such as mice are slow and expensive. You can't interrogate the number of genes that one would like, much less the number of gene combinations. For probably the last 15 years, we've put a lot of effort into developing strategies to either better control gene expression or manipulate gene expression more effectively and also to produce mouse models more quickly. These can involve inactivating genes using RNA interference or, more recently, by genome editing. It can involve transplantation of cells into orthotopic sites to generate tumors more quickly. We've developed inducible systems to look at the function of genes at different times during the course of tumorigenesis, all to get a better handle on how specific mutations or combinations of mutations confer different cancer phenotypes. Ultimately we want to try to find those lesions that are vulnerabilities in cancer cells.

Mirna Kvajo: There is a trade-off in science when creating a model as to whether it gives clear answers to basic cellular processes, or whether the model is pertinent for the disease. When you create your models, do you also think about how close they are to a human disease?

Dr. Lowe: You touch on an important point. No model is perfect, and it's important to understand what any model can teach you and what it can't. In 20 years of using genetically engineered mouse models to study cancer, there has been a lot of debate as to how accurately they reflect human cancer, how predictive they are in terms of therapeutic responses: Different groups would claim more or less often, but often these groups were talking about widely different things. Over the years, we have come to appreciate the advantages of different models. In vitro models are easy; they are simple to study. If you study human cells in vitro, they have the precise genetic configurations that exist in people. On the other hand, you miss the biology of the microenvironment. We've come to appreciate that the immune system is important, but that it can't be captured in cultured cells or even in the typical immunocompromised mouse models that one studies. The genetically engineered mice address many of these caveats, but they are not people.

There are a couple of things I've noticed in our own research that give me hope that these sorts of models are an important part of the present and future. As we characterize the genetic changes that exist in tumors and determine whether they are drivers or passengers, we often look at the data that is coming out of the human genome projects to look at which lesions co-occur in human tumors. Invariably, when we do that and we put those mutations together in mice, we see that we get very potent tumorigenesis. If we use combinations that don't occur together in people, we don't. To me, that's an indication that the mice are weeding out the same biology that occurs in human tumors.

An example from our work involved studying the IDH (isocitrate dehydrogenase) gene, which many groups have shown can block proliferation and differentiation. We made a mouse model of AML (acute myeloid leukemia) that demonstrated that these IDH mutants were oncogenic. Taking advantage of tricks one can only do in mice, we produced a regulated IDH transgene in established leukemia; we could use it to drive the cancer then take it away. What we saw was a paradox: The leukemic cells initially proliferated more, and then they collapsed and differentiated. That was an unexpected response that suggested that inhibiting IDH in leukemia would be a good thing to do, but that there was this paradox. It turns out, that's exactly what is observed in patients when oncologists treat IDH mutant AML patients with small molecule IDH inhibitors. That was a prime example that genetically engineered mouse models can be predictive of the types of pharmacologic responses we might expect in people.

Mirna Kvajo: More recently, you've looked not only at smaller genes but also at larger lesions, which are very common in cancers and which typically affect several genes. These lesions are complex: Several genes are gone and a lot of things are probably dysregulated. Is it a challenge to try to understand what happens when several different genes are gone? How do you address this problem? What kind of questions do you think are going to be answered, especially in regard to human mutations?

Dr. Lowe: Most of the work the community at large has done in terms of trying to understand the genetic changes that exist in human cancers has focused on the single-nucleotide variants: in other words, the point mutations that activate genes such as KRAS or IDH or alternatively inactivate p53. These types of mutations are easy to study. You can make those models quite simply. We have learned an enormous amount about cancer by doing that. However, it's well established that cancers also have, to varying degrees, different changes in gene copy number, regions of amplification, or regions of deletion. These are recurrent, suggesting that they are selected for. In many cases, they are much more common than the single-nucleotide variants. I think it has broadly been assumed that there is one driver in these amplified or deleted regions that promotes cancer or is selected for while the rest of the genes with altered dosage are passenger events that do not contribute to the cancer phenotype.

A number of laboratories have come at copy number variants in different ways. We did it by initially looking at RNAi to suppress different genes in the region where there was a chromosomal region that was recurrently deleted in cancer cells. We were surprised to find that almost always there was more than one gene that, when suppressed, could promote tumorigenesis—for example, in a mouse. That suggested that perhaps it's not a single gene but rather multiple genes in these regions that are driving the selection for a copy-number change.

This concept creates challenges in terms of trying to understand the biology. Using RNAi and different CRISPR approaches to look at the combination of different genes in a region or to produce larger chromosomal alterations provides new ways to do this. The concept also creates challenges in terms of thinking of developing therapies to exploit these types of alterations.

Of course, a difference between normal cells and tumor cells in regions of copy number gain or loss is that not only the gene dosages of the drivers are different but the passengers are as well. People theorized that you could exploit the fact that, say, if a tumor had fewer copies of an essential gene than a normal cell, you perhaps could try to inhibit the essential gene slightly, perhaps killing the tumor (which already had a lower dose) while sparing the normal cells (which had a higher dose). In the case of amplification events, historically scientists have focused on the most obvious driver in the region. If certain amplification events have more than one driver, they may be additional therapeutic targets. It's exciting to think about all the new strategies that might be identified for treating cancer by exploiting copy number alterations in tumors.

A Conversation with Guillermina (Gigi) Lozano

INTERVIEWER: GEMMA ALDERTON

Senior Editor, Nature Reviews Cancer

Guillermina Lozano is Chair of the Department of Genetics at the University of Texas MD Anderson Cancer Center.

Gemma Alderton: Could you give us a bit of background about where we're at in the field of p53 and what's happening with regard to mutations of p53?

Dr. Lozano: All the new, next-generation sequencing data that has been generated over the last few years has basically said that p53 is mutated in almost every cancer that you can think of. We've been working with p53 mutations in animal models, trying to understand how mutant p53 converts a normal, perhaps preneoplastic, cell into something that becomes a very aggressive cell. Mutations in p53 are very common, and these mutations lead to very aggressive tumor types. We've developed some of the animal models that allow us to examine the functions of p53 in vivo.

p53 is a tetramer that binds directly to DNA and activates transcription of probably several hundred genes. There's also a lot of tissue specificity. For example, p53 will activate the p21 cell cycle inhibitor 200-fold in one tissue, but only twofold in another. There's still a lot to learn about its function as a transcription factor. The fascinating part about p53 is the kinds of mutations that occur in human cancers. Most of these mutations occur in the DNA binding domain. Basically, they disrupt the ability of p53 to function as a transcription factor.

Gemma Alderton: These mutations can be loss-of-function or gain-of-function, right?

Dr. Lozano: Yes. There are a few mutations that are truncating mutations, but depending on the tumor type, 70%–80% are missense mutations. Years ago, there was a lot of talk about those missense mutations and why it was so common to have a p53 mutation. One idea was that it was very easy to make a missense mutation in a genomic region. I think the real reason is because these mutant p53 proteins give the cell a growth advantage. The cell becomes able to cope with other changes and they evolve more quickly when it has a missense mutation versus no p53 protein.

Gemma Alderton: Do you see these mutations in specific types of cancer, or is it very broad?

Dr. Lozano: It's very broad. There are five hotspots, all arginines. Four of those arginines directly contact the DNA, so it makes sense that those are common mutations. The other hotspot basically disrupts the whole structure of the DNA binding domain. There are a few differences in different tissues but not enough to really know if there are differences between the kinds of mutations. We've been able to mimic these mutations in the mouse, and we're beginning to see some subtle differences. I think the models will be able to tell us if there is a difference in how these mutations lead to aggressive breast tumors, or lung tumors.

Gemma Alderton: Does the p53 mutation have more to do with initiation or tumor progression?

Dr. Lozano: In ovarian cancer, it's definitely the first mutation that happens. It is very early, and about 95% of high-grade serous ovarian cancers have p53 mutations. Some clinicians suggest that it's really 100%, and that the few percent without p53 mutations are just misdiagnosed tumors. On the other hand, if you look at colorectal cancer, it's the last mutation; it's what takes the lesion from staying in the colon to becoming a metastatic lesion. These are two very different scenarios. In both cases though, once that p53 is mutant (and on top of that, it has to become a stable mutant p53 by signals that the tumor sends out), it contributes to a very aggressive cancer. Timing is important, but it varies between the different cancers.

Gemma Alderton: Are different mechanisms involved depending on whether mutation happens early or late?

Published by Cold Spring Harbor Laboratory Press; doi: 10.1101/sqb.2016.81.031187

Dr. Lozano: There are two answers to your question: One is what happens upstream and when. From studies in animal models, we know that mutant p53 protein is not very stable, in contrast to what we'd thought for years. In most tumors, there are some signals—stress signals, inflammatory signals—that lead to stable mutant protein. Once you have a stable mutant protein, you can have these additional activities, this more aggressive cancer. It's really a two-step process. Perhaps those signals make the mutant stable because the cell is stressed and to proliferate it has to mutate p53, but to progress that mutant protein has to be stabilized.

The second part of the answer involves what these mutant proteins do after they become stable. They bind to numerous proteins, and a lot of them are transcription factors. In those cases, the mutant p53, which has a very potent transcription activation domain, is being dragged to other promoters and is allowing expression of genes the cell would never see normally. Which of those proteins a mutant p53 interacts with will be different in different tissues and will also depend on what else is altered in the cell.

Gemma Alderton: p53 is traditionally associated with regulating processes like metabolism and DNA damage responses. Does mutant p53 serve other roles?

Dr. Lozano: Mutant p53 does a lot of things. Carol Prives published studies showing it affects the sterol synthesis pathway, but we still have to figure out what it means physiologically. If affects integrin recycling (this is from Karen Vousden's laboratory) and the vitamin D receptor (Varda Rotter's laboratory). The field knows that there are a lot of things that mutant p53 does. We need to move from the tissue culture systems where most of these assays have been performed to in vivo systems to ask which mutants, and targets, and pathways it affects.

Perhaps instead of looking at specific genes as markers for the evolution of tumors, we should look at biological processes, pathways, or stress signals. Instead of putting all our hopes into a single gene or gene product to characterize the cancer, we need to really understand the biology of that cancer and how it evolved and how it got to where it is. p53 certainly has a role, but we still have to figure out exactly what that role is. All we really know is that if the tumor cell has a stable mutant p53 protein, it's a much more aggressive tumor.

Gemma Alderton: Where are we at in terms of understanding the role of p53 in promoting metastasis?

Dr. Lozano: The models that Tyler Jacks made a number of years ago basically said that a cell with a mutant p53 protein expressed at high levels is going to be highly metastatic. We're beginning to understand that evolution. The problem is that we currently only have two types of models. In the first, the p53 is mutant everywhere, so the mice develop all kinds of tumors spontaneously. It's very difficult to just study breast cancer, because you get lymphomas, you get sarcomas. Just one model will give you so many different kinds of cancers.

The other model starts out null for p53, so the whole mouse is heterozygous for p53. You can then have a p53 mutation in a single cell, or a single tissue, or organ. That helps a bit because now the mutation is only in a single cell type, but the immune system and stroma is heterozygous for p53. There's quite a bit of data that say that the stroma affects how a tumor develops. One of the new alleles that we made is one in which we can make a mutant p53 basically in one single cell surrounded by a completely normal stroma and immune system. That allows us to do a whole bunch of things. We're still in the early stages, but we're beginning to get some tumors one to two years later. We weren't even sure we were going to get a tumor because we were making p53 mutations in such a small number of cells. What if we needed to hit 10 million cells before we could get a tumor? We're really into uncharted territory.

The other thing that we're doing is that when we mutate the cells—it's a Cre-mediated recombination process—we can label cells with the p53 mutation with red fluorescence. Now you can follow circulating tumor cells; you can actually see a tiny little metastasis. We're hoping that we can examine tumor dormancy in tissues. It's still early, but it's the first time that we've been able to label a mutant p53 cell in a completely normal environment.

Gemma Alderton: You mentioned that you could extract circulating tumor cells from these mice.

Dr. Lozano: It's a very small number, but they're present in some of these models. Comparison of the primary tumor versus what's circulating versus what targets the bone or the liver is going to be really helpful. Metastasis is such a complicated process. We've learned so much from our tissue culture systems, but we need to take that next step and go to physiological systems where we're actually looking at the entire process and the whole organism.

Gemma Alderton: The tumor marker environment is extraordinarily important for metastases. What do we know about p53's role in the stroma?

Dr. Lozano: There's one study published a number of years ago by Terry Van Dyke that looked at the development of prostate cancer. Because of the alleles that we had at the time, the stroma in her model was p53 heterozygous. The stroma and the tumors interacted, and that tumor evolved very rapidly, but the stroma was losing that other p53 allele so that the tumor could get nutrients or something from the stroma. That is one of the best examples showing that what happens in the stroma with regard to p53 affects what happens in the tumor. We're doing similar experiments now in our new model, but it takes time. First, we had to characterize the model to make sure it worked. We can now take out p53 just in the stroma, and look at, say, a *Wnt* tumor in the breast or another kind of tumor and see how it develops in this environment.

The mouse models are fascinating. I started in the field because I went to interview for a postdoc with Arnie Levine and he said we can manipulate the mouse and make a tumor model. That's what really drew me to the field—that you can make a point mutation in the mouse and then study the effects of that mutation. Now with the new CRISPR technologies, we can do so much more.

Gemma Alderton: p53 is part of a family of proteins that are also mutated in cancer.

Dr. Lozano: The two other family members are p63 and p73. There are very few mutations in those proteins in cancers. I'm convinced that p63 has a role in epithelial cancers, but it's usually downmodulated. There's some data that p73 is also downmodulated, but nowhere near the level of p53. I do think they have a role, but it's probably minor.

Gemma Alderton: What are the key challenges in your work going forward?

Dr. Lozano: Ute Moll's laboratory recently published data showing that the lymphoma she was studying in an animal model is "addicted" to having a mutant p53. If you take out that mutant p53, the tumor basically can't deal with it. That's one direction that the field has to go in. If you have a tyrosine kinase expressed at high levels, the drugs you tend to use are aimed at inhibiting this activity. Here you have to reactivate p53, to put it back into the system. That's hard. How do you do that?

If Moll's results turn out to be more broadly applicable to other kinds of cancers, we should be thinking about drugs that disrupt mutant p53 interactions with other proteins that affect metastasis. We're not going to cure cancer. Hopefully, we can make it a stable disease where we can understand enough about what the tumor wants or requires to grow that it can be stuck there in its own little environment and not grow. We need to think more broadly about how to target p53. There are some drugs in the clinic right now that disrupt the interaction between p53 and an important negative regulator called Mdm2, which is an E3 ligase that targets p53 for degradation. In many tumors, you'll have p53 mutations, whereas in others you'll have amplification of Mdm2 at very high levels. Drugs targeting Mdm2 have been around for a long time; some are in clinical studies with mixed results right now. These drugs can disrupt the p53/Mdm2 interaction, but I think you also have to keep these two proteins apart for a longer time. I think the drugs just create a seesaw effect where p53 isn't constantly on, and so you can't tip the balance.

Thanks, Gemma. I enjoyed talking to you.

A Conversation with David Lyden

INTERVIEWER: PAULA KIBERSTIS

Senior Editor, Science

David Lyden is the Stavros S. Niarchos Professor in Pediatrics and Cell and Developmental Biology at Weill Cornell Medicine.

Paula Kiberstis: Your expertise is in metastasis, which is the process by which primary tumor cells journey to a distant site and start growing. Most cancer patients die of metastases, and if we understand how it occurs, we might be able to prevent it or treat it. How did you get into this field?

Dr. Lyden: We focus on translational research in my lab, so that means our investigations start with patient data, then we bring this information for further studies back to the laboratory or the bench. I'm a pediatric neuro-oncologist by training. Although the cancer rate is very low in children compared with adults (10% of all cancers are in children younger than 18 years), what's really significant is that metastasis is very highly noted in all the pediatric cancers, and they have a short latency period for disease progression. That means that when they're diagnosed, if you don't treat the pediatric patient right away, they'll develop metastases very quickly. This is in stark contrast to some adult cancers, such as colon or prostate cancer, where the metastases take years to develop. As a pediatric oncologist, I've a golden opportunity to understand the evolution of the whole metastatic process by using my patients as examples and testing some of the pediatric cancers in the laboratory.

Paula Kiberstis: Metastasis doesn't seem to be a random process. There are certain tumors that prefer to metastasize to certain sites. Prostate cancer will metastasize to bone and liver, whereas pancreatic cancer never metastasizes to bone. One of the central players appears to be a little membrane vesicle called an exosome. Could you tell us about that?

Dr. Lyden: One of the biggest mysteries in cancer is why it spreads to certain organ sites. Pediatric cancers, adult cancers, they all go to certain organ sites, and much of the process was thought to be based on the tumor cell dictating all the steps: The tumor cell leaves the primary tumor, enters the bloodstream, exits the bloodstream, and then dictates its future site, but that seemed too simplified to me.

Cancer is clearly a systemic disease. If you see a cancer patient who's newly diagnosed, there are a lot of complications. They could have clotting, they could have loss of muscle, which we call cachexia; there are a lot of systemic problems. How does the cancer communicate with the body? Originally, I thought it was just soluble growth factors, but when we studied early metastases in our laboratory, we noticed a conditioned microenvironment, a premetastatic niche, at the future sites of the metastases. Even before the tumor gets there, changes are taking place at these organs, making them favorable sites for metastases. We noticed bone marrow–derived progenitor cells clustering in the lung, and extracellular matrix-like scar tissue with collagen deposits, and together they were forming this microenvironment. We noticed that it occurred at the exact site where tumor cells were landing, like a landing pad for metastasis formation. When we studied this site by electron microscopy, we found these bubble-like bodies the size of viruses but they were labeled identically to the original tumor. They were tumor-derived vesicles that we now know as exosomes.

Paula Kiberstis: Do you know what's inside them?

Dr. Lyden: Pretty much whatever we see in cells can be packaged in exosomes: DNA, RNA, and proteins. A cancer patient is secreting billions of exosomes into the bloodstream that package tumor proteins, RNA, and DNA as carriers of tumor information. If they're able to fuse with cells at distant sites and pass on tumor information to normal, healthy cells, that's detrimental to the patient.

Paula Kiberstis: Is there specificity as to how the vesicle finds its way to the organ? Is there something on the outside or within the membrane that helps it get to the right place?

Published by Cold Spring Harbor Laboratory Press; doi: 10.1101/sqb.2016.81.032110

Dr. Lyden: Our laboratory was one of the first to do a protein characterization of adhesion molecules, which stick out from the exosome like antennae. We found a 24-member protein family called integrins. Which member or members of the integrin family is present on exosomes dictates which organ (and which cells in the organ) the exosomes will fuse with. Different integrins explain why some tumor exosomes go to the liver versus the lung versus the brain. This is a start to explaining why cancer spreads to certain organs.

Paula Kiberstis: You pointed out that MET, which is a tyrosine kinase, was within the vesicle. How does that help the tumor cell grow?

Dr. Lyden: The MET oncoprotein is actually the product of an oncogene. When overactive, such a protein promotes cancer growth and spread. Oncoproteins are packaged in exosomes secreted by tumor cells and can be transferred to normal cells, altering their function. C-MET oncoprotein can promote blood vessel development, growth, and migration, and that's exactly what we saw. When that MET protein was transferred, say, to bone marrow cells, we saw migration and survival properties that brought these abnormal bone marrow cells to this premetastatic microenvironment supporting metastases. Basically, it can help transform normal cells at distant sites.

Paula Kiberstis: Is the communication two-way or is it unidirectional from the tumor?

Dr. Lyden: That's a good question. There are other laboratories studying the exosomes in different cell types. Immune and blood vessel cells in your body can also secrete exosomes. They can be normal, but in the cancer environment they can actually give abnormal information back to the tumor and make the tumor grow even faster. We have tumor exosomes going in one direction and all your body's exosomes going back to the tumor, so it's actually a very complex exchange of information.

Paula Kiberstis: What about clinical applications? Where do you see the future for early detection or for using this information to develop a therapy?

Dr. Lyden: There's a lot of potential for studying exosomes and helping clinicians to determine the clinical outcome of a patient. For example, if you have colon cancer and you have a CT scan of your liver, and there's no metastasis in the liver, the surgeon will take out the primary tumor, but we know that 20% of these patients will develop liver metastases in the future. Basically, everyone who has the surgery is sent home, and when those 20% of the patients develop liver metastases, they then go on to the oncologist who will start an aggressive therapy.

Can we do a better job? Can we help predict which patients are in trouble right from the beginning, so you can initiate advancing therapeutics from an early start and prevent metastases? We're hoping we can take a blood sample and collect the exosomes from the plasma by centrifugation. It's complicated by the fact that exosomes occur normally in the body; they're mostly from bone marrow cells. Nevertheless, tumor exosomes are very distinct. If we could recognize them and analyze their protein content and determine if some of the protein signatures are more supportive of metastases, we could make a big impact in how we treat our patients.

If tumor exosomes are missing integrins, they're unlikely to fuse with organs, but the more integrins that you have expressed on the exosomes, the more likely they are to stick to multiple organs. It's not that uncommon to see a lung cancer patient who might have metastases to 10 different organs with up to 15 different integrins expressed on that exosome. We find the more integrins on the exosome surface you have, the worse the prognosis.

A Conversation with Elaine Mardis

INTERVIEWER: RICHARD SEVER

Assistant Director, Cold Spring Harbor Laboratory Press

Elaine Mardis is Robert E. and Louise F. Dunn Distinguished Professor of Medicine and co-Director, McDonnell Genome Institute at Washington University School of Medicine.

Richard Sever: I should welcome you on behalf of the journal, because you're Editor-in-Chief of Cold Spring Harbor's new journal, *Molecular Case Studies*. We've heard a lot about precision medicine and, in particular, "*N* of 1 studies", which is very much a focus of the journal. Could you tell us what "*N* of 1 studies" means in terms of precision medicine?

Dr. Mardis: Many cancer patients find themselves in a situation where they've exhausted all options they might have from FDA-approved drugs or other therapeutic interventions. They're depending on that last look at their cancer genome to reveal some new clues about treatments they might benefit from. Often, these therapies are off-protocol, or not part of a clinical trial, or even require reaching out to pharmaceutical companies to use drugs for which the tissue site of their cancer is not approved by the FDA.

When Carl June described his clinical trials with chimeric antigen receptor T cells, he didn't just report in terms of patient populations; he also broke down the things that he had learned from individual patients. That sort of data goes on to inform how other patients get treated in the future. *Molecular Case Studies* tries to make what we've learned from individual patients widely available to the whole clinical community, not just the cancer realm, so that other people can learn and maybe begin "*N* of 2" or "*N* of more" studies. Individual cases, both where there have been successes but also where there have been failures to respond to a therapy that *should* get a response, are informative to medical care. Most of these studies involve sequencing the genome of the cancer patient tumor cells and comparing it to their normal sequence and then possibly looking at molecules like RNA or proteins for confirmation that this is really the way to go. That forms the basis for a clinical framework where we can identify the right drugs to bring the patient relief from their tumor.

Richard Sever: Someone you know was the focus of one such study. Can you tell us more about that?

Dr. Mardis: This young man, Lukas Wartman, is a colleague of mine. He's a hematologic oncologist. He wrote an important perspective for MCS to describe his experience from the standpoint of being not only a cancer doctor, but also a beneficiary of a very precise therapeutic answer that we found for him by sequencing his cancer genome. Again, this was an "off-label" (i.e., non-FDA-approved) treatment, but by virtue of our finding, he went into remission from his secondary relapse of B-cell acute lymphocytic leukemia. By virtue of being in remission, he was able to receive an unrelated stem cell transplant that saved his life. He's alive and well today. That was in 2011—5 years out from his cancer. He's obviously committed to research, but he's also actively treating patients in the clinic and studying more about heme malignancies. He's even spoken on a personal level as a sort of advocate to patients who've been diagnosed.

Richard Sever: You've also been working on personalized cancer vaccines as immunotherapy against melanoma. How does that work?

Dr. Mardis: Much like an invading bacterium or virus, the proteins that are produced by cancer cells often look different than normal proteins in normal cells. Cancer mutations at the level of the DNA mean changes in the amino acid sequences in the proteins. Those modified peptides, or "neoantigens," that are unique to the cancer cells should look different to the body's defense system. We know that cancers find ways to suppress the immune cells around them so they go undetected, so these mutant proteins go unattacked by immune molecules that would normally target the cell for death.

The vaccine approach uses genomics to tell us which amino acids are changed and of those, which ones are likely to elicit a strong immune response from vaccination. Those sequences are identified by comparing to the nonmutated sequences from the patient in the context of their specific immune molecule sequences

Published by Cold Spring Harbor Laboratory Press; doi: 10.1101/sqb.2016.81.031633

(HLA). The proteins that bind most avidly to the HLA should be the ones that elicit the strongest immune response. The vaccine approach is meant to re-educate the patient's immune system about what it should recognize and to overcome that immune suppression by stimulating it to produce specific T cells that will now identify those tumor-specific proteins. We get a lot of other information out of the genomics component, including identifying specific human leukocyte antigen molecules. Those are the molecules that will identify and bind the mutant proteins. More than anything, it's a computational prediction of those peptides that are most non-self.

Richard Sever: Has it been challenging to develop the ability to do that?

Dr. Mardis: It has, and we haven't solved it completely. At this point, there are more questions than answers. That's not a reason to not try these modalities, especially in patients that have exhausted their possibilities. We're pretty good at choosing neoantigens, but not as good as we need to be.

Richard Sever: How many potential targets do you start with that you have to refine down to a manageable number for use in a vaccine?

Dr. Mardis: Typically, we choose the top ten peptides. That's an arbitrary number because we don't know the right number. Another area of interest is that so far we've chosen single point mutations that result in a single amino acid substitution. That peptide looks different, but imagine how much more different it would look if, instead of a single point mutation, we've shifted to a different reading frame in the DNA. You would likely add a completely unique peptide sequence to its end that isn't present in the normal protein. Another example might be if two genes in the cancer DNA end up fused together: BCR-Abl and the Philadelphia chromosome are nice examples of that. Such modifications could be unique to a patient. They might not be in drivers like the BCR-Abl, but they could make a completely unique peptide that would then elicit a strong immune response.

Richard Sever: Unlike an oncogene driver, it doesn't matter what the peptide does; it's just the fact that it's different. It's a flag saying, "I'm a tumor cell. Come get me."

Dr. Mardis: Exactly. Of course, if any of these neoantigens were also parts of known driver genes, we would definitely target them because you could shut down the cancer cell on multiple levels. Most neoantigens, however, turn out to be passenger mutations. Usually when we try to marry specific targeted therapies to specific mutations, we really go after the known driver genes and tend to think of these passenger mutations as rather mundane. In the context of vaccine therapy, they turn out to be incredibly valuable. It's a different philosophical approach to the genomic data.

Richard Sever: You're validating this with RNA-seq. Can you explain what that means and why?

Dr. Mardis: On a practical level, we don't want to put a peptide in the vaccine that is not actually being produced by the cancer cell. In an ideal world, we would choose our peptides and look for them in protein data, say, from mass spectrometry. Currently, it's not practical to go that far because mass spectrometry is less sensitive than DNA sequencing and requires more tissue that we often have available. Instead, we look at the RNA expression of specific mutations that correspond to the neoantigens we want in our vaccine. This is especially important in high mutation load tumors like melanomas. We have lots of neoantigens to choose from, but by looking carefully at the RNA-seq data, we know that only ~50%–60% of the predicted neoantigens are actually produced as RNA. It's a pretty good intermediate as to whether a gene is actually being expressed and, in particular, whether that altered gene is being expressed in the tumor cells.

Richard Sever: How do you make the vaccine once you've got the peptide sequence?

Dr. Mardis: So far, we've tried synthesizing the peptides in a high-quality GMP [good manufacturing practices]-compliant way and then marrying them with specifically cultured dendritic cells from that patient. The dendritic cells are the parts of the immune system that present these peptides to the major histocompatibility complex and direct the T cells to be produced against the peptides. You basically co-opt the delivery vehicle from that patient, marry it up with your neoantigen peptides, and then infuse it back into the patient as a vaccine. That's one way. In a simpler method, we just produce a peptide or set of peptides that correspond to the neoantigens, dissolve them in an adjuvant much like you would a normal vaccine, and inject intramuscularly. They find their own way and marry up with the various immune molecules, and off the immune system goes.

Richard Sever: You're counting more on the body there than doing it ex vivo. Presumably, that would be more scalable?

Dr. Mardis: Presumably. Although the peptides can be quite expensive, there are commercial entities that are working on rapid GMP-compliant synthesis of all the peptides for each patient together. They just set aside a specific room for that patient's peptide collection of neoantigens, and the turnaround time is about a week. This is a good way of condensing down what can be a rate-limiting step. The dendritic cell approach takes longer because they first have to be isolated from the patient, specifically cultured, married up with the peptide cocktail, and then returned to the patient over a schedule of infusions. It's a bit more involved.

We can also make RNA and package it according to the neoantigens, or we can work completely back upstream and insert DNA corresponding to the different peptides that we want to produce into a vector just behind a human promoter. These get electroporated into the patient's cells and injected back into the patient, so the cells themselves make the neoantigens. There are many forms and flavors to this, all of which are being worked on by a variety of groups. Each approach has its own merits, its own associated costs, and its own associated difficulties or challenges to produce and to scale. If this is ever going to catch on as a way to battle cancer, it has to happen in a clinically relevant time frame. In other words, can we do all of this in a time frame and at a cost that meets the patient's needs in terms of where they're at in their disease fight?

Richard Sever: You're looking at melanoma now. How far can one go with this?

Dr. Mardis: We're focusing on the high-mutation-load cancers because they have more neoantigens to choose from. We're focusing on a second trial in melanoma, and we have a trial open in triple-negative breast cancer. We're also looking at relapse gliomas in children and glioblastomas in adults. Many of these patients after their first diagnoses and removal of their tumor receive a specific chemotherapy called temozolomide that has been shown in ~30% of patients to actually increase the mutation rate significantly when their cancer recurs. This is a special situation that we can take advantage of, because now these patients have the mutation load that gives us lots of things to choose from in our search for neoantigens. We also want to work toward understanding tumors with a lower mutation rate. In essence, how low can we go in terms of the number of neoantigens that are chosen but still have an effective vaccine? That's an important and completely unanswered question.

A Conversation with Sean Morrison

INTERVIEWER: MIRNA KVAJO

Scientific Editor, Cell

Sean Morrison is Mary McDermott Cook Chair in Pediatric Genetics and the Kathryne and Gene Bishop Distinguished Chair in Pediatric Research and Director of the Children's Research Institute at the University of Texas Southwestern Medical Center.

Mirna Kvajo: Stem cells are studied in many contexts, but you're interested in the specific process of self-renewal. Can you tell us about this process and why you decided to study it?

Dr. Morrison: There are many different kinds of stem cells. Self-renewal is the process by which they divide to make more stem cells. This process is fundamental to their ability to persist throughout life in your tissues. We need these in our skin to make new skin cells and in our bone marrow to make new blood cells and bone cells. When I first started my laboratory 17 years ago, I thought the most important question in stem cell biology was to understand the molecular mechanisms that regulate their self-renewal, partly because such self-renewal is necessary for the expansion of stem cells during development during the formation of tissues, but also because it's necessary for the maintenance of stem cells throughout adult life to maintain our tissues.

At the time, there was also a sense that these mechanisms might have something to do with cancer. There were pro-proliferation genes and antiproliferation genes, and it seemed obvious that stem cells would depend on pro-proliferation genes. It turned out to be a lot more interesting than that. As we studied stem cell self-renewal, we learned that the processes were regulated by networks of proto-oncogenes and tumor suppressors. We were bumping into the same genes that regulate cancer cell proliferation. We also learned that stem cell self-renewal is different from the proliferation of restrictive progenitors in the same tissues. We went back and forth between hematopoietic stem cells and neural stem cells, trying to identify key regulators of stem cell self-renewal. We defined this as "gene products that are required for the self-renewal of the stem cells but are not necessarily required for the proliferation of restricted progenitors in the same tissue."

Mirna Kvajo: Do different types of tissue require different regulators?

Dr. Morrison: There are different regulators, but what was more remarkable is that there's a group of regulators that are incredibly conserved across stem cells in many different adult tissues. Those tend to be the genes that get hijacked by cancer cells. Whenever we identified a new regulator of stem cell self-renewal, invariably, it would turn out to be involved in cancer. At least to first approximation, it looks like cancer cells hijack stem cell self-renewal mechanisms, more so than restricted progenitor proliferation mechanisms. Presumably, there's some advantage to cancer cells hijacking molecular mechanisms that enable a long-term capacity to replicate themselves rather than the limited replicative capacity that you typically see in restricted progenitors.

Mirna Kvajo: What is so special about the self-renewal process that could be so advantageous?

Dr. Morrison: Cancer cells need to replicate themselves extensively. Restricted progenitors often have a limited replicative capacity, whereas stem cells are distinguished from restricted progenitors precisely by their ability to replicate themselves extensively. It means that cancer is a disease of dysregulated self-renewal. The mechanisms that we think of as being so important in cancer biology didn't evolve to regulate the proliferation of cancer cells. These networks of proto-oncogenes and tumor suppressors evolved to regulate normal tissue homeostasis. Their role in cancer reflects the ability of cancer cells to hijack those processes to enable a selfish neoplastic proliferation.

Mirna Kvajo: Do you think that this mechanism is general to most cancers?

Dr. Morrison: Many cancers, but probably not all. For example, Iannis Aifantis and others have shown that there are some mechanisms in T-cell acute lymphoblastic leukemia that are borrowed from restricted T-lineage progenitors. T cells are more like stem cells, though, in the sense that they can go in and out of the cycle for

Published by Cold Spring Harbor Laboratory Press; doi: 10.1101/sqb.2016.81.031609

long periods of time and persist throughout life. In one sense, they're restricted progenitors, but on the other hand, their mode of division is a bit more like that of stem cells.

Mirna Kvajo: You've identified a core of conserved regulators, but you're trying to find others, and you've implicated molecules such as estrogen and leptin in stem cell renewal that might also regulate more general systems. What led you to this line of inquiry?

Dr. Morrison: When we started studying self-renewal mechanisms, there was almost nothing known at a molecular level. If we identified a new transcription factor that was required for hematopoietic stem cell self-renewal or an epigenetic regulator like B-cell-specific Moloney murine leukemia virus integration site 1, it was a *Nature* paper or a *Cell* paper. The approaches that we took to identify the genes that were necessary for stem cells' self-renewal are now being used to look for transcription factors, epigenetic regulators, and signaling molecules. We're more interested now in studying aspects of cellular physiology that haven't been studied before in somatic stem cells. Generally, they haven't been studied before because people are used to using 10 million cells, and you can't get 10 million cells if you're studying somatic stem cells. We end up having to scale down assays. As a consequence, we're studying proteostasis in stem cells, we're studying metabolism. We're trying to address the general question of whether stem cells differ from other cells based purely on executive control, where there are a few transcription factors or epigenetic regulators that distinguish stem cells from other cells.

Maybe a lot of the other cellular physiology is the same: the way they segregate chromosomes or metabolize glucose. Maybe most of that is basically the same as in many other cell types. The other possibility is that there are many aspects of cellular physiology that have to be different for a stem cell to create a stable, self-replicating, multipotent state. That's what we're trying to get at. There's an enormous amount of new biology that can be understood knowing how things that we think of as housekeeping functions—whether it's metabolism, or protein synthesis, or whatever—work differently in different kinds of dividing somatic cells.

Mirna Kvajo: What's your take on the importance of specific environments for stem cell function?

Dr. Morrison: Stem cells, we think, do depend on highly specialized microenvironments that we refer to as niches for their maintenance. A major interest in my laboratory is to understand both local signals from the niche that maintain stem cells (particularly blood-forming stem cells), as well as long-range signals. We want to understand the mechanisms that integrate stem cell function across different tissues. When you have a full-body physiological change, like starvation or pregnancy, stem cells in all different tissues have to be reading from the same song sheet. You can't have stem cells making it up as they go along in different tissues. There have to be mechanisms that integrate stem cells function across different tissues with full-body physiological changes. Estrogen and leptin are two examples of factors that do that, at least in certain stem cell systems.

We're also trying to better understand the local microenvironment that maintains stem cells. A lot of the growth factors that we know of—and that are used in regenerative medicine—either are so highly and broadly expressed that you can purify them from urine or are members of a big growth factor family that can be identified by homology with other family members (e.g., BMPs, WNTs, TGF-βs, notch ligands). Even though we've identified scores of growth factors, there are certain things that we still can't control in terms of regenerative medicine. We still can't really grow hematopoietic stem cells in culture, which limits our ability to perform bone marrow transplants in patients. We still only have one FDA-approved anabolic growth factor to promote the formation of new bone in osteoporosis. This is a huge unmet medical need. We're looking at bone marrow to see whether there are additional growth factors that haven't been identified yet, that are highly restricted in their expression to specialized microenvironments like niches, and that explain some of the biology that we can't control. We're in the process of characterizing some of these growth factors now. Hopefully, some of them will be useful in the context of regenerative medicine.

Mirna Kvajo: When you think of stem cells as a potential source of tumor cells, do they also need the right environment to create a tumor?

Dr. Morrison: We go back and forth between trying to understand mechanisms that regulate normal stem cell self-renewal versus the self-replication of cancer cells. On one hand, cancer cells are influenced by their environment in lots of different ways. That can have important consequences on which cancer cells are actually fated to expand and which cells are not and which cancer cells are therapy-resistant. On the other hand, it's important to keep in mind that cancer cells are less factor-dependent than normal stem cells. People talk about niches for cancer stem cells, and there are already some compelling data to support that, but I don't think that cancer cells are going to show the same obligate dependence on highly specialized microenvironments as normal stem cells. In the end, we know that cancer cells are invasive and that they overgrow tissues. When you get a breast cancer cell, or a melanoma cell, or a prostate cancer cell in the bone marrow, they don't only grow in their restricted microenvironments where normal stem cells grow. Eventually, they ablate the structure of the tissues and they grow everywhere. It's clearly worthwhile to study the microenvironmental mechanisms that regulate cancer cells, but it will probably be more of a facultative or situational dependence than what we see with normal stem cells.

Mirna Kvajo: Like most fields, work on stem cells has been limited by the available technology. When you think about what needs to happen to move your research to the next level, do you have a wish for a specific new technique that would make your life easier?

Dr. Morrison: It would be nice to have a PCR for proteins. It would change the world if we could amplify proteins, and not just genes. Beyond that, it's technique by technique. Throughout my career, we've always had to scale down every technique that we work with. It's usually not hard. Usually, people use 10 million cells because they have 10 million cells, not because they need 10 million cells. If you're just a little more careful, it's amazing how easy it is to scale down a lot of things. Right now, we want to study metabolism in small numbers of highly purified cells, and really understand how metabolic pathways are used differently by different kinds of dividing somatic cells in the same tissue. We've collaborated with Ralph DeBerardinis at UT Southwestern to scale down metabolomics. We're doing metabolomics in 10,000 cells and discovering new pathways that regulate stem cell function.

A Conversation with Benjamin Neel

INTERVIEWER: RICHARD SEVER

Assistant Director, Cold Spring Harbor Laboratory Press

Benjamin Neel is a Professor in the Department of Medicine and Director of the Laura and Isaac Perlmutter Cancer Center at New York University's Langone Medical Center.

Richard Sever: Looking back to 1980 and your work identifying *myc* and those early days of oncogenes and how far we've come since, have we come far enough? Have we done everything you'd hoped?

Dr. Neel: Cancer is probably the number one cause of death in the United States by now, so we clearly haven't come far enough. That being said, the road that's been traveled the last 35 years has been amazing. It's hard to believe that in 1980 we hadn't cloned a single human gene yet. We didn't know a single human gene that caused cancer until the year after, when we first heard about *ras*. Now we have a complete map of the whole human genome, and we have an almost complete list of all the genes that are either activated or inactivated in cancer. We know about epigenetic modifications in cancer. We have a much more sophisticated understanding of cell biology and tumor biology and the interaction with all kinds of host cells. For the first time, we really do have, if not a road map, at least a pretty good map of how to go forward to make a major impact on this disease. We're seeing cures in the clinic that are developed rationally, not just by empirical chemotherapy.

Richard Sever: You're optimistic that the molecular dissection that's given us that map is giving us targets for precise therapies that actually work?

Dr. Neel: If I weren't optimistic, I wouldn't let our cancer center put out an advertisement that says we can cure cancer. We already do cure some cancers, and I think we'll see increasing numbers of cancers cured in the next 15 years. More importantly, we'll be able to prevent a lot of cancers, or at least detect them earlier so that they're not as dangerous and don't develop into the full-blown metastatic states we face all too often now. Ironically, I'm less enthusiastic about the use of combination-targeted therapy being curative. There's going to be a role for targeted therapies, but figuring out how to integrate them with other modalities like epigenetic manipulation or tumor immunotherapy is the real key to making major progress.

Richard Sever: One of the landmarks on that map is protein tyrosine phosphatase 1B (PTP1B). Before we get to that, can you disabuse people of the notion that phosphatases are just there to mop up after kinases?

Dr. Neel: Phosphatases are the Rodney Dangerfield of signal transduction; they don't get any respect. Nevertheless, they play a critical role as both positive and negative regulators, and it's important to remember that some kinases can be signaling inhibitors. I think that all of us who work in the signaling field have a more sophisticated understanding than just "plus or minus" science.

Richard Sever: You've looked at this in a human epidermal growth factor receptor 2 (HER2) model of breast cancers where the phosphatase does not shut things down. In fact, it seems that when you shut the phosphatase down, you inhibit the potential for the tumor to survive.

Dr. Neel: That's right, and that was a big surprise. PTP1B is a recent addition to the map for cancer biology. It was best understood as an enzyme that regulates body mass and glucose homeostasis. It's a major regulator of the insulin receptor and of the kinase that's associated with a leptin receptor in the brain that acts as a major controller of body mass regulation. Most of the work on PTP1B up until, say, the last 8 years focused on its role in total body metabolism. The idea of targeting PTP1B was mainly for the possibility of weight reduction and/or diabetes therapy.

About 8 years ago, both our group and Michel Tremblay's laboratory at McGill made the unexpected discovery that PTP1B deficiency actually antagonizes HER2-mediated transformation in mice. We agreed on the biology, but not on the mechanism. That's what led to our studies on human HER2-positive breast cancers that suggested a totally unexpected connection between PTP1B, an unusual, rare disease called Moyamoya syndrome, and

Published by Cold Spring Harbor Laboratory Press; doi: 10.1101/sqb.2016.81.031302

a fascinating class of enzymes called α-ketoglutarate-dependent dioxygenases.

Richard Sever: The phosphatase allows the cells to survive hypoxia, correct?

Dr. Neel: Right. Tumors actually are put in a very stressful situation most of the time. They often are quite hypoxic. They're nutrient-depleted. To survive, they have to buffer those conditions molecularly in some way. The signature molecule that controls hypoxia regulation is hypoxia-inducible factor (HIF), but there are also other mechanisms like the unfolded protein response and the TOR pathway. These three mechanisms control the ability of the tumor to survive at different levels of hypoxia. We found that, at really severe levels of hypoxia, there was a fourth pathway that involved PTP1B and an unusual E3 ligase, RNF213 (ring finger protein 213).

Richard Sever: When one thinks about hypoxia responses, one typically thinks of the von Hippel–Lindau factor and the HIF transcription factor. Does this involve that?

Dr. Neel: No. As far as we can tell, they're different. We actually went to the trouble of getting one of these commercial profiling kits where you look at 98 HIF-regulated genes and there's like a 99% concordance between the response of HIF-regulated genes and PTP1B-depleted cell lines versus PTP-replete cell lines. There's no shared gene. It appears to be a completely different pathway and we think it has a completely different function.

The purpose of the HIF pathway is to turn down mitochondrial consumption of oxygen through oxidative phosphorylation, which is very energy- and oxygen-intensive, and to shift metabolism into a glycolytic state. The second major pathway is the unfolded protein response/ER stress pathway, which comes into play at more severe hypoxia. That pathway seems to control oxygen consumption in the endoplasmic reticulum for protein folding and other ER-oxidative processes. The third major pathway is the AMP-activated protein kinase TOR pathway, the purpose of which is to control excess energy consumption by protein and lipid synthesis.

There's a large family of about 60 to 80 enzymes in mammalian cells that are called α-ketoglutarate-dependent dioxygenases. These enzymes use molecular oxygen, but it's nonmitochondrial oxygen consumption. Some of these enzymes are in the cytoplasm. Some are probably in the mitochondria as well, but they basically use nonmitochondrial mechanisms. They're not involved in oxidative phosphorylation and electron transport. When you get to severe levels of oxygen depletion, you have to make sure that these enzymes aren't active at a very high level. Otherwise, you would consume the oxygen in the local microenvironment and the cells would die. We found that PTP1B-deficient cells have an excess level of nonmitochondrial oxygen consumption that could be suppressed by using inhibitors of α-ketoglutarate-dependent dioxygenase activity. Several biochemical assays showed increased activity of several members of this family, probably including the TET family of demethylases. That was one surprise.

The other, really big, surprise was a connection we still don't understand in molecular detail between PTP1B and the α-ketoglutarate-dependent dioxygenases, which appears to be this large E3 ligase at 600 kDa called RNF213. It's a tyrosine phosphoprotein. We think that PTP1B dephosphorylates a specific site or sites on that protein, but we don't know the exact sites yet. If we deplete RNF213, we can completely rescue the excess nonmitochondrial oxygen consumption and the increased activity of those α-ketoglutarate-dependent dioxygenases. We can restore hypoxia-induced survival in culture and, to a substantial extent, restore tumorigenesis and all the excess hypoxia in the tumor. We don't know the exact mechanism (and that's why it's novel and exciting), but we think that it's a major new pathway.

I want to emphasize that in these cells and tumors, as far as we can tell, all of the other pathways are turned on normally or even precociously. There's no obvious connection to the other pathways. We think it's because this is another set of oxygen-consuming enzymes. Under ambient oxygen or even mild hypoxia, they don't contribute a lot, but under severe hypoxia they probably do become limiting.

Richard Sever: α-Ketoglutarate is part of the citric acid cycle. Is there a metabolic connection there beyond just oxygen?

Dr. Neel: α-Ketoglutarate has a lot of functions, but its role as a substrate in α-ketoglutarate-dependent dioxygenase reactions hasn't been examined much. I think that it's a completely separate role for this metabolite.

Richard Sever: Would a more general connection of PTP1B to metabolism be advantageous when considering it as a target in cancer?

Dr. Neel: There's already discussion about testing PTP1B inhibitors in cancer. I believe that Cold Spring Harbor is starting a trial in HER2-positive breast cancer. The other exciting development in the field is that there are actually some real PTP inhibitors. Interestingly, the only bona fide ones are both allosteric inhibitors. They have actually managed to get around the major limitation for PTP targeting by small molecules, which has been the polar and highly reactive nature of the active site. We're also expecting a report from a pharmaceutical company on a really interesting SHP2 inhibitor (another major interest in my laboratory) that has activity, at least in preclinical models for cancer.

In the twilight of my career, phosphatases are hot again. I've been called by a lot of investors who want to start phosphatase companies. There are already several companies that are working on phosphatase targets now, so it's kind of funny. It's like a rebirth.

A Conversation with Luis Parada

INTERVIEWER: JAN WITKOWSKI

Executive Director, Banbury Center at Cold Spring Harbor Laboratory

Luis Parada is the Director of the Brain Tumor Center, the Albert C. Foster Chair, and an American Cancer Society Research Professor at Memorial Sloan Kettering Cancer Center.

Jan Witkowski: You were a second-year graduate student when you first started working on cancer genes. Did you have any sense of what that would lead to?

Dr. Parada: I wasn't clairvoyant about the future, but the excitement was absolute and instantaneous. As a first-year graduate student at MIT, we didn't rotate through laboratories the way most graduate programs work. Instead, we took course work and then we listened to seminars from the faculty. When an assistant professor named Robert Weinberg got up and said, "We can take DNA from tumor cells and put it into normal cells and they become tumorigenic, and then we can take the DNA and first clip it with one restriction enzyme and retain tumorigenic properties, but when we clip with another enzyme it loses its tumorigenic capacity." I realized, "It's a gene. Cancer is a gene." At that moment, I committed to work with Robert Weinberg. It was the best decision I ever made in my career. Of course, it was rudimentary molecular biology. We had to purify our own restriction enzymes before New England BioLabs started selling them. The primary technique we used was the Southern blot, which again, I think is now a thing of the past. However, the Southern blot permitted us, for the first time, to visualize genomic pieces of DNA. It was the key to my central project as a Ph.D. student, which was to determine if the transforming genes we saw in human cells were related to the oncogenes that were resident in retroviruses that were the heart of the cancer field at that time, if these were innate tumor genes that caused cancer. The answer was, resoundingly, yes.

Jan Witkowski: The other noteworthy thing about that 1982 paper is that there's not a single DNA sequence in it.

Dr. Parada: Nope. The sequence of the amino acid 12 mutation in *ras* was the sequel, performed by my good friend and colleague Cliff Tabin. He sequenced, maybe, 1000 bases, and even that required a collaboration. We couldn't do it all by ourselves.

Jan Witkowski: There was a bit of a race on at the time, wasn't there? There was Mike Wigler's group. There were Geoff Cooper and Mariano Barbacid.

Dr. Parada: We were aware of Geoff Cooper's work, which was orthogonal to ours, in that he wasn't seeing the same things that we were seeing. We were aware of Mike Wigler's work because Mitch Goldfarb was a Bob Weinberg graduate student who did postdoc work with Mike Wigler at the Cold Spring Harbor Laboratories and pursued the same line of inquiry. We weren't aware of Barbacid's work until it came out, so we weren't in discussion with him at the time. It was a funny race, in the sense that we didn't know the answer. It wasn't like the race to, for example, identify the structure of DNA where we knew that there had to be a structure to be identified and we were going to find it. It was more like, once we realized that there was such a thing, we were all trying to figure out how to clone cancer genes.

Jan Witkowski: More recently, you've been working on glioblastomas. One of your key findings is that glioblastomas don't arise through glial cells, but rather through a special set of neural stem cells.

Dr. Parada: Correct. As someone who has always been interested in the developmental biology aspects of disease and the natural history of how tumors evolve, I've been frustrated that so many textbooks claim to tell us where and how cancer arises. There has actually been very little, if any, direct investigation into the origins of cancer. The only really important work of that sort has been done on how hematopoietic stem cells are the root of leukemias. In solid tumors, the origins are really uncharted territory, at least until the last few years. I've spent many years studying the development of the brain and the nervous system, so we decided to tackle the problem in tumors of the nervous system. We generated genetically engineered mouse models that harbored the mutations that were found most frequently in sporadic glioblastoma, an incurable tumor of the brain in humans. We discovered that

Published by Cold Spring Harbor Laboratory Press; doi: 10.1101/sqb.2016.81.032565

when we mutated the genes in the stem cells of the brain, 100% of the mice developed glioblastomas that the neuropathologists at the hospital could not distinguish from the human disease. If we mutated the genes in any other cell in the brain, we never got glioblastoma. We concluded that this is a disease of stem cells, and that—analogous to what is known in many hematopoietic cancers—the mutations actually accumulate in the stem cells. The disease manifests in a cell in a way that resembles some normal product of the stem cell, but that's just, if you will, the "dressing"; it's not the source, the core. The core is actually that there's a unique and important property of stem cells that we refer to as "self-renewal". Self-renewal means that these cells are immortal. As long as we're alive, thankfully, we have stem cells in our brain and in our blood that we need. The analogy to cancer cells is that they also are immortal.

Jan Witkowski: Does that imply that the standard treatments of chemotherapy and radiation aimed at the bulk of the tumor are really misdirected, that what you need to target is just a small number of stem cells?

Dr. Parada: That's a key implication. Chemo- and radiotherapies are designed, on purpose, to kill proliferating cells, cells that divide aggressively. The consequence is that anticancer therapies are not really anticancer therapies. They don't distinguish between a cancer cell and a normal cell; they distinguish between a dividing cell and a nondividing cell. That's why you get all the terrible side effects of chemotherapy: we lose our hair, become anemic, develop gastrointestinal problems. Interestingly, when the chemotherapy is removed, our hair grows back, our blood comes back, and our digestive problems go away. Why? Because the chemotherapy, in addition to targeting the proliferating cells in the tumor, also targets the proliferating cells in our normal stem cell compartments but not the stem cells themselves. That's why we can grow our hair back, and why we recover from anemia, but it's also why the tumors come back: The stem cells were never touched by the therapy.

Jan Witkowski: Are there examples from other solid tumors of these sorts of system?

Dr. Parada: There are emergent examples. There are several groups that have taken the same developmental biology approaches that we have. In my view, they've demonstrated convincingly the stem cell of origin for skin cancer and for colon cancer. There is very good evidence for cancers of the liver and the bladder as being stem-cell-of-origin cancers and moreover, having a cancer stem cell as the driving force in this hierarchical tumor growth, which was really something that was not appreciated until recently.

Jan Witkowski: Does your research provide clues as to what the target in stem cells might be?

Dr. Parada: It does. I cannot emphasize enough how important it was to realize that there is such a thing as a cancer stem cell and that there is a hierarchical method of tumor growth, because we now understand who the enemy is. You have to understand the enemy to isolate it, and then target it. We can do this now with glioblastoma. Our research team is making great advances and we have what we believe are promising strategies that directly target the cancer stem cell and not normal dividing cells.

Jan Witkowski: Do stem cells have particular molecular characteristics that could be used?

Dr. Parada: Yes and no. There are certain common denominator properties held within all stem cells (e.g., embryonic, blood, skin, etc.) that also are found in cancer stem cells. That's not what we want to target, though. When I cure your tumor, I don't want to kill all of your hematopoietic stem cells. Those common denominators are not ideal targets. However, similar to what is seen in the hematopoietic system, cancer stem cells do acquire certain properties that are unique. They are stem cells on steroids, if you will. We have to target the steroids, and conceptually, that's doable.

A Conversation with Sir Michael Stratton

INTERVIEWER: ELIZABETH MCKENNA

Associate Editor, Cancer Discovery

Sir Michael Stratton is Director of the Wellcome Trust Sanger Institute, co-leader of the Cancer Genome Project, and member of the International Cancer Genome Consortium.

Elizabeth McKenna: Can you tell us about the efforts to identify mutational signatures from cancer genome sequences?

Sir Michael: We've known for about half a century that all cancers are caused by somatic mutations: mutations that take place in all cells of our body throughout the course of our lifetime. Remarkably, we know relatively little about the causes of those mutations. We know, for example, that smoking cigarettes damages the DNA of the lung and that damage gets converted into mutations, and those mutations cause lung cancer. Similarly, we know that exposure to ultraviolet light from the sun causes damage in cells of the skin, and that damage gets converted into mutations and can result in skin cancer. However, we have only a rudimentary idea as to what causes the mutations in breast cancer, or glioma, or pancreatic cancer.

One possible way to elucidate this problem is to actually look at the cancer genome itself. A series of studies in the 1980s and 1990s looked at one particular gene, the *p53* gene, across a range of human cancers and found that the pattern of mutations gave insight into the cause of the cancer. If one looked at the *p53* gene in lung cancers, one found a particular type of mutation, a C going to an A being predominant. Conversely, if you looked at melanomas or squamous carcinomas of the skin, you saw C to T mutations predominantly. This made a certain amount of sense, because it was already known that ultraviolet light causes the sort of damage that would result in C to T mutations, and that mutagens in tobacco cause the sort of DNA damage that results in C to A mutations. In other words, the mutations that were being found in the cancers bore the imprint—what we today call the signature—of the exposures that caused them.

Today, we have the genome sequences of somewhere between 25 and 30 thousand cancers. With these, we can now reexamine this question to see if we can get a comprehensive view of the different patterns, the different signatures of mutations, that are present across the face of human cancer, and let that give us a sense of what the different processes are that are causing the mutations that cause cancer.

Elizabeth McKenna: What are some of the other signatures you've identified and what are they associated with?

Sir Michael: Thus far, about 40 different signatures have been found. That's probably more than we were collectively expecting; 40 signatures basically means 40 different mutational processes. As we analyze more cancer cases, the number is increasing. It'll probably hit around 50 or perhaps a bit more. About half of those, we actually have no idea what causes them at all. Basically, half the causes of cancer are, at the moment, opaque to us.

We do have some idea about the other half, or at least some suspicions and in some case, more than suspicions. One of the major findings in the last few years has been the discovery of signatures that are attributable to the activities of the so-called APOBEC enzymes. The normal, day-to-day job of this family of enzymes is to edit DNA. APOBEC enzymes are cytidine deaminases. They remove amine groups from cytosine and convert cytosine to uracil. If DNA replication takes place when there's uracil in the DNA, it gets converted to a thymine, so you end up with a C to T mutation.

These DNA editing enzymes have a normal day job. They do good things for us. They are believed to be involved in innate immunity and protecting us against invaders. For example, when viruses invade cells, the cell senses it. One of the ways that it can protect itself against viruses or other forms of naked DNA is to mutagenize them, to essentially "machine gun" them with mutations, such that the virus can no longer function. That is one of the functions of this group of APOBEC enzymes.

In many cancer types, we find cases with very large numbers of mutations that we believe are caused by these APOBEC enzymes. There are two signatures that we believe are the result of the activity of APOBEC enzymes.

Published by Cold Spring Harbor Laboratory Press; doi: 10.1101/sqb.2016.81.031773

About a third of breast cancers have got a lot of APOBEC mutations, as do almost all bladder cancers, whereas we rarely find them in colorectal cancers. You've got this enormous diversity of exposure to this mutational process that is absolutely overwhelming in the number of mutations that it generates when it happens.

The other mystery is, what is switching on the APOBEC enzymes to create this tidal wave of mutagenesis in the cancers in which it attacks? Here you end up essentially "hand waving," because we don't know the answer. We believe it has to do with the activity of the APOBECs. Perhaps some time in the past, decades ago, a virus—any old virus, really—entered the cell and the cell decided to protect itself against that virus. It decided to mutagenize the virus, but at the same time, it couldn't control its activities and when it deployed the APOBEC against the virus, there was collateral damage on the nuclear DNA as well. That cell would be sitting there for decades, but now with many more mutations than the cells around it. Those mutations would mean that cell was primed to become a cancer cell compared to those around it.

If that idea is true, it gives us a cause of cancer we hadn't anticipated previously: That a variety of different infections that otherwise we had not linked to cancer could be doing this. As a hypothesis, it feels plausible, and it certainly would be interesting, The trouble is, we don't have much evidence in favor of this. For example, we don't find great numbers of APOBEC mutations in liver cancers where hepatitis B and C viruses have been operating and you might have expected them. You do, however, find a lot of APOBEC mutations in almost all cancers of the cervix, and in a subset of cancers of the head and neck. In both those cases, human papilloma virus is known to play a role. You can't dismiss the hypothesis, but at the moment, we also can't prove it.

Elizabeth McKenna: What about other commonly mutated genes, like driver genes? Are they associated with specific signatures?

Sir Michael: Some are. The best examples are the breast cancer susceptibility genes *BRCA1* and *BRCA2*. It's been known for a while that when *BRCA1* and *BRCA2* are inactivated by their mutations, the cell is prone to suffering rearrangements of DNA that are aberrantly repaired, such that one bit of DNA is joined to another bit of DNA when it shouldn't be, and that can inactivate or activate a cancer gene. The mutational signature analyses showed that it's not just these gross rearrangements. There is a substitution mutational signature that is absolutely clearly associated with defective *BRCA1* and *BRCA2* function. That gives us new insight into the biology of *BRCA1* and *BRCA2* deficiency and how the susceptibility is orchestrated. In the future, that signature could also be a good predictor of patients who will be treatable with particular drugs. It's already known that *BRCA1* and *BRCA2* mutant cancers have a deficiency in homologous recombination–based repair. Those cancers are vulnerable to DNA damaging agents like cisplatin and PARP inhibitors. Currently, those patients are identified on the basis of having either a *BRCA1* or a *BRCA2* mutation. However, what is clear from the signature analysis is that although both those groups do have this signature (which we call Signature 3), there is another subset of patients of more-or-less equal numbers that clearly have the signature and have the abnormalities of genome rearrangement too, but you can't find the *BRCA1* or *BRCA2* mutations. If this is correct, then the signature is identifying much more widely the cancers that have a deficiency in homologous recombination repair and therefore could identify with greater sensitivity those patients that are going to be responsive to PARP inhibitors and cisplatin and other drugs. That's to be seen, but I would be optimistic that the signatures will be used as predictors for therapeutic deployment.

Elizabeth McKenna: You envision the signatures being used for patient stratification and things like that?

Sir Michael: Absolutely. It's not so straightforward to get the signatures. Small segments of genome being sequenced don't give you quite as much information as you might need. We might need at least exomes, but possibly whole-genome sequences to be absolutely sure of the presence of a signature. I'm optimistic that, over a period of 5 to 10 years, whole-genome sequencing of cancers to find these signatures associated with DNA repair deficiency will be used to find the patients who are going to be responsive to those drugs.

Elizabeth McKenna: Are there signatures associated with aging, or is it mainly just mutations and things you're exposed to?

Sir Michael: That's an interesting question. There are two signatures that we call 1 and 5. For both, the number of mutations that you find in a cell correlates with the age of the person. These signatures reflect underlying mutational processes which are essentially operating in all of our cells throughout the course of our lives. If one is a 20-yr-old, one has had 20 years for one's cells to acquire mutations of these two signatures, so one will have a certain number. If one is a 40-yr-old, then one will have double the number of mutations in cells of one's body. Both of these mutational signatures are mutational clocks. They are slightly different from each other, however.

Mutational Signature 1 is a very consistent clock; it can't be deflected. We believe that Signature 1 reflects the number of mitoses a cell has been through over a lifetime. We have cells in the tissues of our body that are turning over, day after day, year after year, decade after decade. Each time a cell divides, it goes through DNA replication and a small number of mutations are acquired. Those seem to be Signature 1 mutations. In the future, actually finding the number of Signature 1 mutations will give us great insight into how many divisions a cell has been through. These mutations add up throughout life. If they fall in the wrong genes, they can convert a normal cell into a cancer cell.

Signature 5 is more mysterious. It is also clock-like, and the rates at which it generates mutations are different in different cell types but it differs from the rates at which Signature 1 produces mutations and is not correlated with the number of mitoses a cell has been through. Frankly, we have very little idea what is causing it. We think it can be deflected from keeping time, and there are various things that can cause a cell to acquire more Signature 5 mutations. We believe smoking is one of those things, and there probably are others. Signature 5 is clock-like, but it's a floppy clock: Other factors can push up the rate. I don't know if you can push the rate down.

So, yes, we do have the seeds of our destruction within us. We are acquiring mutations all the time. These are the two major processes that we think are doing so. Indeed, we believe that these two processes, which are operating in all the somatic cells in our bodies, also operate in the germline to give all the mutations that create the differences between us: the phenotypic differences in the way we look and behave and our predispositions to disease. Remarkably, although we do have an idea what is causing it for Signature 1, for Signature 5 (which is about two-thirds of the mutations that occur in a germline), we really have no idea. There's all this interindividual variation, and at the root of it is this mysterious process, and we need to find out what is doing it.

A Conversation with David Tuveson

INTERVIEWER: GEMMA ALDERTON

Senior Editor, Nature Reviews Cancer

David Tuveson is Director of the Cancer Therapeutics Initiative and Deputy Director of the Cancer Center at Cold Spring Harbor Laboratory.

Gemma Alderton: Could give us an overview about your work?

Dr. Tuveson: My laboratory studies pancreatic cancer. We are trying to go beyond the "what" of pancreatic cancer to the "why": Why it makes you sick and why the medicines we have don't work. When I started in the field 20 years ago, we didn't understand what caused cancer. Many people investigated the genes of cancer, found mutations that were associated with that cancer, built mouse models to demonstrate that the gene was sufficient to cause the cancer. More recently, mouse models have been used to show that it was necessary for the cancer. At this stage, we now know the "what."

Knowing what causes pancreatic cancer, why is this still a bad disease? Why don't our medicines work better? Why is it that mice and people are sick when they have pancreatic cancer? To answer those questions, my laboratory collaborated with Hans Clevers using a different, tissue-based, model system called "organoids". This has allowed us to ask "why" questions both with mice and with tissue from human patients. The results have been startling.

Gemma Alderton: The development of organoids improves on traditional cell culture approaches, particularly with regard to cancer, because it takes into account the tumor microenvironment.

Dr. Tuveson: The HeLa cells of the '50s were the start for the cancer field. Today, you can grow those types of cultures for many, but not all, cancers. The value of these tissue-based models—be they organoids, conditionally reprogrammed cells, or related approaches—is that you can grow both normal cells and neoplastic cells. There aren't other methods that allow you to grow normal cells beside ones where you have to put in genes to immortalize the cells. Here, you can compare normal cells to tumor cells within the same patient or experimental animal model. That's a huge advantage. Also, our early efforts suggest that you can indeed add in the microenvironment or at least some of the types of cells. We've focused on the cancer fibroblast; we can add that into the epithelial pancreas organoid cultures. We've learned things that we never would have known to look for if we were starting from tissue samples from patients or from mice. I do think that organoids are transformative for the field. We have some exciting new findings as a result of the hard work and excellence of the postdocs and students in my laboratory.

Gemma Alderton: Do the organoids replicate the physiology that you would see within a tissue?

Dr. Tuveson: They approximate it. For the pancreas, you can grow ducts, but not as tubular structures; they actually grow as spheres. Clevers' group and others have shown that, at least in the intestines, you can grow multiple lineages and get mini-guts. In the stomach, you can make glands that produce acid and exotic things of that nature. We have not been able to make an intact pancreas, but that's not the focus of my laboratory. Developmental biologists are pushing those types of questions because in the field of human pancreatic health, diabetes is the main problem. Growing the endocrine cells of the pancreas would potentially be very beneficial for that line of research. We, on the other hand, focus on cancer. We feel that the duct is very much related to the cells that cause at least a majority of types of pancreatic cancer. It would be an exaggeration to say that we are probing physiology deeply. We probe pathophysiology deeply. That, for our purpose, is sufficient.

Gemma Alderton: Can these organoids also be used translationally?

Dr. Tuveson: We can measure things that we couldn't measure before. That's how you discover new genes, new pathways, etc. Using organoids, we can study metabolism differently than you can in two-dimensional culture, and

Published by Cold Spring Harbor Laboratory Press; doi: 10.1101/sqb.2016.81.031310

the differences between two- and three-dimensional culturing are very evident.

Gemma Alderton: What's the mechanism for that? What causes that difference?

Dr. Tuveson: We don't precisely know. In the three-dimensional organoids, the cells are polarized. There are apical, basal, and lateral surfaces, whereas when you're growing in two dimensions, the cells can't tell up from down or side from side. To survive on plastic you have to be able to proliferate under high tension, which activates many signaling pathways such as the integrins and others. These are cells that are able to exist under harsh conditions. The organoids are less stressful in terms of the actual tension on the cell membrane. I think that the polarization of the cells allows you to look at the pathways as they would exist in vivo. That's a hypothesis we're testing right now.

As for the translational potential of organoids, I'm a cancer doctor. I got into the field because I was trying to think of better ways to help my patients. I spent the last 15 years doing more science than medicine, but with the organoids, we now can identify aspects of pancreatic cancer that are relevant to how we might treat or diagnose the disease that we couldn't examine using cell lines or mice.

For example, you can use organoids to compare normal cells to neoplastic cells in an effort to identify proteins that are expressed specifically in cancer but not by normal tissues. These proteins become "fingerprints" for the cancer. The only normal cells in our system that proliferate are epithelial cells that are dividing because they're developing (as in an embryo) or because they're regenerating in response to trauma. In the pancreas, the trauma would be called pancreatitis. Pancreatitis makes it hard for doctors to figure out if you have a benign or a malignant disease. With the organoids, we can separate that out and very quickly say that these proteins, for example, are in proliferating normal cells (which would only happen in pancreatitis), whereas those proteins are in proliferating cancer cells. By using such comparisons, we could very rapidly distinguish between the two. This very talented scientist in my lab, Dannielle Engle, has used this approach to identify proteins that she thinks are potential biomarkers specific for pancreatic cancer. When she looks for those proteins in the blood of humans who either do or do not have pancreatic cancer, she's diagnosed correctly, I think, 10 out of 11 times. Even from a skeptical standpoint, you have to be encouraged that we can see deeply into problems that we couldn't before.

The other obvious question is whether you could use the organoids to personalize therapies. Could you use it to choose the medicine that might help a patient? A team of workers in my laboratory has taken that approach. We're in the early stages, but so far they've identified a few therapies that look active in the animal models. The organoids appear to be putting us on the right track.

The organoid system has also allowed us to study the microenvironment and the interaction between the cells in the stroma of the tumor as they communicate with the epithelial cells. The tumor microenvironment team in our laboratory has identified a type of stromal cell: a version of a fibroblast that's never been reported and has the right characteristics of something that might promote cancer in a significant way.

Gemma Alderton: There seems to be a controversy as to whether the microenvironment is protective versus promoting in pancreatic cancer. Could you comment on that?

Dr. Tuveson: Think of pancreatic cancer as an oatmeal cookie with raisins in it. The raisins are the cancer cells, and the oatmeal dough is the stroma. There's more dough than there are raisins. We found some time ago that the stroma made it difficult for the vasculature to both develop and function. We hypothesized that because of this, we couldn't deliver drugs into the tumor. When we used various methods to lessen the stroma, the drugs could get in better. In mouse models, it appeared that would be beneficial. You could kill more cancer cells and the mice would live longer. However, this idea did not translate well to the clinic. In fact, the patients did worse when the stromal ablation approach was taken with the *hedgehog* inhibitor.

The dissatisfying part of that was that, as is true of many clinical trials, there was no real scientific angle to the trial. Biopsies were not taken. There was no scientist with an eye on the trial. In the end, when they wanted my advice, I asked them "What do the biopsies show?" They said "What biopsies?" It's frustrating for scientists to see our work not tested properly in the clinical setting.

Nevertheless, we hadn't lost sight of the fact that it failed in the humans even though it looked promising in mice. The organoid coculture allowed us to go back and ask why this happened. We think now that this other type of fibroblast could partially explain why things got rocky for the patients. That's our hypothesis based on the data we have available to us right now. We don't think that the stroma is a friend in this. It clearly can regulate the differentiation of the cancer cells, and well-differentiated tumors tend to behave differently than undifferentiated tumors. The *hedgehog* inhibitor trials made it difficult for the well-differentiated cancers to persist, so you selected for a type of carcinoma that was less differentiated, but we think that was only part of it. The other part was that the type of fibroblast you were left with was one that would promote cancer much worse than the first type.

Gemma Alderton: How would you better do that clinical trial in the future?

Dr. Tuveson: As both a physician and a basic scientist, I feel that we have an obligation to perform early-phase clinical trials—so-called phase 0 trials—on the first few patients treated with a new medicine or evaluated with a new diagnostic technique. These are trials where the end points are molecular and scientific, where there are still the safeguards for the patients' well-being, but the doses of drugs used are limited and the time periods of the experiment are short. Phase 0 trials aren't generally performed for several reasons, the main one being that academic centers don't have such facilities. The second one being that the pharmaceutical companies who are our friends in developing new medicines don't have the patience to conduct these studies nor the bankroll to support them. What we're getting better at, though, is convincing the funding agencies that such studies are important.

I'm hopeful that basic scientists can become part of the early-phase clinical trails process, working with physicians to help interpret the early data. At Cold Spring Harbor, we feel that if we could design these phase 0 studies in a medical setting where you could have, if need be, 24/7 monitoring, you could learn in a few hours or a few days whether or not you were on the right track for most therapies and diagnostics, even immunotherapies. That's our hope. The organoids are expected to be the ex vivo part of this process, but we also need to do in vivo phase 0 studies. When you study a disease where its mortality rate is about equal to its incidence, you're in the red zone. You can try to kick field goals all day long, but you'll never win anything. We have to change our approach. We have to be aggressive, not complacent.

A Conversation with Christopher Vakoc

INTERVIEWER: PAULA KIBERSTIS

Senior Editor, Science

Christopher Vakoc is an Associate Professor at the Cold Spring Harbor Laboratory.

Paula Kiberstis: Could you tell us a little bit about the history of transcription in cancer research and why you feel it's a good target for therapy?

Dr. Vakoc: What many decades of cancer research have taught us is that there are some fundamental hallmarks of cancer biology that I think everyone in the field recognizes. What's clear is that behind every hallmark feature of a cancer cell is an aberrant transcriptional program, which is not surprising. All of biology is driven by changes in gene expression, and cancer biology is no different. There's a kind of obviousness to the role of transcription in cancer.

More compelling, though, is what the sequencing of cancer genomes has taught us about the causes of cancer. There are old examples, like *p53* and *myc*; those heavily mutated genes in cancer are transcription factors. We've known that for ages. There've also been recent discoveries that remind us of the importance of transcription, specifically the discovery that almost all human cancers have mutations of chromatin regulators—the so-called "epigenetic regulators"—involved in transcription, as well. This discovery was made in the last 6 years or so. The new wave of evidence behind the role of transcription is the discovery that there are noncoding mutations in cancer genomes that are incredibly common, which are often missed when you sequence only exomes in protein coding genes. We're realizing that the binding sites of transcription factors are also heavily mutated in different malignancies. You put all these older and more recent lines of evidence together and it's really indisputable that transcription abnormalities act in cancer as a driver mechanism.

Paula Kiberstis: Five years ago, you published an influential paper in which you and a group in Boston independently arrived at the same protein, BRD4 (bromodomain-containing protein 4), as a target for cancer. Could you tell us about that and the story of how that happened?

Dr. Vakoc: If one is aware that aberrant transcription is a driver of cancer, the obvious usefulness of that information is the idea that cancer cells are going to be vulnerable to perturbations of transcription. My whole laboratory has been built on that hypothesis. When I came to Cold Spring Harbor, we went hunting for vulnerabilities in transcription control that cancer cells possess but that normal cells do not. Part of the drive coming to Cold Spring Harbor was to use RNA interference (RNAi) screening that was pioneered in part by Greg Hannon and Scott Lowe here. We set about systematically knocking down every chromatin protein in an aggressive cancer called acute myeloid leukemia. We were asking which chromatin proteins leukemia cells needed that normal cells do not. This led us to BRD4. Targeting BRD4 had a catastrophic phenotype on leukemia and less of an effect on normal cells. It was a discovery based on a genetic screen.

There was undoubtedly a lot of interesting biology behind it, so the therapeutic relevance of this was really off on the horizon, at least in those early days. That's when my encounter with Jay Bradner took my scientific career, and my life, in a totally unexpected direction in the best possible sense. My Ph.D. advisor knew about our BRD4 observation with our RNAi screen and alerted me that Jay Bradner just published a paper in *Nature* in 2010 describing JQ1, the first small molecule I was aware of that directly targets BRD4. This was just the ultimate serendipity, when lightning strikes in the best possible fashion.

Paula Kiberstis: You were not aware of this paper beforehand or you just didn't relate it to what you had discovered?

Dr. Vakoc: Apparently, in Pharma there's been an effort going on in the background that I was unaware of to make molecules against this protein. Jay Bradner was the first to take these molecules from behind the iron curtain of what's happening in secret in Pharma. He brought these

Published by Cold Spring Harbor Laboratory Press; doi: 10.1101/sqb.2016.81.031617

molecules out into the public and started distributing them. He's become very famous for this open source distribution of JQ1. We were one of his first customers. It was a very exciting phone call where I told Jay all our results from our RNAi screen. Literally, the next day he sent us huge quantities of JQ1 and we started treating our leukemia mouse model with this compound, thinking that our RNAi screen predicted that this should be a good target. Remarkably, it was. We just went straight for the killer experiment and it worked. It really just stunned us that this could be true.

Jay has become a very dear friend and one of my closest collaborators. Working together in these animal models of leukemia, we showed that this is actually a respectable, preclinical, drug-like molecule. Defining the mechanism was kind of the cherry on top. It turns out that JQ1 shuts off transcription of c-*myc*, *myc* being one of these elusive oncogenes that everyone recognizes as one of the major drivers of human cancer. This drug was working through this classically "undruggable" target. This was like science moving at the speed of light and with maximal excitement. For me, as a young scientist here, this was a career-changing moment. We published this result in *Nature* in 2011.

Another key detail of all this work is that it's been highly reproducible in independent laboratories. Similar work done in Tony Kouzarides' laboratory at Cambridge using a JQ1 molecule that GlaxoSmithKline had developed showed the exact same findings. Since then Pharma has replicated these results widely, which is probably the most important detail in all of this. This has led to the rapid movement of these molecules into clinical trials, I think purely because of the replicability of the finding.

Paula Kiberstis: Can you tell us a little bit about how this protein works? Can you tell us what a bromodomain is and what it does to affect transcription?

Dr. Vakoc: Bromodomains are protein domains that bind to chromatin histones, but only when they're modified chemically by lysine side-chain acetylation. We call them chromatin reader domains. The fascinating thing is that this is a protein–protein interaction surface. JQ1, this really remarkable chemical probe, works not by inhibiting an enzyme (which is how a lot of cancer drugs work), but by inhibiting how proteins interact with one another. Basically, when cells are pulsed with JQ1, it causes BRD4 to lift off of chromatin by binding competitively to the bromodomain pockets.

Paula Kiberstis: Do you know the mechanism by which it has an antileukemic effect? Also, why would you expect this molecule to selectively affect cancer cells and not normal cells?

Dr. Vakoc: These are the central questions in our laboratory. We're making progress, but we still do not have a solid answer. My background is in biochemical mechanisms and transcription. This project, for me, is really an anomaly because there was this fascinating therapeutic response without a mechanism that's been the source of the excitement. Now, we're working backward to try to figure out why this actually works. The short answer is that we don't really understand this very well. When we target BRD4 with molecules, transcription changes in cancer cells, transcription changes in normal cells. Cancer cells don't like it as much as normal cells do. That is probably the most accurate way I can describe this at this point.

Of course, there are molecular details that are fascinating for how this protein works. We've taken the plunge into trying to decipher every nugget of how BRD4 executes its function in leukemia cells. The gist of it is that the genes that are changed when you target BRD4 are genes like *myc*: a lot of genes that cancer cells are very dependent on. I suspect that JQ1 stresses cancer cells by withdrawing a lot of the key ingredients necessary to maintain a transformed cell state. Nontransformed cells are not as dependent on the genes that JQ1 suppresses. It's more mysterious than it is understandable at this point.

Paula Kiberstis: How is the drug doing in clinical trials?

Dr. Vakoc: I have very little involvement in those trials. We work very much on the basic side of finding targets. Many independent Pharma groups have made molecules like JQ1, but it's not JQ1 itself. Jay Bradner, in his laboratory, made a drug-like derivative of JQ1 and started a small biotechnology company to move these into clinical studies. GlaxoSmithKline and other groups moved into trials starting in 2013. Several of these are in relapsed patients with refractory acute leukemia that have failed prior chemotherapies.

The first trial was led by a company called OncoEthix using a molecule in the same chemical series as JQ1. A few months ago in *The Lancet Haematology*, they reported that the molecule, OTX015, is actually quite well tolerated at certain doses. Furthermore, at those well-tolerated doses there's some initial evidence with a subset of leukemia patients experiencing complete remission of their disease following single-agent therapy with this molecule. We've been enormously excited to see these early lines of evidence. It's also crystal clear though that this molecule is not Gleevec. The responses were not durable; the patients very quickly relapsed, so it's bittersweet in a way.

Paula Kiberstis: Tumor cells are very smart; they always figure out a way to get around whatever you're doing to them. Are there ways to work out the mechanisms of resistance?

Dr. Vakoc: This has become an area of intense interest in the last year. I think everyone knew resistance would occur. The field is at the stage of trying to figure out what the resistance mechanisms will be. Several reports have found mechanisms in various contexts; they're all different. The mechanisms of resistance appear at this stage to be nongenetic; it looks to be a kind of rewiring of gene expression through what seem to be a multitude of different ways.

As one would expect from a target like JQ1, the mechanisms of resistance are equally complicated, if not more so, than the actual mechanism of action. The challenges are many at this stage, but one is to predict which patients might be most sensitive to these drugs. The predictive biomarker is kind of the Holy Grail of this field—that, and the right combination therapies that will synergize and allow these responses to be more durable. This is an area that we, and many others, are pursuing to see whether these molecules can make a real impact on the management of this disease.

A Conversation with Harold Varmus

INTERVIEWER: JAN WITKOWSKI

Executive Director, Banbury Center at Cold Spring Harbor Laboratory

Harold Varmus is the Lewis Thomas University Professor of Medicine at Weill Cornell Medicine and a Senior Associate at the New York Genome Center.

Jan Witkowski: I thought we might start by reflecting on the history of this field. The first Cold Spring Harbor symposium for which the words "cancer" or "tumor" appeared in the title was the Tumor Virus Meeting in 1974. That was largely to do with the biology of the viruses rather than the cancer-causing ability.

Dr. Varmus: Yes, I think that's right in one sense: we knew more about the replication strategies of those viruses than about how they caused tumors. But many, even most, of the people who came to that symposium were interested in cancer; they were drawn to the study of tumor viruses, either those that carried their genes as DNA, like SV40 virus, or as RNA, like Rous sarcoma virus and other retroviruses, because those viruses were able to change a normal cell into a cancer cell and cause frank cancers, solid tumors or leukemias, in animals. For those of us who believed that cancer probably had its origins in genetics, these viruses, which have very few genes but are capable of changing a complex mammalian or avian cell into a cancer cell, offered promise for understanding some fundamental aspects of carcinogenesis at a time before we had molecular cloning and DNA sequencing and all these things that make modern eukaryotic genetics possible. At that time, we had no idea how many genes were actually in a eukaryotic cell. We were looking for simple ways to ask what sort of genes might be able to mediate this dramatic change between normal cell behavior and the excessive growth that is characteristic of cancer cells.

Jan Witkowski: By the time of the next Cancer Symposium in '79, the title of which was "Viral Oncogenes," the field was transformed.

Dr. Varmus: It was and it wasn't. We did know a lot more about viral cancer genes by then, but we also knew something about them in '74. In fact, some of the most important mutations in viral cancer genes—mutations that defined the cancer-causing functions of those genes—had been identified in the late '60s and early '70s. They included, for example, the temperature-sensitive and nonconditional deletion mutants of Rous sarcoma virus, first isolated in the late '60s and early '70s. One of the reasons that viral oncogenes became so much more important in the late '70s is because we knew, in the case of the retroviruses, that every one of those retroviral cancer genes was derived from a normal cellular gene, and we had much greater capacity to study those genes—to figure out what kind of proteins they made, how they became cancer-causing, and how they affected cell functions, like growth—when they were incorporated into viral genomes. At that point, recombinant DNA technology was available, but it had not yet been widely applied to the study of cancer genes from eukaryotic cells because of restrictions on the use of the technology.

Jan Witkowski: Was it a surprise that cells had genes in them that, when mutated, could make them cancerous?

Dr. Varmus: At the symposium in '79, that was still a prediction based on comparisons between viral oncogenes and their cellular progenitors, called proto-oncogenes. Strong evidence that at least some of those genes contributed to human cancers by undergoing mutations—for instance, point mutations or chromosomal translocations—came a few years later. Those discoveries were both surprising and gratifying. Many theories existed about how cancer might arise, some dating back to the early days of the 20th century. One of those ideas, first outlined by the developmental biologist Theodor Boveri, was that normal cellular genes could drive a cell to behave abnormally if they became too numerous or were deleted. What I think everybody was very surprised by was the idea that some of the human genes with that potential could be found by using the genes that were present in tumor viruses of chickens or mice or cats. That was the big surprise.

When I entered the tumor virus field in 1970, we already knew that there was at least one gene in a virus like the Rous sarcoma virus that was required to elicit cancer-

Published by Cold Spring Harbor Laboratory Press; doi: 10.1101/sqb.2016.81.031641

ous behavior in an avian or mammalian cell after infection. Our group—Mike Bishop and I—approached this problem initially without a preconceived idea of where it might lead, thinking, "We're going to be studying a virus that has only a few genes and at least one of those genes we know to be involved in viral carcinogenesis." The idea that the gene would itself be a slightly altered version of a normal cellular gene was certainly not what I expected. I thought it might be similar to a gene that is present in a normal cell. In retrospect, you can view subsequent developments as an early stage of the Human Genome Project: finding out what proved to be the normal versions of human cancer genes by studying the 20 or 30 retroviruses that were known to cause cancer in animals.

Jan Witkowski: That led to a whole raft of similar genes being identified.

Dr. Varmus: It turns out that the Rous sarcoma virus was, for experimental purposes, the best possible virus to work with. It was fully competent to grow (by doing all of those interesting things during the viral life cycle that we talked about at the 1974 symposium). But it also had the ability to transform cells. Many of the other so-called oncogenic or transforming retroviruses were more complicated. The highly oncogenic viruses were defective for replication. Other closely related viruses could replicate, but not transform cultured cells and induced tumors in animals only very slowly if at all. These two kinds of viruses, competent for replication or able to transform cells efficiently, had to grow together to propagate the transforming virus. Rous sarcoma virus was different, simplifying both genetic and molecular studies; it proved to be a great experimental tool and we were lucky to have started with it.

Jan Witkowski: Do you know if Jim Watson was disappointed that his bet on DNA tumor viruses didn't come out?

Dr. Varmus: DNA tumor viruses taught us a lot about DNA synthesis and gene expression in eukaryotic cells, including RNA splicing, in the 60s and 70s; then, in the 80s, they were illuminating about the functions of another set of genes important in human neoplasia, the tumor-suppressor genes. And much of that work was done at Cold Spring Harbor. The route to discovery of tumor-suppressor genes was just a little more complicated, involving chromosomal abnormalities, hereditary cancers (like retinoblastomas), and the binding of viral oncoproteins, T antigens, to cellular proteins like P53. So the role of DNA tumor viruses ended up being incredibly heuristic. Further, Jim was very excited by the discovery of cellular oncogenes and did not seem to regret having placed most of his bets on the DNA tumor viruses. They taught us a lot.

Jan Witkowski: Once people have got hold of genes, you could look at proteins, and you could then begin to look at functions of the normal proteins.

Dr. Varmus: You could. But it was surprisingly difficult to identify the product of the Rous sarcoma virus oncogene, called *src*. The gene had been defined by both temperature-sensitive and deletion mutants around 1970. But it wasn't until 1975 when Joan Brugge, working with Ray Erikson, first showed the gene makes a 60-kDa protein. It took another few years before we understood it was a protein kinase, and another couple of years before Tony Hunter and his colleagues showed that src protein was a kinase of a special kind because it adds phosphate not to the most commonly phosphorylated amino acids, serine and threonine, but to tyrosine. So it took 10 years to go from the identification of this gene as defined by temperature-sensitive mutations to the demonstration that it makes a protein tyrosine kinase that transforms cells when it is not properly regulated. Over the next several years, a community effort revealed that it was only one of what turned out to be roughly 90 tyrosine kinases, many of which are encoded by proto-oncogenes and are now the targets for some of the most effective drug therapies in human cancers driven by mutant versions of those genes.

Jan Witkowski: That led into things like the cell cycle, which was the topic of the '91 symposium.

Dr. Varmus: The proto-oncogenes discovered through the study of retroviruses were actually less central to identification of the cell cycle machinery than were the DNA tumor virus oncoproteins that interact with tumor suppressors. The DNA virus oncogenes that control the activities of the retinoblastoma, E2F, and p53 proteins played a bigger role in the cell cycle symposium than did retroviral oncogenes. The products of proto-oncogenic precursors of retroviral oncogenes have several roles in delivering growth signals from the cell surface to the transcriptional apparatus; but many of the components that govern the cell cycle itself—not all, but many of them—are central to the mechanism by which DNA tumor viruses transform cells.

Jan Witkowski: Presumably, as the genes and the proteins for normal cells were examined, the role of viruses in this research diminished.

Dr. Varmus: You could argue that, because recombinant DNA technology now permits us to isolate or make DNA, RNA, or protein and manipulate the functions of cancer genes in various ways. But I think it is important to remember that tumor viruses, both RNA and DNA, remain very important in several ways. For instance, the work that was done on retroviruses that cause tumors in chickens and mice taught us many things about how retroviruses grow, more than a decade before we knew what caused AIDS. And, of course, AIDS is caused by a retrovirus that is very similar in its growth strategy to other retroviruses, so we were serendipitously well prepared for the discovery of HIV. Retroviruses also remain essential laboratory and clinical instruments—vectors—for introducing genes into cultured cells and patients. I think people have to understand that the investment that was made in tumor viruses because they cause cancer taught us about a class of viruses that ended up being important in many other ways too.

Jan Witkowski: Tell us a bit about what you are doing now.

Dr. Varmus: For the last 20 or 25 years—ever since I became less of a tumor virologist and more of a cancer biologist—my lab has been looking at mechanisms by which normal cells become cancer cells. In a sense, it's still the same general question but it's being looked at in a different way. Our world has been transformed by high-throughput sequencing and the development of highly detailed genetic maps of cancer cells compared to normal cells in patients in which those cancers have arisen. Our community now knows a lot about the complex pattern of genetic mutations and about the physiological changes that are involved in turning a normal cell into a cancer cell. From that rich repertoire of mutations and other kinds of changes, new therapies have evolved. There's discovery of cancer genes; there's novel therapeutics. In between, there's a rich terrain in which many of us are trying to understand how the constellations of mutations, different in every single tumor, end up causing a cell to be cancerous and yet have certain vulnerabilities.

More specifically, we're studying three things about this general situation.

One is how mutations that affect splicing factors and perturb normal RNA splicing patterns help to make cells cancerous.

The second is an effort to understand why certain kinds of mutations are found in certain kinds of cancer. What is it about the cell type in which those mutations are found that make the mutations more "penetrant" in those cells? In other words, why do those mutations give rise to what we call the cancer phenotype in that cell type? What accounts for the behavior and appearance of cells in small cell lung cancer, squamous cell lung cancer, and adenocarcinoma of the lung, three very distinct pathological types of cancer that have rather distinct patterns of mutation, although they all arise in the lung from cells belonging to one developmental lineage?

The third, the one I discussed at this year's symposium, is why certain mutations that are commonly seen—for example, in lung adenocarcinomas—don't occur together. It's been very striking to those who have sequenced literally hundreds of tumors of a certain type that certain genes can be found mutated in perhaps 10%, 20%, or 30% of the tumors, and yet never be found mutated in the same tumor. Why is that? Is it because there's no selective pressure for the second mutations once the first has occurred—for example, because they exert identical physiological effects—or is it because the two together are synthetically damaging? As far as we can tell from the common examples we have studied, the latter is the case. That's interesting because one of the things we're trying to do is to figure out some novel ways to approach therapeutics.

Traditionally, people say, "your cancer is driven by a bad gene, so inhibit or eliminate the protein that bad gene makes." Another way to think about the problem is to say, "When you have a bad gene driving your cancer, it creates a kind vulnerability in the cells such that the cells can't tolerate the loss of some other gene." It is possible there's another kind of vulnerability, a vulnerability to hyperactivity of some other genes. What we're looking for in this situation is a lethal conspiracy that occurs when both an oncogene commonly affected in many types of cancer, the so-called *KRAS* gene, is combined with a mutant epidermal growth factor receptor (EGFR) gene. If we can understand why those two are not tolerated by cells, and we can mimic that in a patient with cancer, we could have a new approach to therapy.

Jan Witkowski: It sounds as though you've got a cancer cell with a *ras* mutation that's the driver of that cancer. You're now going to introduce another bad gene into that stage?

Dr. Varmus: We call them bad, but we only call them bad because the cancers kill the patient. A *ras* mutant is not bad for the cell. It gives the cell a selective advantage. It grows more, it dies less, it drives the cell to success. Success for the cell is becoming a tumor. It's not good for the patient, but it's good for the cell. Surprisingly, if you have two of these bad actors in the same cell, they conspire to make life difficult for the cell. That's potentially a good thing for the patient. What we're trying to do is exploit the toxicity of having two cancer-causing genes in the same cell. Obviously there are probably many occasions in which there are multiple cancer-causing genes in the same cell, but they don't have a toxic effect on the cell. Some combinations do, and one way to identify those is to look through these reams of data for mutations that are common but are mutually exclusive—they do not occur together. If we understand why a combination is bad for the cell, we might be able to mimic that situation with a drug that has little or no effect on normal cells while specifically damaging the cancer cells.

Jan Witkowski: How many such pairs have you found?

Dr. Varmus: We've looked at a few just by using computational methods to look at patterns of mutations. We haven't subjected them all to biological tests ourselves, although other labs have looked at a few of these combinations and have seen some form of cellular damage or growth arrest in those cases. I don't believe that every mutually exclusive pair is going to prove to be damaging, but some will. Other laboratories will need to join us in this endeavor and try to enumerate which of the combinations are synthetically toxic. There's potential for doing something important here if we could understand how that damage is mediated. Is there some switch, some change in behavior of the cell that we can understand and manipulate?

Jan Witkowski: Presumably, you needn't introduce a second mutation if you understand the consequence of that second mutation. There could be something downstream that you can manipulate.

Dr. Varmus: Yes, very likely. But we also need to remember that cells are capable of adapting to escape damage, and those that do learn to escape become the dominant component through an evolutionary process.

For instance, one of the things we're learning as we treat cancers with drugs that inhibit a single oncogenic protein is that cancer cells have a lot of ways to escape the effect of a drug. Drug resistance can most simply occur because of a secondary mutation affecting the target site for the drug, but it can also occur because the cell undergoes a rewiring to work in a different way. Many people are now looking for ways to put one kind of therapy together with another so that cells have a harder time finding a way around the therapeutic inhibitor. The same logic may apply to efforts to induce synthetic lethality. Still, there is great potential here for understanding a cell's wiring system better; for example, there are both negative and positive forces that operate in these signaling networks. Grasping the logic of such signaling systems is likely to yield new ways to think about therapeutics.

Jan Witkowski: Let's spend a couple minutes on a bit of sociology: preprints. You've long been an advocate of open access publishing and data sharing. Are preprint servers a further development of that?

Dr. Varmus: They're a further development in the sense that they represent another change in publication practices in biology. There are two points I would make. First, providing your work as a preprint by putting it online before it's been reviewed by peers is very different from open access publication with peer review. Both methods make work accessible to all at no cost to the reader, but preprint servers are simply inexpensive mechanisms for getting work very quickly to readers. Open access journals, with their review processes have significant costs that have to be met, for example, by charging a fee to the authors. Secondly, the public posting of preprints is not really new. It is only new to biology. In physics, mathematics, astrophysics, and computational sciences, people were doing this through a preprint server called "arXiv" ever since Paul Ginsparg developed it in 1991. So it's been around a long time, and most of the well-known journals that we all use, like *Nature* and *Science*, are perfectly comfortable accepting a paper that was posted as a preprint a year or 2 or 3 years earlier for formal publication after traditional peer review. Preprints therefore don't really break new ground as a publishing practice, but they break new ground in biology. We're all hemmed in by our cultural traditions.

Jan Witkowski: Why do you think biologists are uncomfortable with preprints?

Dr. Varmus: I'm not entirely sure whether it's inherent to the nature of the life sciences. Some people argue that it is, because of the size of our field and the patentability of some of our work. But I don't think so. I think what happened is that we simply didn't adopt it when the physicists did. Perhaps some leaders didn't see a big need for it and didn't adopt it, even though some of them began their careers in physics or math. I'm not sure why we didn't see a need for it when physicists took it up so avidly; maybe because we had a lot of meetings at Cold Spring Harbor and elsewhere to talk about our results.

But now there are other factors. Everyone is worried about the competition for jobs and grants. So much of what determines who's going to advance in their career now depends on publication in certain traditional journals, especially the most highly ranked journals; so people are very nervous about changing their publication practices.

This is a cultural element that we need to fight. For one thing, we should not be evaluating each other based on what journal we publish in; we should be evaluating each other based on the quality of our work. Secondly, we have to remember that as scientists we are paid to spend these highly enjoyable years of our lives working with brilliant colleagues to make discoveries and share them with our colleagues. What better way to do that than to get our work out early in a preprint that you can still publish in a distinguished journal later on. The journals, excepting one or two, are not resistant to this. And we, our colleagues, and the public can all benefit from faster communication, with minimal if any risk or expense.

A Conversation with Karen Vousden

INTERVIEWER: JAN WITKOWSKI

Executive Director, Banbury Center at Cold Spring Harbor Laboratory

Karen Vousden is Director of the Cancer Research UK Beatson Institute.

Jan Witkowski: Over the past few years, there's been a markedly increasing emphasis and interest in the metabolism and biochemistry of cancers. It's moving away from just the genes and proteins. Am I right in thinking that?

Dr. Vousden: Yes, I think you're absolutely right. There's been a resurgence of interest in metabolism. In the middle of the last century a lot of energy was spent on understanding metabolism and biochemists at this time were making huge progress. Many metabolic pathways were worked out in exquisite detail in the '50s and the '60s. Then came the oncogene revolution, and much of the emphasis switched to understanding the regulation of the cell cycle and signal transduction. At this point, metabolism was almost forgotten in cancer biology until about 10 years, when there was a reemergence of interest. Craig Thompson was showing a big poster of all the cancer-signaling pathways, and another with all the metabolic pathways, pointing out that there was not a single gene on one that's also in the other. I think we are fixing that now.

Jan Witkowski: Was there any particular event or finding that has led to this increased interest in biochemistry?

Dr. Vousden: I have to attribute a lot of this to the work of a few labs who led the way in this reemergence of tumor metabolism. This was provoked by the general realization that control of the cell cycle is not enough to explain cancer biology, and that cells need to grow as well as divide. As our understanding of the regulation of the cell cycle grew, the community started to look beyond cell division to the importance of other aspect of proliferation.

Jan Witkowski: Warburg came up with the idea that aerobic glycolysis was a characteristic feature of cancer cells. In modern findings, has that proved true at all?

Dr. Vousden: It has held up. In the end, Warburg's observation is what underlies PET (positron emission tomography) imaging. It certainly is the case that many cancers have a high rate of glucose uptake, which is what we measure when we do functional imaging, or a PET scan.

However, over recent years our understanding of the Warburg effect has become more sophisticated, although we still don't quite understand it. We know now that there probably are changes in the metabolism of almost all cancers. We're interested in understanding that and then asking, "Can we harness any of this information to develop new therapies?"

Jan Witkowski: Can you tell me a little about your current work?

Dr. Vousden: We became very interested in a specific aspect of metabolism, which is how cells use and produce the amino acid serine and why changes in serine metabolism might be particularly important in cancers. Although serine is not an essential amino acid—cells can make their own serine—tumors can be broadly divided into two groups. One group has amplifications in the de novo serine synthesis pathway, so these tumors can make a lot of serine from glucose. The others seem to depend more heavily on exogenous serine, although many of these cancers can switch to making their own serine when the external supply is taken away. This adaptation to serine starvation is a point of vulnerability for those cells, which we are hoping to exploit to deliver a therapeutic advantage.

Jan Witkowski: Why is serine such an important amino acid?

Dr. Vousden: Obviously it's important for the synthesis of proteins and it can also be used to make other amino acids. But it is also a major one-carbon donor to support the folate cycle. The folate cycle is necessary for nucleotide synthesis—purine and thymidine synthesis—and also contributes to the SAM (*S*-adenosyl methionine) cycle, and all the methylation reactions that we are hearing

Published by Cold Spring Harbor Laboratory Press; doi: 10.1101/sqb.2016.81.031930

about. Interestingly, some of the first successful chemotherapeutics work by targeting the folate cycle. We are now exploring additional functions of one-carbon metabolism that might be playing a role in providing ATP, NADH, and NADPH.

Jan Witkowski: Did you start with just serine? Did you just come across it all or was this a carefully thought-out step?

Dr. Vousden: Oh, it was definitely not a carefully thought-out step! My primary interest has been with the investigation of how the tumor suppressor p53 works. We've been very interested in the past in understanding how p53 arrests and kills cells. I can't even remember how this came up, but we did begin to realize that having p53 is not always detrimental to the survival of the cell and that under some circumstances p53 actually helps cells cope with certain types of stress. Those stresses tended to be mild metabolic, nongenotoxic stresses, stresses that you can easily imagine being reversible.

Previous work had shown that cells with p53 deal better with low levels of glucose than cells lacking p53. That was a really interesting observation, because it was paradoxical to the general canonical activities of p53, which are to stop cells from growing by arresting or killing them. We thought, "What about other types of starvation?" Oliver Maddocks, a postdoc in the lab, simply tested the requirement for every amino acid and found that serine starvation made cancer cells grow less well, but that growth was further reduced if they lost p53. That was our starting point. At the same time, several other groups—Matt Vander Heiden, David Sabatini, Lew Cantley—all made observations that showed serine synthesis was going to be important in cancers. It was really great because we were all working on the same problem that I think no one had really considered 5 years previously.

Jan Witkowski: Are you working on enzymes in the serine synthesis pathway?

Dr. Vousden: There are several aspects of this. We're still very interested in the idea that p53 can provide survival functions under the conditions of nutrient stress. We see this not only with serine, but also with glutamine and we are looking at other types of nutrients now. Another interesting thought is around mutant p53. We have found that while cells that don't have any p53 are more vulnerable to this type of stress, some of the p53 mutations that occur in cancer support survival similarly to wild-type p53. This seems to make sense—these tumor-derived mutants have lost the "death" activities of p53 but they retain survival functions. Interestingly, these are the mutations that are associated with the worst prognosis.

In terms of serine synthesis and the one-carbon cycle, we are looking at drugs that target enzymes in serine synthesis pathways, as well as some of the downstream enzymes that are involved in the metabolism of serine. We have been interested in a step that produces mitochondrial NADPH, which may be important for antioxidant activities. But our particular take on this whole area has been to simply say, "Can we see a therapeutic effect by lowering circulating serine levels simply by taking serine out of the diet?"

Jan Witkowski: How easy is it to affect serine levels like that?

Dr. Vousden: We can give mice a completely serine-deficient diet, and we have been using various different iterations of this with diets lacking various combinations of amino acids. Interestingly, and much to my surprise,—Oliver carried on doing this even though I said it wouldn't ever work—you can reduce serum serine levels by $\sim$50% simply by giving mice serine-free diets.

There are examples where in-built errors of metabolism can be rescued by modulating amino acids in the diet in humans. Treatment of phenylketonuria by limiting phenylalanine is the classic example. Then there are syndromes leading to the accumulation of glycine, where treatments to reduce serum glycine levels are used to try to rescue some neurological phenotypes, as well as defects in serine synthesis pathway enzymes that can be rescued by increasing dietary serine. So it's certainly possible, and we are talking to some companies now to see whether we can try to at least formulate a diet to determine whether we can lower serine levels in people.

Jan Witkowski: How much serine is in the typical diet?

Dr. Vousden: Protein-rich foods like meat, cheese, and eggs are all high in serine, and most people won't be serine-deprived. I think that it would be difficult to really limit serine without having an absolutely synthetic diet, which would be, of course, very problematic to administer.

We have clues from our animal models that there is synergy between serine limitation and ROS (reactive oxygen species) activation, leading to an enhanced vulnerability that means cells adapting to serine starvation are more sensitive to increases in oxidative stress. Fortunately, many common chemotherapies induce ROS. With our clinical colleagues we are now considering whether we could carry out a clinical trial where we move patients onto a serine-free diet only over the course of their chemotherapy. In this way they wouldn't go on this diet forever, but we may be able to increase the efficacy of the chemotherapy by using the diet for only a week or two.

Jan Witkowski: That makes a great deal of sense because then it's not such a burden on the patient. I presume these diets are not going to be inexpensive.

Dr. Vousden: There are companies who make these sorts of diets for clinical use, and simply reformulating them to take out serine rather than some of the other amino acids would be relatively straightforward. I don't know what the expense in the end would be, but I think it depends on how effective it becomes.

The other important consideration at the moment is whether there will be any side effects that aren't associated with the tumor. The most common concern I hear is neurological defects. Obviously, serine and glycine are very important in the brain. That, I think, is more of a problem for the inhibition of serine synthesis pathways because the brain is where cells make their own serine. So we are probably less likely to have a neurological effect in response to removing serine from the diet.

Jan Witkowski: Do cells generally prefer to make their own serine?

Dr. Vousden: Most tumor cells, we think, prefer to take it up, and that's why we can have an effect by taking it away from the circulation. But the response is very varied. We are working out now how the genetic constitution of the tumor affects their sensitivity to exogenous serine. Obviously, tumors that have already ramped up serine synthesis won't care, but most of the other cancers are sensitive to some degree, and we would like to understand this more fully, so we can target those cancers that depend on exogenous serine the most.

Jan Witkowski: Are there particular cancers that ramp up their synthesis of serine?

Dr. Vousden: This had been shown in breast cancers, melanoma, and other tumor types. The initial observations were that 10%–20% of breast cancers show evidence of amplification of one of the three serine synthesis pathway enzymes, although this is higher in ER-negative tumors. It's all very interesting because these tumors depend on the serine synthesis pathway, even when they have access to exogenous serine. We're still trying to understand why that might be.

A Conversation with Valerie Weaver

INTERVIEWER: ELIZABETH MCKENNA

Associate Editor, Cancer Discovery

Valerie Weaver is a Professor in the Departments of Surgery, Bioengineering and Therapeutic Sciences, Anatomy, and Radiation Oncology and the Director of the Center for Bioengineering and Tissue Regeneration at the University of California, San Francisco.

Elizabeth McKenna: Can you tell us a little bit about the role of forces in cancer progression?

Dr. Weaver: The way that I view cancer is that, in terms of the mechanics, it's a physical alteration of the tissue on multiple levels. On a gross level, you have a large cell mass surrounded by a dense extracellular matrix that is also highly vascularized. Tumors contain a lot of fluid in part because the vasculature is altered and lymphatic drainage is often compromised. The fibrotic tumor mass and high interstitial pressure can create a milieu where it's really hard to get drugs in. That is certainly an aspect of tumor mechanics that has been studied a lot and what most people think about when they discuss mechanics. What my group has been looking at, and what people haven't really thought about as much, is that tumor cells also alter their behavior in response to the tissues mechanics and the genomic alterations in the tumor cells changes their contractility behavior. Many common oncogenes change both how squishy the cells are and how much the cells can pull on each other and on the matrix around them. A highly contractile tumor cell will remodel the matrix around itself. The contractility of the tumor cell will also alter how the cell responds to exogenous signals like growth factors. It also determines the kind of soluble factors that the tumor cells will secrete and we believe that this is, in part, some of what ultimately will stimulate some of the stromal cells—including infiltrating immune cells—as well as the fibroblasts. The fibroblasts themselves are super-synthetic. They create and secrete a lot of matrix proteins that they lay down and remodel, and these matrix proteins create a denser environment that is a bit stiffer. They also have enzymes that cross-link that matrix and remodel it, and they can even generate aligned matrix bundles that act like migration tracks along which tumor cells, immune cells, and even fibroblasts can migrate. A lot of cytokines are also being constantly secreted within a tumor that can also stimulate the fibroblasts to synthesize and remodel the matrix and these cytokines will attract immune cells to come into the tumor—and they also get in on the action by secreting things like TGF-β, which additionally drives the fibroblasts to produce even more matrix proteins that ultimately create this really dense, stiff, fibrotic tissue. The induction of the stiff, fibrotic tissue has repercussions across the board in terms of the physical properties of the environment and how the tumor cells, the immune cells, and the vasculature are influenced by that stiffness to change their behavior and responsiveness to soluble factors.

What people also don't think about is that, as the tumor mass expands, it's not expanding like a balloon into infinite space. It's expanding into this really dense fibrotic environment. Of course, the dense environment resists. You have a solid stress from the tumor cells as they expand. You have a reciprocal tension from this dense environment. You have a buildup of fluid, so you ultimately have this massive solid stress in the tumor. All of these conspire to create an environment that is really difficult to get drugs into. It creates a hypoxic environment. This also changes the tumor's behavior as well by fundamentally reprograming how that tumor cell behaves.

The key point is that a tumor is not mechanically homogeneous. It's not like one big ball of a stiff tumor. It's a very heterogeneous physical environment. You have little tiny soft niches interspersed with thick matrix bundles that act like little railroad tracks where the cells migrate on these stiff tracks. It's a really complicated and fascinating thing to try to understand how these forces add yet another layer of complexity to the tumor, and how they change the initiation, progression, and aggressiveness of the cancer.

Elizabeth McKenna: You've spoken about intrinsic and extrinsic forces. Can you elaborate about that and the differences between the two types of forces?

Dr. Weaver: Every cell in your body, even if it's not a muscle cell, has myosin. Cells are contractile. That's how I define intrinsic tension. "Force" has a time element and

Published by Cold Spring Harbor Laboratory Press; doi: 10.1101/sqb.2016.81.032128

is often represented as stress which is measured as force over area. It's important to realize that it is an active process. Cells are continuously interrogating their microenvironment. They'll either pull on each other or they'll pull on the matrix. That's intrinsic tension. In a tumor or in a normal cell, if you put them in a stiff environment they tune their tension to the best of their capacity to the stiffness of the environment around them or to the stiffness of the other cells that they're interacting with. In a tumor, we know that there are genetic modifications. What's really interesting is that many of the same genetic pathways that are selected for in cancer are same ones that alter intrinsic tension in the cell. Of course, if you have a big mass of cells and it's pushing out, that's also a type of intrinsic force.

One type of extrinsic mechanical property in a tumor is the stiffness of the extracellular microenvironment. Another type of extrinsic or external force is the shear force that a tumor cell would experience when it enters into the vasculature. Other types of extrinsic forces include interstitial pressure, as well as any long-range stresses such as tissue distortion or stretch. What is important to realize is that extrinsic forces in a tissue cross length scales from a subcellular level to a tissue level. For instance, there is a gross "organ and tissue-level" length scale that affects how the cell and the tissue behaves that can influence the direction of fluid flow or perhaps the alignment of collagen fibers or even the vasculature within the tumor. However, what the cell really cares about is what's happening right around it. That determines how it signals and how it behaves given the various different soluble factors that it's simultaneously experiencing.

Elizabeth McKenna: How do the tumor cells sense the force? What molecules are involved in sensing? How do they propagate the signal to effect changes in the phenotype?

Dr. Weaver: It's not as though you have specialized mechanotransducers, although in some instances you do; some ion channels, for example, have been known for decades as being sensitive to mechanical cues. Cilia are quintessential mechanosensors. Hearing is basically mediated through ion channels because cilia are sensitive to change induced by sound waves and they open up and regulate the ion channels. That's one way to sense mechanical cues. Integrins also are classic mechanotransducers. An integrin responds to the mechanics of the environment, but it also responds to different extracellular matrix proteins, so it's also a biochemical signal. Biochemistry and biomechanics are different sides of the same coin.

Virtually any transmembrane receptor that can change its conformation in response to a strain could be viewed as a mechanoreceptor—in the case of integrins, talin and vinculin are sensitive to force, which favors their open configuration to favor binding to the integrin tail and to each other. For a protein to be considered a mechanosensor, the molecule must be able to respond to force by altering its structure/function, whether this is opening up a cryptic site, enhancing the timing of the opening of an ion channel to allow in ions, or closing an ion channel: any of these things will function as a mechanosensor. Then something has to be able to amplify that response, but even then it doesn't do anything unless it can actually effect a change, either by changing actin polymerization, by changing higher-order chromatin structure so it modifies the next signal, or by just amplifying the signal. All of those things will eventually affect how the cell responds.

Elizabeth McKenna: Do all tumor cells react to forces and interact with their extracellular matrix in the same way, or does it depend on their genetic background?

Dr. Weaver: It depends. It's something that we're just starting to scratch the surface of. A lot of people in this engineering and biophysics field are looking for certain principles that we can be guided by to think about how tissue mechanics can be layered on top of everything that we know about how tumors develop and behave. Having said that, if you think about a glioblastoma versus a pancreatic cell versus a bone, each of these cells has fundamentally different mechanobehaviors to begin with. The kind of mechanical cues and how they're going to change the cell's behavior are probably going to be slightly different. The kind of pathways that would be affected would be slightly different because they're different to begin with, but there could be similar principles that we're working across.

Certainly, our data with the pancreas suggest that if you have a more basal genotype in the tumor you have a much more contractile phenotype and enhanced mechanobehavior: a very unique mechanical signature. We also have some new data on glioblastomas that, if you have an isocitrate dehydrogenase mutation that is known to create a much lower grade or less aggressive tumor, it curbs or curtails the mechanophenotype. Those are two independent things that say that it's different. Triple-negative breast cancers versus luminal breast cancers have fundamentally different mechanics. Maybe it's a continuum with different molecules that affect things differently, but ultimately higher or lower stiffness affecting how the tumor behaves.

Elizabeth McKenna: Do you have any ideas on ways that we could design drugs that could target certain pathways that are involved in mechanotransduction, or should we be thinking about targeting the stroma and affecting the components of the extracellular matrix?

Dr. Weaver: A naïve perspective would be to say, "Oh, we're just going to go in and inhibit mechanics." The problem is, ROCK (ρ-associated protein kinase) inhibitors are contraindicated for treatment of brain tumors because they cause a massive drop in blood pressure. If you just completely ablate the stroma, certain of the tumor cells are really contractile to begin with, so they won't care. That could explain why there have been problems. Plus, the stroma has good things that it does, not just bad things.

We're looking at three or four different approaches. One: Can we exploit aspects of the mechanics and the things that we know about mechanics to enhance patient stratification? If we can identify those patients, for example, that have a very unique mechanical signature, maybe they would be contraindicated for some of the stromal ablation therapies. Or those individuals might have tumors with a higher propensity to metastasize, so those individuals would be treated differently. That's how we're thinking about it for pancreatic cancer.

We're also doing some studies with other people where we're looking at lipophilic statins, which have effects on protein prenylation. They calm down the tension in the tumor. Those individuals that are on these lipophilic statins have a lower incidence, or at least less aggressive, pancreatic cancer. They have a longer expected survival. Those are ways to think about it.

We've also developed mass spectrometry methods to look at the profile of crosslinks in the matrix. Those would then tell us which enzymes are changing the crosslinking. If those are unique in tumors, we could target those. Lysyl oxidase is one, but it's not good for pancreas because the propeptide has anti-Ras effects. It could work for other kinds of tumors, but not for a Ras-dependent tumor. There are also prolyl hydroxylases out there, so that's another way to think about it. I know Sunil Hingorani is leading an effort to reduce solid stress by treating pancreatic patients with hyaluronidase—this is a fantastic approach—although how this truly works remains unclear. Finally, understanding how a mechanical change in a tumor drives its dysfunction could identify unique molecular targets that could be the tumor's Achilles' heel.

Another factor to consider is whether an altered tumor mechanics alters immune function. We have active studies exploring this interesting area—could we enhance the antitumor immune response by normalizing tumor mechanics? Clearly, only one approach will be insufficient, and it will be important to explore this issue using a lot of different angles. We will have to work as a team with our colleagues who are doing fantastic work in immune manipulation. We're also doing some really cool studies with biophysicists to understand how ECM mechanics regulate at the nanolevel, how Ras signaling occurs, and how RTK (receptor tyrosine kinase) signaling is processed in the cell. This collaborative work is yielding some really interesting insights and key targets that we could use. It turns out that the whole dynamics of Ras signaling and effector recruitment and turnover are exquisitely sensitive to the mechanical cues in the environment. The devil is in the details. We need to go after this in a lot of different ways. A really simplistic perspective is not going to work. I would say that's what got us into these problems in the first place. I'd rather use a scalpel than a hatchet.

Elizabeth McKenna: There'll probably be a lot of context specificity and maybe different combination of strategies will be needed.

Dr. Weaver: When I first started doing this, colleagues of mine in cancer biology thought I was a flake. Now I'm pleased to say that—maybe not in the mainstream cancer field, but certainly in the engineering and biophysics—they're all going crazy mad on this stuff. They think that it's going to be one-size-fits-all, and it's all going to be supersimple. I'm coming along, and of course they don't want to hear this, but I think, "No, no, no. It's not like that." It's actually more elegant and more interesting.

A Conversation with Robert Weinberg

INTERVIEWER: JAN WITKOWSKI

Executive Director, Banbury Center at Cold Spring Harbor Laboratory

Robert Weinberg is a Professor of Biology at the Massachusetts Institute of Technology and a Founding Member of the Whitehead Institute for Biomedical Research.

Jan Witkowski: The first Cold Spring Harbor symposium where anything cancer-related appeared in the title was in 1974: the "Tumor Viruses" meeting. That looked very much to do with the biology of viruses rather than much to do with cancer. You were at that meeting.

Dr. Weinberg: Yes. At that time, tumor viruses were studied because they apparently had an interesting way of copying themselves. It was almost incidental that they also caused cancer.

Jan Witkowski: That was only 2 or 3 years after reverse transcriptase?

Dr. Weinberg: Yes, but again, people like me—and there were many others like me—looked at tumor viruses because they would clone genes for you in their own genomes, doing so through their own replicative processes. It was just a sideshow that they also happened to generate tumors in animals. It was almost a distraction for a while.

Jan Witkowski: That changed significantly only 5 years later with the 1979 symposium on viral oncogenes. What led to this transformation in that 5-year period?

Dr. Weinberg: There was the 1975–1976 discovery in the Varmus–Bishop laboratory at the University of California at San Francisco of the astounding fact that a virus that caused sarcomas—connective tissue tumors in chickens—actually had acquired that ability to do so by extracting and remodeling a gene that resided in the normal chicken genome, ostensibly because its role in the normal chicken genome was to help a bird be a bird, therefore having nothing at all to do with the whole process of developing cancer.

Jan Witkowski: It's very hard to put yourself in the mind-set of that period: The idea that it could be a normal cell's genes that, when mutated, could lead to cancer and that viruses could've transduced these.

Dr. Weinberg: Well, it was way out there. It wasn't in the center of everyone's thinking. There was another consequence of that discovery and the discovery that preceded it by several years: that a small number of viral genes could totally reconfigure the whole biology of an infected cell, converting a normal cell into a cancer cell. That had begun to be accepted by that time, but it was hardly obvious. The alternative, more intuitive, hypothesis was that if you asked how many genes needed to be altered in a normal cell to make it into a cancer cell, it might've been dozens. The fact that it was only a single one, at least in these models of virus-induced cancers, totally changed everybody's thinking. Suddenly, one could understand the process of transformation by studying a small number of master genes.

Jan Witkowski: Another important part of that paper was that the *src* gene could be found in the cells of different species. There was an evolutionary conservation.

Dr. Weinberg: The *src* gene is in the genomes of all animals, even in the genome of a sponge, which is a very primitive animal. This suggested in a very compelling way that the *src* gene was important in the biology of a wide spectrum of normal organisms and that its involvement in cancer was only a rare aberration.

Jan Witkowski: Let me read you a quotation from Peter Duesberg: "A critical test for the viral transduction hypothesis would be to isolate a viral transforming gene from a normal cell." So, cue a drumroll for your work where you first isolated a human oncogene.

Dr. Weinberg: Interestingly, the work I got involved in really wasn't that focused on tumor viruses. Much of my interest came from a study by Bruce Ames at the University of California at Berkeley who showed that if you looked at a whole series of chemical carcinogens, their potency in inducing normal cells to become cancer cells was reflected in their ability to damage DNA. In retro-

Published by Cold Spring Harbor Laboratory Press; doi: 10.1101/sqb.2016.81.031625

spect, maybe the correlation was a little overblown, but at least in my mind—and in the minds of many others, I assume—it suggested the inescapable conclusion that within the DNA of a cancer cell, there lay damaged, mutated genes that were responsible for the abnormal behavior of that cell. That really was the take-home lesson from Ames' work.

Jan Witkowski: In hindsight, it seems obvious that if there is a normal cellular equivalent of *src* in a cell, it could undergo mutation that would activate in a way that a viral oncogene could when a virus infects the cell. Ames' work on chemical damage inducing mutations in the cellular gene: That had particular potency for you?

Dr. Weinberg: It did, because already by the mid-1970s retroviral carcinogenesis seemed increasingly implausible as being a major factor in human cancers. On the contrary, chemically induced cancer—for example, cancers induced by cigarette tobacco tar—seemed increasingly to be a more plausible mechanism. Certainly by '77 or '78 I became convinced that retroviruses were interesting but they were not going to lead us directly to understanding how chemically induced human cancers, and indeed many of the spontaneously arising cancers one sees in humans, arose. I thought there might be a much different causal mechanism.

Jan Witkowski: What strategy did you use to clone *ras*?

Dr. Weinberg: We had been working on retroviruses, the sort that Howard Temin had first discovered, the sort that Varmus and Bishop had first studied in great detail. We acquired certain laboratory skills, including the ability to take DNA from one cell and put it into another cell experimentally. The DNA that was moved actually entered the second cell and began to function as if it were now an introduced gene and, ultimately, a native gene of the recipient cell.

By 1978, we knew that we could take genomic DNA from a virus-transformed cell and put it into a second cell that then converted from a normal cell into a cancer cell. Maybe, I speculated, we could do the same thing if we extracted DNA from chemically transformed cells that had never experienced a viral infection. To the extent that there was a cancer-inducing gene in such cells, it was with great likelihood a mutated version of a normal gene that somehow had undergone damage or corruption as a consequence of being exposed to, for example, a chemical carcinogen.

That was the thinking. It was way out there, not because it was conceptually complex, but simply because it was quite challenging experimentally. Starting in 1978, a fellow in my laboratory began the onerous task of extracting DNA from chemically transformed cells and introducing that DNA into previously normal cells, converting the latter into cancer cells. After many months of trial and error, he discovered one that could do that. That ended up being the most important set of experiments in my own career because it revealed a simple and obvious truth: In the DNA of chemically transformed cells—and by extension, human cancer cells—there must lie a single gene or a small number of genes responsible for orchestrating the cancerous behavior of the cells in which they resided.

Jan Witkowski: Coming back to the 1991 symposium, we saw the effects of these genes on controlling the cell cycle. At the '94 meeting, Ed Harlow thought that the elucidation of signaling and signal transduction pathways had become a critical aspect of studies of cancer, arising from the ability to clone cancer genes and examine what their proteins were making.

Dr. Weinberg: Underlying all that was the ability acquired in the late 1970s to actually clone a gene: to extract a single gene from the midst of tens of thousands of genes with which it co-resides within a single cellular genome, the cell being either normal or cancerous. By 1990, we'd also acquired the ability to study the resulting proteins. The earlier work indicated that DNA from a cancer cell could transform a normal cell into a cancer cell but yielded no insight whatsoever into how this introduced cancer gene worked in terms of the mechanisms involved. That is exactly what Ed Harlow was referring to in the early 1990s. He focused on the protein biochemistry of cancer cells, which held the promise of revealing the mechanisms by which cancer-causing genes—oncogenes—could transform a normal cell into a cancer cell.

Jan Witkowski: You've continued to be a leader in the field of the mechanisms by which cancers arise.

Dr. Weinberg: By about the year 2000–2001, I came to the conclusion that we understood, in outline, how a normal cell becomes a cancer cell and forms a primary tumor. We still didn't understand the last step of cancer development, which is how cells from the primary tumor seek to form new cancer colonies in distant sites, the process called metastasis. Starting in 2001–2002, people in my laboratory began to look for metastasis-inducing genes that empowered cancer cells from a primary tumor to disseminate to distant sites in the body.

In 2004, we discovered one such gene, which had been studied previously in the context of early fruit fly development. The genes that convert a normal cell into a primary cancer cell—and ultimately a primary cancer cell into one that invades and metastasizes—have been around for six hundred million years, which is about the distance between us and our distant cousins, the fruit flies. We soon learned that the way by which these metastasis genes worked was to take carcinoma cells, which are cells of an epithelial origin, and convert them into cells that are more like connective tissue, or mesenchymal, cells. This conversion process is called the epithelial/mesenchymal transition (EMT).

Jan Witkowski: How many genes are involved or needed for the EMT to occur in a cell?

Dr. Weinberg: One can take a primary carcinoma cell and force it to go from the epithelial state to an aggressive mesenchymal state with just a single introduced EMT-inducing gene. That cell is now empowered to disseminate to distant sites in the body. That's very satisfying, because it means one can trace the ability to spread from one site to another in the body to a small number of master control genes.

That doesn't solve the whole problem, however. It's the major problem of metastasis that has not yet been attacked successfully. Let's say we start out with a breast cancer cell in a primary tumor and it spreads to the brain, or the liver, or the lungs, or the bone marrow. We kind of understand how it does that (i.e., through the workings of the EMT program). The big challenge is to understand how a disseminated cancer cell, which has now inserted itself into a foreign tissue, makes a living once it gets there. Its site of destination, whether it be the brain or the bone marrow, presents a foreign environment to which the disseminated cancer cell is poorly adapted at the moment of its arrival. How does it learn how to make a living there? We don't really understand that.

Jan Witkowski: It'll be interesting to know how many potentially metastatic cells circulate in the blood and how many actually manage to establish themselves. Do you think that's a very small proportion?

Dr. Weinberg: It's a minute proportion. Between 30% and 35% of women who are diagnosed with a primary breast cancer already have thousands of micrometastases—small clumps of metastatic cells—in their bone marrow on the day of diagnosis, which means the cancer has already dispatched cells to sites throughout the body. You might think this condition dictates that 30%–35% of breast cancer patients will die from metastatic disease, but it's not true. Only about half of those women who initially come into the clinic with disseminated breast cancer cells will ever develop metastatic breast cancer in their lifetimes. What that means is what you just implied: The chance of a cancer cell that has wandered to a distant tissue successfully founding a colony is minute: maybe one in ten thousand, maybe one in a million. We don't really know the inefficiency of the process.

Jan Witkowski: Do metastases from particular cancers like breast or liver have preferred sites of forming metastases, or are different tissues equally good targets for all cancers?

Dr. Weinberg: The cancer cells originating in certain tissues have a predilection to generate metastases in certain distant sites. For example, prostate cancer typically creates metastases in the bone marrow. Colon cancer typically creates metastases in the liver. Breast cancer is less narrowly specialized, creating metastases in bone marrow, lung, and brain. I suspect, but I can't prove it to you, that the reason why these various primary tumor cells exhibit these tendencies is that, for example, if a prostate cancer cell travels through the circulation into the bone marrow, it is readily able to adapt to the foreign microenvironment in which it's landed, whereas it has a very hard time figuring out how to make a living in the liver. Therefore, even though primary prostate cancer cells get spread all over the body, the metastases that arise are testimonials to the ease or difficulty of these cells to make a living in one or another tissue.

Jan Witkowski: This implies that the microenvironment of the metastatic cell in the prostate presumably has some affinity with the microenvironment in the bone marrow.

Dr. Weinberg: I would agree with that, but articulate it slightly differently. It could be that a prostate cancer cell that was grown in the prostate has to make fewer adaptations when it lands in the bone marrow than the number of adaptations it would need to develop if it were to go to the brain. Of course, I use the word "adaptations" with some reluctance because we don't really know how cancer cells from one site of origin make a living in a distant site in the body.

Jan Witkowski: Would "selection" be a better word than "adaptation"? The prostate cell in the bone marrow struggles to survive but might be able to do two or three divisions or something, and then you get selection by the microenvironment?

Dr. Weinberg: It's both, in Darwinian terms. Let's say that a prostate cancer cell ends up in the bone marrow. It could sit there for days, weeks, or years, trying with little success to make a living there. However, one of its progeny might, through happenstance, eventually acquire the adaptive program to enable that cell to start proliferating in the marrow. It's a model, albeit a crude one, of Darwinian evolution.

Jan Witkowski: Do these findings have implications for dealing with metastases clinically?

Dr. Weinberg: I don't know. I could sell what I've just told you as being the answer to preventing metastases, but I wouldn't want to, because I don't see a clear path leading from these discoveries to useful pathways. Moreover, my own scientific trajectory has always been to try to figure out what's going on mechanistically during cancer pathogenesis without being biased or influenced by how useful it's going to be in the clinic. To my mind, curiosity-driven research is much more likely to turn up unexpected insights that could become useful to someone someday than is a highly directed and focused research program that is attempting, for example, to develop a cure for metastatic disease.

A Conversation with Eileen White

INTERVIEWER: ELIZABETH MCKENNA

Associate Editor, Cancer Discovery

Eileen White is a Distinguished Professor of Molecular Biology and Biochemistry at Rutgers University and the Deputy Director and Associate Director for Basic Science at the Rutgers Cancer Institute of New Jersey.

Elizabeth McKenna: Could you explain what autophagy is?

Dr. White: It's a very interesting process where, in response to starvation, cells will make vesicles that will capture components in their cytoplasm—mostly proteins and organelles—and then degrade them. The idea is that this degradation of proteins and organelles provides breakdown products that can be recycled into metabolic pathways. In that way, cells in animals can tolerate starvation for short periods of time. It's a survival pathway first identified in yeast that allows yeast to survive nitrogen starvation, and the use of autophagy to survive starvation is also apparent in mammals. As a matter of fact, we made a mouse where we can systemically, acutely delete an essential autophagy gene. Those mice are okay in the short term, but if you fast them, they die.

Clearly, autophagy's a survival pathway in yeast that's conserved all the way to mammals. What's interesting is that this pathway is normally on only at a very low level under normal conditions, but it's massively induced when cells or animals are starved. What was unusual is that when you look at cancer, even under fed conditions, the autophagy pathway appears to be on. When we interfered with the function of the autophagy pathway, this diminished the growth and survival of cancer cells. What that meant was that autophagy was a survival pathway for normal cells, but cancer cells can usurp or take advantage of that pathway for their own survival.

Elizabeth McKenna: You've been working on the role of autophagy in K-RAS-driven cancers. Can you tell us what you've found? Is there a specific way that autophagy is promoting survival in K-RAS-driven cancers?

Dr. White: I'm really working on two things. One is: What specifically is autophagy doing, and does it in fact recycle intracellular components? If so, what metabolic pathways are supported by this recycling? The first step is, does recycling of intracellular components occur in cancer cells? We did this by demonstrating that labeling components in cancer cells, and then starving the cells, caused the label to start appearing in an autophagy-dependent manner into specific metabolic pathways. We were able to do that. In fact, proteins and organelles are degraded by the process of autophagy, but if you delete an essential autophagy gene that process does not occur efficiently. Cancer cells are recycling macromolecules by autophagy.

The second question is: What are they using it for and how does it promote survival? The short answer is that they need autophagy to prevent fatal nucleotide pool depletion. When you starve RAS-driven cancers, if they don't have autophagy, they undergo an energy crisis and they deplete their pools of nucleotides. What autophagy is doing is supplying substrates to maintain energy, nucleotide synthesis, and nucleotide pools.

Elizabeth McKenna: Does autophagy work to promote survival in the same way in different cancers with different genetic backgrounds, like B-RAF for example, or is it a different mechanism?

Dr. White: I don't know that we know the answer to that. These metabolic interrogations of RAS-driven cancers that were wild type and knocked out for autophagy have only been done in RAS-driven lung cancer. We have not yet done them in other cancers. What we know is that there are probably multiple functions of autophagy that are contributing to cancer. For example, in B-RAF-driven melanoma models, autophagy prevents senescence. That could really be related to a metabolic defect caused by the loss of autophagy, but we'll have to see.

Elizabeth McKenna: Do you suspect that it's a difference based on the genetics or based on just the nature of different tissues?

Dr. White: I think the genetics is going to make a difference, but I think the tissue type might also make a difference. We'll have to see.

Published by Cold Spring Harbor Laboratory Press; doi: 10.1101/sqb.2016.81.031591

Elizabeth McKenna: Other people in the field have found that, in certain contexts, autophagy can actually be tumor-suppressive. Can you comment on that?

Dr. White: We have published papers related to that too. The take-home lesson is that autophagy has a context-dependent role in cancer. In some cancers, autophagy promotes survival like it does in yeast and mammals during starvation. In some cases, autophagy does nothing; there are cancer cells that don't care if autophagy is there or not. That's interesting, because how are these cells dealing with stress if they don't have autophagy? That's an important future area for investigation. In other models, if you knock out autophagy, especially in the liver, you end up with benign hepatomas. It's clear that in mouse models autophagy is suppressing the development of benign liver tumors. One possibility, and I think this is generally what people are thinking, is that autophagy protects cells. It promotes their survival, but in tissues like the liver where chronic damage and cell death leads to chronic inflammation—and chronic inflammation in the liver and pancreas and a few other tissues is a well-known carcinogen—that it is probably the mechanism whereby autophagy is tumor-suppressive. This is occurring in mouse models, and it's not clear whether this occurs in humans.

Elizabeth McKenna: Can you tell us about some of the efforts to characterize this in humans?

Dr. White: One key indicator of anything in cancer is simply to look at its genome. Right now, more than 10,000 human tumors have been sequenced. We have a huge amount of data on genomic alterations in a vast number of genes. Essentially, the genetic footprint of whether or not these genes are involved in cancer can be seen by looking at this data. What we've learned is that, in general, the core autophagy machinery is not mutated in cancer and these genes are often expressed in cancer. What we think is that this pathway is remarkably preserved in most cancers.

Elizabeth McKenna: Can you tell us about efforts to try to drug or treat or target autophagy?

Dr. White: There's a lot of interest in that, and I think their initial attempts involve using hydroxychloroquine, which is a lysosomotropic drug. What happens is that the autophagy pathway captures the cargo and delivers it to the lysosome for degradation. The idea is that if you interfere with the lysosome, you're going to block all lysosome-dependent degradation including that which goes through the autophagy pathway. Hydroxychloroquine is involved in clinical trials to see if interfering with lysosome function might have some anticancer activity. Some of the early results from these trials have been published and are mildly encouraging. Other pharmaceutical companies are making deliberate inhibitors of specific components of the autophagy pathway. There could be something that's more specific other than just targeting the lysosome.

Elizabeth McKenna: Are there specific combination approaches that you foresee might be a good idea based on the literature?

Dr. White: Autophagy generally is induced in response to stress. Not just starvation, but pretty much any kind is stress. There are many papers published that say that autophagy is induced by just about any type of cancer treatment. One possibility is that anything that's damaging or stressful to cells might benefit from combination with autophagy inhibitors. Anything that's interfering with metabolism would probably be predicted to be greatly enhanced by inhibiting autophagy, because autophagy might be a rescue mechanism.

Elizabeth McKenna: Can you talk at all about some of the links between abnormal metabolism in cancer and autophagy? Are there certain triggers or certain pathways that are especially linked to autophagy?

Dr. White: What we've learned is that when autophagy degrades substrates and uses them to support metabolism, the key pathways appear to involve the mitochondria. The substrates are delivered to the mitochondria. They enter in the TCA cycle and keep it going. The TCA cycle maintains mitochondrial respiration and also generates aspartate, for example, to provide carbon for nucleotide synthesis. Everything that we've been able to see is related to autophagy providing substrates for mitochondrial metabolism to maintain respiration and energy homeostasis and to provide carbon for de novo nucleotide synthesis.

Elizabeth McKenna: The field of autophagy has been really hot lately. Where do you see the field going? What do you think the next big unanswered questions are?

Dr. White: I think it would be the connection between autophagy and the immune system. What we have seen, and many other people have also seen, is that when you inactivate autophagy in cancer, you promote an immune response to those tumors. In lung cancer, this is so dramatic that when you have autophagy-deficient lung tumors, even though they're small and benign, the inflammatory response of these tumors is so great that the mice die of pneumonia. It could be that what's happening is you've got sick tumor cells, you've got dead tumor cells, and the immune system is doing what it's supposed to do: go in there and clean things up. With the advances of immune therapy, this presents an intriguing question as to whether autophagy inhibition is going to be an advantage in combination with immune checkpoint inhibition or in immune therapy where T cells are activated to kill tumor cells.

Elizabeth McKenna: Does autophagy play a role in the immune cells themselves?

Dr. White: I think the answer is yes, but the open question is, which comes first? If autophagy is dispensable in the short term for immune cell function, then that doesn't matter.

Author Index

Subject Index